PAUL
HERSHKOWITZ

Y0-AGL-584

Fundamentals of Transportation Engineering

A Multimodal Systems Approach

Fundamentals of Transportation Engineering

A Multimodal Systems Approach

Jon D. Fricker
Robert K. Whitford
School of Civil Engineering
Purdue University

Upper Saddle River, NJ 07458

Library of Congress Cataloging-in-Publication Data on file.

Fricker, Jon D.
  Fundmentals of transportation engineering; a multimodal approach/Jon D. Fricker,
Robert K. Whitford.
    p. cm.
  Includes bibliographical references and index.
  ISBN 0-13-035124-5
  1. Transportation engineering. I. Whitford, Robert K. II. Title.

TA1006.A1F75 2004
629.04--dc22

2004040015

Vice President and Editorial Director, ECS: *Marcia J. Horton*
Acquisitions Editor: *Laura Fischer*
Editorial Assistant: *Andrea Messineo*
Vice President and Director of Production and Manufacturing, ESM: *David W. Riccardi*
Executive Managing Editor: *Vince O'Brien*
Managing Editor: *David A. George*
Production Editor: *Donna King*
Director of Creative Services: *Paul Belfanti*
Art Director: *Jayne Conte*
Cover Designer: *Bruce Kenselaar*
Cover Photo Collage: *Daniel Sandin*
Art Editor: *Greg Dulles*
Manufacturing Manager: *Trudy Pisciotti*
Manufacturing Buyer: *Lynda Castillo*
Marketing Manager: *Holly Stark*

© 2004 Pearson Education, Inc.
Pearson Prentice Hall
Pearson Education, Inc.
Upper Saddle River, NJ 07458

Printed in the United States of America

10 9 8 7 6 5 4 3 2 1

ISBN: 0-13-035124-5

Pearson Education Ltd., *London*
Pearson Education Australia Pty. Ltd., *Sydney*
Pearson Education Singapore, Pte. Ltd.
Pearson Education North Asia Ltd., *Hong Kong*
Pearson Education Canada, Inc., *Toronto*
Pearson Educación de Mexico, S.A. de C.V.
Pearson Education—Japan, *Tokyo*
Pearson Education Malaysia, Pte. Ltd.
Pearson Education, Inc., *Upper Saddle River, New Jersey*

To Karen, Douglas, Laurel,
for their patience and support as this book took shape
and to my recently departed parents, Donald and Beatrice Fricker,
for providing a loving and nurturing home.

J.D.F.

In Memory of Sue Whitford, wife and cherished companion for 46 years and
Ralph A. (Jack) Whitford, father, engineer, and mentor.

R.K.W.

Contents

Preface

Between the two of us, we have taught a course called "Introduction to Transportation Engineering" more than 30 times since the Fall semester of 1984. As we worked together to improve the course, we continued to move farther away from the textbook that we had adopted. We began to assemble a set of course notes that, over time, began to resemble chapters of a textbook. Although we think there are several good transportation engineering books on the market, none of them seemed to fit our style of teaching and, more importantly, the styles of learning that we see in our students. We prefer to encourage classroom discussion, use computer tools to augment lectures, form small groups for in-class discussions or out-of-class problem solving, and base assignments on real-world situations. We see the need to lead students toward improved learning and away from the usual practice of using class time for extensive note taking.

Our course notes finally reached the point where we used them in place of a published textbook. The notes covered all facets of transportation engineering, not just highway modes. Although the emphasis is on highway and air transportation, there are lessons on railroad and waterborne modes of transportation, the systems aspects of transportation economics and program evaluation, public transit, and logistics management. Because we realize that this scope is ambitious for an introductory course, we adopted certain strategies that have proven successful:

1. Integration of topics
2. Clear, enunciated objectives
3. Special discussion boxes
4. A large number of examples that are based on real situations
5. Hundreds of photographs and other illustrations.

These features are continued in this textbook.

SEQUENCE AND FLOW OF TOPICS

An introductory course in Transportation Engineering is susceptible to being simply a series of fragmented topics. We have addressed this problem in several ways:

1. Because a system cannot be operated before it is designed, and a system cannot be designed until it is planned, we have put the chapters in our textbook in an order that reflects this. However, planning in transportation can be rather abstract and mysterious to an undergraduate engineering student. For this reason, we put topics of greatest familiarity to students—traffic—near the front of our textbook.

2. To minimize the fragmented nature of the topics and to emphasize the relationship between the phases of a transportation project:

   A. We have invented the mythical County of Mythaca, with cities and towns of various sizes and various transportation challenges.

   B. We begin each chapter in our textbook with a Scenario. These Scenarios are meant to orient the student to the material about to be covered in the chapter. The material in that chapter is meant to provide the student with knowledge and methods that can be used to address the problems stated in the Scenario.

There is about 20 percent more material than can be covered in a single-semester course at the junior–senior level. The text is structured so that, in most cases, sections within Chapters 6 to 13 can be skipped without major loss of continuity. With a modicum of care, the instructor can change the sequence in which the chapters are covered. The Engineering Economy portions of Chapter 5, however, ought to be covered before the economic analyses in the Transit (Chapter 10) and Air Transportation (Chapter 11) chapters are encountered.

CHAPTER OBJECTIVES

For many years, we have had the following objectives in mind as we taught "Introduction to Transportation Engineering":

- Provide the student an adequate basis for deciding which more specialized transportation courses to take.
- Familiarize the student with the standard terminology and resources involved in transportation engineering, in case the student would need to conduct such an analysis in the future or work with someone else who was doing so. This situation occurs in our capstone senior design course.
- If this would be the only transportation course a student would take, give the student enough of an appreciation of transportation issues to be an informed citizen and professional engineer.

In recent years, we have found it helpful to be more explicit in what we want to accomplish in the course and in its components. We have adopted the Instructional Objectives method advocated in the National Effective Teaching Institute (NETI) Workshop, for several reasons. The instructional objectives for each chapter are presented immediately after the Scenario. In our opinion, the NETI-based objectives:

- Lead to a clear statement of capabilities that a student is expected to acquire.
- Are observable and measurable.
- Form a sort of contract between teacher and learner.

- Impose a measure of discipline on the authors as we seek to present material in a textbook format.

The word "Fundamentals" appears in the title of this book, because topics that every student of transportation engineering should know are emphasized. Also included are some topics that are "nice-to-know" or just interesting. Interspersed in the textbook are items that are meant to reflect the enjoyment that can come from the study of transportation engineering. It is our hope that a student who has read this textbook will never look at a traffic signal, traffic sign, semitrailer truck, or airport runway the same way as before.

TEACHING AND LEARNING

Unfortunately, our experience has been that many students are not easily convinced to come to class prepared. One feature of our text that may help in this regard is the "Think About It" boxes in most sections. These boxed questions are placed in the text wherever a pause to think about the material being presented would be helpful to the student. Often, there may be more than one way to approach the problem being presented. To assist the instructor in leading the discussion in class, an instructor's guide will include the ideas we had in mind when we inserted the "Think About It" boxes.

The most important guidance for this textbook comes from the questions asked by students after class and during office hours. In response to their feedback, we have made numerous changes to the content and style of the text. The result is a presentation style that is more inductive than the traditional textbook.

ACKNOWLEDGMENTS

We gratefully acknowledge the more than 1300 students who have taken CE361 Introduction to Transportation Engineering with us at Purdue University. They helped us try different approaches in the classroom and gave us valuable feedback on the class notes that were the basis for this textbook. Graduate students Sergio Lugo Serrato, Koh Ee Huei, and George Kopcha have been particularly helpful in providing solutions to the end-of-chapter exercises.

Colleague Darcy Bullock reviewed the manuscript, taught from it, and provided many helpful suggestions for its improvement. We also thank colleague Kumares Sinha for his help and encouragement.

To Laura Fisher at Pearson Prentice Hall, we extend our gratitude for her faith in this new approach to Transportation Engineering.

Jon D. Fricker
West Lafayette, IN

Robert K. Whitford
Juneau, AK

About the Authors

Professors Whitford and Fricker bring contrasting backgrounds to this project. Prof. Robert Whitford holds B.S., M.S. and Ph.D. degrees in electrical engineering from Purdue University. Following graduation in 1955, he worked in the aerospace industry, designing and analyzing Missile and Spacecraft systems at TRW Systems (Redondo Beach, California) for 17 years. The last six years at TRW he served as Operations Manager for Guidance and Control (3 years) and Advanced Electronic Systems (3 years). He then became deeply involved in the systems and design aspects of multimodal transportation while serving 6 years as Deputy Director at what is now the US DOT's Volpe National Transportation Systems Center in Cambridge. He joined the automotive transportation-energy research team at Purdue in 1980. In 1982 he began serving as part-time director of Purdue's Public Policy Center, while teaching large scale systems and freight/logistics courses in both civil and industrial engineering. He became full-time on the Civil Engineering Faculty in 1989. In addition to his research interests highway congestion management, logistics systems, and air transportation, he has taught graduate courses in transportation planning, transportation project evaluation, urban planning, and airport planning and design. He was the coordinator of the capstone civil engineering design course for five years. He retired from the Purdue faculty in January 2002 and now lives in Alaska, working part time for the Alaska Department of Transportation.

Prof. Jon Fricker received the S.B.C.E. degree from the Massachusetts Institute of Technology and the M.S.C.E. and Ph.D. degrees from Carnegie-Mellon University. His degrees all came from civil engineering programs that required a solid grounding in all facets of CE. Since joining Purdue's Civil Engineering faculty in 1980, his principal teaching and research interests have been in transportation planning, public mass transportation, and urban planning. Prof. Fricker's service on the local bus company's Board of Directors and the county's Technical Transportation Committee has been a source of many reality-based problems for homework and in-class use.

Professors Whitford and Fricker have interests that overlap in urban planning, economic analysis, and systems analysis. In areas where neither of them had a strong background, they have worked to become competent by attending short courses and seeking the advice of experts. Because their backgrounds are somewhat different, they complement each other well. More importantly, they are in close agreement about what an introductory course in transportation engineering should include and how it should be presented. They think that their ideas, reflected in this textbook, will find acceptance among others teaching a similar course.

Transportation Basics

SCENARIO

Mythaca is not a small town any more. After decades of growth, it is beginning to have the transportation problems of larger cities. No longer can a person drive across town in less than 20 minutes. In fact, traffic congestion is now a common occurrence (Figure 1.1). Travelers spend a greater portion of their waking hours trying to get to their destinations, and households spend a greater share of their budgets on "mobility."

The quality of life in Mythaca is threatened by these and other transportation-related problems. Tailpipe emissions affect air quality. Deteriorating access to raw materials and markets makes it more difficult for Mythaca to attract and retain high-paying manufacturing firms. The airport needs to expand its runways and the public transit agency is asking for increased subsidies. Increasing the capacity of Mythaca's road system will be costly not only in the funds needed but also in the neighborhoods that may be disrupted, the open space that may be lost, and the added traffic that the new roadway lanes will probably attract.

The citizens of Mythaca demand solutions. Elected officials are asking transportation experts to work with citizens to assemble a set of projects and policies that will solve (or reduce) the community's transportation problems at a cost that the taxpayers can afford. The officials need to consider a variety of strategies to improve mobility, but getting citizens to alter their driving habits and try walking, bicycling, and public transit may be difficult. To distinguish among alternatives, decision makers need to have a reliable way to estimate the effectiveness and cost of each proposed project or program. They need to turn to persons

FIGURE 1.1
Mythaca traffic jam. *Source:* Jon D. Fricker.

who can understand and implement the fundamentals of transportation engineering—principles and methods that are introduced in this book.

CHAPTER OBJECTIVES

At the end of this chapter, the student will be able to:

1. Describe the role transportation has played in the development of our communities and our society.
2. List and explain the three "legs" that support the understanding of transportation systems.
3. Develop a broad system specification for a particular transportation system.
4. Calculate and apply performance measures needed to analyze a transportation system.

1.1 THE CALL FOR A NEW TRANSPORTATION ENGINEER

"Moving America" is key to the economic vitality and quality of life in the United States. Too often, we take transportation for granted. Yet we spend about 11 percent of our spendable income on transportation (BTS, 2003, Table 3-12), and transportation is directly responsible for about 16 percent of the U.S. Gross Domestic Product (BTS, 2003, Table 3-12). Nearly every day, items in the news remind us of transportation's vital role in our economy and its significant relationship to our quality of life. Throughout the United States, thousands of citizen groups, such as "Make Mythaca More Livable" (MMML), have been formed. The group is making a difference in the way that both the public and the transportation engineers think about how to solve Mythaca's transportation problems. Mobility is important to the whole community. The journey on which we are about to embark in this book—an exploration of the realm of transportation, with emphasis on key aspects of its engineering and its close relationship to our social and economic lives—can be interesting, even exciting. As we examine transportation problems in Mythaca, we explore methods that can lead to transportation engineering solutions in the real world.

1.1.1 Definitions of Transportation

What is transportation? How do you define your relationship to transportation? Is it only the trips that you make? Or is it the car that you drive? Whether we are considering people or goods, each trip begins at an *origin* and ends at a *destination*. Transportation is everything involved in moving either the person or goods from the origin to the destination. Consider the businessman's trip depicted in Figure 1.2. The trip is from the businessman's home (origin) to a hotel in a distant city (destination).

The trip could begin in his personal automobile, on a public transit vehicle, or in a taxi. This first link of his trip takes him from home to the airport parking garage or to the door of the airport terminal. This first segment is one of several *line-haul* portions of the trip. If he drives his car, he parks it at the airport parking garage, changing from the highway *mode* to the walking mode for a short distance, then taking the shuttle bus to the airport. If he left home by public transit or taxi, he gets dropped off directly at the door to the airport terminal. The places where there is a change of

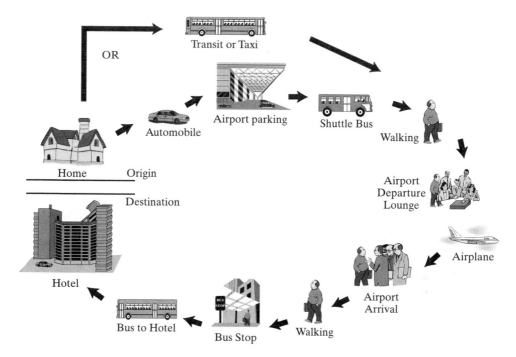

FIGURE 1.2
Tripmaking from origin to destination.

mode are referred to as *intermodal transfer points*. Table 1.1 indicates that this trip has several points where the businessman changes mode. Although the main portion of his trip is by airplane, there are numerous other uses of the transportation system involved.

1.1.2   What Is Transportation Engineering?

Transportation engineering is the application of the principles of engineering, planning, analysis, and design to the disciplines comprising transportation: its vehicles, its physical infrastructure, safety in travel, environmental impacts, and energy usage. Transportation engineering involves the "hard" physical sciences, as the engineer evaluates pavements, geometric design, vehicle design, environmental effects, and the like. It involves thorough analyses of the impact on transportation design and operations from a variety of "soft" or social sciences, such as human behavior, welfare economics, urban planning, and political science. The competent transportation engineer must be capable of integrating the factors found in both the "hard" and the "soft" sciences when searching for the best solution to a given transportation problem.

Take another look at Table 1.1. Each segment of the businessman's trip depended on at least one constructed facility, such as a roadway or a runway. At the intermodal transfer points, constructed facilities such as parking lots or airport terminals

TABLE 1.1   Transportation Functions and Activities for Figure 1.2

Trip Segment	Transportation Function	Activity	Location
START	Origin	Start Trip	Home
1*	Line haul Intermodal transfer point	Drives to the airport Parks car and boards shuttle bus	City streets Airport parking garage
2*	Line haul Intermodal transfer point	Rides shuttle bus to terminal Leaves bus	Airport roadways Airport curbside
3	Line haul Intermodal transfer point	Walks to departure area Boards aircraft	Airport terminal Jetway
4	Line haul Intermodal transfer point	Flies Lands	Airspace Airport airside
5	Line haul Intermodal transfer point	Get bags and walks Gets rental car, or boards taxi or hotel shuttle bus	Airport landside Ground transportation facilities
6	Line haul	Drives car, or rides bus or taxi	City streets
STOP	Destination	End trip	Hotel

*Line haul segments 1 and 2 in Table 1.1 can be replaced by the "Transit or taxi" segment shown in Figure 1.2.

are necessary. Transportation engineers play an important role in planning and designing such facilities. They also need to be aware of the operational capabilities and limitations of the various transportation modes and services that need to be integrated.

Transportation is much more than people making trips. The movement of goods is a critical part of local, regional, and national economies. As goods move from origin to destination, transfer points can be rail yards, truck terminals, warehouses, or distribution centers. Line-haul goods movement will be by rail, truck, water, pipeline, or some combination of these modes. Again, properly designed and operated facilities are essential to an efficient transportation system. If a transportation facility or service is overdesigned, the result may be a waste of resources. If it is underdesigned, bottlenecks that cause delays, lost productivity, or unsafe conditions may arise.

Transportation engineering involves working with the public, with industry, with citizens' groups, with elected officials, and with employees of the agencies of local, state, and federal governments. It involves facing real problems that usually require engineering judgment. It is the county engineer, the city traffic engineer, or the state DOT's engineer who will be asked to propose solutions for citizens and local officials to grapple with in public forums.

To provide an ongoing context, many of the problems presented in this book take place in the mythical County of Mythaca, whose largest city is the City of Mythaca. Along Lake Murdoch, there is also the industrial port of Mazurka and the recreation town of Shoridan. State Road 361 (SR361) between Mythaca and Shoridan is heavily traveled, including weekend traffic to recreation spots. Halfway between the two cities is the little town of Middleville. The map in Figure 1.3 shows the Mythaca area.

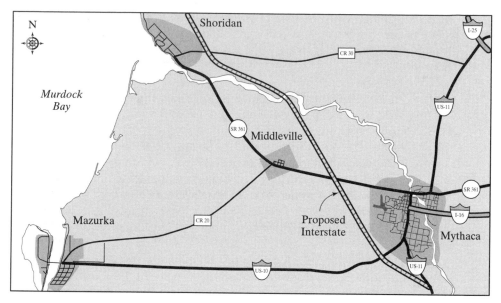

FIGURE 1.3
Map of Mythaca and environs. Map courtesy of R. David Oliver and Rachel Brown, Alaska DOT & PF.

F Y I :   Most interstate, U.S., and state highways are numbered according to the direction that applies to most of the road. A highway that runs mostly east-west will end in an even number. A road that is predominately north-south will end with an odd number. Do the state, U.S., and interstate highways in Figure 1.2 follow this convention? Hint: SR361 starts in the southern part of the state.

1.1.3   Transportation's Role in Shaping Society

Transportation has played a vital role in shaping society, fighting wars, and building cities since humanity's beginnings. Even now, a reformation is taking place, as we become involved in discovering new challenges for transportation in the early years of the 21st century.

Since the beginning of time, moving people and transporting goods and materials has been a necessary part of our societies. Great structures like pyramids and castles were built thousands of years ago. Timber and large rocks were brought to building sites by the sweat of many workers, with some aid from animals providing power and logs used as rollers. Before rollers, materials were dragged across the land on flat plates or makeshift sleds. Later, the power of animals, such as horses, mules, and oxen, were combined with the reduced friction of wheels to ease the burden of long-distance travel. Well-traveled trade routes existed more than 3500 years ago, bringing spices, gems, and other materials from Asia to the Middle East and on to Europe or Africa. The speed of such travel depended on the routes involved, but probably didn't exceed 2 or 3 miles per hour. The Romans built roads, not only for trade, but to help move their great armies.

Movement by sea seemed to advance faster, as sails were used to harness the power of the wind. Oars added human power when it was needed. Speed was anything but fast, with stagecoaches achieving 3 to 5 miles per hour (mph) and the fastest ships about 10 miles per hour in the 1600s (Wilson, 1995).

Critical elements in society's view of transportation are trip time, speed, and access to desirable destinations. After World War II, suburban sprawl resulted when low-cost housing loans for returning veterans were combined with the construction of roads out from the city center. The veterans were able to take advantage of the lower cost of land farther from the city center. Inexpensive transportation has opened many markets to products that could be brought to the consumer cheaply enough to be competitive in price.

Great cities were located near the water. The relative ease and effectiveness of movement in the water, especially across the seas—first for exploration and later for the transport of people and goods—has had a large impact on land use. Because of the convenience of water transport for fostering warfare, imperialism, and movement of resources across the vast oceans, great cities grew up around ocean ports. Ocean, lake, and river travel is still important today, with more than 75 percent of the U.S. population residing in counties along the Great Lakes and sea coasts (53%) or within 35 miles of a navigable river or canal (23%). (U.S. Department of Commerce, 2001).

The 360-mile Erie Canal aided the development of the upper Midwest by connecting the Great Lakes to the Atlantic Ocean in 1825. As shown in Figure 1.4, early canal barges were pulled by animals moving on towpaths next to the water. One canal was built across Indiana to connect the Great Lakes to the Wabash, Ohio, and Mississippi Rivers, resulting in easier movement of goods from the Great Lakes to the Gulf of Mexico. The history of the Wabash and Erie Canal is integral to the history of the State of Indiana. The coming of the canal brought people and goods, and the people built towns where before had only been swamp, forests, and tall grass prairie. However, before the canal could earn enough revenue for its builders to recoup their investment, railroads appeared on the scene. Railroads started carrying many of the grain and coal shipments that the canal had depended on and the Indiana canal could not survive (Fatout, 1985).

Ocean shipping continues to be a large and important part of our society's activity. Transportation continues to use a large amount of imported oil. Automobile parts, appliances, and many other goods come in containers, crossing the ocean daily to meet the needs and wants of the American consumer.

Urban transit predates the steam engine. Early cities developed with all their activities in close proximity to each other. As more and more people moved to the cities, the city had to spread out. The demand for the movement of large groups of people in the urban area became apparent. Throughout the world, large cities began to use horse-drawn streetcars, as shown in Figure 1.5. This "mass transportation" provided important benefits to the growing cities. When it was too far to walk and the

FIGURE 1.4
Mules pull barge along Erie Canal, ca. 1825. *Source*: Photo taken from the ASCE History and Heritage
Collection archive. Reprinted by permission of ASCE.

person did not own a horse, inexpensive transportation was available to take low-wage workers from their homes to the factories to work, to places to shop, and to recreational areas.

As travel speeds changed, travel patterns changed. Things moved at a "snail's pace" until about 200 years ago, when the use of steam to drive rotating machinery found its way into transportation. Example applications are the steamboat in 1786, the steam locomotive in 1813, urban street cars (pulled on ropes connected to pulleys driven by steam rather than horses), and paddlewheel boats for river travel (circa 1807). The use of steam was the beginning of changes in transportation technology that continue today.

Before automobiles, railroads were the premier mode of transportation in the United States for both people and most materials. With the driving of the "golden spike" in Utah in 1869, eastern and western railroads joined, and the country became tied together across land. Previously, settlers of California either took an arduous trip across the country in covered wagons or by ships that took many months to sail around the tip of South America. The California Gold Rush and other movements to seek natural resources in the West were aided by rail. During the late 1800s and early 1900s, railroads were the principal means for moving the agriculture products of the

FIGURE 1.5
Horse drawn streetcar in 1886. *Source*: Neenah Historical Society collection, Neenah, Wiscousin.

Midwest and bringing coal for electric plants and ore to the steel-making foundries in the Midwest. Railroads were also responsible for most intercity passenger travel. After the devastating Civil War, the faster transportation provided by rail helped the country begin to come back together. In fact, the railroads became so strong, especially in areas where there was no competition, that price-gouging and other predatory practices became commonplace. In response, the federal government intervened to promote fair prices and the Interstate Commerce Commission was formed.

In the late 1870s, the automobile appeared on the scene. (See Figure 1.6a.) It did not find itself in widespread use until after about 1925. Before that, there were steam cars (Figure 1.6b), electric cars, and cars using early versions of the internal combustion engine.

As the 20th century began, automobile competition was mainly between the electric automobile (which required large batteries and didn't go very far on a single battery charge) and the internal combustion auto (which had more promise, but required a very difficult manual cranking to start the engine). Ironically, it was an electric invention—the electric starter motor operating from a small battery—that spelled the death-knell for the electric automobile. The advances of the internal combustion car, together with its affordability and the growing level of personal wealth, have made the United States the most automobile-dependent society in the world,

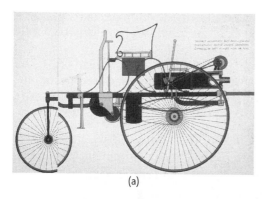

(a)   (b)

FIGURE 1.6

Early Automobiles. (a) The first automobile built by Karl Benz reached a speed of eight miles per hour. (b) An early American steam motor car. *Source*: Getty Images, Inc. – Hulton Archive Photos.

with an average of about 2.1 cars per household. The automobile, when combined with about 4 million miles of roads in the United States, has offered most U.S. citizens a range of personal freedom of movement unsurpassed anywhere in the world. The growth of our automobile driving has become almost an obsession, and Americans who can afford it have come to believe and act as though driving is more of a right than the privilege that it is.

1.1.4   A Present View of Transportation

With just the brief background provided in the previous subsection, it should be clear that society and the cities in which we live have an extremely important linkage to transportation. In the decades after World War II, two further major developments in U.S. transportation took place—Interstate highways and a modern air transportation system. Both began to provide faster and less expensive connections to distant locations. With these developments, the vision of many Americans to easily move across vast expanses had been largely realized.

However, some of the negative side effects of achieving this vision began to appear. Congestion, air pollution, and urban sprawl are now recognized as some of the costs that society is paying for increased mobility. Some modification of our travel behavior may be necessary. Society and transportation have never been more closely connected than they are today. As congestion increases and air pollution remains a concern, citizens are facing a series of difficult choices. Everyone wants pleasant "livable" cities with clean air, but many people also want unrestricted use of their automobiles. Many people see the potential benefits of a more extensive transit system, but few are willing or able to use transit much. Neighbors disagree as to whether "traffic-calming" devices, such as "speed humps" and roundabouts, are a positive safety measure or a nuisance. Some people are asking that local governments control suburban growth and urban sprawl through "Smart Growth" initiatives, whereas others oppose the land use controls and high-density developments they usually involve. Debates continue on how best to seek alternatives to the internal combustion engine.

As these conflicts indicate, even when general goals can be agreed on, the means to achieve them can be very difficult. Associations of citizens and transportation professionals are forming throughout the country to search for new methods of transportation and land use. The Surface Transportation Policy Project (STPP) is a major citizen oriented organization. Its goals are:

- Providing a safe, healthy, and secure transportation environment.
- Supporting sound investments in transportation and other areas to support prosperity and growth.
- Enhancing the quality, livability, and character of all American communities.
- Promoting resource efficiency and conservation. (STPP, 2002a)

STPP undertakes projects that examine a wide variety of transportation changes (e.g., promoting pedestrian travel, measuring congestion, studying the land use-transportation relationship, and exploring non-automobile options).

In Table 1.2, some observations about today's transportation system are summarized. These observations will form the basis for many of the topics to be covered in this book—some directly and others indirectly.

THINK ABOUT IT

What type of trip is likely to have a higher value of *average persons per car*—commuter trips to work or shopping trips? Explain. What factors prevent the commute trip from having a higher average vehicle occupancy rate?

1.1.5   Transportation Goals

The function of transportation is to provide for the movement of persons and goods from one place to another safely and efficiently, with minimum negative impact on the environment. The demand for transportation is derived from the demand for some activity or good. When no particular demand for a specific movement of people or goods exists, then no such transportation for that movement will occur. This could happen because persons do not want to travel to a certain place or because the cost of transport exceeds the value placed on it by the traveler. Sometimes a transportation option is proposed based on marketing studies that suggest that there is a "pent-up" demand for a service that is presently nonexistent. When viewed as a system, transportation has three aspects of use or utility that form the broad goals, namely, Access to Space or Location, Reasonable Travel Time, and Low Cost. These goals are described below.

Space or Location Utility.   Transportation has the function of creating location utility or *access*. As discussed previously, the demand for most automobile and truck transportation

TABLE 1.2   Observations About Today's Transport System

Derived demand for transportation. Transportation usually does not occur unless the movement of someone or some good is desired for some purpose and can be done in a cost-effective and timely manner.

Different modes give different service. With a car, we can generally come and go when and where we please; it is essentially self-service. The other modes do not provide the same basic freedom, i.e., we can use the service only when someone else provides it. As can be seen in Figure 1.7, most of the dollars and passenger miles are consumed in the highway travel mode.

Safety and security are important. Safety has always been an important concern in transportation systems. Since September 11, 2001, security concerns have increased enormously.

Some people are "transit-dependent." These are primarily non-drivers—those who cannot afford to own an automobile, those too young or too old to operate a motor vehicle, and those with disabilities. Equity in mobility must always be a concern of the transportation officials. The average poor person must spend about 35 to 40% of his/her income on transportation. (Bullard and Johnson, 1997)

Congestion costs have several components. Besides the extra time in traffic, fuel is wasted, air quality is degraded, and personal frustration mounts.

Urban sprawl is the primary cause of traffic congestion. According to the USDOT, about 69% of the increase in driving between 1983 and 1990 resulted from our spreadout land use patterns. Average miles per person increased while carpooling and transit use decreased. (BTS, 2003)

Traffic peaks during the day. The commuter trip has resulted in traffic saturation peaks twice a day in the typical large city. (See Figure 1.8.)

We are still building roads. The amount of *highway miles per person* in 76 metropolitan areas showed an increase of more than 10% over the last 16 years. Road miles grew at a faster rate than did population. (STPP, 2001)

Delay is growing much faster than population. Population grew by 22% in the years 1982 to 1997, but the delay experienced by motorists grew by 278% in the same period growth. (BTS, 2003, Table 1-63)

Americans are driving to work alone. The average number of persons in a car is 1.14 during the peak period.

It is often easy to find parking. The availability of parking, with its low cost and ease of access, contributes to our ability to drive nearly anywhere to work, to shop, etc.

The fleet mix is changing. In 1990, the U.S. vehicle fleet included 133.7 million automobiles and 47.5 million light trucks. In 2000, there were still 133.6 million automobiles, but 77.8 million light trucks. (Vans and SUVs are considered light trucks.) (BTS, 2003)

Driving has significant hidden costs. Recent studies (FHWA 2001) show that the real cost of driving most intermediate cars, SUVs, and vans ranges from about 47 cents to 52 cents per mile. Gasoline would cost the consumer $3 to $5 per gallon when the hidden costs of our energy imbalance, pollution, and infrastructure are included.

Household transportation budgets are significant. The average American household spends about 18% of its spendable income for transportation (up from 13% in 1975). This compares to 32.6% for housing and 13.6% for food (FHWA 2001).

---

is to provide access from an *origin*, such as a residence, to a *destination*, such as a given industry, commercial center, or public place. Thus, transportation is closely tied to the manner in which land or space is used. Mythaca is like most cities in the United States. It has many urban streets on which people drive, walk, or ride buses to get from their homes to work or to school. It has major highways that connect it with its neighboring cities. Mythaca also has an airport that permits its residents to go long distances by airplane. It is

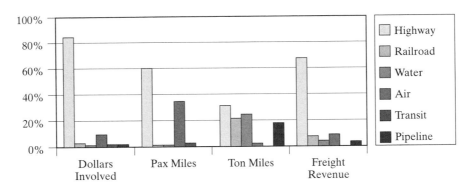

FIGURE 1.7
Comparison of modal activity. *Source*: BTS, 2003.

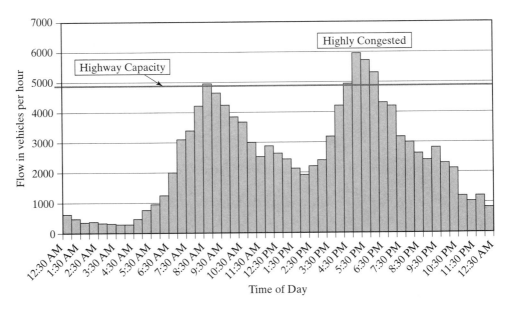

FIGURE 1.8
Typical daily traffic pattern showing periods of high congestion.

also located on a river, which allows its residents and businesses access to boating and limited barge service.

Time Utility. Time utility is related to the *trip speed* of transportation. True *trip speed* is really not the speed of a given vehicle; it is the distance traveled divided by total trip time, including stops, delays, and vehicle changes from *origin* to *destination*.

All forms of ground transportation have limited speed. Although race cars can go more than 200 mph, *safe* travel on our road system generally occurs at speeds less than

65 mph. Railroad trains in Europe (the French TGV and the German ICE) are operating in commercial service at speeds in excess of 250 mph, but passenger rail speeds in the United States are limited to 60 to 100 mph, depending on track conditions.

Speed is also important for longer distance travel, such as by aircraft. Commercial aircraft speeds of about 400 mph are adequate. Faster speeds would be possible with the Supersonic Transport (SST), but the cost and the time for the rest of the trip tends to reduce the overall speed and usefulness for many overseas travelers.

Example 1.1

You are going to travel from Mythaca to Denver (1200 miles). There are two choices: Flight 333 is non-stop and takes 3 hours, while Flight 444 requires that you change planes in St. Louis, with a total time to Denver of 4 hours. The time to change planes in St. Louis is 40 minutes. The difference in flight distance is negligible. Assume that it takes 30 minutes to drive the 20 miles from your home to the airport and that you must arrive at the airport 90 minutes early to check in. When you land in Denver, it takes 20 minutes to get your luggage and an additional 40 minutes to ride to the downtown hotel 30 miles from the airport.

A. What is the percent difference in the air speed for Flights 333 and 444? This excludes the transfer time in St. Louis.

B. What is the percent difference in the speed of air travel for Flights 333 and 444? This includes the transfer time in St. Louis.

C. What is the percent difference in the total origin-to-destination (home-to-hotel) speed for Flights 333 and 444?

Solution to Example 1.1

| | **Flight 333** | **Flight 444** | | |
	Non-Stop Flight	**One-Stop Flight**	**Difference**	**Note**
A. Air speed	1200 mi/3 hr = 400 mph	1200 mi/3.33 hr = 360 mph	10 percent	Time flying only
B. Air travel Speed	1200 mi/3 hr = 400 mph	1200 mi/4 hr = 300 mph	25 percent	Includes ground transfer
C. Total O-D speed	1250 mi/6 hr = 208 mph	1250 mi/7 hr = 179 mph	14 percent	Add ground distance of 50 miles and time of 180 minutes

---

Cost Utility.   Cost-effective transportation is required for both passenger and freight movements to take place. When the cost to move goods produced in one region of the country (*origin*) to a market in another region of the country (*destination*) causes the good to be noncompetitive at the destination, the goods will *not* be shipped. In such cases, companies who want to compete will consider establishing either a production capability or a warehouse (distribution center) in that market area. The distribution center becomes a valid option when the products can be shipped by slower, less costly means (e.g., railroad), thereby reducing the transport portion of the selling price, while still providing timely delivery.

Example 1.2

The airfare for the non-stop Flight 333 in Example 1.1 is $500.00, whereas the ticket for Flight 444 with the transfer at St. Louis is $360.00. Assume that all other costs are the same at $60 and that you value your time at $50 per hour.

- **A.** Which flight is cheaper to you? By how much?
- **B.** Under which conditions would you choose the more expensive flight?
- **C.** At value of time (VoT) would the two flights be equally expensive?

Solution to Example 1.2

- **A.** Add together airfare, ground transportation costs, and value of time:

$$\text{Cost of Flight 333} = \$500 + \$60 + (6 * \$50) = \$860$$

$$\text{Cost of Flight 444} = \$360 + \$60 + (7 * \$50) = \$770$$

Flight 444 is cheaper by $90.

- **B.** (1) Flight 333 does not involve a change of planes, which is extra effort and something else that can "go wrong" during a trip. (2) If Flight 444's scheduled arrival time in Denver is what you consider to be uncomfortably close to the time you *must* be in Denver, and Flight 333 gives you more "breathing room," then Flight 333 may be worth the extra airfare.
- **C.** Set the two equations in Part A equal to each other with value of time (VoT) as the unknown:

$$\$500 + \$60 + (6 * \text{VoT}) = \$360 + \$60 + (7 * \text{VoT})$$

$$\$560 + (6 * \text{VoT}) = \$420 + (7 * \text{VoT})$$

$$\text{VoT} = \$560 - \$420 = \$140/\text{hr}$$

THINK ABOUT IT

What is the value *you* would assign to an hour of travel time? Does your VoT change with circumstances? If it does, give some examples of different circumstances and how your travel choices may be affected as a result.

1.2   UNDERSTANDING TRANSPORTATION SYSTEMS

Figure 1.9 shows a simple stool with three legs. Just as the stool will not stand without all three legs, understanding transportation systems requires knowledge in three major areas. One leg represents the components of the system. A second leg stands for those activities involved in putting a transportation system in place, from planning to operation

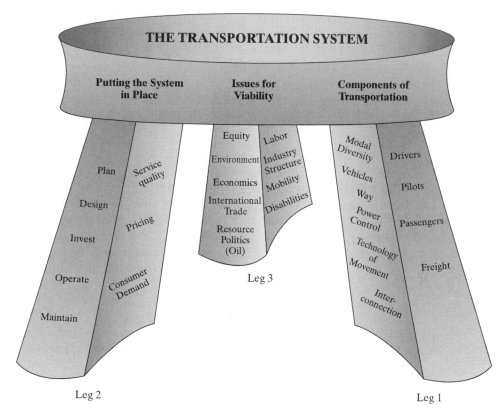

FIGURE 1.9
Three-legged stool of transportation systems.

and maintenance. The third and final leg identifies issues that may not be included in the transportation decision-making processes, although they may be affected by the decisions. Such issues are often identified as "externalities."

1.2.1   Leg 1: The Components (Modes, Movement, and Flow)

The transportation system is very diverse, with differing functional characteristics depending on the medium of movement (ground, air, or water), the particular technology used, and demand for movement in the particular medium. Aspects of the modes are:

- *Vehicle.*   Each mode has particular vehicles that provide carriage and flow, such as automobiles, bicycles, buses, trucks, railroad trains (cars and locomotives), light-rail vehicles, ships, barges, and airplanes. Pipeline is probably the most unusual. The pumps along the pipeline act as the vehicle, imparting energy for the movement of the fluid passing through them and overcoming the resistance to flow along the walls of the pipe. Another often-overlooked mode of transport is the pedestrian mode, which involves the movement of human bodies, sometimes assisted by wheelchairs, crutches, or moving sidewalks.

- *The way.*   Each mode has what is called a highway, guideway, right of way, or simply "the way." For automobiles, trucks, and buses, the way is the highway, road, or street. Other forms of way are railroad tracks, waterways, airways, and pipelines.

- *Control.*   Two forms of control are usually present in transportation. There is the *control of the system*, such as signals and signs for highway traffic, and radar and air traffic control for aircraft. The driver or pilot also must *control the vehicle* through steering, accelerating, and braking.

- *The technology of motion.*   Each mode uses a specific technology to facilitate the movement of the vehicle in its way. For example, ships use the principle of buoyancy in water; railroads use steel wheels on steel rail; automobiles, buses, and trucks use rubber tires on a friction-producing surface; and aircraft use the airfoil wing shape for lift.

- *The technology of power.*   Each mode has a particular type of engine designed to provide the most efficient power source for that mode. Spark-ignition combustion engines operating on gasoline (oil refined for the particular use) are normally used in automobiles. Buses and trucks often use diesel-powered engines. Locomotives and transit cars frequently use traction or series electric motors to drive the wheels. The electricity to operate these motors is either brought to the vehicle through electric wires or buss bars, or can be carried aboard the car using diesel generators as is done with diesel-electric locomotives. Ship propellers are usually driven by gas turbines.

- *Intermodal transfer points.*   Points of connectivity between modes or within a mode are terminals, ports, harbors, airports, bus shelters, and parking lots. These are the places where the passenger or freight changes from one mode to another mode or from one vehicle to another. The function of an airport, for example, is to collect travelers together in one place. Passengers arrive from numerous places and gather in the airport departure lounge to board an airplane. The reverse process occurs when an aircraft lands and each person goes to their individual destinations. Similar functions occur in harbors, bus and rail terminals, warehouses, and distribution centers. (Note: A terminal is not necessarily the terminus of a trip. It is usually the place where the vehicle stops. The traveler usually leaves one mode (or vehicle) for another at the terminal.)

- *Payload.*   For the transportation system to have a purpose, something must be transported—either passengers and/or freight. The nature of the payload is probably the most important factor in the design of the system. Is the good to be transported a perishable item? Does the arrival of the passenger or good need to be within a few minutes of the promised time? Answers to questions such as these provide information for the aspects listed above.

- *Drivers and Pilots.*   The requirements to be placed on a vehicle operator must be part of the design process for a transportation system. Airline pilots must be highly trained to operate complex aircraft. The population of motorists, on the other hand, includes a wide range of capabilities, which the highway designer must take into account.

1.2.2   Personal Transportation

Figure 1.10 shows a hierarchical diagram of modes for the movement of persons. The hierarchy of "people" transportation ranges from nonmotorized modes (walking and bicycling) to high-speed, long-distance highway, rail, and air modes. The shaded box enclosed by dashed lines indicates categories that are usually used for urban transit.

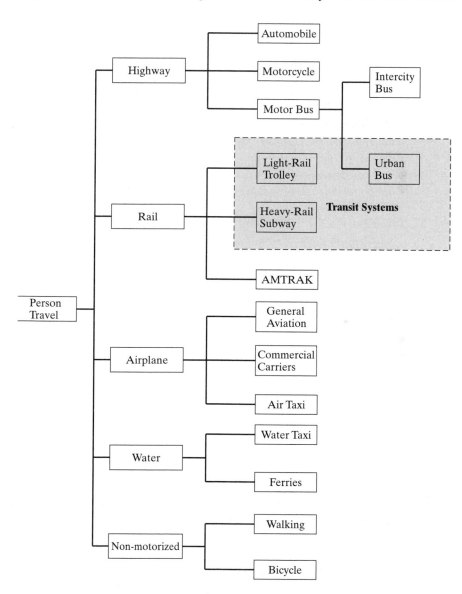

FIGURE 1.10
Hierarchy of personal/people transportation options.

1.2.3   Freight or Goods Movement and Flow

Figure 1.11 shows the hierarchy of freight movement. The freight modes cover all forms of transportation. The pipeline mode is used to transport only liquids (e.g., oil) and various gases under pressure. The shaded portion of the diagram shows an important form of transportation called intermodal. *Intermodal transport* is when two or more modes are involved in the movement of goods from origin to destination. Several of the more common intermodal combinations are:

- *Rail and truck.*   For example, TOFC, which means an over-the-road Trailer (riding) On a railroad Flat Car. (See Figure 1.12.)
- *Water and rail.*   For example, containers that arrive on a ship and are transported by rail from the port.
- *Water and highway.*   For example, ferries that cross a body of water carrying trucks or cars that are driven on and off. The ships are called RORO for "roll-on; roll-off."
- *Water and water.*   A variation of the RORO is the FOFO or "float-on; float-off," where specially designed ocean ships will open so that loaded barges can be floated on and water is pumped out for transit and brought back in to allow the barge to float off at the destination.
- *Air and highway.*   For example, trucks carry airfreight to and from the airport.
- *Drayage.*   The short-distance movement of freight between major modes, such as trucks between port and railhead.

Because each mode has its unique properties and related cost elements, intermodal transportation of freight continues to grow. The container, which can be sealed and can carry a variety of goods, is becoming more and more prevalent as a means of shipping freight.

1.2.4   Leg 2: Putting the System in Place

The various transportation technologies do not operate in a vacuum. Each particular mode has its set of special technical characteristics that result in different design and operational characteristics. As was indicated in Subsection 1.1.2, the system must meet some particular need or demand. Planning and design are critical to its eventual use. Management of a transportation operation—even in the public sector—must satisfy consumer needs or demands.

Planning and Analysis.   Transportation planning aims to predict the needs/demands for a particular service. Several alternatives are analyzed to determine how well each can satisfy the needs/demands.

Investment.   Many transportation options are very expensive to install. The excellent road system in the United States has also required a large investment, primarily financed through gasoline taxes. If there is not likely to be ample return on the investment, the investment will not take place. A recent example is the high-speed rail system. There was insufficient projected demand for ridership to provide adequate return for those who would consider investing capital.

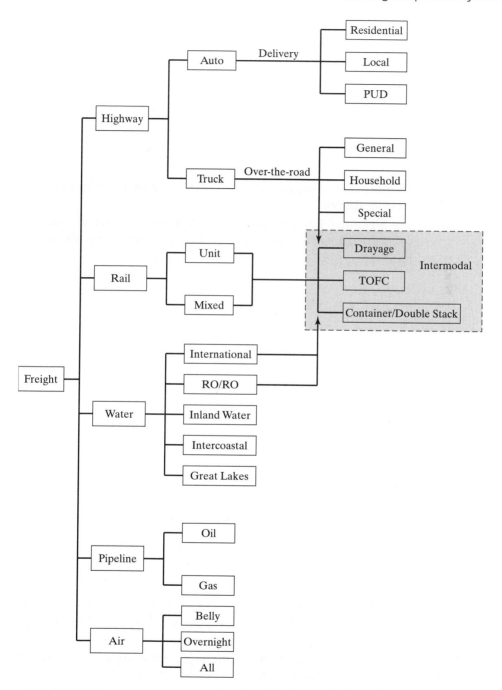

FIGURE 1.11
Means of transporting freight.

FIGURE 1.12
Trailer and container on flat car (TOFC and COFC). Photo: Jon D. Fricker.

Consumer Demand and Pricing. Pricing transport can determine the level of demand and whether the company (or public agency) providing transport services will be profitable (or viable). When the demand is too great, as is the case during the afternoon peak hour on many of the nation's urban freeways, the price of transport is too low. This condition permits *too much* demand, resulting in congestion.

Service Quality. Quality of service is often a major determinant in how freight is moved, as shippers work closely with those carriers that provide reliable, on-time, cost-effective service. Service is also important in providing personal travel. There are persons who will not use a given carrier (such as an airline), because they perceive that its service is poorer than a competitive carrier (airline).

Operations. The manner in which a transportation provider operates may differ according to the operating environment. Is the service expected to cover its expenses and make a profit? In contrast, a bus company may be expected to serve certain neighborhoods, regardless of the financial considerations, so that social objectives can be achieved.

Industry Structure. The industry structure determines the degree to which an enterprise can respond to changes in its operating environment. If customer demands change, competing modes pose a threat, or government changes the rules (through new tax rates or regulations), can the transport provide response effectively? Some analysts say that railroads have lost much business because of their inability to be flexible enough to meet the changing demands of their customer. *Just-in-time (JIT)* delivery of goods to an assembly line requires a carrier that is streamlined enough to meet exacting user requirements. In the JIT case, the carrier is no longer independent of the firm: In fact, the carrier often becomes an integral part of the production line.

Research and Technology.   Industry will not perform research on transportation technologies if it expects that there will not be ample return on its R&D investment. In such cases, the development of technology for transportation must be initiated either through the infusion of government funds or through stringent government regulations, which can only be satisfied by the implementation of new technology. Historically, important advances in safety and environmentally related equipment have been made only when government provides funds for research and/or direction through regulation.

THINK ABOUT IT

What role can transportation engineers play in the various facets of transportation described above?

1.2.5   Leg 3: Issues for Viability

Who has been caught in a traffic jam and has not wondered if there isn't a better way? Who has driven into the reddish-brown haze that we call smog and not wondered why it is there and what can be done about it? What carless poor person has had to depend on unreliable bus service to get to a job, wondering why it isn't any easier? Viability of transportation relates to the ability to meet its overall mobility requirements while meeting the social, health, economic, and access goals of the community.

Mobility/Accessibility.   In simplest terms, mobility is the ability to make trips. Accessibility has the additional qualification that desired destinations can be reached with reasonable effort or cost. Persons dependent on public transit may not be able to reach certain employment opportunities, for example. If vehicle congestion reaches intolerable levels, even automobile owners may find that some destinations cannot be reached with reasonable effort or cost. The way in which a community deals with these issues will say a lot about the community's values.

Equity.   Transportation decisions can have an inequitable effect on poor and underrepresented groups. Subway systems, for example, may provide quick efficient rides to the Central Business District (CBD) from a suburban area but may not serve the poor, whose jobs may still be inaccessible or hard to reach using those same transit routes. Low-income neighborhoods have often been demolished to make way for a new highway or transit line.

Government Decision Making.   Government regulations can affect transportation service and cost, often by economic and physical regulation. Regulatory agencies such as the former Interstate Commerce Commission (ICC) or the Civil Aeronautics Board (CAB) restrained competition by means of both price controls and market entry management. Those "economic" regulations have been largely phased out since 1980. On the other hand, physical product design regulations continue to be issued by the government in the name of maintaining/improving the safety of travel. It is possible that we might not have seat belts, airbags, and car structure reinforcements in every car, if the federal government had not mandated them.

THINK ABOUT IT

How much do car buyers really value safety features? How much more would a customer pay for a better rollbar, side running lights, or a high-mounted rear brake light, if they were options (not government-mandated standard features) on an automobile?

International Trade.   Because it facilitates demand for goods produced in other countries, international trade affects the amount of overseas transportation. For example, the import tariffs on some major goods may lead to reduced market share for goods from another country and hence less demand for transportation. Many airports, ports, cities, and other jurisdictions have created "Free Trade Zones" to promote international trade through that facility or city, thereby enhancing the community's international business. According to the Reuters Financial Glossary [Reuters 2000], a free trade zone is a "designated extra-territorial area within a country in which businesses can operate free of government hindrance, customs duties or currency restrictions. There is usually a tax-free profits element for a set period. Also known as a free trade processing zone and a foreign trade zone."

Transportation Labor.   Transportation labor has played a role in some transportation decisions. The Taft-Hartley Law of 1947 allows the government to step in to avert a strike that it deems a peril to national health or safety. The law has been invoked several times in the transportation sector. A railroad shutdown becomes critical after 10 to 14 days because many electric power plants maintain only a 14-day supply of coal. If manufacturers cannot get raw materials or products to market, unemployment will also occur. The government reserves the right to intervene in such a situation.

The Politics of Oil.   Because transportation is about 95 percent reliant on oil, either a major disruption in oil supply or a significant increase in price would have major impacts on the provision of transportation. The oil embargo of 1973 had a significant effect on short-term transportation demand. However, in the long run, it caused the United States to institute technology that significantly increased the fuel economy of the automobile fleet—from 13.3 miles per gallon in 1973 to an average of 24.3 miles per gallon in 1994 (ORNL 2002, BTS 2003). However, because of the increase in the number of automobiles in the United States and the increase in vehicle miles driven by automobiles, vans, and trucks, the use of gasoline by transportation now surpasses its 1973 pre-embargo level by about 45 percent (BTS, 2003).

Environmental Factors. These factors affect transportation design at the vehicle level (e.g., catalytic converters for pollutant reduction and quiet engine design for aircraft) and at the system level (e.g., the desire to change the ways people use their automobiles).

Driver Behavior. The behavior of drivers influences the set of issues that include safety and vehicle design. The setting of traffic signals, speed limits, and road markings are based on studies of driver behavior.

Safety. A key role of the government is in the area of public safety. Although it is the safest form of travel, the news media treats air transportation as if it were one of the worst. When a plane crash has a large number of fatalities, it is featured in news reports. Table 1.3 shows that, in the 10 years between 1984 and 1993, fatalities occurred more than six times as frequently in automobiles as in air travel. In these calculations, a vehicle occupancy factor of 1.64 persons per automobile is used (NPTS, 1990). Many of the highway design elements that address safety on the highways are presented in Chapter 6.

THINK ABOUT IT

Does 1.64 persons per vehicle seem high or low to you? Do you think average vehicle occupancy varies by time of day, day of week, or trip purpose?

Accessibility for Handicapped Persons. The Americans with Disabilities Act (ADA) mandated *full accessibility* in the transportation industry by standardizing accessible services and establishing requirements for both public and private operators. This 1990 legislation affected highways, transit systems, private transportation providers, airports, and water transportation. A variety of provisions, such as curb cuts and fully accessible highway rest areas, have affected the design of highways and pedestrian sidewalks.

Accessibility to fixed route transit often requires more expensive buses with kneeling capability and wheelchair lifts. The ADA did not mandate retrofitting existing

TABLE 1.3 Relative Safety Between Highway and Air Travel from 1984 to 1993

Mode	Fatalities	Passenger Miles (million)	Deaths per Billion Passenger Miles
Automobile	438,735	34,345,000	13.23
Air carriers	8,424	4,142,000	2.05

Sources: Statistical Abstract 1994 and NPTS 1990.

systems. However, purchases of new buses were to be for "accessible" vehicles. The critical provision of the ADA legislation is that paratransit provide service *comparable* to the fixed route service for persons with disabilities who cannot use the fixed route system. Special vans, such as that shown in Figure 1.13, are dispatched for curb-to-curb service in response to telephone requests from eligible elderly and handicapped persons. Eligibility for ADA transportation services is assessed on the basis of an individual's ability to access the fixed transit routes and use fixed route vehicles. Comparability of service is based on the following criteria:

- service criteria
- service area coverage
- days and hours of service
- fares
- response time to demand

Although they do not need to purchase special wheelchair-accessible vehicles, taxicab operators cannot refuse service, and they must provide assistance in stowing mobility aids, such as wheelchairs. In addition, they may not charge higher fares when serving disabled persons.

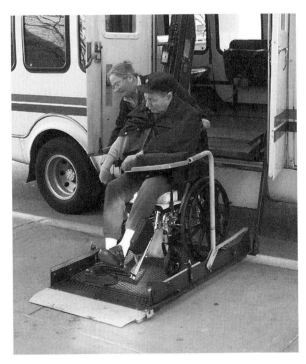

FIGURE 1.13
A paratransit van for wheelchair accommodation. *Source:* Greater Lafayette Public Transportation Corporation, 2002. Used with permission.

1.3   A SYSTEMS VIEW OF TRANSPORTATION

1.3.1   The Systems Challenge

Transportation systems analysis is a way to design or modify a transportation system to meet the needs of the end user(s). It begins with a specification of the system's objectives, and normally includes

- an investigation of the feasibility of the proposed system or its modifications
- an estimate of the costs involved
- an evaluation of the alternative ways to achieve the objectives.

One prominent transportation systems analyst has said:

> The challenge of systems analysis is to intervene, delicately and deliberately, in the complex fabric of a society to use transport effectively, in coordination with other public and private actions to achieve the goals of that society (Manheim, 1979, p. 14).

Note that any intervention must be "careful and deliberate." There will always be persons with an intense interest (or "stake") in any proposed intervention or change. These persons, called stakeholders, are part of the "complex fabric" mentioned above who will need to have a voice in any decision. In this section, the systems analysis process is introduced. The role of stakeholders is covered more fully in Chapter 5.

1.3.2   System Specifications

In specifying a transportation system, there are several important questions that the systems engineer/analyst should ask. The analysis of a system should begin with a clear understanding of its goals, purpose, or mission. The technology required for the system to perform its mission and the system's potential ability to meet the customer demands of accessibility, readiness, reliability, service, and cost must be considered.

Most systems will have a hierarchy of specifications that begin with the system requirements. These requirements are, in turn, used in generating specifications for vehicles, the way, the communications, and other subsystems within the transport system. Finally, these specifications will result in the specification of components that comprise each subsystem. It is the designer's responsibility to ensure that specifications are generated at each level and that, when they are combined, the system will perform according to the overall or "top" system specification. The process of developing an overall specification generally results from answers to the following seven questions. [*To emphasize how the questions relate to a real system, possible answers pertaining to specifying an emergency medical response system for the County of Mythaca are shown in italic font.*]

1. *What is the purpose of this system?*   The first statement in any specification contains a broad indication of the mission or scope of the system.
   [The Emergency Response (ER) system for the county is to provide quick effective medical attention for all 500,000 residents in the county.]

2. *What is the area of coverage for this system's operation?*   The answer to this question generally specifies the range or geographic area of use. It would also contain any special climatic conditions in which the system should work.

   [The ER system is to operate in Mythaca County, with emergency access to hospitals in nearby larger cities. It is to operate fully in 98 percent of all weather conditions.]

3. *What are the technical specifics of the system's intended mission/use?*   These specifics are usually defined in terms of the desired statistics stating the probability of achieving a given mission or by defining the level of service or operational capability to be provided. (Note that these are not design specifications, but they are the first step leading to them.)

   [The system will have a 93 percent probability over the 24-hour day to be on the scene of an emergency situation anywhere in the county in less than 15 minutes from the time the system is requested (usually via a 911 call). It must serve 80 percent of the population within 8 minutes of time of notice.]
   Note: When safety is an important issue, it is often part of this specification.

4. *What is the capacity of the system?*   The demand for the system, especially in the peak periods, is critical. What level of use does it take to make the system ineffective? When is it saturated? What is the performance of the system near saturation?

   [Mythaca's ER system will be able to handle at least six calls simultaneously.]

5. *What is the availability or operational readiness when the system is called into use?*   Sometimes the accessibility for the user to avail himself of the system is included.

   [The Mythaca ER system will have an availability of 96 percent. This includes communication from the caller to the system dispatcher and includes readiness of the vehicle(s) and emergency technicians to be deployed once the call is received.]

6. *What is the reliability of the system in operation?*   A system that is available may fail while it is in use. Reliability is estimated on the probability of functioning properly when in use. The life or wearout of the system may also be specified. If the system can fail while being stored, that reliability must also be specified.

   [The ER system reliability will be greater than 99.5 percent.]

7. *What is the cost-effectiveness of the system?*   Every system needs to be evaluated on the basis of its life cycle cost, which includes the costs of research, development, implementation, investment, operation, and maintenance. Often costs are weighed against the potential benefits of the system. Systems with a "benefit-to-cost ratio" in excess of 1 are usually funded.

   [The Mythaca ER system shall show a benefit-to-cost ratio greater than 1 when the benefit of a life saved is estimated at 1 million dollars and the discount rate is 10 percent.]

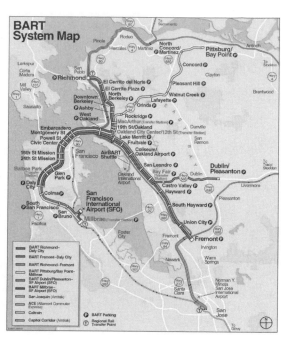

FIGURE 1.14
Bay Area Rapid Transit Subway Train and System Map. *Source*: Bay Area Rapid Transit District.

Example 1.3

The Bay Area Rapid Transit (BART) system is a high-speed rail rapid transit system in the San Francisco-Oakland Bay region (see Figure 1.14). The original BART system was constructed in the period from 1962 to 1975. It has 75 miles of guideway. BART's stated purpose was to relieve congestion by getting residents out of "single-occupant" automobiles, to reduce congestion, and to integrate the Bay Area (Hall, 1982). The BART system was plagued early in its development with cost overruns, safety concerns, lack of financing, and lack of political backing in jurisdictions such as Marin County. This caused the initial system design to be reduced from 123 miles to a length of 75 miles. BART has had many years of safe operation and survived the major earthquake of 1993. It has had a profound impact on the economic development of the area as business areas have grown up around the stations. What is a possible set of system specifications for BART?

Solutions to Example 1.3   A set of sample specifications for BART is shown in Table 1.4.

Example 1.4

The U.S. Coast Guard needs to develop an oil spill response system. How many stations, with what oil spill containment equipment, are necessary to serve the U.S. coastal waters? Where should these stations be located to meet the stringent response time required to reasonably contain a spill? As analysis proceeded on this system, budget limitations had to be traded off for longer response time and the number of coincident spills that could be served by a single station. What specifications might have been presented to the U.S. Coast Guard?

TABLE 1.4   Sample System Specifications for Bay Area Rapid Transit System

1. Purpose/goal	CBD travel, reduced congestion on bridge, spatial integration of the Bay area
2. Coverage	Five counties around San Francisco (123 miles)
3. Technical specs	Avg. Speed = 40 mph, 90-second headway (peak)
4. Capacity	800 passengers per train, 500,000 per week
5. Availability	90% on-time performance, <10-minute walk, wait <5 minutes 98% of time
6. Reliability	99%
7. Cost-effectiveness	Farebox receipts cover 80% of operational cost

Source: Hall, 1982.

TABLE 1.5   Sample System Specifications; Oil Spill Response System

1. Purpose/goal	Effective, quick action response to oil spills
2. Coverage	All navigable waters within 20 miles of U.S. coastline
3. Technical specs	Provide 90% effective cleanup in 2-sigma sea state. (A two-sigma sea state indicates roughness of the sea that is exceeded only 5% of the time.)
Safety	Fatalities <1 in 10,000 hours of operational deployment
4. Capacity	2 major spills that occur within 12 hours of each other
5. Availability	60% of equipment on-site <2 hours
	80% <4 hours
	Full deployment <8 hours
6. Reliability	99%
7. Cost-effectiveness	Benefit/cost >1.2

Source: Discussions at Volpe National Transportation Systems Center, USDOT, 1975.

Solutions to Example 1.4   The original specifications that were discussed with the U.S. Coast Guard are shown in Table 1.5.

1.3.3   Performance Measures as a Prime Tool of Systems Analysis

In performing systems analysis, the analyst must consider the impacts on the user, the operator, and the environment. The use of resources to achieve the system is also of concern. Throughout this text, *performance measures* are used to evaluate a design or assess the impact proposed alternatives

The concept of performance measures is not new; we use some every day. For example, it takes 30 minutes to go from point A to point B, or my automobile gets 22 miles per gallon of gasoline, or the price of gasoline is $1.40 per gallon. The transit operator thinks in terms of operating cost per vehicle mile or farebox revenue per passenger. In design, the engineer needs to consider a number of performance measures to effectively address engineering problems in transportation and to evaluate transportation capability. The performance measures on which transportation investment decisions are based should be chosen carefully. Several of the more pertinent performance measures are shown below. The list below is not exhaustive, but is intended to be illustrative.

- *Average speed and maximum velocity.*   Miles per hour (kilometers per hour)
- *Freight or passenger travel.*   Ton miles or passenger miles per year
- *Operational use/capacity.*   Operations per hour or passenger cars per hour per lane or vehicle miles traveled per year
- *Density of traffic.*   Passenger cars per mile per lane
- *Range.*   Miles
- *Energy use or intensity.*   BTUs per ton mile or passenger mile
- *Acceleration and braking.*   Feet per second per second, or G's
- *Cost of transportation operation.*   Cents per ton mile (or passenger mile)
- *Safety.*   Fatalities or crashes per year or per mile traveled
- *Reliability.*   Failures per unit time or per unit distance traveled; mean time between failures
- *Availability or readiness.*   Percent probability
- *Weather performance.*   Feet or meters of braking distance on wet pavement
- *Chemical emissions.*   Grams of HC per unit of distance or time
- *Noise emissions.*   Decibels
- *Productivity.*   Ton miles delivered per labor hour

Example 1.5

The Mayor of Mythaca has asked you, as City Engineer, to help her prepare for a meeting with the citizens group "Making Mythaca More Livable." In particular, she has asked you to characterize the amount of congestion in Mythaca and trends concerning congestion. What kinds of data would you collect and what analyses do you think would provide useful information for the mayor?

Solution to Example 1.5   Because congestion is worst during the morning and afternoon peak periods and affects the time it takes to commute to work, a survey of commuters may be a good source of information. The survey findings can be combined with traffic measurements made on several sections of the urban freeway, which is 12.3 miles long. (The freeway has six lanes for the 7.7 miles closest to the CBD and four lanes until it becomes a highway.) Although the survey may lack the precision of traffic measurements, it does incorporate the perspective of the commuter. In Table 1.6, census data, survey results, and traffic measurements are summarized.

The second column in the table indicates that the population in the eastern suburbs has grown by 28.3 percent since 1990. At the same time, the number of workers living in the eastern suburbs (column 3) has increased by about 16 percent, mainly young couples with young children. The number of eastern suburb persons with jobs in the CBD (column 4) has increased by about 31 percent. The length of commute trips has increased in distance (column 9) and off-peak

TABLE 1.6 Data for Mythaca Congestion Study—WB Freeway to CBD

(1) Year	(2) Population of East Suburbs	(3) Total Workers Living in E Suburbs	(4) Workers Going to CBD from E Suburbs	(5) Avg. Pk Auto Occup	(6) Pk Hr Commute Time (min)	(7) Avg Off-Pk Trip Time(min)	(8) Pk Hr Vehicle Flow	(9) Avg. Commute Distance (mi)	(10) Lane Miles
1990	32,500	14,000	8,000	1.35	56	16	3500	14.0	42
2000	41,700	16,225	11,500	1.15	76	20	4500	17.5	64.6

trip time (column 7). This reflects the outward growth of the urban area. The growth in traffic (column 8) is the likely explanation for the increase in peak travel time (column 6). Column 5 indicates a significant drop in average vehicle occupancy.

In addition to the observations made directly from the data in Table 1.6, other trends may be found by examining relationships in the data, as demonstrated below.

1. The percentage of workers living in the eastern suburbs with jobs in the CBD has actually increased since 1990 (columns 3 and 4).

$$\frac{8,000}{14,000} = 57 \text{ percent in 1990}; \quad \frac{11,500}{16,225} = 64.7 \text{ percent in 2000}$$

2. Although population in the eastern suburbs has grown by 28.3 percent (column 2) and peak hour traffic flow has increased by 28.6 percent (column 8), the supply of roads in the eastern corridor has increased at about the same pace. The freeway expansion from four lanes to six lanes was completed in 1995, and old county roads and local streets in the eastern suburbs were upgraded so that the number of lane miles in the corridor has increased by about 29 percent.

3. Because of reduced vehicle occupancy, westbound peak vehicle flow (28.6 percent in column 8) has increased faster than westbound peak commuter flow (9.5 percent):

$$\text{Commuters}_{1990} = \text{Vehicle Flow}_{1990} * \text{Occupants per Vehicle}_{1990}$$
$$= 3500 * 1.35 = 4725 \text{ peak commuters}$$

$$\text{Commuters}_{2000} = \text{Vehicle Flow}_{2000} * \text{Occupants per Vehicle}_{2000}$$
$$= 4500 * 1.15 = 5175 \text{ peak commuters}$$

4. Average peak hour commute time has gone from 56 to 76 minutes, but this is over a longer commute distance. How has the average peak period travel speed changed?

$$\text{PeakSpeed}_{1990} = \frac{14.0 \text{ mi}}{(56/60)\text{hr}} = 15.0 \text{ mph};$$

$$\text{PeakSpeed}_{2000} = \frac{17.5}{(76/60)} = 13.8 \text{ mph}$$

This is a decrease of 8 percent.

5. Comparing step 3 to column 4, it appears that the proportion of east suburban commuters to the CBD that use the freeway during the peak hour has dropped considerably.

$$PkHrFwyUse_{1990} = \frac{4725}{8000} = 59.1 \text{ percent};$$

$$PkHrFwyUse_{2000} = \frac{5175}{11500} = 45.0 \text{ percent}$$

They either use the freeway before or after the peak hour or use non-freeway routes.

Certainly, other information could be derived from the Table 1.5 and from the survey. What the brief analysis above shows is that growth of the suburbs and a decline in ridesharing are major contributors to the congestion in the eastern corridor. The mayor and the MMML group can discuss actions to respond to these trends.

SUMMARY

Rarely is a trip taken for its own sake. A trip is made because of the activity to be undertaken at the destination. Goods are shipped because they will be used, consumed, stored, or sold at the receiving end. Despite the apparent secondary role assigned to transportation by these statements, transportation actually plays an enormous role in a region's economic vitality and a community's quality of life. If a community or region enjoys good accessibility, persons have a wider range of opportunities for employment, shopping, and recreation. If a community or region is well connected to other regions, a wider variety of products will be available to its residents at reasonable prices.

Efficient transportation systems are based on appropriate physical structures, adequate funding, and sound operating policies. Among the many possible ways to achieve the desired objectives of a transportation system, a methodology is needed to provide a framework for determining the best designs. This chapter reviewed transportation's historical role in shaping society and then introduced the concepts needed to understand and implement the transportation systems analysis approach in this modern era. In Chapter 2, we begin to introduce the fundamentals of transportation engineering by looking at a topic familiar to everyone—traffic on the highway.

ABBREVIATIONS AND NOTATION

ADA   Americans with Disabilities Act
AKDOT & PF   Alaska Department of Transportation and Public Facilities
ASCE   American Society of Civil Engineers
BTU   British thermal units
CAB   Civil Aeronautics Board
CBD   Central Business District
HC   Hydrocarbons
ICC   Interstate Commerce Commission
JIT   Just-in-time
STPP   Surface Transportation Policy Project
VoT   Value of Time

GLOSSARY

Americans with Disabilities Act: Federal legislation passed in 1990 to make public accommodations, including transportation facilities and services, accessible to individuals with handicaps.

Availability: The percent time the system can be accessed and used.

Capacity: The amount of goods and/or persons a system can handle before reaching saturation.

Coverage: Range or geographic area of system operation.

Demand: The requirement for goods or persons to be moved.

Destination: The place where a trip ends.

Intermodal transfer point: Place where persons or goods shift from one mode to another or one vehicle to another.

Line haul: The movement of persons or goods from one place to another.

Mission: The scope or purpose for a system; includes broad goals.

Mode: The form of transport—highway, air, carpool.

Multimodal: Transport using at least two modes.

Operation: Defines the resources and the manner in which a system functions.

Origin: The place where a trip begins.

Reliability: The percent time the system performs according to its specification.

Safety: The number of fatalities or injuries per unit of operation.

Terminal: The part of a trip where a line-haul segment stops. Because the trip may continue from a terminal, the term *connection* may be more precise.

REFERENCES

[1] Bullard, Robert D. and Glenn S. Johnson, *Just Transportation: Dismantling Race and Class Barriers to Mobility*, New Society Publishers, Stony Creek, CT, 1997.

[2] Burchell, Robert W., *The Costs of Sprawl Revisited Transportation Cooperative Research Project*, Report #30, TRB 1998.

[3] Bureau of Transportation Statistics (BTS) website, U.S. Department of Transportation, www.bts.gov, as downloaded 15 March 2003.

[4] Fatout, P., *Indiana Canals*, Purdue University Press, ISBN 0911198784, 1985.

[5] Federal Highway Administration, *Our Nation's Highways 2000*, U.S.DOT 2001.

[6] Hall, Peter, *Great Planning Disasters*, University of California Press, March 1982.

[7] Mannheim, Marvin L., *Fundamentals of Transportation Systems Analysis*, MIT Press, Cambridge MA, 1979.

[8] McCann, B, *Driven to Spend* Surface Transportation Policy Project and the Center for Neighborhood Technology. 2000.

[9] *Nationwide Personal Transportation Survey*, Federal Highway Administration, 1990, www.cta.ornl.gov/npts/1990/index.html

[10] Oak Ridge National Laboratory, *Transportation Energy Data Book*, Edition 22, September 2002, Chapter 2.

[11] Reuters, *Reuters Financial Glossary*, www.financialminds.com/glossary/definition.asp?term=1598, 2000, as downloaded 15 March 2003.

[12] Surface Transportation Policy Project, *Ten Years of Progress*, STPP Report, 2002.

[13] Surface Transportation Policy Project, *Easing the Burden: A Companion Analysis of The Texas Transportation Institute's 2001 Urban Mobility Study*, STPP Report, 2002b.

[14] *Surface Transportation Policy Project, Why Are The Roads So Congested?*, STPP Report, 2001.

[15] U.S. Department of Commerce, Statistical *Abstract of the United States*, 2001.

[16] Vuchic, Vukan R., *Transportation for Livable Cities*, Center for Urban Policy Research, Rutgers University, NJ, 1999.

[17] Wilson, Anthony, *The MacMillan Visual Timeline of Transportation*, Dorling Kindersley Limited, London, 1995.

EXERCISES FOR CHAPTER 1: TRANSPORTATION BASICS

The Call for a New Transportation Engineer

1.1 Transportation Studies.   Consider the two situations described below. For each one, design a data collection procedure that will provide information to answer the questions stated in the descriptions. Your "design" should include (as appropriate) the types of primary and secondary data that will be sought, how these data will be collected, how many people will be needed, when and where they will be positioned, their specific tasks, what data collection forms will aid them, and so on.

A. Mythaca has four elementary schools—Minton, Nancell, Osprey, and Palatine. Because elementary school enrollments are declining, the school board is considering closing the Nancell School. Parents of the children attending Nancell argue that closing their school would make the distance between their homes and the nearest remaining schools so large, that their children could no longer walk to school. School busing would have to be instituted for the first time in this district. Protectors of other schools say that most Nancell students are driven to school each morning, invalidating walking distance as a reason to keep the Nancell School open. The school board asks you to gather data and draw conclusions regarding this issue of access mode to Nancell School.

B. A research team has been interviewing individuals on the Mythaca State University campus who are known to be members of carpools. Among other things, the researchers have compiled data on carpool size (number of members in each), commuting trip origins (home locations), and driving distances. As the research team prepares a preliminary report to their research sponsor, they are concerned that their campus sample may not be representative of the "population" as a whole. Specifically:

- Is the auto occupancy rate at the university and in Mythaca close to the national average of vehicle occupancy? What is the national average and where did you find the value?
- Is the distribution of carpool sizes (1, 2, 3, ... occupants) in the sample representative of all ridesharing to the university and all ridesharing in the Mythaca area?
- Are the home locations of the subjects in the sample in the same proportion as all commuters who carpool to the college or other Mythaca area work sites?

Hint: As you think through this problem, use your own campus and community to facilitate your thought process.

1.2 Transportation in Magazines.   Visit your school library and find three magazines, each having to do with a different mode of passenger transportation. Select one article (two or more pages in length) from each magazine and write a summary (no more than one page double-spaced) of the article. Indicate the title, author, theme of the article, and the arguments used by the author in presenting the issues. Attach a copy of the article.

1.3 Newspaper Articles. Look in the local newspaper from the most recent week. Cut out (or photocopy) and attach three articles about different topics in transportation. What topic related to transportation did they address? Examples of topics are Safety, Highway Construction, Transportation Planning or Financing, Taxes, Environmental Problems, Technology, and Economic Impacts. Do your clippings represent transportation events that happen frequently? Which article covered a topic that was the most unusual?

1.4 Determining the Value of Time. How valuable is your time, especially as it relates to travel? During the next week, be aware of experiments that you could use to estimate how valuable your time is. For example, imagine a case in which you could spend 75 cents to take a 5-minute bus ride, instead of walking 15 minutes to the same destination. How much would your time be worth if you took the bus? What would be the flaws in such an "experiment"?

1.5 Modal Contribution. Figure 1.7 shows the economic contribution of each mode of transportation. Explain why the "dollars involved" in the highway mode is so much greater than the other modes. You are allowed to consult the source of the figure, the Bureau of Transportation Statistics at www.bts.gov, for the information you need.

Understanding Transportation Systems

1.6 Per Capita Freight Movement. For a recent year, find the total ton-miles of freight shipped in the United States. Approximately how many ton-miles per U.S. resident was that? *Hint:* A good source is the Commodity Flow Survey found at the Bureau of Transportation Statistics website www.bts.gov.

1.7 Containers in Freight Transportation. The use of containers (see Figure 1.12) to transport freight in the United States and worldwide has increased enormously over recent decades. Use resources such as the Internet to determine the extent of this trend. Express your findings in terms of raw numbers and in terms more easily understandable, such as percentages.

1.8 Freight Transportation and Warehousing. Figure 1.11 shows the structure of freight transportation. Because the movement of goods is involved, do you think warehousing should be included in this description? In your answer, include a discussion of the levels of inventory that a business may have to maintain and the implications of just-in-time goods delivery.

1.9 The "Transportation Disadvantaged." Think of an adult you know who does not drive a car. Why is that person a non-driver? How does that person make the trips he/she needs to? Is there anything the transportation engineer can do to increase that person's mobility?

A Systems View of Transportation

1.10 Safety Performance Measures. Using your judgment, propose one or more measures that could be used to compare the relative safety of various transportation modes, such as automobile, public transit, airplane, and intercity rail. What about your measure makes it a fair one?

1.11 Specifications for Trash Collection. The City Council of Shoridan has become concerned about Saturday morning automobile traffic at the City Dump, because each household must dispose of its own trash. There also has been a rash of illegal dumping of trash by the residents in vacant lots and along roads. The city has purchased property in the hills at the edge of the city. That property is suitable for a landfill that will last 30 to 60 years. The council members now want to hire a company to provide curbside trash pickup service and transport it to the landfill. The city will pay for the landfill operations and the road needed to reach the landfill by charging dumpers a "yet to be determined" fee. The private company will also need to provide some recycling services. The council, therefore, has asked you, a consultant, to help them prepare the specification for the service that the prospective bidders will respond to. Using the general criteria for system specifications and your knowledge of the kind of services needed, prepare a set of specifications for the council to use in framing their Request for Proposal.

1.12 Shoridan Beach Front and Visitor Health and Safety. Often large crowds arrive on the weekends in the summer and especially on holidays. The areas around the harbor and the new mall along the water front are very crowded. Last summer there were 23 water-related incidents resulting in 5 drownings and 65 injuries brought into the emergency room, many from unruly beach activity. This finding and the crowds that need to be controlled have caused the city council to want to rewrite the emergency/enforcement code for the city and to apply for a state grant to aid in trying to reduce the number of potentially harmful incidents. Using your knowledge of such a scenario and the data for Shoridan given below, develop comments and criteria for each of the system specifications given in Section 1.3 that might be used in framing a new emergency code. Do not write the code. Simply develop the criteria and consider what performance measures are necessary.

DATA ABOUT THE TOWN OF SHORIDAN

The town of Shoridan is located on a large bay and has a series of fairly rugged hills to the north. Shoridan has existed as a sleepy little resort/retirement community for many years. Its population is about 35,000. The town consists of seven zones, with characteristics indicated below. See also Figure 1.15, which indicates the seven zones on a map.

Zone 1. The Central Business District (CBD), to which a modern office and financial center has been recently added. It is scheduled for further revitalization. Because of current traffic patterns, it takes a long time (15 minutes) to drive through the CBD at rush hour.

Zone 2. Primarily an industrial/commercial complex containing, among other things, the community hospital, which is due for expansion, and the area's Junior-Senior High School complex.

Zone 3. An older residential area located between a marshy area and a relatively poor beach area. This area is not likely to grow much.

Zone 4. An area likely to see considerable expansion. It borders on the best portion of the small boat harbor in the lagoon and contains a popular, sandy beach. It also contains a small popular amusement park operated by the city. There is a mall along the beach drive across the road from the beach.

Zone 5. A largely undeveloped area where there are some expensive hillside homes nestled in the hills. Many of these homes have a panoramic view of the lake. It will be the choice of many who would come to settle in Shoridan.

Zone 6. A small residential and light industrial area close to the small airport.

Zone 7. The main highway into Shoridan runs through this zone. It is the likely site of a future highway interchange. The highway already has a number of commercial establishments, including fast food restaurants and other light shopping areas. A shopping mall is planned for the Shoridan side of the future interchange.

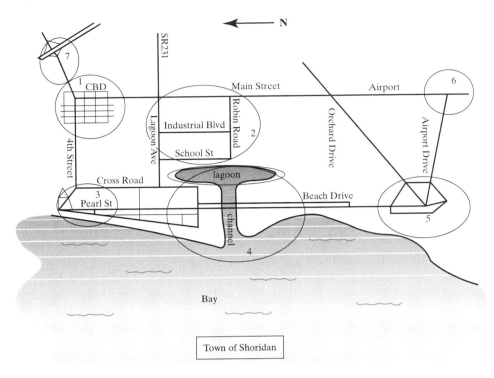

FIGURE 1.15
Map of Shoridan.

Traffic Flow: Theory and Analysis

SCENARIO

Although Interstate Highway 25 is about 15 miles from Mythaca, it serves as an important link to other parts of the state and the country. In recent years, its importance to Mythaca and other cities in the region has resulted in a rapid increase in traffic using I-25. Traffic jams on I-25 are no longer unusual (Figure 2.1). The State Department of Transportation (DOT) has installed equipment to measure the traffic on I-25, so that it can analyze the data and assess the performance of the highway. Because traffic data can often be misunderstood, the DOT must take care to collect and analyze the traffic data according to standard practice. Only then can the need to upgrade I-25 be compared with the needs on other roadways in the state.

CHAPTER OBJECTIVES

By the end of this chapter, the student will be able to:

1. Design and conduct activities to measure traffic stream characteristics.
2. Analyze field data and estimate values for traffic stream parameters.
3. Build a standard traffic flow model to fit observed traffic data.
4. Apply a random model in the analysis of traffic flow when appropriate.
5. Evaluate the performance of a roadway by using standard procedures.

FIGURE 2.1
Interstate traffic jam. Photo: Jon D. Fricker.

INTRODUCTION

Because the automobile is the dominant mode of travel in the United States, everyone is familiar with highway traffic. In this chapter, we introduce the concept of traffic flow and show how it is used in the planning and design of highways. Because the highway system provides a service to its users, it is important to understand how that service is perceived by the consumer (driver). To do so, some fundamental relationships are introduced and explored.

2.1   MEASURING TRAFFIC FLOW AND SPACING

2.1.1   Headway Calculations

A simple way to measure a familiar aspect of traffic flow is to determine the separation between two consecutive vehicles. This separation can be measured in terms of distance or time. *Headway* is the separation between a given point on adjacent vehicles (e.g., from front bumper to front bumper as illustrated in Figure 2.2).

Example 2.1   Headway Data and Computation

In Table 2.1, the results of a 60-second experiment are summarized. At midblock on Green Avenue, an observer began noting at what time after an arbitrary start time vehicles cross a mark on the pavement in the lane nearest the near curb. (We can envision the measurements being made by a person holding a stopwatch, but a more modern method would be by installing a sensor in the pavement that would send signals to a remote data collection device, or by using a video camera to create a videotape with a "time stamp.")

The first vehicle crossed the mark at 6.52 seconds after the arbitrary start time. The second vehicle crossed the mark at 11.26 seconds after the arbitrary start time. (The observer in this example used the front bumper of each vehicle crossing the mark on the pavement as the definition of the event being observed. Alternative definitions could be "rear bumper," "front tires," or "rearmost tires." The choice should be based on the most convenient way to collect the data that meets the needs of the study. The method must be consistent from vehicle to vehicle.)

The time between the two vehicles' arrivals is $11.26 - 6.52 = 4.74$ seconds. This is the time headway for those two vehicles. After 60 seconds, the observer ceases making observations, having seen eight vehicles. What is the average headway for the eight vehicles observed during the 60-second period?

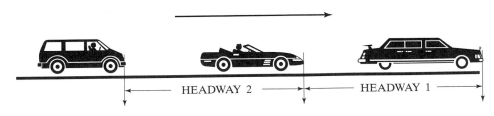

FIGURE 2.2
Vehicle headways.

TABLE 2.1   Data for Headway Calculations

(1) Vehicle Number	(2) Crossed line at (in seconds)	(3) Time headway (in seconds)
1	6.52	
2	11.26	4.74
3	14.59	3.33
4	19.33	4.74
5	28.30	8.97
6	39.93	11.63
7	43.76	3.83
8	58.16	14.40

Solution for Example 2.1   The average headway of the observations in Table 2.1 is the mean value of the seven headway values shown in column 3:

$$58.16 \text{ sec} - 6.52 \text{ sec} = 51.64 \text{ sec}$$

$$51.64 \text{ seconds}/7 \text{ headways} = 7.38 \text{ seconds between vehicles}$$

2.1.2   Definition of Traffic Flow and Volumes

When traffic flow is high enough, it is very difficult to make a left turn from EB (eastbound) Stadium Avenue onto NB Grant Street. Should a traffic signal be installed there? At another intersection a few blocks away, a left-turn arrow does not allow more than seven vehicles per cycle to complete left turns. How often will this left-turn capacity be inadequate? Transportation engineers are often teased about being "car counters," but choosing the proper remedies for the two situations just described depends on having adequate data about traffic flow.

Traffic *counts* can be as simple as recording the number of vehicles that pass a point during a specified time period. Example: 826 vehicles on southbound I-99 in 30 minutes. We could refine this volume count by allocating the counted vehicles into classes (e.g., 663 automobiles, 6 motorcycles, 82 light trucks, and 75 heavy trucks). The process of breaking down the total number of vehicles observed into categories or classes is called *vehicle classification*.

THINK ABOUT IT

Why might vehicle classification be helpful to a traffic engineer?

Traffic volumes are counts made for some specific time period. A way to put various volume counts taken over different lengths of time on a common basis is to compute *traffic flow rates*, expressed in units of vehicles per hour (vph). Thus, a volume count of 826 vehicles in 30 minutes is equivalent to a flow rate of 1652 vehicles per hour.

The flow in vehicles per hour can also be found from headway data as follows:

$$q = \frac{1}{\text{average headway}} = \frac{1}{\bar{h}} \qquad (2.1)$$

where $\bar{h}$ equals the average headway. From the headway data in Table 2.1, the flow q can be calculated as

$$q = \frac{3600 \text{ sec/hr}}{7.38 \text{ sec/vehicle}} = 488 \text{ vph}$$

Example 2.2   Traffic Counts

Immediately after his workday ends, the County Engineer starts driving to his parents' home about 200 miles away for a weekend visit. The truck traffic on I-99 seems very heavy for a Friday evening. A little after 9 PM, the engineer stops for dinner. From his table in the fast-food restaurant, he can see the traffic on I-99 clearly. Because the number of trucks still seems surprisingly high, he decides to count the total number of vehicles and the number of trucks that pass his location in the NB direction during the next 10 minutes. The result of his count is 168 total vehicles, including 56 trucks.

A. What is the NB flow rate during the engineer's 10-minute count?

B. What percent of the NB traffic flow is truck traffic?

Solutions to Example 2.2

A. The engineer's 10-minute count was 168 vehicles. The corresponding flow rate is

$$q = \frac{168 \text{ vehicles}}{10 \text{ min}} \times \frac{60 \text{ min}}{\text{hr}} = 1008 \text{ vph}$$

B. The percentage trucks can be computed from flow rates (vph), but it is easier to use the 10-minute counts on which the flow rates would be based:

$$\% \text{ trucks} = \frac{56 \text{ trucks}}{168 \text{ total vehicles}} \times 100 = 33.3 \text{ percent}$$

THINK ABOUT IT

Does 33.3 percent trucks seem high to you? Based on your intuition or experience, how does this value vary by highway type? How does percent trucks vary by time of day or time of night?

2.1.3   Accounting for Uneven Flows

For many analyses, a simple 10-minute, 30-minute, or 1-hour count may not be what is needed. Rather, the measure of the fluctuation in traffic flow during the period of analysis may be required. As we shall see in Chapter 3, some analyses of roadway capacity are based on the largest of the four 15-minute flow rates during the hour being studied. It is possible to have four equal 15-minute counts during an hour, but it is much more likely that those four counts will vary within the hour. The peak hour factor (PHF) for a roadway (or lane of roadway) is calculated by using the equation

$$PHF = \frac{V}{4\,V_{15}}$$

(2.2)

where V = hourly volume and V_{15} = peak 15-minute count.

Example 2.3   Determining a Peak Hour Factor

Figure 2.3 shows a set of vehicle counts made in 5-minute intervals on an urban freeway. What is the peak hour factor (PHF) for the peak hour shown in Figure 2.3?

Solution to Example 2.3   Figure 2.3 shows traffic flow data for the middle SB lane on I-35W. The peak hour for that lane begins at 7:20 AM. That lane's PHF is calculated by applying Equation 2.2 to the data in Table 2.2. The volume for the hour beginning at 7:20 AM is also the flow rate for that hour, or V = 1623. Of the four 15-minute periods that occur during the hour that began at 7:20 AM, the peak 15-minute period began at 7:35 AM. The count during that peak 15 minutes was 495 vehicles. The flow rate for the hour beginning 7:20 AM is 1623 vph. The maximum 15-minute flow rate in Table 2.2 is 4 * 495 = 1980 vph. Equation 2.2 becomes

$$PHF = \frac{1623}{4 \times 495} = 0.82$$

If only the flow rates (column 3 of Table 2.2) had been provided, it would be a waste of time to recreate the counts from which the flows were computed, just to match the requirements of Equation 2.2. Instead, using the flow rates directly in $PHF = \dfrac{1623}{4 \times (1980/4)} = 0.82$ produces the same result.

THINK ABOUT IT

As traffic flow within an hour becomes more variable (i.e., fluctuates more), does the PHF increase or decrease? Is it ever possible that PHF > 1.00?

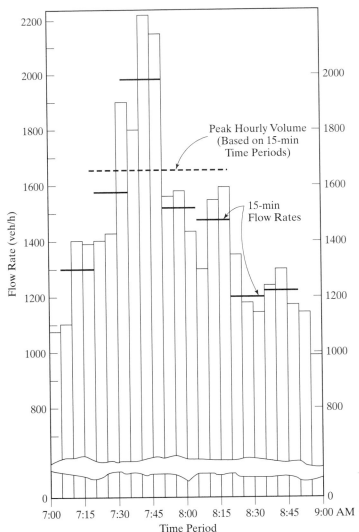

This figure summarizes measurements made August 1983 in the middle lane of three southbound lanes on Interstate 35W in Minneapolis, Minnesota, 4 miles south of the Central Business District. Each 5-minute count has been converted into an equivalent hourly flow rate, with units vehicles per hour (vph). The 12 consecutive 5-minute periods that produce the maximum 1-hour vehicle count (1623 vehicles) begin at 7:20 AM. Within that hour, the 15-minute volumes and flow rates are as shown in Table 2.2. Remember that the measurements are for only one lane.

FIGURE 2.3
Relationship between long-term and short-term flows. *Source*: HCM, 2000, Exhibit 8–10, p. 8–10 (from Minnesota Department of Transportation).

TABLE 2.2   Traffic Volumes and Flow Rates on I-35W

15 Minutes Beginning	Vehicle Count	Flow Rate (vph)
7:20 AM	389	1556
7:35 AM	495	1980
7:50 AM	376	1504
8:05 AM	363	1452
7:20–8:20 AM	1623	1623

2.1.4   Traffic Flow Data Collection Techniques and Technologies

Counting vehicles may seem like a simple concept, and it is—in concept. Traffic counts are also important for several applications. Among the uses are:

A. Documenting locations of congestion

B. Establishing trends in traffic growth as a basis for future investment to increase the capacity of a roadway

C. Determining whether a stop sign or traffic signal is needed at an intersection

D. Supporting the design and evaluation of new or improved traffic signal-timing plans

E. Meeting the data reporting requirements of the federal government or other agencies

For some applications, the data needed may be simple traffic counts taken for a short period of time, such as during peak periods or during special events. In some cases, such as "D" above, even off-peak periods may be of interest, to help make the signal-timing plan as efficient as possible for all times of day. Case "D" also requires that the number of vehicles making each turning movement (left, through, or right) from each approach direction be recorded.

There are many different ways to collect traffic count data, depending on the intended use of the data, the nature of the roadway being studied, and the resources available to the counting agency. For example, a familiar way to do traffic counts is to use a pneumatic road tube stretched across one or more lanes of roadway. (See Figure 2.4.)

FIGURE 2.4
Pneumatic road tubes to count traffic. Photo: Jon D. Fricker.

A tire passing over the rubber tube affixed to the road surface creates a burst of air pressure that activates a ceramic diaphragm in a metal box to record the event. Tube counters can store the number of times a set of tires has activated the tube since it was set out. However, tube counters have the following shortcomings:

1. They count only axles, not vehicles. A vehicle with four axles will register as if two 2-axle vehicles have passed by. An "axle correction factor," equivalent to the average number of axles per vehicle, must be used to convert axle counts into vehicle counts. Typical values for an axle correction factor range from 2.04 for local streets with few heavy trucks to 2.86 on interstate highways.

2. The more advanced tube counters can store counts by 1-hour or 15-minute intervals but cannot store the arrival times of individual vehicle (or axle) arrivals.

3. They often malfunction, are damaged, or are vandalized.

4. Many local governments cannot afford to own them. In such cases, local governments may be able to rent them from private companies or borrow them from other public agencies or from their state's Local Technical Assistance Program (LTAP). Each state has an LTAP, whose mission is to provide technical assistance to local public agencies (cities, towns, and counties) on matters related to transportation.

THINK ABOUT IT

How would you determine the value of the axle correction factor to use on a given roadway? What is the minimum possible value for an axle correction factor?

Several alternatives to road tubes exist. At the low-technology end, persons using clipboards (see Figure 2.5) or electronic countboards (Figure 2.6) can record traffic volumes.

Higher on the technology spectrum than road tubes are the following:

- *Magnetic sensors in aluminum cases (Figure 2.7).*   These portable sensors can be affixed to the pavement or buried in a gravel road to collect traffic data.

- *Loop detectors (Figures 2.8 through 2.10).*   Sreedevi and Black (2001) offer a good description of loop detectors. "Loop detectors operate on the principle of inductance, the property of a wire or circuit element to 'induce' currents in isolated but adjacent conductive media. A detector consists of an insulated electrical wire, placed on or below the road surface, attached to a signal amplifier, a power source, and other electronics. Driving an alternating current (normal operating frequency between 10 and 200 kHz) through the wire generates an electromagnetic field around the loop. Any conductor, such as the engine of a car, which passes through the field will absorb electromagnetic energy and simultaneously decrease the inductance and resonant frequency of the loop. For most

TABULAR SUMMARY OF VEHICLE COUNTS

Observer _____ Date _____ Day _____ City _____   R = Right turn
S = Straight
INTERSECTION OF _____ AND _____   L = Left turn

TIME BEGINS	from NORTH				from SOUTH				TOTAL North South	from EAST				from WEST				TOTAL East West	TOTAL ALL
	R	S	L	Total	R	S	L	Total		R	S	L	Total	R	S	L	Total		

FIGURE 2.5

Top of field summary form for an intersection turning movement count. *Source*: Box and Oppenlander, 1976, p. 29.

FIGURE 2.6

Electronic count board. The models shown are for turning movement counts. *Source*: Jamar Technologies Inc., used with permission.

FIGURE 2.7
Nu-Metrics Hi-Star Magnetic Counter. Photo from Nu-Metrics
used with permission.

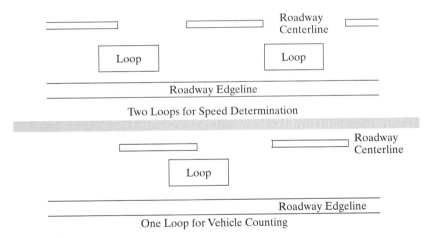

FIGURE 2.8
Loop detector configurations.

conventional installations, when the inductance or frequency changes a preset
threshold in the actuate detector electronics, this indicates that a vehicle has
been detected. Many factors determine loop inductance, including wire size,
wire length, the number of turns, lead length, and insulation."

- *Video methods (Figure 2.11).* By detecting the presence of vehicles at prescribed locations in a video image, in-pavement loops can be mimicked.

- *Infrared, microwave, and ultrasonic methods.* Klein (1997) offers a good introduction to technologies being tested for use in traffic data collection, including
radar and acoustic methods.

FIGURE 2.9
Pavement marked for sawing machine to make cuts for loop detector wire. Photo: Darcy Bullock.

FIGURE 2.10
Loop detector to actuate traffic signal, after wire has been installed and sawcut has been sealed. Photo: Jon D. Fricker.

The more sophisticated technologies can provide greater reliability, accuracy, and detail in the data they collect, although each has its drawbacks and the performance of each must be frequently checked.

THINK ABOUT IT

What factors would you consider when evaluating any technology for use in counting traffic?

2.1.5   Turning Counts into Design Data

Among the many possible examples of collecting and using traffic counts, let us concentrate on the ongoing effort by each State Department of Transportation (DOT) to monitor the amount of traffic that uses each segment of its highway system. The goal of a *traffic monitoring system for highways (TMSH)* is an accurate *annual average daily traffic (AADT)* value for each state highway segment, expressed in vehicles per day that travel either direction on the road segment. For most road segments (those that do not have a permanent traffic counter present), AADT is the result of two principal activities:

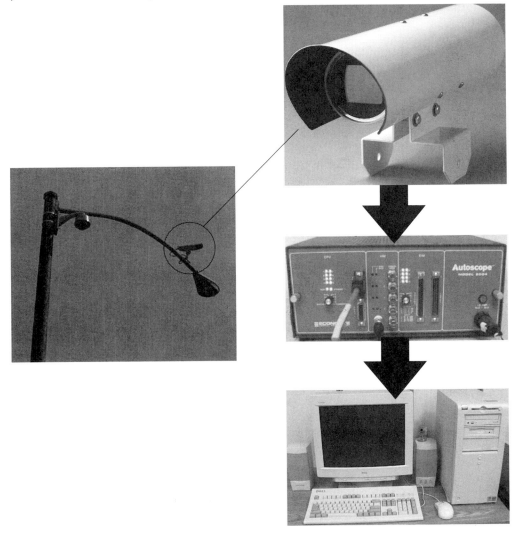

FIGURE 2.11
Autoscope MVP image sensor. *Source:* Photo courtesy of Econolite Control Products, Inc.

1. Periodic 48-hour "coverage" counts taken on the roadway of interest. Ideally, these counts should be taken every year or two, but limited state DOT person-nel, equipment, and budgets, coupled with a DOT's "special count" responsibili-ties, can make the time between coverage counts more like 3 to 6 years. Coverage counts are usually done during a 48-hour period between Monday noon and Friday noon, thereby avoiding the abnormalities of weekend traffic (FHWA, 2001).

2. Seasonal adjustments. Because of limited staff and counting equipment, coverage counts are done on different road segments within a jurisdiction at different times of the year. This means that the coverage counts must be adjusted to account for seasonal fluctuations in traffic levels. The degree of seasonal traffic fluctuation varies by classification of highway (see Table 2.3) and proximity to vacation areas or special event locations. The seasonal adjustment factors are derived from counts made at permanent count stations (PCSs) that are located around the state on a representative sample of roads in each functional class.

TABLE 2.3 Functional Classes of Roadways

Class 1 = Rural Interstate	Class 11 = Urban Interstate
Class 2 = Rural Principal Arterial	Class 12 = Urban Other Freeways and Expressways
Class 6 = Rural Minor Arterial	Class 14 = Urban Principal Arterial
Class 7 = Rural Major Collector	Class 16 = Urban Minor Arterial
Class 8 = Rural Minor Collector	Class 17 = Urban Collector

Source: Highway Performance Management System Field Manual, Chapter IV, FHWA, 30 August 1993.

Design Hourly Volume.   The number of lanes needed on a given roadway is based on the existing or anticipated demand for travel on a particular road segment. Traffic counts are the starting point for the design of roadways. The standard procedure uses the *30th highest hourly volume* in a year as the *design hourly volume (DHV)*. If the road segment happens to have its own permanent count station, all that needs to be done is to rank the most recent year's worth of hourly traffic counts from highest to lowest and then choose the volume ranked number 30. For most roadways, however, a full year of hourly counts are not available. In Indiana, for example, there are approximately 90 permanent count stations, but about 11,000 miles of roadway on the state highway system. If the typical highway segment for traffic count purposes is 2 miles long, only about 1 in every 60 segments has its own PCS. For those segments without a PCS, the design hourly volume could be estimated either by:

(a) Carrying out a "special count" for 1 year on the road segment in question and then ranking the 1-hour volumes so recorded or

(b) Using relationships established by observing other roadway segments with similar characteristics.

THINK ABOUT IT

What are the (dis)advantages of data collection methods (a) and (b) above to determine DHV? Which method do you prefer? Under what conditions would you favor either (a) or (b)?

Method (b) is implemented by using a plot like that in Figure 2.12. In the figure, the highest hourly volume for the year, expressed as a percentage of the average daily traffic (ADT) value, is plotted at the left edge of the graph. For example, if a particular rural road's highest hourly volume for the year was 1044 vehicles per hour (vph), and its ADT was 4777 vehicles per day (vpd), its highest "Traffic as a Percentage of ADT" value would be $(1044/4350)*100 = 21.85$ percent. This calculation could be repeated for an adequate number of rural roads of the same functional class having at least 1 year's worth of hourly count data. From these data, one could produce a median value, a value exceeded by 15 percent of the roads in the sample (called the *85th percentile value*) and a value exceeded by 85 percent of the roads in the sample (called the *15th percentile value*). In Figure 2.12, the median value for the highest hour is 24 percent of the ADT. The highest hour exceeds 30 percent of the ADT for only 15 percent of the roads in the sample; this is the left end of the upper curve. The left end of the lower curve shows that 15.5 percent of the ADT is exceeded by 85 percent of the sampled roads during their highest hour. From the same data, three such percentages could be

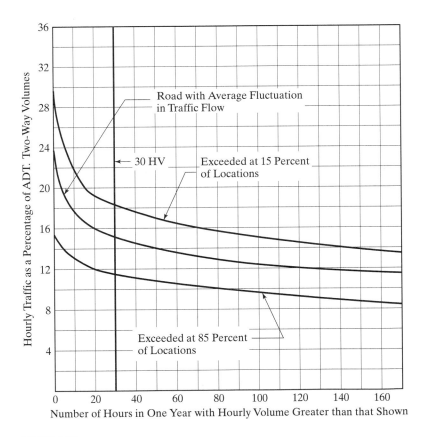

FIGURE 2.12

Relationship between two-way peak hour volume and annual average daily traffic on rural roads. *Source:* AASHTO 2001, Exhibit 2-28.

calculated for any hour rank on the horizontal axis in Figure 2.12. Of greatest interest to us is the 30th highest "Hourly Traffic as a Percentage of ADT" value. This will help the analyst decide whether additional lanes should be considered for the roadway. Example 2.4 demonstrates.

THINK ABOUT IT

Each of the three curves in Figure 2.12 "slopes down" as the "Hours in One Year" value increases. Explain why this makes sense.

Example 2.4   Design Hourly Volume

Mythaca County is studying the possibility of asking the state to increase a rural section of SR43 from two lanes to four lanes. The county engineering staff believes that Figure 2.12 applies to SR43, which has an AADT of 8110 vpd according to the latest state traffic counts. What is the DHV for this section of SR43?

Solution to Example 2.4   For the 30th highest hour in Figure 2.12, the corresponding "% of ADT" value is approximately 15.5 percent. This makes the DHV = 0.155 * 8110 = 1257 vph.

Discussion of Example 2.4.   Under ideal conditions—adequate sight distance, no traffic signals or stop signs, and no heavy vehicles—this DHV would not justify additional lanes. In Chapter 3, a more detailed analysis is introduced to determine the actual need for more lanes. To be thorough, let us also calculate the 85th and 15th percentile DHV values: 0.183 * 8110 = 1484 vph and 0.115 * 8110 = 933 vph. These values help us account for the variability in the sample used to create Figure 2.12 and to recognize that the SR43 segment being studied is not necessarily an "average" member of the rural roads in the sample.

K-Factor.   Besides the peak hour factor and the design hourly volume defined above, another measure of the temporal variation in traffic is the *K-factor*. The K-factor is the proportion of daily traffic at a site that occurs during the peak period.

$$K = \frac{V_{\text{peak}}}{\text{AADT}} \qquad (2.3)$$

If V_{peak} is important to the analysis or design of roadways or intersections, the best situation is to collect peak period data for the sites of interest. In many cases, however, there are too many sites and/or too little time to collect the data directly. In these situations, K-factor values must be assumed or borrowed. Local data may be available for similar facilities with similar demand characteristics. (HCM 2000, p. 8–9) The K-factor value usually falls within the range $0.08 < K < 0.13$ and can vary from site to site in the same city. Table 2.4 shows some of the K-factors developed for use in Florida (Florida DOT, 1995).

TABLE 2.4   Typical K-Factors for Florida Roads

Area Type	K-Factor
Urbanized	0.091
Urban	0.093
Transitioning/urban	0.093
Rural developed	0.095
Rural undeveloped	0.100

Source: HCM, 2000, p. 8–9.

FYI: Make sure you are clear about the difference between the K-factor and the peak hour factor (PHF).

Example 2.5   The K-factor

According to counts taken 3 years ago that are available on the State DOT's web site, SR361 had an ADT value of 17,140 in downtown Mythaca. While waiting for the State DOT to retrieve the hourly traffic counts for SR361 from its archives, the city engineer decided to use the Florida default values to begin his analysis. A short time later, the state provided the requested data—the peak hourly volume was 1779. (a) What K-factor should the city engineer have used in his preliminary analysis while waiting for the State's data? (b) What K-factor did the segment of SR361 actually have, according to the state's data?

Solution to Example 2.5

A. A downtown road segment is in an urban area, so $K = 0.093$ is the best value to borrow from Table 2.4. The peak volume would then be estimated as $V_{peak} = K * \text{AADT} = 0.093 * 17,140 = 1594$.

B. From Equation 2.3,

$$K = \frac{V_{peak}}{\text{AADT}} = \frac{1779}{17,140} = 0.104.$$

Using the default K-factor values in Table 2.4 would have underestimated the peak hour volume by $1779 - 1594 = 185$ vehicles.

The measures of temporal variation in traffic covered in this chapter so far have used two-way volumes as their bases. It is quite common to have traffic flow at a point that is not 50 percent in each direction during the time of interest, especially during the peak period. Traffic data on directional distribution should be collected, borrowed, or assumed, to better assess the ability of the existing or proposed roadway to accommodate the traffic flow in the major direction. This is covered in more detail in Chapter 3.

To this point in Chapter 2, the reader has been introduced to the following:

- Basic traffic characteristics and how to measure them.
- Traffic data collection technology and limitations.
- Examples of converting traffic data for use in roadway design.

Now that the basic definitions of traffic characteristics have been introduced, the focus shifts to how to measure them and how to use them for design. The next sections in Chapter 2 discuss measuring traffic characteristics.

2.2   MEASURING TRAFFIC SPEEDS AND DENSITIES

2.2.1   Speed Calculations

Why collect speed data? Three common applications come to mind:

1. To monitor the quality of traffic flow. Reductions in speed are evidence of congestion. Relationships to be introduced in this lesson will make inferences about flow quality from speed values possible.
2. To monitor speeds with respect to the speed limit or with respect to driving conditions. If speeds on a road segment are excessive, then speed limit enforcement measures there may have to be intensified.
3. To establish the basis for a speed limit. One standard starting point is a legislative standard speed limit for urban or rural road segments. For example, the state may set 55 mph as a default speed limit for all roads in rural areas and 30 mph as the default urban street speed limit. These default values can be adjusted after appropriate engineering studies, such as the one featured in an example later in this chapter.

There are two different ways to compute average speeds in a traffic stream. According to McShane, Roess, and Prassas (1998):

- *Time mean speed (TMS)* is the average speed of all vehicles passing a point on a highway over some specified time period.
- *Space mean speed (SMS)* is the average speed of all vehicles moving over a given section of a highway over some specified time period.

The time mean speed is a point measure, such as that obtained by a radar gun. The space mean speed is ascertained by timing a vehicle over a known distance. The question is: Does the method affect the values that are calculated? Example 2.6 will help.

Example 2.6

Remember the mark on Green Avenue used for headway measurements in Example 2.1? The mark also serves as the beginning of a "speed trap" 100 meters long. The time needed for

TABLE 2.5   Data for Space Mean Speed Calculations

Vehicle Number (i)	Time for 100 meters (sec)	Speed (u_i) (kph)
1	5.95	60.50
2	5.92	60.81
3	5.23	68.83
4	5.04	71.43
5	5.90	61.02
6	5.18	69.50
7	5.45	65.93
8	5.51	65.34

each vehicle to traverse the 100 meters has been measured and stored in Table 2.5, along with the computed value of *space mean speed (SMS)* for each vehicle.

A. Verify that the speed of each vehicle shown in Table 2.5 has been properly computed.

B. Calculate the space mean speed (SMS) for the eight vehicles in Table 2.5.

Solutions to Example 2.6

A. The speed of each vehicle in Table 2.5 is calculated by dividing the distance traveled by the time needed to travel that distance. For Vehicle 1, the space mean speed is

$$u_1 = \frac{100 \text{ m}}{5.95 \text{ sec}} = 16.81 \text{ meters per second (mps)}$$

$$= 60.50 \text{ kilometers per hour (kph)}$$

A single vehicle can have a space mean speed, because its instantaneous speed could have varied over the 100-meter distance, however slightly.

B. To conform to the basis for the space mean speed definition, we cannot simply find the mean value of the eight "speed" values in the rightmost column of Table 2.5. Instead, we must use the total distance covered by the eight vehicles in the sample and the total time it took them (collectively) to cover that distance.

The space mean speed is:

$$\text{SMS} = \frac{8 * 0.1 \text{ km}}{(5.95 + 5.92 + 5.23 + 5.04 + 5.90 + 5.18 + 5.45 + 5.51)\text{sec} * (1 \text{ hr/3600 sec})}$$

$$\text{SMS} = \frac{0.80 \text{ km}}{(44.18 \text{ sec}) * (1 \text{ hr/3600 sec})} = 65.19 \text{ kph}$$

The equation for space mean speed can also be expressed as a *harmonic mean* of the spot speeds:

$$\text{SMS} = \frac{N}{\sum_{i=1}^{N} \frac{1}{u_i}} = \frac{8}{\left(\frac{1}{60.50} + \frac{1}{60.81} + \cdots + \frac{1}{65.93} + \frac{1}{65.34}\right)} = \frac{8}{0.1063} = 65.17 \text{ kph}$$

FIGURE 2.13
Solar mobile speed monitor. *Source*: Decatur Electronics, used with permission.

The two SMS calculations above are equivalent. The small difference is due to round-off error.

THINK ABOUT IT

What is the basis for the first statement in the solution to Part B of Example 2.6? In other words, why not simply find the mean of the values in the rightmost column of Table 2.5?

If a point speed measurement were possible, wouldn't that be a more convenient method, and wouldn't the result be the same? To answer these questions, let us assume that each of the eight vehicles in Table 2.5 maintained precisely the speed given in column 3 for the entire 100 meters of the speed trap. (Normally, a vehicle's speed will vary over the speed trap, even if intentional acceleration or braking is not taking place.) If the constant speed assumption is correct, the time mean speed calculation is:

$$\text{TMS} = \frac{1}{N}\sum_{i=1}^{N} u_i = \frac{60.50 + 60.81 + \cdots + 65.93 + 65.43}{8} = \frac{526.96 \text{ kph}}{8} = 65.87 \text{ kph}$$

The difference between either SMS value and the TMS value is too large to be explained by round-off error. It is the consequence of using a different definition as the basis for computing mean speed in each case.

The difference between SMS and TMS is seldom large enough to worry about. This is good news, because the choice between them is usually the result of which method was used to collect the raw data.

THINK ABOUT IT

In one case, a radar gun is used to record speeds. In a second case, a speed trap is set up with road tubes or loop detectors. In which case would the resulting mean speed be space mean speed? Which would be time mean speed?

2.2.2 Monitoring Speeds on Roadways

One of the considerations in setting speed limits is how fast drivers currently travel on the roadway. A common rule of thumb is to determine what the *85th percentile speed* is on a roadway. This is the speed that is *not* exceeded by 85 percent of the vehicles whose speeds have been recorded in a representative sample of those vehicles that travel the roadway.

Example 2.7

Residents along County Road 750S in rural Mythaca County have been complaining about vehicles traveling at "excessive" speeds on that road in recent months (Figure 2.14). Because no speed limit is posted, the statewide rural default speed limit (55 mph) is in effect on the two-lane road. As part of an engineering study, the County Engineer's office collects speed data for 48 consecutive hours beginning 4 PM Tuesday. During this time, the speeds of 749 vehicles are recorded. The speeds are downloaded from the counting device to a spreadsheet file, where the speeds are sorted in descending order, from fastest (row 1) to slowest (row 749). On which row of the spreadsheet will the 85th percentile speed appear?

Solution to Example 2.7   The 85th percentile speed is the speed that is faster than 85 percent of the other speeds in the database. Row 1 in the sorted spreadsheet is the 100th percentile speed and row 749 is the zero percentile speed. Accordingly, the 85th percentile speed occurs on row X, where $X/748 = (1.00 - 0.85)$. Solving for X gives us $X = 112$. Checking this, $112/748 = 0.1497$, meaning that only 14.97 percent of the speeds are faster than the speed in row 112.

Discussion of Example 2.7.   Let us say that the 85th percentile speed found in Example 2.7 was 57 mph. Does this justify retaining the default speed limit of 55 mph? In the absence of other factors, yes. But there may be other factors that affect the setting of a speed limit. Several of these factors will be introduced in a later chapter.

FIGURE 2.14
Mythaca County Road 750S. Photo: Jon D. Fricker.

THINK ABOUT IT

What factors do you think might have to be considered in setting a speed limit on CR750S?

If, instead of 57 mph, the 85th percentile speed on CR750S was found to be 41 mph, this is evidence that most (85 percent) of the drivers behave as if speeds greater than 41 mph are unwise. The presumption behind the 85th percentile method of setting speed limits is that most drivers know what a safe speed is for prevailing roadway conditions. It may be that only a few drivers on CR750S have been driving at "excessive" speeds. The data, if collected accurately and without changing driver behavior, will show how many excessively fast drivers there are and at what speeds. In the case of a 41 mph 85th percentile speed, a speed limit of 40 mph is justified, subject to consideration of other factors that may reduce the speed limit further.

Sample Size.   Did the county need to get speeds for 749 vehicles in Example 2.7? How many vehicles would have comprised an "adequate sample" of all speeds on the roadway? If 749 was more than the minimum required, of what value was the enlarged sample size? A minimum sample size to achieve a desired degree of statistical accuracy for some percentile speed can be found by using Equation 2.4 (Robertson et al., 1994).

$$N = \frac{S^2 z^2 (2 + U^2)}{2E^2} \tag{2.4}$$

where N = minimum number of measured speeds

S = estimated sample standard deviation

z = a constant corresponding to the desired confidence level (Table 2.6). In most statistics textbooks, this constant is called the *standard normal deviate*.

U = a constant that corresponds to the desired percentile speed (Table 2.7)

E = permitted error in the average speed estimate, mph

How does an analyst set the value of S?

A. Borrow S from previous local speed studies done under similar conditions.

B. If no local data exist, use a default S value of 5.3 mph. (Box and Oppenlander, 1976, p. 80).

C. Using an estimated S from either A or B above, calculate N and collect at least that many speeds, and then calculate S for the actual data collected. If the actual S causes an N value greater than the actual sample size, collect additional data until Equation 2.4 is satisfied.

To use Equation 2.4, we still need a value for the confidence level. This is the probability that the difference between the calculated percentile speed and the actual percentile speed is less than the permitted error E. See Table 2.6 for a range of values. The standard confidence level is 95 percent, in which case $z = 1.96$. The value of E indicates the precision required in estimating the percentile speed. Typical permitted errors range from ± 1.0 to ± 5.0 mph. The corresponding values of E in Equation 2.4 would be $E = 1.0$ and $E = 5.0$.

TABLE 2.6 Constant z That Corresponds to a Specified Level of Confidence

Confidence Level (%)	z
68.3	1.00
86.6	1.50
90.0	1.64
95.0	1.96
95.5	2.00
98.8	2.50
99.0	2.58
99.7	3.00

TABLE 2.7 Constant U That Corresponds to the Desired Percentile Speed

Desired Percentile Speed	U
85th	1.04
50th	0.00

Example 2.8

Answer questions A to D below as if you had been in charge of designing the speed study described in Example 2.7.

A. In the few speed studies done by the county on rural roads in the past year, the standard deviation has been 4.3. What value of S would you have used in Equation 2.4?

B. What value of E would you have used in Equation 2.4?

C. What sample size does Equation 2.4 give you?

D. If the answer to question C turns out to be less than 749, what do you do? If the answer to question C turns out to be greater than 749, what do you do?

Solutions to Example 2.8

A. The $S = 4.3$ value based on recent county experience is based on a few studies; the $S = 5.3$ value in Table 2.7 for rural two-lane roads is based on more studies, but in locations that might be different from our county. The conservative (safe) approach is to use the S value that will increase the required sample size. That value is $S = 5.3$. (It may be wise, however, to try both values in Equation 2.4 and then see how much difference it would make in N. If using the more conservative S in Equation 2.4 would cause considerable extra expense or scheduling the speed study becomes a problem, perhaps the smaller S with adjustments made to other parameters would be acceptable.)

B. Often, the choice of E depends on the reason for doing the speed study. In this case, we want to check the actual vehicle speeds against a 55 mph speed limit. Speed limits are set in 5-mph increments, so an E value of at least 3 mph would seem reasonable. (If the more conservative S value in part A causes problems, we could relax E to be as high as 5 mph.)

C. With $S = 5.3$, $E = 3.0$, and $z = 1.96$, Equation 2.4 becomes

$$N = \frac{(5.3)^2(1.96)^2(2 + (1.04)^2)}{2(3.0)^2} = \frac{332.537}{18} = 18.47$$

This means that only 19 vehicle speeds need to have been observed!

D. If N from Equation 2.4 had been greater than 749, the county should calculate the S value from its collected data and then calculate a new value of N and collect speeds for a number of vehicles equal to the difference between the new N and 749. But part C tells us that a sample size of 19 vehicles would have been statistically adequate for the S and E values used. The county's speed study with $N = 749$ was much more than adequate. But was it wasteful? Perhaps. Because the calculations for part C were set up on a spreadsheet [see column (1) of Table 2.8], calculations using other values of S and E can be easily carried out. In this case, we may want to know, given S from the actual data collected with $N = 749$, and K held at 1.96, how small has our precision value E gotten? Using the Tools/Solver feature in Excel (or "generate and test" otherwise), if actual $S = 3.9$, then $E = 0.347$. [See column (3) of Table 2.8.] This means that the county has collected so much speed data that it can be 95 percent confident of being within 0.347 mph of the actual 85th percentile speed. Unless there is reason to believe that speeds vary significantly by time of day, the county

TABLE 2.8   Calculate Minimum Sample Size for X Percentile Speed

	(1)	(2)	(3)	(4)
$X =$	85	85	85	
$S =$	5.3	4.3	3.9	
$z =$	1.96	1.96	1.96	
$U =$	1.04	1.04	1.04	
$E =$	3.0	3.0	0.3467	
Top $=$	332.537	218.890	180.060	
Bottom $=$	18.0	18.0	0.2	
$N =$	18.47	12.16	749.00	

could have achieved a statistically valid sample with a much smaller data collection effort than it expended.

It is useful to note that, when the median (50th percentile) speed is desired, $U = 0.00$ and Equation 2.4 simplifies to

$$N = \left(S \frac{K}{E} \right)^2 \tag{2.5}$$

In most cases, the median speed and the mean speed are indistinguishable.

Before we leave Examples 2.7 and 2.8, a few more comments may be in order. On the matter of "data collection effort," it may be that there was no significant effort involved in getting those 749 vehicle speeds. Perhaps all it took was placing a speed-sensing device along CR750S at 4 PM Tuesday, checking it once or twice while it was in place and then removing it at 4 PM Thursday. Unlike many radar systems, no personnel had to be present on a continuing basis.

2.2.3   Density

On a recent business trip, the county highway engineer was driving EB (eastbound) on I-80 near Salt Lake City on a Friday night. Because he was surprised at the amount of traffic on the highway, he decided to try to measure just how busy the highway was. A common way to measure traffic magnitudes is to count the number of vehicles that pass a point and compute traffic flow rates (vehicles per hour), but that can't be done from a vehicle in the traffic stream. Another way to measure amounts of traffic is *density*, with typical units being *vehicles per mile (vpm)* or *vehicles per kilometer (vpk)*. If multiple lanes are involved, it might be better to use *vehicles per mile per lane (vpmpl)* as units. One way to compute density is to count the number of vehicles seen on a section of highway of known length, based on an airphoto of the highway. (See Figure 2.15.) The county highway engineer, however, tried to get a rough estimate of the density on I-80 even as he drove his rental car.

FIGURE 2.15
Airphoto showing vehicles on roadway. *Source*: Skycomp, Inc.

Example 2.9

Very quickly, the county highway engineer saw an overpass ahead, glanced at his car's odometer (468.50 miles), counted all the vehicles (13) on EB I-80 between his car and the overpass, and checked the odometer again (468.77 miles) as he drove under the overpass. Eastbound I-80 is a three-lane facility at this point. What is the density (D) of traffic on this section of EB I-80?

Solution to Example 2.9   The density D on three-lane eastbound I-80 is computed as

$$D = 13 \text{ veh}/((468.77 - 468.50 \text{ mi})/3 \text{ lanes}) = (13/0.27 \text{ vpm})/3 \text{ lanes}$$

$$D = 48.15 \text{ vpm}/3 \text{ lanes} = 16.05 \text{ vpmpl}$$

Density is a good measure of the current amount of travel on a highway. So is *flow rate*, but which is easier for the average driver to relate to? Flow rates are computed from traffic counts taken over a period of time at a point on the roadway; density can be observed (over a limited section of roadway) by an individual at one glance. Congestion occurs when traffic flow approaches roadway capacity. (In traffic flow, capacity represents the supply side of the demand–supply relationship.) Capacity can be expressed in terms of flow rate (vphpl) or density (vpmpl). A systematic flow-based approach to capacity estimation and analysis is presented in Chapter 3. At this point, however, we can try to estimate a density-based capacity value, which could be called a *physical capacity*. A common rule of thumb for drivers is to maintain a time interval of 2 seconds between the front of your vehicle and the rear of the vehicle directly in front of you. If this rule of thumb is never violated, the following calculations will produce a density-based capacity estimate.

- If the prevailing vehicle speed is 60 mph, convert that speed to 88 feet per second (fps). (To ease the conversion of miles per hour into feet per second, or vice versa, satisfy yourself that 1 mph = 1.47 fps. Likewise, 1 kilometer per hour = 0.278 meters per second (mps).)
- A two-second interval at that speed is equivalent to 176 feet of space between vehicles.
- Add 15 feet for the length of the typical vehicle.
- The resulting density is (5280 ft/mi)/(176 + 15 ft/veh) = 27.64 vpmpl.

Note that the density-based capacity depends on prevailing speed, the safe interval rule of thumb, and vehicle lengths. If the capacity value just derived above is applied to the county engineer's I-80 case, 16.05/27.64 or 58.1 percent of the roadway capacity (or supply) is being used by the existing demand.

Two ways to measure density have been mentioned: airphoto analysis and observation from a vehicle in the traffic stream. Each method has obvious limitations.

THINK ABOUT IT

What are the practical limitations of each method of measuring density?

2.2.4   Detector Loops and Occupancy Calculations

Another way to determine density is to use sensor equipment to measure a traffic stream characteristic called *occupancy*. Occupancy could be defined as the percent of the roadway that is covered by vehicles. In theory, traffic that is literally bumper-to-bumper would have an occupancy value of 100 percent. Instead, occupancy is measured by detectors in the pavement that indicate whether, at any instant, a vehicle is over the detector. The resulting value is the percent of time in which the detector is "occupied." Because of the electromagnetic aspects of the detection procedure, the vehicle may actually be detected some distance before it is physically over a detector and may continue to be detected a short distance after no longer being directly over the detector. (See Figure 2.16.) For this reason, we refer to the *effective length* (EL) of a detector, rather than its actual length (L). This is important, because a 15-foot-long vehicle that passes over a detector with effective length 8 feet at 46 feet per second will appear to be occupying the detector for (15 + 8 ft)/46 fps = 0.50 seconds. The true occupancy, however, should be based on a detector that has no finite length (i.e., a point). The 15-foot vehicle passing over a point at 46 fps would occupy that point for only 15 ft/46 fps = 0.32 seconds. To get the true occupancy, the analyst must adjust for the EL of the detector. In Figure 2.16, EL = L + B + A.

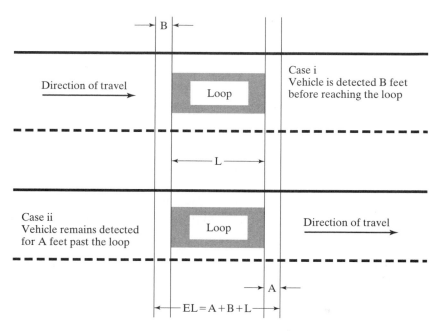

FIGURE 2.16
Vehicle passing over a detector loop.

Example 2.10

A detector with an EL of 6 feet detects a flow rate (q) of 905 vehicles per hour. Other detectors determine that the average speed (S) is 34.6 mph and the average vehicle length (L_V) is 18.2 feet.

A. What percentage of the time will the detector register the presence of a vehicle?

B. Adjust the information provided by the detector in part A to estimate true occupancy.

Solutions to Example 2.10

A. The length of time $t(P)$ that an average vehicle will be detected as "present" by the detector will be

$$t(P) = \frac{\text{distance}}{\text{rate}} = \frac{(L_V + \text{EL})}{S * (5280/3600)} \text{sec/veh} = \frac{(18.2 + 6.0) * 3600}{34.6 * 5280} = 0.477 \text{ sec/veh} \quad (2.6)$$

If the flow rate is 905 vph, the detector will be "occupied" by vehicles for $905 * 0.477 = 431.6$ seconds during that hour. This duration is 12.0 percent of an hour. This is the *apparent occupancy* (O_{app}).

B. The detector senses a vehicle just before it is over the detector, while it is over the detector, and just after it has passed over the detector. The actual length of the detector itself also adds to the "apparent length" of the vehicle. We need to determine what the occupancy value would have been in part A if the detector would be "on" only when a vehicle was directly above a point on the roadway. To do this, we must deduct the effect of the detector's effective length from the "apparent presence time" value $t(P)$ calculated in part A. The actual presence time $t'(P)$ for an average vehicle will be

$$t'(P) = t(P) - \frac{\text{EL} * 3600}{S * 5280} = 0.477 - \frac{6.0 * 3600}{34.6 * 5280} = 0.477 - 0.119 = 0.358 \text{ sec/veh} \quad (2.7)$$

The actual occupancy rate for the flow rate of 905 vph is adjusted as follows:

$$\text{Actual Occupancy } O_{\text{act}} = q * t'(P) = (905 * 0.358)/3600 = 0.090 \text{ or } 9.0 \text{ percent} \quad (2.8)$$

As was the case with density-based occupancy, an occupancy value of 100 percent based on presence time is theoretically possible, if traffic is literally bumper-to-bumper, with no gaps between vehicles. Of course, even the worst traffic jams do not approach 100 percent occupancy. Table 2.9 contains an excerpt of data taken from a section of expressway in the Mythaca area. Each line of data in the table defines the traffic stream characteristics for a period of 20 seconds. In Table 2.9, q represents flow rate and S stands for speed. Note how variable the data can be, even in consecutive 20-second periods. We use the Table 2.9 data in a series of examples and exercises.

THINK ABOUT IT

Why were the data grouped into 20-second intervals, instead of keeping the individual vehicle data?

As was indicated in the previous section, density is a more useful and widely used measure of traffic than occupancy, but the density columns in Table 2.9 are empty. If average L_V is known, *actual occupancy* O_{act} can be converted to density by using Equation 2.9 as follows:

$$D = \frac{O_{\text{act}} * 5280 \text{ ft/mi}}{L_V} = \frac{0.09 * 5280 \text{ ft/mi}}{18.2 \text{ ft}} = 26.1 \text{ vpm} \quad (2.9)$$

If *actual occupancy* has not yet been computed, density can be computed directly from *apparent occupancy* O_{app} without Equations 2.6 to 2.8 by using a variation on Equation 2.9:

$$D = \frac{O_{\text{app}} * 5280 \text{ ft/mi}}{L_V + \text{EL}} = \frac{0.12 * 5280 \text{ ft/mi}}{(18.2 + 6.0) \text{ ft}} = 26.2 \text{ vpm} \quad (2.10)$$

If the value of L_V is not available, it can be determined from Equation 2.7 and

$$L_V = S * t'(P) = 34.6 \text{ mph} * 1.4667 \text{ fps/mph} * 0.358 \text{ sec/veh} = 18.2 \text{ ft} \quad (2.11)$$

TABLE 2.9 Traffic Data for Interstate Highway 25

Time Period	q (veh/hr)	S (km/hr)	Occupancy (%)	Density (veh/km)	Time Period	q (veh/hr)	S (km/hr)	Occupancy (%)	Density (veh/km)
1	1620	91	10		53	1260	20	30	
2	1980	90	11		54	1260	35	18	
3	2160	90	12		55	1260	29	21	
4	1620	95	9		56	1620	32	24	
5	1080	94	7		57	1440	33	20	
6	1980	87	11		58	1440	39	17	
7	1440	86	8		59	1980	44	20	
8	1620	87	10		60	1260	82	7	
9	720	85	3		61	1620	67	12	
10	1980	93	11		62	1620	71	10	
11	1260	93	7		63	1440	78	9	
12	1080	97	6		64	1440	76	9	
13	1800	93	10		65	1620	75	10	
14	1440	89	8		66	1800	75	11	
15	1800	99	9		67	1440	86	9	
16	1800	95	9		68	1440	78	8	
17	1620	43	19		69	1440	76	10	
18	900	14	47		70	1620	63	11	
19	900	13	43		71	540	78	3	
20	900	17	22		72	1620	79	10	
21	1440	41	15		73	900	75	6	
22	1080	82	6		74	1260	72	9	
23	1980	82	11		75	720	85	4	
24	1260	93	7		76	1440	79	9	
25	1260	99	6		77	1260	79	9	
26	2160	87	13		78	1260	81	7	
27	1800	95	9		79	2160	66	16	
28	1620	89	9		80	2160	75	14	
29	1080	85	6		81	1080	89	7	
30	1440	87	9		82	1260	82	7	
31	1620	75	11		83	1080	86	6	
32	1980	76	13		84	1260	82	8	
33	1620	66	13		85	720	79	4	
34	1080	35	16		86	1260	85	7	
35	1620	51	17		87	1080	76	7	
36	1620	32	26		88	1440	78	11	
37	1620	33	26		89	1620	75	11	
38	1800	39	23		90	1620	66	12	
39	1620	28	29		91	1620	66	11	
40	1260	27	21		92	1620	67	11	
41	900	21	21		93	1620	75	10	
42	1440	21	29		94	1800	70	12	
43	1260	31	20		95	1620	70	10	
44	1440	28	30		96	1800	75	12	
45	1440	27	32		97	1620	59	13	
46	1440	25	26		98	1980	74	14	
47	1440	28	24		99	1080	72	8	
48	1440	24	26		100	1440	66	11	
49	1260	20	36		101	1080	76	7	
50	1440	27	25		102	1620	72	11	
51	1260	18	31		103	540	78	3	
52	1620	40	20		104	540	72	4	

If the values of L_V and EL are not available and cannot be estimated with confidence, a simple relationship between speed, density, and flow rate can be used:

$$D = \frac{q}{S} = \frac{905 \text{ vph}}{34.6 \text{ mph}} = 26.2 \text{ vpm} \tag{2.12}$$

This relationship is introduced and explained in the next section of this chapter.

2.3   TRAFFIC MODELS FOR CONTINUOUS FLOW

There are several important models that the transportation engineer should be aware of in the analysis of traffic flow. Because traffic flow can be deterministic or random, and because it can also be continuous (uninterrupted) or sporadic (i.e., with queues), most useful models of traffic fit into the matrix shown in Table 2.10.

TABLE 2.10   Categories of Traffic Flow Models

Flow	Continuous	Sporadic (with queues)
Deterministic	Greenshields Model (Chapter 2.3)	Flow vs. time diagrams (Chapter 3.3)
Random	Poisson Arrival Model (Chapter 2.4)	Stable Queues (Chapter 3.4)

In this chapter, three key traffic stream parameters (and their variations) have been introduced: flow rate, speed, and density. Because these parameters are measuring three different characteristics of the same traffic stream, it is likely that systematic relationships can be identified between them. In this section, some relationships that form the basis for useful models of traffic flow are introduced.

2.3.1   Speed-Density Relationship

In looking for a relationship between speed and density, it would be useful to sketch a speed-density curve. Begin with the free-flow condition of one solitary vehicle being driven at a *free-flow speed* (S_f). As density increases a little, speed decreases hardly at all. At some point, increasing density begins to exert a noticeable impact on speed. As the density of traffic continues to grow, the highway eventually becomes so clogged that all motion stops. The maximum practical density on a roadway is called the *jam density* (D_j). At density $D = D_j$, traffic can no longer move, so we must have speed $S = 0$. Given this narrative, the resulting plot—when speed is plotted on the vertical axis and density on the horizontal axis—would most likely resemble the upper right quadrant of an ellipse. (See Figure 2.17.)

However, as you will find out when doing some of the homework problems, intuition may not match what traffic data seem to show about the relationships between traffic stream characteristics. For a long time, traffic engineers have wanted a better model and have explored a number of them. Our simple "quarter ellipse" in Figure 2.17 can be added to the variety of speed-density hypotheses that have been proposed (Drake et al., 1967).

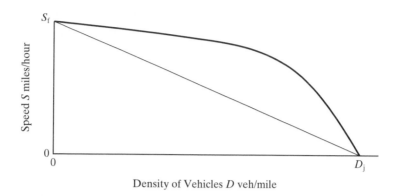

FIGURE 2.17
Two possible speed-density relationships.

The simplest plot to work with is the Greenshields (1934) linear model. The Greenshields linear relationship results from connecting the S_f and D_j points in Figure 2.17 with a straight line. The Greenshields speed-density relationship has the form

$$S = S_f\left(1 - \frac{D}{D_j}\right) \tag{2.13}$$

Regardless of the speed-density relationship that is chosen, the following relationship holds:

$$\text{Flow} = \text{speed} * \text{density} \quad \text{or} \quad q = S * D \tag{2.14}$$

THINK ABOUT IT

Using standard units of measurement for speed and density, verify the validity of Equation 2.14.

Example 2.11

In section 2.2.3, the county highway engineer was estimating traffic density as he drove at 60 mph on I-80 near Salt Lake City. His density estimate was 16.05 vehicles per mile per lane (vpmpl).

A. From this information, can we construct a speed-density model for that section of I-80?

B. What is the approximate flow rate on that section of I-80?

Solution to Example 2.11

A. A speed-density model can be constructed by specifying an equation of the form in Equation 2.13 so that the characteristics of traffic on I-80 are adequately represented.

Although we have only one speed-density point (D, S), we can make reasonable assumptions about one of the remaining two parameters (D_j) and then see what value for the other parameter (S_f) results. We can assume that, at jam density, vehicles are effectively bumper-to-bumper (with only a little space between vehicles). If the average vehicle is 15 feet long and the average space between vehicles at jam density is 15 feet, we can estimate jam density as

$$D_j = \frac{5280 \text{ ft/mi}}{(15 + 15) \text{ ft/veh}} = 176 \text{ vpmpl}$$

We can rearrange Equation 2.13 to be

$$S_f = \frac{S}{1 - \dfrac{D}{D_j}} \tag{2.15}$$

For our county highway engineer's case, $S_f = \dfrac{60 \text{ mph}}{1 - \dfrac{16.05 \text{ vpm}}{176 \text{ vpm}}} = 66.02 \text{ mph}$. This value

of free-flow speed seems reasonable for an urban expressway.

B. Finding the flow rate is simply a matter of using Equation 2.14.
In each lane, $q = S * D = 60 \text{ mph} * 16.05 \text{ vpmpl} = 963 \text{ vpmpl}$.

THINK ABOUT IT

In Example 2.11, should we accept $D_j = 176$ vpm and $S_f = 66.02$ mph as reasonable parameters to use on the section of I-80 observed by the county highway engineer? Are there any adjustments or caveats we should consider before adopting this "model"?

Example 2.12

If the density value (16.05 vpm) estimated for I-80 by the county highway engineer in Example 2.11 at 60 mph is accurate, what jam density value will yield a typical highway design speed of 70 mph? What will be the corresponding space between vehicles at jam density, if the average vehicle is 15 feet long?

Solution to Example 2.12

A. The design speed can be used as the free-flow speed, $S_f = 70$ mph. Rearranging Equation 2.13 to isolate D_j gives us

$$D_j = \frac{S_f D}{S_f - S} = \frac{70 * 16.05}{70 - 60} = 112.35 \text{ vpm}$$

B. The average space between the front bumper of a following car and rear bumper of the leading car must account for the number of vehicles in a lane (112.35 per mile) and the space taken up by each vehicle (15 feet). The space taken up by the vehicles is $112.35 * 15 = 1685.25$ feet of each lane-mile. By dividing the remaining space among the 112.35 vpm, the average space between vehicles is

$$\frac{5280 - 1685.25}{112.35} = \frac{3954.75 \text{ ft}}{112.35 \text{ veh}} = 32.00 \text{ ft}$$

This spacing is much larger than the 15 feet we assumed in Example 2.11.

THINK ABOUT IT

Does the 32-foot spacing found in Example 2.12 seem rather large for jam density conditions? Was the 15-foot spacing assumed for jam density in Example 2.11 too small? What may explain the difference in the two values?

2.3.2   Speed vs. Flow

Consider the case of a vehicle being the only vehicle on a roadway. Its driver can operate the vehicle at any speed he/she decides, influenced only by roadway geometrics (sight distance, cross streets, parked cars, etc.) and the speed limit. Because there is no other traffic to constrain the speed, the speed chosen by the driver is called *free-flow speed*. Based on the data plotted in Figure 2.18, satisfy yourself that the free-flow speed on the San Diego freeway is about 60 mph. As a few other vehicles enter the roadway, the original driver must now share the road with them, but his/her speed is not affected very much. In Figure 2.18, the free-flow speed can be maintained even when flow is at 1000 vehicles per hour per lane (vphpl). As the flow rate (demand) continues to increase with respect to the roadway's fixed capacity (supply), the driver's speed will be noticeably affected. This appears to take place

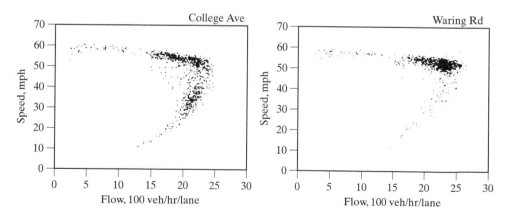

FIGURE 2.18

Observed speed-flow relationship at two locations 1.3 miles apart on the San Diego freeway (Data in 30-second sampling intervals). *Source*: Banks, 1989, p 56, used with permission of Transportation Research Board.

somewhere between 1000 and 1500 vphpl in Figure 2.18. Even at capacity (where max flow rate is about 2400 vphpl), vehicles continue to move along the roadway. After all, capacity is expressed in terms of vehicles per hour, so the vehicles must be moving. Any other point on the speed-flow curve in Figure 2.18 must have a flow rate that is less than the capacity. For a flow rate like 2400 vphpl to be sustained, a "reasonable" speed must be maintained. Still, it is possible for lower speeds to occur. In fact, it is possible for the case in which speed drops to zero (all traffic is stopped) to exist. This is represented by the lower part of the speed-flow curve, which consists of points leading toward the flow = 0, speed = 0 point in Figure 2.18. At zero speed, there will be zero flow. The highway will be jammed up with vehicles, none of which will be moving.

In Figure 2.18, note how the individual data points are scattered around a curve that resembles a parabola, although its shape is not quite symmetric. In Figure 2.18, the capacity (maximum flow) "q_{max}" has a value of approximately 2400 vph. If one more vehicle is added to a traffic stream that has reached q_{max}, the flow rate cannot increase, but the density can. The result must be a decrease in speed. The lower half of the speed-flow curve in Figure 2.18 illustrates how the same flow rate can occur at two different speeds. For example, a curve drawn through the data points in Figure 2.18 would have two values of speed for flow = 1500 vphpl—about 13 mph on the lower part of the curve and about 57 mph on the upper part of the curve. The difference between the upper and lower points is the density. The point upper point has lower density and higher speed (57 mph). Refer to Equation 2.14 for the pertinent relationship. The lower point has the lower speed (12 mph), but enough density to produce the same flow rate (1500 vph). (The lower half of the speed-flow curve represents what traffic flow analysts call the *unstable flow regime*.) Given the approximately parabolic shape of the curve, we now have enough information with which to build a mathematical relationship for speed vs. flow.

Using Equations 2.13 and 2.14, we can derive an equation to use as a model for the speed-flow relationship.

$$S = S_f - \frac{S_f}{D_j} \frac{q}{S} \tag{2.16}$$

or

$$q = [D_j * S] - \frac{D_j}{S_f} S^2 \tag{2.17}$$

or

$$q = D_j \left(S - \frac{S^2}{S_f} \right) \tag{2.18}$$

THINK ABOUT IT

Does Equation 2.16, Equation 2.17, or Equation 2.18 allow the user to build an entire speed-flow curve like that shown in Figure 2.18, assuming that values for D_j and S_f are available?

2.3.3   The Flow-Density Relationship

The county highway engineer found it possible to estimate density while driving his rented car on I-80. There is no practical way to directly measure flow rate while in the traffic stream, but he was able to use Equation 2.14 to convert density and speed into an estimate of flow rate. A roadside observer can easily count vehicles and convert that count into a flow rate. It would be very convenient if one were able to convert a flow value directly into a density or vice versa. We could try the intuitive approach again, but that may be difficult. We could collect lots of flow and density data and then try to fit a curve to those data, but we'll save that for homework problems.

THINK ABOUT IT

Can a roadside observer estimate the traffic density on a section of roadway? If you think so, describe how it could be done.

Equations 2.13 and 2.14 can be rearranged into a flow-density relationship, as shown below:

$$q = S_f\left(D - \frac{D^2}{D_j} \right). \qquad (2.19)$$

Equation 2.19 has a parabolic form that approximates the data plotted in Figure 2.19.

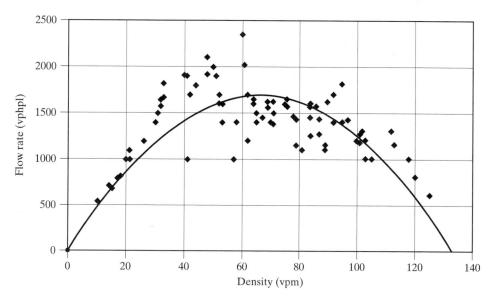

FIGURE 2.19
The standard flow-density relationship. Based on Drake, Schofer, and May (1967, p. 77).

Example 2.13

What is the approximate flow rate for the traffic observed by the county highway engineer on I-80?

Solution to Example 2.13   If the engineer adopts the parameters estimated in Example 2.11, Equation 2.19 becomes

$$q = 66.02 \text{ mph} * \left(16.05 - \frac{(16.05)^2}{176} \right) \text{vpm} = 66.02 * 14.586 = 963.0 \text{ vph.}$$

You may recall that the county highway engineer's data are on a per lane basis. In Table 2.9, we saw that a freeway lane's capacity can exceed 2000 vph. In Figure 2.19, the parabola's highest flow value is about 1700 vph, but the highest single data point is at about 2350 vph. Therefore, $q = 963$ vph seems reasonable. *Always remember to do a reasonableness check on the results of your calculations.*

THINK ABOUT IT

Do you agree that $q = 963$ is a reasonable value for this case? What, if anything, about the county highway engineer's description of the conditions on I-80 causes you to agree or disagree?

The data plotted in Figures 2.18 and 2.19 exhibit some scatter, but those plots are close enough to parabolic in shape to permit us to proceed with the equations proposed in the last two sections. The next section will use the proposed equations to develop relationships and test them against traffic data, such as that contained in Table 2.9.

2.3.4   The Greenshields Model

Earlier in this chapter, a number of basic relationships regarding speed, density, and flow were developed. These relationships were expressed graphically and in equations. Within those relationships are specific cases of interest to the builder and user of traffic flow models. One point of interest is the peak of the parabola in Figure 2.19. This is the point of maximum flow, and it occurs at about half the value of free-flow speed. If the data in Figure 2.18 were to be modeled as a symmetric parabola (as is often done), the peak flow would occur at about half the value of jam density. These simplifying assumptions make it possible to say that the maximum flow point can be expressed as a special form of Equation 2.14:

$$q_{max} = S * D = \frac{S_f}{2} * \frac{D_j}{2} = \frac{S_f D_j}{4} \tag{2.20}$$

Example 2.14

Check one of the speed-flow relationships in Equations 2.16–2.18 against the data observed by the county highway engineer on I-80. Also estimate the maximum flow for that section of I-80.

Solution to Example 2.14

A. In Example 2.11, the prevailing speed was 60 mph and the free-flow speed was been estimated to be 66.02 mph. The jam density was calculated to be 176 vpm, if average vehicle length was 15 feet and the space between vehicles was 15 feet. Equation 2.18 becomes

$$q = 176 \text{ vpm}\left(60 - \frac{60^2}{66.02} \right) \text{mph} = 176 * 5.47 = 962.9 \text{ vph.}$$

This result is consistent with Examples 2.11 and 2.13.

THINK ABOUT IT

Is it just a coincidence that Examples 2.13 and 2.14 end up with essentially the same flow rate?

B. Equation 2.20 produces $q_{max} = \dfrac{S_f D_j}{4} = \dfrac{66.02 * 176}{4} = 2905$ vph for the same input values used in part A.

THINK ABOUT IT

Is the answer $q_{max} = 2905$ vph in Example 2.14B reasonable? What is the basis for your answer?

The three fundamental relationships studied in this section so far are summarized here:

- Speed-density:

$$S = S_f\left(1 - \frac{D}{D_j} \right) \tag{2.13}$$

- Flow-density:

$$q = S_f\left(D - \frac{D^2}{D_j} \right) \tag{2.19}$$

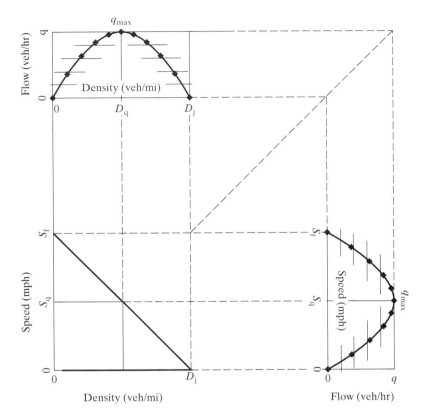

FIGURE 2.20
The interrelationships of the fundamental traffic flow relationships.

- Speed-flow:

$$q = D_j\left(S - \frac{S^2}{S_f}\right) \tag{2.18}$$

Equation 2.13 is linear, and Equations 2.18 and 2.19 have a parabolic shape when plotted. Collectively, we refer to these equations as the Greenshields Model of traffic flow. The three equations can be shown to have a consistent relationship with each other. See Figure 2.20. The explanation for this can be as simple as the fact that each relationship is defined in terms of free-flow speed, jam density, and/or roadway capacity. If you have data such as those shown in Table 2.9, you can try to build a set of models for the three fundamental traffic relationships. The "models" are actually Equations 2.13, 2.18, and 2.19 with values for D_j, S_f, and q_{max} estimated from the observed traffic stream.

Usually the easiest place to start is with the simplest relationship. Because Equation 2.13 is linear, we can use simple linear regression (Walpole and Myers,

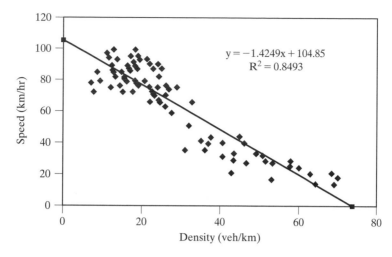

FIGURE 2.21
Fitting data in Table 2.9 to speed-density Equation 2.13.

1998) to fit a straight line through the data points. Table 2.9 contains occupancy data, but we need density data to use Equation 2.13. Fortunately, we can rearrange the simple relationship in Equation 2.14 to calculate entries for the "density" column that had been left empty in Table 2.9, by using the equation $D = q/S$. If the data in Table 2.9 are in spreadsheet format, inserting the density values is a very simple task. Note also that the data in Table 2.9 have been collected in metric units. For example, in the first 20 seconds, $D = 1620$ vph/91 kph = 17.80 vpk. When the resulting density values are plotted against speed, the best-fit regression line (Figure 2.21) is $S = -1.4249D + 104.85$. Note that the R^2 value in Figure 2.21 is 0.8493. This is probably lower than values you may be accustomed to seeing in laboratory experiments involving the behavior of materials, but here we are examining the behavior of people (drivers).

The speed-density plot in Figure 2.21 resembles the straight-line Greenshields plot in Figure 2.17. The best linear fit may give good intercept values for S_f and D_j, but be sure to check the results for reasonableness. For the regression line in Figure 2.21, $S_f = 104.85$ kph and $D_j = S_f/-m = 104.85/1.4249 = 73.58$ vpk are the intercepts. (The parameter **m** is the slope of the regression line.) The free-flow speed $S_f = 104.85$ kph (65 mph) seems reasonable for an urban freeway. The jam density value $D_j = 73.58$ vpk (118.4 vpm) translates into about 44.6 feet per vehicle (fpv), which is reasonable for a freeway of two or more lanes.

Example 2.15

Continue fitting Table 2.9 data to the models implicit in speed-density Equation 2.13 and flow-density Equation 2.19.

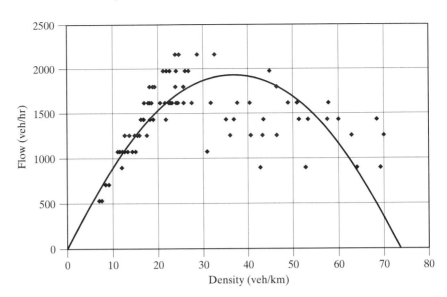

FIGURE 2.22
Fitting data in Table 2.9 to flow-density Equation 2.19.

Solutions for Example 2.15 Now that the density values can be added to Table 2.9, the flow-density plot can be created. See Figure 2.22. The plot does appear to have the inverted "U" shape of the standard plot in Figure 2.19.

Using the S_f and D_j results from the speed-density line fit, Equation 2.19 becomes

$$q = 104.85\left(D - \frac{D^2}{73.58}\right)$$

Note that this equation for q is plotted by using the S_f and D_j values found by fitting a straight line to the data in Figure 2.22. The curve was not fitted to the flow-density data points in Figure 2.22. To do so would be to create another set of values for S_f and D_j, when only one set of values can apply to the traffic that was observed.

A point of special interest in Figure 2.19 or Figure 2.22 is the maximum flow point q_{max}. If we differentiate Equation 2.16 and recognize that the slope of the speed-density curve at q_{max} is zero, we get

$$\frac{dq}{dD} = S_f\left(1 - \frac{2D}{D_j}\right) = 0 \tag{2.21}$$

Because S_f is not zero, the expression within the parentheses in Equation 2.21 must equal zero:

$$D_{\max q} = \frac{D_j}{2} \tag{2.22}$$

The subscript "max q" is added to D in Equation 2.22 to remind us that this derivation was to find the specific value of D that lies at the top of the curve in Figure 2.22 that is based on the

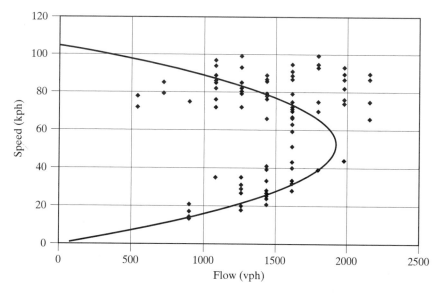

FIGURE 2.23
Fitting data in Table 2.9 to speed-flow Equation 2.18.

flow-density equation. In other words, it is the density at maximum flow, or at "max q". In a similar fashion, let us look for the q_{max} point on the curve in Figure 2.23 based on the speed-flow equation. Substituting Equation 2.22 into Equation 2.13 produces

$$S_{max\,q} = S_f\left(1 - \frac{D_j}{2D_j}\right) = \frac{S_f}{2} \tag{2.23}$$

where $S_{max\,q}$ stands for "speed at maximum flow," not "maximum speed." Equation 2.20 enables us to estimate the maximum flow for a curve fitted to the Table 2.9 data:

$$q_{max} = \frac{(104.85)(73.58)}{4} = 1929 \text{ vph}$$

Note in Fig. 2.23 that 11 of the 104 observations in Table 2.9 exceed this q_{max} value, with 4 of those 11 greater than 2000 vph.

THINK ABOUT IT

Does the fact that 11 values in Table 2.9 exceed q_{max} = 1929 vph invalidate a model based on that value? Explain.

Figure 2.20 and Equations 2.13, 2.18, and 2.19 seem to indicate strong connections among the speed, density, and flow traffic parameters. However, there are some warnings that must be heeded.

- Actual traffic data may be scattered along the curves drawn in Figure 2.20, or they may not even resemble the plots at all. Local conditions, unusual events, or data collection errors may affect the data being used to show the relationships.
- The data collected may not cover a wide enough range of conditions to build the complete relationship. A common mistake is to try to fit a complete speed-flow curve using data that were taken from conditions best described by the top half of the curve.

"Real" data may not be close approximations to the models that are used in an attempt to explain the underlying phenomena. As far back as 1967, Drake, Schofer, and May (1967) were considering numerous alternatives to the well-behaved forms in Figure 2.20. More than 30 years of data collection since then have led to increasing acceptance of *revised* forms of the standard curves. In particular:

- The flow-density curve is close to linear for uncongested conditions (Hall et al., 1992). Notice how the portion of Figure 2.19 for which density is less than 60 vpm is nearly linear.
- The upper portion of the speed-flow curve also tends to be more linear than parabolic, as the top portion of Figure 2.18 indicates.

In Figure 2.22, the data points tend to lie along the model curve based on Equation 2.19. In Figure 2.23, there is more scatter about the model curve. In Figure 2.23, the data points illustrate the warning made earlier that the data may not cover a wide enough range of conditions to build the complete relationship. Because the data in Table 2.9 came from only about 35 minutes of observations, the data are unlikely to reflect a wide range of traffic characteristics. The data in Table 2.9 have been used to show how to fit traffic data to build traffic flow models such as Equations 2.13, 2.18, and 2.19. A more prudent way to build traffic flow models for the expressway segment would be to gather traffic data for all hours of the day and then repeat the processes illustrated in this section on those data points.

2.4   THE POISSON MODEL FOR CONTINUOUS RANDOM TRAFFIC FLOW

2.4.1   The Poisson Process

Siméon Denis Poisson was a French mathematician (1781–1840) who has made a significant impact on civil engineering analysis. His work on the probabilities associated with infrequent events has been applied to analyses of how often earthquakes and floods occur. The Poisson model can answer the question, "How many events will likely take place during a time interval of a specified length?" Can the arrival of vehicles in the traffic stream be treated in much the same way as earthquake or flood occurrences? The answer is "yes," if the assumptions inherent in the Poisson Process are satisfied:

1. *The events are random. The number of events that occur in one time interval is independent of the number of events that occur in any other time interval. There is no pattern, no memory, and no interdependence between events.* Events that occur only a few times a decade (such as earthquakes and floods) are easier to view as random. However, even these events in nature may be interdependent, such as when an unusually erratic rainy season causes repeated flooding. Vehicles may pass a point on the roadway every few seconds but could still be considered as random events. Fortunately, we do not have to rely on intuition to decide whether this assumption is met. Systematic tests exist to help us make that decision by using data about the traffic stream being analyzed.

2. *The event rate remains constant for the duration of the analysis.* If the traffic flow rate in the traffic stream fluctuates during the desired period of analysis, the analysis cannot take advantage of the simplicity of the Poisson equations. More complicated procedures would have to be used.

3. *The probability that a single event will occur during a short time interval is proportional to the length of the time interval.* As the time interval used for the analysis is decreased, the probability of an event occurring also decreases.

FYI: "Poisson" is pronounced "Pwa-SOHN." It doesn't sound anything like "poison."

The Poisson process is a good one for a traffic stream that is made up of free-flowing traffic. It is not a good choice for bumper-to-bumper traffic or for traffic flowing shortly downstream from a traffic signal. The basic Poisson equation is

$$P(n) = \frac{(\lambda t)^n e^{-\lambda t}}{n!}$$

(2.24)

where t = duration of time interval

$P(n)$ = probability that exactly n vehicles arrive in time interval t

λ = average flow rate or arrival rate (vehicles/unit time) (also "q" or "V")

FYI: A very important property of the Poisson distribution is that both the mean and the variance of the Poisson distribution have the same value: λt.

Example 2.16

The Mythaca County Highway engineer and his daughter Donna (age 5) were sitting on lawn chairs in their front yard, enjoying an August evening (Figure 2.24). One-half block to the

FIGURE 2.24
Front yard data collection point. Photo by Karen Springer.

east is Steak Street, a moderately busy street that they can easily see from their front yard. By now, Donna had learned how to count cars—even to classify them by vehicle type. On this particular evening, they decided to record (using the stopwatch/split mode on the father's sports watch) the arrival times of 15 vehicles going in one direction. Donna picked the direction (uphill or downhill) and told her father when to start his watch. To himself, the county highway engineer decided to stop recording arrivals at the next multiple of 15 seconds after the 15th vehicle was seen. They repeated the process for the other direction on Steak Street. The results of their data collection are shown in the "Direction One" and "Direction Two" columns of Table 2.11.

   A. What is the value of λ for the Direction One data?

TABLE 2.11   Vehicle Arrival Times and 15-Second Patterns on Steak St.

Direction One			Vehicle Number	Direction Two			Direction One 15-sec Patterns	
Min.	Sec.	h(sec.)		Min.	Sec.	h(sec.)	Period	Cars
0	1.96		1	0	8.46		1 0 to 15	3
0	4.76	2.80	2	0	11.10	2.64	2 15.01 to 30	0
0	7.05	2.29	3	0	15.80	4.70	3 30.01 to 45	2
0	40.86	33.81	4	0	19.00	3.20	4 45.01 to 60	3
0	44.35	3.49	5	0	21.93	2.93	5 60.01 to 75	4
0	47.58	3.23	6	0	25.39	3.46	6 75.01 to 90	0
0	55.64	8.06	7	1	8.84	43.45	7 90.01 to 105	1
0	57.38	1.74	8	1	24.90	16.06	8 105.01 to 120	0
1	3.75	6.37	9	1	27.30	2.40	9 120.01 to 135	2
1	6.00	2.25	10	2	29.16	61.86		
1	8.71	2.71	11	2	32.17	3.01		15
1	13.77	5.06	12	2	35.90	3.73		
1	38.87	25.10	13	2	38.32	2.42		
2	5.42	26.55	14	2	39.94	1.62		
2	8.56	3.14	15	2	59.13	19.19		

B. What is the probability that zero, one, two, three, four, or five Direction One vehicles arrive in any one 15-second time interval?

C. What is the probability that more than five Direction One vehicles arrive in any one 15-second time interval?

Solutions to Example 2.16

A. The average flow or arrival rate λ has units *vehicles per unit time*. (The inverse of these units is *time per vehicle*, which is the units for *headway*.) Donna and her father saw 15 vehicles in Direction One over a time period of 2 minutes and 15 seconds. Therefore, the average arrival rate in Direction One was 15 vehs/135 sec = 0.111 vps = 6.67 veh/min = 400 vph. As we shall soon see, the units for arrival rate must be consistent with the rest of the terms in Equation 2.24.

B. Because t is given in seconds, we must use the veh/sec version of λ. Equation 2.23 for $n = 0$ then becomes

$$P(0) = \frac{(0.111 \text{ vps} * 15 \text{ sec})^0 \, e^{-0.111 \text{ vps} * 15 \text{ sec}}}{0! \text{ veh}} = \frac{(1)(0.189)}{1} = 0.189$$

$$\text{When } n = 1, P(1) = \frac{(0.111 \text{ vps} * 15 \text{ sec})^1 \, e^{-0.111 \text{ vps} * 15 \text{ sec}}}{1! \text{ veh}} = \frac{(1.67)(0.189)}{1} = 0.315$$

$$\text{When } n = 2, P(2) = \frac{(0.111 \text{ vps} * 15 \text{ sec})^2 \, e^{-0.111 \text{ vps} * 15 \text{ sec}}}{2! \text{ veh}} = \frac{(2.78)(0.189)}{2} = 0.262$$

$$\text{When } n = 3, P(3) = \frac{(0.111 \text{ vps} * 15 \text{ sec})^3 \, e^{-0.111 \text{ vps} * 15 \text{ sec}}}{3! \text{ veh}} = \frac{(4.62)(0.189)}{6} = 0.146$$

$$\text{When } n = 4, P(4) = \frac{(0.111 \text{ vps} * 15 \text{ sec})^4 \, e^{-0.111 \text{ vps} * 15 \text{ sec}}}{4! \text{ veh}} = \frac{(7.71)(0.189)}{24} = 0.061$$

$$\text{When } n = 5, P(5) = \frac{(0.111 \text{ vps} * 15 \text{ sec})^5 \, e^{-0.111 \text{ vps} * 15 \text{ sec}}}{5! \text{ veh}} = \frac{(12.86)(0.189)}{120} = 0.020$$

Note that, although the calculations appear tedious, only the $(\lambda t)^n$ and $n!$ terms change as n changes. This is an easy set of calculations to do in a spreadsheet. The relative frequency histogram in Figure 2.25 was produced by a spreadsheet. Each 15-second observation had as its outcome the number of vehicles that arrived. Three times out of the nine 15-second intervals, zero vehicles arrived in Direction One. According to the Direction One mean arrival rate of 0.111 vps, $9 * P(0) = 9 * 0.189 = 1.70$ is the expected frequency of zero arrivals. Each pair of vertical bars in Figure 2.25 shows the relative values for the actual data and the expected value as calculated by using Poisson Equation 2.24.

THINK ABOUT IT

Why not simply define P(0), P(1), P(2), etc. in terms of the lengths of the headways observed? That is, just keep a tally of how many times n events took place during a specified time period t.

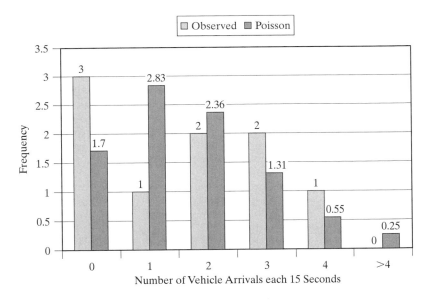

FIGURE 2.25
Frequency histogram for Steak Street Direction One (Actual vs. Theoretical).

Regardless of the method you use, however, be sure to check the resulting $P(n)$ values for reasonableness. Almost always, there is a peak value for $P(n)$, with other $P(n)$ values diminishing as n values get more distant from the mode. (The most frequently occurring value in a distribution is called the *mode*.) In this example, $n = 1$ was the mode, because $P(1)$ had the highest $P(n)$ value. (See the right-hand vertical bars in Figure 2.25.) Look for this kind of peaking pattern in any Poisson distribution. Another check comes with a little experience in using the Poisson model. In this example, we have $\lambda = 6.67$ veh/min. It seems that one or two vehicles every 15 seconds would be typical. The two largest $P(n)$ values we calculated were $P(1)$ and $P(2)$.

C. In this example, $P(n)$ values for $n > 3$ are getting smaller and smaller as n increases. How many more $P(n)$ calculations for $n > 5$ do we have to do before $P(n)$ gets "sufficiently close to zero" (however that is defined) for us to justify stopping? Actually, we need do no more $P(n)$ calculations. We already have the basis for stating $P(n \leq 5)$, the probability that five or fewer vehicles will arrive in any 15-second interval. If we recognize that the sum of all possible $P(n)$ is 1.00, then

$$P(n > 5) = 1.00 - \sum_{n=0}^{5} P(n)$$
$$= 1.00 - (0.189 + 0.315 + 0.262 + 0.146 + 0.061 + 0.020)$$
$$= 1.00 - 0.993 = 0.007.$$

THINK ABOUT IT

Does the choice of t influence the values of $P(n)$? If so, how does one choose t?

Example 2.17

Look at the data in Table 2.11 for Direction Two on Steak Street and reorganize the data into 10-, 15-, and 20-second intervals. (See Table 2.12.) Will the appearance of the frequency distribution change as the interval width used to combine the data is changed? Does the Poisson model change in any important way as the choice of interval width changes?

Solution for Example 2.17   The observations for Direction Two lasted 3 minutes (180 seconds). The average arrival rate λ is 15 vehicles in 180 seconds, or 0.0833 vps. When the interval width is 10 seconds, Equation 2.24 is used to find the probability that *no* vehicles will be observed in an interval, as follows:

$$P(0) = \frac{(0.0833 \text{ vps} * 10 \text{ sec})^0 \, e^{-0.0833 \text{ vps} * 10 \text{ sec}}}{0! \text{ veh}} = \frac{(1)(0.435)}{1} = 0.435$$

This is the first entry in the "0" row in Table 2.12. There are eighteen 10-second intervals in the 3-minute observation period, so 0.435 * 18 or 7.82 intervals can be expected to have zero vehicles observed. We can represent the expected number of times n observations will occur as $E(n)$. In this case, $E(0) = 7.82$. When the interval is 15 seconds long, $P(0) = 0.287$ and only 3.44 of the twelve 15-second intervals are expected to have zero vehicles observed. The remaining calculations are completed in a similar fashion, as summarized in Table 2.12. Note how the mode value becomes less pronounced as the interval gets wider.

THINK ABOUT IT

Does it make sense that $P(0)$ and $E(0)$ are smaller for the 15-second interval than for the 10-second interval? Why?

Figure 2.26 shows the "Actual frequency" distribution for each of the three interval widths in Table 2.12. Each entry in an "Actual freq" column in Table 2.12 stands for the number of times that n vehicles were observed during an interval. This is different from the frequency tally in the two rightmost columns in Table 2.11.

TABLE 2.12   Direction Two Frequency Distributions for Various Interval Widths Used in Example 2.17 (15 vehicles in 180 seconds $=0.8333$ veh/sec).

	10-Second Intervals			15-Second Intervals			20-Second Intervals		
Events per Interval, n	Probability of n Events, $P(n)$	Expected frequency of n, $E(n)$	Actual frequency of n, $A(n)$	Probability of n Events, $P(n)$	Expected frequency of n, $E(n)$	Actual frequency of n, $A(n)$	Probability of n Events, $P(n)$	Expected frequency of n, $E(n)$	Actual frequency of n, $A(n)$
0	0.435	7.82	10	0.287	3.44	5	0.189	1.70	3
1	0.362	6.52	4	0.358	4.30	3	0.315	2.83	2
2	0.151	2.72	1	0.224	2.69	1	0.262	2.36	2
3	0.042	0.75	2	0.093	1.12	1	0.146	1.31	0
4	0.009	0.16	1	0.029	0.35	2	0.061	0.55	1
5+	0.002	0.03	0	0.009	0.11	0	0.028	0.25	1
Sums	1.001	18.00	18	1.000	12.00	12	1.001	9.00	9

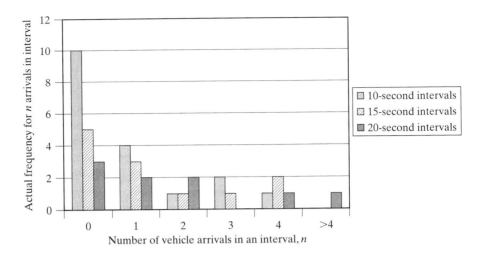

FIGURE 2.26
Frequency distributions for various interval widths in Direction Two.

THINK ABOUT IT

Are you clear about the distinction between "the number of vehicles observed during the interval" and "the number of intervals that had n vehicles observed"?

Although the appearance of the frequency distribution clearly changes with the choice of interval width, does the choice affect calculations using Equation 2.24? No, because no matter how you group data, the calculations using Equation 2.24 can be based on $\lambda = 0.0833$ vps for a period of analysis of any duration. Because the Direction Two arrival rate is five vehicles per minute, let us try computing $P(5)$ for a period of 1 minute. Equation 2.24 becomes

$$P(5) = \frac{(0.0833 \text{ vps} * 60 \text{ sec})^5 \, e^{-0.0833 \text{ vps} * 60 \text{ sec}}}{5! \text{ veh}} = \frac{(3124.4)(0.006751)}{120} = 0.176$$

If λ had been defined in terms of minutes, $\lambda = 15 \text{ veh}/3 \text{ min} = 5.0 \text{ veh/min}$ and the Equation 2.23 calculation becomes

$$P(5) = \frac{(5.0 \text{ veh/min} * 1 \text{ min})^5 \, e^{-5.0 \text{ veh/min} * 1 \text{ min}}}{5! \text{ veh}} = \frac{(3125)(0.006738)}{120} = 0.175$$

Any difference in the two computations would be due to rounding. The lesson: As long as you keep your units consistent (e.g., seconds vs. minutes), Poisson calculations such as $P(n)$ can be done in several ways.

THINK ABOUT IT

In the second $P(5)$ calculation above, $n = \lambda$. Wouldn't you expect $P(\lambda)$ to be much larger than 0.175? In fact, wouldn't you expect $P(\lambda)$ to be the mode of the probability distribution?

Do you trust the Poisson model yet? Are you comfortable with it? Let's try some more experiments with it.

Example 2.18

In using Equation 2.24, is there any difference between using one 15-second time interval and using 15 one-second time intervals?

Solution to Example 2.18 In Example 2.16, $P(0) = 0.189$ when $t = 15$ seconds for Direction One. Because λ had the units "vehicles per second," the $(\lambda t)^n$ term was actually combining 15 one-second intervals, viz., $(0.111 * 15)^n = (1.667)^n$. If λ had been defined in terms of 15-second intervals, $\lambda = 1.667$ vehs/inerval and $t = 1$ interval for the same case, leading to $(\lambda t)^n = (1.667 * 1)^n$. The result is the same, because the time units were kept consistent.

Example 2.19

Let's try to fool the Poisson model. Let us put aside the Steak Street data and invent a case in which vehicles arrive at the middle of every 5-second time period. We could call it the "conveyor belt" case, or think of the vehicles as being towed by a lead vehicle with cables of equal length between them. What would be the parameter values to use in Equation 2.24? What would be the values of $P(0), P(1), \ldots, P(5)$?

Solution to Example 2.19 If a vehicle arrives every 5 seconds, the arrival rate is 0.2 vps = 12 veh/min = 720 vph.

Using Equation 2.24 with $n = 0$, we get

$$P(0) = \frac{(0.2 \text{ vps} * 15 \text{ sec})^0 \, e^{-0.2 \text{ vps} * 15 \text{ sec}}}{0! \text{ veh}} = \frac{(1)(0.050)}{1} = 0.050$$

Note that a value of λ larger than in Example 2.16 has produced a smaller $P(0)$ value.

$$\text{When } n = 1, P(1) = \frac{(0.2 \text{ vps} * 15 \text{ sec})^1 \, e^{-0.2 \text{ vps} * 15 \text{ sec}}}{1! \text{ veh}} = \frac{(3.0)(0.050)}{1} = 0.150$$

$$\text{When } n = 2, P(2) = \frac{(0.2 \text{ vps} * 15 \text{ sec})^2 \, e^{-0.2 \text{ vps} * 15 \text{ sec}}}{2! \text{ veh}} = \frac{(9.0)(0.050)}{2} = 0.225$$

$$\text{When } n = 3, P(3) = \frac{(0.2 \text{ vps} * 15 \text{ sec})^3 \, e^{-0.2 \text{ vps} * 15 \text{ sec}}}{3! \text{ veh}} = \frac{(27.0)(0.050)}{6} = 0.225$$

$$\text{When } n = 4, P(4) = \frac{(0.2 \text{ vps} * 15 \text{ sec})^4 \, e^{-0.2 \text{ vps} * 15 \text{ sec}}}{4! \text{ veh}} = \frac{(81.0)(0.050)}{24} = 0.168$$

When $n = 5$, $P(5) = \dfrac{(0.2 \text{ vps} * 15 \text{ sec})^5 \, e^{-0.2 \text{ vps} * 15 \text{ sec}}}{5! \text{ veh}} = \dfrac{(243.0)(0.050)}{120} = 0.101$

Again, notice the "modal" nature of the results. Only this time there are two modes: $n = 2$ and $n = 3$. With an arrival rate of three vehicles every 15 seconds, we could expect the mode to be $n = 3$, but not $n = 2$. In fact, with the "conveyor belt" pattern of arrivals we have created in our data, we would expect to observe all values of $P(n)$ to be zero except $P(3)$.

THINK ABOUT IT

Do the results of Example 2.19 invalidate the use of the Poisson model for traffic flow?

A few comments about interval size—the t in Equation 2.24—are in order here. In Poisson Assumption 3, "a short time interval" is mentioned. It is possible to choose a time interval that is too long. As Example 2.18 showed, one interval that is 3600 seconds long is equivalent to 3600 intervals that are each 1-second long. However, with one-second intervals there will be a large number of $P(n)$ values to compute over a 1-hour traffic analysis period, and each would be very small. For example, $n = 765$ is just one of many reasonable outcomes for the number of vehicle arrivals in an hour, but $P(765)$ would be a very small value. It is much more efficient to analyze shorter time periods. As for a time interval being too short, a good rule of thumb is to avoid the case where $P(0)$ is larger than any other $P(n)$, $n > 0$. In most cases, if you choose a time interval that meets the needs of your analysis, your results should be satisfactory. For example, if an analysis is focused on the number of vehicles that arrive during a certain existing or proposed length for a red traffic signal phase, that phase length would be a logical basis for choosing the time interval's length. In section 2.6, we present a test to see if the data satisfy the conditions that the Poisson process requires.

2.4.2   The Time Between Vehicle Arrivals

The Poisson process deals with a random variable n that takes on only integer values. What if we want to analyze the time that elapses between the arrival of two consecutive vehicles? That *interarrival time* is measured as a continuous variable. The Poisson Equation 2.24 can be used to derive an equation for this purpose. By definition, the time between two consecutive vehicles has no other vehicles arriving. Let us define X to be the time to the first event after the previous event. The probability that the length of time until the first event will exceed x is the same as the probability that no events will occur in x. According to Equation 2.24,

$$P(X \geq x) = P(0) = \frac{(\lambda x)^0 \, e^{-\lambda x}}{0!} = e^{-\lambda x} \tag{2.25}$$

The cumulative distribution function (CDF) for X is given by

$$P(0 \leq X \leq x) = 1 - e^{-\lambda x} \tag{2.26}$$

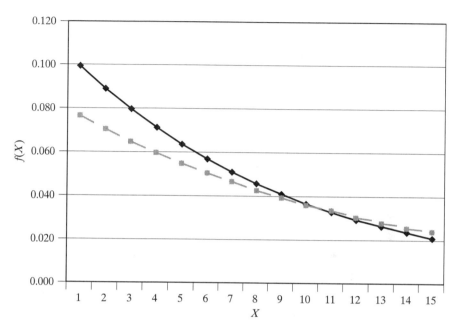

FIGURE 2.27
Plot of negative exponential distribution fitted to Steak Street data.

To get the desired probability density function (PDF) that we need to analyze interarrival times, we take the first derivative of the CDF and end up with the probability density function (PDF)

$$f(X) = \lambda e^{-\lambda x} \qquad (2.27)$$

Equation 2.27 is known as the *negative exponential* distribution. Figure 2.27 shows the shape of the PDF for each direction of traffic on Steak Street.

Because the PDF slopes downward, the PDF indicates that the Poisson model assumptions lead to shorter interarrival times being more likely than longer headways. For practical purposes, the CDF form in Equation 2.26 is more useful than the PDF for analyzing traffic flow, as the next two examples will show.

Example 2.20

The county highway engineer believes that a driver on a side street needs at least a 5-second gap in traffic on Steak Street before feeling safe enough to make a right turn onto Steak Street. If the Direction One data in Table 2.11 represent typical traffic on Steak Street, what is the probability that the next gap on Steak Street will be large enough?

Solution to Example 2.20   Equation 2.24 applies directly: $P(X \geq 5 \sec) = e^{-0.111 \text{ vps} * 5 \sec} =$ 0.574. A side street driver is more likely than not to see a gap greater than 5 seconds.

Example 2.21

The data in Table 2.11 are from actual traffic flow on Steak Street. A current safety rule of thumb is for a driver to leave at least 2 seconds of space between his/her vehicle and the one directly in front. Assume that the vehicles on Steak Street were traveling at 30 mph.

A. How many vehicles in Table 2.11 seemed to be violating the 2-second rule of thumb?

B. Assuming that the data in Table 2.11 represent well the traffic on Steak Street, use the Poisson model (or its relative, the negative exponential model) to determine the probability that any given vehicle will be violating the 2-second rule of thumb.

Solution to Example 2.21   As a preliminary step, we must translate the time between consecutive vehicle arrivals into the time between one vehicle's rear bumper and the front of the next vehicle's bumper. If the average passenger car is about 15 feet long, that distance can be overed in $\dfrac{15 \text{ ft}}{30 \text{ mph} * 1.47 \text{ (fps/mph)}} = 0.34$ seconds at 30 mph. This means that the 2-second rule must be increased to 2.34 seconds to account for the difference between "distance between vehicles" and "time between vehicle arrivals."

A. Three vehicles in Direction One and one vehicle in Direction Two violated the rule of thumb, because they arrived less than 2.34 seconds after the previous vehicle.

B. To find $P(X < 2.34 \text{ seconds})$, use Equation 2.26 as follows:

$$P(X < 2.34 \sec) = 1 - P(X \geq 2.34 \sec) = 1.00 - \frac{(\lambda x)^0 \, e^{-\lambda x}}{0!} = 1.00 - e^{-\lambda x}$$

In Example 2.16, we calculated λ to be 0.111 vps. For Equation 2.25, $x = 2.34$ seconds.

$$P(X < 2.34 \sec) = 1.00 - e^{-0.111 \text{ vps} * 2.34 \sec} = 1.00 - 0.771 = 0.229.$$

Because λ is different for Direction Two, $P(X < 2.34 \text{ seconds})$ will have to be calculated separately for that direction.

Design of Left-Turn Lanes.   The performance of a left-turn lane can be determined if one knows the signal timing and the arrival rate of traffic that is likely to want to turn left during the peak hour. If the traffic is fully deterministic (nonrandom), the length of the lane can be exactly determined, but under the concept of Poisson arrivals, it will be good to make sure that the length of the lane will be adequate some specified percentage of time. When there is not room for a car in the left-turn lane, that car will obstruct the through lanes, creating both congestion and a safety hazard.

Example 2.22

A left-turn lane has a storage capacity of seven cars, which is deemed adequate for an average arrival rate of 60 cars per minute with an average of 10 percent of the cars wanting to turn left. The light cycle is 60 seconds long. The green turn signal will always accommodate the first six cars waiting to turn. However, if we assume Poisson arrivals and that the left-turn lane is emptied in each cycle, what is the probability that there will be more than the seven cars for which the lane is designed?

Solution to Example 2.22 The average arrival rate to turn left is 0.10 * 60 = 6 cars per minute. Using Poisson Equations 2.24 and 2.26 (see the table), the number of cars that arrive in each 1-minute interval will need to be seven or fewer for the lane to work as designed. Because 74.4 percent of the time there will be seven or fewer cars, there will be more than seven cars 25.6 percent of the time. This may be too often for safe operating conditions. If the supervisor wants the left-turn lane to be long enough for 90 percent of the cycles, the lane will need to have a capacity of nine cars.

Number of Cars Arriving in a 60-Second Interval	Probability of That Number of Cars Arriving	Total Probability of $Pr(n < N)$
0	0.002	
1	0.015	0.017
2	0.045	0.062
3	0.089	0.151
4	0.134	0.285
5	0.161	0.446
6	0.161	0.606
7	0.138	0.744
8	0.103	0.847
9	0.069	0.916
10	0.041	0.957

2.5 MEASURING ROADWAY PERFORMANCE

2.5.1 Capacity and Level of Service

The concept of *highway capacity* is actually quite simple. A straightforward definition is:

> the maximum hourly rate at which persons or vehicles can reasonably be expected to traverse a point or uniform section of a lane or roadway during a given time period under prevailing roadway, traffic, and control conditions (HCM, 2000, p. 2-2).

Capacity is an important value to estimate. Some reasons are as follows:

- It provides the service rate value μ for queueing analysis (sections 3 and 4 in Chapter 3).
- It is an essential input value for link performance functions in route choice models (section 5 in Chapter 4).
- It serves as a basis for *level of service* evaluation (defined later in this section and discussed more in Chapter 3).
- It determines whether an existing facility (roadway or intersection) is adequate to meet the demands placed upon it by traffic flows.
- It will indicate whether an expanded or new roadway is required to accommodate traffic demands under specified future conditions.

Despite the simplicity of the definition of capacity, there are some complications involved in determining the number to use for capacity. For example: Does it matter what types of vehicles are in the traffic flow? Clearly, a roadway that has just enough capacity for 1740 passenger cars per hour could never handle 1740 semitrailer trucks or other large vehicles in 1 hour. Perhaps some kind of weighting factor can convert larger vehicles into passenger car equivalents. A way to address this complication is to classify each vehicle by length or by weight at permanent vehicle classification or weigh-in-motion stations. Portable data collection equipment for these purposes also exists.

A concept related to capacity is *level of service*. The flow (demand) at many roadways and intersections never reaches the capacity of the facility. At other facilities, capacity may be reached only a small percentage of the time. However, congestion can be present well before demand approaches capacity. In fact, it is possible to distinguish varying degrees of congestion by relating existing flow to a facility's capacity. For example, we could establish a ratio of flow to capacity below which no congestion is considered to exist. As flow increases with respect to capacity, congestion begins and grows in intensity. The level of service (LOS) of a roadway is much like a grade given in an academic course: A is best, F is worst, with grades B, C, D, and E in between. "LOS A describes free-flow operations. Free-flow speeds prevail. Vehicles are almost completely unimpeded in their ability to maneuver within the traffic stream. LOS F describes breakdowns in vehicular flow. [B]reakdown occurs when the ratio of existing demand to actual capacity or of forecast demand to estimated capacity exceeds 1.00" (HCM 2000, p. 13-8 and 13-10). As the ratio of flow rate to capacity on a roadway increases, level of service degrades from LOS A toward LOS F. The photographs in Figure 2.28 show how the traffic stream looks on a freeway under the various LOS conditions.

The principal source for analyses pertaining to capacity and level of service is the *Highway Capacity Manual* (HCM, 2000). The current HCM was released in the year 2000. The HCM is the product of extensive data collection and the result of considerable analytical work by numerous committees sponsored by the Transportation Research Board. The HCM contains procedures to analyze capacity and level of service for a variety of roadways, such as two-lane highways, freeways, and multilane highways. It also provides techniques for analyzing signalized and unsignalized intersections. There are also HCM chapters on urban streets and freeway ramps. At the beginning of Chapter 3 of this book, the HCM procedures for Two-Lane Highways and Basic Freeway Sections are introduced.

2.5.2 Highway Performance Monitoring

Federal law requires each state to record and analyze data on traffic that flows on the highways in that state. The federal data reporting program is called the Highway Performance Monitoring System (HPMS) (FHWA, 1993). The HPMS attempts to estimate information (traffic volume, annual vehicle distance traveled, annual average daily traffic, vehicle classification, and truck weights) for all public roads in the United States by having each state DOT use a specified statistical sampling program. Let us first look at the nature of the traffic data that must be gathered.

Traffic data gathering can be thought of as a three-dimensional process, as shown in Figure 2.29. Along each of the three axes, the data are segmented to provide for analysis with the required accuracy. Each axis reflects a different type of data to be

(a) LOS A

(b) LOS B

(c) LOS C

(d) LOS D

(e) LOS E

(f) LOS F

FIGURE 2.28
Illustrations of levels of service. *Source*: HCM, 2000, Exhibits 13-5 through 13-10.

gathered. The data gathered are to cover variations that occur in highway traffic during a day and over the course of a year.

Volume or Flow Data.   The first element of data to be gathered is the flow or volume count. See section 2.1.5 for a description of coverage counts and seasonal

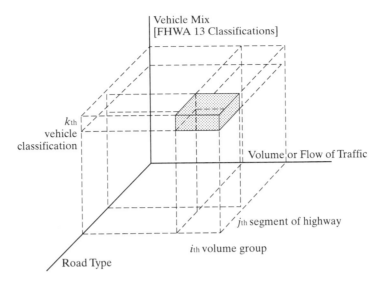

FIGURE 2.29
Traffic space for highway performance monitoring. *Source:* Whitford, 1998.

adjustments. As we have already discussed in section 2.1.2, there are various ways to determine flow. The data are segmented into groups of highways that have similar volumes called *volume groups*.

Road Functional Classification. The flow for each volume group has been segmented according to the functional class of the road (called "Road Type" in Figure 2.29). Table 2.3 indicates the various functional classes of highway. The HPMS requires that data be gathered on enough segments of each classification to provide a good statistical sample of all roads in the state.

Vehicle Classification. Although vehicle counts and speed may be the most familiar types of traffic data to be collected, there is considerable value in knowing about the number and size of heavier vehicles. As we will see when analyzing the performance of two-lane roads and freeways, heavier vehicles have a detrimental effect on the quality of traffic flow. They also affect the life of pavements, which is discussed in Chapter 8. In addition, if vehicle classification can be automated in "real time," electronic toll collection for all vehicles becomes possible.

Just as the federal government has set up standard functional classifications for roads (Table 2.3), the FHWA has established thirteen vehicle classes. (See Table 2.13 and Figure 2.30.) As part of the HPMS sampling program, state DOTs must adopt technologies that identify vehicles reliably and economically. The most common approach is to attempt to identify the axle configurations of vehicles and then match them against the FHWA classes in Table 2.13 and Figure 2.30. The standard Class 9 truck (see Figures 2.30 and 2.31) has five axles.

The only practical way for a state DOT to collect enough data for a statistically acceptable sample is to use a device that detects, interprets, and stores information

TABLE 2.13   The 13 FHWA Vehicle Classes

Class 1 = Motorcycles	Class 8 = 4 or fewer axle combination trucks
Class 2 = Passenger cars	Class 9 = 5-axle combination trucks
Class 3 = Other 2-axle, 4-tire single-unit vehicles	Class 10 = 6+ axle combination trucks
Class 4 = Buses	Class 11 = 5-axle multi-trailer trucks
Class 5 = 2-axle, 6-tire single-unit trucks	Class 12 = 6-axle multi-trailer trucks
Class 6 = 3-axle, 6-tire single-unit trucks	Class 13 = 7+ axle multi-trailer trucks
Class 7 = 4+ axle single-unit trucks	

Source: Traffic Monitoring Guide, FHWA, 2001.

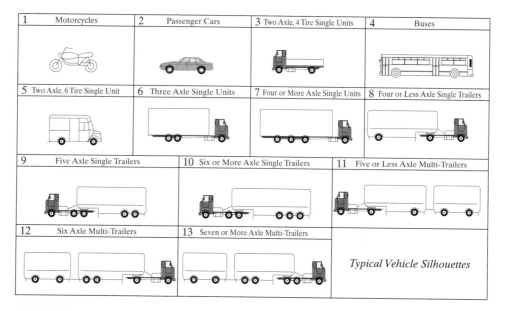

FIGURE 2.30
FHWA vehicle classifications. *Source:* AASHTO, 1992.

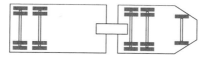

FIGURE 2.31
Axle configuration for Class 9 tractor-semitrailer.

about each vehicle as it passes over the device. As was the case with speed detection and traffic volume counting, the ideal situation is to have a large number of permanent data collection points. Unfortunately, the task of weighing and classifying vehicles while they are moving along a roadway is more difficult than recording speeds and counts. Accordingly, automated vehicle classification (AVC) and weigh-in-motion (WIM) stations are more expensive and less reliable than speed and count sites.

An AVC system consists of lane sensor devices that record the physical characteristics of vehicles and a processing unit, which aggregates input from the various sensor devices and interprets this input to assign a class to each vehicle passing through the lane. Some of the equipment types available to determine vehicle classification are listed below (ETTM On The Web, 1996).

- *Inductive loops.* These are wires placed in channels cut into the road bed and are used to detect vehicle presence by sensing the metallic mass of the vehicle. See also section 2.1.4 for a description of "loop detectors."
- *Treadles.* These are pressure-sensitive devices placed in frames installed in the roadbed and are used to determine the number of axles, number of wheels, and direction of a vehicle crossing the treadle. A parallel series of sensor devices detects the direction of axle movement by the sequence of sensor activation. Treadles can be classified into several types based on the physical principle used to convert the pressure of a vehicle's wheel into electrical signals recognized by the logic units of the treadles.
- *Electromechanical treadles.* These devices are simple electromechanical devices and are in widespread use for low-speed applications; however, they are reported to be inaccurate at speeds over 55 mph. They also have a high maintenance cost.
- *Resistive rubber treadles.* These devices are similar to electromechanical treadles but use resistive rubber rather than metal for contact closure. They are specified to operate accurately at speeds from 2 to 80 mph. They also have a lower maintenance cost than metallic units.
- *Optical treadles.* These devices use infrared beams inside a tube. The beam is broken, and an electrical signal is generated when an axle crosses the treadle. These devices are reported to be accurate at higher speeds and have a long life and low-cost maintenance. They can be installed in standard treadle frames.
- *Piezoelectric treadles.* These devices use special material inside a tube. The material generates an electric current when subjected to pressure caused by axle crossing the treadle. The devices are reported to be quite accurate at speeds greater than 5 mph, and recent developments have made them accurate even in the 0 to 5 mph range.

THINK ABOUT IT

Given one or more axle detectors, describe a procedure whereby you could record the distance between consecutive axles. Once you have solved that problem, decide how you would know which axles belonged to which vehicles.

2.6 TESTING THE PROPOSED MODEL

2.6.1 Is Poisson the Right Model?

Donna and her father were looking at a plot (Figure 2.25) of the Steak Street data (Table 2.11) they had collected. They were trying to decide if the Poisson model was a "good enough" representation of the actual arrival patterns they had observed, when Donna's brother Darren came home. Darren was an engineering major who was about to enter his senior year of college. Having taken a course in Probability and Statistics, Darren recalled seeing a way to test a proposed model against the data it was supposed to represent. He went to his room to look for his Statistics text.

In section 2.4, we introduced the Poisson assumptions. The activity we are studying must have the properties expressed by those assumptions reflected in the data we collect about the activity. In Example 2.19, we intentionally violated the first Poisson Assumption (random events) by using a "conveyor belt" case, but using Poisson Equation 2.24 gave us plausible $P(n)$ values anyway. *They were "plausible" because they were not obviously wrong. However, "plausible" values may not be values that conform to a proposed model.*

We need a reliable systematic way to test the data for compliance with Poisson properties. Such a method is the chi-square goodness of fit test (Ang and Tang, 1975, section 6.3.1; Walpole and Myers, 1998, section 10.14). The chi-square steps are listed below. We use them as a checklist in a subsequent example.

1. Group the possible values of the random variable into intervals. It is preferable to have the number of intervals, k, be at least five. List these values or ranges of values as entries in column one of a goodness of fit test (GFT) table that will eventually have five columns, plus a "column zero" that will contain the row or interval number i, $1 \leq i \leq k$.

2. Enter the number of times each value is observed in the data. Create a column (2) in the GFT table for this purpose. Call these entries o_i, where **i** represents each interval defined in step 1.

3. Calculate the probability that each value will occur on the basis of the assumed probability distribution. In this chapter, we are testing the Poisson model as represented by Equation 2.24, which is repeated here:

$$P(n) = \frac{(\lambda t)^n e^{-\lambda t}}{n!} \tag{2.24}$$

Place the calculated probabilities in the appropriate row of column three in the GFT table.

4. Multiply each probability by the total number of events observed in the data and put these theoretical frequencies e_i into column four of the GFT table. Note: If any $e_i < 5$, for $i < k$, then consider making the intervals larger by returning to step 1 or by increasing the amount of data. If the last interval has too few events, then combine the last two intervals until $e_i \geq 5.0$. If combining original intervals at the bottom of the table causes $k < 5$ to result, you probably have too few observations and your goodness of fit results may be invalid.

5. For each interval i, square the difference between o_i and e_i and divide this squared difference by e_i. Place these χ_i^2 values in column five of the GFT table.

$$\chi_i^2 = \frac{(o_i - e_i)^2}{e_i} \tag{2.28}$$

6. Sum the χ_i^2 values and place the result at the bottom of column five in the GFT table.

7. Calculate the degrees of freedom. The degrees of freedom requires that you subtract out the number of places v that you used the data. (For example, if we used the data to determine the Poisson parameter λ, and our data had been separated into six intervals, $f = k - v - 1 = 6 - 1 - 1 = 4$. Insert the degrees of freedom value at the bottom of column 1 in the GFT table.

8. Choose a value for α, which determines the confidence level of the analysis. The standard value is $\alpha = 0.05$, which means that we would expect the data to satisfy the Poisson distribution 95 percent of the time. The parameter α is called the *tail area*. Insert the α value at the bottom of column 2 in the GFT table.

9. Look up the critical value of χ_α^2 for the number of degrees of freedom f and tail area α found in the two previous steps. Tables with these critical values can be found in almost any Probability and Statistics text. (See Table 2.14.) The critical value χ_α^2 is the value of χ^2 that has an area α to the right of it under the chi-square curve in Figure 2.32. Enter the critical value of χ_α^2 at the bottom of column 3 in the GFT table.

10. Add up the χ_i^2 values in column five of the GFT table and place the sum at the bottom of that column. If

$$\sum_{i=1}^{k} \frac{(o_i - e_i)^2}{e_i} < \chi_\alpha^2,$$

then the observed data match the theoretical frequencies e_i sufficiently well to be represented by the assumed theoretical distribution. (In this case, the assumed distribution is the Poisson.) If not, the proposed model must be rejected.

THINK ABOUT IT

Does it make sense that the chi-square goodness of fit test rejects a proposed model if $\sum_{i=1}^{k} \frac{(o_i - e_i)^2}{e_i} > \chi_\alpha^2$ and accepts the proposed model if $\sum_{i=1}^{k} \frac{(o_i - e_i)^2}{e_i} < \chi_\alpha^2$? Explain in your own words.

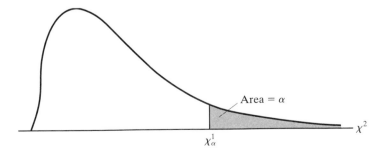

FIGURE 2.32
Critical value of test statistic χ_α^2 for a specified tail area α.

TABLE 2.14   Critical Values for Test Statistic χ^2_α in Chi-Square Goodness of Fit Test

Degrees of Freedom	Upper-Tail Area α						
	.30	.20	.10	.05	.02	.01	.001
1	1.074	1.642	2.706	3.841	5.412	6.635	10.827
2	2.408	3.219	4.605	5.991	7.824	9.210	13.815
3	3.665	4.642	6.251	7.815	9.837	11.345	16.268
4	4.878	5.989	7.779	9.488	11.668	13.277	18.465
5	6.064	7.289	9.236	11.070	13.388	15.086	20.517
6	7.231	8.558	10.645	12.592	15.033	16.812	22.457
7	8.383	9.803	12.017	14.067	16.622	18.475	24.322
8	9.524	11.030	13.362	15.507	18.168	20.090	26.125
9	10.656	12.242	14.684	16.919	19.679	21.666	27.877
10	11.781	13.442	15.987	18.307	21.161	23.209	29.588
11	12.899	14.631	17.275	19.675	22.618	24.725	31.264
12	14.011	15.812	18.549	21.026	24.054	26.217	32.909
13	15.119	16.985	19.812	22.362	25.472	27.688	34.528
14	16.222	18.151	21.064	23.685	26.873	29.141	36.123
15	17.322	19.311	22.307	24.996	28.259	30.578	37.697
16	18.418	20.465	23.542	26.296	29.633	32.000	39.252
17	19.511	21.613	24.769	27.587	30.995	33.409	40.790
18	20.601	22.760	25.989	28.869	32.346	34.805	42.312
19	21.689	23.900	27.204	30.144	33.687	36.191	43.820
20	22.775	25.038	28.412	31.410	35.020	37.566	45.315
21	23.858	26.171	29.615	32.671	36.343	38.932	46.797
22	24.939	27.301	30.813	33.924	37.659	40.289	48.268
23	26.018	28.429	32.007	35.172	38.968	41.638	49.728
24	27.096	29.553	33.196	36.415	40.200	42.980	51.179
25	28.172	30.675	34.382	37.652	41.566	44.314	52.620
26	29.246	31.795	35.563	38.885	42.856	45.642	54.052
27	30.319	32.912	36.741	40.113	44.140	46.963	55.476
28	31.391	34.027	37.916	41.337	45.419	48.278	56.893
29	32.461	35.139	39.087	42.557	46.693	49.588	58.302
30	33.530	36.250	40.256	43.773	47.962	50.892	59.703

Example 2.23

Darren finds his Probability and Statistics textbook. (See the first paragraph of this section.) In the text, he locates the chi-square goodness of fit test just described in the 10 steps above. Darren is ready to determine whether the data Donna collected with her father for Direction One on Steak Street (see Table 2.11) can be represented by the Poisson model.

Solution to Example 2.23   Darren follows the 10 steps in the chi-square goodness of fit test and builds the GFT table as described above.

1. The basis for our table is the choice of 15 seconds for our time interval. There are nine such time intervals in the 2 minutes and 15 seconds during which Direction One data were observed. For each 15-second time interval, the number of events (vehicle arrivals) has been tallied. During one 15-second time interval, four events were observed. In all other time intervals, three or fewer vehicles arrived. In three of the remaining eight time intervals, no vehicles were observed. The intervals in Table 2.15 are defined by the frequency of events—0, 1, 2, 3, and 4—occurring in any 15-second time interval. Note that there are five such intervals in the GFT table (Table 2.15). This is the minimum value of k that is desirable.

2. There were 15 vehicles observed in Direction One on Steak Street, but they have been placed into five intervals in Table 2.15, based on the number of vehicle arrivals in each 15-second time interval. There were nine 15-second time intervals, meaning that there were nine observations, each with its own number of arrivals. The number of times zero arrivals occurred was three, the number of times one arrival occurred was one, etc. Enter these values in column 2 of the GFT table. These are the o_i values for each interval i.

3. We have already calculated the probability that each value will occur on the basis of the assumed Poisson probability distribution. We did this in Example 2.16. Although there were no *observed* cases in Direction One with $n \geq 5$, the probability exists: $P(5+) = 1.00 - \Sigma_{n=0}^{4}P(n) = 1 - 0.973 = 0.027$.

4. If $P(0) = 0.189$ according to the Poisson model and there are nine time intervals of 15 seconds each, what is the expected number of times (out of nine), that zero vehicles will arrive in a 15-second time interval? The first entry in column 4 of the table becomes $e_1 = P(0)*9 = 0.189*9 = 1.70$. After computing each column 4 entry, *we can see that none of the entries meets the $e_i \geq 5.0$ criterion for the chi-square test. With*

TABLE 2.15   Chi-Square Goodness of Fit Test for Example 2.23

(0)	(1)	(2)	(3)	(4)	(5)
Interval i	Frequency n	Observed o_i	Probability $P(n)$	Theoretical $P(n)*N = e_i$	Equation 2.28 χ_i^2
1	0	3	0.189	1.70	0.994
2	1	1	0.315	2.83	1.186
3	2	2	0.262	2.36	0.055
4	3	2	0.146	1.31	0.361
5	4	1	0.061	0.55	0.376
6	5+	0	0.027	0.25	0.248
		$\Sigma o_i = 9 = N$	$\Sigma = 1.000$	$\Sigma e_i = 9.00$	3.220

Degrees of Freedom	Tail Area Probability	Critical Value	
$f = k - 1 - 1 = 4$	$\alpha = 0.05$	$\chi_\alpha^2 = 9.488$	from Table 2.14

$N = 9$, even two intervals would be too many to meet this requirement. Darren notes this shortcoming to his sister and father, but he proceeds with the test to demonstrate its application. In any case, you should check your calculations by verifying that $\Sigma e_i = 9.00 = \Sigma o_i = 9 = N$.

5. The first entry in column 5 will be calculated by using Equation 2.28 as follows:

$$\chi_1^2 = \frac{(o_1 - e_1)^2}{e_1} = \frac{(3 - 1.70)^2}{1.70} = \frac{1.69}{1.70} = 0.994$$

6. The sum of the χ_i^2 values in column 5 = 3.220. We call this sum the *test statistic*.

7. Degrees of freedom $f = k - v - 1 = 6 - 1 - 1 = 4$

8. Use the standard value of $\alpha = 0.05$ for the tail area

9. The *critical value* for χ_α^2 such that the tail area = 0.05 can be found in Table 2.14. In the row for "Degrees of Freedom = 4" and the column "Upper-Tail area = 0.05," the entry "9.488" is found. (See also Figure 2.33.) Enter the values from steps 5–8 at the bottom of the GFT solution table (Table 2.15).

10. If the test statistic is smaller than the critical value, there is no reason to reject the hypothesis that the Poisson model is a good model for the phenomenon represented by the data. In this example, the test statistic (3.220) is smaller then the critical value (9.488). It appears that the proper decision is to accept the Poisson model as appropriate. However, as Darren has pointed out, the $e_i \geq 5.0$ criterion has been violated. If a serious analysis of Steak Street arrival patterns is desired, we ought not to base a decision on such a small amount of data collected in such an impromptu way.

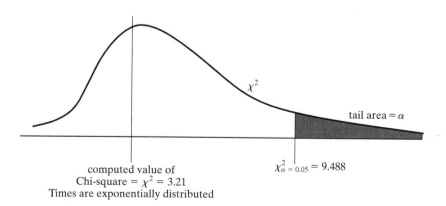

FIGURE 2.33
The critical value of χ_α^2 for Example 2.23.

Example 2.24

Donna didn't pay much attention when Darren showed their father how to conduct a chi-square goodness of fit test, but she thinks counting cars is fun. She prevails upon her big brother to return to their front yard and get more data. Darren agrees, remembering that $k > 5$ and $e_i > 5$ are conditions that lead to a more reliable test. With these conditions in mind, Darren collects data with Donna for 6 minutes. This gives them 24 intervals, each 15 seconds long. Darren organizes the data as shown in Table 2.16. Do the data fit a Poisson model?

Solution for Example 2.24

Step 1: A frequency distribution based on the vehicle arrivals observed by Donna and Darren can be entered in column 4 of a GFT table (Table 2.17).

Step 2: Note that four of the 24 cells in Table 2.16 are empty, meaning that four of the 15-second intervals had no vehicle arrivals. Thus, "4" is the first entry in column 4. Likewise, count the number of intervals in which one vehicle arrived, two vehicles arrived, etc., and enter the results in column 4 of Table 2.17.

Step 3: Because 48 vehicles were observed, the arrival rate λ can be expressed as eight vehicles per minute, two vehicles per 15-second time interval, or 0.133 vehicles per second. Use Equation 2.24 to compute the $P(n)$ values for column 2 of the GFT table. Using a spreadsheet can make these calculations and those that follow much more convenient and error-free.

THINK ABOUT IT

Experiment and convince yourself that the entries in column 2 will be the same, no matter which of the three arrival rates λ given above are used.

TABLE 2.16 New Set of Data on Steak Street for Example 2.24

Seconds	Minutes					
	0	1	2	3	4	5
0.0 to 15.0	1.8 6.3	1.7 7.2 13.2	2.7 8.3			1.2 12.3 3.7 9.6
15.1 to 30.0	15.2 24.6 17.8 26.9 20.4 29.5	25.1	15.1 23.7 17.3 27.6 19.9	17.6 25.5	16.1 23.4 20.2 27.2	
30.1 to 45.0	36.1 42.2		33.4	39.3 43.7	32.3 33.2 40.2	
45.1 to 60.0	52.2		49.2	45.9 58.2 48.6 52.7	46.3 46.6 51.3 58.9 57.3	
Cars	11	4	9	8	9	7

TABLE 2.17 Chi Square Goodness of Fit Test for Example 2.24

(0)	(1)	(2)	(3)	(4)	(5)
Interval	N = Cars/ Interval	$P(n)$	Frequency Expected $e = 24*P$	Observed Frequency	Equation 2.28 χ^2
1	0	0.1353	3.25	4	0.174
2	1	0.2707	6.50	6	0.038
3	2	0.2707	6.50	7	0.039
4	3	0.1804	4.33	3	0.409
5	4	0.0902	2.17	2	0.013
6	5	0.0361	0.87	1	0.021
7	6	0.0120	0.29	1	1.752
8	7	0.0045	0.11	0	0.11
		Σ	Σ	Σ	Σ
		1.0000	24.00	24	2.556

Degrees of Freedom	Tail Area Probability	Critical Value
$f = k - 1 - 1 = 6$	$\alpha = 0.05$	$\chi_\alpha^2 = 12.592$ from Table 2.14

Step 4: Use the equation $e_i = P(n)*N$, with $N = 24$, to fill in each row in column 3.

Step 5: Use Equation 2.28 to complete the χ^2 entries in column 5.

Step 6: Find the sum of the column 5 entries.

Step 7: Degrees of freedom $f = k - 1 - 1 = 8 - 1 - 1 = 6$.

Step 8: Use the standard tail area probability $\alpha = 0.05$.

Step 9: From Table 2.14, $\chi_\alpha^2 = 12.592$.

Step 10: Because the column 5 sum 2.556 is less than the critical value 12.592, the hypotheses that the Poisson model can represent the data can be accepted.

Example 2.25

In Example 2.19, the "conveyor belt" arrivals produced a theoretical frequency distribution of $P(n)$ that resembled a Poisson model, but we didn't subject the results to a goodness of fit test. Let us apply our chi-square test to 6 minutes of "conveyor belt" data now.

Solution to Example 2.25 The $P(n)$ results from Example 2.19 can be inserted into column 2 of Table 2.18. Actually, Table 2.18 is a spreadsheet, which accounts for the "precise" entries in columns 2 and 3. Column 3 indicates that the observations lasted 6 minutes, or 24 intervals of 15 seconds each. Note in column 4 that every 15-second time interval had exactly three vehicle arrivals. For the fourth interval, in which $n = 3$, Equation 2.28 gives us

$$\chi_4^2 = \frac{(o_4 - e_4)^2}{e_4} = \frac{(24 - 5.377)^2}{5.377} = \frac{346.816}{5.377} = 64.500$$

TABLE 2.18   Goodness of Fit Table for Example 2.25

(0) Interval	(1) N = Cars/ Interval	(2) $P(n)$ Equation 2.24	(3) Frequency Expected 24	(4) Observed Frequency	(5) Equation 2.28 x^2
1	0	0.049787	1.194890	0	1.195
2	1	0.149361	3.584669	0	3.585
3	2	0.224042	5.377003	0	5.377
4	3	0.224042	5.377003	24	64.500
5	4	0.168031	4.032753	0	4.033
6	5	0.100819	2.419652	0	2.420
7	6	0.050409	1.209826	0	1.210
		Σ 0.9989	Σ 23.97	Σ 24	Σ 82.320

Degrees of Freedom	Tail Area Probability	Critical Value
$f = k - 1 - 1 = 5$	$\alpha = 0.01$	$\chi_\alpha^2 = 20.517$

to enter in column 5, row 4. For all the other intervals, Equation 2.28 leads to

$$\chi_i^2 = \frac{(o_i - e_i)^2}{e_i} = \frac{(0 - e_i)^2}{e_i} = e_i$$

The column 5 sum is rather large, so large that no value in the $f = 5$ row of Table 2.18 would permit us to accept the Poisson model as a reasonable representation of the conveyor belt data.

THINK ABOUT IT

The e_i values in Example 2.25 and in column 3 of Table 2.18 are based on the conveyor belt arrival rate being used as input to Poisson Equation 2.24. Why didn't the conveyor belt data pass the goodness of fit test in Example 2.25?

SUMMARY

This chapter begins the introduction to the fundamentals of transportation engineering with a familiar topic—traffic. Traffic can be measured in terms of its basic characteristics: speed, density, and flow rate. Standard methods and equipments for carrying out these measurements are described in this chapter. In addition, the fundamental relationships between the basic traffic flow characteristics are presented, along with procedures whereby raw traffic data can be converted into approximations of the proposed traffic flow model relationships.

The random behavior of traffic flow that is not influenced by traffic signals or congestion is introduced. Cases in which random behavior cannot be assumed are covered in Chapter 3. Standard procedures for evaluating the performance of a roadway are also introduced in this chapter, forming the basis for more specific analyses in subsequent chapters.

As will be the case throughout this book, situations that the reader is likely to encounter as an engineer or as a member of a mobile society have been used to illustrate essential principles in transportation engineering. Whenever possible, familiar situations will be used to help the reader see those situations with an increased understanding of the underlying principles at work. This approach will also help the reader to analyze such situations for purposes of evaluation, mitigation, and (re)design.

ABBREVIATIONS AND NOTATION

Traffic Flow: Theory and Analysis

AADT	annual average daily traffic
ADT	average daily traffic
AVC	automated vehicle classification
D	density, vehicles per mile or vehicles per kilometer
DHV	design hourly volume
E	permitted error in the average speed estimate, mph
EL	the effective length of a loop detector
fps	feet per second
fpv	feet per vehicle
h	headway
HPMS	Highway Performance Monitoring System
K	the proportion of daily traffic that occurs during the peak hour at a given location, Equation 2.3
mph	miles per hour
mps	meters per second
pcphpl	passenger cars per hour per lane
PHF	peak hour factor, Equation 2.2
q	traffic flow rate, vph
S	speed
S	estimated sample standard deviation
sec	seconds
SMS	space mean speed
TMS	time mean speed
TMSH	traffic monitoring system for highways
U	constant that corresponds to a desired percentile speed
V or v	volume
V_{15}	peak 15-minute count
veh	vehicle
veh/min	vehicles per minute
vps	vehicles per second
vpd	vehicles per day
vph	vehicles per hour

vphpl	vehicles per hour per lane
vpk	vehicles per kilometer
vpm	vehicles per mile
vpmpl	vehicles per mile per lane
WIM	weigh-in-motion
z	standard normal deviate that corresponds to a desired confidence level

GLOSSARY

Axle correction factor: The average number of axles per vehicle.

Classification: The process of breaking down the total number of vehicles observed into categories or classes.

Coverage counts: 48-hour traffic counts made by using portable equipment.

Design hourly volume: The hourly volume used by roadway designers to determine the capacity needed.

Density: A measure of traffic congestion, the number of vehicles per unit length of a roadway.

Effective length: The distance between the point at which a vehicle begins to be detected by a loop and the point at which it is no longer detected.

Flow: Traffic volume converted to a rate per unit of time, most commonly vehicles per hour.

Free-flow speed: Speed that can be maintained on a given roadway when no other vehicles are present.

Gap: The space between consecutive vehicles, rear bumper to front bumper.

Headway: Time or distance between consecutive vehicles.

Jam density: The maximum possible density on a roadway.

Occupancy: The percent of time in which a point in the roadway has a vehicle that is directly over it.

Seasonal adjustments: Adjustments made to coverage counts to account for seasonal variations in traffic flow at the site of the count.

Space mean speed: The average speed of all vehicles *moving over a given section* of a highway over some specified time period.

Time mean speed: The average speed of all vehicles *passing a point* on a highway over some specified time period.

Volume: A count of traffic past a point made for some specified time period.

85th percentile speed: The speed that is exceeded by 85 percent of the vehicles whose speeds have been recorded in a representative sample of those vehicles that travel the roadway.

REFERENCES

[1] AASHTO. American Association of State Highway and Transportation Officials, *Guidelines for Traffic Data Programs*, Washington DC, 1992, p. 19.

[2] AASHTO. American Association of State Highway and Transportation Officials, *A Policy on Geometric Design of Highways and Streets*, Fourth Edition, Washington DC, 2001.

[3] Ang, Alfredo H-S. and Wilson H. Tang, *Probability Concepts in Engineering Planning and Design, Vol. I Basic Principles*, John Wiley & Sons, 1975.

[4] Autoscope, Freeway Applications, December 04, 2000. Retrieved 3 June 2002 from www.hadlandphoto.com.au/autoscop.htm

[5] Banks, J.H., "Freeway Speed-Flow-Concentration Relationships: More Evidence and Interpretations," *Transportation Research Record 1225*, Transportation Research Board, 1989, p. 53–60.

[6] Box, Paul C. and Joseph C. Oppenlander, *Manual of Traffic Engineering Studies*, Fourth Edition, Institute of Transportation Engineers, Arlington, VA, 1976.

[7] Drake, J.S., J.L. Schofer, and A.D. May, "A Statistical Analysis of Speed-Density Hypotheses," *Transportation Research Record 154*, Transportation Research Board, 1967, p. 78–79.

[8] ETTM On The Web, 1996-present. Retrieved 2 June 2002 from www.ettm.com/

[9] FHWA. Federal Highway Administration, *Highway Performance Management System Field Manual*, Chapter IV, Washington DC, 30 August 1993. [www.fhwa.dot.gov/ohim/hpmspage.htm]

[10] FHWA. Federal Highway Administration, Office of Highway Policy Information, *Traffic Monitoring Guide*, 2001.

[11] FHWA. Federal Highway Administration, *1996 Highway Statistics*, 1997.

[12] Florida Department of Transportation, *Florida's Level of Service Standards and Guidelines Manual for Planning*, Systems Planning Office, Tallahassee, 1995, as cited in HCM 2000, p. 8–9.

[13] Gerlough, D. and M. Huber, *Traffic Flow Theory: A Monograph, Special Report 165*, Transportation Research Board, National Research Council, Washington DC, 1975.

[14] Greenshields, B., "A Study of Traffic Capacity," *Proceedings of the Highway Research Board, Vol. 14*, Transportation Research Board, National Research Council, Washington DC, 1934.

[15] Hall, F., V. Hurdle, and J. Banks, "A Synthesis of Recent Work on the Nature of Speed-Flow and Flow-Occupancy Relationships on Freeways," *Transportation Research Record 1365*, Transportation Research Board, 1992.

[16] HCM 2000. *Highway Capacity Manual*, Transportation Research Board, National Research Council, Washington DC, 2000.

[17] Jamar Technologies, Inc., Traffic Data Collectors. Retrieved from www.jamartech.com/prod02.htm#DB-100 2 June 2002.

[18] Klein, Lawrence A., *Vehicle Detector Technologies for Traffic Management Applications, Part 1*, ITS Online, 1997. Retrieved 2 June 2002 from http://www.itsonline.com/detect_pt1.html

[19] Lapin, Lawrence L., *Probability and Statistics for Modern Engineering*, Second Edition, PWS-Kent Publishing, Boston, MA, 1990.

[20] McShane, William R., Roger P. Roess, and Elena S. Prassas, *Traffic Engineering*, Second Edition, Prentice Hall, 1998.

[21] Nu-Metrics, Product Information, Hi-Star Portable Traffic Sensor. Retrieved from http://www.nu-metrics.com/ 2 June 2002.

[22] Robertson, H. Douglas, editor, *Manual of Traffic Engineering Studies*, Prentice Hall, Englewood Cliffs, NJ, 1994.

[23] Skycomp, Inc., www.skycomp.com/

[24] Sreedevi, Indu and Justin Black, *Inductive Loop Detectors*, ITS Decision Report, last update: 02/01/01. Retrieved from www-path.eecs.berkeley.edu/~leap/TTM/Incident_Manage/Detection/loopdet.html 2 June 2002.

[25] Walpole, Ronald E. and Raymond H. Myers, *Probability and Statistics for Engineers and Scientists*, Fourth Edition, MacMillan Publishing, New York, 1998.

[26] Whitford, Robert K, *The WIM Plan for Alaska*, Alaska DOT&PF, 1998.

EXERCISES FOR CHAPTER 2: TRAFFIC FLOW THEORY AND ANALYSIS

Measuring Traffic Flow and Spacing

2.1 Headways. Six vehicles were observed during a 40-second period of observation passing through a 100-foot segment of expressway. The observations started at an arbitrary time $t = 0$ and are tabulated below. What is the average time headway for the six vehicles observed?

Vehicle Number	Time of Arrival (sec)
1	3.74
2	12.12
3	17.95
4	24.19
5	26.58
6	36.16

2.2 Headways. During a 45-second period of observation that started at an arbitrary time $t = 0$, the arrival times of vehicles were entered into the table below. Calculate the average time headway for these vehicles. Calculate the mean time headway between vehicles over the 45-second period.

Vehicle number:	1	2	3	4	5	6	7
Time of arrival (sec)	7.53	8.04	17.68	27.43	32.93	36.73	42.98

2.3 Safe Interval Rule of Thumb. A current safety rule of thumb is for a driver to leave at least 2 seconds of space between his/her vehicle and the one directly in front. (*Note:* We are talking about space between the front bumper of the following car and rear bumper of the leading car. This is not headway.) More than 20 years ago, the rule of thumb was: One car length for every 10 miles per hour of current speed. At 50 mph, Which rule of thumb is more conservative—the old distance rule of thumb or the newer *time* rule of thumb? Assume a vehicle length of 18 feet.

2.4 Traffic Flow and Volumes. Traffic counts were taken on a minor arterial street near a factory for four consecutive 15-minute periods during the street's peak hour. The 15-minute counts were 264, 204, 357, and 305.

(a) What was the traffic flow rate for each of the 15-minute periods?
(b) What was the peak hour factor for the minor arterial?

2.5 Traffic Volumes and Flow Rates. Vehicle counts were made on one direction of a two-lane roadway three different times last Thursday. The person who made the counts produced the data that appear in the table below for your analysis. Convert the counts in the table into flow rates and average headways for each of the three time periods.

Count ID	Start Time H:M:S	End Time H:M:S	Number of Vehicles	Flow Rate (vph)	Average Time Headway
A.	7:00:00	7:45:00	511		
B.	17:06:45	17:33:30	481		
C.	14:16:45	14:46:45	249		

2.6 Vehicle Classification. Guess at the percent trucks on a major highway near you. Then form a team and collect data to see how close your guess was. Each team member could count total vehicles and trucks for a 30-minute period, keeping 5-minute subtotals to see how the values vary within the 30-minute period. What definition of a "truck" would you use when doing your vehicle classification counts? *Note*: Whenever collecting traffic data, exercise common sense and extreme care. Never put yourself in a dangerous location or situation.

2.7 Peak Hour Factor.   The flow in each 5-minute interval from 4:30 to 6 PM is given in the table below. Determine the when peak hour begins and ends. What is the peak hour factor for the peak hour?

Time	4:30–4:35	4:35–4:40	4:40–4:45	4:45–4:50	4:50–4:55	4:55–5:00
Volume	1200	1350	1490	1780	1720	1540
Time	5:00–5:05	5:05–5:10	5:10–515	5:15–5:20	5:20–5:25	5:25–5:30
Volume	1670	1650	1700	1400	1500	1700
Time	5:30–5:35	5:35–5:40	5:40–5:50	5:45–5:50	5:50–5:55	5:55–6:00
Volume	1300	1300	1550	1350	1780	1450

2.8 Traffic Data Collection.   The City of Mythaca has a Traffic Commission, whose members include the City Traffic Engineer and several citizens. The commission's purpose is to identify and discuss traffic problems in the community. At last month's meeting, a citizen raised a pedestrian safety issue at the Mythaca State University campus. The citizen member complained about how difficult it is to cross Wildcat Avenue between the central campus and a large parking garage. The crossing is midway between two signalized intersections that are about 1500 feet apart. Although there have been no vehicle-pedestrian incidents at the midblock pedestrian crossing, the citizen wants to have a traffic signal installed there. The MSU faculty member on the commission suggested that some observations be made before any action is taken. He volunteers to have some of his students make observations, conduct an analysis, and report back to the commission.

Pick a midblock location that is similar to the situation described above. At what times of day are pedestrian crossings likely to be most difficult? What kind of observations do you think will be useful in assessing the extent of the problem? What data would you collect?

2.9 Turning Movement Counts.   At a fairly busy signalized intersection near you, do a count of turning movements for at least 20 minutes. Use a form patterned after Figure 2.5. Use one line of the form for each signal cycle. Count those vehicles that move through the intersection during a particular cycle but also make some sort of distinctive mark if more vehicles were about to make the movement when the signal turned red. (*Note*: Although these are called turning movement counts, you must also count vehicles that proceed straight through the intersection.) Summarize the data you collect by computing the total volume on each approach and the percent of each turning movement on each approach. For example, what percent of vehicles from the north turned left, went straight, and turned right? Which green phases in the traffic signal cycle (if any) appear to need to be lengthened to accommodate the traffic in a particular turning movement? Save the counts you make for use in other problems.

2.10 Correction Factor.   A pneumatic tube was placed across the WB lane of SR361 last week. During the peak hour, 2429 axles were detected. A set of wheels is detected whenever tires compress the tube.

(a) What is the maximum number of WB vehicles that could have been on SR361 during that peak hour? Explain.

(b) If all of the vehicles had been Class 9 trucks (see Figure 2.31), how many WB vehicles were on SR361?

(c) The table below summarizes a vehicle classification survey carried out a few years ago on WB SR361. What was the average number of wheels per vehicle observed in that survey?

Wheel Sets per Vehicle	2	3	4	5
Frequency	662	119	55	153

(d) Use the results of part C to estimate the number of WB peak hour vehicles traveling on SR361 last week.

2.11 Speed Distributions.   Although the speeds in Table 2.9 are average speeds for vehicles observed in 20-second time intervals, let us use them to demonstrate how to find the 85th percentile speed.

(a) Sort the speeds in the speed column of Table 2.9 and determine the 85th percentile speed for the speeds in your sorted list.

(b) Because we used a collection of average speeds instead of the speeds of all individual vehicles, do you think the 85th percentile speed found in part A will be higher or lower than the actual 85th percentile speed? Explain.

2.12 Design Hourly Volume.   A rural two-lane road has a two-way ADT of 6915 vehicles per day. Transportation planners forecast that this road will carry 9660 vpd 10 years from now. What will be the road's DHV 10 years from now? Will more lanes need to be added by that future year? Assume that a rural road has a capacity of 2000 vphpl.

2.13 Design Hourly Volume.   Is it possible—even in theory—for a road to have a "K" value greater than 1.00? Give an example to support your answer.

2.14 Design Hourly Volume.   The AADT on CR750N is 472. What are the average DHV, the 15th percentile DHV, and the 85th percentile DHV for CR750N? Use the 30th highest hour criterion.

2.15 Design Hourly Volume.   Mythaca County's Highway Engineer is not sure that the standard 30th highest hourly volume is the best criterion on which to base road improvements. He thinks of four roads in the county:

- County Road 800E, which is near the university's football stadium. Traffic is very high on that road only 1 or 2 hours before and after each of the team's five or six home games.
- State Highway 361, which is a major route through the city of Mythaca, with heavy traffic every weekday morning and evening, and midday on Saturdays. The engineer is especially concerned about the approaches to the city on SR361.
- County Road 650E, which serves an industrial park just outside the city limits whose factories operate on a 7 AM to 3:30 PM work shift.
- County Road 200S, which runs past a 20-field youth soccer complex and the campus on which a high school and junior high are located at the edge of town.

For each of these roads, decide whether the 30th DHV from Figure 2.12 would overestimate or underestimate the actual 30th highest hourly volume on the particular roadway. State your reasons.

2.16 Design Hourly Volume.   Repeat Problem 2.14, but apply a criterion that uses the 50th highest hour of the year. How much difference does that choice of criterion make?

Measuring Traffic Speeds and Densities

2.17 Units of Speed.   It is not unusual for an analyst to need to convert units of speed. In each row of the table below, convert the unit speed to its equivalents in the same row. For example, in row one, 1 foot per second = _____ miles per hour = _____ kilometers per hour. Show your calculations.

1 fps	_ mph	_ kph
_ fps	1 mph	_ kph
_ fps	_ mph	1 kph

2.18 Speeds and Average Speeds.   What is the speed of each vehicle listed in the table below? Express these speeds in units of both mph and kph. What is the space mean speed of the six vehicles observed in mph? What is the space mean speed in kph?

Vehicle Number	Time of Arrival (sec)	Time to Travel 100 ft (sec)	Speed (mph)	Speed (kph)
1	3.74	1.10		
2	12.12	1.03		
3	17.95	1.14		
4	24.19	1.25		
5	26.58	0.97		
6	36.16	0.87		

2.19 Speed Monitoring.   Wallace Elementary School is located on Laxford Street, a principal arterial street in Mythaca. Parents of Wallace students are upset over what they perceive as excessive speeds by motorists on Laxford Street during times when children are walking to and from the school. The Mythaca Police Chief says he is not able to divert personnel from other important duties to check speeds on Laxford Street, but he could set up a newly acquired solar mobile traffic monitor near Wallace School. Based on radar, the monitor tells drivers what their speeds are as they pass the monitor. The hope is that _____. If that doesn't work, then _____. Complete the last two sentences above by filling in the blanks.

2.20 Setting a Speed Limit.   What factors do you think might have to be considered in setting a speed limit on a roadway?

2.21 Density and Flow Rate on I-80.   On one section of I-80 that has three EB lanes, a driver counts 40 other vehicles between her rented car and an overpass ahead. It turned out to be 0.78 mile to the overpass from the point at which she started her vehicle count. EB I-80 had three lanes open at that point and the vehicles were moving at approximately 52 mph at the time.

(a) Calculate the vehicle density (including the driver's vehicle) on that stretch of EB I-80.
(b) Assuming the traffic was distributed evenly over the three EB lanes, what was the flow rate?

2.22 Measuring Lane Occupancy.   A truck that is 31.8 feet long passes over a loop detector that has an effective length of 9 feet. If the truck is traveling at 46.9 mph,

(a) How long (in seconds) will the truck keep the loop detector activated?
(b) How long (in seconds) was the truck over any single point of the loop detector?

2.23 Using Lane Occupancy.   A section of road has a loop detector that indicates that apparent occupancy O_{app} is 8.2 percent during the AM peak hour. This occupancy value is based on a cumulative apparent presence time value $t(P) = 294.7$ seconds. The loop detector has an effective length of 8.7 feet. The mean speed for the 997 peak vehicles was 51.9 mph.

(a) What is the actual presence time (cumulative or per vehicle) for the peak hour traffic?
(b) What is the average length of vehicle for the peak hour traffic?

2.24 Using Traffic Measures.   Prof. B of Mythaca State University took his family to see a children's program in an adjacent state. The family arrived at the site of the event at 8:15 AM, hoping to get four of the remaining 3000 tickets. When they arrived, they joined a queue of vehicles (mostly minivans) that held ticket seekers. The queue was 2.5 miles long. Prof. B's family members were worried that they would not get tickets to the event.

(a) Explain how Prof. B could conduct a rough analysis in his vehicle to assess his family's chances of getting tickets to the event. State clearly any assumptions he would have to make to supplement the information given above.
(b) Given the information available to him at the time and the assumptions made in Part A, should Prof. B have been confident of getting tickets? Show your calculations clearly.

2.25 Loop Detectors.   If a loop detector can indicate the presence of a vehicle, develop a procedure whereby signals from one or more loops can be used to determine vehicle speeds and lengths.

2.26 Speed Study.   Several parameter values in Example 2.8 had to be chosen by the analyst. The initial choices were $S = 5.3$, confidence level $= 95$ percent, and $E = 3.0$. As the solution to the example was presented, other reasonable values of S (4.3 or 3.9) and E (5.0) were mentioned. (a) Set up a spreadsheet to carry out the calculations in Equation 2.4 for a range of values for S and E. An appraisal of how much the output variable (N for the 85th percentile speed) changes as one or more input variables change is called a *sensitivity analysis*. It indicates how sensitive the output is to changes in the input. A table to summarize a sensitivity analysis is provided below. (b) Is the value of N more sensitive to changes in S or changes in E over the ranges specified in the summary table? Support your answer by citing specific values in the table.

N for 85th Percentile Speed	S = 2.0	S = 3.9	S = 4.3	S = 5.3	S = 10.0
E = 1.0					
E = 3.0			12.16	18.47	
E = 5.0					
E = 10.0					

2.27 Density.   Explain how you would use the airphoto in Figure 2.15 to determine the density for a section of the main roadway shown in the photo. The section of roadway of interest is the three WB through EB lanes of the freeway at the top of Figure 2.15, with its east end at the right edge of the photo and its west end where it passes over another highway. Do not include the vehicles in the rightmost WB lane about to exit using the offramp. What would you measure, and how would you measure it? What counts would make from the photo, if any? In Figure 2.15, 1 mm = 13 feet.

2.28 Occupancy.   Based on the material presented in this chapter, does occupancy appear to be a good proxy for density? Explain.

2.29 Occupancy and Density.   Use Equation 2.12 to estimate the density for each 20-second period in Table 2.9 and then plot occupancy vs. density for each 20-second period. Do the plots indicate that occupancy is a good proxy for density? Explain.

Traffic Models for Continuous Flow

2.30 Vehicle Lengths for Speed-Density Relationship.   To check the average vehicle length assumed in the solution to Example 2.11, gather data on the typical lengths of passenger cars, buses, semitrailers, and other vehicles commonly seen in traffic. What range of lengths applies to each vehicle type? Where did you find such information?

2.31 Speed-Density Relationship on SR361.   The speed-density relationship for a lane on SR361 is believed to be:

$$u + 2.60 = 0.001 \times (k - 240)^2$$

where u has units mph and k has units vehicles per mile. Find (a) the free-flow speed, (b) the jam density, (c) the lane capacity, and (d) the speed at capacity.

2.32 Speed-Flow Relationships.   The speed-flow relationship for a lane on SR361 is believed to be $q = 273u - 70u (\ln u)$, where speed u has units mph and flow rate q has units vph. Find

(a) the free-flow speed to the nearest 0.1 mph
(b) the maximum-flow speed
(c) the maximum flow rate
(d) the density at maximum flow

2.33 Flow-Density Relationship.   Traffic is flowing at its maximum capacity of 3000 vehicles per hour at 30 miles per hour. Determine its density using the Greenshields model.

2.34 Greenshields Model.   Complete the following table using the Greenshields model of traffic flow.

Number	Design Free Flow Speed (mph)	Jam Density (pcpmpl)	Average Speed (mph)	Average Density (pcpmpl)	Average Flow (pcphpl)	Capacity (pcphpl)	V/C
A	75	250	40				
B		250		150	3000		
C	60	250		50			
D	72	250					0.5

2.35 Greenshields Model.   A single lane of a freeway with a design speed of 68 miles per hour has a capacity of 2550 passenger cars per hour per lane. There is a flow of 1000 passenger cars per hour per lane. Using the Greenshields model, what is the jam density? What is the average speed with the present flow?

2.36 Traffic Flow Models. Only two reliable data points exist for a two-lane westbound section of I-25:

(a) Speed = 31 mph, Density = 95 vpmpl
(b) Speed = 50 mph, Density = 43 vpmpl

Using those two data points in the Greenshields traffic flow model, estimate q_{max}, the capacity of that section of I-25.

2.37 Greenshields Equations. Plot the speed-occupancy and flow-occupancy relationships for the I-25 data in Table 2.9. Use Equation 2.12 to estimate the density for each 20-second period in Table 2.9 and then plot the speed-density and flow-density relationships for the I-25 data. Do the plots resemble Figure 2.17 (linear form), Figure 2.19, or Figure 2.20?

2.38 Developing Greenshields Models. Following the procedures illustrated by Figures 2.21 through 2.23, develop equations to describe the relationships that you plotted in the previous question. Assume that the Greenshields Equations 2.13, 2.18, and 2.19 are valid for the I-25 data, and estimate values for D_j, S_f, and capacity q_{max} from the data.

2.39 Linear Flow-Density Relationships. The points plotted on the left half of Figure 2.22 appear to have a relationship that is more linear than parabolic. For the points that lie to the left of the maximum-flow point of the flow-density plot created for problem 2.37, fit a straight line through those points and the origin. Compare the resulting equation of the straight line with the form of Equation 2.12. Does the slope of the fitted line have any physical meaning?

2.40 Speed-Density-Flow Relationships. Select 30 consecutive time periods from the 104 periods shown in Table 2.9. These data cover the three EB lanes of I-25, which is northeast of Mythaca.

(a) Convert the I-25 data into a scatter plot showing the relationship between speed and density in a format like Figure 2.21. (*Hint*: When needed, use Equation 2.12 to derive density from other data in a time period.) Fit a linear function (Equation 2.13) to the data for I-25. Show this linear function on the scatter plot and determine the values for free-flow speed and jam density. Is the scatter plot consistent with the smooth curve? Explain.
(b) Convert the I-25 data into a scatter plot showing the relationship between flow and density in a format like Figure 2.22. Use Equation 2.19 with the parameters found in part A to generate a smooth curve and show this on the scatter plot. Show a manual calculation of q for $D = 30$ vpk. Is the scatter plot consistent with the smooth curve? Explain.
(c) Convert the I-25 data into a scatter plot showing the relationship between speed and flow in a format like Figure 2.23. Use Equation 2.18 with the parameters found in part A to generate a smooth curve and show this on the scatter plot.

2.41 Greenberg's Model. An alternative to the Greenshields model of traffic flow is one developed by Greenberg (Drake et al., 1967). The Greenberg relationship between speed u and density k is

$$S = S_{\max q} \ln\left(\frac{D_j}{D}\right)$$

The Greenberg relationship between flow rate q and density D is

$$q = S_{\max q} * D * \ln\left(\frac{D_j}{D}\right).$$

Fit the Greenberg-based equations for speed vs. density and flow vs. density to the data in Table 2.9. Plot the fitted curves on the respective scatter plots.

The Poisson Model for Continuous Random Traffic Flow

2.42 Poisson Model.   The table below shows the arrival times of vehicles over a 2-minute period on SR361.

Vehicle No.	Arrival Time (sec.)	Vehicle No.	Arrival Time (sec.)	Vehicle No.	Arrival Time (sec.)	Vehicle No.	Arrival Time (sec.)
1	0.5	11	15.6	21	51.2	31	84.3
2	3.1	12	17.8	22	52.3	32	85.7
3	4.2	13	21.7	23	53.4	33	89.2
4	6.1	14	24.8	24	54.9	34	96.5
5	8.3	15	25.9	25	60.5	35	98.3
6	9.6	16	29.0	26	71.2	36	102.7
7	11.1	17	32.5	27	73.8	37	109.1
8	12.2	18	43.8	28	79.5	38	112.3
9	13.5	19	46.9	29	80.1	39	117.1
10	14.7	20	50.1	30	82.5	40	119.6

(a) What is the average arrival rate during the 2 minutes of observations?

(b) Using 5-second intervals and the format of Table 2.12, determine the frequency distributions for both the observed data and the Poisson model.

(c) How well does the Poisson distribution fit the data? Plot the actual and Poisson frequencies, as was done in Figure 2.25. Comment on the results.

2.43 Poisson Model.   Repeat the previous problem using 10-second intervals. How do the solutions change?

2.44 Poisson Calculations.   The SR361 bridge across the Mythaca River between Middleville and Shoridan is a toll facility. The table below shows the SEB (southeastbound) bridge volume counts by 20-minute periods for a recent weekday morning. If the arrivals during any 20-minute period shown in the table are assumed to follow a Poisson Distribution, what is the probability that at least two vehicles will arrive during any 30-second period within each 20-minute period? Enter the results in the last row of the table below.

Time period (beginning):	7:00	7:20	7:40	8:00	8:20	8:40	9:00
Vehicle arrivals:	55	133	202	193	129	104	76
$Pr(n \geq 2)$:							

2.45 Left Turn Bay.   A traffic engineer is trying to design a left-turn bay on an EB approach to an intersection. The signal timings for left turns will have 10 seconds of "left turn only" green time and 50 seconds of red time each signal cycle. The green time will be sufficient to allow three left turns to be made each signal cycle. An average of 2.5 left-turning vehicles are expected to approach the intersection each signal cycle. Space is limited, so the engineer wants to minimize the storage capacity of the turn bay, but insufficient capacity will cause the excess vehicles waiting for a green light to block through traffic. If the turn bay has capacity for only three vehicles, what percentage of the time will the through lane adjacent to the left-turn bay be blocked?

2.46 Bicycle Crossing.   Traffic on a county road has been counted at 465 vehicles during the afternoon peak hour. A bicycle path crosses this road about midway between two intersections that are one-half mile apart.

(a) What is the mean time between the arrivals of vehicles at the bicycle crossing?
(b) It has been determined by observation that the average bicyclist needs at least 7.3 seconds between vehicles before he/she will attempt to ride across the road. What is the probability that the next time between vehicles will allow the average bicyclist enough time to cross the road?
(c) Find the time between vehicles t such that $P(T \geq t) = 0.90$. To what vehicle flow rate does this time correspond?

2.47 Poisson Events.   A house along a hilly rural road has its narrow driveway enter that road just east of the crest of a hill. The eastbound ADT on the road is 649 vehicles per day (vpd). It takes 9 seconds for the family's car to be backed out of the driveway onto the road. This time includes what is needed to accelerate to the speed limit in the EB direction. Find the probability that no EB vehicle will come over the crest of the hill while the family's vehicle is backing out of the driveway.

2.48 Time Between Events.   Even at the most dangerous intersections, collisions between vehicles can be considered "rare events," so that they can be considered independent of each other. The intersection of Bodly and Equinox Streets had one of the highest collision counts in the county last year, with 37. If that rate were to continue, what is the probability that at least one collision will occur at that intersection during the next week?

2.49 Vehicle Ferry Planning.   Rich and powerful residents of Abalone Island want the state to build a bridge to help them reach their summer residences. Instead, the state DOT staff proposes to increase the frequency of the existing ferryboat service, so that the probability of having any vehicle bound for Abalone Island left behind because of a full ferryboat will be minimized. Each ferryboat now in service operates on a 30-minute cycle. Each ferryboat can carry seven vehicles.

(a) If an average of 9.8 vehicles per hour want to go to Abalone Island, what is the probability that more than 14 vehicles will arrive at the ferryboat dock in a 60-minute period? What is the probability that more than seven vehicles will arrive at the ferryboat dock in a 30-minute period?
(b) If the state DOT wants to ensure that the probability of being left at the dock because the departing ferryboat had insufficient capacity is no more than 0.05, how many seven-vehicle boats should it employ? (*Hint:* Use a spreadsheet to facilitate the repetitive calculations that need to be done in this analysis, but show at least one calculation done by hand.) If only one ferryboat can be put into service, how many vehicles will it have to be able to carry to meet the 0.05 service standard?

2.50 Poisson Processes.   Vehicles leave a shopping center parking lot randomly at an average rate of 152 vehicles per hour. The adjacent major street has a random flow rate of 1200 vehicles in the lane nearest the parking lot exit. If a vehicle exiting the parking lot needs 5 seconds between consecutive major street vehicles to complete a right turn, what is the probability that the next gap on the major street will be greater than 5 seconds?

Measuring Roadway Performance

2.51 HPMS Inputs.   Explain why all three dimensions of traffic data collection in Figure 2.29 are needed as input to the HPMS. (*Hint*: The HPMS data are based on a sample of all roadways in the United States.)

2.52 Axle Count Adjustments.   How do traffic counts taken over a 48-hour period by axle detectors need to be adjusted to determine a value of annual average daily traffic? What adjustments to the raw axle counts need to be made and why?

Highway Design for Performance

SCENARIO

Kara's question: A mother of two children sees the Mythaca County Highway engineer before the start of a 10-kilometer running race. The mother asks him a question that her daughter Kara had asked her a few days earlier. As she and Kara were creeping along in their car on I-25, Kara asked her mom, "Why is there always a traffic jam on this highway?" (See Figure 3.1.)

CHAPTER OBJECTIVES

By the end of this chapter, the student will be able to:

1. Determine the level of service and capacity for a two-lane highway under specified traffic characteristics.
2. Determine the level of service on a basic freeway segment under specified traffic characteristics.
3. Calculate delay-related performance measures using queueing analysis.
4. Answer "Kara's question" clearly and concisely.

INTRODUCTION

In Chapter 2, methods for measuring traffic flows and assessing roadway performance were introduced. Those techniques are important in helping the traffic engineer to keep the roadways operating as efficiently as possible. Another important capability would be to design a roadway expansion in response to existing or anticipated traffic conditions. A later chapter covers designing

FIGURE 3.1
Traffic jam on an urban expressway. Photo: Jon D. Fricker.

highways for safety. This chapter addresses highway design for efficiency. The reader will learn how to use portions of two major chapters in the *Highway Capacity Manual* (2000). In addition, we explain why traffic jams occur and what can be done to prevent them or reduce their severity.

3.1   CAPACITY AND LEVEL OF SERVICE FOR TWO-LANE HIGHWAYS

3.1.1   Level of Service on Two-Lane Highways

There are almost 1 million miles of roadways eligible for federal funding in the United States. Of these "federal aid" roads, more than 83 percent are two-lane highways (FHWA, 1997). (See Table 3.1.) Because they are so prominent and are located in a variety of locations and environments, two-lane highways are a good subject for introducing capacity and LOS analysis procedures (Figure 3.2).

A new method of *operational analysis* for two-lane highways is presented in the Year 2000 edition of the *Highway Capacity Manual*. "The objective of operational analysis is to determine the level of service (LOS) for an existing or proposed facility operating under current or projected traffic demand. Operational analysis also may be used to determine the capacity of a two-lane highway segment, or the service flow rate that can be accommodated at any given LOS" (HCM, 2000, p. 20-1).

Although the new method is entirely different from the method published in previous editions of the *HCM*, it still begins by considering what we might call the "perfect" two-lane road. The "perfect" two-lane road has the following:

- Lanes that are 12 feet wide with paved shoulders at least 6 feet wide
- Level, straight terrain (with no no-passing zones)
- Equal traffic in both directions
- No access points, such as side streets and driveways
- No trucks, buses, or recreational vehicles.

TABLE 3.1   Percent of Federal Aid Roadway Miles That Have Two Lanes

Road Type	Two-lane	Total	Two-Lane (%)
Rural interstate	1,087	32,818	3.3
Rural other principal arterial	72,461	98,131	73.8
Rural minor arterial	130,343	137,359	94.9
Rural major collector	427,588	432,118	99.0
Urban interstate	56	13,218	0.4
Other urban freeways and expressways	607	9022	6.7
Urban other principal arterial	20,003	52,973	37.8
Urban minor arterial	64,566	89,022	72.5
Urban collector	80,384	87,918	91.4
All federal-aid roads (total)	797,095	952,579	83.7

Source: 1996 Highway Statistics (FHWA, 1997).

FIGURE 3.2
Rural two-lane highway. Photo: Jon D. Fricker.

Obviously, there are no perfect highways. To conduct a level of service analysis for a section of two-lane road, we must reduce the section's free-flow speed by using factors that account for the "imperfections" in the highway. (*Level of service* was defined in Section 2.5.) The LOS on a two-lane highway can be determined after the *Average Travel Speed (ATS)* and *Percent Time Spent Following (PTSF)* are estimated. See Figure 3.3 and Table 3.2.

Note that the LOS criteria apply to Class I highways. These are highways that provide mobility between cities and towns. Class II highways primarily serve local traffic. Because drivers on Class II highways have lower expectations with respect to roadway performance, a separate Class II analysis method has been developed (*HCM*, 2000). Only the method for Class I will be covered in this book.

The *Highway Capacity Manual (HCM)* provides a worksheet (Figure 3.4) to summarize the calculations made during the two-lane highway LOS analysis. The worksheet also serves as a "roadmap" or checklist to guide the analyst through the steps in the analysis. The worksheet is presented twice. Figure 3.4 is a blank copy that can be reproduced for the students' examples and homework. Figure 3.5 is a completed version that shows the results of an analysis of a two-lane roadway with a two-way volume of 526 vehicles per hour on rolling terrain. This case is used to demonstrate two-lane highway LOS analysis in the following paragraphs.

In the "Average Travel Speed" and "Percent Time Spent Following" sections of the worksheet, reference is made to Exhibits 20-5 through 20-21. These exhibits are displayed in Tables 3.3 to 3.10, just as they appear in the *HCM*. They contain values that enable the user to assess the impacts of a roadway's departures from ideal conditions. For example, the first line in the "Average Travel Speed" section of the worksheet

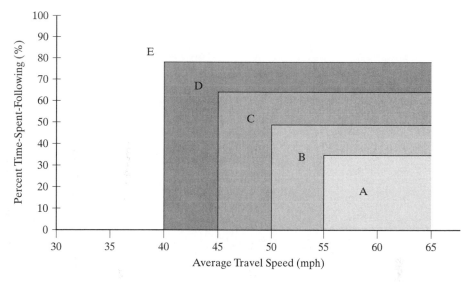

FIGURE 3.3
LOS criteria for two-lane highways in Class I. *Source: HCM*, 2000, Exhibit 20-3.

TABLE 3.2   LOS Criteria for Two-Lane Highways in Class I

LOS	Percent Time Spent Following	Average Travel Speed (mph)
A	≤ 35	>55
B	$>35–50$	$>50–55$
C	$>50–65$	$>45–50$
D	$>65–80$	$>40–45$
E	>80	≤ 40

Note: LOSF applies whenever the flow rate exceeds the segment capacity.

Source: HCM, 2000, Exhibit 20-2.

(Figure 3.4) refers to Exhibit 20-7. (See Table 3.5.) This exhibit estimates the effect on ATS of the terrain over which the roadway is built. (Later, Exhibit 20-8 will provide *Grade Adjustment Factor* (f_G) value to adjust Percent Time Spent Following; for rolling terrain, the two f_G values will be different.) If the roadway being analyzed is in rolling terrain and has a two-way volume of 526 vehicles per hour, an f_G of 0.71 is entered on that line of the worksheet for subsequent use.

THINK ABOUT IT

Do you think that rolling terrain has a bigger negative impact on the level of service of a highway at higher or at lower flow rates?

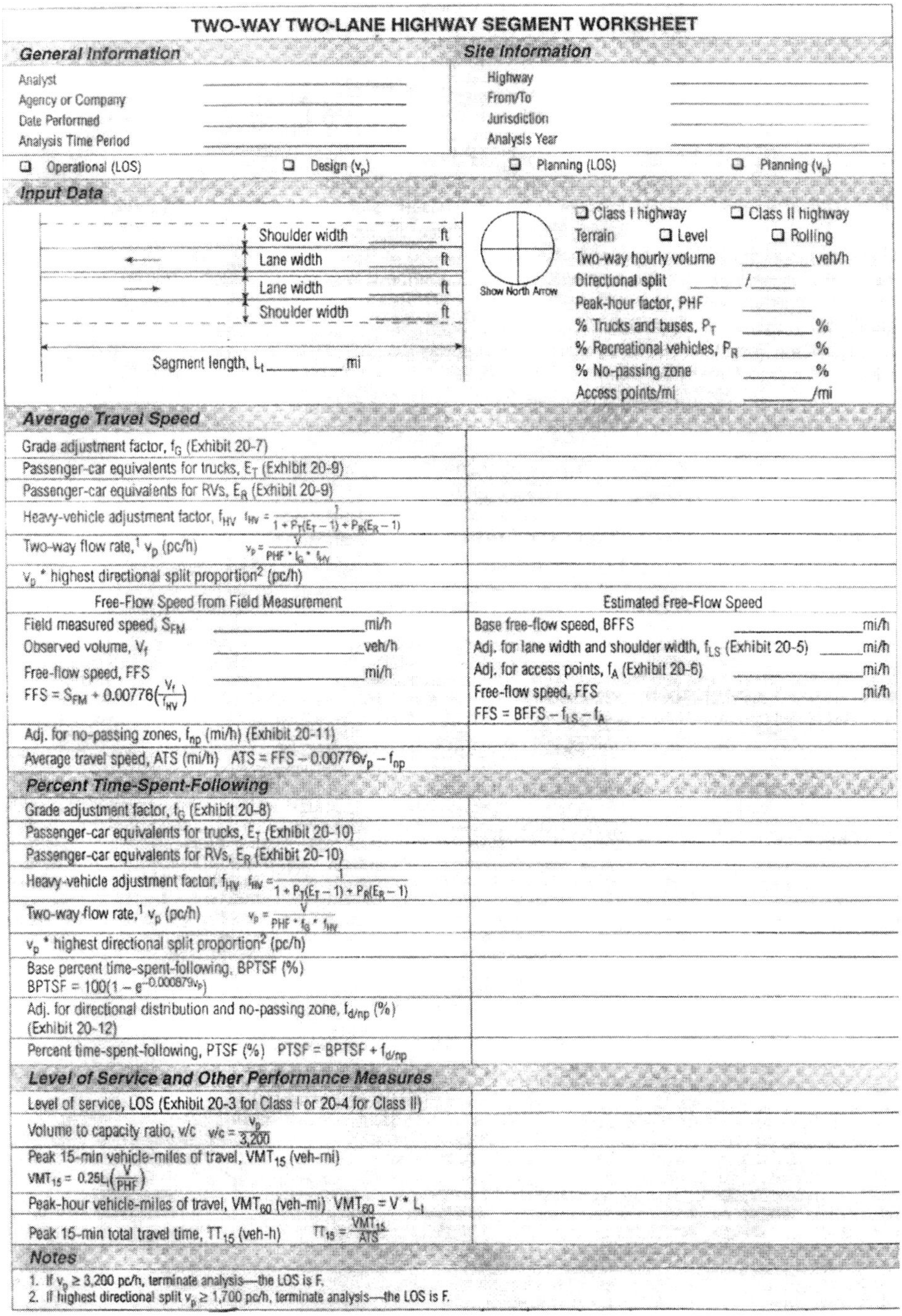

TWO-WAY TWO-LANE HIGHWAY SEGMENT WORKSHEET

General Information

Analyst	_____
Agency or Company	_____
Date Performed	_____
Analysis Time Period	_____

Site Information

Highway	_____
From/To	_____
Jurisdiction	_____
Analysis Year	_____

☐ Operational (LOS) ☐ Design (v_p) ☐ Planning (LOS) ☐ Planning (v_p)

Input Data

Shoulder width _____ ft
Lane width _____ ft
Lane width _____ ft
Shoulder width _____ ft

Segment length, L_t _____ mi

Show North Arrow

☐ Class I highway   ☐ Class II highway
Terrain ☐ Level   ☐ Rolling
Two-way hourly volume _____ veh/h
Directional split _____ / _____
Peak-hour factor, PHF _____
% Trucks and buses, P_T _____ %
% Recreational vehicles, P_R _____ %
% No-passing zone _____ %
Access points/mi _____ /mi

Average Travel Speed

Grade adjustment factor, f_G (Exhibit 20-7)	
Passenger-car equivalents for trucks, E_T (Exhibit 20-9)	
Passenger-car equivalents for RVs, E_R (Exhibit 20-9)	
Heavy-vehicle adjustment factor, f_{HV} $f_{HV} = \dfrac{1}{1 + P_T(E_T - 1) + P_R(E_R - 1)}$	
Two-way flow rate,[1] v_p (pc/h) $v_p = \dfrac{V}{PHF \cdot f_G \cdot f_{HV}}$	
v_p * highest directional split proportion[2] (pc/h)	

Free-Flow Speed from Field Measurement	Estimated Free-Flow Speed
Field measured speed, S_{FM} _____ mi/h	Base free-flow speed, BFFS _____ mi/h
Observed volume, V_f _____ veh/h	Adj. for lane width and shoulder width, f_{LS} (Exhibit 20-5) _____ mi/h
Free-flow speed, FFS _____ mi/h	Adj. for access points, f_A (Exhibit 20-6) _____ mi/h
$FFS = S_{FM} + 0.00776\left(\dfrac{V_f}{f_{HV}}\right)$	Free-flow speed, FFS _____ mi/h $FFS = BFFS - f_{LS} - f_A$

Adj. for no-passing zones, f_{np} (mi/h) (Exhibit 20-11)	
Average travel speed, ATS (mi/h) $ATS = FFS - 0.00776v_p - f_{np}$	

Percent Time-Spent-Following

Grade adjustment factor, f_G (Exhibit 20-8)	
Passenger-car equivalents for trucks, E_T (Exhibit 20-10)	
Passenger-car equivalents for RVs, E_R (Exhibit 20-10)	
Heavy-vehicle adjustment factor, f_{HV} $f_{HV} = \dfrac{1}{1 + P_T(E_T - 1) + P_R(E_R - 1)}$	
Two-way flow rate,[1] v_p (pc/h) $v_p = \dfrac{V}{PHF \cdot f_G \cdot f_{HV}}$	
v_p * highest directional split proportion[2] (pc/h)	
Base percent time-spent-following, BPTSF (%) $BPTSF = 100(1 - e^{-0.000879v_p})$	
Adj. for directional distribution and no-passing zone, $f_{d/np}$ (%) (Exhibit 20-12)	
Percent time-spent-following, PTSF (%) $PTSF = BPTSF + f_{d/np}$	

Level of Service and Other Performance Measures

Level of service, LOS (Exhibit 20-3 for Class I or 20-4 for Class II)	
Volume to capacity ratio, v/c $v/c = \dfrac{v_p}{3,200}$	
Peak 15-min vehicle-miles of travel, VMT_{15} (veh-mi) $VMT_{15} = 0.25L_t\left(\dfrac{V}{PHF}\right)$	
Peak-hour vehicle-miles of travel, VMT_{60} (veh-mi) $VMT_{60} = V * L_t$	
Peak 15-min total travel time, TT_{15} (veh-h) $TT_{15} = \dfrac{VMT_{15}}{ATS}$	

Notes

1. If $v_p \geq 3,200$ pc/h, terminate analysis—the LOS is F.
2. If highest directional split $v_p \geq 1,700$ pc/h, terminate analysis—the LOS is F.

FIGURE 3.4

Blank two-way two-lane highway segment worksheet. *Source: HCM*, 2000, p. 20–50.

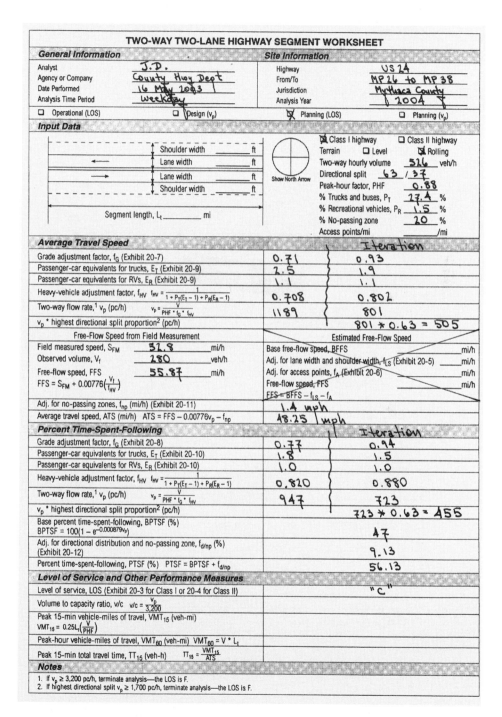

TWO-WAY TWO-LANE HIGHWAY SEGMENT WORKSHEET

General Information

Analyst	J.D.
Agency or Company	County Hwy Dept
Date Performed	16 May 2003
Analysis Time Period	Weekday

Site Information

Highway	US 2A
From/To	MP 26 to MP 38
Jurisdiction	Mythaca County
Analysis Year	2004

☐ Operational (LOS)   ☐ Design (v_p)   ☒ Planning (LOS)   ☐ Planning (v_p)

Input Data

Shoulder width _____ ft
Lane width _____ ft
Lane width _____ ft
Shoulder width _____ ft

Show North Arrow

Segment length, L_t _____ mi

☒ Class I highway   ☐ Class II highway
Terrain ☐ Level   ☒ Rolling
Two-way hourly volume **526** veh/h
Directional split **63 / 37**
Peak-hour factor, PHF **0.88**
% Trucks and buses, P_T **17.4** %
% Recreational vehicles, P_R **1.5** %
% No-passing zone **20** %
Access points/mi _____ /mi

Average Travel Speed

		Iteration
Grade adjustment factor, f_G (Exhibit 20-7)	0.71	0.93
Passenger-car equivalents for trucks, E_T (Exhibit 20-9)	2.5	1.9
Passenger-car equivalents for RVs, E_R (Exhibit 20-9)	1.1	1.1
Heavy-vehicle adjustment factor, f_{HV} $f_{HV} = \frac{1}{1 + P_T(E_T - 1) + P_R(E_R - 1)}$	0.708	0.802
Two-way flow rate,[1] v_p (pc/h) $v_p = \frac{V}{PHF \cdot f_G \cdot f_{HV}}$	1189	801
v_p * highest directional split proportion[2] (pc/h)		801 * 0.63 = 505

Free-Flow Speed from Field Measurement		Estimated Free-Flow Speed	
Field measured speed, S_{FM} **52.8** mi/h		Base free-flow speed, BFFS _____ mi/h	
Observed volume, V_f **280** veh/h		Adj. for lane width and shoulder width, f_{LS} (Exhibit 20-5) _____ mi/h	
Free-flow speed, FFS **55.87** mi/h		Adj. for access points, f_A (Exhibit 20-6) _____ mi/h	
$FFS = S_{FM} + 0.00776\left(\frac{V_f}{f_{HV}}\right)$		Free-flow speed, FFS _____ mi/h	
		$FFS = BFFS - f_{LS} - f_A$	

Adj. for no-passing zones, f_{np} (mi/h) (Exhibit 20-11)	1.4 mph
Average travel speed, ATS (mi/h) $ATS = FFS - 0.00776v_p - f_{np}$	48.25 mph

Percent Time-Spent-Following

		Iteration
Grade adjustment factor, f_G (Exhibit 20-8)	0.77	0.94
Passenger-car equivalents for trucks, E_T (Exhibit 20-10)	1.8	1.5
Passenger-car equivalents for RVs, E_R (Exhibit 20-10)	1.0	1.0
Heavy-vehicle adjustment factor, f_{HV} $f_{HV} = \frac{1}{1 + P_T(E_T - 1) + P_R(E_R - 1)}$	0.820	0.880
Two-way flow rate,[1] v_p (pc/h) $v_p = \frac{V}{PHF \cdot f_G \cdot f_{HV}}$	947	723
v_p * highest directional split proportion[2] (pc/h)		723 * 0.63 = 455
Base percent time-spent-following, BPTSF (%) $BPTSF = 100(1 - e^{-0.000879v_p})$		47
Adj. for directional distribution and no-passing zone, $f_{d/np}$ (%) (Exhibit 20-12)		9.13
Percent time-spent-following, PTSF (%) $PTSF = BPTSF + f_{d/np}$		56.13

Level of Service and Other Performance Measures

Level of service, LOS (Exhibit 20-3 for Class I or 20-4 for Class II)	"C"
Volume to capacity ratio, v/c $v/c = \frac{v_p}{3,200}$	
Peak 15-min vehicle-miles of travel, VMT_{15} (veh-mi) $VMT_{15} = 0.25L_t\left(\frac{V}{PHF}\right)$	
Peak-hour vehicle-miles of travel, VMT_{60} (veh-mi) $VMT_{60} = V \cdot L_t$	
Peak 15-min total travel time, TT_{15} (veh-h) $TT_{15} = \frac{VMT_{15}}{ATS}$	

Notes

1. If $v_p \geq 3,200$ pc/h, terminate analysis—the LOS is F.
2. If highest directional split $v_p \geq 1,700$ pc/h, terminate analysis—the LOS is F.

FIGURE 3.5

Completed two-way two-lane highway segment worksheet. *Source: HCM*, 2000, p. 20–50.

TABLE 3.3   (Exhibit 20-5) Adjustment (f_{LS}) for Lane Width/Shoulder Width

Lane Width	Reduction in FFS (mph)			
	Shoulder Width (feet)			
	0 to <2	2 to <4	4 to <6	≥6
9 to <10	6.4	4.8	3.5	2.2
10 to <11	5.3	3.7	2.4	1.1
11 to <12	4.7	3.0	1.7	0.4
≥12	4.2	2.6	1.3	0.0

Source: *HCM*, 2000, Exhibit 20-5.

TABLE 3.4   (Exhibit 20-6) Adjustment (f_A) for Access Point Density

Access Points per Mile	Reduction in FFS (mph)
0	0.0
10	2.5
20	5.0
30	7.5
40	10.0

Source: *HCM*, 2000, Exhibit 20-6.

TABLE 3.5   (Exhibit 20-7) Grade Adjustment Factor (f_G) to Determine Speeds on Two-Way and Directional Segments

Range of Two-Way Flow Rates (pc/h)	Range of Directional Flow Rates (pc/h)	Type of Terrain	
		Level	Rolling
0 to 600	0 to 300	1.00	0.71
>600 to 1200	>300 to 600	1.00	0.93
>1200	>600	1.00	0.99

Source: *HCM*, 2000, Exhibit 20-7.

TABLE 3.6   (Exhibit 20-8) Grade Adjustment Factor (f_G) to Determine Percent Time Spent Following on Two-Way and Directional Segments

Range of Two-Way Flow Rates (pc/h)	Range of Directional Flow Rates (pc/h)	Type of Terrain	
		Level	Rolling
0 to 600	0 to 300	1.00	0.77
>600 to 1200	>300 to 600	1.00	0.94
>1200	>600	1.00	1.00

Source: *HCM*, 2000, Exhibit 20-8.

TABLE 3.7 (Exhibit 20-9) Passenger Car Equivalents for Trucks (E_T) and RVs (E_R) to Determine Speeds on Two-Way and Directional Segments

Vehicle Type	Range of Two-way Flow Rates (pc/h)	Range of Directional Flow Rates (pc/h)	Type of Terrain	
			Level	Rolling
Trucks E_T	0 to 600	0 to 300	1.7	2.5
	>600 to 1200	>300 to 600	1.2	1.9
	>1200	>600	1.0	1.5
RVs E_R	0 to 600	0 to 300	1.0	1.1
	>600 to 1200	>300 to 600	1.0	1.1
	>1200	>600	1.0	1.1

Source: *HCM*, 2000, Exhibit 20-9.

TABLE 3.8 (Exhibit 20-10) Passenger Car Equivalents for Trucks (E_T) and RVs (E_R) to Determine Percent Time Spent Following on Two-Way and Directional Segments

Vehicle Type	Range of Two-Way Flow Rates (pc/h)	Range of Directional Flow Rates (pc/h)	Type of Terrain	
			Level	Rolling
Trucks E_T	0 to 600	0 to 300	1.1	1.8
	>600 to 1200	>300 to 600	1.1	1.5
	>1200	>600	1.0	1.0
RVs E_R	0 to 600	0 to 300	1.0	1.0
	>600 to 1200	>300 to 600	1.0	1.0
	>1200	>600	1.0	1.0

Source: *HCM*, 2000, Exhibit 20-10.

TABLE 3.9 (Exhibit 20-11) Adjustment (f_{np}) for Effect of No-Passing Zones on Average Speed on Two-Way Segments

Two-Way Flow Rates v_p (pc/h)	Increase in Average Travel Speed (mph)					
	No-Passing Zones (%)					
	0	20	40	60	80	100
0	0.0	0.0	0.0	0.0	0.0	0.0
200	0.0	0.6	1.4	2.4	2.6	3.5
400	0.0	1.7	2.7	3.5	3.9	4.5
600	0.0	1.6	2.4	3.0	3.4	3.9
800	0.0	1.4	1.9	2.4	2.7	3.0
1000	0.0	1.1	1.6	2.0	2.2	2.6
1200	0.0	0.8	1.2	1.6	1.9	2.1
1400	0.0	0.6	0.9	1.2	1.4	1.7
1600	0.0	0.6	0.8	1.1	1.3	1.5
1800	0.0	0.5	0.7	1.0	1.1	1.3
2000	0.0	0.5	0.6	0.9	1.0	1.1
2200	0.0	0.5	0.6	0.9	0.9	1.1
2400	0.0	0.5	0.6	0.8	0.9	1.1
2600	0.0	0.5	0.6	0.8	0.9	1.0
2800	0.0	0.5	0.6	0.7	0.8	0.9
3000	0.0	0.5	0.6	0.7	0.7	0.8
3200	0.0	0.5	0.6	0.6	0.6	0.7

Source: *HCM*, 2000, Exhibit 20-11.

TABLE 3.10   (Exhibit 20-12) Adjustment ($f_{d/np}$) for Combined Effect of Directional Distribution of Traffic and Percentage of No-Passing Zones on Percent Time Spent Following on Two-Way Segments

Two-Way Flow Rates v_p (pc/h)	Increase in Percent Time Spent Following (%)					
	No-Passing Zones (%)					
	0	20	40	60	80	100
Directional split = 50/50						
≤200	0.0	10.1	17.2	20.2	21.0	21.8
400	0.0	12.4	19.0	22.7	23.8	24.8
600	0.0	11.2	16.0	18.7	19.7	20.5
800	0.0	9.0	12.3	14.1	14.5	15.4
1400	0.0	3.6	5.5	6.7	7.3	7.9
2000	0.0	1.8	2.9	3.7	4.1	4.4
2600	0.0	1.1	1.8	2.0	2.3	2.4
3200	0.0	0.7	1.1	1.1	1.2	1.4
Directional split = 60/40						
≤200	1.6	11.8	17.2	22.5	23.1	23.7
400	0.5	11.7	16.2	20.7	21.5	22.2
600	0.0	11.5	15.2	18.9	19.8	20.7
800	0.0	7.6	10.3	13.0	13.7	14.4
1400	0.0	3.7	5.4	7.1	7.6	8.1
2000	0.0	2.3	3.4	3.6	4.0	4.3
≥2600	0.0	0.9	1.4	1.9	2.1	2.2
Directional split = 70/30						
≤200	2.8	13.4	19.1	24.8	25.2	25.5
400	1.1	12.5	17.3	22.0	22.6	23.2
600	0.0	11.6	15.4	19.1	20.0	20.9
800	0.0	7.7	10.5	13.3	14.0	14.6
1400	0.0	3.8	5.6	7.4	7.9	8.3
≥2000	0.0	1.4	4.9	3.5	3.9	4.2
Directional split = 80/20						
≤200	5.1	17.5	24.3	31.0	31.3	31.6
400	2.5	15.8	21.5	27.1	27.6	28.0
600	0.0	14.0	18.6	23.2	23.9	24.5
800	0.0	9.3	12.7	16.0	16.5	17.0
1400	0.0	4.6	6.7	8.7	9.1	9.5
≥2000	0.0	2.4	3.4	4.5	4.7	4.9
Directional split = 90/10						
≤200	5.6	21.6	29.4	37.2	37.4	37.6
400	2.4	19.0	25.6	32.2	32.5	32.8
600	0.0	16.3	21.8	27.2	27.6	28.0
800	0.0	10.9	14.8	18.6	19.0	19.4
≥1400	0.0	5.5	7.8	10.0	10.4	10.7

Source: HCM, 2000, Exhibit 20-12.

Passenger Car Equivalents.   In the next two lines of the worksheet, vehicles in the traffic stream that are not passenger cars are converted into *passenger car equivalents (PCE)*. Exhibit 20-9 converts trucks (and buses) and recreational vehicles into PCE values for purposes of adjusting Average Travel Speed. (As is the case for the Grade Adjustment Factor f_G, there is a separate exhibit—Exhibit 20-10—with PCEs that are used to adjust Percent Time Spent Following.) For a two-way flow rate of 526 passenger cars per hour (pc/h) over rolling terrain, the PCE value for trucks and buses, E_T, is 2.5; for recreational vehicles, $E_R = 1.1$. In the "Input Data" section of the worksheet, the "% Trucks and buses, P_T" and the "% Recreational vehicles, P_R" are entered. Let us say that those values are $P_T = 27.4$ percent and $P_R = 1.5$ percent. Those values must be converted from percentages to proportions before being used in Equation 3.1 to compute the *Heavy-Vehicle Adjustment Factor* (f_{HV}):

$$f_{HV} = \frac{1}{1 + P_T(E_T - 1) + P_R(E_R - 1)} \tag{3.1}$$

where

P_T = proportion of trucks in the traffic stream, expressed as a decimal

P_R = proportion of RVs in the traffic stream, expressed as a decimal

E_T = passenger-car equivalent for trucks, obtained from Exhibit 20-9 or Exhibit 20-10

E_R = passenger-car equivalent for RVs, obtained from Exhibit 20-9 or Exhibit 20-10.

For the illustration using the flow rate of 526 pc/h, the f_{HV} would be

$$f_{HV} = \frac{1}{1 + 0.274(2.5 - 1) + 0.015(1.1 - 1)}$$

$$= \frac{1}{1 + 0.411 + 0.0015} = \frac{1}{1.4125} = 0.708$$

Next in the worksheet is the adjustment made to the "*Two-way hourly volume*" (*V*), which has the units "vph" in the "Input Data" section of the worksheet. Up to this point, we have been assuming that all the vehicles in the traffic stream are passenger cars. That is why we used 526 pc/h as the input to Exhibits 20-7 and 20-9. At this point, we will apply the f_G, the f_{HV}, and the PHF to determine an adjusted *two-way flow rate* (v_p) that reflects these characteristics.

$$v_p = \frac{V}{PHF * f_G * f_{HV}} \tag{3.2}$$

where

v_p = passenger car equivalent flow rate for peak 15-minute period (pc/h)

V = demand volume for the full peak hour (vph)

> **FYI:** In cases where (a) 15-minute counts within the peak hour have not been taken and (b) the roadway being analyzed is in a rural area, a default value of PHF = 0.88 can be used (*HCM*, 2000, p. 12-19).

In our ongoing illustration, the conditions in the FYI box apply. A default value of PHF = 0.88 will be used. Using Equation 3.2, v_p would be

$$v_p = \frac{V}{\text{PHF} * f_G * f_{HV}} = \frac{526}{0.88 * 0.71 * 0.708} = \frac{526}{0.442} = 1189 \text{ pc/h}$$

By applying the adjustment factors called for in the worksheet, the counted hourly volume of 526 vph has been converted to 1189 passenger car equivalents per hour. This PCE value reflects fluctuations in traffic within the hour, the terrain, and the vehicle mix. It also is a more accurate representation of the traffic on the roadway. However, note that 1189 exceeds the 600 pc/h upper bound of the range in which $V = 526$ lay. This means that the values taken from Exhibits 20-7 and 20-9 earlier using $V = 526$ must be replaced by using values from the next higher range. (The *HCM* calls this an *iterative computation*, because the first set of calculations has determined whether a second set was needed.) The new values are $f_G = 0.93$, $E_T = 1.9$, and $E_R = 1.1$. The Heavy Vehicle Adjustment Factor f_{HV} from Equation 3.1 would become

$$f_{HV} = \frac{1}{1 + 0.274(1.9 - 1) + 0.015(1.1 - 1)}$$

$$= \frac{1}{1 + 0.246 + 0.0015} = \frac{1}{1.2475} = 0.802$$

and Equation 3.2 now leads to

$$v_p = \frac{V}{\text{PHF} * f_G * f_{HV}} = \frac{526}{0.88 * 0.93 * 0.802} = \frac{526}{0.656} = 801 \text{ pc/h}$$

Note that we are still using the original $V = 526$ veh/hr value, but we are using adjustment factors taken from the range of values that correspond to the $v_p = 1189$ pc/h found in the first interation. The value $v_p = 801$ pc/h is "acceptable" because it lies in the same range (>600 to 1200) as the v_p value used to determine f_G, E_T, and E_R. According to the first footnote in the worksheet, if the iterated v_p value exceeds 3200 pc/h, the roadway's LOS is F and the analysis can be terminated. Otherwise, the next line in the worksheet uses the higher flow direction to convert the two-way flow rate into the major direction flow. If 63 percent of the two-way flow is WB and 37 percent is EB, the higher proportion (0.63) is used in

$$v_p * \text{highest directional split proportion} = 801 \text{ pc/h} * 0.63 = 505 \text{ pc/h} \qquad (3.3)$$

If this value exceeds 1700 pc/h, LOS F has been found, and the analysis need not continue.

Free-Flow Speed from Field Measurements.   Although somewhat tedious to do by hand, the steps up to this point have depended simply on available data and following the rules prescribed in the *HCM*. The next steps in the Class I method are critical because they establish the free-flow speed on the highway segment to be analyzed. All other steps have the effect of reducing the speed on the highway as the shortcomings with respect to the "perfect" highway are identified. After adjustments, the resulting ATS will provide one of the two criteria by which to assess the LOS of the roadway. If the free-flow speed estimate is inaccurate, the entire analysis will be affected. Two methods for setting the free-flow speed are as follows:

1. Field measurements of speeds on the highway to be analyzed. This, of course, is possible only for existing roadways under low-flow conditions. It may take some time and expense, but it can capture the conditions of the roadway as they exist. If there is a highway very similar to the one being designed, it may be possible to measure the traffic on the surrogate highway and use those characteristics to determine FFS.

2. Estimates of free-flow speed using adjustment factors provided in the *HCM*. This option can be applied to cases for which there is insufficient time or resources to collect data, where the benefits of having actual data are not deemed that critical to a particular analysis, or when hypothetical changes to a roadway need to be evaluated.

Both methods have been given space in the lower half of the "Average Travel Speed" section of the worksheet in Figure 3.4. If speeds are measured in the field, it should be done in periods of low flow (i.e., when two-way flow is less than 200 vph). If field measurements cannot be performed at such low-flow conditions, speeds taken at high-flow rates can be converted into approximate free-flow speeds by using

$$\text{FFS} = S_{\text{FM}} + 0.00776\frac{V_{\text{f}}}{f_{\text{HV}}} \tag{3.4}$$

where

$\quad$ FFS = estimated free-flow speed (mph)

$\quad S_{\text{FM}}$ = mean speed of traffic measured in the field (mph)

$\quad V_{\text{f}}$ = observed flow rate for the period when field data were obtained (vph)

$\quad f_{\text{HV}}$ = heavy-vehicle adjustment factor, determined as shown in Equation 3.1.

If the mean speed in our sample was S_{FM} = 52.8 mph when V_{f} = 280 vph, then Equation 3.4 can be used to determine FFS. But which f_{HV} value should be used in that equation?

(a) The f_{HV} = 0.802 found in the second iteration for V = 526 vph in the analysis period?

(b) The f_{HV} = 0.708 that follows from the V = 280 vph for the time the speeds were observed?

Because the free-flow speed data are being sought by field measurements during some off-peak period, it makes no sense to use the peak period f_{HV} value. Option (b) is the correct choice.

FYI: When the authors used a software package to check their calculations for this problem, they found that the software incorrectly used Option (a). Fortunately, it made a difference of only 0.37 mph in the determination of FFS, but it is a further reminder to check your calculations and to get to know the strengths and weaknesses of any software you use.

With $f_{HV} = 0.708$, Equation 3.4 becomes

$$\text{FFS} = S_{FM} + 0.00776\frac{V_f}{f_{HV}} = 52.8 + 0.00776\frac{280}{0.708} = 52.8 + 3.07 = 55.87 \text{ mph}$$

Exhibit 20-11 in Table 3.9 gives us the speed adjustment for no-passing zones. If our roadway has 20 percent no-passing zones and two-way $v_p = 801$ pc/h, $f_{np} = 1.4$ mph.

Computing Average Travel Speed. Finally, ATS (mph), can be computed in the worksheet as

$$\text{ATS} = \text{FFS} - 0.00776\,v_p - f_{np} \tag{3.5}$$

$$\text{ATS} = 55.87 - (0.00776 * 801) - 1.4 = 55.87 - 6.22 - 1.4 = 48.25 \text{ mph}$$

THINK ABOUT IT

How would you determine the "percent no-passing zones" value for a two-lane highway?

The ATS value of 48.25 mph fits into the LOS C range of Table 3.2, but we still need to determine *Percent Time Spent Following (PTSF)* to check against the possibility that LOS D or E might apply in Figure 3.3. Before doing the PTSF analysis, let us look at the alternate way to determine free-flow speed when field measurements are not available.

Free-Flow Speed Without Field Measurements. Using the "Estimated Free-Flow Speed" lines of the worksheet, begin with an assumption of a *base free-flow speed (BFFS)*. According to the *HCM* (2000, p. 20-5), the BFFS

> reflects the character of traffic and the alignment of the facility. Because of the broad range of speed conditions on two-lane highways and the importance of local and regional factors that influence driver-desired speeds, no guidance on estimating the BFFS is provided. Estimates of BFFS can be developed based on speed data and local knowledge of operating conditions on similar facilities. The design speed and posted speed limit of the facility may

be considered in determining the BFFS; however, the design speeds and speed limits for many facilities are not based on current operating conditions."

The free-flow speed of a two-lane highway is typically in the 45 to 65 mph range. If the speed limit on that road segment is 50 mph, but drivers at low-flow periods tend to exceed the speed limit by 5 mph, a good initial estimate of BFFS might be 55 mph. According to the worksheet, Exhibit 20-5 should be consulted next. If the lanes are 12 feet wide and the shoulders are an average of 5 feet wide, the exhibit indicates an adjustment for lane and shoulder width, $f_{LS} = 1.3$ mph. *Access points* are intersections and driveways on both sides of the roadway. If there are 43 access points along a 1.6-mile stretch of road, the access point density is $43/1.6 = 26.9$ per mile. Exhibit 20-6 translates that density into an adjustment factor f_A by interpolating between $f_A = 5.0$ for density $= 20$ and $f_A = 7.5$ for density $= 30$. The result is $f_A = 6.7$ mph. The free-flow speed found without field measurements is

$$FFS = BFFS - f_{LS} - f_A \qquad (3.6)$$
$$FFS = 55 - 1.3 - 6.7 = 47.0 \text{ mph}$$

Applying Equation 3.5 with this FFS,

$$ATS = FFS - 0.00776v_p - f_{np} = 47.0 - (0.00776*801) - 1.4$$
$$= 47.0 - 6.22 - 1.4 = 39.38 \text{ mph}.$$

This ATS value based on estimated FFS falls in the LOS E range of Table 3.2, whereas the ATS value based on field-measured FFS fell in the LOS C range. It is interesting to note that the two methods of determining free-flow speed in the worksheet have very little in common. The presumption is that the adjustments for lane and shoulder width and for access point density used in the "Estimated Free-Flow Speed" method are reflected in the speeds collected during field measurements for the "Field Measurements" method. It should also be clear by now how important (and difficult) the choice of BFFS is. A difference of a few miles per hour in judging the value of BFFS can shift the resulting FFS value into a different LOS range in Table 3.2 and Figure 3.3.

Compute Percent Time Spent Following.   The Percent Time Spent Following (PTSF) section of the worksheet requires use of exhibits and equations similar to those used to compute ATS. The grade adjustment factor f_G for $V = 526$ vph in rolling terrain is 0.77. (Note that because the adjustment factors for PTSF may be different from the adjustment factors for ATS, the iterative process must be followed again. We cannot presume that the second iteration needed in the ATS steps will be needed in the PTSF steps.) The PCE values are not the same as they were in the ATS steps. For the PTSF steps, they are $E_T = 1.8$, $E_R = 1.0$. Equation 3.1 produces

$$f_{HV} = \frac{1}{1 + 0.274(1.8 - 1) + 0.015(1.0 - 1)} = \frac{1}{1 + 0.219 + 0.0} = \frac{1}{1.219} = 0.820$$

Because the f_G and f_{HV} factors have changed, the v_p value from Equation 3.2 will change:

$$v_p = \frac{V}{PHF*f_G*f_{HV}} = \frac{526}{0.88*0.77*0.820} = \frac{526}{0.556} = 947 \text{ pc/h}$$

Because this v_p exceeds the upper bound (600) of the flow range currently being used, the next higher range (>600 to 1200) must be invoked in another iteration of calculations: $f_G = 0.94, E_T = 1.5, E_R = 1.0$.

$$f_{HV} = \frac{1}{1 + 0.274(1.5 - 1) + 0.015(1.0 - 1)} = \frac{1}{1 + 0.137 + 0.0} = \frac{1}{1.137} = 0.880$$

$$v_p = \frac{V}{PHF * f_G * f_{HV}} = \frac{526}{0.88 * 0.94 * 0.880} = \frac{526}{0.728} = 723 \text{ pc/h}$$

Now we can complete the PTSF section of the worksheet. Repeating Equation 3.3,

$$v_p * \text{highest directional split proportion} = 723 \text{ pc/h} * 0.63 = 455 \text{ pc/h}$$

Again, if this value exceeds 1700 pc/h, LOS F has been found, and the analysis need not continue.

A value for *Base Percent Time Spent Following (BPTSF)* can be estimated from

$$BPTSF = 100(1 - e^{-0.000879 v_p}) \tag{3.7}$$

$$BPTSF = 100(1 - e^{-0.000879 v_p}) = 100(1 - e^{-0.000879 * 723}) = 100(1 - 0.530) = 47.0$$

Checking Table 3.10 for an adjustment factor ($f_{d/np}$) *for directional distribution* (63/37) *and percent no-passing zones* (20) requires some linear interpolation. An "interpolation table" using values from the 20 percent no-passing column in Table 3.10 may help. The italicized entries in the table below have been interpolated, first in the top and bottom rows and then in the middle column.

$f_{d/np}$	60/40 Directional Distribution	63/37 Directional Distribution	70/30 Directional Distribution
$v_p = 600$	11.5	*11.53*	11.6
$v_p = 723$		*9.13*	
$v_p = 800$	7.6	*7.63*	7.7

The last line in the PTSF section of the worksheet in Figure 3.5 is

$$PTSF = BPTSF + f_{d/np} \tag{3.8}$$

$$PTSF = 47.0 + 9.13 = 56.13$$

Determining LOS.   By itself, a PTSF value of 56.13 percent places it in the LOS C range of Table 3.2. Recall that the LOS of a Class I two-lane highway has two criteria in Figure 3.3. The combination of ATS = 48.25 mph and PTSF = 56.13 percent is a point in the LOS C band in Figure 3.3. See Figure 3.6.

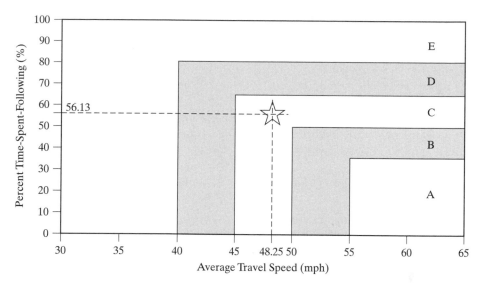

FIGURE 3.6
LOS of two-lane highway based on ATS and PTSF.

Example 3.1

SR361 is a Class I two-lane highway in rolling terrain. An operational analysis is needed for the AM peak hour, during which a two-way count of 1275 vehicles—including 157 trucks, 2 buses, and no RVs—has been made. Fifty-eight percent of the traffic was headed EB. The section of SR361 of particular interest is 1.9 miles long, with 12-foot lanes and 4-foot shoulders. Four intersections and nine driveways are found along the segment. Thirty-eight percent of the segment has no-passing zones. A base free-flow speed of 55 mph will be used. What is the LOS of this section of SR361 during the AM peak? Use a PHF of 0.88.

Solutions for Example 3.1   The operational analysis will follow the steps laid out in the worksheet in Figure 3.4. The values found in the exhibits, the basic calculations, and any assumptions or judgments are presented in sequence.

Steps for Average Travel Speed

f_G (Exhibit 20-7) = 0.99 for V = 1275 vph in rolling terrain. Note that we are using a two-way vehicle count of vehicles at this point. The conversion from vph to passenger car equivalents "pc/h" comes next.

Exhibit 20-9: E_T = 1.5, E_R = 1.1 for V = 1275 pc/h in rolling terrain.

$$P_T = \frac{157 + 2}{1275} = 0.125, P_R = 0.00$$

$$f_{HV} = \frac{1}{1 + 0.125(1.5 - 1) + 0.00(1.0 - 1)} = \frac{1}{1 + 0.0625 + 0.0} = \frac{1}{1.1125} = 0.941 \quad (3.1)$$

$$v_p = \frac{V}{PHF * f_G * f_{HV}} = \frac{1275}{0.88 * 0.99 * 0.941} = \frac{1275}{0.820} = 1555 \text{ pc/h} \quad (3.2)$$

This v_p < 3200 pc/h, so the analysis continues. It is also in the same range as the original V, so no iterative calculation is needed. With 58 percent of the traffic EB, v_p* highest directional split proportion = 1555 * 0.58 = 902.

This value is less than 1700 pc/h, so the analysis continues.

No speed data were provided, so the "Estimated Free-Flow Speed" will be based on BFFS = 55 mph.

f_{LS} (Exhibit 20-5) = 1.3 mph for 12-foot lanes and 4-foot shoulders. Note that in Exhibit 20-5, 4-foot shoulders do not belong in the "2 to <4" range, but rather in the "4 to <6" range.

If there are 4 intersections along the segment, that means there are 8 roads entering the highway. Add the 9 driveways, and there are 17 access points in 1.9 miles, for an access point density of 17/1.9 = 8.9 per mile. f_A (Exhibit 20-6) = 2.24 mph by interpolation.

$$FFS = 55 - 1.3 - 2.24 = 51.5 \text{ mph} \tag{3.6}$$

With 38 percent no-passing zones and 1555 pc/h, f_{np} (Exhibit 20-11) = 0.82 mph by linear interpolation

$$\begin{aligned} ATS &= FFS - 0.00776\, v_p - f_{np} = 51.5 - (0.00776 * 1555) - 0.82 \\ &= 51.5 - 12.07 - 0.82 = 38.6 \text{ mph} \end{aligned} \tag{3.5}$$

Steps for Percent Time Spent Following

f_G (Exhibit 20-8) = 1.00 for V = 1275 pc/h in rolling terrain.

Exhibit 20-10: E_T = 1.0, E_R = 1.0 for V = 1275 vph in rolling terrain, so f_{HV} = 1.0

$$v_p = \frac{V}{PHF * f_G * f_{HV}} = \frac{1275}{0.88 * 1.00 * 1.00} = \frac{1275}{0.88} = 1449 \text{ pc/h} \tag{3.2}$$

This v_p < 3200 pc/h, so the analysis continues. With 58 percent of the traffic EB, v_p* highest directional split proportion = 1449 * 0.58 = 840.

This value is less than 1700 pc/h, so the analysis continues.

$$BPTSF = 100(1 - e^{-0.000879 v_p}) = 100(1 - e^{-0.000879 * 1449}) = 100(1 - 0.280) = 72.0 \text{ percent} \tag{3.7}$$

$f_{d/np}$ (Exhibit 20-12) = 5.08 mph by multiple linear interpolations shown in the table below.

	20% No Passing; Directional Distribution			38%	40% No Passing; Directional Distribution		
	50/50	**58/42**	**60/40**		**50/50**	**58/42**	**60/40**
v_p = 1400	3.6	3.68	3.7		5.5	5.42	5.4
v_p = 1449		3.56		5.08		5.25	
v_p = 2000	1.8	2.2	2.3		2.9	3.3	3.4

$$PTSF = BPTSF + f_{d/np} = 72.0 + 5.1 = 77.1 \text{ percent} \tag{3.8}$$

Plotting ATS = 38.6 mph and PTSF = 77.1 percent in Figure 3.3 results in a point lying in the LOS E band. During the time observed, SR361 was operating at LOS E.

None of the steps in the operational analysis described in this section is difficult, but there are many table lookups and calculations involved. Fortunately, software called *Highway Capacity Software (HCS) 2000* has been produced to carry out the steps in the HCM. *HCS 2000* relieves the analyst of the tedium involved in the two-lane highway operational analysis, but the analyst must understand what *HCS 2000* is capable of and what its limitations are.

3.1.2   Highway Capacity

During the operational analysis presented in the previous section, a two-way capacity of 3200 pc/h and a one-way capacity of 1700 pc/h were mentioned as points beyond which the LOS on a two-lane highway is "F." However, the definition of *highway capacity* in the *HCM* is:

> the maximum hourly rate at which persons or vehicles can reasonably be expected to traverse a point or uniform section of a lane or roadway during a given time period under prevailing roadway, traffic, and control conditions (*HCM*, 2000, p. 2-2).

This definition indicates that the capacity of a lane or roadway depends on several conditions. A level roadway has a greater capacity than a roadway on rolling terrain. A roadway with no heavy vehicles can accommodate more vehicles than a roadway used by trucks, buses, and/or RVs. A roadway with limited access has more capacity than a road with numerous intersections and driveways. It is possible to describe a two-lane highway in terms of its terrain, vehicle mix, and degree of access control, and so on, then determine its capacity. This can be done by determining the V value at which LOS E becomes LOS F. Recall that, in the previous section, V was a count of vehicles without distinguishing vehicle types, whereas v_p was the result of converting V into passenger car equivalents (PCE). It is the v_p value that should be compared with the standard capacity value to calculate a *volume-to-capacity ratio (v/c)*:

$$v/c = \frac{v_p}{c} \tag{3.9}$$

where v/c = volume-to-capacity ratio

c = two-way segment capacity—normally 3200 pc/h for a two-way segment and 1700 for a directional segment

v_p = passenger car equivalent flow rate for peak 15-minute period (pc/h).

Because v_p has PCE units, so must the capacity value c. Just below the Level of Service line in the worksheet in Figure 3.4 is a line for the v/c ratio. Although the LOS ratings A through E offer a familiar "grading scheme," the v/c ratio offers a more quantitative assessment of a roadway's performance.

Example 3.2

For the roadway analyzed in Example 3.1,

A. What value of V would cause $v/c = 1.00$?

B. What value of ATS results from the value of V found in part A, and where does it lie when plotted in Figure 3.3?

Solution to Example 3.2

A. The volume-to-capacity ratio $v/c = 1.00$ when $v_p = 3200$ pc/h. In the ATS part of Example 3.1,

$$v_p = \frac{V}{\text{PHF} * f_G * f_{HV}} = \frac{1275}{0.88 * 0.99 * 0.941} = \frac{1275}{0.820} = 1555 \text{ pc/h}$$

By rearranging Equation 3.2, $V = 3200 * 0.820 = 2624$ vph.

B. Because $v_p = 3200$ pc/h, f_{np} (Exhibit 20-11) $= 0.59$ mph by linear interpolation. Using Equation 3.5,

$$\text{ATS} = \text{FFS} - 0.00776\, v_p - f_{np} = 51.5 - (0.00776 * 3200) - 0.59$$
$$= 51.5 - 24.83 - 0.59 = 26.1 \text{ mph}$$

In Figure 3.3, ATS = 26.1 lies beyond the boundary of LOS E. This means that the LOS is F, which makes sense when $v/c = 1.00$.

THINK ABOUT IT

Although interpolation between values in the *HCM* exhibits is acceptable, it is not always worth the effort. Finding a precise value for $f_{d/np}$ in Example 3.1 took a three-stage interpolation. Look at Exhibits 20-11 and 20-12. Try to develop a guideline for when a careful linear interpolation is not necessary.

3.1.3   Using Level of Service Analysis for Design

In section 3.1.1, an operational analysis of a two-lane highway was demonstrated. It would be helpful to be able to determine which roadway and traffic characteristics could be controlled to improve the roadway's performance. For example, what would be the impact of reducing the access points along the roadway? Or, what would be the impact of restricting heavy vehicle use during certain hours? These kinds of questions are part of design analysis. According to the *HCM* (2000, p. 20–32),

> The objective of design analysis is to estimate the flow rate in passenger cars per hour given a set of traffic, roadway, and FFS conditions. A desired LOS is stated and entered in the worksheet. Then a flow rate is assumed and the procedure for operational (LOS) analysis is performed. This computed LOS is then compared with the desired LOS. If the desired LOS is not met, another flow rate is assumed. This iteration continues until the maximum flow rate for the desired LOS is achieved.

It must be emphasized that the iterative nature of this analysis is unavoidable. It may be tempting to simply try to solve an equation involving v_p, but v_p cannot be isolated that easily. As v_p changes, so do several adjustment factors taken from exhibits that have v_p as input. Even in Example 3.2, the value of f_{np} had to be changed, because v_p had been changed.

In this context, *design* means trying alternative strategies on an existing or proposed roadway to meet certain LOS standards. The design (or redesign) of the roadway could range from closing access points to improving terrain from rolling to level. Operational strategies could include restricting heavy vehicles during certain times of day.

Example 3.3

The LOS E on SR361 found in Example 3.1 is a cause of concern to highway officials in Mythaca. The highway department may want to consider measures to improve the level of service on SR361. For example, will prohibiting trucks during the peak period described in Example 3.1 cause the operation on SR361 to improve to LOS C?

Solution to Example 3.3 Removing 157 trucks from SR361 will reduce the two-way vehicle count **V** from 1275 to 1118. With only two buses remaining, $P_T = 2/1118 = 0.18$ percent $= 0.0018$. These changes will affect the value of f_G taken from Exhibit 20-7 (0.93 vs. 0.99 in Example 3.1) and E_T from Exhibit 20-9 (1.9 vs. 1.5). The computed value of f_{HV} becomes 0.998, leading to $v_p = 1368$. Because this v_p value exceeds 1200 pc/h, which is the upper bound on the flow rate ranges used in Exhibits 20-9 and 20-7, a second iteration is needed. In the second iteration, the ranges used in Exhibits 20-9 and 20-7 are the same as those ranges used for Example 3.1. The updated computations are $f_{HV} = 0.999$ and $v_p = 1284$. ATS is a function of v_p, but its "truck-free" value of 38.6 mph is not much faster than ATS $= 36.58$ mph in Example 3.1. The LOS would still be E, despite the enforced absence of trucks.

The PTSF portion of the analysis will be carried out to determine how sensitive that criterion is to the removal of trucks from the traffic stream. In Exhibit 20-8, $V = 1118$ vph corresponds to $f_G = 0.94$. In Exhibit 20-10, the E_T value to use is 1.5. These values produce $f_{HV} = 0.999$ and $v_p = 1353$. As in the ATS portion of the analysis, the computed v_p value exceeds 1200 pc/h, so a second iteration is needed with $v_p = 1353$, $f_G = 0.94$ in Exhibit 20-8 and $E_T = 1.5$ in Exhibit 20-10. The recomputed $f_{HV} = 0.999$ and $v_p = 1272$, which confirms the flow range found at the end of the first iteration. Because BPTSF is a function of v_p, the lower v_p causes BPTSF to drop to 67.3 percent from 72.0 percent. With $v_p = 1272$ pc/h, another three-stage interpolation in Exhibit 20-12 produces $f_{d/np} = 6.34$ mph, which is a slight increase from the 5.08 mph value found in Example 3.1 when $v_p = 1449$ pc/h. The resulting PTSF $= 73.6$ percent is not much lower than the 77.1 percent value found in Example 3.1. LOS based only on PTSF is "D," just as it was in Example 3.1. LOS based on ATS is still "E," which means that abolishing truck traffic during the PM peak hour will not have much positive impact on the operations on SR361.

THINK ABOUT IT
Can you explain why, in Table 3.10, $f_{d/np}$ usually increases as v_p decreases?

In this section, an analysis of a two-way two-lane highway segment has been introduced. Performance measures for the two-way segment methodology apply to both directions of travel combined. Two-way segments may include longer sections of two-lane highway that have homogeneous cross sections, relatively constant demand volumes, and relatively constant vehicle mixes over the length of the segment. Two-way segments may be located in level or rolling terrain. If the terrain is mountainous, or with grades of 3 percent or more for lengths of 0.6 mi or more, the segment must be separated into specific upgrades or downgrades. Any roadway segment can be evaluated with the directional segment procedure, but separate analysis by direction of travel is particularly appropriate for steep grades and for segments containing passing lanes. The *HCM* has provisions for such segments, which are not covered in this book.

THINK ABOUT IT

Why does it make sense to separate a two-way highway segment on a steep grade into two separate directional segments—uphill and downhill?

3.2   CAPACITY AND LEVEL OF SERVICE FOR FREEWAY DESIGN

Remember Kara's question at the start of this chapter? As she and her mother were creeping along in their car on I-25, Kara asked her mom, "Why is there always a traffic jam on this highway?" In this section, we find out more about what causes traffic jams, even on freeways like the one shown in Figure 3.7.

3.2.1   Freeway Definitions

A *freeway* may be defined as a divided highway that provides uninterrupted flow. This quality of flow is achieved by the physical control of access—permitting vehicles to enter only from ramps designed to facilitate merging with traffic already on the freeway. There are no signalized or stop-controlled intersections. Opposing directions of traffic flow are continuously separated by a barrier or median strip.

FIGURE 3.7
"Wall of Trucks" on Borman Expressway. *Source:* Jon D. Fricker, 7 July 2000.

TABLE 3.11 Fatality Rates by Roadway Class

Roadway Class	Rural Freeways	Urban Freeways	All Freeways	Rural Non-Freeways	Urban Non-Freeways	All Non-Freeways
Fatalities in 1998	3095	3555	6650	21,557	12,468	34,025
VMT (10^6)	251,062	374,407	625,469	782,248	1,217,650	1,999,898
Fatalities per 10^8 VMT	1.233	0.950	1.063	2.756	1.024	1.701

Source: *Highway Statistics*, 1998, FHWA, November 1, 1999, FHWA-PL-99-017, p. V48-V53.

Freeways are the safest of all roadway types. (See Table 3.11.) Traffic safety problems on freeways tend to result from differences in vehicle speeds and *weaving* at on/off ramps. Of course, rain, snow, fog, and/or poor lighting can also be the cause of some traffic incidents. The smooth flow of traffic can also be disrupted by toll collection operations.

As can be seen in Figure 3.8, the freeway consists of three component parts:

1. Basic freeway sections.
2. *Weaving areas*, defined as "the crossing of two or more traffic streams traveling in the same general direction along a significant length of highway without the aid of traffic control devices. Weaving segments are formed when a merge area is closely followed by a diverge area, or when an on-ramp is closely followed by an off-ramp and the two are joined by an auxiliary lane" (*HCM*, 2000, p. 13-13).
3. Ramp junctions, which permit access to and from the freeway.

The basic freeway section is treated in this lesson. Treatments of the weaving and ramp components can be found in the *Highway Capacity Manual* (*HCM*, 2000).

The methods for determining capacity and level of service on freeway sections appear in Chapter 23 of the *Highway Capacity Manual* (*HCM*, 2000). Many of the methods

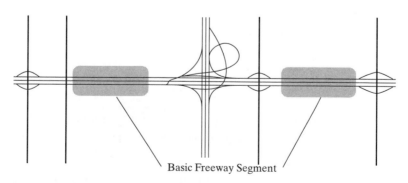

Basic Freeway Segment

FIGURE 3.8
Basic freeway sections. *Source: HCM*, 2000, Exhibit 13-1.

described in *HCM* Chapter 23 use the same paradigm as that used in *HCM* Chapter 20 for two-lane highways, i.e., start with an ideal situation and then convert specified non-ideal traffic and roadway conditions into adjustment factors that reduce the ideal value of speed or flow to a value that can be compared against a table of LOS standards.

3.2.2 The "Perfect Freeway" for Maximum Flow

Just as we did for rural two-lane highways, a set of ideal conditions exists as the starting point that will provide for maximum flow. The conditions are as follows:

- Lanes at least 12 feet wide
- Right-shoulder lateral clearance between the edge of the travel lane and the nearest obstacle or object influencing traffic behavior at least 6 feet (minimum median lateral clearance is 2 feet)
- Traffic stream consists of passenger cars only
- Ten or more lanes (in urban areas only)
- Interchanges spaced every 2 miles or more
- Level terrain, with grades no greater than 2 percent
- Driver population dominated by regular and familiar users of the facility.

These ideal conditions represent the highest type of basic freeway section, one with a free-flow speed of 70 mph or greater. The photo in Figure 3.7 was taken from a car traveling westbound (WB) on the Borman Expressway early on a Friday afternoon in July. The Borman Expressway is I80/94 in Northwest Indiana just east of Chicago. The traffic conditions are certainly not "perfect," as that term has been defined above.

 THINK ABOUT IT
What elements of the "perfect freeway" seem to be violated in Figure 3.7?

Note in Figure 3.7 how a long series of trucks occupies the middle of the three WB lanes heading toward Chicago. In fact, the middle lane for oncoming (EB) traffic is also predominately trucks. A traffic stream with 30 percent or more trucks is generally considered to have a lot of truck traffic.

3.2.3 Freeway Performance Measures

The service offered by a basic freeway section can be characterized by three performance measures:

- Density in terms of passenger cars per mile per lane (pc/mi/ln)
- Speed in terms of mean passenger car speed
- Volume-to-capacity ratio.

Each of these measures is an indication of how well (or how poorly) traffic flow is being accommodated by the freeway. These three measures (speed, density, and *v/c*) are interrelated—if two of these measures are known, the third can be determined.

TABLE 3.12 Density Ranges for Level of Service

Level of Service	Density Range (pc/mi/ln)
A	0–11
B	>11–18
C	>18–26
D	>26–35
E	>35–45
F	>45 *

*Based on the collective professional judgment of the members of the Committee on Highway Capacity and Quality of Service of the Transportation Research Board, the upper value shown for LOS E (45 pc/mi/ln) is the maximum density at which sustained flows at capacity are expected to occur.

Source: HCM, 2000, p. 23-3.

One overall measure of performance, called Level of Service (LOS), was introduced in section 2.5 of this book. However, if *density* is the basic performance measure being used, then Table 3.12 will indicate the freeway's level of service. Table 3.13 gives the LOS criteria for free-flow speeds from 75 mph to 55 mph. In past editions of the *HCM*, the designer would have used a capacity of 2000 passenger cars per hour per lane (pcphpl) for freeway design. Recent data, however, indicate that a freeway truly designed for a free-flow speed of 70 mph will have an ideal capacity of about 2400 pcphpl. The design speed used for most rural and suburban freeways is 70 mph. Freeways in urban areas are often designed with lower speeds in mind. In Figure 3.9, note

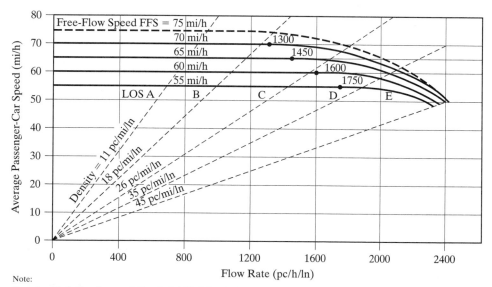

Note:

Capacity varies by free-flow speed. Capacity is 2400, 2350, 2300 and 2250 pc/h/ln at free-flow speeds of 70 and greater, 65, 60, and 55 mi/h, respectively.

FIGURE 3.9

Average car speed vs. flow rate for differing free-flow speeds with LOS indicated. *Source: HCM*, 2000, Exhibit 23-3.

TABLE 3.13 (Exhibit 23-2) LOS Criteria for Basic Freeway Segments

Criteria	LOS				
	A	B	C	D	E
FFS = 75 mph					
Maximum density pc/mi/ln	11	18	26	35	45
Minimum speed (mph)	75.0	74.8	70.6	62.2	53.3
Maximum v/c	0.34	0.56	0.76	0.90	1.00
Maximum service flow rate (pc/hr/ln)	820	1350	1830	2170	2400
FFS = 70 mph					
Maximum density pc/mi/ln	11	18	26	35	45
Minimum speed (mph)	70.0	70.0	68.2	61.5	53.3
Maximum v/c	0.32	0.53	0.74	0.90	1.00
Maximum service flow rate (pc/hr/ln)	770	1260	1770	2150	2400
FFS = 65 mph					
Maximum density pc/mi/ln	11	18	26	35	45
Minimum speed (mph)	65.0	65.0	64.6	59.7	52.2
Maximum v/c	0.30	0.50	0.71	0.89	1.00
Maximum service flow rate (pc/hr/ln)	710	1170	1680	2090	2350
FFS = 60 mph					
Maximum density pc/mi/ln	11	18	26	35	45
Minimum speed (mph)	60.0	60.0	60.0	57.6	51.1
Maximum v/c	0.29	0.47	0.47	0.88	1.00
Maximum service flow rate (pc/hr/ln)	660	1060	1560	2020	2300
FFS = 55 mph					
Maximum density pc/mi/ln	11	18	26	35	45
Minimum speed (mph)	55.0	55.0	55.0	54.7	50.0
Maximum v/c	0.27	0.44	0.54	0.85	1.00
Maximum service flow rate (pc/hr/ln)	600	990	1430	1910	2250

Note: The exact mathematical relationship between density and *v/c* has not always been maintained at the LOS boundaries because of the use of rounded values. Density is the primary determinant of LOS. The speed criterion is the speed at maximum density for a given LOS

Source: HCM, 2000, Exhibit 23-2.

that capacity varies by free-flow speed. The capacities (or flow at $v/c = 1.00$) indicated by the right end of the curves for design speeds of 55, 60, 65, and 70 mph in Figure 3.9 are 2250, 2300, 2350, and 2400 pcphpl, respectively.

THINK ABOUT IT

Is it possible for an incident on one side of a freeway median to cause delays on the opposite of the median? If so, how can this happen?

3.2.4 Applications

The *Highway Capacity Manual* contains worksheets for the analysis of basic freeway sections. (See Figure 3.10 for the top section of the Basic Freeway Section Worksheet.) The approach is similar to the two-lane highway analysis. The six possible types of basic freeway segment analysis (or "applications") are given in the box to the right of the flow-speed curves in Figure 3.10. For example, the standard operational analysis uses free-flow speed (FFS), number of lanes (N), and hourly flow rate (v_p) as inputs to find the LOS on a freeway section, with secondary outputs speed (S) and density (D). The "Design (N)" application is a check on the adequacy of the number of lanes (actual or proposed) for a basic freeway segment, given the volume or flow rate and LOS goal. Another design method calculates achievable flow rate v_p as its primary output. This "Design (v_p)" analysis requires an LOS goal and a number of lanes as inputs and estimates the flow rate that will cause the highway to operate at an unacceptable LOS. In planning analyses, most or all of the input values come from estimates or default values, whereas the operational and design analyses tend to use field measurements or known values for most or all of the input variables. In each of the six analyses, FFS, either measured or estimated, is required as an input.

The hourly flow rate is determined from Equation 3.10:

$$v_p = \frac{V}{(\text{PHF})(N)(f_{HV})(f_p)} \tag{3.10}$$

where PHF = the peak hour factor

N = number of lanes

f_{HV} = heavy-vehicle factor

f_p = driver population adjustment

The overall formula for the heavy-vehicle adjustment factor uses Equation 3.1, which was introduced in section 1 of this chapter, where the passenger car equivalents are given in Table 3.14.

$$f_{HV} = \frac{1}{[1 + P_T(E_T - 1) + P_R(E_R - 1)]} \tag{3.1}$$

The "driver population adjustment" factor in Equation 3.10 attempts to account for the degree of familiarity drivers have with the roadway. When most drivers are regular users (e.g., daily commuters) of the roadway, a value $f_p = 1.00$ is appropriate. If the traffic stream has numerous infrequent users, such as visitors or vacationers who use the roadway less knowledgeably, a value as low as 0.85 can be used. When in doubt, use $f_p = 1.00$ (*HCM*, 2000, p. 23-12).

Highway-Specific Free-Flow Speed. The speed of vehicles on the freeway will be affected by a number of factors in the design of the particular freeway, such as right

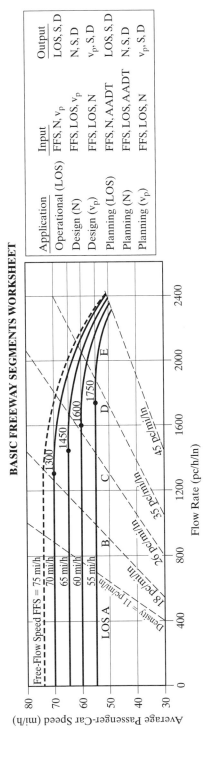

BASIC FREEWAY SEGMENTS WORKSHEET

Application	Input	Output
Operational (LOS)	FFS, N, v_p	LOS, S, D
Design (N)	FFS, LOS, v_p	N, S, D
Design (v_p)	FFS, LOS, N	v_p, S, D
Planning (LOS)	FFS, N, AADT	LOS, S, D
Planning (N)	FFS, LOS, AADT	N, S, D
Planning (v_p)	FFS, LOS, N	v_p, S, D

FIGURE 3.10

Top of worksheet for analysis of basic freeway section. *Source: HCM*, 2000, p. 23–33.

TABLE 3.14   (Exhibit 23-8) Passenger Car Equivalents (E_T and E_R)for Extended General Freeway Segments *

	Type of Terrain		
Category	Level	Rolling	Mountainous
E_T for trucks and buses	1.5	2.5	4.5
E_R for recreational vehicles	1.2	2.0	4.0

* See Exhibits 23-9, 23-10, and 23-11 in the *Highway Capacity Manual* for tables to analyze specific upgrades and downgrades.

Source: HCM, 2000, Exhibit 23-8.

shoulder lateral clearance, lane width, number of lanes, and the density of interchanges. Tables 3.13 to 3.20 show the extent to which the ideal (or base) free-flow speed is reduced by less-than-ideal design characteristics. The equation for the reduction in free-flow speed (FFS) is

$$\text{FFS} = \text{BFFS} - f_{LW} - f_{LC} - f_N - f_{ID} \qquad (3.11)$$

where FFS = estimated free flow speed

BFFS = base free-flow speed, 70 mph (urban) or 75 mph (rural)

f_{LW} = adjustment for lane width from Table 3.15

f_{LC} = adjustment for right-shoulder lateral clearance from Table 3.16

f_N = adjustment for number of lanes from Table 3.17

f_{ID} = adjustment for interchange density from Table 3.18

TABLE 3.15   (Exhibit 23-4) Adjustment Factors for Lane Width (f_{LW})

Lane Width	Reduction in Free Flow Speed
≥12	0.0
11	1.9
10	6.6

Source: HCM, 2000, Exhibit 23-4.

THINK ABOUT IT

Why would any roadway—especially a freeway—be designed to less-than-ideal specifications?

TABLE 3.16 (Exhibit 23-5) Adjustment Factors for Right Shoulder Lateral Clearance (f_{LC})

Right Shoulder Lateral Clearance (ft)	Reduction in Free Flow Speed (mph) with Lanes in One Direction		
	2	3	4
≥6	0.0	0.0	0.0
5	0.6	0.4	0.2
4	1.2	0.8	0.4
3	1.8	1.2	0.6
2	2.4	1.6	0.8
1	3.0	2.0	1.0
0	3.6	2.4	1.2

Source: HCM, 2000, Exhibit 23-5.

TABLE 3.17 (Exhibit 23-6) Adjustment Factors for Number of Lanes (f_N) on Urban Freeways

Number of Lanes in One Direction	Reduction in Free-Flow Speed, f_N, in That Direction
≥5	0.0
4	1.5
3	3.0
2	4.5

Note: For rural freeways, $f_N = 0.0$

Source: HCM, 2000, Exhibit 23-6.

TABLE 3.18 (Exhibit 23-7) Adjustment Factors for the Density of Interchanges (f_{ID})

Interchanges per Mile	Reduction in Free-Flow Speed, f_{ID}
≤0.50	0.0
0.75	1.3
1.00	2.5
1.25	3.7
1.50	5.0
1.75	6.3
2.00	7.5

Source: HCM, 2000, Exhibit 23-7.

3.2.5 Determination of Level of Service

The LOS on a basic freeway section can be determined directly from Figure 3.10 on the basis of the free-flow speed and flow rate. The steps of the procedure—operational

(LOS) in Figure 3.10—are as follows:

Step 1:   Define and segment the freeway sections as appropriate. Each defined segment should have physical and operational characteristics that are reasonably uniform.

Step 2:   Based on the measured or estimated free-flow speed on the freeway segment, construct an appropriate speed-flow curve of the same shape as the typical curves shown in Figure 3.9. If the FFS value is not close to any of the four FFS values for which solid curves have been drawn, it is acceptable to sketch a curve between the solid curves, so that the curve will intercept the y-axis at the desired free-flow speed.

Step 3:   Using the flow rate, v_p, found by using Equation 3.10, read up to the free-flow speed curve identified in step 2 and determine the average passenger car speed and level of service corresponding to that point.

Step 4:   Determine the density of flow as $D = \dfrac{v_p}{S}$, where

D = density in pc/mi/ln
v_p = flow rate in pcphpl
S = average passenger car speed

Step 5:   Determine the level of service using the density ranges in Table 3.12.

Example 3.4

The four NB lanes on an existing urban freeway have a peak hour volume of 2800 vehicles. One particular 8-mile section of the freeway has five interchanges, 11-foot lane widths, a lateral clearance of 2 feet, and a peak hour factor of 0.9. The highway has a posted 60 mph speed limit. Nine percent of the peak hour traffic is heavy trucks and 5 percent is buses.

A.   Assuming a base free-flow speed of 70 mph, what is the free-flow speed (FFS) for this freeway section?

B.   What is the freeway's LOS (level of service) during the peak hour?

Solution to Example 3.4

A.   See Tables 3.15 to 3.18 for the adjustment factors that pertain to the freeway conditions specified. Table 3.19 summarizes the factors found for this example. Unless otherwise specified, an urban freeway is assumed to be on level terrain.

TABLE 3.19   Factors in Speed Determination for Example 3.4

Condition:	Lane Width	Lateral Clearance	Number of Lanes	Interchange Density
Value:	11 feet	2 feet	4	5/8 = 0.625 per mile
Table used:	3.18	3.19	3.20	3.21
Adjustment factor:	$f_{LW} = 1.9$ mph	$f_{LC} = 0.8$ mph	$f_N = 1.5$ mph	By linear interpolation, $f_{ID} = 0.65$ mph

Equation 3.11 produces FFS = $\text{FFS}_i - f_{LW} - f_{LC} - f_N - f_{ID} = 70 - 1.9 - 0.8 - 1.5 - 0.65 = 65.15$ mph.

B. To find LOS, follow the five-steps procedure outlined just before this example.

1. The freeway section being analyzed is presumed to be uniform.
2. Use the free-flow speed found in part A—65.15 mph, rounded to 65 mph. The speed-flow curve in the worksheet below begins with a free-flow speed of 65 mph on its left end is the third-highest curve in the figure. There is no need to sketch in a curve between the solid curves in the figure. The 65 mph curve indicates a capacity of 2350 pcphpl.
3. To use Equation 3.10 to compute the flow rate v_p, we must first calculate f_{HV} from Equation 3.1:

$$f_{HV} = \frac{1}{[1 + P_T(E_T - 1) + P_R(E_R - 1)]}$$

$$= \frac{1}{[1 + 0.14(2.5 - 1) + 0.00(2.0 - 1)]} = \frac{1}{1.21} = 0.826$$

This makes $v_p = V/(\text{PHF})(N)(f_{HV})(f_p) = 2800/(0.9)(4)(0.826)(1.0) = 941$ pcphpl in Equation 3.10. Drawing a vertical line up from this v_p value intersects the 65 mph curve in Figure 3.9 in the LOS "B" zone of the figure. The 65 mph free-flow speed curve is still horizontal at this flow rate, which indicates an average passenger car speed $S = 65$ mph.

4. The density can be computed as $D = \dfrac{v_p}{S} = \dfrac{941 \text{ pcphpl}}{65 \text{ mph}} = 14.5$ pc/mi/ln.

5. In Table 3.14, the density range 11.0 to 18.0 defines the bounds on LOS "B." This confirms the LOS found in step 3 using Figure 3.9.

Note that this was an "Operational (LOS)" application, according to the categories listed at the top right of the Basic Freeway Segments Worksheet (Figure 3.11). The worksheet serves as a checklist of values needed as inputs, steps to follow, equations to use, and adjustment factors to retrieve from *HCM* Exhibits. The *HCM* Exhibit numbers are given in the Title and "Source" portions of Tables 3.13 to 3.18. Using the *HCS* software yields slightly different intermediate values–$f_{ID} = 0.6$ from interpolation and $S = 65.2$ mph without rounding off. The overall result, however, is $v_p = 941$ pc/h, $D = 14.4$ pc/mi/ln, and LOS "B."

THINK ABOUT IT

In Example 3.4, the same LOS was found in two different ways—by using the speed-flow curves in Figure 3.11 and by computing density for use with Table 3.7. (a) Will the two methods always produce the same LOS for a given set of conditions? Explain your answer. (b) Which method do you prefer? Why?

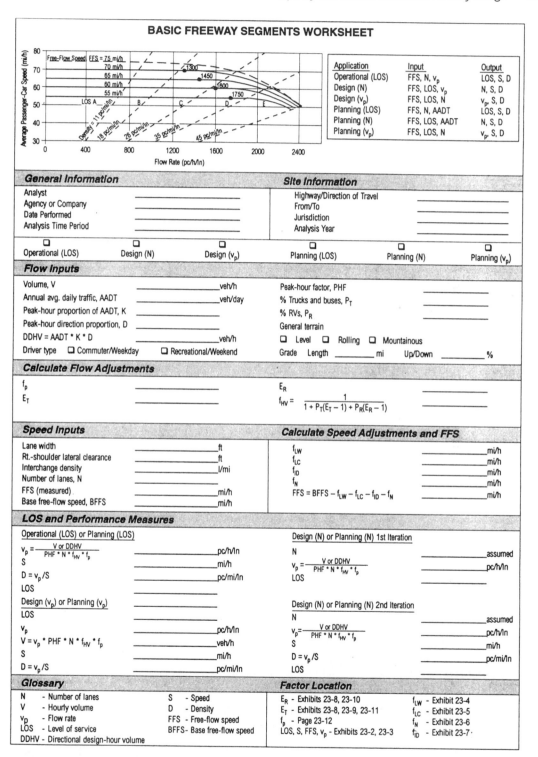

FIGURE 3.11

Basic freeway segments worksheet. *Source: HCM, 2000, p. 23–31.*

Determination of Number of Lanes

When planning or designing a new freeway section, one basic decision is the number of lanes it should have in each direction. When an expansion of freeway capacity is being considered, the designer needs to determine how many additional lanes are needed. In either case, the key inputs are expected flow rate and desired level of service. This type of analysis is "Design (N)" in Figure 3.11. The steps in this analysis are [HCM 2000, p. 23-15]:

Step 1:   Find f_{HV} using Table 3.14 for E_T and E_R, and Equation 3.1.

Step 2:   Try 2 lanes in each direction, unless it is obvious that more lanes will be needed.

Step 3:   Convert volume (vph) to flow rate (pcphpl), v_p, for the current number of lanes in each direction, using Equation 3.10.

Step 4:   If v_p exceeds capacity, add one lane in each direction and return to Step 3.

Step 5:   Compute FFS using Equation 3.11.

Step 6:   Use Table 3.13 to determine the LOS for the freeway with the current number of lanes being considered. If the LOS is not good enough, add another lane and return to Step 3.

Example 3.5 will demonstrate these steps.

Designing for Level of service C is a reasonable starting point for a new road. LOS B would probably result in over-design and LOS D means that when the freeway opens it will already be very crowded. Traffic will be attracted to the freeway once it has been constructed. One important factor in planning is the demand calculation for future traffic. This forecast comes from planning efforts described in Chapter 4.

Example 3.5

A new suburban freeway, cutting through the rolling hills of a fast-growing section west of Mythaca, is being planned. The forecasted peak-period traffic for 20 years from now is 4000 vph in each direction. How many lanes in each direction will be required if LOS C (or better) is to be maintained? The future peak-period traffic mix is expected to be 15 percent trucks/buses and 5 percent RVs. Other important factors are:

- PHF = 0.85
- 1.00 interchange per mile
- Lane width and lateral clearance will be designed to "ideal" standards, that is, 12 foot lanes and >6 foot lateral clearance.

Solution for Example 3.5   Figure 3.12 summarizes the data and the solution process, which is for the "Design (N)" application. The steps are explained here.

Step 1:   Using the Rolling Terrain column in Table 3.14, $E_T = 2.5$ and $E_R = 2.0$. Equation 3.1 becomes

$$f_{HV} = \frac{1}{[1 + 0.15(2.5 - 1) + 0.05(2.0 - 1)]} = \frac{1}{[1 + 0.225 + 0.05]} = \frac{1}{1.275} = 0.784$$

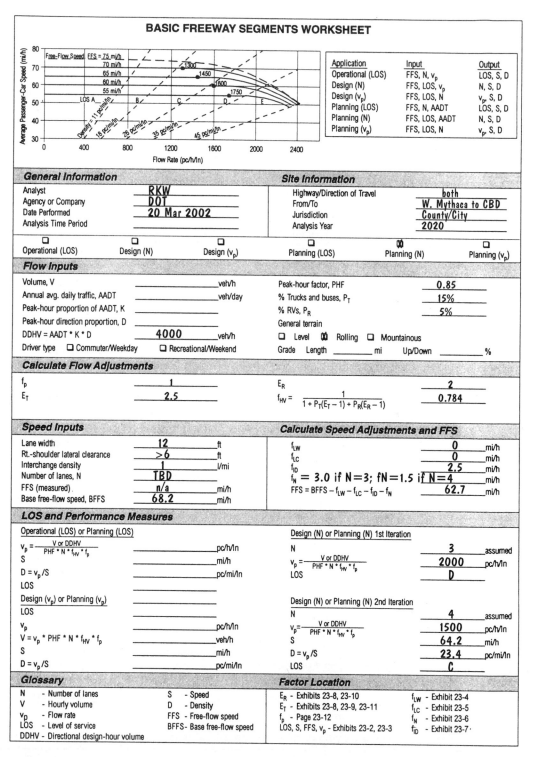

BASIC FREEWAY SEGMENTS WORKSHEET

Application	Input	Output
Operational (LOS)	FFS, N, v_p	LOS, S, D
Design (N)	FFS, LOS, v_p	N, S, D
Design (v_p)	FFS, LOS, N	v_p, S, D
Planning (LOS)	FFS, N, AADT	LOS, S, D
Planning (N)	FFS, LOS, AADT	N, S, D
Planning (v_p)	FFS, LOS, N	v_p, S, D

General Information

Analyst	RKW
Agency or Company	DOT
Date Performed	20 Mar 2002
Analysis Time Period	

Site Information

Highway/Direction of Travel	both
From/To	W. Mythaca to CBD
Jurisdiction	County/City
Analysis Year	2020

☐ Operational (LOS)   ☐ Design (N)   ☐ Design (v_p)   ☐ Planning (LOS)   ☒ Planning (N)   ☐ Planning (v_p)

Flow Inputs

Volume, V		veh/h
Annual avg. daily traffic, AADT		veh/day
Peak-hour proportion of AADT, K		
Peak-hour direction proportion, D		
DDHV = AADT * K * D	4000	veh/h
Driver type ☐ Commuter/Weekday	☐ Recreational/Weekend	

Peak-hour factor, PHF	0.85	
% Trucks and buses, P_T	15%	
% RVs, P_R	5%	
General terrain		
☐ Level ☒ Rolling ☐ Mountainous		
Grade Length ___ mi Up/Down ___ %		

Calculate Flow Adjustments

f_p	1
E_T	2.5

E_R	2
$f_{HV} = \dfrac{1}{1 + P_T(E_T - 1) + P_R(E_R - 1)}$	0.784

Speed Inputs

Lane width	12	ft
Rt.-shoulder lateral clearance	>6	ft
Interchange density	1	l/mi
Number of lanes, N	TBD	
FFS (measured)	n/a	mi/h
Base free-flow speed, BFFS	68.2	mi/h

Calculate Speed Adjustments and FFS

f_{LW}	0	mi/h
f_{LC}	0	mi/h
f_{ID}	2.5	mi/h
$f_N = 3.0$ if N=3; fN=1.5 if N=4		mi/h
FFS = BFFS − f_{LW} − f_{LC} − f_{ID} − f_N	62.7	mi/h

LOS and Performance Measures

Operational (LOS) or Planning (LOS)

$v_p = \dfrac{V \text{ or } DDHV}{PHF * N * f_{HV} * f_p}$		pc/h/ln
S		mi/h
D = v_p/S		pc/mi/ln
LOS		

Design (v_p) or Planning (v_p)

LOS		
v_p		pc/h/ln
V = v_p * PHF * N * f_{HV} * f_p		veh/h
S		mi/h
D = v_p/S		pc/mi/ln

Design (N) or Planning (N) 1st Iteration

N	3	assumed
$v_p = \dfrac{V \text{ or } DDHV}{PHF * N * f_{HV} * f_p}$	2000	pc/h/ln
LOS	D	

Design (N) or Planning (N) 2nd Iteration

N	4	assumed
$v_p = \dfrac{V \text{ or } DDHV}{PHF * N * f_{HV} * f_p}$	1500	pc/h/ln
S	64.2	mi/h
D = v_p/S	23.4	pc/mi/ln
LOS	C	

Glossary

N - Number of lanes	S - Speed
V - Hourly volume	D - Density
v_p - Flow rate	FFS - Free-flow speed
LOS - Level of service	BFFS- Base free-flow speed
DDHV - Directional design-hour volume	

Factor Location

E_R - Exhibits 23-8, 23-10	f_{LW} - Exhibit 23-4
E_T - Exhibits 23-8, 23-9, 23-11	f_{LC} - Exhibit 23-5
f_p - Page 23-12	f_N - Exhibit 23-6
LOS, S, FFS, v_p - Exhibits 23-2, 23-3	f_{ID} - Exhibit 23-7

FIGURE 3.12

Step 2:   Try $N = 2$ lanes.

Step 3:   Convert volume to flow rate (pcphpl) using Equation 3.10:

$$V_p = \frac{V}{(\text{PHF})(\text{N})(f_{\text{HV}})(f_p)}.$$

It is reasonable to assume that most peak period drivers will be commuters or will be otherwise familiar with the freeway, so $f_p = 1.0$ can be used.

$$V_p = \frac{4000 \text{ vph}}{(0.85)(2)(0.784)(1.0)} = \frac{4000}{1.260} = 3000 \text{ pcphpl}.$$

Step 4:   3000 pcphpl is far in excess of the 2400 pcphpl capacity given in Table 3.13 for an ideal highway section; so repeat step 3 with $N = 3$.

Step 3:   (1st iteration) $N = 3$ in Equation 3.10 produces

$$V_p = \frac{4000 \text{ vph}}{(0.85)(3)(0.784)(1.0)} = \frac{4000}{1.999} = 2000 \text{ pcphpl}.$$

Step 4:   Because this value of v_p is less than 2400 pcphl, the question is whether three lanes is sufficient to provide LOS C with the forecasted traffic flow.

Step 5:   To use Equation 3.11, the area type (urban and rural) must be determined. Because a suburban freeway is more nearly urban than rural (and likely to eventually become urban), a base free flow speed of 70 mph is adopted. (Rural freeways are designed for 75 mph.) Using Table 3.17, the free-flow speed is reduced by 3 mph, because the number of lanes in each direction is 3. In Table 3.18, 1.0 interchange per mile causes a further reduction of 2.5 mph. This reduces the FFS to $70.0 - 3 - 2.5 = 64.5$ mph.

Step 6:   In Figure 3.9, the point along the imaginary 64.5 mph curve (between the 60 mph and 65 mph curves shown) at flow rate $v_p = 2000$ pc/h/ln falls near the right-hand boundary of the LOS D range, with a traffic density of about 34 pc/mi/ln. Another way to compute the density is to note that the Average Passenger-Car Speed at this point in Figure 3.9 is just less than 60 mph. Using $S = 60.0$ mph in $D = v_p/S$ leads to a density of 33.3 pcphpl. Table 3.13 indicates that a density below 26 is required for LOS C. The density of 33.3 is too high. Add a fourth lane and return to Step 3.

Step 3:   (2nd iteration) $N = 4$ in Equation 3.10 produces

$$V_p = \frac{4000 \text{ vph}}{(0.85)(4)(0.784)(1.0)} = \frac{4000}{2.666} = 1500 \text{ pcphpl}.$$

Step 4:   Because this value of v_p is less than 2400 pcphl, go to Step 5.

Step 5:   In Table 3.13, FFS = 70.0 mph is reduced by the number of lanes (4; reduction factor 1.5 mph) and the number of interchanges per mile (1 interchange per mile; reduction factor of 2.5). This reduces the FFS to 66.0 mph.

Step 6:   The imaginary 66.0 mph curve at the top of the worksheet at 1500 pcphpl lies in the LOS C range. Because the curve is horizontal at this point, use $S = 66.0$ mph in $D = v_p/S$ to compute the density is about 22.7 pcphpl.

Table 3.12 indicates that density of 26 pcphpl is the upper limit for LOS C. Four lanes in each direction are needed to maintain LOS C.

Table 3.20 summarizes the calculations carried out in this example.

TABLE 3.20 Factors in Speed Determination for Example 3.5

Number of Lanes	v_p pcphpl	FFS mph	Density pc/mi/ln	LOS
2	3000			V > C, add lane
3	2000	64.5	33.3	LOS D
4	1500	66.0	22.7	LOS C

> **THINK ABOUT IT**
>
> In the paragraph above is the sentence "Some of those factors can change from day to day." Which factors are referred to in that sentence? What about hour to hour changes? What will this mean for planning?

3.3 QUEUEING SYSTEMS

3.3.1 Introduction

When the toll plaza at the Shoridan end of the bridge over the Mythaca River was being planned, the assumption was made that traffic would approach the toll plaza in a random pattern (Figure 3.13). The size and layout of the toll plaza was designed accordingly. After a few years, the increased traffic using the bridge displayed a tendency to come in "platoons." Long delays were common, as drivers waited to pay the toll. A state legislator has demanded an investigation. The lawmaker wants to know what could have been done differently in characterizing the traffic that was expected to use the toll bridge. A consultant will be hired to provide expert advice. That consultant could be you.

FIGURE 3.13
Toll plaza on Florida expressway. *Source:* Jon D. Fricker.

3.3.1   Queueing Systems

Queueing systems are common in our daily lives. Waiting in line is a familiar, but unpleasant, experience. The traffic backups at the approach to the above-mentioned tollbooth is certainly one example. Another classic transportation example is vehicles approaching a work zone. Some other examples are not so obvious but would still be familiar to you, such as the sequence in which you receive homework assignments and then decide the order in which you complete them. How do we know when *queueing* will help us perform an analysis (Figure 3.14)? A good place to begin is by defining a *queueing process*:

- Customers arrive at a service facility.
- They wait in line if all servers are busy.
- They (immediately or eventually) receive service.
- They leave the facility.

If you have already thought of one or more examples of a queueing process from your experience, congratulate yourself for reading actively; then see if your examples have the elements of a *queueing system* listed below:

- A set of customers. (It isn't interesting until there are "many.")
- A set of servers. (One or many—this is always interesting.)

An order in which customers arrive and the order in which they are processed.

The *state of a queueing system* is defined in terms of the *number of customers* in the facility at any specified time.

FYI: There is some dispute over how to spell *queuing/queueing*. You may not care, but some are passionate about it. See the letter in Figure 3.14.

Because of Dr. Allen's logic, and because we think the word looks better with the "extra e," we will use the *queueing* version of the spelling in this book.

—The Authors

Now that the basic definitions have been presented, let us prepare to analyze a queueing system by identifying its components:

1. *Arrival patterns.*   Do customers arrive on a uniform, predictable schedule (as if on a conveyor belt) or randomly? If randomly, is there a model that can be used to describe the random arrival patterns?
2. *Service patterns.*   Does each service activity have the same duration, or does service time vary from customer to customer?
3. *Number of servers.*   Must all customers in a queue be served by a single server, or may a customer be served by "the first available" among several servers?

LETTERS

Word Held Dear

To the Editor:
I am writing to protest your vicious attack on one of the most beautiful words in the English language in the August 1992 issue of *OR/MS Today*. I speak of the wonderful word queueing — the only word in the English language with five vowels in a row, and a word all of us who take queueing theory seriously hold dear. You have insulted this word grievously by misspelling it as (ughh) "queuing," even in an article, "Tools of the Trade," whose co-author, Carl M. Harris, is the co-author of "Fundamentals of Queueing" (Second Edition, John Wiley, 1985, Donald Gross co-author) in which, naturally, the word is spelled correctly. By correctly I mean the way queueing theory writers spell it. Those who write books and articles about queueing theory certainly know how to spell the word. I believe it is time to give this beautiful word the respect it deserves. Please tell your editors to allow queueing to be spelled correctly in your journal.

Dr. Arnold O. Allen
Roseville, Calif.

Editor's note: Webster's Ninth New Collegiate Dictionary lists two spellings for the word in question, in this order: *queuing* or *queueing*. In editing circles, it is customary to use the dictionary's first spelling when two or more spellings of a word are given. However, since *queueing* is the preferred spelling in the scientific community, we will see that our editors spell it "correctly" from now on. Incidentally, along with his letter, Allen included a list of 21 books and articles. Twenty of the books and articles spell the word *queueing* in the title. The exception is Kishor Shridharbhai Trivedi's book, "Probability and Statistics with Reliability, *Queuing*, and Computer Science."

FIGURE 3.14
Queueing or queuing?
Source: OR/MS Today, october 1992.

Another factor is whether two or more servers, each acting as part of a sequence, are needed to provide the service requested by the customer.

4. *System capacity.*   This is defined as the maximum number of customers that are permitted to be in the service facility at one time. If a facility has a finite capacity, customers that arrive at a full facility must leave without being served. It is possible, however, to have a facility whose capacity is effectively (although not actually) infinite. A long road would be such a facility.

5. *Queue discipline.*   What is the order in which customers are served? There are four basic rules that can govern the order in which customers are served:

   A. *First in, first out (FIFO).*   This is the analyst's way of saying "First come, first served." It is the most common queue discipline, especially for persons waiting for service at a bank or airport ticket counter.

   B. *Last in, first out (LIFO).*   Sometimes circumstances or physical layouts make this discipline necessary. A good example might be rail cars stored on a siding. The only way to get the first car stored there out is to take out the last car, then the next-to-last car, etc. Of course, we would have to first define *service* to see if this were a true example of LIFO.

   C. *Random.*   The next "customer" is chosen for service without regard for protocol or fairness. By now, it must have occurred to you that a "customer" need not be a human being. For example, a bin full of spare parts (or nuts and bolts) would be accessed and refilled by grabbing and dumping, not by any careful ordering process.

D. *Priority.* Regardless of the order of arrival, the "most deserving" customer is served next. The classic case is *triage* in battlefield medical operations, where wounds that are relatively minor or almost certainly beyond medical help have lower priority than other cases.

THINK ABOUT IT

Think of at least one example of a queueing system for each of the four queue disciplines (or rules) listed above.

Having introduced the various aspects of queueing, we still need to see how it can all fit together. Figure 3.15 provides a good overview of how queueing systems can vary by the number of queues and/or the number and orientation (in series or parallel) of servers.

THINK ABOUT IT

Where do the four (or more) examples that you thought of fit in Figure 3.15? Do you have any that don't seem to fit in any of the four queueing systems shown in Figure 3.15? In the likely situation that your examples did not cover all four queueing systems shown in Figure 3.15, try to fill in the missing cases.

3.3.2 Types of Queueing Models

The shorthand notation to describe a type of queueing system is $x/y/z$, where

- x = the distribution of interarrival times
- y = the distribution of service times
- z = the number of servers

The most common values for x and y are

- D = interarrival or service times are constant; "D" stands for "deterministic" or "degenerate," because there is no variability in the "distribution."
- M = a negative exponential distribution (introduced in Chapter 2.4); "M" stands for "Markovian" or "memoryless," based on the Poisson assumptions introduced in Chapter 2.4.
- G = a general distribution, for cases in which a probability density function other than the negative exponential distribution applies.

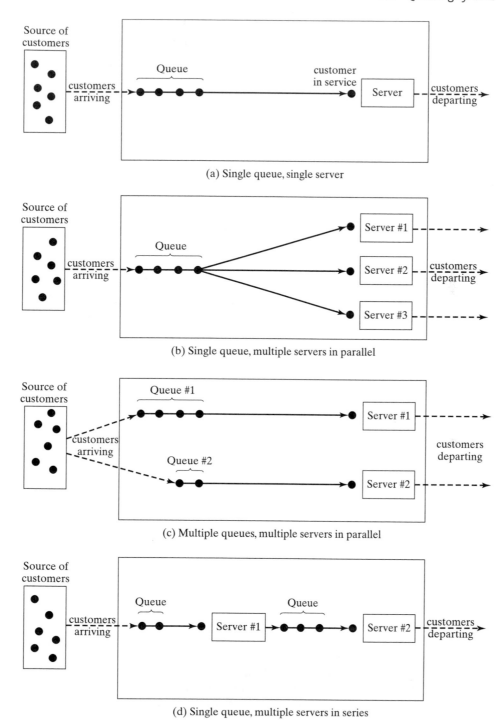

FIGURE 3.15
Queueing systems. *Source*: Bronson, 1982.

For any queueing system, the terms most commonly used are:

- λ = the arrival rate (customers arriving per unit time) at a queueing system
- μ = the service rate (customers served per unit time) by a single server
- $\overline{Q}$ = average number of customers waiting for service
- $\overline{W}$ = average waiting time for each vehicle in the queue
- $\overline{t}$ = average time spent in the queueing system = waiting time + service time
- p_n = steady-state probability that exactly n customers are in queueing system.

3.3.3   The Queueing Diagram

Queueing systems can exist in two different situations. The first is when $\lambda < z\mu$, where z is the number of servers. This means that the arrival rate does not exceed the collective service rate (or capacity) of the z servers. Such situations can be analyzed with formulas that are presented later in this chapter. However, there are situations in which $\lambda > z\mu$, at least temporarily, and these must be analyzed. A common example is the temporary blockage of one or more lanes of a multilane expressway. In the case of a lane closure, the original service rate (capacity) of the roadway is reduced (because the number of servers, or lanes, has been reduced), whereas the arrival rate (flow rate) is unchanged. Another example is where the arrival rate (flow rate) temporarily increases beyond the service rate (capacity) of the roadway. In either of these examples, *reasonable questions* are:

- How long will the queue become until the temporary condition $\lambda > z\mu$ is removed?
- How much delay will drivers have to endure because of the temporary $\lambda > z\mu$ condition?
- How long will it take to dissipate the queue after the $\lambda > z\mu$ condition is removed?

If $\lambda > z\mu$ is the case for a period of analysis, *the equations to be presented in section 4 of this chapter do not apply*. The analyst has two recourses, however:

1. Use simulation to model the situation and generate the desired performance measures, such as those contained in the three questions above. This involves using commercial software or writing your own computer code.
2. Use queueing diagrams or software based on queueing diagrams.

In this section, the fundamentals of queueing diagrams will be introduced. The queueing diagram is a graphical technique that enables an analyst to "see" the results of the situation in which $\lambda > z\mu$. If done by hand, a queueing diagram can be drawn on graph paper. The horizontal axis in Figure 3.16 measures time from some specified start of the analysis. The vertical axis measures the cumulative number of vehicles that arrive at the queueuing system or depart from it. Each queueing diagram has two "curves"—an *arrival curve* and a *departure curve*. In Figure 3.16, the arrival curve is a straight line, having a slope equal to the arrival rate λ in vehicles per hour. The arrival curve is linear, because the arrival rate stays constant throughout the analysis. The departure (or service) curve in Figure 3.16, however, is piecewise linear.

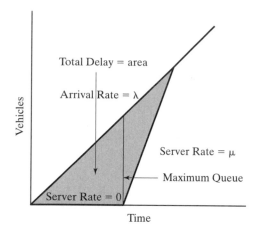

FIGURE 3.16
Generalized diagram for analyzing queues.

This is because the server is shut off for a period (i.e., slope = 0) and then begins serving later at a rate μ. If the server is a roadway, a horizontal departure curve represents the situation in which the roadway is blocked. The area between the arrival curve and the departure curve is the total vehicle delay and the longest vertical line between the two curves represents the maximum length of the queue.

Let us now look at the queueing diagram in more detail. In Figure 3.17, the *arrival curve* "AC1" illustrates a constant arrival rate of 2715 vph. The *departure curve* represents the ability of the roadway to allow traffic to proceed, that is, its capacity. If the departure rate exceeds the arrival rate, a lasting queue will not form and no departure curve needs to be drawn. If, for some reason, one lane of roadway is blocked at time t_1, such that the capacity is reduced to 1340 vph, the departure curve "DC1" will show the new capacity. If, as is shown in Figure 3.17, the slope of "DC1" (1340 vph) is not as steep as the arrival curve (2715 vph), a queue will begin to form. Clearly, we would expect the queue to grow by (2715 − 1340)/60 = 22.9 vehicles during each minute the capacity is reduced to 1340 vph. A queueing diagram will do a good job of answering the "reasonable questions " posed in the bullet list above. Example 3.6 will demonstrate.

Example 3.6: Only Departure Rate Varies

Because of a minor collision, one of the two SB lanes on Freeway 16 is blocked at 8:13 AM. The normal freeway capacity of 3600 vph (60 veh/min) is reduced to 1340 vph (22.33 veh/min). The SB flow rate on Freeway 16 at this time of day is 2715 vph (45.25 veh/min). The blockage is removed after 15 minutes.

A. Did a queue form? If so, what was its maximum length?

B. What was the longest time any single vehicle was in the queue?

C. At what time did the queue clear?

D. What was the total delay to traffic because of the lane blockage?

E. If a vehicle entered the queue at 8:25 AM, how many vehicles would be ahead of it in the queue and how long would the driver have to wait in the queue?

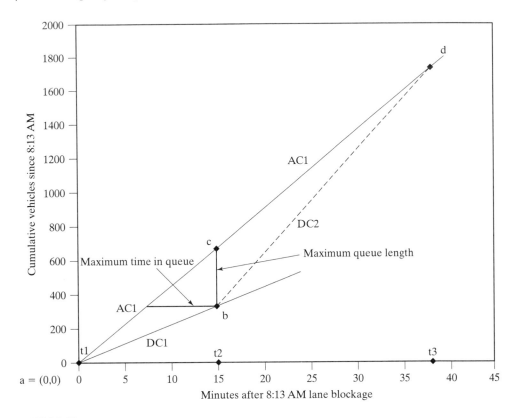

FIGURE 3.17
Queueing diagram when only departure rate varies. (Example 3.6)

Solution to Example 3.6

A. Maximum queue length. Because the departure rate (capacity) during the lane block-age was less than the arrival rate (flow rate), a queue formed. The departure curve "DC1" was in effect until 8:28 AM (time t_2 in Figure 3.17); then full capacity was restored. The restored capacity is represented in Figure 3.17 by the dashed departure curve "DC2," beginning at time t_2 at point b. Whenever a departure curve in a queueing diagram is below the arrival curve, a queue is present. In fact, the number of vehicles in the queue at any time t is the length of the vertical line between the arrival and departure curves. Upon inspecting the "right triangle" in Figure 3.17, it is clear that the queue grows until time t_2 and then dissipates until the queue clears at time t_3. (Whenever a departure curve would appear above an arrival curve in a queueing diagram, it does not have to be drawn.) Because of the relative slopes of the arrival and departure curves between times t_1 and t_3, the longest vertical line between the two curves will occur at time t_2. To determine the actual length of this vertical line, establish a coordinate system within the queueing diagram and use some algebra. (1) Let the coordinates (x, y) of the arrival curve at time t_1 be $(0, 0)$, where $x =$ minutes since t_1 and $y =$ number of vehicles arriving since t_1. (2) At time t_2, the coordinates of the arrival curve will be $x = 15$ minutes and $y = 15$ minutes $*45.25$ veh/min. $= 678.75$ vehicles. At time t_2, the coordinates of the departure curve will be $x = 15$ minutes and $y = 15$ minutes $*22.33$ veh/min. $= 335.0$ vehicles. vehicles. Therefore, the maximum queue length was $678.75 - 335 = 343.75$ vehicles.

B. Maximum time in queue. Whenever a departure curve in a queueing diagram is below (and to the right of) the arrival curve, the time spent in the queue by any single vehicle is the length of the horizontal line drawn from the arrival curve at the time that vehicle arrives to the departure curve. Just as the maximum queue length was reached at time t_2, the maximum time in queue can be found by drawing a horizontal line from the arrival curve, such that it reaches the "DC2" curve at time t_2. Because $y = 335.0$ at t_2, this means the 335th vehicle would be getting past the bottleneck on the freeway 15 minutes after the blockage occurred. When did the 335th vehicle enter the queue? If the arrival rate is 45.25 veh/min, then the 335th vehicle arrived 335.0/45.25 = 7.4 minutes after the blockage began. Because the blockage was removed at $t_2 = 15$ minutes, the maximum time in queue was 15.0 − 7.4 = 7.6 minutes.

C. When does the queue clear? After time t_2, the departure curve has a steeper slope than arrival curve "AC." Eventually, at time t_3 in Figure 3.17, the two curves will converge. At that time, the queue will have been cleared. To find that point in the queueing diagram, develop equations for the two curves and then determine when they intersect. The equation for the arrival curve is $y_{AC} = 45.25x$ and the equation for the departure curve "DC1" starting at x,y coordinates (15.0,335.0) is $y_{DC1} = 335.0 + 60.0(x - 15)$. To check these equations, verify part A of this example using them:

$$y_{AC} - y_{DC1} = 45.25(15) - [335 + 60(15 - 15)] = 343.75 \text{ vehicles}$$

When the two curves intersect, $y_{AC} = y_{DC1}$:

$$45.25x = 335 + 60(x - 15); x = \frac{900 - 335}{60 - 45.25} = \frac{565}{14.75} = 38.3 \text{ minutes}$$

Check this result in the original equations:

$$y_{AC} = 45.25(38.3) = 1733 \text{ vehicles and } y_{DC1} = 335 + 60(38.3 - 15) = 1733 \text{ vehicles}$$

D. Total delay. A nice feature of a queueing diagram is that the units along its axes are time (minutes) and flow (vehicles). If we calculate the area between an arrival curve and a departure curve to its right, the units associated with that area will be vehicle-minutes, which is a good measure of delay. The area of triangle abc in Figure 3.17 is found by defining a new point e at coordinates (15.0, 0.0), which will form the lower right vertex of right triangles aec and aeb.

$$A_{abc} = A_{aec} - A_{aeb} = \frac{1}{2}(15.0 - 0.0)(678.75 - 0.0) - \frac{1}{2}(15.0 - 0.0)(335 - 0.0)$$
$$= 5090.625 - 2512.5 = 2578.125 \text{ veh-min} = 42.97 \text{ veh-hrs.}$$

A shortcut also works:

$$A_{abc} = \frac{1}{2}(15.0 - 0.0)(678.75 - 335.0) = 2578.1 \text{ veh-min} \frac{hr}{60 \text{ min}} = 42.97 \text{veh-hrs}$$

The area of triangle bcd can be found by defining a new point f at coordinates (15.0, 1733), which will form the upper left vertex of right triangles bdf and cdf.

$$A_{bcd} = A_{bdf} - A_{cdf} = \frac{1}{2}(38.3 - 15.0)(1733 - 335) - \frac{1}{2}(38.3 - 15.0)(1733 - 678.75)$$
$$= 16286.70 - 12282.01 = 4004.69 \text{ veh-min} = 66.74 \text{ veh-hrs}$$

Again, the shortcut also works:

$$A_{bcd} = \frac{1}{2}(38.3 - 15.0)(678.75 - 335.0) = 4004.69 \text{ veh-min} = 66.74 \text{ veh-hrs}$$

Together, A_{abc} and A_{bcd} represent the total delay on Freeway 16 because of the collision:

$$42.97 + 66.74 = 109.71 \text{ veh-hrs}$$

So, during the 38.3 minutes in which the queue builds and dissipates, 109.71 veh-hrs of delay is endured by drivers on SB Freeway 16.

E. Queue position and waiting time for a specific vehicle. A vehicle that arrives at 8:25 AM does so 12 minutes after 8:13 AM, when the queue started to build. In part C, we developed equations for curves "AC" and "DC1." Because 8:25 AM occurs when curve "DC2" is in effect, we need the equation for that departure curve: $y_{DC2} = 22.33x$. At 8:25 AM, $x = 12$, $y_{DC2} = 268$ vehicles, and $y_{AC} = 45.25x = 543$ vehicles. This means that the queue length at 8:25AM is $y_{AC} - y_{DC2} = 543 - 268 = 275$ vehicles. Strictly speaking, there would be 274 vehicles ahead of the vehicle that arrives at 8:25 AM. The length of time this vehicle will spend in the queue is equal to the length of the horizontal line drawn from the 8:25 AM point on curve "AC" to the "DC1" curve to its right in Figure 3.17. Using the "DC1" equation developed in part B, $y_{DC1} = 335.0 + 60.0(x - 15) = 543$ vehs. Solving for x gives us the time the 8:25 AM arrival will leave the queue:

$$x_{dep} = \frac{543 + 900 - 335}{60} = \frac{1108}{60} = 18.47 \text{ minutes.}$$

Recall that x is measured from 8:13 AM, so this vehicle will leave the queue just before 8:31:30 AM. Because the vehicle entered the queue 12 minutes after the queue formed, the waiting time for this vehicle will be $18.47 - 12 = 6.47$ minutes.

Example 3.6 concerned a freeway segment that was temporarily blocked. Queueing diagrams can also be applied to traffic signal cases, which have alternating red phases (when $\mu = 0$ and queues form) and green phases (when queues are cleared if $\lambda < \mu$).

Example 3.7: Traffic Signal Cycles

A proposed traffic signal is expected to have a cycle time of 67 seconds. If EB traffic arrives at the rate 1584 vph and the capacity of the EB approach is 3600 vph when the EB signal is green, determine the minimum length of the green phase so that the average intersection approach queues can be cleared 4 seconds before the end of the green phase.

Solution for Example 3.7   Draw the queueing diagram shown in Figure 3.18. The overall cycle time of 67 seconds consists of R seconds of red and $G + 4$ seconds of green. Red time plus Green time to clear the queue = 63 seconds. In the remaining 4 seconds, any arriving vehicle will be able to clear the intersection immediately.

$$\text{Arrival rate} = 1584 \text{ vph} = 0.44 \text{ veh/sec}$$

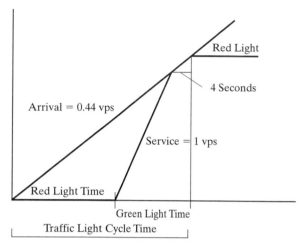

FIGURE 3.18
The queueing diagram with variable server rates for
Example 3.7.

Vehicles arriving in 63 seconds = 0.44 * 63 = 28 vehicles.

At one vehicle per second, the server (the green phase at the intersection) will take 28 seconds to clear the 28 vehicles that joined the queue during the previous red phase and the current green phase.

This means that the red phase can be no longer than 67 − 4 − 28 = 35 seconds, given that the green phase must be at least 28 + 4 = 32 seconds long.

In Example 3.7, the word "average" was used. This is because it is highly unlikely that the same number of vehicles will approach the intersection during each cycle.

Example 3.8: Only Departure Rate Varies

A "gaper's block" occurs when traffic in one direction slows to look (or gape) at an incident on the opposite side of the roadway's median. In Figure 3.19, the gaper's block reduces roadway capacity from 100 vpm to 30 vpm for 45 minutes. If the traffic arrival rate stays at 60 vpm,

    A. How long after the gaper's block starts will the queue be cleared?
    B. What is the total delay to traffic because of the gaper's block?
    C. What is the average delay per vehicle?
    D. What is the maximum length of the queue?
    E. What is the longest time any vehicle spent in the queue?

Solution for Example 3.8

    A. The time at which the queue clears is determined by writing equations for the arrival curve (60 vpm) and the 100 vpm portion of the departure curve. These equations are

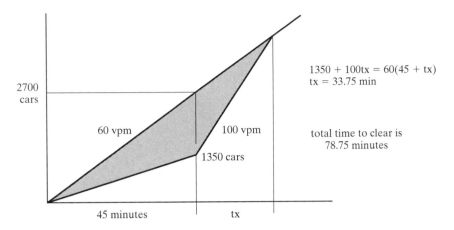

FIGURE 3.19
The queueing diagram with variable server rates for Example 3.8.

shown in Figure 3.19. The queue clears 33.75 minutes after the gaper's block is ended. This is 78.75 minutes after the gaper's block began.

B. The total vehicle delay is represented by the shaded area in Figure 3.19. This area is calculated as

$$\frac{1}{2}(60*78.75*78.75) - \frac{1}{2}(45*1350) - \left[\frac{1350 + (60*78.75)}{2}*33.75\right] = 53{,}146 \text{ veh-min}$$

C. The average delay per vehicle is $\dfrac{53{,}146}{60*78.75} = 11.25$ minutes/vehicle.

D. The maximum length of the queue is the length of the longest vertical line drawn in the delay area:

$$(45 \text{ min}*60 \text{ vpm}) - (45*30 \text{ vpm}) = 1350 \text{ vehicles}$$

E. The longest time any vehicle spent in the queue is represented by the longest horizontal line that can be drawn in the delay area:

$$45 \text{ min} - \frac{1350 \text{ veh}}{60 \text{ vpm}} = 22.5 \text{ minutes.}$$

Example 3.9:   Only Arrival Rate Varies

Interstate I-25 normally has two lanes open in each direction, but a resurfacing project will require that the NB side of the median be closed for 2 weeks while those lanes are resurfaced. During that time, the two SB lanes will be converted to two-way operation—one lane for NB traffic and one lane for SB traffic. One lane in either direction provides enough capacity for most hours of the week, except for certain weekend hours. For example, from 4 PM to 6 PM on a typical Sunday, NB traffic flow is about 1500 vph. The single NB lane is expected to have a capacity of 1340 vph. At 6 PM, NB traffic suddenly drops to 850 vph.

A. How long will the maximum length of the queue be?

B. What is the longest time any single vehicle will spend in the queue?

C. When will the queue be cleared?

D. What is the total delay that drivers will have to endure, from 4 PM until the queue dissipates?

E. What is the average time spent in the queue?

Solution to Example 3.9   The queueing diagram in Figure 3.20 differs from Figure 3.17 in two important respects:

1.   In Figure 3.17, the arrival curve had a constant slope; in Figure 3.20, the arrival curve slope is different before and after 6 PM.

2.   In Figure 3.17, the departure curves had different slopes, depending on whether a freeway lane was blocked; in Figure 3.20, there is only one departure curve.

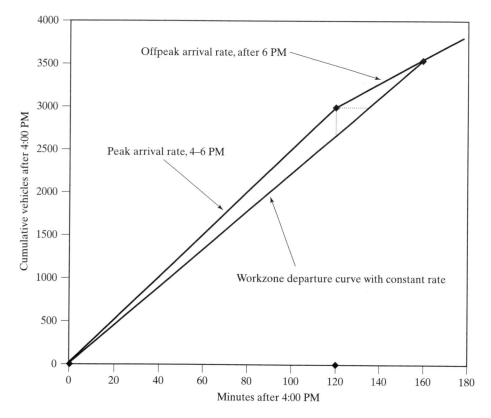

FIGURE 3.20

Queueing diagram when only arrival rate varies. (Example 3.9)

A. The maximum queue length will occur at 6 PM, when the arrival curve switches from a slope of 1500 vph to a slope of 850 vph. At 6 PM, 3000 vehicles will have arrived since the queue started to build at 4 PM. At 6 PM, 2680 vehicles will have passed through the work zone. The length of the vertical line between the arrival and departure curves in Figure 3.20 is greatest at 6 PM. This vertical line's length is the maximum queue length: $3000 - 2680 = 320$ vehicles.

B. The maximum time in queue is shown by the longest horizontal line between the arrival and departure curves in Figure 3.20. By looking at the slopes of the arrival and departure curves, it is clear that a vehicle entering the queue at 6 PM will have the longest wait. The coordinates of the arrival curve at 6 PM are (120, 3000). The x-coordinate of the departure curve when $y = 3000$ is

$$x = \frac{3000 \text{ veh}}{1340 \text{ vph}} * \frac{60 \text{ min}}{\text{hr}} = 134.3 \text{ min.}$$

The time in queue for the vehicle that arrives at 6 PM is therefore $134.3 - 120 = 14.3$ minutes.

C. The queue will clear when the arrival curve intersects with the departure curve. The departure curve equation is $y_{DC} = 1340x$. The equation of the arrival curve after 6 PM is $y_{AC2} = 3000 + 850(x - 2)$. (Note that the units of x are hours in these two equations.) The two curves intersect when $y_{DC} = y_{AC2}$: $1340x = 3000 + 850(x - 2)$.

$$x = \frac{3000 - 1700}{1340 - 850} = \frac{1300}{490} = 2.65 \text{ hr} = 159.2 \text{ minutes after 4 PM.}$$

This means that the queue will clear just after 6:39 PM.

D. Total vehicle delay is represented by the area between the arrival and departure curves. The area between the curves before 6 PM is $A_L = \frac{1}{2}(2 \text{ hr})(320 \text{ vehicles}) = 320$ veh-hrs. The area between the curves after 6 PM is $A_R = \frac{1}{2}(0.65 \text{ hr})(320 \text{ veh}) = 104$ veh-hrs. The total area is 424 veh-hrs, which is the total delay that drivers had to endure because traffic flows exceeded the capacity of the work zone.

E. Average time in queue. A total delay of 424 veh-hrs may not be a value that is immediately useful to an analyst. That value could be the result of a small delay to many vehicles or a large delay to a few vehicles. The average delay time is usually much easier to understand. The number of vehicles that experienced any queueing delay at the I-25 work zone equals those who arrived between 4 PM and 0.65 hr after 6 PM. Between 4 and 6 PM, 3000 vehicles arrived. Between 6 PM and queue clearance, 0.65 hr * 850 veh/hr = 552.5 vehicles arrived. Therefore, the average queueing delay is

$$\overline{W} = \frac{424 \text{ hrs}}{3552.5 \text{ veh}} = 0.119 \text{ hr/veh} * 60 \text{ min/hr} = 7.16 \text{ min/veh.}$$

Those in charge of work zone traffic control can now make the decision whether this average delay is acceptable, or whether measures need to be taken.

FYI: If a typical freeway lane has a capacity of 1800 vph, two freeway lanes would have a capacity of 3600 vph. However, when one of those two lanes is closed for construction work adjacent to the traffic, or to accommodate traffic in the opposite direction as in Example 3.9, studies (Dudek and Richards, 1982) have shown that a more accurate capacity to use is 1340 vph for the single remaining lane.

THINK ABOUT IT

Why is it that, when one of two lanes is closed on a freeway, the remaining lane has a capacity of 1340 vph, not the usual 1800 vph?

Of course, we can have situations where both the server and arrival rates vary. The diagrams can still be used, but it requires a more careful analysis. We consider that case in an end-of-chapter exercise.

Queueing diagrams have been criticized (Nam and Drew, 1998) and defended (Lovell and Windover, 1999). They can be misused and misinterpreted (Lovell and Windover, 1999). Queueing diagrams are, at their best, a way to estimate *approximate* delay when $\lambda > z\mu$. They allow the analyst to "see" how and where queues build up and dissipate. If the problem being analyzed is too complex, or performance measures other than those that can be computed directly from a queueing diagram are needed for an analysis, then a special-purpose simulation computer package should be used.

3.4 SYSTEMS WITH STABLE QUEUES

Introduction

The Mythaca County Highway Engineer and his wife were visited by 10 relatives over Labor Day weekend. On Friday evening, the engineer's wife asked him to get some corn oil from the grocery store. He quickly found it and went to the checkout area. There were four checkout clerks on duty—one at the express lane and three others at "regular" checkout lanes. There were seven people lined up at the express lane and two people (with their carts) at each of the other active cash registers. Which line should the engineer join?

In this section, delay-related performance measures for queueing systems are calculated. The methods demonstrated in this section may help the engineer at the supermarket and on the job.

3.4.1 Types of Queueing Models

The essential elements of queueing systems were introduced in section 3.3.2. Those elements that are needed for this section are repeated here for the convenience of the reader. The shorthand notation used to describe a type of queueing system is $x/y/z$, where

- x = the distribution of interarrival times
- y = the distribution of service times
- z = the number of servers

The most common values for x and y are

- D = interarrival or service times are constant; "D" stands for "deterministic" or "degenerate," because there is no variability in the "distribution."
- M = a negative exponential distribution; "M" stands for "Markovian" or "memoryless," based on the Poisson assumptions introduced in section 3 of this chapter.

- G = a general distribution, for cases in which a probability density function other than the negative exponential distribution applies.

The value of z can be 1, 2, or any integer that indicates the number of servers active in the queueing system. For example, if a toll road exit ramp has a single toll booth, and the driver of each car can throw the quarter toll into the basket in about the same time, the arrivals are probably negative exponentially distributed, the service time is approximately constant, and the most appropriate queueing model is type M/D/1. If the grocery store patronized by the county engineer has determined that service time at a typical non-express checkout lane is normally distributed, the best queueing model notation for such a checkout lane is M/G/1. If there are 10 non-express checkout lanes, the overall queueing model for those lanes would be represented by M/G/10.

The performance of any queueing system can be measured by certain essential values. For most applications, the most important of these performance measures are as follows:

- $\overline{Q}$ = average number of customers waiting for service
- $\overline{W}$ = average waiting time for each vehicle in the queue
- $\overline{t}$ = average time spent in the queueing system = waiting time + service time
- p_n = steady-state probability that exactly n customers are in queueing system

To compute these performance measures, the required input values are:

- λ = the arrival rate (customers arriving per unit time) at a queueing system
- μ = the service rate (customers served per unit time) by a single server

The ratio $\dfrac{\lambda}{\mu}$ is called the *utilization ratio* (ρ). The most important thing about ρ is that it must be *less than* 1 for the equations listed in Table 3.21 to be valid. The condition $\rho < 1$ means that $\lambda < \mu$ (i.e., the arrival rate is less than the service rate). If $\lambda > \mu$, the eventual result will be a queue of increasing length. We could call the situation in which $\lambda > \mu$ a *persistent queue*.

If $\lambda < \mu$, a steady-state condition will result in terms of the performance measures listed above. The queue's length will fluctuate, perhaps even dissipating before a new queue forms. A queue of this type (i.e., with $\lambda < \mu$) is called a *stable queue*. When there are z servers ($z > 1$), the condition $\lambda < \mu$ must be amended to be $\lambda < z\mu$.

For the reader's convenience, the equations that are most useful in transportation applications are grouped in Table 3.21. For a much more extensive set of equations, see Appendix C in Allen (1978). A way to dispense with many of the equations in Table 3.21 is to derive them from Equations 3.12 and 3.13 for the M/G/1 model, using particular values for the standard deviation of the service times:

- $\sigma = 0$ for a M/D/1 model
- $\sigma = \dfrac{1}{\mu}$ for a M/M/1 model.

THINK ABOUT IT

For any M/y/1 queueing system with $\lambda > 0$ and $\mu > 0$, there will always be some delay. Why is that true?

TABLE 3.21   Performance Measure Equations for Queueing Models

Performance Measure	Queueing Model		
	M/G/1		**M/D/1**
$\overline{Q}$	$\overline{Q} = \dfrac{\rho^2 + \lambda^2\sigma^2}{2(1-\rho)}$ (3.12) σ^2 = variance of service time		$\overline{Q} = \dfrac{\rho^2}{2(1-\rho)}$ (3.15)
$\overline{W}$	$\overline{W} = \dfrac{\overline{Q}}{\lambda} = \dfrac{\rho^2 + \lambda^2\sigma^2}{2\lambda(1-\rho)}$ (3.13)		$\overline{W} = \dfrac{\rho(\frac{1}{\mu})}{2(1-\rho)} = \dfrac{\rho}{2\mu(1-\rho)}$ (3.16)
$\bar{t}$	$\bar{t} = \overline{W} + \dfrac{1}{\mu}$ (3.14)		$\bar{t} = \dfrac{2-\rho}{2\mu(1-\rho)}$ (3.17)
p_n			
	M/M/1		**M/M/z**
$\overline{Q}$	$\overline{Q} = \dfrac{\rho^2}{1-\rho}$ (3.18)		$\overline{Q} = \dfrac{p_0\rho^{z+1}}{z!z}\left[\dfrac{1}{(1-\frac{\rho}{z})^2}\right]$ (3.22)
$\overline{W}$	$\overline{W} = \dfrac{\lambda}{\mu(\mu-\lambda)} = \dfrac{\rho}{\mu-\lambda}$ (3.19)		$\overline{W} = \dfrac{\rho+\overline{Q}}{\lambda} - \dfrac{1}{\mu}$ (3.23)
$\bar{t}$	$\bar{t} = \dfrac{1}{\mu-\lambda}$ (3.20)		$\bar{t} = \dfrac{\rho+\overline{Q}}{\lambda}$ (3.24)
p_n	$p_n = (1-\rho)\rho^n$ (3.21)		$p_0 = \left[\displaystyle\sum_{n=0}^{z-1}\dfrac{\rho^n}{n!} + \dfrac{\rho^z}{z!(1-\frac{\rho}{z})}\right]^{-1}$ (3.25a)
			$p_n = \dfrac{\rho^n p_0}{n!}$ for $n < z$ (3.25b)
			$p_n = \dfrac{\rho^n p_0}{z^{n-z}z!}$ for $n \geq z$ (3.25c)

Sources: Allen, 1978; Larson and Odoni, 1981; Hillier and Lieberman, 1986

Table 3.22 summarizes the differences between a single deterministic server and a single server with negative exponential (i.e., Poisson) characteristics.

To clarify and reinforce some points, take a closer look at Equation 3.17 in Table 3.21 or Table 3.22. The value sought is the average time spent by a customer in the queueing system. This time includes both the average time spent in the queue waiting

TABLE 3.22 Comparing Deterministic and Negative Exponential (Poisson) Server Models

Performance Measure	M/D/1 System	M/M/1 System	Comparison of Poisson (M) and Deterministic (D) models
$\overline{Q}$ (Average length of queue)	$\overline{Q} = \dfrac{\rho^2}{2(1-\rho)}$ (3.15)	$\overline{Q} = \dfrac{\rho^2}{1-\rho}$ (3.18)	M model estimates average queue that is twice as long.
$\overline{W}$ (Average wait time per vehicle in queue)	$\overline{W} = \dfrac{\rho\left(\frac{1}{\mu}\right)}{2(1-\rho)} = \dfrac{\rho}{2\mu(1-\rho)}$ (3.16)	$\overline{W} = \dfrac{\lambda}{\mu(\mu-\lambda)} = \dfrac{\rho}{\mu-\lambda} =$ $= \dfrac{\rho}{\mu(1-\rho)}$ (3.19)	M model estimates average vehicle wait or delay time that is twice as long.
$\bar{t}$ (Average total time per vehicle in queue)	$\bar{t} = \dfrac{2-\rho}{2\mu(1-\rho)}$ (3.17)	$\bar{t} = \dfrac{1}{\mu-\lambda}$ (3.20)	D model estimates less total time for $\lambda < \mu/2$; more for $\lambda > \mu/2$, and the same for $\lambda = \mu/2$

for service $(\overline{W})$ and the average time being served $\left(\frac{1}{\mu}\right)$. For example, if the average service *rate* is 20.27 customers per hour (cust/hr), the average service *time* is

$$\frac{1}{\mu} = \frac{1}{20.27 \text{ cust/hr}} = 0.049 \text{ hr/cust} * \frac{60 \text{ min}}{\text{hr}} = 2.96 \text{ min/cust}$$

Equation 3.17 comes from Equation 3.16 and the average service time:

$$\bar{t} = \overline{W} + \frac{1}{\mu} = \frac{\rho}{2\mu(1-\rho)} + \frac{1}{\mu}.$$

Multiplying the $\dfrac{1}{\mu}$ term by $\dfrac{2(1-\rho)}{2(1-\rho)}$ produces

$$\bar{t} = \frac{\rho}{2\mu(1-\rho)} + \frac{1}{\mu}\frac{2(1-\rho)}{2(1-\rho)} = \frac{\rho+2-2\rho}{2\mu(1-\rho)} = \frac{2-\rho}{2\mu(1-\rho)}.$$

Example 3.10

During afternoon peak periods, so much traffic tries to enter Freeway 16 at the Cheddar Street on-ramp that both safety and efficiency have been compromised. A signal has been installed on the ramp to restrict the number of vehicles entering the freeway. This is called *ramp metering*.

The ramp from Cheddar Street has space for about 10 vehicles. The ramp metering signal is controlled by a sensor that looks for gaps in the freeway traffic. The result is that no more than 500 vph may enter the freeway from the on-ramp. During the typical weekday afternoon peak hour, 400 vehicles attempt to enter the freeway from Cheddar Street.

A. Which queueing model x/y/z best fits this problem?

B. What is the average queue length for the situation described?

C. What is the average time a driver will have to wait in the queue?

D. How long will the average driver spend waiting on the ramp?

E. What is the probability that the on-ramp will be full at any time?

Solutions to Example 3.10

A. For arrival pattern, the usual choice is between M (Poisson/negative exponential) and D (deterministic/constant). In traffic, the arrival pattern is usually M, unless the assumption of random arrivals is invalidated by conditions such as a traffic signal upstream. The service pattern for this onramp is probably M, because the ramp meter operation is tied to traffic on the freeway, and the freeway arrivals are probably Poisson (random). Let us use the M/M/1 model in this example, saving other models for separate exercises.

B. The average queue length in a M/M/1 queueing system is found by using Equation 3.18. Because Equation 3.18 uses the utilization ratio ρ, it would be convenient to first calculate $\rho = \lambda/\mu = 400$ vph/500 vph $= 0.8$. Substituting this value of ρ into Equation 3.18 gives us

$$\overline{Q} = \frac{\rho^2}{1 - \rho} = \frac{(0.8)^2}{1 - 0.8} = 3.2 \text{ vehicles.}$$

Note that, even though $\lambda < \mu$, the average queue length is not zero. The randomness of arrivals will cause queues to form and dissipate, but the average queue length is estimated to be 3.2 vehicles, using the M/M/1 equation.

C. If there were never a queue on the ramp, the driver would only have to wait to be "served" by the ramp metering signal. However, there is sometimes at least one other vehicle ahead waiting for the signal to turn green. Equation 3.19 estimates the average time a driver will have to wait in the queue.

$$\overline{W} = \frac{\lambda}{\mu(\mu - \lambda)} = \frac{400 \text{ vph}}{500 \text{ vph}(500 - 400 \text{ vph})} = 0.008 \text{ hr/veh} * \frac{3600 \text{ sec}}{hr} = 28.8 \text{ sec/veh}$$

It is a good idea to develop the habit of showing the units at each step in your calculations. The relationships are not that difficult in the calculation above, but in other problems, embarrassing errors can be averted by keeping the units in order.

D. The total time in the queueing system known here as the Cheddar Street on-ramp can be estimated by Equation 3.20 if the M/M/1 model is used.

$$\overline{t} = \frac{1}{\mu - \lambda} = \frac{1}{(500 - 400)\text{vph}} = 0.01 \text{ hr/veh} * 3600 \text{ sec/}hr = 36.0 \text{ sec/veh}$$

A simple, but useful, check is to verify that $\overline{t} \geq \overline{W}$. Here, 36.0 seconds > 28.8 seconds, so nothing appears to be wrong.

E. Equation 3.21 will allow us to estimate the probability that any particular number of vehicles will be in the queueing system (on the on-ramp) at any given time. For example, the probability that no vehicles will be on the on-ramp is $p_0 = (1 - \rho)\rho^0 = (1 - 0.8)(0.8)^0 = (0.2)(1) = 0.2$. Unfortunately, the question is about $n > 10$. We could recognize that $P(n > 10) = 1 - \sum_{n=0}^{10} p_n$ and calculate by hand all the p_n values for $n = 0, 1, 2, \ldots, 10$. Instead of 10 manual applications of Equation 2.23 for $n = 0, 1, 2, \ldots, 10$, a spreadsheet was used to generate the p_n values. The results (Table 3.23) are worth examining. First of all, the most common value of n is zero. Second, although $\overline{Q} = 3.2$ vehicles, $p_3 = 0.102$. The modal value (section 2.4.1) is not near the $\overline{Q}$ value. Finally, $P(n > 10) = 1.00 - 0.914 = 0.086$.

TABLE 3.23 The $p(n)$ Calculations for Example 3.10

lambda =	400
mu =	500
rho =	0.8
n	$p(n)$
0	0.200
1	0.160
2	0.128
3	0.102
4	0.0825
5	0.066
6	0.052
7	0.042
8	0.034
9	0.027
10	0.021
sum, 0-10	0.914

THINK ABOUT IT

In Example 3.10, the choice of an M/M/1 queueing model is not obvious, because the arrival pattern might not be "sufficiently random." What should an analyst do in such a case?

Example 3.11

Travelers on WB SR361 from Mythaca to Shoridan must cross the Mythaca River using a toll bridge. At present, there are only two tollbooths for each direction of traffic. The average service time for each tollbooth has a negative exponential distribution, with a mean value of 8.8 seconds. Approach traffic follows the Poisson model.

A. Compute the first three performance measures $\overline{Q}, \overline{W}$, and $\overline{t}$ for the WB toll plaza when the arrival rate is 718 vph.

B. How many tollbooths would have to be active to guarantee $\overline{W} < 15$ seconds?

Solutions to Example 3.11

A. When $z = 2$, Equation 3.22 in Table 3.19 becomes

$$\overline{Q} = \frac{p_0 \rho^{2+1}}{2!2} * \left[\frac{1}{(1 - \frac{\rho}{2})^2} \right] = \frac{p_0 \rho^3}{4} * \left[\frac{1}{(1 - \frac{\rho}{z})^2} \right]$$

To compute ρ for use in Equation 3.22, all we need are $\lambda = 718$ vph and $\mu = 3600$ sec/hr$/8.8$ sec/cust $= 409.1$ cust/hr. However, remember that $\mu = 409.1$ cust/hr applies to each server. There are two servers, so

$$\frac{\rho}{z} = \frac{\rho}{2} = \frac{\lambda}{2\mu} = \frac{718 \text{ vph}}{2(409.1) \text{ vph}} = 0.878.$$

In computing p_0 to use in Equation 3.22, use Equation 3.25a with $\rho = 718$ vph$/409.1$ vph $= 1.756$ for a single-server case:

$$p_0 = \left[\sum_{n=0}^{z-1} \frac{\rho^n}{n!} + \frac{\rho^z}{z! \left(1 - \frac{\rho}{z} \right)} \right]^{-1}$$

$$= \left[\sum_{n=0}^{2-1} \frac{(1.756)^n}{n!} + \frac{(1.756)^2}{2!(1 - 0.878)} \right]^{-1} = [1 + 1.756 + 12.633]^{-1} = 0.065.$$

At last, we are ready to use Equation 3.22 to find the average length of the queue:

$$\overline{Q} = \frac{p_0 \rho^3}{4} * \left[\frac{1}{(1 - \frac{\rho}{z})^2} \right] = \frac{(0.065)(1.756)^3}{4} * \left[\frac{1}{\left(1 - \frac{1.756}{2} \right)^2} \right] = 0.088 * [67.186] = 5.91 \text{ vehicles}$$

We now need to determine the average waiting time in the queue, using Equation 3.23:

$$\overline{W} = \frac{\rho + \overline{Q}}{\lambda} - \frac{1}{\mu} = \frac{1.756 + 5.91}{718} - \frac{1}{408.9} = 0.010680 - 0.002446$$

$$= 0.008235 \text{ hr} * 3600 \text{ sec/hr} = 29.6 \text{ sec}$$

and the average time spent in the queueing system, using Equation 3.24

$$\overline{t} = \frac{\rho + \overline{Q}}{\lambda} = \frac{1.756 + 5.91}{718} = 0.010677 \text{ hr} * 3600 \text{ sec/hr} = 38.44 \text{ sec}$$

Notice that, in the $\overline{W}$ and $\overline{t}$ equations, λ and μ have units "customers per hour," so the units of $\overline{W}$ and $\overline{t}$ will be hours. It is often necessary to convert hours to seconds to make the results easier to interpret.

B. How many tollbooths would have to be active to guarantee $\overline{W} < 15$ seconds? In part A, $\overline{W}$ was found to be 29.6 seconds with two tollbooths active. A natural approach to this problem would be to repeat the calculations of part A using $z = 3$ and then check to see if the resulting $\overline{W}$ is less than 15 seconds. This process can be expedited by using a spreadsheet. A good way to start this approach is to replicate with a spreadsheet the calculations in part A that led to $\overline{W} = 29.6$ seconds. In this way, you can check both the calculations in part A against the formulas you have entered into the spreadsheet. You could also use the *solver* feature of the spreadsheet, noting that in this problem the solution (the number of tollbooths) must be integer. If $z = 3$, the spreadsheet solution is $\overline{W} = 8.00$ seconds, which easily meets the criterion $\overline{W} < 15$ seconds. Using the solver in Excel to find z such that $\overline{W} < 15$ seconds, the solution $z = 2.19$ emerged. Only one additional tollbooth needs to be provided to meet the $\overline{W} < 15$ seconds criterion.

> **THINK ABOUT IT**
>
> Most toll plazas now have three ways for drivers to make payment: (1) paying the exact toll at an automatic lane, (2) getting change at a manual lane, and (3) placing a toll tag or transponder in your vehicle for electronic toll collection. Think of ways in which such toll plazas compromise the *efficiency* and *safety* of the roadway to collect tolls.

There are several other considerations regarding queueing systems that ought to be addressed here, in the interest of completeness. We define the relevant terms here:

- *Balking.*   A customer refuses to enter the queue. For example, driver may see that the left-turn lane he/she wanted to use has a long backup and decides to take a different route.
- *Reneging.*   A customer, having joined a queue, leaves before being served. A driver may leave the left-turn lane after having chosen it.
- *Preemptive service.*   A server may suspend service to one customer, because a higher-priority customer (e.g., an emergency vehicle) has arrived.

Each of these complications to the models described by the equations in Table 3.21 can occur in transportation cases.

Example 3.12

In this section's introduction, the county engineer was faced with a decision: join the 7-person express checkout line at the grocery store or enter one of the three 2-customer non-express queues. Although the engineer must make a quick decision based on prior experience, the time he will have to wait will depend on the following values that he can only guess at:

- *The express lane.*   The time between customers arriving at the express checkout lane is about 1 minute. The average service time is 43 seconds.

- *A non-express checkout lane.*   The interarrival time for non-express customers (all three open lanes) is about 2 minutes, 10 seconds. The average service time is 5.7 minutes.

Before evaluating the engineer's alternatives, let us evaluate the grocery store checkout queueing system and its components, ignoring for now its current *state* (number of customers in line).

A. *The express lane.*   Compute the average time an express customer will have to wait for service in the express line.

B. *A non-express checkout lane.*   How long will the average wait be in a non-express line, assuming that the non-express customers will allocate themselves evenly to the three open non-express lanes?

C. *The non-express lanes as a system.*   The county engineer's choice is really between the express checkout lane and *any* of the non-express lanes. Which of the systems shown in Figure 3.16 best describes his situation with respect to the non-express lanes? Use the M/M/z model to compute average wait time for a non-express customer.

Solutions to Example 3.12   This grocery checkout problem is similar to a standard toll plaza problem, in which an approaching driver must choose between manual, exact change, and (perhaps) electronic tollbooths. The service times depend on the type of tollbooth, as do the arrival rates. In the grocery store checkout problem, we must first verify that the grocery store management has kept enough checkout lanes open, such that the condition $\lambda < \mu$ is maintained. In the absence of any better information, we make the further assumption that the best queueing model to use is M/M/z. That is, we assume customers arrive randomly at the checkout area and the service times have a negative exponential distribution. The value of z will depend on the number of checkout lanes observed (or assumed) to be open.

A. The express lane. There is only one express lane, so $z = 1$. The arrival rate at the express (x) lane is $\lambda_x = 1.0$ customer per minute. The express service rate is

$$\mu_x = \frac{60 \text{ sec/min}}{43 \text{ sec/cust}} = 1.395 \text{ cust/min.}$$

The condition $\lambda < \mu$ is met. Using the M/M/1 equation in Table 3.21 for average wait time:

$$\overline{W}_x = \frac{\lambda}{\mu(\mu - \lambda)} = \frac{1.0}{1.395(1.395 - 1.0)} = 1.815 \text{ min}$$

B. A non-express checkout lane. If each of the three open non-express (nx) lanes gets one third of the non-express customers,

$$\lambda_{nx} = \frac{1 \text{ cust}}{130 \text{ sec}} * \frac{60 \text{ sec}}{\text{min}} * \frac{1}{3} = 0.154 \text{ cust/min.}$$

The non-express service rate is

$$\mu_{nx} = \frac{1}{5.7 \text{ min/cust}} = 0.175 \text{ cust/min.}$$

Again, the condition $\lambda < \mu$ is satisfied. Because $z = 1$, the M/M/1 equation still applies:

$$\overline{W}_{nx} = \frac{\lambda}{\mu(\mu - \lambda)} = \frac{0.154}{0.175(0.175 - 0.154)} = 41.9 \text{ min}$$

C. The non-express lanes as a system. There are three non-express lanes in operation, so the M/M/3 version of Equation 3.23 applies. Needed input values are

$$\lambda_3 = \frac{1 \text{ cust}}{130 \text{ sec}} * \frac{60 \text{ sec}}{\text{min}} = 0.46 \text{ cust/min},$$
$$\mu_3 = 0.175 \text{ cust/min},$$

and

$$\rho_3 = \frac{\lambda_3}{z\mu_{nx}} = \frac{0.46 \text{ cust/min}}{3 * 0.175 \text{ cust/min}} = 0.876,$$

as well as $\overline{Q}$ from Equation 3.22. The spreadsheet set up for this problem is shown below as Figure 3.21. Recall that λ in the M/M/z equations is for all arriving customers, but μ is for *each* checkout lane. Likewise, ρ is computed by using a single-server value of μ, not the μ value for the *collection* of servers. In cell N34 of the spreadsheet,

		M	N	O	P
7	Queueing calcs for M/M/3 model				
8					
9	Input:				
10	**lambda**	27.692		
11	**mu**	10.5		
12	**#servers, z**	3		
13					
14	rho	2.631		
15	rho/z	0.877		
16					
17	n.le.z − 1	0	1	2
18	rho^n	1.000	2.631	6.921
19	n!	1	1	2
20	(rho^n)/n!	1	2.631	3.460
21	sum	7.091		
22					
23	rho^z	18.207		
24	z!	6		
25	1 − (rho/z)	0.123		
26	term2	24.656		
27					
28	**p0** Eq 3.25a	0.031		
29					
30	rho^(z + 1)	47.900		
31	term1	0.084		
32	(1 − (rho/z))^2	0.015		
33					
34	**Q bar (cust)**	5.534		
35					
36	**W bar (min)**	11.989		
37					
38	**t bar (min)**	17.689		

FIGURE 3.21
Three non-express lanes as a queueing system in Example 3.12C.

$\overline{Q} = 5.534$ customers. $\overline{W}_3$ in cell N36 is 11.989. The average total time in the system (cell N38) is 17.689 minutes.

THINK ABOUT IT

Shouldn't the answers to parts B and C in Example 3.12 be the same? If they are different, use Figure 3.15 to help explain why they are different.

A member of the Mythaca County Engineer's staff asserted that, if a stream of vehicles approaching a toll plaza with z tollbooths possesses Poisson characteristics, then simply analyzing the queueing behavior at one tollbooth with arrival rate λ/z will be a valid procedure. However, the system does not work that way. One cannot take a stream of traffic coming to six tollbooths and divide the arrival rate by 6. The Poisson distribution may still apply before a tollgate is chosen by each driver. However, because each driver can make a determination of which toll line to enter, the pattern will change. In the limit, each tollgate will have the same number of vehicles in its queue, regardless of the arrival pattern. See the solution to Example 3.12B for some insight into this issue.

SUMMARY

The simplest answer to Kara's question is that backups on a roadway occur whenever the demand (traffic flow) on the roadway exceeds its supply (capacity). However, we have seen in the sections on rural two-lane roads and basic freeway sections that several factors go into determining the capacity of the roadway. Some of those factors can change from day to day (or hour to hour). Or delays can be the result of an event that happens a significant distance ahead of the section being analyzed, such as a breakdown or collision that blocks part of the roadway. In Chapter 2, some basic traffic flow relationships were presented that help explain how such traffic behaves. In the last two sections of this chapter, methods for quantifying delay were introduced. Even experienced traffic engineers, when caught in a "traffic jam," have a hard time determining the cause as they proceed along the roadway. Given the information in Chapters 2 and 3 and the ability to make direct measurements and observations, traffic engineers can usually explain the cause of traffic congestion. More importantly, they can propose remedies. Remedies can be the result of better planning, design, operations, or a combination of these fundamental activities.

ABBREVIATIONS

ATS	average travel speed
EB	eastbound
E_R	passenger car equivalent for RVs, obtained from Exhibit 20-9 or Exhibit 20-10
E_T	passenger car equivalent for trucks, obtained from Exhibit 20-9 or Exhibit 20-10

FFS	free-flow speed (mph)
$f_{d/np}$	Adjustment factor for directional distribution and no-passing zones
f_G	grade adjustment factor, for terrain
f_{HV}	heavy-vehicle adjustment factor, determined by in Equation 3.1
f_{ID}	Interchange density adjustment factor
f_{LC}	Adjustment factor for lateral clearance
f_{LW}	Adjustment factor for lane width
f_{LW}	Adjustment factor for number of lanes
f_{up}	Adjustment factor for no-passing zones
HCM	*Highway Capacity Manual*
NB	northbound
P_n	steady-state probability that exactly n customers are in queueing system.
PCE	passenger car equivalents
pcphpl	Passenger cars per hour per lane
P_R	RVs as proportion of the traffic stream, expressed as a decimal
P_T	trucks as proportion of the traffic stream, expressed as a decimal
PTSF	percent time spent following
$\overline{Q}$	average number of customers waiting for service
SB	southbound
S_{FM}	mean speed of traffic measured in the field (mph)
$\bar{t}$	average time spent in the queueing system = waiting time + service time
v/c	Volume/capacity ratio
V	volume count for a specified time period
V_f	observed flow rate (vph) for the period when field data were obtained
v_p	flow rate expressed in terms of PCE, pc/h/ln or pcphpl
veh	vehicle(s)
$\overline{W}$	average waiting time for each vehicle in the queue
WB	westbound
z	the number of servers in a queueing system
λ	the arrival rate (customers arriving per unit time) at a queueing system
μ	the service rate (customers served per unit time) by a single server
ρ	utilization ratio

GLOSSARY

Balking: A customer refuses to enter the queue. For example, driver may see that the left-turn lane he/she wanted to use has a long backup and decides to take a different route.

Base free-flow speed: The speed a driver would choose to maintain on a particular roadway, after that roadway's characteristics have been taken into account.

Federal aid road: Roadways eligible for federal funding in the United States

Free-flow speed: The speed a driver would choose to maintain on a particular roadway type, in the absence of other traffic and roadway characteristics that cause the driver to reduce his/her speed

Iterative computation: A calculation repeated because the result of the previous calculation indicates that the input to the calculation must be changed.

Passenger car equivalent: A result of converting a truck, bus, or recreational vehicle to its equivalent number of passenger cars in the traffic stream, to reflect how its size and operating characteristics affect the traffic flow.

Persistent queue: A queueing system with $\lambda > \mu$, such that the eventual result is a queue of increasing length

Preemptive service: A server may suspend service to one customer, because a higher-priority customer (e.g., an emergency vehicle) has arrived.

Reneging: A customer, having joined a queue, leaves before being served.

Stable queues: A queueing system in which $\lambda < \mu$, such that a queue will not grow indefinitely.

REFERENCES

[1] Allen, Arnold O., *Probability, Statistics, and Queueing Theory with Computer Science Applications*, Academic Press, Appendix C, 1978.

[2] Bronson, Richard, *Theory and Problems of Operations Research*, McGraw-Hill Book Company, New York, 1982, p. 267.

[3] Dudek, Conrad L. and Stephen H. Richards, "Traffic Capacity Through Urban Freeway Work Zones in Texas," *Transportation Research Record* 869, 1982, p. 14–18.

[4] FHWA, *Highway Statistics 1996*, Report No. FHWA-PL-98-003, Office of Highway Information Management, Federal Highway Administration, U.S. Department of Transportation, Washington DC 20590, November 24, 1997. Data for the most recent year are available at www.fhwa.dot.gov/pubstats.html.

[5] Fricker, Jon D. and Huel-sheng Tsay, "Airborne Traffic Advisories: Their Impact and Value," *Transportation Research Record* 996, 1984, p. 20–24.

[6] Hillier, Frederick S. and Gerald J. Lieberman, *Introduction to Operations Research*, Fourth Edition, Holden-Day, Inc., Section 16.7, 1986.

[7] Larson, Richard C. and Amedeo R. Odoni, *Urban Operations Research*, Prentice-Hall, Section 4.6.2, 1981.

[8] Lovell, David J. and John R. Windover, Discussion of "Analyzing Freeway Traffic under Congestion: Traffic Dynamics Approach," *Journal of Transportation Engineering*, Vol. 125, No. 4, Jul./Aug. 1999, p. 373–375, American Society of Civil Engineers.

[9] Nam Do H. and Donald R. Drew, "Analyzing Freeway Traffic under Congestion: Traffic Dynamics Approach," *Journal of Transportation Engineering*, Vol. 124, No. 3, May/Jun. 1998, p. 208–212, American Society of Civil Engineers.

[10] HCM 2000. *Highway Capacity Manual*, Transportation Research Board, National Research Council, Washington DC, 2000.

EXERCISES FOR CHAPTER 3: HIGHWAY DESIGN FOR PERFORMANCE

Capacity and Level of Service for Two-Lane Highways

3.1 LOS on Two-Lane Highways.   The poor LOS found in Example 3.1 was because of a low average travel speed. Besides the choice of BFFS, can you identify which other traffic or roadway characteristic(s) had the largest detrimental effect on the ATS value in Example 3.1?

3.2 LOS on Two-Lane Highways.   In Example 3.1, the ATS criterion fell in the LOS E range.

(a) What "percent trucks" would cause the PTSF-based LOS to improve to LOS D?

(b) With "percent trucks" as in Example 3.1, what "access point density" would cause the ATS-based LOS to improve to LOS D?

3.3 LOS on Two-Lane Highways.   Repeat Example 3.1 after all access points have been eliminated. By how much has the highway's LOS improved?

3.4 LOS on Two-Lane Highways.   Repeat Example 3.1 after changing the terrain from rolling to level. By how much has the highway's LOS improved? In practice, how could such a change in terrain be achieved? In your opinion, is this likely to be an expensive strategy?

3.5 River Road Capacity Analysis.   A two-lane roadway extends about 25 miles along the Snake River. Because the road follows the river, the terrain is fairly level, but the road has many curves. As a result, about 60 percent of the roadway has no-passing zones. The road's design speed is about 50 mph. The directional distribution of traffic is 80/20 during the most heavily traveled times. The lanes are 11 feet wide and have an average shoulder width of 2 feet. Because it is also the road to the region's landfill, about 25 percent of the traffic is heavy trucks. It has several RV campgrounds along the river, so the RV traffic can be as high as 10 percent. There are no buses. During the peak hour, the two-way traffic is 640 vehicles per hour, with a peak 15-minute volume of 180 vehicles. There are five access points per mile. What is the level of service and approximate average trip time over the 25-mile stretch?

3.6 Highway Capacity and Level of Service.   SR361 southeast of Mythaca is a two-lane rural road through rolling terrain. For most of its 10-mile length from Mythaca to Eastpoint, it has the following characteristics:

> 12-foot lanes, 3-foot shoulders, and 15 percent no passing zones.
> Peak period flow = 410 vph with a 60/40 directional distribution and 2 percent RV, 1 percent bus, 18 percent trucks.
> 15 access points over 10 miles.

(a)   Do the peak period conditions on this section of SR361 correspond to LOS C or better?

(b)   Unfortunately, a bridge on SR361 over White Creek has 10.5-foot lanes and 1.5-foot shoulders. If the bridge is considered to be the bottleneck that determines SR361's capacity, can LOS C be maintained during the peak period?

3.7 Capacity and Level of Service.   Southeast of Mythaca, SR361 is two-lane rural road that goes through the small settlement of Doby. At this point, SR361 has 12-foot lanes with 6-foot shoulders. Although the terrain is level in Doby, no passing is allowed within the settlement. Despite a posted speed limit of 45 mph, the residents of Doby say that many vehicles on SR361 pass through their settlement at much higher speeds. The residents have asked the state DOT to install a "neckdown" at both approaches to Doby. The road would narrow to two 9-foot lanes with no shoulder. You are asked to evaluate the "neckdown" as a "traffic calming" strategy.

(a)   If an average of 285 vph (8 percent trucks) pass through Doby with a 50/50 directional distribution, what is the existing LOS?

(b)   If the proposed "traffic calming" strategy is implemented, will the existing LOS be maintained?

Capacity and Level of Service for Freeway Design

3.8 Adjustment Factors.   Using the view provided in Figure 3.7, attempt to develop a rough estimate of the percent trucks for both the EB and WB directions.

3.9 Weaving Section.   A common example of a *weaving section* is a freeway off-ramp being located immediately after an on-ramp. Make a sketch of the weaving section with which you are most familiar. Where is this section located? Why do you think this roadway section was designed that way? Can you propose any remedies?

3.10 Maintaining a Specified Level of Service on a Freeway.   A four-lane freeway (two lanes in each direction) is located on rolling terrain and has a design speed of 70 mph, 12-foot lanes, and no lateral obstructions within 8 feet of the pavement edges. The traffic stream consists of cars and trucks only (no buses or recreational vehicles). A weekday peak hour volume of 1800 vehicles is observed, with 500 arriving in the most congested 15-minute period. If a level of service no worse than C is desired, determine the maximum number of trucks that can be present in the peak hour traffic volume.

3.11 Freeway Capacity Analysis.   Cars and trucks (14 percent heavy trucks) arrive at a place on a rolling freeway where, because of highway repair, three lanes of the freeway 11 feet wide with 3 feet clearance on either side become two lanes 10 feet wide with no shoulder for a distance of 3000 feet. Later, the traffic opens up to the full three lanes again.

(a) What service flow on the highway entering the repair zone will just avoid a traffic jam within the repair zone?
(b) What level of service does that flow correspond to for the three-lane freeway?

3.12 Freeway Level of Service.   A four-lane urban freeway (two lanes in each direction) was built in part using an abandoned canal. The design speed is 60 mph. The lanes are 10 feet wide without any shoulder on either side. If the traffic flow is 2500 vph with 25 percent trucks in the peak hour, what level of service is the freeway operation if there is an interchange every 0.33 mile? What is the density?

3.13 Freeway Level of Service.   A four-lane freeway (two lanes in each direction) is located on rolling terrain and has a design speed of 70 mph, 12-foot lanes, and no lateral obstructions within 8 feet of the pavement edges. The traffic stream consists of cars and trucks only (no buses or recreational vehicles). A weekday peak hour volume of 1800 vehicles is observed, with 500 arriving in the most congested 15-minute period. If a level of service no worse than C is desired, determine the maximum number of trucks that can be present in the peak hour traffic volume.

3.14 Freeway Level of Service.   An eight-lane interstate highway (freeway) on level terrain has four 10-foot lanes in each direction, with a 3-foot shoulder on the right side and no shoulder on the median side. The highway normally carries 22 percent heavy trucks and has no RVs or buses. The level of service most of the time is "B." (SF = 1120 pcphpl) Most of the drivers are commuters and know the road well, so f_p = 1. PHF = 0.9. What is the directional flow rate of traffic on this highway? How many heavy trucks are there per hour?

3.15 Specifying Design Speed.   The speed limit for the section analyzed in Example 3.4 was posted as 60 mph, yet the design speed of 65 mph was used. Why?

3.16 Adjustment Factors.   Do you think it is true that some of the adjustment factors in Freeway LOS Analysis can change from day to day? Which factors are referred to in this question? Are hour-to-hour changes possible?

Queueing Systems

3.17 Queueing at Amusement Park Vehicle Entrance.   Vehicles begin arriving at an amusement park 1 hour before the park opens, at a rate of four vehicles per minute. The gate to the parking lot opens 30 minutes before the park opens. If the total delay to vehicles entering the parking lot is 3600 vehicle-minutes, (a) how long after the first vehicle arrival will the queue dissipate and (b) what is the average service rate at the parking lot gate? Assume a D/D/1 queueing system at the parking lot gate.

3.18 Queueing.   A rural section of SR361 has two WB lanes with capacity 1400 vph per lane. During the PM peak period, the flow rate is typically 1750 vph. If one lane of WB SR361 must be closed for 30 minutes during the PM peak period, estimate how much total vehicular delay will result.

3.19 I-96 Incident and Queueing Analysis. A truck overturned at 11:57 AM near milepost 138 on NB I-96, completely blocking that highway. Fortunately, the incident site is just beyond an overpass, between an off-ramp and an on-ramp. This means that most vehicles will see the blockage and exit I-96 at the off-ramp, avoiding a long backup and long delay. This also makes the detour of through vehicles simply a matter of using these ramps to go around the incident site. For the first 10 minutes after the truck's mishap, the ramp capacities were governed by the stop sign at the end of the off-ramp and the priority given to cross traffic, which did not have a stop sign. The ramp's service rate for detouring traffic was approximately 325 vph. After 10 minutes, state police began controlling traffic at the end of the off-ramp, increasing the ramp's service rate for detouring traffic to 650 vph. At exactly 1:00 PM, NB I-96 (capacity = 3600 vph) was reopened to through traffic. If the NB I-96 flow rate at this time of day is 1550 vph:

(a) Draw a queueing diagram that shows the buildup and dissipation of the queue.
(b) At approximately what time (to the nearest 2 minutes) does the queue dissipate? Show this event on your diagram.
(c) What was the longest vehicle queue? What was the longest vehicle delay? Show the longest queue and longest delay on your diagram.

3.20 Queueing in "Real Life." You (and several thousand other music lovers) are driving to the Vulgaris concert at Fawn Creek Music Center (FCMC) just off I-25 north of Mythaca. Heavy traffic on the off-ramp from I-25 to the county roads leading to FCMC is causing a backup onto I-25. Describe how you would try to estimate the value of μ for this off-ramp as part of a queueing system without leaving your car, which is stuck in the off-ramp traffic. Use hypothetical but plausible numbers to clearly demonstrate your method.

3.21 Times Measured in a Queueing System. In Example 3.9E, the solutions used the symbol $\overline{W}$ to represent average wait time at the I-25 work zone. Shouldn't the symbol $\bar{t}$ have been used instead? Why or why not?

Systems with Stable Queues

3.22 Equations for Stable Queues. Derive Equations 3.15 and 3.16 for the M/D/1 model and Equations 3.18 and 3.19 for the M/M/1 model from the M/G/1 Equations 3.12 and 3.13, using $\sigma = 0$ for a M/D/1 model and $\sigma = \frac{1}{\mu}$ for a M/M/1 model.

3.23 Review the Solution. Think about the results in Example 3.12E. Do they make sense, or are they indications of an error in the analysis? If there is an error, where could it lie?

3.24 Poisson Calculations. The SR361 Bridge across the Mythaca River between Middleville and Shoridan is a toll facility. The table below shows the SEB bridge volume counts by 20-minute periods for a recent weekday morning.

Time Period (beginning at):	7:00	7:20	7:40	8:00	8:20	8:40	9:00
Vehicle Arrivals:	55	133	202	193	129	104	76

(a) If the arrivals between 7:00 and 7:20 AM are assumed to follow a Poisson Distribution, what is the probability that at least two vehicles will arrive during any given 30-second period?
(b) Adopt a M/M/1 queueing regime and treat the parallel toll booths as a composite service channel with $\mu = 450$ vph. Calculate the average length of queue, average waiting time (seconds) in the queue, and average time (seconds) spent in the system for drivers arriving in the 40 minutes beginning 7:00 AM.

(c) For the consecutive time period(s) in which cumulative $\lambda > \mu$, draw and label clearly a queueing diagram patterned after Figure 3.17 or 3.20 that shows the buildup and dissipation of the queue.

(d) Using the queueing diagram you constructed in the previous subproblem...

- When will the queue dissipate?
- How long (according to the diagram) was the longest queue and when did it occur?
- During the period in which a queue existed, what was the approximate total vehicular delay and average delay per vehicle?

3.25 Left-Turn Lane Analysis Using Poisson Equations.   The intersection of Coliseum Avenue and Wakefield Street near Mythaca State University has a signal with a 60-second cycle length. At the start of each cycle, a "protected left-turn phase" is provided for EB traffic on Coliseum that is turning left onto NB Wakefield. This left-turn (LT) phase is long enough to allow only 7 vehicles to turn left. Two conditions prevent an easy solution at the intersection. (1) Traffic near the university fluctuates during each hour, depending on the times at which classes begin or end. Between 15 and 25 minutes past each hour on Monday, Wednesday, and Friday, an average of 9.7 drivers per minute try to enter the LT lane. During the rest of the hour, 5.7 drivers per minute want to turn left. (2) Other approach volumes are too large to permit additional time to be given to the left turns made from the EB approach.

(a) What is the probability that the seven-vehicle capacity of the LT lane will be exceeded during any cycle within the 10-minute peak each hour?

(b) What is the probability that the seven-vehicle capacity of the LT lane will be exceeded during any cycle during the rest of the hour (the "off-peak" period)?

3.26 Left-Turn Lane Analysis Using Queueing Equations.   The county staff thinks that the LT traffic from Coliseum onto Wakefield (see previous problem) follows an M/D/1 queueing regime. Assuming they are correct, show how to carry out the following calculations for the off-peak arrival rates:

(a) Average length of the LT queue

(b) Average waiting time in the LT queue

(c) Average time spent in the system

3.27 Fast-Food Queueing Systems.   Students at Mythaca State University were interested in comparing service at two nearby fast-food restaurants. The two restaurants used different queue structures—one used a single server in each of several parallel queues, whereas the other employed two servers in series in a single queue. When the students collected data on the "peak hour" service rates at the two restaurants, they were surprised to find the hourly service rates were 71.46 and 69.76, respectively. Because these values were so similar, the students decided to adopt a μ value of 70 customers per hour for both restaurants. Assuming a λ value of 558 fast-food customer arrivals per hour, use the appropriate equations in Table 3.21 to determine:

(a) The minimum number of "channels" or queues a fast-food restaurant would need to meet the service quality standard that the average waiting time in the queue does not exceed 1 minute? Call this result k_{min}.

(b) The average queue length with k_{min} channels.

(c) The probability that the system will be empty (to five decimal places).

(d) What assumptions must you make to allow you to use Table 3.21?

3.28 Queueing. Taco Terrace, a new fast-food restaurant, is opening in Middleville. The current plan to serve drive-up customers involves two servers. The first server window will take the customer's order with a mean service time of 38.5 seconds. The second server window will collect the money and give the order to the customer with a mean service time of 60.5 seconds. The expected arrival rate is 50 drive-up customers per hour. Both interarrival times and departure time patterns are exponentially distributed.

(a) If you neglect the "move-up" time from the first window to the second, what is the average time a Taco Terrace drive-up customer will spend in the system?

(b) How many car lengths of space need to be provided between the two server windows such that, under "average" conditions, the first server window is not blocked by a backup at the second window?

3.29 Data Collection for Queueing Analysis. If the county engineer wanted to verify or update the service times for express and non-express checkout lanes at his favorite grocery store, how should he collect and process the data he needs?

3.30 Grocery Checkout, Revisited. Actually, the county engineer's decision in Example 3.12 does not depend on the calculations carried out for parts A to C. Fortunately, if the information given in the example about checkout service times is reasonably accurate, the decision is quite simple. Which lane should the county engineer join?

Modeling Transportation Demand and Supply

SCENARIO

State Route 361, which runs through Mythaca, is becoming congested—or at least the citizens of Mythaca think so. SR361 not only carries most of the traffic destined for downtown Mythaca from the north and south, it is also the main route *through* the city. The trips through downtown include trips between the south part of the county and the campus of Mythaca State University. Because SR361 is a state highway, the local authorities must prove to the state's Department of Transportation (DOT) that the problem is (or will become) serious enough to justify spending some of the state's limited highway funds on a solution. Because Mythaca County is served by the Mythaca Regional Planning Commission (MRPC), local officials call on the MRPC staff to study the matter. The MRPC Executive Director agrees to have the MRPC staff document the current situation, forecast future trends, propose several possible solutions, and recommend the best course of action.

For the state DOT to accept any study of this sort, the MRPC must follow procedures that are well documented, supported by data, and can be replicated by the DOT. Over the past 40 years, certain procedures to carry out travel demand forecasting have become widely accepted. Although researchers have looked at other procedures—and a major national effort to completely replace the current methods is now underway—this chapter introduces the main features of each step in the current "state of the practice."

CHAPTER OBJECTIVES

By the end of this chapter, the student will be able to:

1. List the four steps in the traditional travel demand modeling process.
2. Explain the two-way relationship between land use and travel.
3. Estimate the number of trips that will be generated to and from a specified area.
4. Calculate the number of trips that can be expected to go to any particular destination from a specified origin.
5. Determine the proportion of travelers who will choose each transportation mode from a set of available modes.
6. Explain equilibrium and use that concept to calculate the flow patterns that satisfy equilibrium conditions.
7. Discuss the strengths and limitations of standard travel demand models as the basis for major public investment decisions.

4.1 BASIS FOR TRANSPORTATION PLANNING

4.1.1 Anticipating Future Network Needs

Transportation planning is a process that involves the analysis of current travel patterns, the forecasting of future travel patterns, the proposal of transportation infrastructure and services, and the evaluation of proposed alternative projects. The result of the process is a *plan*—a set of improvements in the transportation system to be considered by decision makers for implementation.

Figure 4.1 shows part of an urban street network in the year 1999. Traffic counts for important street links in the network for that base year are shown in the diagram. The current path of SR361, mentioned in this chapter's scenario, is highlighted. The first label on SR361 as it enters the diagram from the south is "8000," which represents two-way 24-hour *annual average daily traffic (AADT)* volume of 8000 vehicles per day. The AADT for SR361 increases to 22,000 and then drops off to 14,000 and 15,800 as it

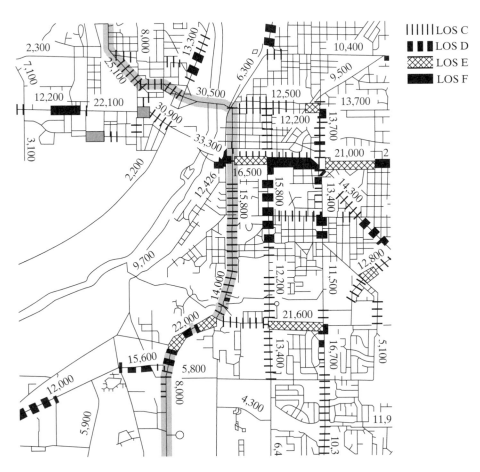

FIGURE 4.1
AADT and LOS on links in base year urban network. *Source:* APC, 2001, p. 61.

heads north toward downtown. Much of this traffic is heading for the university across the river, located in the northwest corner of Figure 4.1. The two bridges that cross the river have AADT values of 33,500 and 30,500. The base year AADT values in Figure 4.1 have been converted to level of service (LOS). Links with LOS C or worse have special markings in Figure 4.1. If some of the forecast year LOS values are poor, planners and decision makers should start *immediately* to consider strategies to alleviate the congestion. It is not unusual for a road construction project to take 10 years to go from decision to completion. In Figure 4.1, it doesn't make sense for the traffic that comes from the south and goes to the university to clog the downtown streets. What can be done?

THINK ABOUT IT

A state highway that runs through a city is causing delays to the drivers traveling through the city and congestion at intersections within the city. What options are available to transportation decision makers to alleviate this problem? Be sure to include some options that do not require major construction.

The MRPC staff has created a model of the Mythaca street network. The model permits them to simulate the way trips are being made in the area being modeled. If the staff can replicate the flow patterns in the base year, they may be able to predict how travel patterns will respond in a future year to a change in the transportation network or the service provided on it. Figure 4.2 is the result of such a travel demand modeling process. The proposed change in the network is the construction of a bypass route for SR361 that includes a third bridge across the river. The proposed SR361 bypass is highlighted in Figure 4.2. The AADT predictions in Figure 4.2 are for the year 2010, if the SR361 bypass is built. The new route diverts traffic away from the old SR361 route. The new bridge is expected to carry 38,500 vehicles per day. Compare the link AADTs in the southwest corner of Figure 4.1 with the AADTs on the same links in Figure 4.2. How did the MRPC staff build their travel demand model? How reliable is the model to use as a basis for deciding whether to invest millions of public dollars in the SR361 bypass project? These questions are considered in this chapter.

THINK ABOUT IT

Notice the 2010 AADT values on the two "old" bridges in Figure 4.2. Compare these values with the 1999 AADT values for the same two bridges in Figure 4.1. Develop an explanation for why the AADT values do not change much. Does this observation support an argument for the SR361 bypass project, or against it?

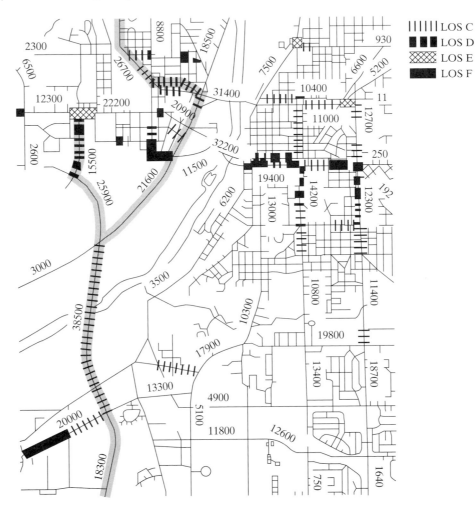

FIGURE 4.2
AADT and LOS on links in forecast year urban network. *Source:* APC, 2001, p. 71.

The MRPC travel demand model's primary output is the amount of traffic that is expected to be using each link in the network in some "horizon" (forecast) year, given certain assumptions about such inputs as population, employment, land use patterns, and network structure in that year. To conduct an analysis of travel patterns, the transportation planner must consider what causes trips to be made and what decisions are made by a tripmaker. The reader is invited to think about these issues as they relate to trips that he or she makes.

The standard travel demand modeling methods currently in use are based on several concepts. First, most trips are not made for the sake of traveling, but to do something at the destination—work, shop, study, play, etc. This is known as the *derived demand* for transportation. Second, travel patterns are influenced greatly by land use patterns. The nature of the activity on a given plot of land will determine the amount of travel to and from that location. Finally, the various elements of individual decision-making regarding trips may be made subconsciously, more or less simultaneously, based on either careful thought or very little rational analysis, out of habit, and may be subject to change during the trip. Nevertheless, most practicing travel demand modelers follow a four-step procedure that assumes a tripmaker considers the following four questions in the following sequence:

1. Should I make a trip?
2. What should be my destination?
3. What mode of transportation should I use?
4. What route or path should I take?

THINK ABOUT IT

Think back to the last four or five trips you made. Did you ask yourself any or all of the four questions above? If so, in what order did you consider the questions?

Obviously, most individuals do not think about trips in this sequential structure. But remember: A travel demand modeler is seeking to describe (and explain) the travel decisions of hundreds—even thousands—of tripmakers. To keep the analysis—and the collection of data to support it—manageable, most travel demand modelers have adopted a standard method that is based on the four questions listed above. As each of the four steps in the process is introduced in the lessons that follow, each question above will be restated in a way that recognizes that many persons may be making the decision implied by the question.

4.1.2   Land Use and Tripmaking

A 500-acre undeveloped site lies adjacent to State Highway 361 at the edge of Mythaca (Figure 4.3). The land is suitable for either residential or industrial development. The Mythaca County land use planners are concerned about the patterns of land use and how they will affect the quality of life in the area. The transportation planners want to be sure that the area's transportation system—roadway and public transit—will be adequate to accommodate the new tripmaking associated with the new development.

A person's tripmaking is the result of a decision to go somewhere and do something there. Tripmaking is the response to landowners' decisions as to how to use their land. Transportation facilities and services are also a response to land use, either in anticipation of the traffic that will result or in response to the tripmaking that is already taking place. The relationship also works in reverse (i.e., land use decisions are affected by existing and promised transportation facilities and services), but it is usually not as

FIGURE 4.3
Vacant land awaiting development. Photo: Jon D. Fricker.

strong. A decision to build a factory at a particular site, for example, may depend on several factors in addition to transportation, such as

- The availability of skilled or trainable employees.
- Tax abatements or other concessions from the local and state governments.
- The suitability of the soil at the site for the intended development.

THINK ABOUT IT

Make a list of the factors that influenced your choice of housing for the current semester. Will these factors be different when you choose your first permanent residence after getting your degree(s)?

The asymmetric nature of the land use-transportation relationship makes land use forecasting more difficult than transportation forecasting, but transportation forecasting is difficult enough. To see how land use can be translated into trips, a few specific examples will be helpful.

People make trips to engage in some activity at the destination. The amount of tripmaking to a plot of land is due in large part to the intensity of use at that location. At one extreme of land use intensity is a vacant lot. Such a "land use"—or lack of it—will normally attract very few trips. If neighborhood children use the vacant lot as a playground, their trips to the lot will typically be made by walking or by bicycle, not by motor vehicle. An example of an intense use of land is a fast-food restaurant, to which a large number of vehicle trips will be made, especially at typical mealtimes. It would be very helpful to a transportation planner to be able to predict how many trips will be made to a particular type of land use at a particular location.

Example 4.1

Consider the above-mentioned 500-acre undeveloped site. The transportation planners at the MRPC want to predict how many trips will be made to and from this site if the development is (a) residential or (b) industrial in nature.

Solution for Example 4.1 A reasonable way to estimate the number of trips that will be made to the 500-acre site would be to base that estimate on the nature and intensity of land use on the site. Fortunately, tripmaking data of that sort have been collected and published by the Institute of Transportation Engineers (ITE) in its *Trip Generation* report (ITE, 1997).

> **FYI:** Many individuals call the ITE *Trip Generation* report a "manual," but they are wrong. The report is simply a compilation of trip generation studies done by consultants, government agencies, and ITE student chapters across the United States. Unlike a manual, the report does not prescribe procedures or set standards. The data in *Trip Generation* must be used with care.

If the 500-acre site is developed as a subdivision with single-family housing on lots of one-third acre each, and 10 percent of the total land area is devoted to streets and other public space, as many as 1350 homes will occupy the remaining land. Using the average trip rate of 9.57 vehicle trips per dwelling unit per weekday found under "Average Rate" in Figure 4.4(a), the transportation planner can estimate that the new subdivision will cause $1350*9.57 = 12,920$ vehicle trips per weekday to be "generated." Using the "Fitted Curve Equation" at the bottom of Figure 4.4(a) leads to an alternative calculation:

$$\text{Ln}(T) = 0.920\,\text{Ln}(1350) + 2.707 = (0.920*7.20786) + 2.707 = 6.631 + 2.707 = 9.338$$

$$T = \exp(9.338) = 11,364 \text{ vehicle trips per weekday}$$

Even though the two estimates (12,920 and 11,364) are fairly close, the planners would be wise to recognize that these values are estimates based on data collected in 348 studies conducted in other cities in unspecified years. Using Figures 4.4(b to d) to corroborate these estimates should be considered.

> **FYI:** Do not let the label of the vertical axis in Figure 4.4(a) confuse you. It says "T = Average Vehicle Trip Ends." The word *average* applies to each point in the figure's scatter plot, because each point may be the result of several observations at that location. A better label might be "Average Total Vehicle Trip Ends" at a site for the time period being studied—in this case, weekday.

Single-Family Detached Housing
(210)

Average Vehicle Trip Ends vs: Dwelling Units
On a: Weekday

Number of Studies: 348
Avg. Number of Dwelling Units: 198
Directional Distribution: 50% entering, 50% exiting

Trip Generation per Dwelling unit

Average Rate	Range of Rates	Standard Deviation
9.57	4.31–21.85	3.69

Data Plot and Equation

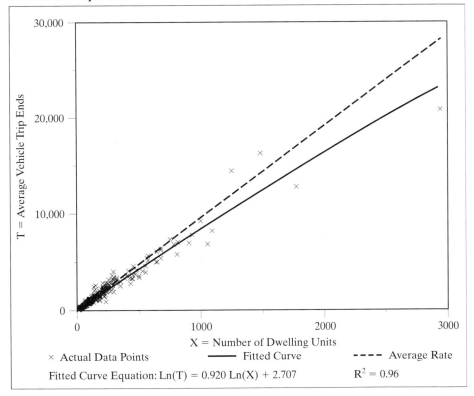

Fitted Curve Equation: $\text{Ln}(T) = 0.920\ \text{Ln}(X) + 2.707$ $R^2 = 0.96$

FIGURE 4.4(a)
ITE *Trip Generation* data page for single-family detached housing. *Source*: *Trip Generation, 6th edition*, 1997, p. 263. © 1997 Institute of Transportation Engineers. Used by permission.

Single-Family Detached Housing
(210)

Average Vehicle Trip Ends vs: Persons
On a: Weekday

Number of Studies: 185
Average Number of Persons: 557
Directional Distribution: 50% entering, 50% exiting

Trip Generation per Person

Average Rate	Range of Rates	Standard Deviation
2.55	1.16–5.62	1.69

Data Plot and Equation

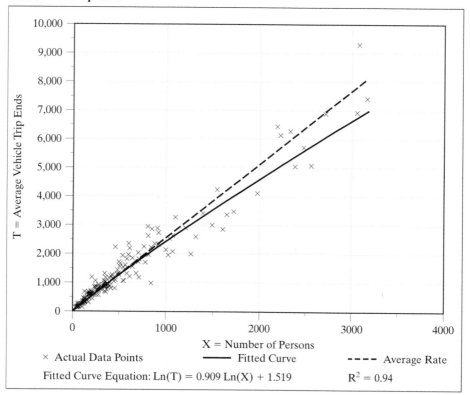

FIGURE 4.4 (b)
ITE *Trip Generation* data page for single-family detached housing. *Source: Trip Generation, 6th edition,* 1997,
p. 272. © 1997 Institute of Transportation Engineers. Used by permission.

Single-Family Detached Housing
(210)

Average Vehicle Trip Ends vs: Vehicles
On a: Weekday

Number of Studies: 120
Average Number of Vehicles: 257
Directional Distribution: 50% entering, 50% exiting

Trip Generation per Vehicle

Average Rate	Range of Rates	Standard Deviation
6.02	2.69–9.38	2.77

Data Plot and Equation

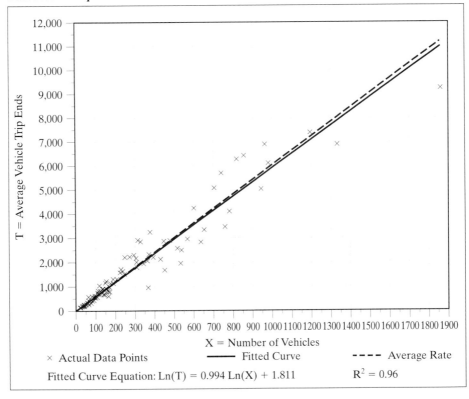

X = Number of Vehicles

× Actual Data Points   —— Fitted Curve   - - - - Average Rate

Fitted Curve Equation: $Ln(T) = 0.994\ Ln(X) + 1.811$ $R^2 = 0.96$

FIGURE 4.4(c)
ITE *Trip Generation* data page for single-family detached housing. *Source: Trip Generation, 6th edition,* 1997, p. 281. © 1997 Institute of Transportation Engineers. Used by permission.

Single-Family Detached Housing
(210)

Average Vehicle Trip Ends vs: Acres
On a: Weekday

Number of Studies: 144
Average Number of Acres: 70
Directional Distribution: 50% entering, 50% exiting

Trip Generation per Acre

Average Rate	Range of Rates	Standard Deviation
26.04	3.17–84.94	19.62

Data Plot and Equation

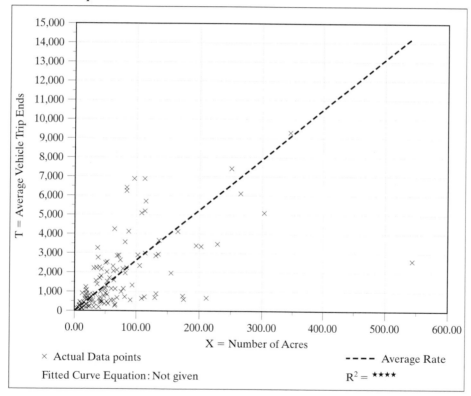

$\times$ Actual Data points

Fitted Curve Equation: Not given

- - - - Average Rate

$R^2 = $ ★★★★

FIGURE 4.4(d)

ITE *Trip Generation* data page for single-family detached housing. *Source: Trip Generation, 6th edition*, 1997, p. 290. © 1997 Institute of Transportation Engineers. Used by permission.

In contrast to a proposed residential subdivision, if the 500 acres were to be devoted to a manufacturing facility, Figure 4.5(a) from the ITE *Trip Generation* report could be used:

Average trip rate: $T = 38.88 * 500$ acres $= 19,440$ vehicle trips per weekday

A fitted curve equation for land use 140 in Figure 4.5(a) is not given, because its R^2 value is too low. If it is known that the proposed manufacturing facility will have 2.5 million square feet of floor space, Figure 4.5(b) could be used to estimate vehicle trips per weekday:

Average trip rate: $T = 3.82 * \dfrac{2.5 * 10^6 \text{ sq ft}}{1000 \text{ sq ft}} = 9550$ vehicle trips per weekday

Fitted curve equation: $T = 3.881 * \dfrac{2.5 * 10^6 \text{ sq ft}}{1000 \text{ sq ft}} - 20.702 = 9682$ vehicle trips per weekday

If the number of employees was predicted to be 3000, Figure 4.5(c) could be used:

Average trip rate: $T = 2.10 * 3000 = 6300$ vehicle trips per weekday

Fitted curve equation: $T = (1.740 * 3000) + 229.975 = 5450$ vehicle trips per weekday

The results of these calculations for manufacturing use of the land are summarized in this table:

Variable	Average Trip Rate	Fitted Curve Equation
$X = 500$ acres	$T = 19,440$	No curve given
$X = 2.5$ million square feet	$T = 9550$	$T = 9682$
$X = 3000$ employees	$T = 6300$	$T = 5450$

THINK ABOUT IT

Which of the five T values computed for the manufacturing land use would you adopt? Why?

The large range of T values—from 5450 to 19,440—that were computed in Example 4.1 using Figures 4.5(a to c) is not unusual. (In fact, the input for the manufacturing land use in this example comes from an actual facility.) After all calculations are done with the ITE method, the planner must decide which single value or range of values is most appropriate. Buttke (1990) offers some guidance in making this decision, but it also depends heavily on the judgment and experience of the planner.

The ITE procedures can be applied to any land use for which data are available. A common use of this procedure is to predict the impact on nearby streets of a new land use, from something as small as a fast-food restaurant or convenience store to a development as large as a regional shopping center or an automotive assembly plant. However, transportation planners often must deal with an entire metropolitan area. It would be extremely tedious to attempt to model tripmaking for even a small town on a site-by-site basis. Instead, transportation planners break up a study area into *traffic*

Manufacturing
(140)

Average Vehicle Trip Ends vs: Acres
On a: Weekday

Number of Studies: 56
Average Number of Acres: 35
Directional Distribution: 50% entering, 50% exiting

Trip Generation per Acre

Average Rate	Range of Rates	Standard Deviation
38.88	2.54–396.00	41.93

Data Plot and Equation

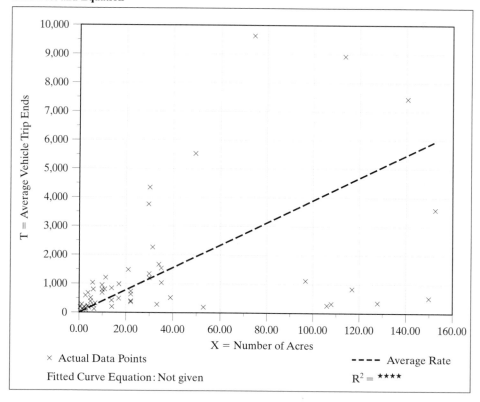

FIGURE 4.5(a)
ITE *Trip Generation* data page for manufacturing land use. *Source: Trip Generation, 6th edition,* 1997, p. 179.
© 1997 Institute of Transportation Engineers. Used by permission.

Manufacturing
(140)

Average Vehicle Trip Ends vs: 1000 Sq. Feet Gross Floor Area
On a: Weekday

Number of Studies: 62
Average 1000 Sq. Feet GFA: 349
Directional Distribution: 50% entering, 50% exiting

Trip Generation per 1000 Sq. Feet Gross Floor Area

Average Rate	Range of Rates	Standard Deviation
3.82	0.50–52.05	3.07

Data Plot and Equation

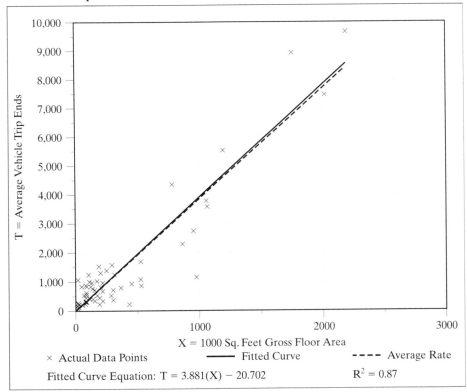

Fitted Curve Equation: T = 3.881(X) − 20.702 $R^2 = 0.87$

FIGURE 4.5(b)
ITE *Trip Generation* data page for manufacturing land use. *Source: Trip Generation, 6th edition,* 1997, p. 170.
© 1997 Institute of Transportation Engineers. Used by permission.

Manufacturing
(140)

Average Vehicle Trip Ends vs: Employees
On a: Weekday

Number of Studies: 61
Avg. Number of Employees: 641
Directional Distribution: 50% entering, 50% exiting

Trip Generation per Employee

Average Rate	Range of Rates	Standard Deviation
2.10	0.60–6.66	1.65

Data Plot and Equation

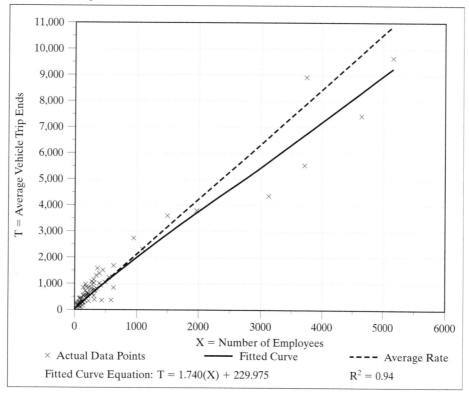

Fitted Curve Equation: $T = 1.740(X) + 229.975$ $R^2 = 0.94$

FIGURE 4.5(c)
ITE *Trip Generation* data page for manufacturing land use. *Source: Trip Generation, 6th edition,* 1997, p. 161.
© 1997 Institute of Transportation Engineers. Used by permission.

analysis zones (TAZs). It is with this zonal basis that we will proceed with the standard four-step travel demand modeling methodology in the sections that follow.

THINK ABOUT IT

Begin a site-by-site analysis of any urban area you choose by dealing first only with one land use type, say, restaurants. How many restaurants are there in that urban area? How do you determine how many there are? Describe the steps you would have to take to estimate the number of trips made to and from all restaurants in that urban area, using data from the ITE *Trip Generation* report. Is this an efficient way to do trip generation in an urban area?

4.2   TRIP GENERATION

To anticipate the future transportation needs in Mythaca County, the Mythaca Regional Planning Commission staff is asked to predict how many vehicle trips will be generated by residents and visitors as they go about their daily activities. The MRPC staff needs to identify and adopt a method that can replicate current tripmaking levels and make reliable forecasts of future tripmaking. For example, how many trips are made on a typical day to a large discount store? (See Figure 4.6.) Without a satisfactory trip generation procedure, the rest of the four-step travel demand modeling process cannot be successful.

4.2.1   Trip Generation Concepts

How many trips will begin or end in each traffic analysis zone?
Instead of attempting to study tripmaking on the level of detail of individual persons or specific origins and destinations, travel demand modelers normally assign the location

FIGURE 4.6
Parking lot in front of "big box" discount store. Photo: Jon D. Fricker.

of each trip end (origin and destination) to a traffic analysis zone (TAZ). In densely populated neighborhoods, a TAZ may consist of only a few city blocks. In more rural areas, a TAZ may cover almost 1 square mile. A TAZ's boundaries are usually set to coincide with the edges of census tracts or census blocks.

When grouping or aggregating tripmaking by TAZs, the first question in the four-question sequence described earlier in this chapter is converted from the individual's "Should I make a trip?" to the collective "How many trips are made from (or to) this zone?" A common preliminary step is to recognize that each trip is made for a particular purpose. Standard trip purposes are work, shopping, school, and recreation. Another important assumption in the modeling process is that the amount of tripmaking from a given zone is a function of certain measurable characteristics of the zone or of the people who live in it. For example, it is generally accepted that a family or household will make more vehicle trips per weekday if it has

- more members of driving age
- more people with jobs
- higher income
- more vehicles to use.

4.2.2   Regression Models for Trip Generation

Based on a list of household characteristics that could be expected to influence trip-making, a planner could collect data on an adequate sample of households in each zone for these variables and build a linear regression model of the form

$$T = a_0 + a_1X_1 + a_2X_2 + \cdots + a_nX_n \tag{4.1}$$

where X_i is a factor (usually demographic) that explains the level of tripmaking and a_i is a coefficient or constant that converts these factors into number of trips T.

Example 4.2   Household-Based Regression for Trip Generation

Whenever a new subdivision is proposed in Mythaca County, the Mythaca Regional Planning Commission staff is asked to predict how many vehicle trips will be generated by the development. Because the ITE method has not been providing good forecasts, one MRPC staff member proposes a method based on certain household characteristics. Among the characteristics that the staff thinks might influence the number of trips is the number of persons in the household. Census data and surveys of a sample of households from recently completed subdivisions have been compiled. A part of that dataset is shown in Table 4.1 below. The data in the table are sorted by household size [persons per *household* (HH) in column 1]. Column 2 contains the number of trips reported by each household when the head of household was asked about the previous day's travel activity. Is it possible to develop a "model" from these data that will help the MRPC staff predict tripmaking levels from similar land uses (i.e., new residential subdivisions)?

Solution to Example 4.2   A glance at the data in columns 1 and 2 in Table 4.1 reveals that, in general, larger households make more trips. This relationship is expected. Can it provide the basis for a good forecasting tool? By entering columns 1 and 2 into a spreadsheet, we can use a feature like "Tools/Data Analysis/Regression" in Excel to produce the best straight-line fit to the

TABLE 4.1   Data for Example 4.2

Household Travel Survey Data		
(1) Persons/HH	(2) Vehicle Trips/Day	(3) T (model)
1	3	5.4
1	2	5.4
2	3	6.8
2	5	6.8
2	10	6.8
2	1	6.8
2	7	6.8
3	6	8.3
3	10	8.3
3	12	8.3
3	11	8.3
4	11	9.7
4	14	9.7
4	10	9.7
5	12	11.1
6	18	12.5
8	12	15.3
8	16	15.3
9	15	16.8
10	16	18.2

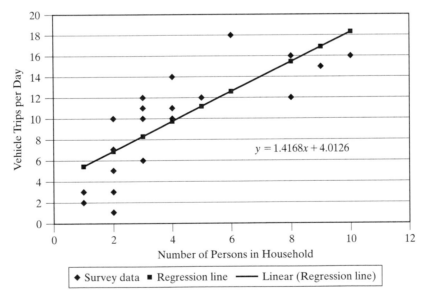

FIGURE 4.7
Linear fit of household vehicle trip data.

data. The resulting line is $y = 4.0126 + 1.4168x$ with an adjusted r^2 value of 0.575. Using this equation, the expected number of vehicle trips per day for each household size can be computed. These expected values are shown in the "T (model)" column in Table 4.1. The plots of the original data and the fitted regression line are shown in Figure 4.7.

THINK ABOUT IT

In Example 4.2, the relationship between tripmaking and household size was considered. Think of at least three other factors that might help explain levels of household tripmaking.

Examples of trip generation models built using linear regression are (FHWA, 1967)

$$\text{Trips produced in a zone} = 36.03 + (5.09 * \text{cars/zone}) \tag{4.2}$$

$$\text{Trips produced per household} = 0.69 + (1.39 * \text{persons/HH}) + (1.94 * \text{cars/HH}) \tag{4.3}$$

$$\text{Shopping trips produced in a zone} = 30.98 + 1.03(\text{HHs/zone}) + 2.20(\text{cars/100HHs}) - 5.39(\text{cars/100 persons}) \tag{4.4}$$

For most trip types, trips are produced at the origin and attracted to the destination. When one end of the trip is at home, however, that end is always the production end and the other end must be the attraction end. This means that *home-based work (HBW)* trips could have a production equation like Equation 4.2 above for the traveler's home zone and an attraction equation for the employment zone such as (NCHRP365, 1998)

$$\text{HBW attractions} = 1.45 * \text{Total Employment} \tag{4.5}$$

4.2.3 Cross-Classification Models for Trip Generation

Although regression models are computationally convenient, their use of zonal totals or zonal averages can mask important variations between households within a zone. An alternative to using zone location to put households into groups is to use important characteristics such as household size and auto availability. This approach assumes that a household with three members and one car has tripmaking characteristics similar to other three-member, one-car households, whether or not they are located in the same

TABLE 4.2 Average Daily Person Trips by Household Size and Autos Owned Urbanized Area Population: 50,000–200,000

Autos Owned	Persons per Household					Weighted Avg.
	1	2	3	4	5+	
0	2.6	4.8	7.4	9.2	11.2	3.9
1	4.0	6.7	9.2	11.5	13.7	6.3
2	4.0	8.1	10.6	13.3	16.7	10.6
3+	4.0	8.4	11.9	15.1	18.0	13.2
Weighted avg.	3.7	7.6	10.6	13.6	16.6	9.2

Source: NCHRP365, 1998.

zone. Table 4.2 is the result of collecting data on the tripmaking frequency of households for all reasonable combinations of household characteristics. Table 4.2 is called a *cross-classification* table. If forecasts can be made of the number of each household type in each zone for some future year, the cross-classification table can be used to compute the total trip productions for each zone. Note, however, that the entries in Table 4.2 are for *person* trips, not *vehicle* trips. The entries may include trips by public transit or by non-motorized travel modes.

THINK ABOUT IT

Look at the trends in the trip rates shown in Table 4.2, as the values of the household characteristics increase or decrease. Do these trends make sense?

Example 4.3

Based on recent surveys, the Mythaca Regional Planning Commission (MRPC) staff has estimated the number of households in Mythaca that fall in each household size/vehicle ownership category. These estimates are shown in Table 4.3. The MRPC does not have the resources (money, personnel, or time) to develop its own cross-classification trip rate table like Table 4.2, so the MRPC staff decides to use the rates in Table 4.2 until better trip rate information can be obtained.

A. How many person trips per day will be produced by a five-person household that owns two vehicles?

B. Using the cross-classification trip generation method, calculate the total number of person trips per day that will be produced by Mythaca households, if the data in Table 4.2 and the "Number of Households" Table 4.3 are reliable.

C. If 94 percent of the person trips calculated in part B used private automobiles (not public transportation) and the average auto occupancy is 1.96, how many vehicle trips were produced by Mythaca households in part B?

TABLE 4.3   Numbers of Households, Each Household Category

	Persons per Household					
Autos Owned	1	2	3	4	5+	
0	1005	403	213	114	105	1840
1	2909	2038	875	526	369	6717
2	408	3915	1802	1425	1406	8956
3+	111	1075	1375	1383	860	4804
Column total	4433	7431	4265	3448	2740	22,317

Solutions to Example 4.3

A. According to the MRPC staff, there are 1406 households in Mythaca that have five or more persons and own two vehicles. This total includes households with exactly five members and two vehicles. In Table 4.2, each of these households produces an average of 16.7 person trips per weekday.

B. Begin with the first cell in the cross-classification tables (i.e., the household category for one member and no auto owned). There are 1005 such households, each of which generates an average of 2.6 person trips per day. To find the number of trips produced by all such households, multiply 1005 by 2.6 to get 2613 person trips per weekday. To compute the number of person trips for all households, simply repeat this calculation for all cells in the "Number of Households" Table 4.3. Multiply the value in each such cell by the trip rate in the corresponding cell in Table 4.2 and then add the products.

$$(1005*2.6) + (2909*4.0) + \cdots + (1406*16.7) + (860*18.0) = 209{,}871 \text{ person trips per day}$$

The cell-by-cell calculations were performed with a spreadsheet, as shown in Table 4.4.

C. Of the 201,574 person trips calculated in part B, 94 percent use private vehicles:

$$0.94*209{,}871 = 197{,}279 \text{ person trips by private vehicle}$$

If the average private vehicle carries 1.96 persons, the number of vehicle trips produced by Mythaca households on an average day is

$$\frac{197{,}279 \text{ person trips}}{1.96 \text{ persons/vehicle}} = 100{,}652 \text{ vehicle trips}$$

TABLE 4.4   Total Trips per Weekday, Each Household Category

	Persons per Household					
Autos Owned	1	2	3	4	5+	Row Total
0	2613	1934	1576	1049	1176	8348
1	11,636	13,655	8050	6049	5055	44,445
2	1632	31,712	19,101	18,953	23,480	94,877
3+	444	9030	16,363	20,883	15,480	62,200
Column total	16,325	56,331	45,090	46,934	45,192	209,871

4.2.4   Beginning the Middleville Case Study

Middleville is a small but fast-growing city between Mythaca and Shoridan. Its current population is estimated to be 5000. To anticipate and plan for the transportation impacts of future growth in the Middleville area, the MRPC has begun to apply travel demand modeling techniques to the area. The MRPC staff has identified four sections of Middleville, each of which is devoted exclusively to a particular activity. These sections will be called *traffic analysis zones*. Zone 1 is a manufacturing zone, zone 2 is a retail zone, and zones 3 and 4 are residential. (See Figure 4.8.)

The data that describe the amount of activity in each zone are summarized in Table 4.5.

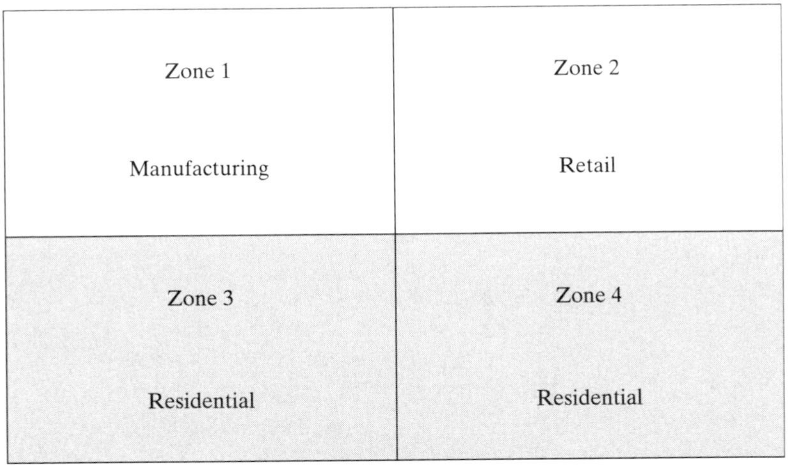

FIGURE 4.8
Four-zone study area.

TABLE 4.5   Middleville T/G Data

TAZ	pop	HH	vehs	empl
1	0	0	0	1000
2	0	0	0	1500
3	3000	1100	1400	0
4	2000	900	1600	0
Totals	5000	2000	3000	2500

TAZ = traffic analysis zone; pop = population in zone; HH = households in zone; vehs = vehicles owned by HHs in zone; empl = employment = jobs in zone.

Example 4.4   Trip Generation in Middleville

Because the MRPC staff does not have the time or resources to develop its own trip generation model, it chooses to borrow a model developed for cities similar to Middleville, with the hope that it is suitable for Middleville. The borrowed model consists of two equations:

$$\text{Trip productions: } P = 93 + 4*\text{vehs} + 0.1*\text{HH} + 0.7*\text{empl}$$

$$\text{Trip attractions: } A = 327 + 2.2*\text{empl} + 1.3*\text{HH}$$

According to the borrowed equations, how many trips per day will be produced and attracted by each TAZ?

Solution to Example 4.4   Applying the borrowed equations to each zone i results in the productions $P(i)$ and attractions $A(i)$ shown in Table 4.6 below. For example, $P(1) = 93 + (4*0) + (0.1*0) + (0.7*1000) = 793$ and $A(2) = 327 + (2.2*1500) + (1.3*0) = 3627$. This is the first step in the four-step travel demand modeling process that is applied to Middleville in this chapter.

TABLE 4.6   Middleville T/G Results

TAZ i	$P(i)$	$A(i)$
1	793	2527
2	1143	3627
3	5803	1757
4	6583	1497
Total	14,322	9408

THINK ABOUT IT

In Example 4.4, the total daily productions (14,322) in Middleville do not equal the total daily attractions (9408). Is this a problem? If so, what are the possible causes? What options does the MRPC staff have?

In practice, both cross-classification and regression are used to compute zonal trip productions, depending on modeler preference, computer software capabilities, and data availability. Trip attractions for each zone are almost always calculated using regression equations.

Two words used in the previous paragraph deserve further mention: *data* and *software*. A large part of the effort involved in travel demand modeling is the collection and analysis of data. These data must be based on a representative sample of the trip-makers being studied. The analysis of these data must be statistically valid and suitable for travel demand modeling purposes. Over the years, many computer packages have been developed to help the transportation planner carry out the calculations involved in the modeling process. Although most software packages are designed to implement the standard four-step methodology, many of them have capabilities and limitations that influence what specific data and modeling techniques a planner can employ.

4.3   TRIP DISTRIBUTION

There are four cities in the region around Mythaca: (1) Econoly, (2) Mythaca, (3) Shoridan, and (4) Middleville. Each city has an increasing number of retail stores. Residents of Mythaca now have many more shopping opportunities competing for their dollars. Each resident must make a choice as to the shopping locations he/she will choose as his/her destination. For shopping trips, work trips, and any other trip type, a planner must be able to estimate the number of travelers at any given origin location (such as in zone 2 in Figure 4.9) who will choose any particular destination (from among zones 1 to 4 in Figure 4.9).

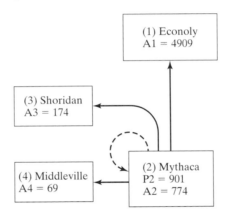

FIGURE 4.9
Schematic diagram of shopping trips in Mythaca region.

4.3.1   Trip Distribution Concepts

How many trips that begin at a given origin will end at a given destination?
Each trip has two ends: an origin and a destination. In the trip generation step, only the number of trip ends in each zone was calculated. In the trip distribution step, trip ends are connected to form *trip interchanges* (T_{ij}), which represent the number of trips produced in zone i and attracted to zone j. Once the production end and attraction end of each trip have been determined, a *trip matrix* can be established. A trip matrix has one row and one column for each zone. The ith row and jth column of the matrix contains the number of trips that are produced in zone i and are attracted to zone j. Another name for a trip matrix is *trip table*.

Table 4.7 summarizes the peak hour shopping trips made between the four communities in or near Mythaca County, according to data collected by the Mythaca Regional Planning Commission. Because Econoly is the largest city in the region and has the most stores, it is likely to attract a lot of shopping trips from the other three cities.

TABLE 4.7   Productions and Attractions for Shopping Trips in Mythaca Region

Cities	(1) Econoly	(2) Mythaca	(3) Shoridan	(4) Middleville
Productions	4724	901	193	108
Attractions	4909	774	174	69

However, shoppers tend to favor their own town, if there are enough shopping opportunities to satisfy their needs, because it saves travel time. The MRPC wants to know how many shoppers make trips to cities other than their own during the peak hour.

The standard procedure for converting zone-by-zone production and attraction totals into trip interchanges T_{ij} is called the *Gravity Model*. The Gravity Model builds on two principal assumptions:

- A trip produced in zone i is more likely to be attracted to an attraction zone that has a higher number of attractions.
- A trip produced in zone i is more likely to be attracted to an attraction zone that is closer to zone i.

The Gravity Model is an analogue of Newton's law of gravity. The gravitational force between two bodies with masses m_1 and m_2 is

$$F_{12} = \frac{m_1 * m_2}{d_{12}^2} \tag{4.6}$$

where d_{12} is the distance between those bodies. In the 1920s, the Swedish investigator Pallin used a form of Equation 4.6 to determine traffic flows between cities (BPR, 1963). He used productions in city 1 in place of m_1 and attractions in city 2 in place of m_2. Since then, the travel demand modeling form of Equation 4.6 has become

$$T_{ij} = P_i \frac{A_j F_{ij}}{\sum_k A_k F_{ik}}. \tag{4.7}$$

T_{ij} is the number of trips that go from zone i to zone j. F_{ij} is a function of the separation between zones i and j and is usually called the *traveltime factor* or the *friction factor*. Normally, the separation between zones is measured in terms of travel time t_{ij}. The most common form of F_{ij} is

$$F_{ij} = \frac{1}{t_{ij}^2} = t_{ij}^{-2} \tag{4.8}$$

The general form of Equation 4.8 is $F_{ij} = at_{ij}^b$.

Example 4.5

To illustrate the relationships contained in the two principal Gravity Model assumptions stated above, consider the regional shopping trips summarized in Table 4.7. How well does the Gravity Model explain the way those trips distribute themselves from each production zone to

TABLE 4.8 Average Travel Times in the Mythaca Region (Minutes)

Origin Cities	Destination Cities			
	(1) Econoly	(2) Mythaca	(3) Shoridan	(4) Middleville
(1) Econoly	7	35	45	40
(2) Mythaca	35	5	20	12
(3) Shoridan	45	20	3	8
(4) Middleville	40	12	8	2

the available attraction zones? Begin with the 901 shopping trips that are produced in Mythaca (zone 2 in Table 4.7). How many of the shopping trips that start in Mythaca will be attracted to stores in Econoly, Mythaca, Shoridan, and Middleville? The average travel times in the region are summarized in Table 4.8. Note that intrazonal travel times t_{ii} are not zero. The diagonal entries are the average travel times for vehicle trips that stay within the same zone.

Solution to Example 4.5 We can use Equation 4.7 to find T_{21}, T_{22}, T_{23}, and T_{24}. First use Equation 4.8 to convert travel times t_{21}, t_{22}, t_{23}, and t_{24} into friction factors F_{21}, F_{22}, F_{23}, and F_{24}:

$$F_{21} = 35^{-2} = 0.000816, F_{22} = 5^{-2} = 0.0400, F_{23} = 20^{-2} = 0.0025, F_{24} = 12^{-2} = 0.006944$$

Because it is a bit awkward to deal with numbers so small, scale up each F_{ij} by a factor of 1000:

$$F_{21} = 0.816, F_{22} = 40.00, F_{23} = 2.5, \text{ and } F_{24} = 6.944$$

THINK ABOUT IT

What is it about the structure of Equation 4.7 that allows us to simply multiply each value by an arbitrary scalar and not affect the T_{ij} values that result from Equation 4.7?

The A_j values for this problem are the trip attraction totals for each city in Figure 4.9:

$$A_1 = 4909, A_2 = 774, A_3 = 174, \text{ and } A_4 = 69$$

Now calculate T_{21} by using Equation 4.7:

$$T_{21} = P_2 \frac{A_1 F_{21}}{\sum_k A_k F_{2k}} = P_2 \frac{A_1 F_{21}}{A_1 F_{21} + A_2 F_{22} + A_3 F_{23} + A_4 F_{24}}$$

$$T_{21} = 901 \frac{4909 * 0.816}{(4909 * 0.816) + (774 * 40.0) + (174 * 2.5) + (69 * 6.944)} = 101$$

This result tells us that only 101 of Mythaca's 901 shopping trips (produced in zone 1) will be attracted to Econoly (zone 2). Despite the high number of shopping trip attractions in Econoly—which is an indication of many shopping opportunities there—most Mythaca shoppers prefer closer destinations. The calculations for T_{21}, T_{22}, T_{23}, and T_{24} can be made easier by using a spreadsheet such as that shown in Table 4.9. In this table, parameter **a** is the scalar applied to

TABLE 4.9   Gravity Model Calculations for Example 4.5

(1) Zone j	(2) $A(j)$	(3) $t(1j)$	(4) $F(2j)$	(5) $A(j)F(2j)$	(6) $AF(j)/\text{sum}(AF)$	(7) $T(2j)$
1	4909	35	0.816	4007.3	0.112	101
2	774	5	40.000	30960.0	0.863	777
3	174	20	2.500	435.0	0.012	11
4	69	12	6.944	479.2	0.013	12
	5926			35881.5	1.000	901

$P = 901$ from Zone 2; $a = 1000$, $b = -2.00$

Equation 4.8 to get the column 4 values, columns 5 and 6 contain the intermediate steps in Equation 4.7, and column 7 shows the results for T_{21}, T_{22}, T_{23}, and T_{24}.

The form of Equation 4.8 seems to make sense when auto travel between cities is involved. Short trips are more likely to made by auto than long trips. Example 4.6 illustrates.

Example 4.6

What does the friction factor of Equation 4.8 look like when b in the general form of Equation 4.8 takes on values between -0.5 and -2.0?

Solution for Example 4.6   Figure 4.10 shows how the value of b controls how quickly F_{ij} diminishes as t_{ij} increases. The lowest of the three curves plotted in Figure 4.10 is the standard version of Equation 4.8, with $b = -2.0$. If the T_{ij} values calculated using Equation 4.8 with

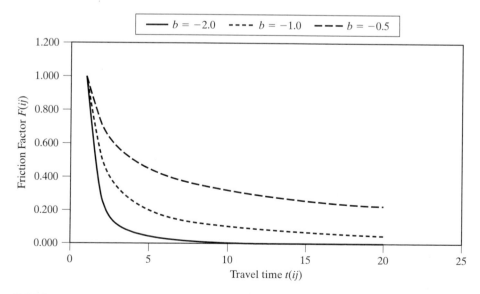

FIGURE 4.10
Friction factor plots as parameter b varies, with $c = 0$.

$b = -2.0$ clearly overstate (or understate) the number of short trips, adjusting the value of b should be considered.

THINK ABOUT IT

If the results in Example 4.5 understate the number of peak hour shopping trips from Mythaca to Econoly, while overstating the number that stay in Mythaca, what should be done with the value of **b** in Equation 4.8?

There may be cases in which short trips are not the most common trip lengths in the travel data being analyzed. Consider, for example, the study area in Example 4.7.

Example 4.7

Consider a small town that has only four zones. (See Figure 4.11.) In this study area, 100 trips are produced in zone 1. How many of those trips will find destinations in zones 1, 2, 3, and 4?

Solution to Example 4.7   In column 3 of Table 4.10, the travel time from Zone 1 to each zone in the study area is given. If travel time did not matter, most of zone 1's 100 trips would go to

Zone 1	Zone 2
$P_1 = 100, \ A_1 = 50$	$A_2 = 200$
Zone 3	Zone 4
$A_3 = 75$	$A_4 = 675$

FIGURE 4.11
Four-zone study area.

TABLE 4.10   Gravity Model Calculations for Four-Zone Town

(1) Zone j	(2) $A(j)$	(3) $t(1j)$	(4) $F(1j)$	(5) $A(j)F(1j)$	(6) AF(j)/sum(AF)	(7) $T(1j)$
1	50	2	125.000	6250.0	0.515	52
2	200	5	20.000	4000.0	0.330	33
3	75	10	5.000	375.0	0.031	3
4	675	15	2.222	1500.0	0.124	12
	1000			12125.0	1.000	100

$P = 100$ from zone 1; $a = 500$; $b = -2.00$

zone 4, because it has by far the most attractions. If travel time is the most important consideration, most trips produced in zone 1 will not go to zones other than zone 1. If the parameter values $a = 500$ and $b = -2$ are used in Equation 4.8, the friction factor function is $F_{ij} = 500\,t_{ij}^{-2}$. Table 4.10, which has the same structure as Table 4.9, summarizes the Gravity Model calculations for the four-zone town.

The results of Example 4.7 are cause for concern. In Table 4.10, $T_{11} > A_1$, but it is not possible to have more trip interchanges than trip attractions for any zone. Assuming the P and A totals in Figure 4.11 are correct, the Gravity Model must be overestimating short trips and underestimating long trips. One recourse is to adjust the b parameter, using lessons from Figure 4.10, but short trips would still be favored, only to a lesser degree. For example, if $b = -1$, T_{11} would be 21.

THINK ABOUT IT

What is the shape of the expected trip length distribution for trips made by automobile? Are very short trips (a few blocks) more common than trips of about 1 to 2 miles? Are trips of more than 5 miles more common than trips of about 3 miles in an urban area? Make a rough sketch of a trip length frequency distribution for an urban area, using the format of a probability density function.

4.3.2 Other Forms of the Trip Length Distribution

It has long been known that very short trips are normally not the most frequent vehicle trips. See Figure 4.12.

A friction factor function of the form

$$F_{ij} = at_{ij}^{b}e^{ct_{ij}} \tag{4.9}$$

can create a trip length distribution like that shown in Figure 4.12. Parameters $a, b,$ and c are used to fit the model's trip table (or trip length distribution) to the one observed for the study area. Equation 4.9 offers the analyst much more flexibility than Equation 4.8 in matching the Gravity Model to travel patterns observed in a study area. Because of J.C. Tanner's (1961) work on this subject, Equation 4.9 is often called the *Tanner function.*

Example 4.8

What does the friction factor of Equation 4.9 look like when $a = 1, b = +2,$ and c takes on values between -1 and -0.5?

Solutions for Example 4.8 Figure 4.13 shows how a proper combination of b and c values can replace the monotonically decreasing function in Figure 4.10 with a function that increases as t_{ij}

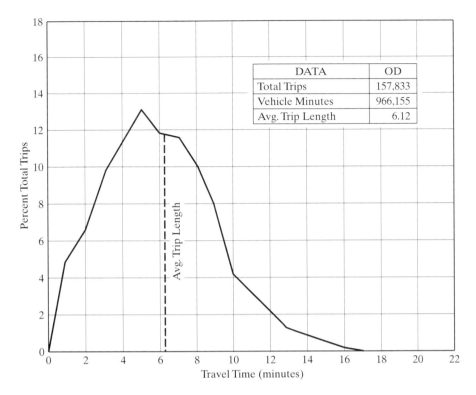

FIGURE 4.12
Trip length frequency distribution for an urban area. *Source:* BPR, 1963.

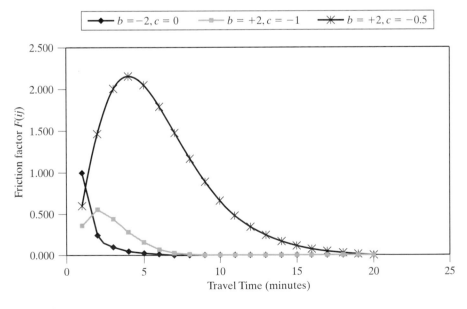

FIGURE 4.13
Friction factor plots for Equation 4.9.

increases from very low values and then decreases as t_{ij} gets large. A special form of Equation 4.9, in which $a = 1$, $b = -2$ and $c = 0$, is actually Equation 4.8. In Figure 4.13, the plot marked by diamond-shaped points represents Equation 4.8. The other two plots in the figure show how the frequency of short trips and long trips can be reduced, with trips of medium length being the most common.

Example 4.9

In Table 4.10, the Gravity Model results for Example 4.7 with $b = -2.0$ indicated that more than half the 100 zone 1 productions would stay within that zone. As Figure 4.12 shows, it is often the case that the shortest trips are not as common as slightly longer trips. Based on Figure 4.13, use the Tanner function (Equation 4.9) with $b = +2$ and $c = -0.5$ in the Gravity Model to compute T_{1j} for $j = 1$ to 4. Compare the results with those in Table 4.10.

Solutions for Example 4.9   The spreadsheet (or manual) calculations that led to Table 4.10 are modified to use friction factor Equation 4.9 instead of Equation 4.8. The results, shown in Table 4.11, now send most trips to the second-closest zone. The number of trips to the two most distant zones does not change much. Most of zone 2's new trips have been redistributed from zone 1. Because zone 4 has such a large number of attractions, it attracts the second most trips from zone 1.

TABLE 4.11   Gravity Model Calculations for Example 4.9

(1) Zone j	(2) $A(j)$	(3) $t(1j)$	(4) $F(1j)$	(5) $A(j)F(1j)$	(6) $AF(j)/\text{sum}(AF)$	(7) $T(1j)$
1	50	2	14.715	735.8	0.119	12
2	200	5	20.521	4104.2	0.664	66
3	75	10	6.738	505.3	0.082	8
4	675	15	1.244	840.0	0.136	14
	1000			6185.4	1.000	100

$P = 100$ from zone 1; $a = 10$; $b = 2.00$; $c = -0.50$

THINK ABOUT IT

Another difference between Table 4.10 and Table 4.11 is the value of the ***a*** parameter. Will the friction factor values in column 4 or the trip interchange values in column 7 change if the value of ***a*** is changed?

If reliable data are available on the frequency distribution of trip lengths in a study area, the Tanner function can be adjusted to reflect the travel patterns. Statistical methods exist for estimating the best values for ***b*** and ***c***. As in Example 4.5, ***a*** is just a scalar.

The Tanner function offers the analyst an opportunity to build a trip distribution model that matches trip length travel data. Sometimes, *intrazonal* trips T_{ii} (trips that have both ends inside the same TAZ) do not fit the tripmaking behavior that the Gravity Model attempts to explain. If intrazonal trips cannot be properly included in a single application of the Gravity Model, the intrazonal trips can be treated separately. The Gravity Model can be applied for interzonal trips only, leaving each $T_{ii} = 0$ until a separate intrazonal analysis can be carried out.

Regardless of the form of the friction factor and whether intrazonal trips are included, the trip distribution step continues until trips have been distributed from each production zone. In Examples 4.7 and 4.9, the distribution of trips from production zones other than zone 1 still needs to be done. The result will be a trip matrix (or trip table) that summarizes all T_{ij} trip interchange values. The intrazonal trips T_{ii} that would fill the diagonal cells in the matrix could be done by a separate process, could be included in the Gravity Model calculations, or could be left out of the trip distribution step entirely, depending on the objectives of the modeling task.

4.3.3   The Middleville Case Study, Continued

Because the MRPC planners have more faith in their production equations than their attraction equations, they decided to balance the Ps and As that came out of Example 4.4 by multiplying each A_j value by $\Sigma P / \Sigma A = 14322/9408$. The resulting A_j values are $A_1 = 3847$, $A_2 = 5521$, $A_3 = 2675$, $A_4 = 2279$, so that $\Sigma A = \Sigma P = 14,322$. Now that both trip ends are accounted for, the planners can attempt to build a trip matrix.

Example 4.10   Trip Distribution for Middleville

The MRPC planners decide to use Equation 4.8 with the traditional parameter value $b = -2.0$ to compute each interzonal T_{ij} trip interchange value. Intrazonal trips T_{ii} will be treated in a separate analysis. The average auto travel times between the zones in Middleville have been estimated and are shown in the upper left quadrant of Table 4.12 below. What will the resulting interzonal trip matrix look like?

Solution for Example 4.10   With $b = -2.0$ and $a = 100$, Equation 4.8 becomes $F_{ij} = 100 * t_{ij}^{-2.0}$. For trips produced in zone 1 and attracted in zone 2, $F_{12} = 100 * t_{12}^{-2.0} = 100 * (11.0)^{-2.0} = 0.83$. For trips produced in zone 4 and attracted in zone 2, $F_{42} = 100 * t_{42}^{-2.0} = 100 * (5.5)^{-2.0} = 3.31$. Although average interzonal travel times are not always symmetric $(t_{ij} = t_{ji})$, they are symmetric here. Consequently, the friction factors are also symmetric $(F_{ij} = F_{ji})$ in the upper right quadrant of the Table 4.12. In the lower left quadrant of Table 4.12, the intermediate calculations for $A_j F_{ij}$ are stored. For example, $A_2 F_{32} = 5521 * 1.83 = 10,083$. In the denominator of Equation 4.6, $\sum_k A_k F_{ik}$ appears. This is simply the row sum of the $A_j F_{ij}$ values. In row 2 of the lower left quadrant of Table 4.12, we see that $3179 + 0 + 4884 + 7534 = 15,597$. Finally, each T_{ij} value can be calculated by using Equation 4.6: $T_{41} = P_4(A_1 F_{41})/ (\sum_{k=1,4} A_k F_{4k}) = 6583 * (8831)/(29,813) = 1950$. The lower right quadrant of Table 4.12 shows the results for all the interzonal T_{ij} calculations. This is the trip matrix (or trip table) for the Middleville study area.

TABLE 4.12   Calculations for Example 4.10

| −2.00 = friction factor exponent | | | | | 100 = friction factor multiplier | | | | |

auto t(ij)	1	2	3	4		F(ij)	1	2	3	4
1	2.4	11.0	8.2	6.6		1	17.36	0.83	1.49	2.30
2	11.0	2.1	7.4	5.5		2	0.83	22.68	1.83	3.31
3	8.2	7.4	3.4	9.9		3	1.49	1.83	8.65	1.02
4	6.6	5.5	9.9	2.3		4	2.30	3.31	1.02	18.90

A(j)F(ij)	1	2	3	4	sum	T(ij)	1	2	3	4	Tot. P
1	0	4563	3978	5232	13,773	1	0	263	229	301	793
2	3179	0	4884	7534	15,597	2	233	0	358	552	1143
3	5721	10,083	0	2325	18,129	3	1831	3227	0	744	5803
4	8831	18,253	2729	0	29,813	4	1950	4030	603	0	6583
						Tot. A	4014	7521	1190	1598	14,322

Note: In Table 4.12, row labels (1 to 4) indicate origin zones; column labels (1 to 4) indicate destination zones. For example, the average auto travel time from zone 2 to zone 4 is 5.5 minutes.

THINK ABOUT IT

Although the travel time and friction factor portions of Table 4.12 are symmetric, the resulting trip matrix is not. Why is that?

An experienced transportation planner would not be satisfied with the trip matrix produced in Example 4.10. He or she would notice that we started with zonal attraction totals $A_1 = 3847$, $A_2 = 5521$, $A_3 = 2675$, and $A_4 = 2279$ and ended up with $A_1 = 4014$, $A_2 = 7521$, $A_3 = 1190$, and $A_4 = 1598$. The Gravity Model used in Example 4.10 has changed the zonal attraction totals for zones 2, 3, and 4 quite a lot. The remedies most likely to help are:

- Adjust the friction factor exponent in Equation 4.8 if trip lengths are not being represented correctly.
- Use the friction factor Equation 4.9, instead of Equation 4.8, if the actual trip length distribution is not monotonically decreasing.
- Look for another variable besides travel time to explain how travelers choose between multiple destination zones.

These are topics more suitable for coverage in a separate course in transportation planning.

4.4   MODE CHOICE

The Mythaca Bus Company (MBC) is experiencing financial difficulties. Ridership is falling as operating costs continue to increase. Other travel modes—drive alone auto and ridesharing—are increasing their mode shares (Figure 4.14). Three strategies for off-peak service are under active consideration by MBC:

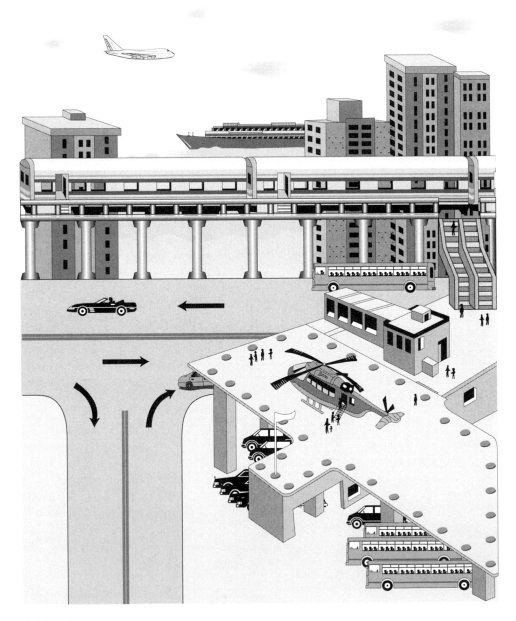

FIGURE 4.14
Five passenger modes in one view.

A. Increase fares from 75 cents to $1.00, in hopes of increasing revenues.

B. Decrease service frequency from four times per hour to twice per hour, to reduce operating costs.

C. Increase service to six times per hour, in hopes of attracting more ridership and more revenue.

Which of these alternate strategies would help MBC's financial situation the most?

4.4.1 Mode Choice and Utility

How many travelers will choose each mode of transportation?
In many urban areas, public mass transportation is not a major factor. In fact, nationwide, the percentages of urban commuting trips made by the various modes in 1999 were (HUD et al., 2000):

- 78.2 percent alone in privately operated vehicles
- 9.4 percent carpool
- 4.9 percent by bus and rail transit
- 3.1 percent walking
- 2.8 percent work at home
- 1.6 percent by bicycle, taxi, other modes.

Many persons ride public transit because they have to. They may be too young or too old to operate a motor vehicle. They may not be financially able to own and operate a motor vehicle, or they may not be physically able to do so. These persons are called *captive* transit riders, to distinguish them for modeling purposes from *choice* riders. Choice riders use transit because they have compared the attributes of all available travel modes and have decided that public transit is the best alternative for their trips. This decision to use public transit is most likely to occur in large cities, where traffic congestion and parking costs make using a private vehicle sufficiently unpleasant and expensive, and where public transit service is available and of adequate quality. Although this chapter continues to use examples applicable to an urban context, the reader should note that models similar to those introduced here could be applied to choices of travel modes (air, bus, private vehicle, or rail) for intercity trips.

The current *state of the practice* in mode choice modeling is the *multinomial logit (MNL) model*. The MNL model incorporates the notion that a traveler with a choice tends to choose the travel mode that has the greatest *utility* to him/her, but it also recognizes that (a) the utility may be difficult for the modeler to measure and (b) individual travelers may perceive the same mode choice alternatives in different ways. The result is a *utility function* of the form

$$U_m = a_{0,m} + a_{1,m}x_{1,m} + \cdots + a_{n,m}x_{n,m} + \varepsilon \qquad (4.10)$$

The utility function includes measurable variables $x_{i,m}$ that help explain the likelihood that a traveler in category i would choose mode m. Common examples of $x_{i,m}$ are the travel time and out-of-pocket cost for each mode. The $a_{i,m}x_{i,m}$ terms in Equation 4.10 comprise the *measurable utility* (V_m). The random variable ε accounts for factors that may influence a tripmaker's mode choice decision but that are not easily measured or observed. For instance, two otherwise identical individuals with the same available travel mode alternatives may place a different value on the ability to read the newspaper while commuting. As a result of these variations, each mode's utility to a particular individual will often be higher or lower than represented by the equation.

THINK ABOUT IT

Think of at least two other personal factors (like wanting to be able to read the newspaper) that might influence your choice of travel mode but does not directly involve mode attributes such as travel time or cost. Would your personal factors be observable and measurable?

The values of the coefficients $a_{i,m}$ in Equation 4.10 are the result of surveys taken of individuals as they consider real or hypothetical mode choices. The choices made by the persons surveyed and the values of the measurable variables $x_{i,m}$ for the available alternatives are fed into a statistical analysis. The software that carries out this analysis normally uses *maximum likelihood estimation (MLE)* techniques. Research has indicated that treating ε in Equation 4.10 as a random variable with a Gumbel distribution and an expected value of zero leads to the easy-to-use logit equation:

$$P_m = \frac{e^{V_m}}{\sum\limits_{j=1,k} e^{V_j}} \tag{4.11}$$

where P_m = the probability that an individual will choose mode m from a specified set of alternatives.

Example 4.11

Discouraged because only 6 percent of the workers at the new office park at the edge of Mythaca use the express bus service from a certain white-collar neighborhood, the Mythaca Bus Company asks the Mythaca Regional Planning Commission to conduct a survey of persons who are commuting to the new development. The MRPC finds that two factors affect commuter mode choice the most: *out-of-pocket cost (OPC)* and *total travel time (TTT)*. The MRPC computations for a MNL model result in a utility function $V_m = a_o - (0.47 * OPC_m) - (0.22 * TTT_m)$, where $a_{0,auto} = 0.73$, $OPC_{bus} = \$0.75$, $TTT_{auto} = 10.5$ minutes, and $TTT_{bus} = 18$ minutes. All other values are zero.

A. Does the MNL model developed by MRPC replicate the actual bus mode share?

B. According to the MNL model, what would the bus mode share be if the MBC reduces the bus fare to zero?

Solution to Example 4.11

A. Entering the coefficient and variable values provided by MRPC into the measurable utility part of Equation 4.10 leads to

$$V_{auto} = 0.73 - (0.47 * 0.00) - (0.22 * 10.5) = -1.58$$

$$V_{bus} = 0.00 - (0.47 * 0.75) - (0.22 * 18.0) = -4.31$$

Note that all of the $a_{i,m}$ coefficients in this version of Equation 4.10 are negative, and that both utilities are negative. Entering these V_m values into Equation 4.11 produces

$$P_{auto} = \frac{e^{V_{auto}}}{e^{V_{auto}} + e^{V_{bus}}} = \frac{e^{-1.58}}{e^{-1.58} + e^{-4.31}} = \frac{0.2060}{0.2060 + 0.0134} = 0.939$$

$$P_{bus} = \frac{e^{bus}}{e^{V_{auto}} + e^{V_{bus}}} = \frac{e^{-4.31}}{e^{-1.58} + e^{-4.31}} = \frac{0.0134}{0.2060 + 0.0134} = 0.061$$

According to the MRPC model, about 94 percent of the commuters would choose auto and only 6 percent would choose bus. This matches the observed mode share, so the MRPC model seems to be correctly specified. *Always check these mode share calculations. Find the sum of the P_m values that are calculated by using Equation 4.11. The sum should always be equal to 1.00.*

B. If MBC reduces its fares for this express service to zero, the auto utility equation remains unchanged, but the bus utility equation becomes $V_{bus} = 0.00 - (0.47 * 0.00) - (0.22 * 18.0) = -3.96$. When the P_m calculations using Equation 4.11 are redone, the mode shares are now

$$P_{auto} = \frac{e^{V_{auto}}}{e^{V_{auto}} + e^{V_{bus}}} = \frac{e^{-1.58}}{e^{-1.58} + e^{-3.96}} = \frac{0.2060}{0.2060 + 0.0191} = 0.915$$

$$P_{bus} = \frac{e^{bus}}{e^{V_{auto}} + e^{V_{bus}}} = \frac{e^{-3.96}}{e^{-1.58} + e^{-3.96}} = \frac{0.0134}{0.2060 + 0.0191} = 0.085$$

Even when no bus fare is charged, more than 91 percent of the commuters are expected to drive to the office park. Looking at Equation 4.10 to explain this strong preference for auto, only two reasons can be found. One is the 7.5-minute difference in total travel time. This difference probably includes walking to the bus stop and waiting for the bus—both activities that many people dislike. The other factor might be in the first term of Equation 4.10, the *mode-specific constant* $(a_{0,m})$. In this example $a_{0,bus} = 0.00$, but $a_{0,auto} = 0.73$. Unlike the ε term in Equation 4.10, which attempts to explain individual variations around the measurable utility V_m, the $a_{0,m}$ term reflects a general degree of preference for mode m that is not measured by the other $a_{i,m}\ x_{j,m}$ terms in Equation 4.10. The flexibility and/or comfort of driving one's own vehicle may give the auto mode an advantage that cannot otherwise be measured in the utility function, except in the mode-specific constant $a_{0,m}$. If the MRPC mode share model is correct, introducing fare-free service to the office park will not have a major impact on bus mode share.

THINK ABOUT IT

Notice that both utilities computed in Example 4.11 had negative values. Does it make sense for something to have negative utility? If so, offer an explanation.

4.4.2   Logit Model Attributes

When only two choices are involved, the structure of the logit model is easy to see. Consider two mode choices, A and B. A and B can be any modes, not necessarily auto and bus. According to Equation 4.11, mode A's share of the travelers will be

$$P_A = \frac{e^{V_A}}{e^{V_A} + e^{V_B}}.$$

Mode B's share will be

$$P_B = \frac{e^{V_B}}{e^{V_A} + e^{V_B}} \quad \text{or } P_B = 1 - P_A.$$

The P_A equation can be rewritten by dividing both the numerator and the denominator of its right-hand side by e^{V_A} to get

$$P_A = \frac{1}{1 + e^{V_B - V_A}} \qquad (4.12)$$

Note in Equation 4.12 that, when $V_B = V_A$, $P_A = 0.50$. This means that, when the utilities of two alternatives are equal, their mode shares are equal. In the case of an individual, he/she is equally likely to choose alternative A or alternative B. Figure 4.15

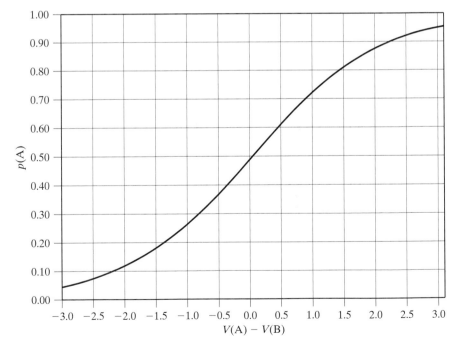

FIGURE 4.15
Plot of logit function for the two-alternative case.

shows the relationship between $V_A - V_B$ on the horizontal axis and P_A on the vertical axis. When $V_A - V_B = 0$, $P_A = 0.50$. If V_A becomes larger than V_B, then $V_A - V_B > 0$ and P_A will become larger than 0.50. If V_A is smaller than V_B, then $V_A - V_B < 0$ and P_A will be less than 0.50.

THINK ABOUT IT

In Example 4.11(a), the two modal utilities were -1.58 and -4.31. The difference in utilities was ± 2.73. Find the solution to Example 4.11(a) on the logit plot in Figure 4.15. If the two modal utilities in Example 4.11(a) had instead been -11.58 and -14.31, what would the mode shares have been? Find this point in the logit plot in Figure 4.15. What have you discovered?

Example 4.12

In this section's opening paragraph, three strategies for off-peak service were under active consideration by the Mythaca Bus Company:

A. Increase fares from 75 cents to $1.00, in hopes of increasing revenues.

B. Decrease service frequency from four times per hour to twice per hour, to reduce operating costs.

C. Increase service to six times per hour, in hopes of attracting more ridership and more revenue.

Which of these alternate strategies would help MBC's financial situation the most? Using data from the last mode choice survey done by the Mythaca Regional Planning Commission, MBC has developed the MNL mode choice model shown here: $U_m = a_o - (0.41 * OPC_m) + (0.24 * FREQ_m) - (0.68 * TTT_m)$. In this model, $FREQ_m$ = frequency per hour of service with mode **m**, and OPC and TTT have the same definitions as in Example 4.11. MBC wants to try the model in a corridor that is served by a toll road whose toll for *drive-alone (DA)* motorists is $0.50. Vehicles that have at least two occupants *[ridesharing vehicles, (RS)]* pay no toll. Buses operate in the toll road's median on an exclusive busway and pay no toll. In the MRPC mode choice model, $a_{0, DA} = 1.56$, $a_{0, RS} = 0.96$, $FREQ_{DA} = 12$, $FREQ_{RS} = 6$, $TTT_{DA} = TTT_{RS} = 25.3$ minutes, and $TTT_{bus} = 21.8$ minutes. The current bus mode share of the 545 off-peak travelers through the corridor in a typical off-peak hour is 0.206 (20.6 percent), which translates into 112 bus passengers. Use the MPRC model to evaluate the three strategies with respect to fare revenue changes.

Solution to Example 4.12 The corridor has three competing travel modes: drive-alone (DA) auto, ridesharing (RS), and express bus. This means that the utility equation must be applied three times:

$$V_{DA} = 1.56 - (0.41 * 0.50) + (0.24 * 12) - (0.68 * 25.3) = -12.969$$

$$V_{RS} = 0.96 - (0.41 * 0.00) + (0.24 * 6) - (0.68 * 25.3) = -14.804$$

$$V_{bus} = 0.00 - (0.41*0.75) + (0.24*4) - (0.68*21.8) = -14.172$$

$$P_{bus} = \frac{e^{V_{bus}}}{e^{V_{DA}} + e^{V_{RS}} + e^{V_{bus}}} = \frac{e^{-14.172}}{e^{-12.969} + e^{-14.804} + e^{-14.172}} = \frac{7.0*10^{-7}}{34.0*10^{-7}} = 0.206$$

The fare revenue from $0.206*545 = 112$ passengers is $112*\$0.75 = \$84.00$.

A. If bus fares rise to $1.00, $V_{bus} = 0.00 - (0.41*1.00) + (0.24*4) - (0.68*21.8) = -14.274$ and

$$P_{bus} = \frac{e^{V_{bus}}}{e^{V_{DA}} + e^{V_{RS}} + e^{V_{bus}}} = \frac{e^{-14.274}}{e^{-12.969} + e^{-14.804} + e^{-14.274}} = \frac{6.32*10^{-7}}{33.4*10^{-7}} = 0.190$$

Bus ridership would drop to $0.190*545 = 104$, which would produce $104*\$1.00 = \$104$ in fare revenue. Despite the ridership drop, the fare revenue increases $\$104 - \$84 = \$20$.

B. If bus service is reduced from four times in the hour to twice, the bus utility function changes to

$$V_{bus} = 0.00 - (0.41*0.75) + (0.24*2) - (0.68*21.8) = -14.65 \text{ and}$$

$$P_{bus} = \frac{e^{V_{bus}}}{e^{V_{DA}} + e^{V_{RS}} + e^{V_{bus}}} = \frac{e^{-14.65}}{e^{-12.969} + e^{-14.804} + e^{-14.65}} = \frac{4.33*10^{-7}}{31.4*10^{-7}} = 0.138$$

Bus ridership would decrease to $0.138*545 = 75$, which would produce $75*\$0.75 = \$56.25$ in fare revenue. Because the ridership drop is not accompanied by a fare increase, total fare revenues in the corridor will decrease.

C. If bus service is increased to six times per hour, the bus utility function becomes

$$V_{bus} = 0.00 - (0.41*1.00) + (0.24*6) - (0.68*21.8) = -13.69 \text{ and}$$

$$P_{bus} = \frac{e^{V_{bus}}}{e^{V_{DA}} + e^{V_{RS}} + e^{V_{bus}}} = \frac{e^{-13.69}}{e^{-12.969} + e^{-14.804} + e^{-13.69}} = \frac{1.13*10^{-6}}{3.84*10^{-6}} = 0.295$$

Bus ridership would increase to $0.295*545 = 161$ passengers. The resulting fare revenue would be $161*\$0.75 = \$120.75$. This is $36.75 more than the original fare revenues.

These revenue changes must be compared with the changes in operating costs that are associated with each strategy

---

THINK ABOUT IT

The coefficient for service frequency in the utility function in Example 4.12 had a positive sign. Why is that?

Example 4.13

In recent years, Murdoch Bay has become a popular place for wealthy people to spend their weekends. In recent years, some less familiar travel modes (viz., helicopter and hovercraft) have

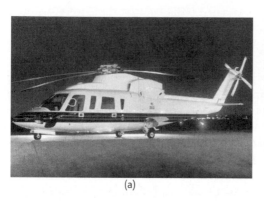 

(a)   (b)

FIGURE 4.16
Less familiar passenger modes. (a) Helicopter and (b) Hovercraft. *Sources:* (a) Photo courtesy of Avpro Inc., (b) Photo courtesy of Lynden Alaska Hovercraft.

attracted a surprisingly large number of vacationers who want to reach the far side of Murdoch Bay from Shoridan in a novel manner (Figure 4.16).

Local transportation planners have collected data about these services. The MNL model they developed for trips across Murdoch Bay from Shoridan is shown in the equation below.

$$V_m = a_m - 0.009 \, \text{Cost} - 0.057 \, \text{IVTT} - 0.061 \, \text{OVTT}$$

The modal attributes that are expected to be effective next summer are summarized in Table 4.13. What will be the mode shares P_m for each mode m next summer?

TABLE 4.13 Modal Attribute Data for Example 4.13

m	Mode	a_m	Cost ($)	IVTT (min)	OVTT (min)
1	Helicopter	0	85.40	5.35	33
2	Hovercraft	0.40	88.16	15.7	34

Note: *IVTT* = in-vehicle travel time; *OVTT* = out-of-vehicle travel time

Solution to Example 4.13   Substituting the appropriate values into Equation 4.10, we get

$$V_{\text{helicopter}} = 0 - 0.009(85.40) - 0.057(5.35) - 0.061(33)$$

$$= 0 - 0.7686 - 0.3050 - 2.0130 = -3.0866$$

$$V_{\text{hovercraft}} = 0.40 - 0.009(88.16) - 0.057(15.7) - 0.061(34)$$

$$= 0.40 - 0.7934 - 0.8949 - 2.0740 = -3.3623$$

Equation 4.11 allows us to convert these measurable utilities into the probability that each mode will be chosen:

$$P_{\text{helicopter}} = \frac{e^{-3.0866}}{e^{-3.0866} + e^{-3.3623}} = \frac{0.0457}{0.0457 + 0.0347} = \frac{0.0457}{0.0803} = 0.5685$$

$$P_{\text{hovercraft}} = \frac{e^{-3.3623}}{e^{-3.0866} + e^{-3.3623}} = \frac{0.0347}{0.0457 + 0.0347} = \frac{0.0347}{0.0803} = 0.4315$$

Note that $P_{\text{helicopter}} + P_{\text{hovercraft}} = 0.5685 + 0.4315 = 1.0000$, because these were the only two choices analyzed. In theory, the logit equation can be applied to three modes or to as many modes for which reliable data are available with which to estimate the coefficients in the utility equation.

Before moving on to another example, look again at the table of input values in Example 4.13. Both modal alternatives use the same coefficients in computing utilities, but the hovercraft's a_m value was 0.40, whereas the helicopter's a_m value was 0.00. The a_m term is called the *mode-specific constant*. It is produced by the same statistical software that estimates the best values of the coefficients for the variables $x_{i,m}$. The a_m term accounts for modal attributes (as perceived by the traveler) that are difficult or impossible to measure, such as comfort, convenience, and feelings of security.

Example 4.14

Consider the introduction of a third travel mode for trips across Murdoch Bay from Shoridan. To illustrate the strengths of the multinomial logit model, let the new mode be one that may appear to be frivolous—a hot air balloon (Figure 4.17). This new mode is used here to demonstrate that the MNL model only "cares" about the attributes of a mode, not its name or technology. If the attributes for the helicopter and hovercraft modes in Example 4.13 do not change, what will be the mode shares for the three modes now available to travelers? (See Table 4.14.)

FIGURE 4.17
Hot air balloons. Photo by Gary Tomlin, ABQ Photography; courtesy of Albuquerque International Balloon Fiesta®.

TABLE 4.14   Attributes of Third Mode in Example 4.14

m	Mode	a_m	Cost ($)	IVTT (min)	OVTT (min)
3	Hot air balloon	1.1	70.59	38.0	42

Solution to Example 4.14   Applying the MNL model begins with using the utility equation in Example 4.13:

$$V_{balloon} = a_{balloon} - 0.009 \text{ Cost} - 0.057 \text{ IVTT} - 0.061 \text{ OVTT}$$

$$V_{balloon} = 1.1 - 0.009(70.59) - 0.057(38.0) - 0.061(42)$$

$$V_{balloon} = 1.1 - 0.635 - 2.166 - 2.562 = -4.263$$

Nothing about the helicopter and hovercraft attributes has changed, so $V_{helicopter} = -3.0866$ and $V_{hovercraft} = -3.3623$, as in Example 4.13. In the denominator of Equation 4.11,

$$\sum_{j=1,3} e^{V_j} = e^{-3.0866} + e^{-3.3623} + e^{-4.263} = 0.0457 + 0.0347 + 0.0098 = -0.0902.$$

Using Equation 4.11, the predicted mode shares are

$$P_{helicopter} = \frac{0.0457}{0.0902} = 0.507, P_{hovercraft} = \frac{0.0347}{0.0902} = 0.385, \text{ and } P_{balloon} = \frac{0.0098}{0.0902} = 0.109$$

The sum of the three mode shares is 1.001, so the calculations seem to be correct. More importantly, the mode with the highest utility value has the highest mode share, and so on. As usual, it is wise to check the calculations for reasonableness.

THINK ABOUT IT

What percent of the non-hot air balloon travel market is captured by the helicopter mode in Example 4.14? How does this compare with the helicopter mode share in Example 4.13?

   If a fourth mode were to be considered for trips across Murdoch Bay from Shoridan, the attributes of the three previously existing modes (and their utilities) would remain as before. The attributes and utility of the fourth mode would be calculated, and Equation 4.11 would again be used to "divvy up" the customers. This ability to leave the prior modes' attributes and utilities unchanged is called the *independence of irrelevant alternatives* (*IIA*) property of the MNL model. It holds as long as the competing modes are truly independent of (i.e., distinct from) each other. For example, if one set of transit buses ran on diesel, and another set of buses ran on an alternative fuel, but each bus type had identical service attributes, it would be better to combine the two bus types into one mode.

4.4.3 Middleville Case Study

As in most small and medium-sized cities, the automobile mode in Middleville is dominant. However, some people must rely on modes other than auto, and the Mythaca Bus Company provides reasonable transit service in this community. It would be helpful to the MBC if it had a mode choice model with which to test some proposed changes in fares and service.

Example 4.15 Mode Choice in Middleville

The MRPC planners have assembled a simple MNL model that relies primarily on travel time as its basis. The utility function for trips between any pair of zones i and j is $U_m = a_m - (0.22 * t_{ij})$, where $a_{auto} = 0.9$ and $a_{transit} = 0.0$. Average auto travel times between zones were displayed in the upper left quadrant of Table 4.12 in the section on trip distribution. Average transit travel times between zones are displayed in the upper left quadrant of Table 4.15. What percent of trips between each zone pair will use the transit mode, if the only alternate mode is automobile?

Solution for Example 4.15 Using the MNL model is a two-step process: compute utilities for each mode using Equation 4.10 and then compute mode shares using Equation 4.11. This two-step process must be carried out for each zone pair for this example. For trips between zones 1 and 2, the utilities are

$$V_{1,2,auto} = 0.9 - (0.22 * t_{1,2,auto}) = 0.9 - (0.22 * 11.0) = 0.9 - 2.42 = -1.520 \text{ and}$$

$$V_{1,2,transit} = 0.0 - (0.22 * t_{1,2,transit}) = 0.0 - (0.22 * 22.8) = 0.0 - 5.016 = -5.016$$

The auto utilities for all zone pairs are stored in the lower left quadrant of Table 4.15; the transit utilities for all zone pairs are stored in the upper right quadrant of Table 4.15. The probability that transit will be used to make a trip from zone 1 to zone 2 is found by using Equation 4.11 again:

$$P_{transit} = \frac{e^{-5.016}}{e^{-5.016} + e^{-1.520}} = \frac{0.006631}{0.006631 + 0.218712} = \frac{0.006631}{0.225343} = 0.029$$

TABLE 4.15 Data and Calculations for Example 4.15

Transit $t(ij)$	1	2	3	4	Transit $V(m)$	1	2	3	4
1	8.1	22.8	15.0	18.2	1	−1.782	−5.016	−3.300	−4.004
2	22.8	8.2	13.4	14.9	2	−5.016	−1.804	−2.948	−3.278
3	15.0	13.4	7.5	13.2	3	−3.300	−2.948	−1.650	−2.904
4	18.2	14.9	13.2	7.1	4	−4.004	−3.278	−2.904	−1.562

Auto $V(m)$	1	2	3	4	$p(tr)$	1	2	3	4
1	0.372	−1.520	−0.904	−0.552	1	0.104	0.029	0.083	0.031
2	−1.520	0.438	−0.728	−0.310	2	0.029	0.096	0.098	0.049
3	−0.904	−0.728	0.152	−1.278	3	0.083	0.098	0.142	0.164
4	−0.552	−0.310	−1.278	0.394	4	0.031	0.049	0.164	0.124

Note: In Table 4.15, row labels (1 to 4) indicate origin zones; column labels (1 to 4) indicate destination zones. For example, the average transit travel time from zone 2 to zone 4 is 14.9 minutes.

This probability appears in the p(tr) portion (the lower right quadrant) of Table 4.15. The remaining p(tr) calculations were completed using a spreadsheet.

Transforming the Model Output. Before we proceed to the fourth, and final step, of the travel demand modeling process, two other tasks must be performed. The first task is to convert mode shares into person trips made using each mode in Middleville. This is accomplished by simply multiplying the $p(tr)$ transit mode share matrix in Table 4.15 by the $T(i,j)$ trip interchange matrix in Table 4.12. For example, the number of persons using transit from zone 3 to zone 1 can be computed as follows: $p_{31}(tr)*T(31) = 0.083*1831 = 152$ transit users. The person trips by auto from zone 3 to zone 1 is calculated as follows: $(1 - p_{31}(tr))*T(31) = (1 - 0.083)*1831 = 1679$. These values appear in the (3,1) locations in Table 4.16. [Because Table 4.16 is taken from a spreadsheet, the hand-calculated answers (152 and 1679) do not exactly match the spreadsheet computations (153 and 1678).] Using a spreadsheet, this first task of computing person trips can be finished quickly.

The second task involves converting person trips by auto into vehicle trips by auto. This recognizes that each automobile can carry more than one person. If the average vehicle occupancy is estimated by MRPC to 1.29 persons per vehicle, dividing each person trip entry in the auto half of Table 4.16 will produce the auto vehicle trip values in Table 4.17.

THINK ABOUT IT

What is an efficient way to estimate average vehicle occupancy? Do you think the value might vary by time of day and/or by zone pair?

TABLE 4.16   Calculating Interzonal Person Trips by Each Mode in Middleville

Auto Person Trips $T(i,j)$	1	2	3	4		Transit $T(ij)$	1	2	3	4	
1	0	255	210	292		1	0	8	19	9	
2	226	0	323	525		2	7	0	35	27	
3	1678	2911	0	622		3	153	316	0	122	
4	1890	3833	504	0	13,270	4	60	197	99	0	1052

Note: In Table 4.16, row labels (1 to 4) indicate origin zones; column labels (1 to 4) indicate destination zones. For example, the number of auto person trips from zone 2 to zone 4 is 525.

TABLE 4.17   Auto Vehicle Trips T_{ij} Between Middleville Zones

Origin Zones **i**	Destination Zones **j**				
	1	2	3	4	
1	0	198	163	226	
2	175	0	250	407	
3	1301	2257	0	482	
4	1465	2972	390	0	10,287

Now that the Middleville travel data have been converted into vehicle trips by travel mode, we can turn our attention to how travelers choose routes to reach their desired destinations.

4.5 TRIP ASSIGNMENT

Five or six times each autumn, crowds in excess of 50,000 people travel to Mythaca State University's football stadium to see a college football game (Figure 4.18). Under their normal configuration, the streets of Mythaca do not have the capacity to accommodate all the traffic that heads for the stadium. The University asks the Mythaca Regional Planning Commission to use its computer models to test some ideas its Athletic Department staff has about rerouting and controlling traffic. That is certainly a better way to decide on traffic control strategies than "trial and error."

4.5.1 Trip Assignment Concepts

What route or path will be taken by each tripmaker?

After three steps of the four-step method, we have established T_{ij}^m, the number of trips that go from zone i to zone j by mode m. What remains to be determined for each T_{ij}^m is which set of links in the transportation network will be used to reach zones j from zone i. The assumption on which most route choice models are based is that travelers will choose the origin-destination path that has the shortest travel time. (Common variations on travel time are distance, out-of-pocket cost, number of traffic signals, or a combination of such factors.) The simplest route choice model is called the *all or nothing (AON)* assignment model. It assigns *all* trips from origin zone i to destination zone j to the links along the path that has the shortest travel time. The AON method is simple but is not very realistic. Quite often, there are several different reasonable paths

FIGURE 4.18
Football stadium on game day. Photo courtesy of Purdue University Marketing Communications.

from zone i to zone j. This is true for transit trips in very large cities with well-developed transit service and for automobile trips in cities of even modest size. Furthermore, even when one path is clearly the fastest while traffic volumes are very low (free-flow conditions), if an increase in traffic causes this path to become congested, other routes may become attractive as alternatives. Accounting for increases in travel time as traffic flow increases is known as capacity restrained traffic assignment.

THINK ABOUT IT

Pretend that, as you read this, you suddenly remember that you had promised to meet a friend at a _____ across town. (Fill in the blank with a specific store, theater, restaurant, or other location across town from where you are now.) You need to drive to that location as soon as possible. Write down the streets you would use to reach that destination. Would everyone else take the same route as you would?

At low-traffic flow rates, travel time increases slowly. However, as traffic flow approaches the capacity of the links on the paths being used, travel time increases quickly. The most common function to represent the travel time effects of congestion is written as

$$t = t_0\left[1 + a\left(\frac{V}{C}\right)^b\right]$$
(4.13)

where t = travel time on link
  t_0 = free-flow link travel time
  C = capacity of link
  V = current   flow rate on link
  a and b are parameters.

The functional form in Equation 4.13 is known as the BPR or FHWA *volume delay function (VDF)*, because it was published in 1964 by the Bureau of Public Roads (BPR), a government agency whose responsibilities were later taken over by the Federal Highway Administration (FHWA). The 1964 BPR report used the values $a = 0.15$ and $b = 4.0$, and many transportation planners still use these parameter values in their models. In recent years, however, modelers have begun referring to VDFs as *link performance functions (LPFs)* and have begun to use different values for parameters a and b. An increasingly popular idea is to fit Equation 4.13 to the speed-flow curves in the *Highway Capacity Manual* (Feng and Gion, 1995; Singh, 1995), but an even better idea would be to fit the LPF parameters to the data observed in the study area being modeled (Fricker, 1989; Fricker and Moffett, 1993).

The FHWA LPF in Equation 4.13 should be applied to each link in a network. The capacity term "C" in the FHWA LPF is normally defined in terms of the upper limit of level of service (LOS) "C" and is called the "practical capacity." The link's LOS "E" (or "absolute") capacity is commonly taken from tables that provide default values for link types by functional class, location within the urban area, spacing of traffic signals, etc., but it could also be determined for each link more precisely by using techniques in the *HCM*. Although one could also determine the link's LOS "C" capacity using the *HCM*, it is usually sufficient to simply multiply the link's LOS "E" capacity by 0.75 to estimate the LOS "C" capacity.

Equation 4.13 can be rewritten as

$$\frac{t}{t_0} = \left[1 + a\left(\frac{V}{C}\right)^b\right]. \tag{4.14}$$

Using the traditional values $a = 0.15$ and $b = 4.0$, the resulting LPF looks like the curve in Figure 4.19 that has diamond-shaped markers. A few variations on the standard LPF are also shown in the figure.

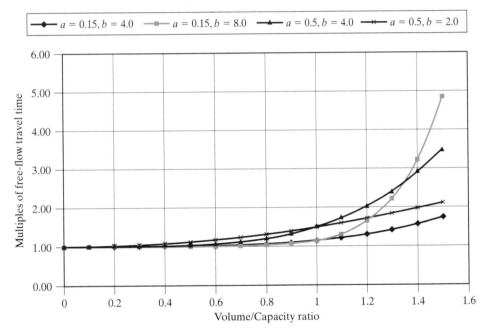

FIGURE 4.19
Link performance functions as parameters *a* and *b* vary.

Note that the form of Equation 4.13 allows travel times to be computed for capacity values that are greater than those physically possible. This is actually a desirable feature, because a route choice model's initial "guess" may put too much traffic on some links. As long as the LPF penalizes a link with a very large travel time for having had more traffic assigned to it than is actually possible, further iterations of the model will correct for this.

Example 4.16

An urban principal arterial that has a capacity of 1400 vphpl and a length of 1.0 mile.

A. What is the LOS "C" capacity of the principal arterial?

B. If the link's free flow speed is 45 mph and the standard values of **a** and **b** are used in Equation 4.13, what is the link travel time for traffic flow rates $V = 0$, $V = 500$, $V = 1000$, and $V = 1500$? Plot these points and the curve through them.

C. Recent research (Feng and Gion, 1995) has suggested that $a = 0.76$ and $b = 5.1$ be used for an urban arterial. Repeat part B with these values. How do the two LPFs differ?

Solutions for Example 4.16

A. LOS "C" capacity $= 0.75 * $ LOS "E" capacity $= 0.75 * 1400 = 1050$ vph

B. The link's free-flow travel time t_0 is 1.0 mile/45 mph $* 3600$ sec/hr $= 80.0$ seconds. A spreadsheet was used to apply Equation 4.13 and produce the table of link travel times below (Table 4.18) for each of the flow rates of interest, for both this part and part C. The values are summarized in Table 4.18 and plotted in Figure 4.20.

C. The suggested revisions to parameters **a** and **b** make little difference at low flow rates, but they produce significantly larger travel times as flow rates approach practical capacity (i.e., $V/C = 1.0$). Beyond that capacity, the travel times increase much more rapidly than for the standard FHWA LPF.

TABLE 4.18 Link Travel Time (seconds) as Values of LPF Parameters **a** and **b** Change

a	b	V = 0	V = 500	V = 1000	V = 1500
0.15	4.0	80.00	80.62	89.87	129.98
0.76	5.1	80.00	81.38	127.41	454.89

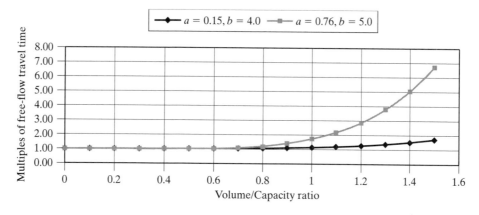

FIGURE 4.20
Two LPFs for Example 4.16.

THINK ABOUT IT

Does the general shape of the LPFs in Figures 4.19 and 4.20 remind you of a curve you have seen in Chapter 2 of this text? Which curve is it? Explain the similarities and differences.

4.5.2   User Equilibrium

Having established the role of the LPF, one more important concept in route choice modeling can be introduced: *user equilibrium*. To keep the first example of equilibrium simple, let us consider a corridor with just two reasonable paths, each with a linear LPF:

$$t_A = 8.8 + 0.4\,V_A; \quad t_B = 7.0 + 0.8\,V_B \tag{4.15}$$

In Equation 4.15, the units of flow are thousands of vehicles per hour. According to a minimum-time route choice model, the first driver (when flow $V = 1$) will choose path B because its free-flow travel time is 7.0 minutes, compared to 8.8 minutes on path A. In fact, all subsequent drivers will choose path B, until $t_B = 8.8$, which is path A's free-flow travel time. To determine the number of such drivers, find V_B in the equation $t_B = 7.0 + 0.8\,V_B = 8.8$. The solution, $V_B = 2.25$, means that the first 2250 drivers will choose path B as the shorter time path, but the 2251st driver will find that path B is no longer the shortest time path. Any subsequent driver (according to the route choice model) will choose the path that has the shorter travel time for the flow levels that prevail at the time of the decision. In other words, once both paths are in use, the user equilibrium condition states that the travel time on both paths will remain essentially equal.

Example 4.17

In the two-path corridor described above, 3800 vph are making the trip. What will be the values of V_A and V_B, if the two paths are found to be in equilibrium? What are the corresponding travel times t_A and t_B?

Solutions for Example 4.17   Use $V_A = 3.8 - V_B$ in the LPF for path A, then set that LPF equal to path B's.

$$8.8 + 0.4(3.8 - V_B) = 7.0 + 0.8\,V_B;$$

$$V_B = \frac{8.8 + 1.52 - 7.0}{0.8 + 0.4} = 2.767; V_A = 3.8 - 2.767 = 1.033$$

$$t_A = 8.8 + 0.4\,V_A = 8.8 + 0.4(1.033) = 9.21 \text{ minutes}$$

$$t_B = 7.0 + 0.8\,V_B = 7.0 + 0.8(2.767) = 9.21 \text{ minutes}$$

The path time calculations for t_A and t_B not only verify that the path flow calculations were done correctly, they also verify that the two paths in the corridor are in equilibrium.

The graphic representation of what was analyzed in Example 4.17 is shown in Figure 4.21. Route B has a lower free-flow travel time (when $V = 0$) than does route A, but after $V_B = 2.25$, the travel time on route B becomes greater than route A's free-flow travel time. Normally, when two lines cross in a plot, it has special meaning. In Figure 4.21, all it means is that both routes have the same travel time and the same flow rate. For equilibrium to be in effect, all used routes must have the same travel time. In most cases, the flows on the routes in use will not be the same. See the results for Example 4.17 to verify

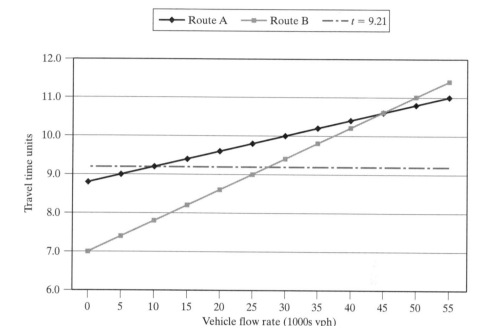

FIGURE 4.21
Seeking equilibrium traffic assignment for two routes.

this statement. A dashed horizontal line at $t = 9.21$ has been added to Figure 4.21 to illustrate what user equilibrium means. According to the principle of user equilibrium, travelers choose routes so that no change of routes will improve their travel times. The result is the same travel time on each used route. A vertical line drawn down from each point where the $t = 9.21$ horizontal line crosses an LPF will intersect the horizontal axis at a flow rate for a route. The sum of the flow rates found this way must equal the total flow using the routes. An analyst could draw horizontal lines in Figure 4.21 by trial and error and then compute the corresponding route-specific flows and total flows until the actual total flow was duplicated, but the mathematical approach in Example 4.17 is more convenient and efficient.

An important concept in user equilibrium is whether a particular route is being used. In Example 4.17 and Figure 4.21, it is clear that route A will not be used until the travel time on route B reaches at least 8.8 time units. Computing the flow on route B at which this situation occurs is not very difficult. It may not be so easy, however, when three or more routes are involved. Fortunately, a systematic procedure has been developed to assist in such cases (Lewis, 1993). The Lewis method is a way to check which routes are in use when more than two routes are involved. In the steps below,

n = number of possible routes between origin A and destination B

q_{OD} = total flow between origin O and destination D

The Lewis Method to Determine Route Use

1. Identify the route $\mathbf{r'}$ between origin A and destination B that has the worst free-flow travel time $t_r^0; r = 1, n$.
2. For each other route $\mathbf{r} \neq \mathbf{r'}$, determine $V_r(t_{r'}^0)$. This is the flow rate on a route that would have the same travel time as the free flow travel time found in step 1.
3. If $\sum_{\forall r \neq r'} V_r(t_{r'}^0) > q_{OD}$, then route r' will not be used. This means that there is not enough existing flow to make the routes with longer free-flow travel times competitive. In such cases, return to step 1 with $n = n - 1$. Otherwise, all remaining routes will be used, so there is no need to continue the route use check.

Example 4.18 will demonstrate the Lewis method.

Example 4.18

As travel increases in the corridor described in Example 4.17, other paths are being tried. A county road that used to have a gravel surface has recently been paved with asphalt. Its free-flow travel time through the corridor is 6.0. Because the county road, even with its new surface, is easily congested, its LPF is $t_C = 5.8 + 1.7\, V_C$.

A. If the flow rate in the corridor is the same as it was in Example 4.17 (3800 vehicles per hour), will all routes get used?

B. How much flow will be on each used route?

Solutions to Example 4.18

A. Apply the Lewis method to routes A, B, and C in the corridor.

1. Route A has the worst free-flow travel time of the three routes. $t_A^0 = 8.8$.
2. $V_B(t_A^0)$: Find the flow rate on route B, such that $t_B = t_A^0 = 8.8$.

$$t_B = 7.0 + 0.8\, V_B = 8.8;\ V_B = \frac{8.8 - 7.0}{0.8} = 2.25 \text{ when } t_B = t_A^0 = 8.8$$

$V_C(t_A^0)$: Find the flow rate on route C, such that $t_C = t_A^0 = 8.8$.

$$t_C = 5.8 + 1.7\, V_C = 8.8;\ V_C = \frac{8.8 - 5.8}{1.7} = 1.765 \text{ when } t_C = t_A^0 = 8.8$$

3. $\sum_{\forall r \neq r'} V_r(t_{r'}^0) = V_B(t_A^0) + V_C(t_A^0) = 2.25 + 1.765 = 4.015$. The total flow in the corridor would have to be 4105 vph for route A to have a travel time competitive with routes B and C. However, the flow in the corridor now, $q_{OD} = 3.8 * 10^3$, is not that high. Route A will not be used. What about the remaining routes? Return to step 1.

1. Route B has the worst free-flow travel time of the two remaining routes. $t_B^0 = 7.0$.
2. $V_C(t_B^0)$: Find the flow rate on route C, such that $t_C = t_B^0 = 7.0$.

$$t_C = 5.8 + 1.7\, V_C = 7.0;\ V_C = \frac{7.0 - 5.8}{1.7} = 0.706 \text{ when } t_C = t_B^0 = 7.0$$

3. $\sum_{\forall r \neq r'} V_r(t_r^0) = V_C(t_A^0) = 0.706$. The total flow in the corridor would have to be 706 vph for route B to have a travel time competitive with route C. Because the flow in the corridor now, $q_{OD} = 3.8 * 10^3$, exceeds that level, route B will be used. So will any other route with a free-flow travel time better than route B's.

B. Now that we know only routes C and B will be used at the current flow levels in the corridor, we can use $V_C = 3.8 - V_B$ in the LPF for path C, then set that LPF equal to path B's.

$$5.8 + 1.7(3.8 - V_B) = 7.0 + 0.8V_B;$$

$$V_B = \frac{5.8 + 6.46 - 7.0}{0.8 + 1.7} = 2.104; V_C = 3.8 - 2.104 = 1.696$$

$$t_C = 5.8 + 1.7 V_C = 5.8 + 1.7(1.696) = 8.68 \text{ minutes}$$

$$t_B = 7.0 + 0.8 V_B = 7.0 + 0.8(2.104) = 8.68 \text{ minutes}$$

Again, we calculate the path times t_A and t_B to "check the math" and to verify that the two paths in the corridor are in equilibrium. Note that the equilibrium travel time (8.68 minutes) is below the free-flow travel time on unused route A (8.80 minutes). See Figure 4.22 for a plot of the solution.

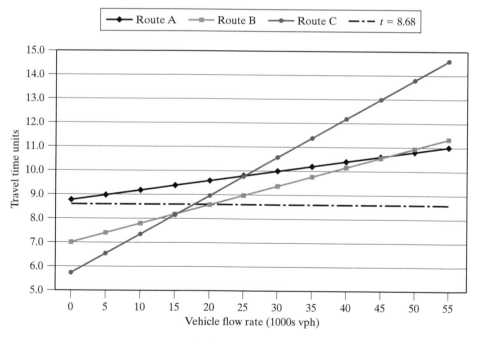

FIGURE 4.22
Seeking equilibrium traffic assignment for three routes.

THINK ABOUT IT

In Example 4.17, the equilibrium travel time for the two-route case was 9.21 minutes. In Example 4.18, a third route was added. Route A became so inferior at the prevailing flow rate that it was no longer used, yet the equilibrium travel time fell to 8.68 minutes. Does this make sense?

To this point, we have examined traffic assignment using linear LPFs. Sometimes it is possible to use a spreadsheet to help solve a problem where the travel time function is nonlinear such as in Equation 4.13. What can be done is to set the LPFs up and find the traffic flow on each route, such that the travel times on all used routes are equal.

Example 4.19

Two possible routes from an origin O to a destination D are shown in Figure 4.23. The northern route has a shorter distance, but it has less capacity. The link performance functions that govern the north and south routes are

$$t_N = 16 * \left[1 + 0.76 \left(\frac{5000 - V}{2400} \right)^5 \right] \text{ and } t_S = 20 * \left[1 + 0.76 \left(\frac{V}{3200} \right)^5 \right].$$

5000 cars leave the origin O at the speeds indicated in the figure. How many of the 5000 vehicles will use the north route and how many will use the south route, such that travel times on the north and south routes will be equal?

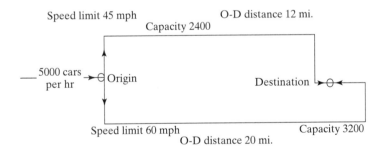

FIGURE 4.23
Trip assignment equilibrium with non linear LPFs.

Solution to Example 4.19 The table below is the spreadsheet set up to solve by trial and error the flow on each route such that the total O-D flow equals 5000 vehicles and the two route travel times match. A convenient alternative is to use the Solver feature in the spreadsheet. Both solution methods show that $V_S = 2680$ vehicles will choose the longer distance (but faster speed) southern route and 2320 cars will take the northern route. The equilibrium travel time for either route is about 26 minutes.

Trial V_S	t_S	t_N
2000	21.44958	53.10938
3000	31.00779	20.88683
2700	26.49999	25.82916
2650	25.92002	26.94502
2670	26.14681	26.48714
2680	26.26279	26.26402

Traffic Assignment in Current Practice

The traffic assignment examples given so far in this lesson are, of course, greatly simplified. For one thing, travelers probably use other criteria besides travel time in choosing their routes. Even if travel time is the principal criterion, different travelers may perceive the same travel time differently. Although tolls and other factors can be combined to replace travel time with a generalized cost value, most analysts are satisfied that travel time gives a result that meets their requirements. The other complication is that any link in a street network is likely to have traffic on it that is coming from many different origins and going to many different destinations. The examples in this section implied that all travelers in the corridor had the same origin and destination, or that the origin and destination did not matter. Fortunately, efficient algorithms have been incorporated in standard travel demand modeling software that can implement user equilibrium traffic assignment for an entire transportation network.

The results of the traffic assignment step are critical for two reasons. (1) The link flows in the results for the existing network are usually the best indication of how well the travel demand model has replicated existing travel patterns. Traffic counts on links are much easier to collect and are more reliable than the data collected for trip generation and trip distribution. If there are major discrepancies, the modeler must look throughout the four-step process for input and modeling errors. (2) The flow patterns produced by the model's forecasts usually form the basis for public investment decisions involving large amounts of public funds. If the forecast shows future demand far exceeding present capacity in some portion of the transportation system, a decision regarding major increases in capacity or measures to reduce travel must be made. The LOS values in Figure 4.1 were based on a four-step process that ended with trip assignment.

Traffic assignment also serves as the basis for other important analyses, as Example 4.20 will illustrate.

Example 4.20

Carrying out a full-scale traffic assignment on the Middleville road network is beyond the scope of this text. If traffic assignment had been completed for the Middleville study area, travel time and speeds could be estimated for roads that connect the zone pairs. Table 4.19 shows such interzonal speeds. State and federal environmental agencies require that the

TABLE 4.19 Average Interzonal Auto Speeds (mph)

Origin Zones	Destination Zones			
	1	2	3	4
1	13	19	17	17
2	30	15	36	27
3	32	24	17	22
4	20	26	27	19

MRPC provide estimates of the amount of travel that takes place in the Middleville study area. Two ways to measure the amount of travel are *vehicle hours of travel (VHT)* and *vehicle miles of travel (VMT)*. Calculate the VHT and VMT totals for interzonal trips in the Middleville study area.

Solutions for Example 4.20 VHT estimates: In the absence of congested interzonal travel times from a completed traffic assignment, let us use the free-flow travel times in Table 4.12 and the vehicle trips by auto in Table 4.17. For auto trips from zone 3 to zone 4, the free-flow travel time is 9.9 minutes. There were 482 auto trips from zone 3 to zone 4. Therefore, the VHT for this zone pair is

$$\frac{9.9 \text{ minutes}}{60 \text{ minutes/hour}} * 482 \text{ vehicles} = 80 \text{ veh-hrs.}$$

VMT estimates: The VMT estimate for the $(3, 4)$ zone pair uses the 22 mph average speed found in Table 4.19. The VMT estimate for $(3, 4)$ is 80 vehicle hours $* 22$ mph $= 1750$ VMT. The total interzonal VMT for the study area is 30,738 vehicle miles. The total VHT is 1217 vehicle hours. See Table 4.20 for a summary of the VHT and VMT calculations.

TABLE 4.20 VHT and VMT Calculations for Middleville Zone Pairs

	Interzonal VHT Using Auto $t(ij)$				Total VHT		Interzonal VMT Using Auto $t(ij)$ and Speed (ij)				Total VMT
	1	2	3	4			1	2	3	4	
1	0	36	22	25		1	0	689	378	423	
2	32	0	31	37		2	964	0	1111	1007	
3	178	278	0	80		3	5690	6680	0	1750	
4	161	272	64	0		4	3224	7082	1739	0	
					1217						30,738

THINK ABOUT IT

What is the average travel speed for all interzonal trips in the network?

4.5.4   Beyond the Current Practice

Having been introduced to the standard four-step travel demand modeling process, the reader is asked to consider again the notion that a typical person (even subconsciously) makes travel decisions in such a sequential manner. Some researchers, bothered by this drastic simplification, have looked into trying to combine some of the steps. Evans (1976) investigated the merging of the trip distribution and assignment steps, and Hicks et al. (1985) looked into combined mode and route choice models. Other modelers appreciate the strengths of the logit model as applied to the mode choice problem so much, that they prefer treating the trip distribution step as a destination choice phenomenon (Mannering and Kilareski, 1998).

Another innovation in travel demand modeling stems from the recognition that many trips are not simply from point A to point B and back to point A. Increasingly, persons are linking their trips together into trip chains. A common example is the personal errand, such as stopping at a store, on the way home from work. This person's Home-Work-Shop-Home trip chain would be treated as three separate trips under the standard four-step methodology. In theory, each trip in the chain would involve a fresh look at the mode choice question, whereas in reality, the mode used by the traveler in the next trip would be enormously dependent on the mode used in the previous link in the trip chain. This approach to the travel decision problem is called *Activity-Based Modeling* (Kitamura, 1988).

Although the four-step method has been an acceptable basis for transportation planning for several decades, the desire for a fresh approach that addresses transportation planners' frustrations, incorporates planning lessons learned, and takes advantage of today's considerable computing power has led to the Travel Model Improvement Program (TMIP) sponsored by the U.S. Department of Transportation. A major element of TMIP is the TRANSIMS microsimulation. TRANSIMS "tracks every car, every driver, every stoplight, acceleration, deceleration, braking average and turning ... for one second intervals" (Shunk and Bass, 1995). That is quite a departure from the grouped data used in most models today.

Despite these developments, the four-step process will be the standard for the foreseeable future. Applied with adequate data, skill, and judgment, it will continue to be a reasonable basis for sound transportation planning.

SUMMARY

The transportation demand modeling process attempts to replicate the tripmaking decisions made by individuals under observable circumstances, in the hope that a forecasting tool can be developed. If certain inputs pertaining to present tripmaking can be converted into the travel patterns that are taking place today, perhaps those same relationships can be used to predict how future conditions can affect the travel patterns of the future. The "future" can be very short-term, in the case of "What happens if that bridge is closed next week?" More often the future is on the distant horizon, as in "Where will congestion be the worst in 25 years if current growth continues and no transportation projects are undertaken?" Having reliable travel models gives decision makers the ability to see—at least approximately—the consequences of major public investments. This ability helps decide between competing projects. It also can illustrate the consequences of not taking certain action. Travel demand modeling is an important element of the transportation-planning process, but it is not the only component. The broader view of transportation planning is presented in the next chapter.

ABBREVIATIONS AND NOTATION

$a_{0,m}$ or a_m	mode-specific constant in MNL utility equation
$a_{i,m}$	coefficients in MNL utility function
A	trip attractions
AADT	annual average daily traffic
AON	all or nothing trip assignment
BPR	Bureau of Public Roads
DOT	Department of Transportation
ε	the random variable in the total utility function that accounts for factors that may influence a tripmaker's mode choice decision, but that are not easily measured or observed
empl	employment = jobs in zone
FHWA	Federal Highway Administration
F_{ij}	friction factor function for Gravity Model
G/M	Gravity Model
HBW	home-based work trips
HH	households in zone
IIA	independence of irrelevant alternatives
ITE	Institute of Transportation Engineers
IVTT	in-vehicle travel time
LOS	level of service
LPF	link performance function
MLE	maximum likelihood estimation
MNL	multinomial logit model
MRPC	Mythaca Regional Planning Commission
M/S	modal split or mode choice
O-D	origin-destination
OPC	out-of-pocket cost
OVTT	out-of-vehicle travel time
P	trip productions
P_m	probability that mode m would be chosen; proportion of travelers who would choose mode m
pop	population in zone
T	trip ends per specified time unit (weekday, site peak hour, etc.) at a particular land use
T/A	traffic assignment or trip assignment
TAZ	traffic analysis zone
T/D	trip distribution
T/G	trip generation
T_{ii}	intrazonal trips
T_{ij}	trip interchanges
T_{ij}^m	the number of trips that go from zone i to zone j by mode m
t_{ij}	travel time from zone i to zone j
TMIP	Travel Model Improvement Program

t_0	free-flow travel time
U_m	total utility of mode m
VDF	volume-delay Function
vehs	vehicles owned by HHs in zone
VHT	vehicle hours of travel
V_m	measurable utility of mode m
VMT	vehicle miles of travel
X	independent variable in ITE *Trip Generation* equations
X-E	external-internal
X-I	external-internal
X_i	a factor (usually demographic) that explains the level of tripmaking
$x_{i,m}$	measurable independent variables that help explain the likelihood that a traveler in category i would choose mode m.
X-X	external-external

GLOSSARY

Base year: The year used as the starting point for travel demand forecasts; usually a recent year for which data are available.

Capacity restrained traffic assignment: Accounting for increases in travel time as traffic flow increases

Cross-classification: A trip generation method that organizes household trip rates in accordance with certain household characteristics

Derived demand: The recognition that a trip is made because of the activities to be undertaken at the end of the trip.

Free-flow travel time: The time that an average driver would take to traverse a link or path if no other vehicles were present.

Friction factor: A means of converting travel times or other measures of separation between a trip's origin and possible destinations for use in the Gravity Model.

Gravity Model: A trip distribution method that is based on the relative spatial separation of traffic analysis zones and the relative amount of activity in the destination zones.

Horizon year: The specified year for which a forecast is made; usually 5, 10, or 20 years into the future.

Link performance function: A function that estimates how travel time on a link will vary as the flow on that link varies.

Multinomial logit model: A mode choice model that uses the utility of competing modes to estimate the share of tripmakers that each mode will earn.

Traffic analysis zone: A geographic subset of a study area.

Traffic assignment: A procedure by which traveler route choice in a street or highway network is simulated.

Trip assignment: The generic form of traffic assignment, to include transit assignment if the transit network offers more than one reasonable way to reach a destination.

Trip distribution: A procedure to describe and explain how travelers choose their destinations.

Trip generation: A procedure to estimate how many trips are made to and from certain locations, such as households, individual sites, or TAZs.

Trip interchanges: The number of trips that go from zone i to zone j

Trip matrix: A matrix that summarizes the number of trips that go between zone i (represented by the ith row of the matrix) and zone j (represented by the jth row of the matrix).

Trip table: Another name for a trip matrix.

User equilibrium: Travelers choose routes so that no change of routes will improve their travel times.

Volume-delay function: The older term for link performance function.

REFERENCES

[1] Area Planning Commission, Tippecanoe County, Indiana, *Transportation Plan for 2025, Greater Lafayette Area Transportation and Development Study*, May 2001.

[2] BPR. Bureau of Public Roads, U.S. Department of Commerce, *Calibrating & Testing a Gravity Model for Any Size Urban Area*, July, 1963.

[3] BPR. Bureau of Public Roads, U.S. Department of Commerce, *Traffic Assignment Manual*, June, 1964. Retrieved from www.bts.gov/publications/tsar/2000/

[4] BTS 2000. Bureau of Transportation Statistics, Transportation Statistics Annual Report 2000, Chapter 4 - Mobility and Access to Transportation, Table 1-Mode of Travel to Work: 1999. Retrieved from www.bts.gov/publications/tsar/2000/ 27 October 2002.

[5] Buttke, Carl H., "Guidelines for Using Trip Generation Rates or Equations," *ITE Journal*, p. 14–16, August, 1990.

[6] Evans, Suzanne P. "Derivation and Analysis of Some Models for Combining Trip Distribution and Assignment," *Transportation Research*, Vol. 10, 1976, p. 37–57.

[7] Feng, Li-Yang and Lisa Gion, "Traffic Assignment: BPR to HCM (Part I)—Denver's Application," *Compendium of Papers, Fifth National Conference on Transportation Planning Methods Applications—Volume I*, p. 3–5 to 3–14, April, 1995. Transportation Research Board.

[8] FHWA, *Guidelines for Trip Generation Analysis*, Federal Highway Administration, U.S. DOT, 1967. Reprinted April 1973.

[9] Fricker, Jon D., "Two Procedures to Calibrate Traffic Assignment Models," *Proceedings, Second Conference on Application of Transportation Planning Methods*, Orlando, FL, April, 1989. Transportation Research Board.

[10] Fricker, Jon D. and David P. Moffett. "Traffic Assignment Model Calibration When Precision is Essential," *Compendium of Technical Papers*, 1993 Annual Meeting, Institute of Transportation Engineers, The Hague, The Netherlands, September, 1993.

[11] HCM, *Highway Capacity Manual*, Special Report 209, Transportation Research Board, 2nd and 3rd editions, 1985 and 1994.

[12] Hicks, James E., David E. Boyce, and Lars Lundqvist. "Application of Combined Models of Mode and Route Choice in Stockholm, Sweden," presented at the Meeting of the Operations Research Society of America, Atlanta, GA, November, 1985.

[13] HUD et al. 2000. U.S. Department of Housing and Urban Development and U.S. Department of Commerce, U.S. Bureau of the Census, American Housing Survey for the United States: 1999, H150/99 (Washington, DC: 2000), as cited in BTS 2000.

[14] ITE, *Trip Generation*, Informational Report, Institute of Transportation Engineers, 6th edition, Washington, DC, 1997.

[15] Kitamura, R., "An Evaluation of Activity-Based Travel Analysis," *Transportation*, Vol. 15, Nos. 1–2, 1988, p. 9–34.

[16] Lewis, Michael A., Class discussion, CE361 Introduction to Transportation Engineering, Purdue University, Fall Semester, 1993.

[17] Mannering, Fred L. and Walter P. Kilareski, *Principles of Highway Engineering and Traffic Analysis*, 2nd ed., John Wiley and Sons, 1998.

[18] NCHRP167, *Transportation Planning for Small Urban Areas*, National Cooperative Highway Research Program Report 167, Transportation Research Board, 1976, p. F-4.

[19] NCHRP365, *Transportation Estimation Techniques for Urban Planning*, National Cooperative Highway Research Program Report 365, Transportation Research Board, 1998.

[20] Shunk, Gordon A. and Patricia L. Bass, *Travel Model Improvement Program Conference Proceedings*, Fort Worth, TX, August 14–17, 1994, Report DOT-T-95-13, 1995.

[21] Singh, Rupinder, "Beyond the BPR Curve: Updating the Speed-Flow and Speed-Capacity Relationships in Traffic Assignment," *Compendium of Papers, Fifth National Conference on Transportation Planning Methods Applications–Volume II*, April 1995, p. 3–17 to 3–26.

[22] Tanner, J.C., Factors Affecting the Amount of Travel, Department of Scientific and Industrial Research, *Road Research Technical Paper No- 51*, London, 1961. As cited in BPR (1963).

EXERCISES FOR CHAPTER 4: MODELING TRANSPORTATION DEMAND AND SUPPLY

Land Use and Tripmaking

4.1 Trip Generation Using the ITE Report. (*Note*: This problem requires access to the ITE *Trip Generation* report or copies of the appropriate pages in the report.) Three firms whose activities could be categorized as "general light industry" have plans to build new facilities on adjacent parcels of land on the outskirts (viz., in zone 10) of Mythaca. The materials testing laboratory will have about 52 employees, the book-binding company will employ about 378 people, and the electronic equipment firm will have a workforce of 896. All firms plan to have one work shift, from 8 AM to 5 PM. Using both the "Average Trip Rate" and "Fitted Curve Equation" approaches, estimate the number of average weekday vehicle trip ends that will be generated by each of the three firms. If the two results for any given site are not in close agreement, explain how you would choose a value of T to use.

4.2 Trip Generation. The ITE *Trip Generation* report was used to convert three different independent variables into forecasts of average vehicle trip ends per weekday that would be generated by a new automobile assembly plant. The results are shown in the table below. A colleague in your consulting firm says, "Be conservative. Always use the highest value." What is a good, concise argument against such a philosophy?

Independent variable	T
Acres of land	33,800
Employees	3360
Gross floor area	8900

4.3 Trip Generation by ITE Method. An automobile company is planning to build a 1.8 million square foot factory in zone 1 in Middleville. Estimate the expected number of trip ends at the factory on an average weekday.

4.4 Transportation and Land Use

(a) How does transportation planning influence land use?
(b) How does land use influence transportation planning?

Trip Generation

4.5 Zone-Based Regression for Trip Generation—Shoridan.   The new census figures are available. The number of vehicles owned in each zone in the Shoridan area has been entered into the table below, along with base year estimates of the number of trips made in a typical day by all households in each TAZ. Use regression analysis to create an equation that can be used to forecast trips by households in a zone in a future year, given an estimate of vehicle ownership in that zone.

Trip Generation Data for Exercise 5					
TAZ	**Trips by HHs**	**Vehicles**	**TAZ**	**Trips by HHs**	**Vehicles**
1	3860	427	6	4578	527
2	13,338	1187	7	9519	931
3	4043	710	8	8861	929
4	6468	672	9	9337	734
5	2434	373	10	3920	571

4.6 Zone-Based Regression for Trip Generation—Winder.   The table of data below reflects the home-based trips plus other demographic data from the small city of Winder. The city is divided into eight zones for the purpose of trip generation. Determine the coefficients of the best regression equation using the data if the equation is of the form:

$$\text{TRIPS} = a + b*(\text{income}/1000) + c*(\text{persons/HH}) + d*\text{Log(Density)}$$

(a) Determine a, b, c, and d. Set it up on a spreadsheet.
(b) What is the coefficient of determination?
(c) What is the coefficient of correlation?

Zone	Trips/DU	Avg HH Income	Persons/HH	log (DU/Acre)
1	5.6	19,600	3.3	1.62
2	7.5	30,600	2.9	0.76
3	6.3	24,900	3.3	1.30
4	10.0	41,500	3.6	0.50
5	9.1	37,800	3.4	0.90
6	7.1	29,800	2.7	1.55
7	12.1	42,000	3.9	0.30
8	4.7	21,000	3.4	1.75

4.7 Trip Generation by Regression.   An automobile company is planning to build a 1.8 million square foot factory in zone 1 in Middleville. The factory is expected to employ 2500 workers. Using the production and attraction equations that were "borrowed" for Example 4.4, calculate new values for productions and attractions for zone 1. How many new trip ends are attributable to the new factory?

4.8 Trip Generation by Cross-Classification.   Zone 8 in Mythaca is largely residential. Your client has provided the zone's expected household composition in the horizon year to you in the table below. Use the household trip rates in Table 4.2 to estimate the number of home-based trips per day that will be produced by zone 8.

Persons per HH	Vehicles per HH		
	0	1	2+
1	100	300	150
2	110	250	50
3	90	250	50
4	150	210	60
5+	20	50	0

4.9 Trip Generation

(a) Calculate the average trip rate of the zone 8 households considered in the previous problem, if "Persons per HH" is ignored and only "Automobile ownership" is thought to be important. In other words, find a "trips per household" value for each "Auto ownership" value (0, 1, 2, or more).

(b) If T = trips per household and X = automobile ownership, would it be satisfactory to use a linear regression model of the form $T = b X$ to represent the results found in part A of this problem?

4.10 Trip Generation.   Identify the closest relationships between elements in adjacent columns. For example, if expression A is most often associated with "Cross-Classification Analysis," write the pair "A, D". Do not try to connect elements in column 1 with elements in column 3.

(1)	(2)	(3)
A. 6.17 trip ends per day to city park for each parking space provided.	D. Cross-classification analysis	G. Attractions
B. $T = a_1(\text{GFA}) + a_2(\text{employees}) + a_3(\text{seats})$	E. ITE report	H. Productions
C. 2 veh/HH, 4 persons/HH → 11.98 person trips per HH per day?	F. Multiple regression	

Trip Distribution

4.11 Gravity Model Calculation.   Example 4.5 used the friction (or travel time) function in Equation 4.8, which uses $b = -2.0$. Repeat Example 4.5, this time using $b = -2.8$. Your work must be presented in the same format used in Table 4.9.

4.12 Trip Lengths and Friction Factors.   If the use of Equation 4.8 with $b = -2.0$ causes the average trip length in the resulting trip matrix to be too high should the value of b be increased or decreased?

4.13 Trip Distribution by the Gravity Model.   Using acceptable trip generation procedures, the recent Middleville Area Transportation Study (MATS) has produced the *horizon-year* (*H-yr*) production and attraction totals for each zone, as shown in Table 4.21a below. MATS has also determined that

$$F_{ij} = 1000 \, t_{ij}^{-2.4}$$

The H-year t_{ij} values are given in Table 4.21b. (∞ = infinity) Using the new F_{ij} equation and the format of Table 4.9, calculate the predicted values of T_{21}, T_{23}, and T_{24}.

TABLE 4.21a   MATS H-Year Productions and Attractions

Zone	1	2	3	4
P_i	1100	300	1600	1000
A_j	1400	600	500	1500

TABLE 4.21b   H-Year Interzonal Travel Times (minutes)

T_{ij}	1	2	3	4
1	∞	13	18	13
2	13	∞	13	18
3	18	13	∞	16
4	13	18	16	∞

4.14 Trip Distribution.   An automobile company is planning to build a 1.8 million square foot factory in zone 1 in Middleville. In anticipation of the new factory, Middleville receives funding from the state for an expressway that runs between zones 1 and 3. The new expressway reduces auto travel time between zones 1 and 3 to 6.1 minutes. Use the Gravity Model with $F_{ij} = t_{ij}^{-2}$, where t = auto travel time (minutes), to allocate the 5803 trips produced in zone 3 after the factory and expressway are in operation. Use $A(1) = 5770$, $A(2) = 4370$, $A(3) = 2380$, and $A(4) = 2980$ for this problem. Use the tabular format shown in Table 4.9. *Note*: Do not include $T(3, 3)$ in your calculations; assume $T(3, 3) = 0$.

4.15 Trip Distribution.   The sketch map below illustrates the potential for shoppers to go from Lafayette to Indianapolis, the Chicago Loop, or the (local) Tippecanoe Mall. The Gravity Model is used to apportion the shopping trips with the friction factor equal to

$$F_{ij} = 0.8t^{-2}.$$

The attraction to shoppers is proportional to the population. If the production of shopping trips in Lafayette on a given day is 8000, how many shopping trips will end in each place?

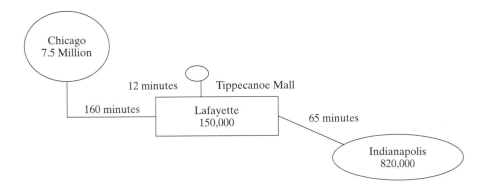

4.16 Trip Distribution by Gravity Model.   The Gravity Model calculations in Table 4.9 "predicted" where the 901 shopping trips from zone 2 would go if $b = -2.0$. However, the MRPC has acquired survey data that indicate the actual trip interchange values for shopping trips from zone 2. These "target values" are:

$$T(2, 1) = 491, T(2, 2) = 367, T(2, 3) = 27, T(2, 4) = 16$$

The MRPC wants to adjust its Gravity Model to match the survey data.

(a) Build a spreadsheet that duplicates the structure shown in Table 4.9, and add the $T(2j)$ survey data as a new column 8. Instead of "-2" in Equation 4.8, try different values of b until the values in column 7 match the values in column 8. (*Note:* In real life, you would never be able to match survey data exactly, but this is a HW problem.) What value of b allows you to match the target values?

(b) What does this result say about the *trip length distribution (TLD)* of shopping trips from zone 2 based on the survey data, as opposed to the original TLD in column 7 of Table 4.9? Explain how changing b made the match possible, making reference to Figure 4.10.

Mode Choice

4.17 Mode Choice.   An express bus service is introduced on the new expressway between zones 3 and 1, making the total transit travel time only 7.1 minutes, instead of 15.0 minutes. Estimate the transit and auto mode shares between zones 3 and 1 after the expressway and express bus service are in operation. Use the MNL model and the auto and bus utility functions that were used in Example 4.15.

4.18 Multinomial Logit Mode Choice Model.   A calibration effort resulted in the following utility equation:

$$U_m = a_m - 0.025\ X_1 - 0.032\ X_2 - 0.015\ X_3 - 0.002\ X_4$$

where X_1 = access plus egress time in minutes
X_2 = waiting time in minutes
X_3 = line-haul time in minutes
X_4 = out-of-pocket costs in cents

The trip distribution forecast for a particular interchange was a horizon year volume of T_{ij} = 5000 person trips per day. During the horizon year, trip makers between zones i and j will have a choice between modes A and B, with the following horizon year attributes:

Attribute:	X_1	X_2	X_3	X_4
Mode A	5	0	20	100
Mode B	10	15	40	50

(a) Assuming that the calibrated mode-specific constants are $a_A = -0.12$ and $a_B = -0.56$, apply the multinomial logit model to estimate p_A and p_B for the horizon year and then convert to the total number of trips that will occur between zones i and j by each mode.

(b) Did both modes need to have a non-zero mode-specific constant? What if $a_A = 0.44$ and $a_B = 0.00$? Would the results found in part A of this problem change? Explain.

4.19 Modal Split.   The red cruiser bus line presently runs 30 express buses from a suburban community (capacity 45 passengers per bus) that is 80 percent loaded during the rush hour.

Headways are 2 minutes (i.e., average wait time is 1 minute). The buses go to the CBD 10 miles away, where they discharge the passengers at a single bus terminal. This means that an individual passenger walks on the average of 9 minutes to his/her destination. The buses operate on a reverse flow portion of the expressway (limited to buses) without tolls, which reduces the line-haul portion of their trip from the auto trip time of 20 to 13 minutes. The pickup time, however, increases the in-vehicle time for the bus rider an average of 7 minutes, whereas the automobile time to reach the expressway is 4 minutes. No bus rider has to walk more than 3 minutes from home. The average time from the expressway to parking for an auto in the CBD involves 4 minutes IVTT. The average walk from the parking garage to the place of work is 3 minutes OVTT. The bus fare is $1.00 per trip. The auto is subjected to a $6.00 per day parking charge. Other out-of-pocket costs for auto are $1.00 for gas and oil and $1.20 for tolls per auto trip. A survey has indicated that the individual choice utility is determined by

$$U_m = B_m - 0.2*IVTT_m - 0.4*OVTT_m - 0.15*OPTC_m$$

Mode	B	IVTT	OVTT	OPTC in $
AUTO	1	Min	Min	
BUS	0	Min	Min	

Because it is a major urban area, the average automobile has 1.8 persons riding.

(a) What is the mode split during the rush hour?

(b) Based on the bus ridership, how many automobiles come from this community?

4.20 MNL Mode Choice Model. A group of teachers in the Mythaca School District have agreed to stop commuting by motor vehicle. Instead, each of these teachers will choose each morning between walking and bicycling to school, depending on the weather. The utility functions for the two non-motorized modes for these teachers are:

$$U_{bike} = 0.0 - 0.5\,t_{bike} - c_3 W_{bike}$$
$$U_{walk} = +2.9 - 0.5\,t_{walk}$$

where W is a weather-related variable and t is travel time in minutes. $W = 0$ in good weather; $W = 1$ in bad weather.

(a) When the weather is good, what is the probability that a "non-motorized" teacher with a choice between a 15-minute walk and a 6-minute bike ride will choose the bicycle mode?

(b) At what value of the weather coefficient c_3 will the teacher in part A be equally likely to choose walk and bicycle in bad weather?

4.21 Mode Choice. The multinomial logit choice model determines the mode shares. A consultant for Shoridan is using the following utility model to predict the H-year modal split between zones 3 and 4:

$$V_m = B_m - 0.05\,TT_m - 0.25\,cost_m$$

where B_m is the mode-specific constant, TT_m is travel time, and $cost_m$ is the travel cost per person.

The modal attributes are:

Mode	B_m	TT_m	$Cost_m$
Auto (A1)	2.1	16 min	$2.50
Carpool*	1.2	21 min	$0.80
Transit (T)	0.0	30 min	$0.75

*Assumes an average of 3.2 persons per carpool

(a) What is the mode share for autos, vanpools, and buses between zones 3 and 4?

(b) The industries that attract the travel into zone 3 need more space. They want to use the land they own more productively. The only place to get the space they need is from their large parking lots. The companies decide to enter into a massive program that encourages vanpool and transit use. Employees who arrive in a high-occupancy vehicle (HOV) will have to pay a parking fee of $1.00 per day per vehicle. Employees who enter a company lot in a single-occupant vehicle (SOV) will have to pay a parking fee of $3.00 per day per car in addition to fuel. The vans will hold an average of seven persons, each of whom will pay $1.00 per 5-day week to cover administrative costs. The company will provide the fuel. Those who ride transit will be able to ride the buses free, because the companies will buy transit passes for their employees. The average travel time by vanpool on surface streets will be 9 minutes longer than for SOV travel time, because of the time needed for pickup and/or rendezvous. Vanpool and bus travel time on the expressway will, however, be an average of 7 minutes shorter than SOV travel time because of an HOV lane that is being installed for vans/cars with five or more occupants. The modal constant for the van is $B_m = 1.0$ and, because the bus company made some adjustments in its schedule and routes, B_m becomes 0.3 for bus riders. If the present total attractions over the entire morning commuting period from zone 4 to zone 3 are 7000 person trip ends and each vehicle consumes an average of 400 square feet in the parking lot, how many acres of land can be recovered? *Hint:* You will need to use the answer in part a above to calculate the number cars expected in the lot. You will then need to spread those employees into the vehicles based on the modal split with the vanpool. From that modal split plus the ridership, the number of vehicles can be obtained (1 acre = 43,560 sq ft).

Trip Assignment

4.22 Equilibrium Traffic Assignment.   There are two travel paths in use between towns A and B. Path 1 has a link performance function $t_1 = 1.4 + 0.8 x_1$, whereas path 2's LPF is $t_2 = 3.7 + 0.3 x_2$, where t = travel time (minutes) and x = flow (1000s vph). If the total flow rate between A and B is 3184 vph, what are the equilibrium flows and travel times on paths 1 and 2?

4.23 Route Choice.   A new expressway has an LPF $t(X) = 6.1 + 3.3 V(X)$, where $V(X)$ is in 1000s of vehicles per hour in a given direction. The old arterial streets between zones 3 and 1 are still available, with a LPF of $t(A) = 8.2 + 17.0 V(A)$.

(a) If all drivers from zone 3 to zone 1 want to minimize their individual travel times, at what flow rate $V(X)$ will drivers begin to divert back to the arterial route?

(b) If $T(3, 1) = 2265$ vph during the peak hour, find the equilibrium travel time from zone 3 to zone 1 and calculate $V(X)$ and $V(A)$ for the peak hour.

4.24 Trip Assignment. There are two routes from a rural origin to the CBD destination. The most direct route A is 21 miles with an average free flow speed of 45 mph and a capacity of 2000 cars per hour. The alternate route B is 30 miles with an average free flow speed of 60 mph and a capacity of 4000 cars per hour. As traffic increases, the time to traverse the routes is given by the following functions:

$$ t_A = \frac{1}{\left(1 - \dfrac{V}{C}\right)} \text{ and } t_B = \frac{1}{\left[1 - \left(\dfrac{V}{C}\right)^2\right]} $$

How would you anticipate that 3000 cars would divide themselves between routes A and B?

4.25 Traffic Assignment Fundamentals

(a) Do you think that route choices in your local urban area are made such that the user equilibrium condition exists in the area's road network? Support your answer by citing examples.

(b) Is free-flow speed defined as the fastest speed at which one can drive on a given highway?

4.26 Link Performance Functions and Delay Estimates. It is 9:40 PM as the airport shuttle van is taking you on SB US281 to your downtown San Antonio hotel. Up ahead, you see a flashing blue light. It turns out to be a police car blocking the left lane of three lanes, because of an accident. You are alert enough to

(a) Determine the current vehicle density

(b) Convert the density to a vehicle flow rate V (2700 vph)

(c) Remember the standard LPF for a 1-mile segment of freeway with a design speed of 70 mph:

$$ t_{seg} = 51.4 \left[1 + \left(\frac{V}{C}\right)^{5.4} \right] \text{ sec} $$

where $C = 1800$ vph per lane times the number of lanes. Using the data and the LPF above, how much delay can be expected if SB US281 is restricted from three lanes to two lanes for 1 mile? Show your calculations to 0.1 second.

4.27 Highway Link Flows. Three inbound lanes of an urban interstate (I-16) have a capacity of 1800 vehicles per hour per lane. It has been determined that the following link travel time equation describes the congestion effects along a particular 1-mile segment of I-16:

$$ t = t_o \left[1 + 0.65 \left(\frac{V}{C}\right)^4 \right] $$

(a) During the morning peak hour, it takes traffic an average of $1.75\, t_o$ minutes to travel that 1-mile segment of I-16. What is the peak hour volume (vph/lane) on this segment?

(b) Assume that the current three-lane volume on this segment is 4800 vph and that $t_o = 1$ minute. If one of the three lanes were to be converted to a high-occupancy vehicle (HOV) lane and 25 percent of the current vehicles would be eligible to use it, what would be the travel time in the HOV lane over the 1-mile segment? What would be the travel time in the two remaining lanes?

4.28 Trip Assignment. One of the big events in Mythaca County each year is the Murdock Bay Regatta, which takes place in and around Shoridan. Immediately after the last Regatta event, however, thousands of vehicles leave Shoridan and head inland to (and through) Mythaca. There are five possible routes from Shoridan to Mythaca. The link performance functions for these five routes are:

$$t_1 = 18 + 4.5\, x_1;\ t_2 = 21 + 7.0\, x_2;\ t_3 = 26 + 4.9\, x_3;\ t_4 = 29 + 3.4\, x_4;\ t_5 = 34 + 0.3\, x_5$$

where t is in minutes and x is in 10^3 vph.

(a) It is expected that 8300 vehicles will be making the Shoridan-Mythaca trip in the maximum-flow hour after the Regatta. Will all five routes be used?

(b) A Mythaca County planner has tried to convert the expected total flow and the LPFs above into route flows. His solution includes $x_1 = 3.50$; $x_2 = 2.82$; $x_3 = 1.58$. Do these values seem correct to you?

Planning and Evaluation for Decision Making

SCENARIO

In the Scenario for Chapter 4, the MRPC staff was asked to study the perceived congestion on SR361 as that highway runs through the City of Mythaca (Figure 5.1). As the MRPC staff collected data and began to apply the four-step travel demand model to the SR361 corridor, several citizen groups were proposing solutions to the congestion.

A. Developers advocate the relocation of SR361 to an alignment that bypasses the city entirely. This alternative would involve the construction of a bridge across the river, acquisition of prime farmland as right-of-way west of the city, and modifications to the roads at the north end of the new bridge.

B. Downtown merchants urge the state DOT to expand two-lane SR361 to four lanes in the city. This alternative would require that some homes, trees, and commercial property along the urban segments of SR361 be taken and demolished.

C. Environmental interests want increased transit service offered along the existing SR361 in an elevated monorail from South Mythaca, through downtown, and across the river to the university campus. Some land would be taken but far less than for alternative B.

D. Taxpayer watchdog groups and neighborhood preservationists favor the "do

FIGURE 5.1
SR361 passes through a residential neighborhood. Photo: Jon D. Fricker.

nothing" approach. These people say that increasing road capacity only encourages more driving and that investment in expensive transit technology is never justified.

It is almost certain that other projects will be proposed. Eventually, someone will have to decide what course of action to take.

CHAPTER OBJECTIVES

By the end of this chapter, the student will be able to:

1. Explain how travel demand modeling fits into the transportation-planning process.
2. Explain how the transportation-planning process is used to help make public investment decisions.
3. List stakeholders who should be involved in the planning process.
4. Perform a benefit-cost analysis on alternative transportation projects.
5. Rank projects with non-quantifiable benefits or costs.

INTRODUCTION

The travel demand modeling and forecasting procedures discussed in Chapter 4 are not done in a vacuum. As the Scenario for this chapter indicates, there is more involved in transportation planning than applying the four-step travel demand model and computing vehicle hours of travel (VHT) and vehicle miles of travel (VMT) values. Each of the four alternatives in the Scenario—even the "do nothing" alternative—will affect different people in different ways. Each alternative will confer benefits on some groups and cause some groups to bear costs. Some of these benefits and costs will be measurable or quantifiable; some will be impossible to assess in economic terms. Ideally, the choice of alternative should be based on an objective and rational assessment of each alternative. This chapter describes processes and techniques that can be used to develop a recommended course of action.

THINK ABOUT IT

For each alternative in the Scenario, name at least one benefit and one cost that results. Identify the groups that will receive the benefit and bear the costs.

5.1   THE TRANSPORTATION-PLANNING PROCESS

According to a *FHWA Briefing Notebook* (FHWA, 2001) on the subject, "Transportation planning in metropolitan areas is a collaborative process, led by the metropolitan planning organization (MPO) and other key stakeholders in the regional transportation system." The process can be complicated and time-consuming, as Figure 5.2 implies. Federal law makes many of the activities shown in Figure 5.2 necessary. For example, the Transportation Equity Act for the 21st Century (TEA-21) requires consideration of seven broad areas:

1. Support the economic vitality of the metropolitan area.
2. Increase the safety and security of the transportation system for motorized and non-motorized users.

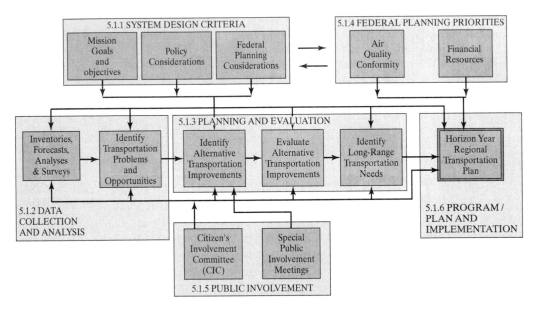

FIGURE 5.2
A typical structure for the transportation planning process. Based on AMATS, 2002.

3. Increase the accessibility and mobility options available to people and for freight.

4. Protect and enhance the environment, promote energy conservation, and improve quality of life.

5. Enhance the integration and connectivity of the transportation system, across and between modes, for people and freight.

6. Promote efficient system management and operation.

7. Emphasize the preservation of the existing transportation system.

The seven areas are mandatory (not optional), which means that the planning process must be comprehensive and well-documented.

As the lead agency in the process, the MPO must carry out five core functions (FHWA, 2001):

- *Establish a setting.*   Establish and manage a fair and impartial environment for effective regional decision making in the metropolitan area.

- *Evaluate transportation alternatives.*   Using well-documented and accepted procedures, carry out the technical analyses needed to support the planning process.

- *Maintain a Long-Range Transportation Plan (LRTP).*   The LRTP is a document that defines the vision for the region's transportation systems and services. The plan indicates all the transportation improvements scheduled for funding over the next 20 years.

- *Develop a Transportation Improvement Program (TIP).*   A TIP is a document prepared by an MPO that lists projects to be funded with FHWA/FTA funds for the next 1- to 3-year period.
- *Involve the public.*   Involve the general public and all the significantly affected subgroups in the four essential functions listed above.

As the principal elements of the transportation-planning process are described below, consult Figure 5.2 to see how an MPO could carry out the necessary activities.

5.1.1   System Design Criteria

The transportation-planning process needs a context. The process can react to transportation problems as they arise. Preferably, the process can anticipate problems, prevent them, or implement a long-range vision for the future of the community. In reality, some of each—react, anticipate, prevent, implement—is part of the planning process. In most situations, a variety of actions are possible. Choosing what is best for a community may depend on the viewpoints of those making those choices.

Mission Goals and Objectives.   The goals of a region's planning process may begin with actions or outcomes prescribed by government legislation. Even when federal requirements are not driving the process, the community needs a sense of direction within which to make plans and implement them. The "vision" for an area's plan can come from various segments of the community—political leaders, transportation and planning professionals, business leaders, special interest groups, and citizens. Ideally, all such community segments should be involved. Citizens can contact their elected and appointed leaders directly, and they can serve on citizens' advisory committees. The results of these activities are *goals*—the general statement of community values and the desired ends of the process. It is premature at this stage to state specific objectives; they emerge as specific projects are proposed and evaluated.

Policy Considerations.   There are often local policies that have been established either by tradition or by the political process through the action of mayors, councils and other political forces in the community, sometimes this happens through a referendum where the will of the voters is sought.

Federal Planing Considerations.   Federal mandates exist for air quality, mobility for handicapped persons, and use of seatbelts. Urban areas or states must comply with these mandates or face sanctions such as loss of federal transportation funds. The federal government is a source of funding and oversight for many projects. Any local and state government contemplating a project should examine the federal statutes pertaining to that project. The Federal Highway Administration (FHWA), Federal Transit Administration (FTA), and the National Highway Traffic Safety Administration (NHTSA) are funding and regulatory sources of which the planner must be aware.

5.1.2   Define the Needs

It is important to define as clearly as possible the transportation needs. In this part of the process, shown in Figure 5.2, the key is the collection and analysis of data.

Data Collection

- Inventory and inspection data that can be used to assess directly the condition of the transportation infrastructure, the relationship of travel and capacity, the quality of service by a given mode, and the impacts of specified transportation activity. The extent to which bridges and pavements need to be repaired or replaced, travel volumes approach system capacity, transit (or air) service is inadequate, and traffic noise affects adjacent neighborhood are examples of situations that need to be considered in the transportation-planning process.
- Survey data that can be used to describe what citizens think about transportation issues in the community or how trip decisions are being made by individuals. The first kind of survey is another attempt to get citizen input. The second survey is an attempt to understand how tripmakers behave in response to a variety of situations.
- Travel data that can be used in standard, well-established models to analyze existing transportation systems and predict the impact of proposed changes. How well does each element of a street network perform, in terms of congestion and vehicle emissions? A reliable model using accurate data can help identify trouble spots.

Transportation Problems and Opportunities: In the process of collecting and analyzing data it is appropriate to improve on the definition of the transportation problem(s) to be solved. This analysis may lead to the need to collect more or better data and refine the goals and objectives developed in Section 5.1.1.

At this stage of the transportation-planning process, the general goals can be transformed into more specific, measurable objectives. Later in this chapter, measures of effectiveness are illustrated.

5.1.3   Generate and Evaluate Alternatives

This is the heart of the planning process. It is the subject of Sections 5.3 and 5.4.

Identify Alternative Transportation Improvements. Generally, there may be more than one solution to achieve the transportation goals set out in Element 5.1.1. There may be a variety of routes, modes, differing technologies, and funding schemes to be considered. Each alternative must be identified according to system specifications, as indicated in Section 1.4.

Evaluate Alternative Transportation Improvements. Having identified alternatives, the process turns to a major analysis described in Section 5.3. There, the alternatives are ranked against a set of criteria to determine which ones best solve the problem.

Identify Long-Range Transportation Needs. Any solution to a transportation problem must address not just "today," but also "tomorrow." The analyst must develop a forecast of future demand for the transportation facility in question.

5.1.4 Federal Planning Priorities and Constraints

The two boxes in this element require a higher level of analysis than that implied by "Federal Planning Considerations" under Element 5.1.1.

Air/Water Quality Conformity. There are numerous federal guidelines and standards that will govern the process of design. The impact of a transportation change on ambient air quality or water quality will be very important in the evaluation of transportation alternatives.

Financial Resources. The federal government is one source of funding for many transportation projects. Many transportation projects receive a federal share of between 50 and 90 percent of the project cost. Sometimes, projects are selected on the basis of the federal share, which reduces the local share that must be raised by bonds and taxes. Often, projects proceed or get halted, based on a vote of the citizens. For instance, a majority of citizens in one county near San Francisco voted not to incur the bonding debt necessary for the extension of BART into their county.

5.1.5 Involving the Public and Others (Stakeholders)

The transportation-planning process is ultimately a public process. To be successful, it must involve a variety of stakeholders, each of whom has his/her own interests. A stakeholder is a person, a group of persons, a company, or an organization that has a stake in the decisions being made. Transportation investment decisions, such as building a new road, improving port facilities, or extending an airport runway, can have a widespread effect. Some of these impacts are positive; some impacts will be harmful to some groups. A person who owns (or works in) a convenience store that relies on pass-by traffic to stay in business may be irreparably harmed by a project that diverts that traffic to another route. The store's neighbors may be glad to see the traffic go elsewhere but may regret the loss of the store and its jobs. A person handling baggage at an airport may welcome the job security from the increased air traffic that a runway extension will make possible, but the airport's neighbors will object to the noise associated with larger aircraft and their more frequent takeoffs and landings.

An effective early planning strategy is to identify (and notify) all possible "stakeholders" (i.e., those who may have a stake in the outcome of a project). Including stakeholders in the planning and development of the project is not only a requirement in most cases, it usually leads to a better result. Even where community *consensus* cannot be reached, a sufficient level of *consent* to a particular solution means that a satisfactory outcome has been obtained.

Example 5.1

List all the stakeholders who need to be involved in the planning process for the SR361 project described in this chapter's Scenario.

Solution to Example 5.1   Table 5.1 contains a list of potential stakeholders and the issues that might concern each. Some stakeholders may not accept an invitation to be involved. Other stakeholders may be identified during the notification process. Some may take on very prominent roles. Other stakeholders may participate simply as observers.

TABLE 5.1   Stakeholders for Example 5.1

Local community—Public agencies

Mayor's office/city council	Overall project—effect on community
County councils and boards	Overall project—effect on community, taxation
Taxation agencies (federal, state, city)	Funding plan, bond issues, tax implications, taxation plan
Visitor information bureau	Project information and timeline, public information program
Utilities	Routing and temporary power/water
Emergency services coordinator	Fire/ambulance/police routes during construction
Local police	Traffic control during construction and requirements for future enforcement
City engineer, design engineer, Consulting (staff/consulting)	All aspects of physical design
Transit agencies	Effects on bus system, changes in bus routes
Regulatory agencies (Environmental, economic, zoning)	Establish and manage control systems
Airport officials	Changes in access to the airport

Local community—Citizens' groups

Chamber of commerce	Overall project—Effect on business in the community, valuation of land, attraction of new industry
Environmental groups	Land utilization, impact on wetlands, noise
Local neighborhood groups	How project will affect their neighborhoods
Employees and local union(s)	Employment practices
Community improvement	How project affects the group's mission
Sports stadium owners	Potential impact on sporting events
Basketball arena and convention facility operators	Potential impact on events

Local businesses and individuals affected

Persons with business along SR361	Change in customer base, downtime during construction
Hotels and restaurants depend on 361 access	Access to their businesses, duration of disruption
Construction companies	Construction design, bidding process
Banks and developers	Financing, magnitude of operation, land acquisition, zoning/land use
Material supplier-to-contractor for construction and maintenance	Specifications and available supply of products
Truckers using SR361	Alternate routing, duration of construction, project benefits
Real estate investor	Changes in property values, commercial investment demands
Drivers who use SR361 daily	Duration of construction, alternate routes, project benefits
Persons who live or own land along the ROWs	Impacts on property values and quality of life during project implementation and after completion.

THINK ABOUT IT

Are there any stakeholders that you think should be added to the list? Which of the stakeholders on the above list are likely to be the most powerful and influential? Which of the stakeholders on the above list are likely to have the least "clout" and therefore need some assistance in getting their concerns heard?

As was demonstrated in this chapter's Scenario, ideas for projects can be advanced by any individual or group. The alternatives are formalized by transportation professionals for analysis. In cases where complex networkwide impacts are being assessed, the transportation-planning staff will conduct a computer-based study of proposed alternatives and report the results to a technical committee. This committee typically consists of representatives of the various jurisdictions in the area—city, county, law enforcement, public transit, airport, and citizens. The comments and recommendations of the technical committee are sent to the planning commission, which is usually made up of elected officials and citizens.

5.1.6   Programming and Implementation

The term ***programming*** is used to describe the selection and scheduling of projects. Transportation projects typically go through conceptual planning, preliminary engineering, land acquisition, design, and construction phases. It is common to schedule the phases of each project for appropriate years, depending on the availability of funding, the expected duration of each phase, and the urgency of one project with respect to the others. The resulting schedule is published as a Transportation Improvement Program (TIP) for the urban area. Each urban area's TIP becomes part of the state TIP. With surprising frequency, it is sufficient to have a full discussion at an open meeting of the technical committee to establish the priorities and sequence of projects in a TIP. A complex priority-setting algorithm is not needed, unless the number of projects is great and reaching consensus within the technical committee is too difficult. The next two sections in this chapter describe methods to conduct a formal technical analysis, if one is needed. The technical committee's recommendations are forwarded to the planning commission level, where political and other non-technical factors are considered. It is at this stage that information needed to update the community vision can be collected for feedback to stage 1 in the transportation-planning process.

Project implementation involves the actual construction of any proposed facility and/or the management and operation of the transportation facility or service that has been created or improved.

5.1.7   Trends in the Transportation-Planning Process

The four phases or elements of the transportation-planning process just described must address the seven broad areas listed at the start of section 5.1. There is increasing emphasis on an explicit consideration of freight transportation in the planning process.

Planning agencies need to be transformed into multimodal agencies that can accommodate and encourage the interface between highway, rail, and air freight modes. Intelligent Transportation Systems (ITS) technologies offer possible solutions to transportation problems that were not available just a few years ago. The integration of ITS into the transportation-planning process is one of the requirements of TEA-21. MPOs must develop an ITS element and integrate ITS planning into their transportation-planning processes. The metropolitan-planning process need not follow exactly the structure of Figure 5.2, but it must be a cooperative, continuous, and comprehensive framework for making transportation investment decisions in metropolitan areas.

5.2   BRIEF REVIEW OF ENGINEERING ECONOMICS

This section covers only those aspects of Engineering Economics that the transportation engineer needs to rank and rate specific projects that are being planned or evaluated. For coverage of related topics, such as depreciation, taxes, and inflation, the student is encouraged to consult one of the many textbooks on the subject of engineering economics. Several texts with which the authors are familiar are given in the reference list.

5.2.1   What Is Engineering Economics?

Engineering economics is a particular way of looking at the economic (or financial) side of engineering decisions. Engineering economics can be applied in both the private and the public sectors of the economy. An economic analysis determines a proposed project's benefits to the public and the costs that must be expended to achieve those benefits over the life of the project. A life cycle analysis is often done for several possible alternative public projects to determine which one has the most worth (value) to the public. Economic analysis is also done to examine some more complex private sector projects, where the benefit is usually expressed in terms of profit or future sales.

Sometimes, it is not obvious whether it is the public or the private sector that should undertake a certain project. For example, who should provide a shuttle service from a downtown hotel district to the airport outside the city? The general criteria are:

- If private benefits > private costs, the private sector will find the project financially worthwhile. The private sector provider will be able to charge a fee that will cover the cost of the project.
- If public benefits > public costs, the public sector can justify carrying out the project. The project will be paid for by tax revenues, user fees, or a combination of the two.
- If public benefits > private costs and private costs > private benefits, government regulations or incentives can make the project a reality. In the case of auto safety features, such as the third rear brake light or safety belts, government regulations were necessary. The development of automobiles with lower emissions may be hastened by government subsidies.

Although economic analysis can be applied to any of these cases, the emphasis in this text is on the second item above.

5.2.2   Time-Oriented Value of Money

Once money is spent on a project, it becomes a *sunk cost*. The opportunity to earn interest on those particular funds is lost. In simple terms, the use of money costs more money. Therefore, project costs or benefits that will occur 5 years from now are worth less in terms of today's dollar.

The rate at which the value of a dollar amount diminishes with time is called the *discount rate*. (*Note*: We are not talking about inflation when comparing projects, but rather the *cost* of money. Inflation rates can be factored in later, if necessary.) The term used in economic analysis when the discount rate is applied over the life of a new project is the *net present value (NPV)*. This approach takes the costs for a multiple-year project and discounts the funds to be spent each year to develop the NPV of the project.

In evaluating a project, it is usually helpful to draw a cash flow diagram that shows the "stream" of costs and benefits. In Figure 5.3, each upward arrow B_i indicates a benefit in (or at the end of) year i and each downward arrow C_i indicates a cost in (or at the end of) year i. The project is shown to have an expected life of N years. Using a discount rate of d, the present value (PV) of a stream of costs $C_0, C_1, C_2, C_3, \ldots, C_N$ is

$$\text{PV}_{\text{Costs}} = C_0 + \frac{C_1}{(1+d)} + \frac{C_2}{(1+d)^2} + \frac{C_3}{(1+d)^3} + \cdots + \frac{C_N}{(1+d)^N} \qquad (5.1)$$

Likewise, the PV of the benefits is

$$\text{PV}_{\text{Benefits}} = \frac{B_1}{1+d} + \frac{B_2}{(1+d)^2} + \cdots + \frac{B_N}{(1+d)^N} \qquad (5.2)$$

The *net present worth (NPW)* of a project is $\text{PV}_{\text{Benefits}} - \text{PV}_{\text{Costs}}$, or

$$\text{NPW} = -C_0 + \frac{B_1 - C_1}{1+d} + \frac{B_2 - C_2}{(1+d)^2} + \cdots + \frac{B_N - C_N}{(1+d)^N} \qquad (5.3)$$

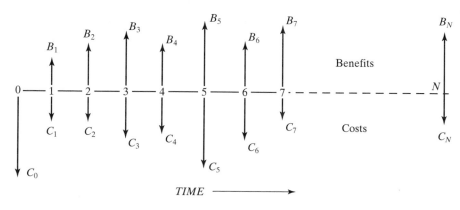

FIGURE 5.3
Stream of costs and benefits for an N-year project.

Any problem can be solved by using these simple relationships. Spreadsheets are ideal to use when the annual values are not all equal. Special relationships also exist when the values of $C_1, C_2, C_3, \ldots, C_N$ are identical or increase or decrease in either a geometric or an arithmetic manner. These and other cash flow relationships are summarized in Figure 5.4.

When $C_1 = C_2, = C_3, = C_N = C_i$, Equation 5.1 simplifies to

$$PV_{\text{Costs}} = C_0 + C_i \sum_i \frac{1}{(1 + d)^i},$$

Type	Factor and Formula	Relation	Simple Cash Flow Diagram
Single amount	$[F|P, i, n] = (1 + i)^n$	$F = P * [F|P, i, n]$	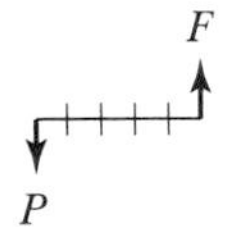
	$[P|F, i, n] = \dfrac{1}{(1 + i)^n}$	$P = F * [P|F, i, n]$	
Uniform series	$[P|A, i, n] = \dfrac{(1 + i)^n - 1}{i(1 + i)^n}$	$P = A * [P|A, i, n]$	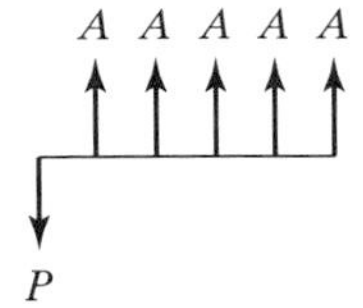
	$[A|P, i, n] = \dfrac{(1 + i)^n}{(1 + i)^n - 1}$	$A = P * [A|P, i, n]$	
	$[F|A, i, n] = \dfrac{(1 + i)^n - 1}{i}$	$F = A * [F|A, i, n]$	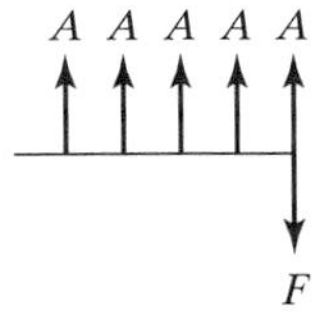
	$[A|F, i, n] = \dfrac{i}{(1 + i)^n - 1}$	$A = F * [A|F, i, n]$	
Arithmetic gradient	$[P|G, i, n] = G\left[\dfrac{(1 + i)^n - in - 1}{i^2 (1 + i)^n}\right]$	$P = G * [P|G, i, n]$	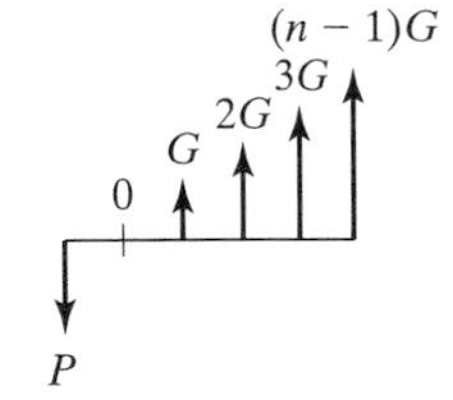
	$[A|G, i, n] = G\left[\dfrac{1}{i} - \dfrac{n}{(1 + i)^n - 1}\right]$	$A = G * [A|G, i, n]$	
Geometric gradient	$[P_g|A, i, g, n]$	$P_g = \dfrac{A_1\left[1 - \left(\dfrac{1 + g}{1 + i}\right)^n\right]}{i - g}$ if $g \neq i$	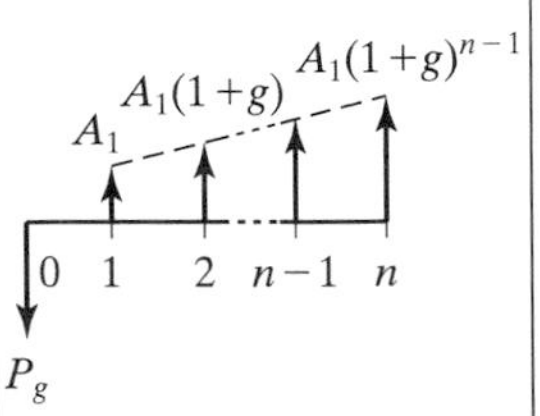
		$P_g = A_1\left(\dfrac{n}{1 + i}\right)$ if $g = i$	

FIGURE 5.4
Cash flow relationships with end-of-period compounding.

where C_i is the constant (or uniform) cost each year i. The sum of the series $\sum_i (1/(1 + d)^i)$ is often denoted as $(P|A, d, N)$, a factor that converts the uniform annual A values (here, $A = C_i$) into the present worth of the uniform series.

$$\text{PV}_{\text{Costs}} = C_i(P|A, d, N) = C_i \frac{(1 + d)^N - 1}{d(1 + d)^N} \qquad (5.4)$$

As before, d is the discount rate and N the number of years.

The notation $P|A$ stands for the present value, given the uniform annual amount. The factor $A|P$ means the uniform annual amount, given the present value. It is the reciprocal of $P|A$. When the annual benefits are uniform throughout the project and the annual costs are also uniform, then the NPW becomes

$$\text{NPW} = -C_0 + (B - C)(P|A, d, N) \qquad (5.5)$$

5.2.3   Discount Rate

When dealing with life cycle costs, the discount rate reflects the time value of money. How does a 20-year project A that has a low initial investment (e.g., $2000) and costs $700 per year to operate compare with project B, whose initial investment is $5000 but costs $450 per year to operate? For project A with $d = 5\%$, Equation 5.4 produces the present value of the annual costs:

$$\text{PV}_{\text{Costs}} = \$700(P|A, 0.05, 20) = \$700\frac{(1 + 0.05)^{20} - 1}{0.05(1 + 0.05)^{20}} = \$700[12.4622] = \$8723.55$$

Adding that value to the initial investment of $2000 leads to the $10,725 entry in Table 5.2.

In Table 5.2, project A has a higher PV_{Costs} value than does project B until the discount rate is increased. At a 5% discount rate, the projects have almost the same present value. At 10%, project A becomes better. A higher discount rate gives less weight to benefits and costs in the more distant future. As a result, project A's extra $300 per year in operating costs will not offset its lower purchase price, if the discount rate is high enough. A brief discussion of how discount rates are established is given later in this section.

Varying the length of the analysis can also affect the outcome. In Table 5.3, the PV_{Costs} values for projects A and B using a 5% discount rate were computed for project lifetimes of 10, 20, and 30 years. As might be expected, the higher annual value of project A begins to dominate when the project lasts longer.

TABLE 5.2   Comparison of Two Projects with Various Discount Rates

| Project | Investment | Annual Cost | Present Value of Costs | | |
			$d = 0\%$	$d = 5\%$	$d = 10\%$
A	2000	700	16,000	10,725	7,960
B	5000	400	14,000	10,610	8,830

TABLE 5.3   Net Present Value for Different Project Lifetimes

			Present Value of Costs		
Project	Investment	Annual Cost	$N = 10$ years	$N = 20$ years	$N = 30$ years
A	2000	700	7405	10,725	12,760
B	5000	400	8090	10,610	11,150

In a comparative economic analysis, all values must be discounted with the same discount rate. Applying another discount factor may alter the selection, depending on the makeup of the costs and benefits. Repeating the analysis with other reasonable values of the discount rate is a good way to determine how sensitive the results are to the choice of discount rate. Example 5.2 presents one example of sensitivity analysis.

Example 5.2

A proposed project can be accomplished by using either of two possible technologies. Version "A" is accomplished by investing $8 million. Each year, the *operation and maintenance (O&M)* costs are estimated to be $680,000, until the project reaches its lifetime of 20 years with no salvage value. Version "B" involves investing in a much higher quality system at $14 million and sustaining much lower O&M costs of $100,000 per year. Version B will have a residual (salvage) value of $3 million at the end of 20 years. If the discount rate used is 10%, which version should be chosen?

Solution to Example 5.2   The present value of costs for version "A" is

$$PV_{cost}^{A} = 8,000,000 + \left[\frac{(1.1)^{20} - 1}{0.1(1.1)^{20}} \times 860,000 \right] = 8,000,000 + (8.51 \times 860,000) = \$15,320,000$$

The version "B" approach to accomplishing the project has a present value of costs of

$$PV_{cost}^{B} = 14,000,000 + (100,000 \times 8.51) - \frac{3,000,000}{(1.1)^{20}} = \$14,457,000$$

Version B has the lower present value of costs, but other factors may enter into the choice of technologies. Because public sector budgets are usually so limited and because the public sector is influenced by the time to the next election, choosing version A with its lower initial cost may be preferable to elected officials, given that lifetime costs for the two alternatives are so close.

5.2.4   Projects with Gradients

Often, a cost or a benefit changes with time in a predictable manner.

Constant Arithmetic Growth or Decline.   The case in which the increase or decrease G is constant each year is shown in Figure 5.5.

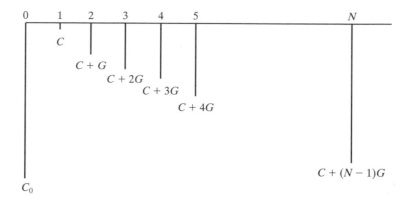

FIGURE 5.5
Cash flow diagram for uniform gradient.

There are several ways the present value of these costs can be computed:

$$\text{PVC} = C_0 + \frac{C_1}{1+d} + \frac{C_2+G}{(1+d)^2} + \frac{C_3+2G}{(1+d)^3} + \cdots + \frac{C_N+(N-1)G}{(1+d)^N}$$

where $C = C_1 = C_2 = \cdots = C_N$

or $\text{PVC} = C_0 + \left\{G * \left[P|G, d, N\right]\right\} + \left\{C + \left[P|A, d, N\right]\right\}$

Figure 5.4 shows the terms and equations used for the most common cash flow problems in engineering economics. In Figure 5.4, i = discount rate and n = number of time periods.

Example 5.3

The Mythaca County Highway Engineer wants to buy a new software package to track and analyze pavement maintenance costs. The software has a price tag of $5000. Other county engineers who have been using the same software like it, but they warn their Mythaca colleague to set aside an additional $600 the first year for technical support. This extra cost drops to about $400 the second year, $200 the third year, and nothing after that.

What is the present value of the cost of the software, including technical support? Use a discount rate of 5.2 percent.

Solution to Example 5.3   If the technical support stayed at $600 per year for 3 years, the present value of that uniform series would be (using Equation 5.4)

$$\text{PV}_{\text{Costs}} = C_i(P|A, d, N) = 600 * \left[\frac{(1.052)^3 - 1}{0.052(1.052)^3}\right]$$

$$= 600 * \left[\frac{1.1643 - 1}{0.052 * 1.1643}\right] = 600 * 2.7131 = 1627.84$$

However, the cost decreases $200 each year over the second and third year. Using the $P_G|G$ equation in Figure 5.4 with $i = d = 0.052$, $n = N = 3$, and $G = -\$200$,

$$[P_G|G] = \frac{(1 + d)^N - (d*N) - 1}{d^2*(1 + d)^N} = \frac{(1.052)^3 - (0.052*3) - 1}{(0.052)^2*(1.052)^3}$$

$$= \frac{1.1643 - 0.1560 - 1}{0.002704*1.1643} = \frac{0.008253}{0.003148} = 2.6214$$

and $P_G = G* -200*[P_G|G] = 2.6214 = -\$524.28$. Combining the uniform component and the negative gradient component, the present value of the 3-year technical support costs is $P = \$1627.84 + (-\$524.28) = \$1103.56$. Add that to the "year zero" purchase price of the software, and the total present value of the software cost is $\$5000 + \$1103.56 = \$6103.56$.

Constant Annual Percentage Growth.   When the growth is a geometric gradient g and the discount rate is d, the following cash flow relationship occurs.

$$NPV = C_0 + C_1\left(\frac{1}{1 + d}\right) + C_2\frac{1 + g}{(1 + d)^2} + C_3\frac{(1 + g)^2}{(1 + d)^3} + \cdots + C_N\frac{(1 + g)^{N-1}}{(1 + d)^N}$$

If $C_1 = C_2 = C_3 = \cdots = C_N = A$, then the $[P_g|A, d, g, n]$ equation in Figure 5.4 applies.

5.2.5   Annual Costs

Frequently, the choice among alternatives is made on the basis of the lowest *equivalent uniform annual cost (EUAC)*. The EUAC is the investment (minus discounted salvage, if any) multiplied by the $(A|P)$ factor. In the two projects in Example 5.2, $(A|P) =$

$$EUAC_A = \frac{\$8,000,000}{8.51} + \$860,000 = \$1,800,000 \text{ per year}$$

and

$$EUAC_B = \frac{\$14,000,000 - \dfrac{\$3,000,000}{1.1^{20}}}{8.51} + \$100,000 = \$1,693,000 \text{ per year}$$

The EUAC method allows the comparison of alternative projects that have different expected lifetimes. The primary assumption is that each alternative project being evaluated will be replaced at the end of its useful life with another project just like it. If that assumption can be made, the project with the smallest EUAC value has the lowest cost.

Example 5.4

In Example 5.3, if the county can spread the cost of the software over 3 years at a discount rate of 5.2 percent, what uniform amount should the county set aside each of the 3 years to pay for the software and technical support?

Solution to Example 5.4 There are two reasonable ways to find the EUAC of the software. One way is to compute $5000*[A|P] + \$600 + (-\$200*[A_G|G])$. Because we have already calculated the present value of all the software costs in Example 5.3, why not simply convert that value into its uniform annual equivalent?

$$[A|P, d = 0.052, N = 3 \text{ yr}] = \frac{d(1 + d)^N}{(1 + d)^N - 1} = \frac{0.052*(1.052)^3}{(1.052)^3 - 1} = \frac{0.0605}{0.1643} = 0.3686$$

$$A = P*[A|P] = \$6103.86 * 0.3686 = \$2249.80.$$

5.3 ECONOMIC EVALUATION OF TRANSPORTATION ALTERNATIVES

5.3.1 Introduction

The responsible engineer is obligated to search for the best solution(s) to specified transportation problems. The solution must explicitly meet the system specifications, demands, or needs for the system. (See Chapter 1.3.) Sometimes this is not an easy task.

Many times, it is possible to make a recommendation regarding competing transportation systems or options based purely on cost or net benefits (benefits minus cost). In justifying such recommendations, the engineer should use the principles of engineering economics. Each candidate project should be evaluated over its entire expected life. This brings the time value of money into the problem.

There are three types of projects that could be considered:

1. A project needs to be done. Which alternative has the lowest "life cycle" cost?
2. Several alternatives that have equal quantified benefits. Which alternative has the lowest cost?
3. The candidate alternatives have different quantifiable benefits and costs. How can the benefits and costs be compared in some rational way?

We examined demand for travel in Chapter 4. In the Trip Assignment section of Chapter 4, we assigned travel to various routes, thereby identifying routes that would be congested if no changes to the network were made. The next step is to examine the question of what new supply of transportation should be developed to meet the expected demand. For example, a decision must be made about the timing of a new investment in highways (increase in supply) or adding an alternative transportation service such as new or expanded bus routes to satisfy the projected demand. As transportation planning occurs, we need to continually think about supply and demand as interactive elements.

Because of the large number of public concerns in transportation (e.g., safety, pollution, and equity), the project to be evaluated is frequently a government project or a project where the major funding comes from government. Sometimes it is a local municipal project; other times it could involve federal money or a combination of funds from local, state, and federal sources. In these types of projects, the benefits usually accrue to the users of the system, although sometimes the operator also benefits. The costs, on the other hand, will usually be costs paid from public funds, such as taxes, an existing trust fund, or a bond issue. In most cases, taxpayers fund these systems, whether they use them or not. (One exception for automobiles is a toll road.)

Example 5.5:   Bridge Repainting

The state DOT wants to repaint some highway bridges that pass over I-25. To do this, they must close one or more lanes of traffic, causing delays during peak periods of travel. The DOT has two choices:

A. Do the bridge repainting only during normal daylight work hours. This requires that the job site be set up and torn down each day. The normal bridge repainting job done this way takes 3 days.

B. Work continuously on the bridge repainting job until it is finished. Working into the night means that the job can be completed within 24 hours, but local work rules require that nighttime labor be paid at a rate 1.5 times the standard daytime rate.

What are the advantages and disadvantages of each option, and who is affected?

Solution to Example 5.5   Under option A, at least 3 days of peak period traffic must endure the delays resulting from one or more lanes being closed. Under option B, two cycles of setup and teardown are eliminated, but nighttime work is not as productive or safe as working with daylight. Option B will probably be much more expensive to the state DOT than option A, but who will receive the benefits from those increased costs? The drivers on I-25 who will experience much less delay! If the DOT routinely adopts the more expensive option B, it will be able to paint fewer bridges with its annual paint budget. Option B is an example of one party paying more, so that another party can realize the benefits. In some cases, it may be the best thing to do!

5.3.2   Evaluation Process Using Engineering Economics

As illustrated in the previous section, engineering economics includes the *time value of money*. The decision to spend money eliminates the possibility of investing it instead. Therefore, the value of money is discounted over time. (Figure 5.4 shows a summary of the most common cash flow relationships.) This is not *inflation*, which is a measure of the rise in prices over time. The time value of money accounts for the effects of using money over time, rather than investing it in a bank or in an activity that promises a return on the investment. Projects are then judged on the basis of some economic criterion such as their net present worth or their net benefits, including all the costs and benefits that exist.

The evaluation process begins by establishing specifications for the system(s) being evaluated, defining each particular alternative, and then performing an alternative preliminary design on each. (*Note*: One alternative is frequently the "do nothing" alternative.) With the design options in mind, costs of the alternative approaches to design can be estimated. Options frequently evaluated are:

- The approximate or exact location of a new road.
- The number of new lanes to be added.
- The technologies to use.
- Additional safety features that may be required.
- Operating strategies to be employed.
- Traffic control methods to adopt.

The comparison of alternatives is always based on a comparison of *measures of effectiveness (MOEs)*. In this section, we are concerned with those comparisons for which cost is the only MOE. The MOEs are based on the choice of criteria to be used to distinguish alternatives. These criteria are discussed in the next section.

5.3.3   Criteria for Comparing Alternatives

The criteria most often used in the evaluation of transportation projects are (1) net present worth (NPW), (2) *benefit-cost ratio (BCR)*, and (3) equivalent uniform annual cost (EUAC). It is often helpful to apply more than one of the criteria before making the project selection.

- Select the project that has the highest NPW. Determine and compare the net present value or net present worth or net present benefits. This is calculated by determining the benefits accrued over the life cycle of the project and subtracting the investment plus the annual costs over the life of the project, less any anticipated salvage value. The life cycle costs are usually discounted by using a rate d.

$$\text{NPW} = -C_0 - \sum_{i=1}^{N} \frac{C_i}{(1 + d)^i} + \sum_{i=1}^{N} \frac{B_i}{(1 + d)^i} + \frac{S}{(1 + d)^N} \tag{5.6}$$

where

C_i = amount of cost in time period i,

B_i = amount of benefit in time period i,

d = discount rate

S = salvage value

- Select the project that has the highest BCR. The benefit-cost ratio is calculated by determining the benefits over the life cycle of the project and dividing by the costs (investment plus annual costs, less any anticipated salvage) determined over the life of the project. The life cycle costs are usually discounted by a discount rate of **d**. Assuming the annual benefits and costs do not vary from year to year, the BCR is shown in Equation 5.7.

$$\frac{\text{Benefit}}{\text{Cost}} = \frac{[P|A, d, N] \times B}{C_0 + [P|A, d, N] \times C - \text{Salvage} \times [P|F, d, N]} \tag{5.7}$$

where P = present value

A = annual value

F = future value

$[P|A, d, N]$ is the present worth factor, given the annual amount A over N years at an annual discount rate d. (See Figure 5.4.)

$[P|F, d, N]$ is the present worth factor, at an annual discount rate d, of an amount F that occurs N years into the future. (See Figure 5.4.)

- Select the project with the lowest EUAC or highest *equivalent uniform annual benefit (EUAB)*. In this method, the present values are transformed into the equivalent uniform annual cost or benefit. In the case of costs,

$$\text{EUAC} = \text{PV}_{\text{cost}} * [A|P, d, N] = \frac{\text{PV}_{\text{cost}}}{[P|A, d, N]} \tag{5.8}$$

This method is especially helpful when comparing projects that have different lifetimes. It also helps with projects whose costs are made up of several distinct pieces.

$$\text{Total EUAC} = \text{EUAC}_1 + \text{EUAC}_2 + \text{EUAC}_3 + \cdots + \text{EUAC}_N \tag{5.9}$$

Example 5.6:   Projects with Different Lifetimes

Compare projects A, B, and C, each with a different lifetime, using discount rates of 10 percent and 5 percent.

P	Investment	Life (yr)	Annual Cost	Salvage	Annual Benefit	EUAB-EUAC at $d = 5\%$	at $d = 10\%$
A	$3,000,000	20	100,000	30,000	700,000	$460,000	$148,000
B	$5,000,000	30	200,000	100,000	1,000,000	$476,000	$270,000
C	$10,000,000	15	300,000	2,000,000	1,500,000	$329,000	−$52,000

Solution to Example 5.6   The best way to handle the discrepancy between natural project lifetimes is to convert benefits to EUAB and costs to EUAC. In the equations that follow, the annual benefit is the first term, and the annual cost is the last term. In the numerator of the term in brackets, the present value of the salvage value is subtracted from initial investment, the result of which is a net cost. The denominator in the bracketed term is the equation for $[P|A, d, N]$. As seen in Equation 5.8, dividing by $[P|A, d, N]$ is equivalent to multiplying by $[A|P, d, N]$. This means that the bracketed term represents the uniform annual equivalent of the present value found in the numerator.

$$\text{Project A}_{5\%} \quad \text{EUAB} - \text{EUAC} = 700,000 - \left[\frac{3,000,000 - \dfrac{30,000}{(1.05)^{20}}}{\dfrac{(1.05)^{20} - 1}{0.05 \times (1.05)^{20}}} \right] + 100,000$$

$$= 700,000 - \left[140,000 + 100,000 \right] = \$460,000$$

$$\text{Project B}_{5\%} \quad \text{EUAB} - \text{EUAC} = 1,000,000 - \left[\frac{5000,000 - \dfrac{100,000}{1.05^{30}}}{\dfrac{1.05^{30} - 1}{0.05 \times (1.05)^{30}}} \right] + 200,000$$

$$= 1,000,000 - \left[324,000 + 200,000 \right] = \$476,000$$

$$\text{Project } C_{5\%} \quad \text{EUAB} - \text{EUAC} = 1,500,000 - \left[\frac{10,000,000 - \dfrac{2,000,000}{1.05^{15}}}{\dfrac{(1.05)^{15} - 1}{0.05 \times (1.05)15}} + 300,000 \right]$$

$$= 1,500,00 - \left[871,000 + 300,000 \right] = \$329,000$$

When the discount rate is increased to 10%, the answers are:

$$\text{Project } A_{10\%} \quad \text{EUAB} - \text{EUAC} = \$148,000$$
$$\text{Project } B_{10\%} \quad \text{EUAB} - \text{EUAC} = \$270,000$$
$$\text{Project } C_{10\%} \quad \text{EUAB} - \text{EUAC} = -\$52,000$$

Project B is the best at both discount rates.

Example 5.7:   Transit AVL

A transit automated vehicle location (AVL) system has an initial cost of $1 million, with annual costs of $100,000 and no salvage value at the end of a 15-year lifetime. If the benefits are $260,000 per year and the discount rate is 7%, determine (A) the NPW and (B) the BCR. (C) Is the AVL project worth pursuing?

Solution to Example 5.7

A. Bring the value of the annual costs and benefits back to their equivalent present values.

$$\text{NPW} = -C_0 - C_{\text{annual}}([P|A, 7\%, 15]) + B_{\text{annual}}([P|A, 7\%, 15])$$

$$[P|A, 7\%, 15] = \frac{(1.07)^{15} - 1}{.07(1.07)^{15}} = 9.1$$

$$\text{NPW} = \text{Benefits} - \text{Costs} = -\$1,000,000 - \$100,000(9.1) + \$260,000(9.1) = \$456,000$$

B. The BCR is also based on the present worth of benefits and costs.

$$\frac{\text{PW (Benefits)}}{\text{PW (Costs)}} = \frac{260,000 \times 9.1}{1,000,000 + (100,000 \times 9.1)} = 1.24$$

C. For the project to be worthy of further consideration, the necessary conditions are NPW > 0 and B/C ratio > 1.0. The closer to zero the NPW is, and the closer to 1 the BCR is, the more likely the decision will be based on other criteria, if no other projects are competing for available funds. Based on its NPW and BCR values calculated above, the AVL project is worthwhile for implementation, or it can continue to be considered in comparison with any other proposed projects.

Example 5.8:   Railroad Section

A section of railroad is composed of the elements listed in the table below. Determine the EUAC for each element when a 10% discount rate is used. $[P|A, d, N]$ is the present value, given the annual value (at the discount rate and number of years). Also note that $[A|P]$, which is what is really wanted for the EUAC, is simply the reciprocal of $[P|A]$.

Element	Investment	Annual Cost	Life (yr)	Salvage Value	$[P|A, 10\%, N]$	EUAC
Land	500,000	1,200	100	500,000	9.999	51,207
Grading	65,000	0	50	20%	9.915	6,567
Railbed	180,000	12,000	30	40%	9.427	31,532
Track	200,000		20	5%	8.513	23,667
Signals	40,000	5,000	10	0	6.145	11,510
Total						124,483

Solution to Example 5.8

Because "Salvage Value" is treated as a discounted reduction in the initial Investment, the $500,000 salvage value is discounted back to Year 0 using $[P|F, 0.10, 100] = 1/(1.10)^{100}$. The resulting Year 0 net cost is annualized over 100 years using $[A|P, 0.10, 100]$ and added to the $1,200 annual Land costs:

$$\text{EUAC}_{\text{Land}} = \left(\$500,000 - \frac{\$500,000}{(1.10)^{100}} \right) * [A|P, 0.10, 100] + \$1,200$$

$$= \left(\$500,000 - \frac{\$500,000}{(1.10)^{100}} \right) * \left[\frac{1}{9.999} \right] + \$1,200 = \$51,207$$

For the Railbed element, the computations are similar:

$$\text{EUAC}_{\text{Railbed}} = \left(\$180,000 - \frac{0.40 * \$80,000}{(1.10)^{30}} \right) * [A|P, 0.10, 30] + \$12,000$$

$$= \left(\$180,000 - \frac{0.40 * \$80,000}{(1.10)^{30}} \right) * \left[\frac{1}{9.427} \right] + \$12,000 = \$31,532$$

These EUAC values appear in the rightmost column of the table above. The EUAC method allows project elements with different lifetimes to be compared on a common basis.

5.3.4 Multiple Projects

When an engineer has many projects to decide between, one method of evaluating them is to examine the "efficiency frontier" [Tyner et al., 1981]. Proposals to use the BCR suffer from the difficulty that the project with the highest benefit-cost ratio is not always the one with the best NPW. The project with the greatest net benefit (benefit − cost) is often one of the most expensive projects. For the practicing engineer, some judgment is appropriate. That judgment is exercised by plotting the efficiency frontier and then selecting the project(s) near the knee of the efficiency frontier curve. This becomes especially important when there are several competing projects and when some or all of them may be implemented at several levels.

Example 5.9

There are four different alternative projects (1, 2, 3, and 4) for which a decision must be made with limited funds. The engineer can choose one of them, several of them, or all of them. In each project, there are two implementation scenarios—A and B. A gives the costs and benefits of doing a partial project, while B gives the benefits and costs for doing the entire project. C is the do-nothing alternative. The data below show the benefits and costs for each combination of project and implementation strategy. There are $3^4 = 81$ possible combinations, which are plotted in Figure 5.6. Note that, for all projects except C, the individual BCR is greater than one.

Benefits and Costs of Multiple Projects ($Millions) for Example 5.5

Benefits	1	2	3	4	Costs	1	2	3	4
A	6	5.1	8	8.3	A	3	2.5	5	4.1
B	9	8.8	12	9.8	B	5	6.5	7	6.2
C	0	0	0	0	C	0	0	0	0

Solution to Example 5.9   For each combination of projects, compute the net benefit = benefit − cost and plot it on a graph against the Cost as shown in Figure 5.6. The project combinations with the best overall net benefits for the lowest cost will appear on the left edge of the plotted points. This is the "efficiency frontier." Pick the points that lie along the frontier. In this case, let us examine points P1, P2, P3, and P4. The projects related to these four points are given in the table below.

Point	Implementation Level				NB	Cost	B/C
	Project 1	Project 2	Project 3	Project 4			
P1	A	A	C	A	9.8	9.6	2.02
P2	A	A	B	A	14.8	16.6	1.89
P3	B	A	C	A	10.8	11.6	1.93
P4	A	A	A	A	12.8	14.6	1.87

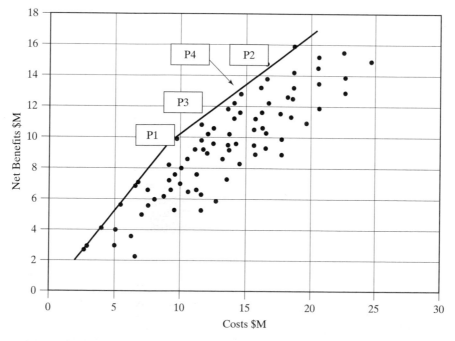

FIGURE 5.6
Plot of Net Benefits versus Total Costs for several competing projects.

The net benefits (NB) for any point is the sum of the net benefits for each constituent project at its specified level. For example, P1 represents the A-level (partial) implementation of Projects 1, 2, and 4, with no implementation of Project 3. The net benefit of this combination is

$$NB(P1) = NB(\text{Partial Proj. 1}) + NB(\text{Partial Proj. 2})$$
$$+ NB(\text{Partial Proj. 3}) + NB(\text{Partial Proj. 4})$$
$$NB(P1) = (6 - 3) + (5.1 - 2.5) + (0 - 0) + (8.3 - 4.1) = \$9.8 \text{ million.}$$

The total cost of P1 is $Cost(P1) = 3.0 + 2.5 + 0 + 2.1 = \$9.6$ million, and its BCR is $(9.8 + 9.6)/9.6 = 2.02$.

The best point along the efficiency frontier is at its knee (near P1). In Figure 5.6, Point P1 is closest to the knee. Almost as close to the knee is Point P3, representing a combination in which Projects 2 and 4 would still be implemented at level A, but Project 1 is now at level B (full implementation). The BCR for Point P3 is a little lower, at 1.93, and the NB is $10.8 million. It is wise to examine other points on or near the frontier, like P2 and P4. P2 is the set of projects with the highest cost, because it brings in Project C at partial implementation. P4 brings in Project 3 at the A level. At both points P2 and P4, the BCR, while still good, is less than at Points P1 and P3. From the information given, the best choice if there are adequate funds would be strategy represented by Point P1—partial implementation of Projects 1, 2, and 4, or Point P3—full implementation of Projects 1, and not to do Project 3 at all. This method of plotting the full array of project possibilities is easily done using a computer. It gives the manager a basis for eliminating some project combinations at the outset of the process, while identifying promising competing alternatives for further evaluation.

5.3.5 Evaluating Public Projects

Investments in private business and government are made to increase the efficiency of delivering goods or services. When new efficiencies are instituted, as they almost always are in a competitive environment, the cost of producing the goods or services is reduced. These savings, sooner or later, will be passed on to the consumer. In private business, the benefits received from investments that improve efficiency are often in the form of increased profits. In the public sector or government, the benefits from investment that improve efficiency are almost always in the form of an increase in service and/or safety, which results from a reduction in the provider's costs. Unfortunately, as indicated in the bridge repainting case in Example 5.5, the benefits from government projects are usually not in the form of direct monetary payments.

In assessing the benefits of a purchase to a consumer (called *consumer surplus*), begin by considering consumers who value the product much more highly than the equilibrium price p_1 shown in Figure 5.7. These consumers would be willing to pay a higher price than the p_1 currently paid by the q_1 customers. If, for some reason, the price drops to p_2, Figure 5.7 indicates that sales will increase to q_2 units. All those customers who were willing to buy q_1 units of product at price p_1 will receive a surplus benefit at the new price p_2. The increase in benefits to the former consumers is shown as the darker rectangular area A in Figure 5.7. The new customers also receive a surplus benefit from being able to buy the product at price p_2. The amount of this consumer surplus also depends on the shape of the demand curve and is shown as

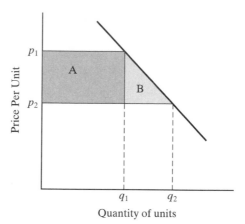

FIGURE 5.7
Consumer surplus.

Quantity of units

triangular area B in Figure 5.7. The total consumer surplus benefit then is found in the area of the trapezoid: $A + B$.

$$\text{Benefit} = (p_1 - p_2) \times \frac{1}{2}(q_1 + q_2) \tag{5.10}$$

Example 5.10

Anything but That Ferry! Years ago, when the Mythaca County Engineer was traveling with friends in Europe, they made the unfortunate decision to take the car ferry across the English Channel from Oostende, Belgium to Dover, England. The fare was equivalent to approximately \$40. The trip took 6 hours, the Channel was rough, and almost every passenger became ill. On the return trip, the group decided that, *no matter what the cost*, they would take the hovercraft from Dover to Calais, France (Figure 5.8). To their astonishment, the hovercraft fare was also about \$40. The hovercraft trip took only 27 minutes. Certainly, the group had *some* limit on how much they would have paid for the hovercraft ride. How much would *you* have been willing to pay to ride the hovercraft under those conditions?

FIGURE 5.8
Hovercraft used for channel crossings. *Source*: Photo Transport (U.K.) website,
www.photo-transport.co.uk/ferries/hovercraft/hovercraft.htm.

Solution to Example 5.10   Whatever amount you choose in excess of $40 is the surplus benefit you would have derived from the hovercraft ride.

In the private sector, investments are made by a firm with the intent to receive benefits as a result. When evaluating public projects, such as the widening or rerouting of a highway, there is the likelihood that the investment is made with public funds through a public agency, like a state department of transportation, which bears the investment and operations cost. However, as we saw in the bridge repainting in Example 5.5, the primary benefits are reduced user costs, such as reduced driver time and fuel use. For such a case, the engineer can compare the benefits and costs according to Equation 5.11.

$$\frac{\text{Benefit}}{\text{Cost}} = \frac{\sum \text{UserCosts}_{\text{Present}} - \left(\sum \text{UserCosts}_{\text{New}} + \text{ConsumerSurplus}\right)}{\text{Investment}_{\text{new}} + \text{O\&M}_{\text{new}} - \text{O\&M}_{\text{present}} + \text{Salvage}_{\text{present}} - \text{Salvage}_{\text{new}}}$$

$$(5.11)$$

In Equation 5.11, "O&M" stands for operations and maintenance.

Example 5.11:   New Road to Industrial Park

A new road is being proposed to create a shortcut from a major residential area to an industrial complex 15 miles away (as the crow flies). The current drivers [two-way *annual average daily traffic (AADT)* of 10,000 vehs/day] travel 27 miles each direction on a curvy, hilly road to the industrial park. Many do not go to the shopping center near the industrial park because of the present substandard road. The new alignment would put a new bridge over the river and cross some wetlands to shorten the distance to 16 miles. The traffic is expected to increase to 12,000 AADT. The cost of the new highway will be $72,000,000, including the bridge. Some rehabilitation (estimated at $5,000,000) would have to be done on the existing road by the state DOT in 10 years, if the new road is not built. If the new road is built, the old road will be turned over to the local governments. The O&M costs will decrease from $300,000 per mile on the old road to $250,000 per mile on the new road. When the cost of gasoline and time are included, the cost to the present users is $7.50 per trip; when the new road opens up, it will be $5.50. Because of the new road, the residential area is expected to grow faster. Traffic is expected to increase by 200 new users per year. The road will be designed to last 20 years. The DOT uses a 5% discount rate to evaluate projects such as this one. Use the three criteria in section 5.3.2 (viz., Equations 5.6 to 5.8) to evaluate the proposed project.

Solution to Example 5.11

Costs to the Public Agency.   The $72,000,000 construction cost is assigned to year 0. The present worth of avoiding a $5 million rehab project in year 10 is:

$$\$5,000,000 * [P|F, 0.05, 10 \text{ yr}] = \$5,000,000 * \left[\frac{1}{(1.05)^{10}}\right] = \$5,000,000 * [0.614] = \$3,070,000$$

The annual O&M costs on the existing road over the next 20 years will be $300,000/mi * 27 mi = $8,100,000.
The annual O&M costs on the new road over the next 20 years would be $250,000/mi * 16 mi = $4,000,000.

The annual saving in O&M costs from building the new road and relinquishing the old road to local government would be $4,100,000. The present worth equivalent of this annual O&M saving over 20 years is

$$\$4,100,000 * [P|A, 0.05, 20 \text{ yr}] = \$4,100,00 * \left[\frac{(1.05)^{20} - 1}{0.05(1.05)^{20}}\right]$$

$$= \$4,100,000 * [12.46] = \$51,086,000.$$

Therefore, the present worth of the project's costs is

$$PWC = \$72,000,000 - \$3,070,000 - \$51,086,000 = \$17,844,000$$

Note that these are all costs (or reduced costs) associated with the state DOT.

Benefits to Users of the Road.   The existing travelers to the industrial park benefit by the decrease in cost, put on a yearly basis, discounted and summed. Assuming 365 days per year, the annual reduction in user costs is ($7.50 − $5.50) per trip * 10,000 trips per day * 365 days per year = $7,300,000 per year. The same $[P|A, 0.05, 20]$ factor used in the cost analysis above can be applied here to calculate the present worth of benefits to current users.

$$PWB_A = \$7,300,000 * [P|A, 0.05, 20] = \$7,300,000 * [12.46] = \$90,958,000$$

The new road will attract 200 new users each year. Each new user will benefit from the $2 per trip saving each day of the year. The present worth of the benefits "enjoyed" by the increasing number of new users can be calculated using the $[P_G|G]$ factor in Figure 5.4, because the gradient is constant (+200 per year), not geometric (**g** percent per year).

$$PWB_G = (\$2/\text{trip} * 200 \text{ trips/day} * 365 \text{ days/year}) * [P_G|G, 0.05, 20]$$

$$PWB_G = \$146,000 * \left[\frac{(1.05)^{20} - (0.05 * 20) - 1}{(0.05)^2 * (1.05)^{20}}\right] = \$146,000 * [98.488] = \$14,379,308$$

The sum of the benefits to existing and new travelers is

$$PWB = PWB_A + PWB_G = \$90,958,000 + \$14,379,308 = \$105,337,308$$

Using the criteria introduced in section 5.3.3:

1. Net present benefit = $PW_{\text{benefits}} - PW_{\text{costs}} = \$105,337,308 - \$17,844,000 =$ $87,493,308.

2. $BCR = \dfrac{PWB}{PWC} = \dfrac{105,337,308}{17,844,000} = 5.90$

3. Equivalent uniform annual benefits and costs. First convert PWB to EUAB and convert PWC to EUAC, using $[A|P, 0.05, 20] = \left[\dfrac{0.05 * (1.05)^{20}}{(1.05)^{20} - 1}\right] = 0.080$. Note that $[A|P] = \dfrac{1}{[P|A]} = \dfrac{1}{12.46} = 0.080$ and that the PWC already include the gradient in benefits.

$$EUAB = PWB * [A|P] = \$105,337,308 * 0.080 = \$8,426,985$$
$$EUAC = PWC * [A|P] = \$17,844,000 * 0.080 = \$1,427,520.$$

The net uniform annual benefits are $8,426,985 − $1,427,520 = $6,999,465

In Example 5.10, the benefits associated with reduced user costs were much greater than the millions of dollars spent on construction and maintenance of the new road. Although this may be surprising, it is not unusual. It does make the estimation of user costs (especially if the value of travel time saved is included) the most important and difficult value to determine in the analysis.

5.3.6 Projects with Multiple Quantifiable Benefits

The same methods of benefit-cost comparison that were applied in the previous sections are applicable to problems where there are benefits from different sources (e.g., fuel use and safety) that are quantifiable in monetary terms. The benefits need to be determined separately and then combined as shown in Example 5.11.

Example 5.12: Proposed Overpass

An overpass will eliminate the heavy traffic congestion and reduce the crash rate at a busy intersection. The cost of the construction is $12,000,000, including land acquisition and the new interchange. It is assumed that the overpass will provide a higher level of service during the 4-hour peak period over the 2-mile design length. It is also assumed that no appreciable benefits will accrue during the off-peak periods, so that an analysis covering the 250 weekdays per year will be satisfactory. A discount rate of 8.5 percent is to be used. Other pertinent information is summarized in the table below. What is the net present worth of the project?

	Existing	**Proposed**
Vehicular operating speed	16 mph	40 mph
Peak period flow rate	4500 vph	6000 vph
Crash rate	500 per year	25 per year
Cost (time and fuel)	30 cents/VMT	22 cents/VMT
Average cost/accident (all property)	$4,000	$4,000
Annual O&M costs	$300,000	$400,000
Design life		15 years (no salvage)

Solution to Example 5.12 The direct benefits from the overpass are determined by using the concept of consumer surplus. (See Figure 5.6.)

Existing cost (price) $= 30$ cents $*2$ miles $= 60$ cents

Proposed cost (price) $= 22$ cents $*2$ miles $= 44$ cents

Existing traffic (quantity) $= 4500$ vph $*4$ hours $*250$ days/year $= 4.5$ million vehicles per year

Proposed traffic (quantity) $= 6000$ vph $*4$ hours $*250$ days/year $= 6$ million vehicles per year

Existing drivers $+$ new (consumer surplus benefit)

$= \frac{1}{2} * ((4.5 + 6)$ million$) * ($0.60 - $0.44) = $840,000$/year

Safety savings $= 4000/crash $* (500 - 25)$ crashes/year $= $1,900,000$ per year

Investment cost $= $12,000,000$

Increased net annual operating cost = $400,000 − $300,000 = $100,000 per year

Present worth factor = $[P|A, 8.5\%, 15] = \left[\dfrac{(1.085)^{15} - 1}{0.085*(1.085)^{15}}\right] = 8.304$

Total annual user benefits = $840,000 + $1,900,000 = $2,740,000

NPW of the project = $−$12,000,000 − ($100,000*[P|A]) + ($2,740,000*[P|A]).
Costs are shown as negative cash flow; benefits take on a positive value.
NPW = −$12,000,000 − ($100,000*[8.304]) + ($2,740,000*[8.304]) = $9,922,560

Example 5.13:   Road Widening

A $3.5 million road widening project over a 10-mile stretch is proposed. The project will not induce any increase in the 3-hour peak period traffic volume (currently 5000 vehicles). Use 250 as the number of days in a year having peak period traffic. The average speed of the existing traffic will increase from 35 to 45 mph. Energy use will go from an average of 23 to 24.5 mpg. The average annual cost of maintenance and operations is $123,000. If time is worth $12 per hour and gasoline costs $1.25 per gallon, determine the project's net present worth and benefit-cost ratio with a design life of 20 years and a discount rate of 8 percent.

Solution to Example 5.13   Calculate $[P|A, 8\%, 20] = \dfrac{(1.08)^{20} - 1}{.08(1.08)^{20}} = 9.82$. The NPW of the project's life cycle cost is then

$$\text{NPW}_{\text{LCC}} = \$3,500,000 + \$123,000*[P|A, 8\%, 20] = \$4,707,860$$

The NPW of each component of the benefits is calculated as follows.

- *Time saved per day.*   10 mi/45 mph − 10 mi/35 mph = 0.0635 hours saved per vehicle per day.
- *Annual value of time saved.*

    5000 vpd × 250 commuting days per year × 0.0635 hr/vehicle
    × $12 per hr = $952,500 per year

- *Energy cost reduction.*

    Gallons saved per day.

    10 mi/23 mpg − 10 mi/24.5 mpg = 0.4347 gal − 0.4181 gal

    = 0.0166 gal per vehicle

    Annual value of gallons saved (250 days of commuting)

    = 5000*250*0.0166*$1.25/gal = $25,750 per year

The annual benefit beginning in year 1 = $952,500 + $25,750 = $978,250 (assumes no "inflation").

The discounted present value of uniform annual benefits of 20 years = $978,250*9.82 = $9,606,415.

The NPW of the project is $9,606,415 − $4,707,860 = $4,898,555.

The BCR is $9,606,415/$4,707,860 = 2.04.

Example 5.14

For Example 5.13, assume that the traffic grows at the rate of 2 percent per year. Calculate the new net present worth and benefit-cost ratio for the project.

Solution to Example 5.14   Because traffic grows at $g = 2$ percent per year, the annual benefit will grow according to the geometric gradient in Figure 5.4. The discount rate $d = 0.08$; this is i in Figure 5.4. Because $g \neq i$, use the first $[P_g|A]$ equation in Figure 5.4.

$$[P_g|A, 0.08, 20] = \left[\frac{1 - \left(\frac{1+g}{1+i}\right)^n}{i - g} \right] = \left[\frac{1 - \left(\frac{1.02}{1.08}\right)^{20}}{0.08 - 0.02} \right]$$

$$= 11.353; \ P_g = \$978,\!250 * [11.353] = \$11,\!106,\!278$$

This is larger than the $9,606,415 found in Example 5.13 under the assumption of no traffic growth. The NPW of the project becomes $11,106,278 − $4,707,860 = $6,398,418 and the benefit-cost ratio becomes $11,106,278/$4,707,860 = 2.36.

5.4   RANKING TRANSPORTATION ALTERNATIVES

5.4.1   Introduction

Some citizens of Mythaca County have questioned how the county highway engineer has been spending the county's limited road maintenance budget. The citizens' group claims that more roads are in bad shape than ever before, and its members suspect that the engineer makes his maintenance decisions on a political basis (Figure 5.9). The engineer points out that he has only $365,000 per year with which to maintain more than 1000 miles of county roads. He admits that he does not use a formal system to decide which roads should be patched or resurfaced in any given year. Instead, he uses his judgment. After meeting with the citizens' group, the engineer agrees to ask for advice from the state's Local Technical Assistance Program (LTAP) at Mythaca State University. The LTAP offers technical assistance to local public agencies (cities, towns, and counties) in the state on matters related to transportation.

Have you ever had to make an important decision, but the choice between two or more options was not an easy one? The county engineer remembers looking for an apartment, as he was about to begin graduate school in a city new to him. After visiting several apartments, he had narrowed down his choices to two—one in the Bloomfield neighborhood and the other in East Liberty. Having been trained to think systematically, the engineering student made a list of the attributes favoring each apartment.

THINK ABOUT IT

What attributes are you likely to consider in choosing a place to live? Once you have made your initial list, try to rank your attributes from most important to least important.

FIGURE 5.9
Gravel county road. Photo: Jon D. Fricker.

When the engineer-to-be saw his lists of attributes, arranged in two columns, guess what he saw? He saw two columns of almost exactly the same length. In this case, if the choice was so difficult that the engineering student felt compelled to make a structured list, the result of having made such a list was likely to be inconclusive. Well, the future county engineer made a decision, and it was a decision with which he turned out to be very satisfied. As he looked back over this experience, what surprised the engineering student was that the attributes that made his choice so satisfying were not even on his original list!

In the apartment choice example, the engineering student had identified only two alternatives as worthy of further, detailed evaluation: the apartments in Bloomfield and East Liberty. To keep the situation simple, let us say that only two attributes mattered to the student: rent and distance to campus. These factors have the advantage of being easily quantified [e.g., using units of dollars per month and miles (or kilometers) to campus]. The quantities used to assess factors are often called measures of effectiveness (MOE). If the two factors are not equally important to the student, he might choose to apply weights to their values. Normally, weights are assigned so that their sum equals 100 percent or 1.00. In the apartment case, rent might be given a weight of 60 percent or 0.60, and distance a weight of 40 percent or 0.40. These values are summarized in Table 5.4.

TABLE 5.4   Apartment Choice Problem

Factor	Weight	Measure of Effectiveness	
		Bloomfield	East Liberty
Rent per month	0.60	$270	$180
Distance to campus	0.40	2.1 miles	3.9 miles

THINK ABOUT IT

If rent and distance to campus really are the only factors that matter, which apartment in Table 5.4 would you choose? Why? How much higher would the rent have to be for the apartment you chose before you would choose the other apartment? Suppose one apartment appeals to you more than the other, how might you weight appeal?

The apartment choice case described a personal decision. It was likely to have significant impacts only on the landlord, the renter, and perhaps the renter's neighbors. In the public sector, the situation is likely to be much different. As custodians of public budgets, public officials are entrusted with making decisions regarding programs or projects that could affect thousands of people. Because of this, public officials need to make decisions that are consistent and defensible. They need a system. That system can be referred to as the multiattribute effectiveness method, which is described in the next section.

5.4.2   Multiattribute Effectiveness Analysis Process and MOEs

Figure 5.10 illustrates the process for setting up the multiattribute effectiveness analysis. The analysis begins by identifying all the relevant alternative projects that should be considered. Then, the analyst determines those attributes that will likely have the greatest impact on the outcome of the evaluation (e.g., environmental impact, investment cost, energy use, and ease of construction). From these attributes, a set of MOEs can be developed. Given the MOEs, the analyst identifies factors that are important to making the decision and chooses performance measures that will be used in determining the weights to be applied to the MOEs.

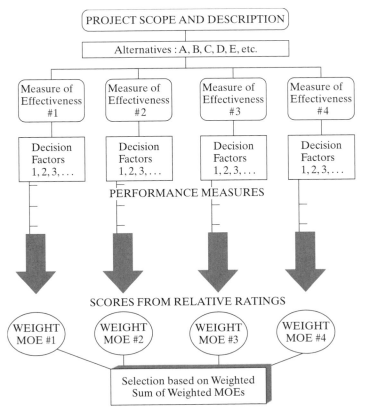

FIGURE 5.10
General representation of multiattribute effectiveness analysis.

The choice of MOEs can determine the outcome of the analysis; they should be chosen in an objective manner that adheres to the characteristics listed below:

- Comprehensive—include all the important aspects of the problem.
- Relevant—useful in differentiating between alternatives.
- Well-defined—easily understood by all.
- Non-redundant—avoid double counting of alternatives' attributes.

When "appeal" (a personal preference) is added to the apartment example, the whole analysis may be changed. This factor may be impossible to quantify by using standard units, such as dollars or distance. The result may be switching to an evaluation method that accounts for a ranking of alternatives.

The following set of MOEs might be considered for a public project. Each MOE is listed with some of its possible decision factors.

- *Minimize the impact on or preserve the environment.*   Impact on wetlands, endangered species, air quality, and water quality.
- *Maximize the economic return to stakeholders.*   Optimize initial investment, minimize life cycle costs, examine individual benefits and costs to stakeholders, maintain system reliability and accessibility.
- *Improve or maintain the quality of life of the residents.*   Enhance mobility, maximize employment potential, minimize construction disruption, increase land values.
- *Ability to implement.*   Time to implement, political and public support, legal challenges, existing infrastructure.

The fundamental objectives and their decision factors will vary from analysis to analysis. The list above is just one sample from a larger set of decision factors.

5.4.3   MOE Decision Factors and Performance Measures

After the MOEs are fully developed and the decision makers agree that they represent the most important issue(s) involved in the system, the analysis proceeds by evaluating each alternative in accordance with the MOEs. For example, if *cost* is an MOE, then the analyst makes a determination of the costs of each system, perhaps including a sensitivity analysis by analyzing the problem and comparing results by using a range of discount rates. Often, two or three different MOE values are developed from the sensitivity analysis for each alternative, giving the decision makers a range of values to consider.

Performance measures must be developed to determine how well the alternative projects perform in comparison to the objectives. Some variables (e.g., air pollutants, energy use, and noise) can be measured directly. Such direct measurements can be used in lieu of dollar amounts to determine the impact of the variable. If pollution were an important element, analysis of the emissions or health effects from each alternative would indicate to the analyst the relative merit of each alternative on the basis of pollution. Comparison of carbon monoxide, volatile organic compounds (usually hydrocarbons), nitric oxide, and small particulates for each alternative system will be made. It may be that an alternative will significantly decrease ozone but adds a large amount of small particulates. How that system is evaluated requires an understanding of the health risks of each.

How nonquantifiable items like aesthetics, neighborhood opinion, and equity will be handled in the analysis must be discussed with those who will be the decision makers. In particular, if is there is a safety advantage of one system over another, how will a reduced number of crashes or a life saved be valued?

THINK ABOUT IT

What factors should go into the estimate of the value of a human life for purposes of evaluating alternatives that will have an impact on number lives lost or saved?

Even if MOEs are quantifiable, their values for each alternative will probably need to be transformed. For example, in Table 5.4, would you really want to treat dollars of rent and miles of distance without transforming the units into a common basis? Subsection 5.4.6 shows how this can be done. Subsection 5.4.5 demonstrates another method—a relative ranking scheme that works for both quantifiable and non-quantifiable MOEs. This second method rates the MOEs on a scale of 1 to 5 or 1 to 10. For example, if the MOE related to cost is better (lower-cost) for one alternative than the other, then the better alternatives might receive an MOE ranking of 5, with the higher-cost alternative receiving a rating of 3 or 4. Care must be taken to be consistent in assigning ratings to the MOEs. *Higher* mobility and economic development MOE values should receive *higher* ratings; *higher* costs and pollution MOE values should receive *lower* ratings.

5.4.4   MOE Weights

The next step in the analysis of alternatives is to determine the relative importance of each MOE. For example, which is more important, protecting the environment or attracting employers to the community? (Giving equal weight to all the MOEs is normally not appropriate.) This is a critical step, one in which consensus (or consent) among those involved in the decision is necessary.

The weight given to each measure of effectiveness usually depends on the policies or dictates of the decision makers or agency looking at the alternatives. Sometimes it is inherent in the reason for proposing the project. For example, "We need a project to improve access to the new sports stadium." If, for that project, limited funds makes investment cost very important, travel time is fairly important, with safety and pollution less important, then the weights could be 40 percent on investment cost, 30 percent on travel time and 15 percent on each of the other two. The sum of the weights will always be 100 percent.

The transportation engineer/planner needs to find an approach by which the weights can be fairly and effectively established. Some of the mechanisms for assigning weights to MOEs include:

- Public involvement, such as a survey or a series of public meetings to get citizen input.
- A consensus of community leaders consulted by the planner.
- Consulting the agency funding the project.

The weight assigned to each MOE is a subjective process. It sometimes helps to vary the weighting criteria before presenting the finding to the decision maker. This is done to understand the sensitivities of the project (or the analysis) when examined from different perspectives.

5.4.5 Weighted MOE Sum for Each Alternative

The final step in the process is the application of the weights to the MOEs. Example 5.15 illustrates the steps discussed in subsections 5.4.2 through 5.4.4.

Example 5.15: The New Mazurka Bridge

The residents who live on the island across Mendel Bay from Mazurka and the developers who want to develop the island would like to replace the ferry service with a bridge. (See Figure 5.11.) Two bridge sites were proposed—one across the north bay along (or near) an existing rail bridge and a shorter bridge over the south bay.

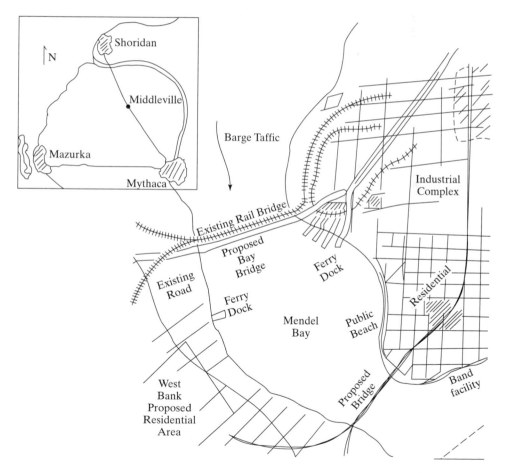

FIGURE 5.11
Downtown Mazurka and west side of Mendel Bay.

A. Identify the MOEs that should be used to analyze the two options.

B. Determine the MOE values for each alternative.

C. Assign weights to each MOE.

D. Compute the weighted sums for each alternative.

Solution to Example 5.15   The engineer designer/analyst must consider bridge length and height, traffic forecasts, soil conditions, foundation and pier design, and environmental impacts. With that information, developing the least cost bridge design can be accomplished. However, maximizing societal benefits is more complex and is seldom amenable to a mathematical model. Assessing some of the impacts and casting them as MOEs allows the analysis of the problem to begin.

A. Among the many MOEs that could be considered, those that will most impact the decision are:

1. *Accessibility and travel time.*   Which of the solutions will improve accessibility to destinations in Mazurka the most?

2. *Neighborhood impacts.*   How will the bridge affect the residents and businesses at the ends of the bridge?

3. *Roads and streets.*   To what extent will the roads in the area need to be upgraded to handle the expected increase in traffic?

4. *Cost of the bridge and its maintenance.*

B. Although it would be possible to estimate improved accessibility in terms of travel time savings and estimate the costs of road upgrades, these and the other MOE values were assigned relative values on a scale of 1 to 5:

MOE	Northern Bridge	Southern Bridge
Accessibility and travel time	4	3
Neighborhood impacts	5	1
Roads and streets	2	4
Cost of the bridge/maintenance	2	4

The committees that decided on these ratings offered the following rationale.

- *Accessibility and travel time.*   The Northern Bridge offers better accessibility to key destinations in Mazurka.
- *Neighborhood impacts.*   (N) The Northern Bridge will feed into the same industrialized commercial area that the ferry does today. Its impact on the neighborhood will not be great. (S) The greatly increased car and truck traffic through the residential neighborhood at the east end of the southern bridge will be disruptive. Noise and pollution will increase.
- *Roads and streets.*   (N) The existing streets in the neighborhood at the northern end of Mazurka are narrow, with sharp turns and pavement designed for light vehicular traffic. A new arterial would have to be fashioned to provide a road with a reasonable level of service to traffic from the island to the commercial district and CBD. Estimated cost is $6.7 million. (S) The industrial and commercial sectors of town would benefit from improved access to their suppliers and to existing (and prospective) employees via the southern bridge. A refurbished highway will need to be built to handle the traffic, especially the added truck weight. However, land is

available to construct well-designed entry and exit ramps for the bridge. Estimated cost is $3.8 million.

- *Cost of the bridge and its maintenance.*   (N) The cost of the Northern Bridge is estimated at $16 million, with maintenance at $400,000 per year. (S) The cost of the Southern Bridge is estimated to be $8 million, with annual maintenance of $200,000.

C. Assigning weights to each MOE took into account the original purpose for building a bridge to Mazurka. Also considered were the opinions of the citizens with a stake in the project, either as beneficiaries or as those adversely affected.

- *Accessibility and travel time.*   Reduction in travel time will cause the traffic to increase, spur development, and increase the tax base of the city. Weight = 40 percent.
- *Neighborhood impacts.*   With concerns about increased traffic through neighborhoods, especially trucks, the weight of this MOE was 30 percent.
- *Roads and streets.*   The bridge will mean that there will be a need for wider streets, with a new road through the community near the southern bridge. The assigned weight was 15 percent.
- *Cost of the bridge and its maintenance.*   This received a weighting of 15 percent. Although the funds are available for the project, the cost of each proposed bridge is considered in the name of fiscal responsibility.

D. The MOE weights are multiplied by the MOE value for each alternative. These products are then summed for each alternative as shown in Table 5.5.

According to this evaluation, the Northern Bridge would be the recommended alternative to improve the supply of transportation to the island across Mendel Bay.

TABLE 5.5   Solution to Mazurka Bridge Example

MOE	Weight (%)	Northern Bridge	(Weighted)	Southern Bridge	(Weighted)
Accessibility and travel time	40	4	160	3	120
Neighborhood impacts	30	5	150	1	30
Roads and streets	15	2	30	4	60
Cost of the bridge/maintenance	15	2	30	4	60
Weighted total			370		270

THINK ABOUT IT

In Table 5.5, the ratio of the biggest weight (40%) to the smallest (15%) is almost 3:1. What chance do the two MOEs with the small weights have to affect the results?

In some analyses, the weighted sums of the MOEs do not differ by much. When two alternatives are nearly equal, other considerations (usually qualitative in nature) may determine the selection. Example 5.15 involved only two alternatives. Example 5.16 presents a four-alternative case.

Example 5.16

Several transportation projects are being considered to help meet air quality standards in a freeway corridor that connects a major residential area with the industrial/commercial section of a city. The corridor is about 12 miles long. The freeway has become extremely congested, and the source of irritating smog must be reduced. There four projects being suggested by transportation planners are as follows:

- Build a fourth lane for traffic.
- Put in a light-rail transit (LRT) system in the median of the existing freeway.
- Convert the third lane into a lane for high-occupancy vehicles (HOVs) only.
- Add a busway without altering the present freeway.

Each of these alternative solutions has its pluses and minuses. The MRPC staff must analyze the alternatives and identify (and defend) the "best" alternative to local public officials.

Solution to Example 5.16   To evaluate the pluses and minuses of these alternative solutions, the planning agency has developed a list of MOEs. The procedure uses the following steps:

1. Set up a simple, yet realistic, model of the system within its proposed environs.
2. Determine the relative importance of each MOE. Assign weights to the each MOEs. Table 5.6 presents these data.

   - *Reduction in congestion.*   Congestion causes pollution, but it also affects productivity and quality of life.
   - *Reduction of air pollution.*   At present, stop-and-go traffic during periods of congestion creates vehicle emissions that exceed federal air quality standards.
   - *Access.*   Maintaining or improving access to businesses in the CBD is a key factor for some stakeholders.
   - *Investment cost.*   The funding available is constrained.
   - *Operational cost.*   The system operation will take tax dollars away from other activities.
   - *Community economic development.*   The region needs to attract more businesses.

3. Assign to each project alternative a rating for each MOE. In Table 5.7, the ratings are in the range 1 though 5 (5 = best), but 1 through 10 or any other range is acceptable. This rating is based on an analysis of each MOE for each of the alternative systems.
4. Apply the weights to the various performance ratings. The weighted results are then summed and compared. In Table 5.8, the LRT system alternative has the highest weighted sum, which suggests that the LRT system offers the best option for the city.

TABLE 5.6   Weighting the MOEs

MOE #	Measures of Effectiveness	How Measured	Weighting
A	Reduction in congestion	Improved travel time	20
B	Reduction of air pollution	Reduced ozone and particulates	35
C	Access	Ease of using the system	5
D	Investment cost	Dollars	20
E	Operational cost	Maintenance and labor	10
F	Community economic development	Land use and attractiveness to investors	10

TABLE 5.7   Ranking the Alternatives Relative to Their MOE Performance (5 = best)

MOE #	Measures of Effectiveness	Add Lane	LRT	HOV	Busway
A	Reduction in congestion	3	5	2	3
B	Reduction of air pollution	2	5	3	3
C	Access	5	2	1	4
D	Investment cost	1	2	5	2
E	Operational cost	4	2	4	3
F	Community economic development	1	4	1	2

TABLE 5.8   Applying the Weights to the Rankings

MOE #	Measures of Effectiveness	Weight	Add Lane	LRT	HOV	Busway
A	Reduction in congestion	20	60	100	40	60
B	Reduction of air pollution	35	70	175	105	105
C	Access	5	25	10	5	20
D	Investment cost	20	20	40	100	40
E	Operational cost	10	40	20	40	30
F	Community economic development	10	10	40	10	20
	Total		225	385	300	295

5.4.6   Ranking Alternatives with Transformed MOE Values

At the start of this section (Section 5.4), the Mythaca County Highway Engineer wanted to improve the way in which he chooses road segments for maintenance and repair. To test any proposed ranking system, a list of 11 road segments that are eligible for road maintenance activity in the next fiscal year has been prepared. Three factors have been identified as the basis by which the "neediest" road segments must be selected: pavement condition [*pavement serviceability rating (PSR)* = 5.0 is best, or least needy], daily traffic volumes (lower AADT is less needy), and a *hazard index* (HAZ = 0) is least needy). The measures of effectiveness for each factor on each road segment are given in Table 5.9. Note that the units for each MOE are different. So are the range of values. Example 5.17 shows how such a situation can be addressed in a systematic, rational way by using a procedure called the *percentile method*.

THINK ABOUT IT

Using only your personal judgment, choose and rank the three neediest road segments in Table 5.9.

TABLE 5.9   Roadway Data for Priority Setting

Segment	PSR	AADT	HAZ	Length	$/mile
A	1	366	0	2.3	79,000
B	3	448	0	2.5	18,000
C	2	5704	0	6.6	61,000
D	2	106	2	1.2	75,000
E	3	263	1	1.5	31,000
F	5	359	0	2.6	0
G	4	278	0	2.0	11,000
H	2	125	1	1.9	85,000
I	3	119	0	3.2	20,000
J	1	672	0	1.2	65,000
K	2	98	0	0.5	60,000

Segment: road identifier
PSR: pavement condition (5 = best)
AADT: annual average daily traffic
HAZ: index of traffic safety hazards (0 = safest)
Length: segment length (miles)
$/mile: cost to remove pavement deficiencies
Source: Shaffer, 1987.

Percentile Method.

Example 5.17:   Ranking Alternative Projects Using the Percentile Method

If the weight of each factor in Table 5.9 is 1/3 and if the engineer had about $600,000 to allocate, which of the 11 projects should he fund?

Solution to Example 5.17   Most likely, you have taken an aptitude test. The Scholastic Aptitude Test (SAT) is an example. Along with your raw score (out of 1600), you probably were informed of your relative standing among all individuals who took the SAT. This relative standing is expressed as a percentile score. If your raw SAT score put you in the 89th percentile, this means that your score was better than 89 percent of all scores recorded on the SAT. We can apply this normalization technique to the roadway data in Table 5.9. By definition, the roadway segment with the highest AADT value (segment C) is the neediest roadway in terms of the traffic volume factor and should be assigned a percentile score 100. Likewise, segment K has the lowest AADT, which is the least needy of the roadway segments and ought to have a percentile score of zero. This indicates that segment K is needier than no other segment when the factor being considered is traffic volume. Unless other road segments have AADT values that are tied for first or last place, they will have to be assigned percentile scores between 0 and 100. In the case of road segment A, there are three other segments with AADT values greater than segment A's value of 366 and seven other segments with AADT less than 366. This means that the percentile score for segment A in Table 5.9 is

$$\text{AADT Percentile (A)} = \frac{L}{\text{Total}} * 100 = \frac{7}{10} * 100 = 70.00 \tag{5.12}$$

where

L = number of segments less needy than segment A

Total = total number of segments other than A.

In the general form of this equation, the words "less needy" may have to be replaced by words that reflect the reason for the ranking process. For example, if a county was considering several road graders, the county could base its percentile definition on how many graders are "worse" than a specific grader with each respect to each factor. Another modification to Equation 5.12 may be in order, in cases where two or more alternatives have the same factor values. This is often the case when factor values are integer and have few possible values. In Table 5.9, PSR can take on integer values between 1 and 5, and HAZ has integer values between 0 and 2. Clearly, segment F should have a PSR percentile score of 0, because it (and only it) has a PSR factor value that is less needy than any other segment. Segments A and J share the neediest PSR factor value and deserve a PSR percentile score of 100. But how can we calculate a percentile score for segments that have the same factor value as other segments, but the factor value is not at the high or low extreme? In the case of road segment B, PSR = 3. There are six road segments with PSR values that are lower (needier), two with higher (less needy) values, and two that are the same as that of segment B. A reasonable way to make Equation 5.12 more general is to exclude those segments whose factor values that are the same:

$$\text{Factor percentile score}(i) = \frac{L}{D} * 100 \qquad (5.13)$$

where

L = Number of alternatives with a "less needy" factor value than alternative i

D = Number of alternatives with factor values different from alternative i

Using Equation 5.13, segment B's PSR percentile score would be:

$$\text{PSR percentile score (B)} = \frac{2}{10 - 2} * 100 = 25.0$$

Segment B's AADT percentile score is $(8/10)*100 = 80.0$ and segment B's HAZ percentile score is $(0/3)*100 = 0.0$.

Once these factor-specific percentile scores have been calculated for a segment, that segment's *composite percentile index (CPI)* can be computed:

$$\text{CPI(Alt.i)} = \sum_{j=1}^{n} w_j * P_j \qquad (5.14)$$

where w_j is the weight assigned to the jth factor and P_j is the segment's jth percentile score for the jth factor. If the three factors in Table 5.9 have equal weights, segment B's CPI will be:

$$\text{CPI(B)} = (1/3)(25.0) + (1/3)(80.0) + (1/3)(0.0) = 8.33 + 26.67 + 0.0 = 35.00.$$

After all the CPI values for all segments have been computed by using Equation 5.14, they are entered into Table 5.10.

TABLE 5.10   Priority Ranking of Highway Segments

Segment	PSR	AADT	HAZ	CPI	Dollars ($)	Cumulative ($)
J	100	90	0.0	63.3	78,000	78,000
A	100	70	0.0	56.7	181,700	259,700
H	43	30	89.0	54.0	161,500	421,200
D	43	10	99.0	51.7	90,000	511,200
E	25	40	89.0	51.3	46,500	557,700
C	43	100	0.0	47.7	402,600	960,300
B	25	80	0.0	35.0	45,000	1,005,300
G	11	50	0.0	20.3	22,000	1,027,300
F	0	60	0.0	20.0	0	0
I	25	20	0.0	15.0	64,000	1,091,300
K	43	0	0.0	14.3	30,000	1,121,300

The segments can be ranked in order of need, so that the limited county road budget can be allocated to the neediest road segments. Segments J, A, H, D, and E receive the five highest CPI scores. Together, those five road projects would cost $557,700. Adding the sixth neediest project (segment C) would exceed the $600,000 budget specified in the problem. The percentile method is one ranking method (among many) that the county highway engineer might adopt as a rational system that can be replicated by anyone having the same road segment data.

THINK ABOUT IT

The percentile method used in Example 5.16 was called a *rational* method? Does this mean it is an *objective* method? Explain your answer.

Range Index Method.   Note that, in Table 5.9, the busiest road (segment C) had an ADT value much greater than that of the second busiest road (segment J). In Table 5.10, however, segment C's percentile value for ADT was only a little higher than segment J's. If the *cardinal* values of factors are just as important as their *ordinal* values, an alternative to the percentile method to consider is the range index method (Shaffer and Fricker, 1987). It normalizes the raw factor values like this:

$$\text{Factor RI(Alt.i)} = \frac{x}{\text{Range}} * 100 \tag{5.15}$$

Equation 5.15 computes the *range index (RI)* for Alternative i with respect to a specified factor. Figure 5.12 helps explain Equation 5.15. For a given factor, the values for each alternative are ranked from worst to best, neediest to least needy, and so on, depending on the intent of the ranking process. In Table 5.9, segment C has the "worst" AADT value, because segment C is the "neediest" segment in terms of traffic volume. Therefore, $f_w = 5704$. The best (least needy) AADT value in Table 5.9 is 98 for segment K, so $f_b = 98$.

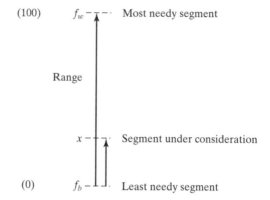

FIGURE 5.12
The Range Index Method. *Source:* Shaffer and Fricker, 1987.

In Equation 5.15, range $= f_w - f_b = 5704 - 98 = 5606$. For any given segment i with factor value f_i, $x = f_i - f_b$. For the factor AADT and segment B, $x = 448 - 98 = 350$ and AADT RI(B) $= (350/5606)*100 = 6.24$. Applying Equation 5.15 to segment B's other factor values, PSR RI(B) $= ((3 - 5)/(1 - 5))*100 = 50.00$ and HAZ RI(B) $= ((0 - 0)/(2 - 0))*100 = 0.00$. As was done with the percentile scores, the range indices for segment B can be combined into a composite range index by using appropriate weights:

$$\text{CRI}(\text{Alt.}i) = \sum_{j=1}^{n} w_j * \text{RI}_j \qquad (5.16)$$

For segment B, CRI(B) $= (1/3)(50.00) + (1/3)(6.24) + (1/3)(0.00) = 18.75$. After all the CRI values for all segments have been computed by using Equation 5.16, the segments can be ranked in order of need, with the highest CRI as the neediest.

THINK ABOUT IT

Based on what you have seen so far about the percentile and range index methods of normalizing factor values, what are the pros and cons of each method? Which method do you prefer?

5.4.7   Complications

Don't Make the Weights too Heavy.   There is a common tendency to give the more important factor a weight that is too high with respect to the less important factors. Under most ranking or priority setting schemes, assigning weights so that the highest weight is more than three times the lowest weight often causes a *preemptive* situation. What this means is that the higher-weighted factor so dominates the low-weighted factor, that the low-weighted factor has little or no influence on the outcome of the ranking procedure. If this is the case, why even bother collecting data for the low-weighted factor? This is one reason why the methods presented in this lesson call for allocating factor weights so that their sum is 1.00 or 100 percent. It tends to be much easier to assign weights like "5" and "1," than to assign weights like "83" and "17." In any case, try to keep the ratio of highest factor weight to lowest factor weight smaller than 3.0.

MOEs that Are Unpleasant or Nearly Impossible to Quantify.   We return to the problem of placing a value on human life. A great many elements could go into arriving at such a figure. The effort would be immense, controversial, and even emotional. The result would eventually be a number to put into an analysis such as those described earlier in this chapter. One strategy that may avoid much of this difficulty is to estimate the values for all the other (presumably less controversial) factors and then determine the value of human life that would just barely justify the project being undertaken. In many cases, the "break-even" value will be so low or so high as to make the decision obvious (or nearly so). Even in cases where the break-even value is in the "gray area," where well-intentioned analysts will disagree, the dispute has been reduced to a decision about a single number: Is the break-even value of human life high enough? Yes or no?

5.4.8   More on Ranking Methods

Many ways to rank alternatives have been developed (Adler and Posner, 2000; Collier, 1982; Merkhofer, 1987). Some are rather complicated and may require special data, special software, and the services of an expensive consultant. The LTAP advisor who was sent (without charge) by Mythaca State University to meet with the county highway engineer and the citizens' group suggested that they develop several different ranking procedures of their own, test them, then decide which one(s) to adopt. Each of the methods turned out to be quite simple, because the county engineer and the citizens' group preferred "transparent" methods they could understand to "black boxes" designed by outsiders.

The choice of the ranking approach often depends on the kind of projects being evaluated. When the nature of the projects is the same, such as road maintenance in Example 5.16, ranking using a method such as percentiles is usually best. Similar approaches, such as range indexing or normalizing, may also be used (Shaffer, 1987). However, when the projects are quite different, a multiattribute effectiveness analysis (like the method used in Example 5.14) is often used.

Project evaluation, ranking, and selection is a difficult, but necessary process, especially for transportation engineers who have limited budgets funded by tax revenues. Whatever method is used, it must be consistent and reflect appropriate factors and their measures of effectiveness.

5.5   THE EXISTING INSTITUTIONAL STRUCTURE

The existing public institutional structure fostering the provision of transportation has evolved over a considerable period of time since the late 1800s when the first railroad legislation was enacted. For years, a variety of agencies handled the diverse federal responsibilities for transportation, including the U.S. Department of Commerce. Since 1967, however, the federal level responsibility for most aspects of transportation was placed under a new cabinet level department, the United States Department of Transportation (U.S. DOT).

5.5.1 The United States Department of Transportation

In the legislation that established the United States Department of Transportation, the department's role is to:

- Provide general leadership in identifying and solving transportation problems.
- Develop and recommend transportation policies and programs.
- Achieve transportation objectives considering the needs of the public, users, carriers, industry, labor, and national defense.

A 1990 policy statement from U.S. DOT, laying out an agenda for transportation in the 1990s, builds on the original themes.

1. Maintain and expand the nation's transportation [system].
2. Foster a sound financial basis for transportation.
3. Keep the transportation industry strong and competitive.
4. Ensure that the transportation system supports public safety and national security.
5. Protect the environment and the quality of life.
6. Advance U.S. transportation technology and expertise for the 21st century.

Since the events of September 11, 2001, item 4 (transportation security) has taken on added significance.

At present, U.S. DOT is organized along modal lines, with the Office of the Secretary of Transportation providing the integration between modes. The modal agencies are the Federal Highway Administration (FHWA), National Highway Transportation Safety Administration (NHTSA), Federal Transit Administration (FTA), Federal Railway Administration (FRA), Federal Aviation Administration (FAA), and Federal Maritime Administration (FMA). The Coast Guard was part of U.S. DOT until it was moved to the new Department of Homeland Security (DHS). (In time of war, the FAA is transferred to the Department of Defense.) Most modal administrations are further organized geographically into regions and districts. There is one quasi-private/public agency called the St. Lawrence Seaway Development Corporation. Other federal agencies that play a role in the provision of transportation include the Forest Service and Bureau of Land Management, which provides roads on public lands. The U.S. Army Corps of Engineers is responsible for construction and operation of the inland waterway system and development and maintenance of some parts. U.S. DOT also provides for pipeline safety, hazardous material movement, emergency transportation, transportation statistics, and multimodal systems analysis.

5.5.2 State and Local Government

The states play the primary role in the provision of major highways. Each state has an organization responsible for the design, construction, maintenance, and operation of a state highway system. Originally, these organizations were almost universally known as state highway departments. During the 1970s and 1980s, many states broadened the role of these organizations to include responsibility for other modes of transportation, and many of them changed their names to state DOTs. The state role in the provision of modes other than highways includes establishment of policy and, in some cases, funding. However, in most states, the bulk of the DOT personnel and the major transportation funding are still in support of the state's highway system.

The involvement of local governments (cities and counties) in the provision of transportation varies from jurisdiction to jurisdiction. In most cases, local governments are responsible for setting design standards for local streets and roads and for maintaining and operating them. This includes providing traffic control and snow removal. Many communities have some form of mass transportation that may be provided by either a public authority or by a private, for-hire carrier. Many communities have a bus system supplemented by private taxis or vans for transporting elderly. In addition, communities may be directly involved in the design and construction of major transport facilities. Many local streets in urbanized areas, even if provided initially by land developers, are maintained by local government. Local government often works in partnership with developers to ensure adequate access and proper traffic control. In addition, some local governments are directly involved in the provision of transit services or port facilities.

There are a variety of local agencies such as special districts and authorities. These organizations are publicly owned but operate to some extent like private businesses. They commonly operate transit systems, airports, seaports, toll roads, and toll bridges. One example is the Port Authority of New York and New Jersey. Working across the two states, the Port Authority has responsibility for the harbor, ports, the four airports serving the New York City metropolitan area, the toll bridges, and the intercity bus lines between New York and New Jersey.

Finally, transportation planning is primarily the responsibility of agencies that exist at the level of the metropolitan region. Such organizations are known as metropolitan planning organizations (MPOs). MPOs are most commonly associations of local governments known as councils of governments (COGs). Some MPOs are statutory organizations created by state legislatures or organizations created by interstate compacts. In all cases, however, the governing body of the MPO is composed of representatives of local governments.

In addition to the public institutional structure involved in the provision of transportation, there are a number of private organizations. The most conspicuous of these are the carriers, such as airlines, truck lines, barge lines, and railroads that provide fleet operation for the freight and commercial intercity passenger systems. In addition, some transit services are provided by private firms under contract to public agencies. In addition, the maintenance of local streets is often done by the private sector under contract from the city.

5.5.3   The Financing of Transportation

The system for financing transportation has also evolved over time. Sources of funds include user charges, public general funds, private investment, and cross-subsidization among various levels and types of systems. User charges include direct charges such as fares and tolls and, more importantly, indirect sources such as fuel taxes. In either case, however, the major share of the financing comes from users and is to some degree proportional to the extent of their use of the system. Over the years, the user taxes have gone into major development funds such as the Highway Trust Fund and the Airport and Airway Development Fund. These funds are held by and disbursed by the federal government to the states, based on a formula that accounts for the amount collected from each state in user taxes and each state's need.

Table 5.11 indicates that road funds account for the largest share of transportation spending, often in excess of 70 percent. This should not be surprising, considering the large number of road miles that exist within the United States. Although most of that money is used for road construction and maintenance, a portion goes for improved highway safety. Airports consume about 15 percent, or about half, of the remaining funding. Overall, the federal government puts up only about 27 percent of the total money spent on transportation projects, whereas the other levels of government expend about 73 percent.

Private investment occurs in the parts of the transportation system that are normally under private control, such as the carriers. Furthermore, there are instances where public agencies will issue operating franchises, as in toll facilities or in the purchase of transportation services under contract.

Cross-subsidization occurs when revenues collected from users of one type of transportation system are used to finance some other type of system. A common example in recent years has been the option some communities have of using the grant money from the Highway Trust Fund to help finance new public transit systems, such as the light-rail system in Cleveland.

TABLE 5.11 Expenditures by U.S. Government Entities for Transportation Facilities and Services in 1993

Mode	Federal Government (%)	State/Local Governments (%)	Total (%)
Highway	56	77	71
Air	30	10	15
Transit	6	10	9
Marine	5	2	3
Railroad	3	1	1
Total	100	100	100
1993 dollars (millions)	$32,500	$88,500	$121,000
Percent	27	73	

Source: Transportation in America, ENO Foundation, 1994.

It is commonly the case in the United States that the higher levels of government provide a significant portion of the funding in partnership with the state and local government. The local government, often operating through a politically chosen authority, usually oversees the project. State and federal highways are managed by the state highway component of the state DOT, although much of the funding is provided by the federal government. Because of insufficient fare box revenue, public transit systems, which are managed and operated locally, receive funding from all levels of government. The federal share is usually the largest contributor. Distribution of transit funding in any state usually occurs in accordance with some formula based on such quantities as population, amount of service provided, and transit ridership.

In the public expenditure process, there are normally two stages of decision making. The first is the authorization of expenditures. When Congress or state legislatures pass "transportation bills," such as the Intermodal Surface Transportation Efficiency Act of 1991 (ISTEA) or the Transportation Equity Act for the 21st Century (TEA-21), they normally set up programs and authorize expenditures for several years into the future. The authorizations provide some indication of the level at which the legislature intends to fund the programs. They normally establish ceilings for funding. Funds are then usually appropriated on an annual basis by separate bills.

SUMMARY

The provision of transportation involves much more than designing and building a facility or service. The needs and preferences of the community must be determined. This may be difficult, because of the different viewpoints held by various stakeholders in the community. Once problems and needs have been identified, alternative solutions can be generated and evaluated. The evaluation of each alternative involves an assessment of benefits and costs. Whenever possible, the benefits and costs are expressed in monetary terms, and appropriate economic analysis tools can be applied. However, some criteria are difficult or impossible to convert to dollars. This chapter covers both cases and presents standard procedures by which competing projects can be ranked with respect to specified criteria.

The transportation planning process is a public one. Members of the public must be involved in the process. The process is also greatly influenced by government entities, whose regulatory, funding, and administrative activities help determine which transportation projects become reality.

ABBREVIATIONS AND NOTATION

AADT	annual average daily traffic
BCR	benefit-cost ratio
CPI	composite percentile index
EUAB	equivalent uniform annual benefit
EUAC	equivalent uniform annual cost
FTA	Federal Transit Administration
HAZ	hazard index for roadway section
ITS	intelligent transportation systems
LRTP	long-range transportation plan
MOE	measure of effectiveness
MPO	metropolitan planning organization

NPV	net present value
NPW	net present worth
O&M	operation and maintenance
PSR	pavement serviceability rating
RI	range index
TEA-21	Transportation Equity Act for the 21st Century
TIP	Transportation Improvement Program

GLOSSARY

Consumer surplus: The aggregate amount of value enjoyed by consumers who purchase a good or service at a price lower than they were willing to pay.

Discount rate: The rate at which the value of a dollar amount diminishes with time.

Goals: A general statement of the direction a program should go.

Intelligent Transportation Systems: The application of technologies to improve the efficiency and/or safety of transportation facilities and services

Life-cycle cost analysis: An analysis that includes all costs and benefits that are expected during the complete life of a project, including salvage value.

Net present value: The discounted value of a series of cash benefits or costs in a future year, brought back to "Year Zero."

Objectives: A more specific list of activities designed to meet goals. Objectives should be measurable or otherwise permit a clear appraisal of the extent to which they have been achieved.

Sensitivity analysis: Varying inputs to see how much the outputs vary. In this way, the sensitivity of a process to the variability of certain inputs can be ascertained.

Sunk cost: An expenditure that is committed to and, therefore, will not affect the cost analysis.

Transportation Improvement Program: Published schedule of all transportation projects for an urban area or state.

REFERENCES

[1] Adler, M. and E. Posner, eds., *Cost-Benefit Analysis: Legal, Economic, and Philosophical Perspectives*, University of Chicago Press, 2000.

[2] AMATS (Akron Metropolitan Area Transportation Study), 2025 Regional Transportation Plan, Figure 1-1, May 2002, http://ci.akron.oh.us/AMATS/RTP_2025/, as viewed August 2002.

[3] Collier, C. and W. Ledbetter, *Engineering Cost Analysis*, Harper and Row, 1982.

[4] FHWA, The Metropolitan Transportation Planning Process: Key Issues, A Briefing Notebook for Metropolitan Planning Organization Board Members, A Publication of the Metropolitan Capacity Building Program, Federal Highway Administration, November 2001. Available on the Internet at www.mcb.fhwa.dot.gov/Documents/BriefingBook/BBook.htm#2BB

[5] Merkhofer, M., *Decision Science and Social Risk Management: A Comparative Evaluation of Cost-Benefit Analysis, Decision Analysis, and Other Formal Decision-Aiding*, D Reidel Pub Co; ISBN: 9027722757, February 1987.

[6] Shaffer, Joseph L., "A Methodology for Determining and Prioritizing County Highway Network Needs," MSCE thesis, School of Civil Engineering, Purdue University, May 1986.

[7] Shaffer, Joseph L. and Jon D. Fricker, "Simplified Procedures for Determining County Road Project Priorities," *Transportation Research Record 1124*, Transportation Research Board, 1987, p. 8–16.

[8] Tyner, W., J. Binkley, M. Matthews, and R. Whitford, *Transportation Energy Futures—Paths of Transition, Volume 2: Benefits and Costs*, Purdue University, DOT-ATC-81-14, Report for U.S. Department of Transportation, November, 1981.

EXERCISES FOR CHAPTER 5: PLANNING AND EVALUATION FOR DECISION MAKING

The Transportation-Planning Process

5.1 Metropolitan Planning Organization. Is the city in which you live located within an MPO's jurisdiction? If so,

(a) What is the MPO's official name?

(b) How are the cities, counties, and unincorporated areas within the MPO's jurisdiction represented on the MPO's governing body?

(c) What is the current horizon year for the MPO's long-range transportation plan?

(d) Does the MPO have responsibility for anything besides transportation planning and land use controls in its jurisdiction?

If the place in which you live is not located within an MPO's jurisdiction, choose the MPO nearest to where you live or go to school and answer the questions above.

5.2 Stakeholders. Apex Industries, a major employer in the Eastern Mythaca area, is threatening to move its operations to a city in another state. Apex cites deteriorating transportation infrastructure and services as a major factor in the decision to leave Mythaca. Its business depends heavily on the port facilities at Mazurka and, without a circumferential bypass highway, they have to face the very heavily traveled freeway through the CBD every day. A meeting of major stakeholders in Mythaca was convened. The result was a decision to hire you as a consultant to document the nature and extent of the transportation problems in the Mythaca area. Name five major stakeholders other than Apex and describe why they are key "players." List five kinds of data that you would want to collect to document the nature and extent of the transportation problems that plague Apex so much that its management seeks to relocate.

5.3 Stakeholders. Because of the governor's desire to provide more recreational opportunities for the state's residents, the state is proposing to build an interstate highway connecting Mythaca County to New Cyberon, a large city to the north. The proposed route will pass within 3 miles of the Shoridan Central Business District. (A description and map of Shoridan is given in Chapter 1.) Because of its undeveloped resort potential, Shoridan's City Council cannot reach consensus on whether to favor having an interstate highway so close to their small town. There would be significantly improved access to the community, but that is not viewed as a good thing by some individuals and groups. Use your knowledge and judgment about communities and the forces that are at work as the local and state governments consider this "better access" question. List at least 12 stakeholders (local banks and environmentalists would be two). For each stakeholder, indicate which side of the issue you would expect them to be on and write a short statement that outlines the arguments they would present to validate their position at a public meeting on the proposed location of the Interstate.

5.4 Stakeholders. There are a number of pilots who fly small, single-engine aircraft into Shoridan regularly. However, the 2200-foot gravel airstrip limits the number and kind of aircraft that can safely land and take off at Shoridan. There is an air taxi company that would like to fly regular daily service from Shoridan to the nearest major city in a nine-passenger aircraft that requires 3700-foot paved runway and improved lighting. Doing this would mean a whole new airport, whose location would be about a mile from the existing one. The proposed airport relocation would involve building a runway across some wetlands. Also, a radio tower would need to be moved to provide the require safe approach angle. Develop a list of three stakeholders most likely to be in favor of the new airport and a list of three stakeholders most likely to be opposed. Indicate for each stakeholder their main argument(s) for their position.

5.5 Stakeholders.   This questions is based on a 75-minute video "Keep America Moving" (Eno Foundation, 1990). In the video, the conflict between various stakeholders is portrayed. The management of the large company "Worldbusters" has decided to move its operations to another state. Apparently, it has become too expensive for Worldbusters to get their products to market. Of course, the state (Franklin) and its major city (Benjamin) want the company to stay. Worldbusters has a large employment base and its presence provides an important economic base for the state. Various transportation options are discussed in the video in an attempt to keep Worldbusters from moving out of state.

(a) What is the relationship of transportation and economics (growth, stability, development)?
(b) Who are the major stakeholders in a transportation investment decision, such as the move of a major industry?
(c) What are the interfaces between the public and private sectors in transportation?
(d) Why is transportation planning so difficult for public officials?

Economic Evaluation of Transportation Alternatives

5.6 Net Present Value.   Compute the NPV of the following three highway projects, analyzed over a 30-year period, with the local social discount rate of 8.76 percent discount rate. Note that when the service life is up for those not lasting 30 years, you will need to spend the money to do it again. The cost is considered to be the same as the initial cost. You must, however, take the cost forward in time.

Project	Initial Investment ($M)	Salvage Value ($M)	Service Life	Annual Benefits ($M)	Reduction of Travel Costs
A	10	0.50	10 years	1.80	
B	19	2.50	15 years	2.30	(inc. @ 3%/yr)
C	33	3.60	30 years	3.70	(dec. @ 4%/yr)

M = millions.

5.7 Benefit-Cost.   Two possible new road routes are proposed to improve the traffic flow into Econoly from the nearby "bedroom community." One road is 15 miles long and will cost $12 million to construct. The maintenance of the road is expected to cost about $0.5 million per year. The second road is 24 miles long and will cost $20 million to construct with a maintenance cost of about $700,000 per year. Both roads will last 20 years and have a salvage of 25%. The benefit to the residents using the first road in fuel expense and time saving is fixed at $2 million per year. The second road will have increasing benefits. They begin at $3 million and grow at 2 percent per year (*Hint*: Use a new rate that includes the discount rate and the growth rate). The county commissioners who must approve the road use a discount rate of 8.8 percent and prefer to look at BCRs.

- What are the BCRs for each?

5.8 Annual Cost. A proposed option for the Mythaca airport capacity improvement will cost $15 million and have a lifetime of 12 years with no salvage value. The increase in O&M costs will be negligible. Air traffic is growing at the rate of 3 percent per year. The benefits ($3.5 million) at the time the runway improvement comes on line are significant, but they will decline proportional to the traffic increase. The social discount rate is 6.5 percent.

(a) What is the equivalent uniform annual cost of construction?
(b) What is the dollar value of the benefit in year 12?
(c) What is the discounted value of the benefit in year 12?
(d) Determine the BCR over the 12-year life of the project. *Hint*: The decline in benefits will look just like a larger discount rate.

5.9 Road Improvement. A road is to be built to accommodate traffic between two industrial sites and a major metropolitan area that is located between them (not on a straight line). There are two specific routes A and B that have been proposed. Route A on the north would be built over very hilly and curvy terrain, whereas B, the southern route, is considerably longer but flat, level, and on good soil. Traffic is expected to be about 7 million vehicles per year, with 9.3 percent trucks.

The major benefits include travel time, safety and fuel as indicated below.

- Automobile time is valued at $12.00 per automobile hour saved.
- Fuel: Average auto fuel economy on the hilly terrain it is 16 mpg, whereas on the flat terrain it is 24 mpg.
- Fuel costs are $1.25 per gallon.
- Convenience/service/safety is valued as $(v/c)^{-1.5}$ times $1 per auto trip.

Truck time and truck fuel can be valued at $50 per minute.

(a) Determine the equivalent annual cost including the O&M of each A and B using an 8.5 percent discount rate.

Project	Length (mi)	Investment (M)	Annual O&M (M)	Service Life (yr)	Salvage (m)	Auto Velocity (mph)	Truck Velocity (mph)	v/c
A	48	$80	$4	25	$4.8	45	30	.5
B	68	$140	$8	50	nil	65	55	.8

5.10 Value of Time. Why is it important to try to estimate the *value of travel time* in transportation analysis?

Ranking Transportation Alternatives

5.11 Evaluation of Alternatives. Mythaca Airport is becoming an increasingly important part of the regional air transportation system. Several major airlines are interested in offering service to Mythaca. But Mythacans are fussy. They want only the best airlines to be considered. A local consumer protection group has acquired performance data on 10 major airlines. Note that a high value for "Percent of on-time flights" is good, but a high value for "Complaints per 100,000 passengers" and "Mishandled baggage per 100,000 passengers" is not.

(a) Explain how you will modify these data for use in a ranking procedure and use a spread-sheet to implement your modifications.

	Percent of on-time flights			Complaints (per 100,000 passengers)			Mishandled baggage (per 100,000 passengers)	
1.	TWA	87.9	1.	Southwest	0.10	1.	America West	3.44
2.	Delta	86.5	2.	US Airways	0.68	2.	US Airways	3.58
3.	Northwest	85.9	3.	Alaska	0.69	3.	TWA	3.58
4.	Continental	84.1	4.	Delta	0.78	4.	Delta	3.72
5.	US Airways	83.2	5.	United	1.24	5.	Continental	4.01
6.	Southwest	81.5	6.	TWA	1.33	6.	Southwest	4.05
7.	American	77.4	7.	Northwest	1.35	7.	American	4.32
8.	United	76.9	8.	American	1.47	8.	Northwest	4.56
9.	Alaska	76.5	9.	Continental	1.59	9.	Alaska	5.54
10.	America West	69.4	10.	America West	2.50	10.	United	6.46
	Average	81.7		Average	1.07		Average	4.39

In parts b to d below, use a spreadsheet to rank the 10 airlines using the following weights:

0.40 for "Percent of on-time flights"
0.30 for "Complaints"
0.30 for "Mishandled luggage"

and the following ranking methods:
(b) The Range Index method
(c) The percentile method. *Note*: Excel has a "PERCENTRANK" feature.
(d) At this point, which three airlines (in order of performance) would you recommend Mythaca Airport to invite? Explain your recommendation.

5.12 Comparing Alternatives.   Various county council members and citizens' groups have been demanding that work be done on certain road segments in Mythaca County. Of course, there is not enough money to work on all of them. The county engineer's staff has prepared a table of factor values for each of the segments nominated for repair, resurfacing, and so on. Apply the Range Index method to only 1 of the 12 segments listed below. (Choose a segment that never has either the neediest or least needy value for any of the factors.)

(a) Which segment did you choose?
(b) What Range Index value did you compute for that segment?
(c) Show your method clearly.
(d) Does anything about the values in the table above bother you? If so, what is a potential problem?

Factors: Peak V/C ratio (Wt = 36); Project Cost ($1000) (Wt = 52); Crash rate (Wt = 12)

Road Segment	MOE Adj. V/C	MOE Adj Cost	MOE Adj. Crash Rate	Range Index
A	0.55	39.9	1.84	
B	0.18	49.2	9.79	
C	0.93	82.0	17.82	
D	0.36	76.8	15.38	
E	0.36	62.4	13.68	
F	0.87	43.6	17.40	
G	0.81	28.7	3.70	
H	0.90	20.7	12.74	
I	0.74	14.0	8.70	
J	0.28	79.8	2.12	
K	0.19	45.1	13.74	
L	0.70	18.1	3.03	

5.13 When Do Weights Get too Heavy?   Repeat the County Road Segment (Table 5.9) and Mythaca-Shoridan Corridor (Table 5.7) examples with increasingly disparate maximum and minimum weights, until you think you have reached a "preemptive" set of weights. Did the 3:1 rule of thumb work? If not, what ratio limit should be applied to these examples?

Safety on the Highway

CHAPTER 6

SCENARIO

The headline in the Mythaca Daily News was "5 KILLED, 3 INJURED IN FIERY CRASH ON COUNTY ROAD" (Figure 6.1). The story tells of a young couple and their three children who were killed instantly in a head-on crash with a Sports Utility Vehicle. The SUV was driven by a 28-year-old man, who had his fiancée and a friend as passengers. The driver of the SUV had a blood alcohol content of 0.14 and apparently ran the stop sign at the intersection of County

FIGURE 6.1
Fatal crash on county road. *Source: Post-Bulletin Online*, Red Wing MN, 5 September 2000.

Roads 500 East and 220 South. The condition of the fiancée is critical; the driver and friend suffered minor injuries. What can a transportation engineer do to improve safety on the highway?

CHAPTER OBJECTIVES

By the end of this chapter, the student will be able to:

1. Conduct a highway safety improvement program.
2. Use human factors in the design and analysis of highways for safety.

3. Evaluate and design roadway sections for safe stopping sight distance.

4. Apply prescribed standards in the use of roadway signs and markings.

INTRODUCTION

The death rate on U.S. highways continues to be around 40,000 per year. Because the number of miles traveled continues to increase, the apparent rate of safety is improving. This does not change the fact that the highway death toll is the equivalent of one major plane crash killing more than 100 persons *each day*. This chapter deals with the aspects of highway design, automobile performance, and human/driver behavior, which together form the basis for assessing safety on our highways.

> **FYI:**   Over the 10 years from 1984 to 1993, six times as many deaths occurred in U.S. automobile crashes as in airplane-related incidents.
>
> *Source*: NHTSA, 1998

6.1   HIGHWAY SAFETY—DATA AND ANALYSIS

Despite his limited budget, the Mythaca County Highway Engineer is expected to keep the county's roads in safe operating condition. Hardly a week goes by without an impassioned call or a letter to the area newspaper about some dangerous intersection or stretch of road. Why can't the county highway engineer do something about the obviously hazardous situation? If the engineer is wise, he has a system in place to (a) identify locations within the county that are hazardous, (b) determine which remedies are appropriate, and (c) establish the sequence in which the improvement projects will be accomplished. The system should be as objective as possible, using the limited funds as effectively as possible.

6.1.1   History and Perspective

In the middle 1960s, the death toll on the nation's highways became a matter of national attention and concern. In 1967, the National Highway Traffic Safety Administration (NHTSA) was formed in the new Cabinet-level Department of Transportation. NHSTA was given broad regulatory powers and made responsible for most of the 16 safety goals that Congress had promulgated in the National Highway Act of 1967. Figure 6.2 shows how much highway safety has improved in exposure (fatalities per vehicle mile travel) and fatalities per registered car since the formation of NHTSA. The data show that even occupant fatalities have declined. However, the number of vehicle miles traveled, which is a common measure of *exposure*, has more than doubled in the same period.

> **FYI:**   In the summer of 1997, NHTSA began using the term "crashes" instead of the traditional term "accidents." This change was in recognition of the fact that most crashes have a cause and are not simply the result of uncontrollable circumstances.

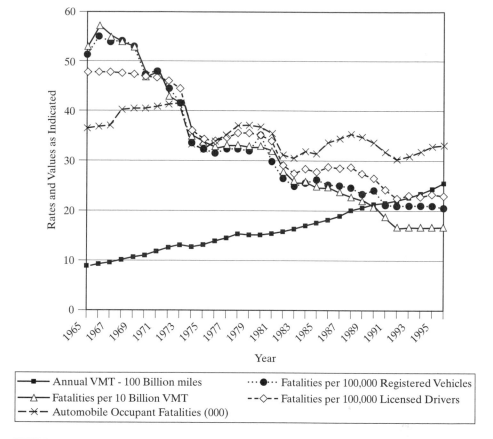

Annual VMT - 100 Billion miles
Fatalities per 10 Billion VMT
Automobile Occupant Fatalities (000)
Fatalities per 100,000 Registered Vehicles
Fatalities per 100,000 Licensed Drivers

FIGURE 6.2
Highway fatality data since the formation of the NHTSA. *Sources*: Statistical Abstract of U.S., 2000 and National Personal Transportation Survey of 1990.

6.1.2   Factors that Contribute to Highway Safety

The major elements that have improved highway safety include (1) vehicle safety regulatory standards for items such as roof design, side door supports, motor mounting, headlights, windshields, and wiper blades; (2) the design of highways, especially the interstate; (3) the social pressure to reduce drunk driving; (4) enhanced driver training; (5) the installation of seat belts and airbags; and (6) imposing a national speed limit.

THINK ABOUT IT

It has been speculated that returning the authority for setting speed limits on interstate highways to the states will cause the death rate to rise. Do you tend to agree with that viewpoint? What has been the experience to date?

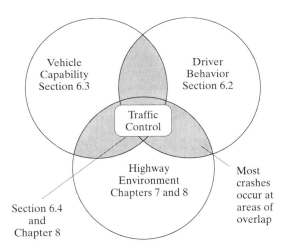

FIGURE 6.3
Venn diagram showing the causal factors of
highway safety.

Law enforcement and public education efforts continue to make sure that seat belts are being worn and that drunk driving is eliminated. When it was learned that passenger side airbags could cause death of small adults and children, the use of air bags in certain situations was suspended.

Highway crashes generally have more than one causal factor, as suggested in the Venn diagram of Figure 6.3. Important items pertaining to the design and performance aspects that affect safety are discussed for each area of the diagram. Traffic control markings and devices are usually used to help the driver control the vehicle. Each of these areas is the subject of a section in this chapter, and the highway environment is covered in Chapters 7 and 8.

6.1.3   Highway Safety Improvement Programs

Despite the progress illustrated in Figure 6.2, highway safety remains a primary concern to highway engineers and public officials. For many years, the U.S. federal government has enacted legislation to encourage (and sometimes require) states to adopt programs "to reduce traffic accidents and the resulting deaths, injuries, and property damage" (HSIP, 1981). The Highway Safety Improvement Program (HSIP) was defined in a federal publication in 1979. The HSIP remains to this day a good framework for planning, implementing, and evaluating safety programs and projects. Figure 6.4 shows the basic components that comprise the HSIP, along with the processes that make up each component. In this section, methods commonly used to carry out the processes are introduced and demonstrated.

Planning Component, Process 1: Collect and Maintain Data.   If collected properly, crash data can help identify the scope and nature of traffic safety problems in a community. Traffic crash data depend on collisions being reported to law enforcement agencies, so that the events can be documented. Once a crash report is filled out and filed, it becomes the basis for an analysis of that crash and the compilation of summary

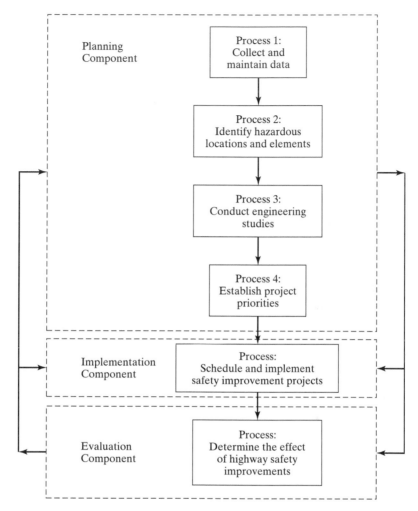

FIGURE 6.4
Highway Safety Improvement Program at the process level. *Source*: FHWA, 1981.

data. The right-hand side of Figure 6.5 is the top section of the back page of a crash report that was filled out at the crash scene by a police officer. These locally created reports eventually become part of a national database. See, for example, the Fatality Analysis Reporting System at www-fars.nhtsa.dot.gov. From these data, the analyst can compute crash rates, look for patterns in the history of crashes at a particular location, propose solutions, and evaluate their expected effectiveness.

THINK ABOUT IT

Have you ever been involved in a traffic collision? If so, did a police officer fill out a crash report? If not, why not?

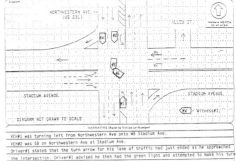

The passenger of a Toyota Camry involved in a two-vehicle accident around 4:30 p.m. Monday at the intersection of Stadium Street and Northwestern Avenue removes her belongings from the wreck before it was towed. No injuries were reported in the accident that occurred after a northbound Chrysler LeBaron driven by ███████ 17, of West Lafayette collided with the southbound Toyota, driven by ███████ University student ███████

CHRIS PICKARD/CHIEF PHOTOGRAPHER

FIGURE 6.5

A crash and its crash report. *Sources: Purdue Exponent*, Tuesday 21 September 1999, p. 1 and West Lafayette, Indiana Police Department.

Planning Component, Process 2: Identify Hazardous Locations and Elements.   Although there are more than 6 million crashes each year on U.S. roadways (NHTSA, 1998), a traffic collision is considered a "rare event."

THINK ABOUT IT

If you were asked to go to a busy intersection near your residence and wait for the next collision to occur, how long would you have to wait?

To help determine how "dangerous" a roadway section or intersection is, a local agency could refer to the crash reports on file and then determine the *total number of crashes* at that location. Another way to assess the safety performance of a location is to compute its *crash rate*. For an intersection, the standard measure is crash *rate per million entering vehicles (RMEV)*. The key "ingredients" are (a) the number of crashes in a given year and (b) the *average daily traffic volume (ADT)* on all approaches to the intersection.

$$\text{RMEV} = \frac{\text{crashes/year}}{\text{approach ADT} * \text{days/year}} * 10^6 \qquad (6.1)$$

Example 6.1

Several intersections in Mythaca had an apparent increase in collisions last year. One such intersection—Fisk at Kissimmee—may need some special attention. It had 13 crashes. The four legs of that intersection had two-way ADT values of 9671, 2893, 9506, and 2611 vehicles per day last year. Calculate the intersection's crash rate, so that it may be compared with the rate for other similar intersections.

Solution to Example 6.1 Equation 6.1 uses *approach* ADT, not two-way ADT. That part of a road's ADT that departs from the intersection must be excluded.
A reasonable way to estimate approach ADT is to take half of the two-way ADT. Therefore, the intersection crash rate at Fisk and Kissimmee last year was

$$\text{RMEV} = \frac{13}{0.5*(9671 + 2893 + 9506 + 2611)*365}*10^6 = \frac{13}{0.5*24{,}681*365}*10^6 = 2.886$$

THINK ABOUT IT
Why must we exclude the departing ADT from the calculation of RMEV?

If a roadway section is not "near" (usually defined as within 100 feet of) an intersection, a crash *rate per hundred million vehicle miles* (*RHMVM*) is computed.

$$\text{RHMVM} = \frac{\text{crashes/year}}{\text{ADT}*\text{days/year}*\text{miles in section}}*10^8 \qquad (6.2)$$

Example 6.2

A new state highway safety program offers funds to improve local roads with RHMVM values greater than 100. A 6.1-mile section of Tyler Road in Mythaca County had six crashes last year. The two-way ADT was 755 vehicles per day. Does Tyler Road qualify for the state program?

Solution to Example 6.2 Unlike the intersection crash rate, the roadway section crash rate uses the two-way ADT.

$$\text{RHMVM} = \frac{6 \text{ crashes}}{755 \text{ vpd}*365 \text{ days}*6.1 \text{ miles}}*10^8 = 356.9$$

Tyler Road had an average of only one crash per mile last year, but the corresponding crash rate was high enough to qualify for state funds.

THINK ABOUT IT
Why does RHMVM use two-way ADT?

Does the crash rate found in Example 6.1 "prove" that Fisk and Kissimmee is a dangerous intersection? Did the state set its RHMVM threshold for funding local roads at the right level? A common way to answer either question is to consider a representative sample of intersections or roadways and then determine whether the crash history of any particular location is "extreme." A critical rate analysis begins by identifying locations with similar characteristics (geometrics, ADT, traffic control). Crash rates for the comparison group of sites are calculated by using several years of data and either Equation 6.1 or 6.2, as appropriate.

THINK ABOUT IT

Why are several years of data preferable to using just last year's data when computing RMEV or RHMVM?

The mean crash rate $\bar{x}$ and sample standard deviation σ_S for the comparison group is calculated. Any rate greater than C in Equation 6.3 below can be considered to be "extremely high."

$$C = \bar{x} + (Z * \sigma_S) \qquad (6.3)$$

Because of the variability of the data being analyzed, a confidence level needs to be specified. Most often, a 95 percent confidence level is chosen. The Z value that corresponds to a 95 percent confidence level is 1.96.

Example 6.3

The intersection of Wyckliffe Boulevard and Kolfax Avenue has had 9.35 collisions per million entering vehicles (MEV) during the last 3 years. The crash rate for rear-end collisions is 5.7 per MEV. Eight other intersections in Mythaca have characteristics similar to Wyckliffe and Kolfax. Their crash rates for the last 3 years are summarized in Table 6.1. Does Wyckliffe and Kolfax have an unusually high crash rate for total collisions?

TABLE 6.1   Critical Rate Analysis for Example 6.3

Comparison Site	Total Collisions Rate	Rear-End Collisions Rate
1	7.89	2.67
2	9.24	2.56
3	7.09	2.64
4	8.67	4.00
5	7.20	4.36
6	7.93	2.58
7	9.94	4.77
8	6.68	1.97
Mean, $\bar{x}$	8.08	3.19
Sample standard deviation, σ_S	1.129	1.025

Solution to Example 6.3   The mean crash rate $\bar{x}$ for the comparison group is 8.08 crashes per MEV. The sample standard deviation $\sigma_s = 1.129$ is computed by using

$$\sigma_S = \sqrt{\frac{\sum_{i=1}^{n}(\text{RMEV}_i - \bar{x})^2}{n - 1}},$$

where n is the number of comparison group members. The denominator uses "$n - 1$," because the comparison group is treated as a sample of "all" intersections like those in the comparison group. Using Equation 6.3 with $Z = 1.96$, $C = \bar{x} + (Z*\sigma_S) = 8.08 + (1.96*1.129) = 8.08 + 2.21 = 10.29$. Although Wyckliffe and Kolfax has a crash rate higher than all but one of the intersections in the comparison group, its crash rate does not exceed the critical rate set by Equation 6.3. Therefore, it does not qualify as a hazardous intersection based on total collisions.

Planning Component, Process 3: Conduct Engineering Studies.   The HSIP (FHWA, 1981) listed 24 types of studies that could be used to provide information needed to determine safety deficiencies at locations that had been identified in the previous process as hazardous. The best single method for our purposes is the collision diagram (see Figure 6.6). The diagrams drawn as part of each crash report (see the right-hand side of Figure 6.5) for a given intersection are transferred to a single graphic representation of that location. In Figure 6.6, 15 crashes for the year 1975 are depicted with arrows. Except for the events involving northbound vehicles, each collision's arrow is labeled with a date and time. Some also have conditions ("nite," snow, wet) noted. Also important is the symbol that represents the *type* of collision: rear-end, head-on, sideswipe, out of control, left turn, right angle. It is often possible to detect a pattern in the collisions shown in the diagram. If so, specific countermeasures can be proposed for evaluation. Common categories by which trends are classified are:

- *Type of collision*
- *Severity.*   Fatal, personal injury, or property damage only
- *Contributing circumstances.*   DUI, reckless driving, equipment failure, and so on
- *Environmental conditions.*   Weather, roadway surface, lighting conditions
- *Time of day.*   Daylight, night, dawn, dusk

If any of this information is not present on the collision diagram, the original crash report should be checked. After one or more recurring characteristics have been identified, a list of possible causes can be developed. If, for example, left-turn head-on collisions are prevalent, possible causes are (U.S. DOT 1981, p. 101):

- Restricted sight distance
- Yellow phase too short
- Absence of separate left turn phase
- Approach speeds too high.

For each of these possible causes, one or more countermeasures can be proposed. For a summary of the procedure in concise form, see Table 6.2.

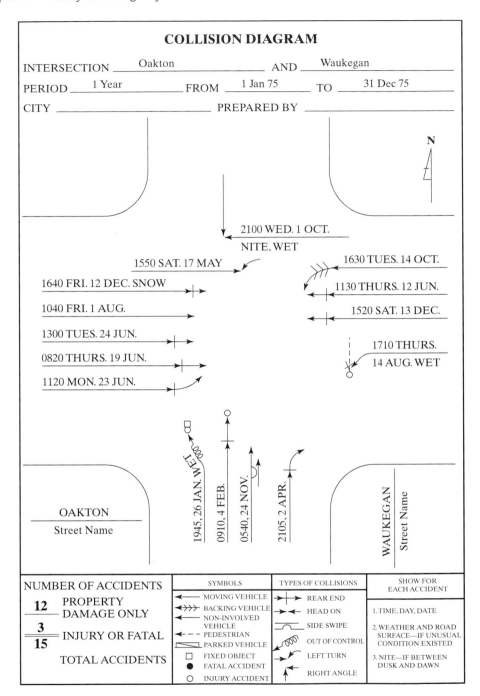

FIGURE 6.6
Collision diagram. *Source*: Box and Oppenlander, 1976, p. 62.

TABLE 6.2   Sample Accident Pattern Table

Accident Pattern	Probable Cause	General Countermeasure
Right-angle collisions at an unsignalized intersection	Restricted sight distance	Remove sight obstructions. Restrict parking near corners. Install stop signs. Install warning signs. Install/improve street lighting. Reduce speed limit on approaches. Install signals. Install yield signs. Channelize intersection.
	Large total intersection volume	Install signals. Reroute through traffic.
	High approach speed	Reduce speed limit on approaches. Install rumble strips.

Source: U.S. DOT, 1981, p. 117.

Planning Component, Process 4: Establish Project Priorities.   After countermeasures have been proposed, they must be evaluated. The basic evaluation procedure is to estimate the effectiveness of a particular countermeasure and then compare that estimate against the countermeasure's cost. Estimating a countermeasure's effectiveness is not easy. A common method is to determine by how much the crash rate will be reduced and then convert that reduction into a benefit in economic terms. There are two serious problems with this method.

1. It is far from certain how effective a certain countermeasure will be. If, in Table 6.2, "install stop signs" is chosen as the countermeasure, by how much will "right-angle collisions at an unsignalized intersection" be reduced? Numerous attempts have been made to develop crash reduction factors for this purpose (Creasey and Agent, 1985; Ermer et al., 1992).

2. It is difficult to translate any given estimated reduction in crashes into benefits. How much is a human life worth, if its loss because of a traffic collision can be avoided? How much is a personal injury worth? How much should be spent to prevent each property damage only crash from happening?

THINK ABOUT IT

What method would you use, if you were required to estimate the value of a human life and justify it? Answer the question again, this time for a personal injury crash or a property damage only crash.

Table 6.3 shows values published in a NHTSA report (Blincoe, 1994). The "economic costs" result from goods and services that must be purchased or productivity

TABLE 6.3   Sample Costs for Collision Countermeasure Analysis

Crash Severity	Economic Costs ($) (Blincoe, 1994)	Comprehensive Costs ($) (Miller et al., 1991)
Property damage only	1,440	
Minor personal injury		10,840
Moderate personal injury		133,700
Serious personal injury		472,290
Severe personal injury		1,193,860
Critical personal injury	706,000	2,509,310
Fatal	830,000	2,854,500

that is lost as a result of motor vehicle crashes. They do not represent the more intangible consequences of these events to individuals and families, such as pain, suffering, and loss of life. The estimates of "comprehensive costs" (Miller et al., 1991) combine both economic costs and values for "intangible" consequences. The costs someone might be willing to pay to prevent the crash can be much higher than the economic costs of injuries (Blincoe, 1994). Currently, most authors seem to agree that the value of fatal risk reduction lies in the range from $2 million to $5 million per life saved.

If the effectiveness of a particular countermeasure is known, the following equation can be used to determine its benefit.

$$\text{crashes prevented} = EC * CRF * \frac{\text{forecast ADT}}{\text{base ADT}} \tag{6.4}$$

where
- EC = expected number of crashes over a specified time (usually a year) if the countermeasure is not implemented and the traffic volume remains the same
- CRF = crash reduction factor = percent reduction in crashes if the countermeasure is implemented
- base ADT = traffic volume per day before countermeasure is implemented
- forecast ADT = traffic volume per day for specified time after countermeasure is implemented

Note that there may have to be a separate CRF for each degree of crash severity. A particular countermeasure might be expected to reduce fatal crashes by 12 percent, *personal injury (PI)* crashes by 33 percent, and *property damage only (PDO)* crashes by 24 percent. Or, for each crash prevented, the severity of the prevented crash could be drawn from a probability distribution (e.g., 41 percent fatal, 40 percent PI, and 19 percent PDO). A further complication is that a countermeasure may not prevent a crash but may instead reduce its severity. The basis for the CRF value and the use of Equation 6.4 would have to be adjusted accordingly. In Example 6.4, we use Equation 6.4 in a relatively simple form.

Example 6.4

The Mythaca County Engineer's staff believes that installing stop signs at a previously un-controlled intersection will reduce crashes by 26 percent. In the base year (last year), there were 11 right-angle collisions at the intersection, whose approach volume was 3273 vehicles per day. Ten years from now, the approach ADT is forecasted to be 4000 vehicles per day. How many crashes will be prevented 10 years from now, if the stop signs are installed?

Solution to Example 6.4   Using Equation 6.4, the estimate of right-angle crashes prevented

$$= 11 * 0.26 * \frac{4000}{3273} = 3.50 \text{ per year.}$$

Example 6.4 tells only part of the story. A more complete analysis would include the following:

A. The crashes prevented in the years between the base year and the 10th year.
B. Years beyond the 10th year, if the life of the countermeasure extended that far.
C. An estimate of the cost of the project, spread out over the life of the project.

Let us extend the period of analysis to 15 years. In those 15 years, traffic is expected to increase at the same rate as in the 10-year analysis. Because $(1 + r)^{10} = 4000/3273 = 1.222$, the annual *traffic growth rate* $(r) = 2.0$ percent. In the absence of any counter-measure, CRF = 0.0 in Equation 6.4 and the number of right-angle crashes will also increase 2.0 percent per year. So, for any year k,

$$\text{crashes prevented in year k} = CP(k) = (EC)(1 + r)^k * CRF \tag{6.5}$$

During year 1, $CP(1) = 11 * (1.02)^1 * 0.26 = 2.917$ crashes prevented. During year 2, $CP(2) = 11 * (1.02)^2 \, 0.26 = 2.975$ crashes prevented. Over a 15-year period, the total number of crashes prevented would be calculated as follows:

$$\sum_{k=1}^{n} CP(k) = \sum_{k=1}^{n} EC * (1 + r)^k * CRF \tag{6.6}$$

For EC = 11, $r = 0.02$, $n = 15$, and CRF = 0.26, a total of 50.45 crashes would be prevented. See the calculations in the "CP(k)" column of spreadsheet Table 6.4.

The spreadsheet in Table 6.4 has also been used to estimate the benefit of the crashes prevented. If one half of 1 percent of the crashes at the intersection are fatal and the "comprehensive cost" of a fatal crash in Table 6.3 is used, the corresponding value of the countermeasure in the first year is

$$\text{Fatal benefit} = 2.917 \text{ crashes prevented} * 0.005 \text{ fatal} * \$2{,}854{,}500/\text{fatal crash}$$
$$= \$41{,}636$$

TABLE 6.4   Crashes Prevented Over Life of Countermeasure Project

11 EC			15 years of project life		
0.02 Traffic growth rate			0.26 CRF		
0.005 fatals			$2,854,500	Fatal unit benefit	
0.195 PI			$472,290	PI unit benefit	
0.800 PDO			$1,440	PDO unit benefit	

Year k	CP(k)	Fatal Benefit	PI Benefit	PDO Benefit	Tot. Benefits
1	2.917	$41,636	$268,664	$3,361	$313,660
2	2.976	$42,468	$274,037	$3,428	$319,934
3	3.035	$43,318	$279,518	$3,496	$326,332
4	3.096	$44,184	$285,108	$3,566	$332,859
5	3.158	$45,068	$290,811	$3,638	$339,516
6	3.221	$45,969	$296,627	$3,710	$346,306
7	3.285	$46,889	$302,559	$3,785	$353,233
8	3.351	$47,826	$308,611	$3,860	$360,297
9	3.418	$48,783	$314,783	$3,937	$367,503
10	3.486	$49,759	$321,078	$4,016	$374,853
11	3.556	$50,754	$327,500	$4,097	$382,350
12	3.627	$51,769	$334,050	$4,179	$389,997
13	3.700	$52,804	$340,731	$4,262	$397,797
14	3.774	$53,860	$347,546	$4,347	$405,753
15	3.849	$54,937	$354,497	$4,434	$413,868
Total	50.448	$720,024	$4,646,120	$58,117	$5,424,260

A similar calculation is done for the proportion of crashes that are expected to be personal injury (PI) and property damage only (PDO), year by year for all 15 years. The estimate of total benefits from the proposed countermeasure over the next 15 years is $5,424,260.

THINK ABOUT IT

Do the relative values of the total benefits in Table 6.4 for each crash type seem reasonable to you?

Now that the benefits side of the analysis has been completed, the cost estimates for the countermeasure must be made. If it costs $85 to install each of four stop signs and an average of $15 per year to maintain or replace each sign, the present worth (at $i = 3.0$ percent) of the cost to implement the stop sign countermeasure for 15 years is

$$\text{PWC} = P_o + A\left[\frac{(1 + i)^n - 1}{i(1 + i)^n}\right] = (4 * \$85.00) + (4 * \$15.00)\left[\frac{(1 + 0.03)^{15} - 1}{0.03(1 + 0.03)^{15}}\right]$$

$$\text{PWC} = 340.00 + 60.00\,[11.938] = \$1056.28.$$

In this case, the benefits of the proposed countermeasure far exceed the costs. Often, the analysis is not so clear-cut. Some analyses are very sensitive to the choice of the value of human life, the CRF, and so on. An analyst must be clear in stating his/her assumptions and may find sensitivity analyses helpful. Even when such crucial values can be agreed on, there are usually more projects proposed than can be funded in a given budget cycle. For that reason, a system for ranking competing projects must be adopted. Chapter 5 covered ways to rank alternatives.

Implementation Component: Schedule and Implement Safety Improvement Projects. The scheduling of projects is more of a management and budgeting activity, but the proper implementation of the projects depends on good design and construction practices.

Evaluation Component: Determine the Effect of Highway Safety Improvements. Notice that a feedback loop exists on the left side of Figure 6.4. The loop begins at the evaluation component of the HSIP. By monitoring the performance of the implemented projects, the agency can ensure that maximum benefits are derived from the projects and the agency can collect information to help make future decisions regarding highway safety improvement projects.

6.1.4   Traffic Conflict Analysis

Although there are too many collisions, they are still rare events at any given location. However, some locations have the potential to be dangerous or to be more dangerous than they already are. To assess the potential for actual collisions, a procedure has been developed to collect data on traffic conditions that are conducive to dangerous interactions between vehicles. This procedure is the *traffic conflict technique*.

As originally defined by Perkins and Harris (1996), a traffic conflict was any evasive action taken by a driver to avoid a collision. Classic cases are (a) sudden application of the brake to avoid "rear-ending" a vehicle in front that has slowed to turn and (b) sudden swerving to avoid a rear-end collision. During the 1970s and 1980s, the definition was expanded and refined to include time-distance and severity elements. "A traffic conflict occurs when two or more road users approach the same point in time and space, and at least one road user takes successful evasive action to avoid a collision within a predefined minimum time to collision" (Hamilton Associates, 1996).

Figure 6.7 provides a structure for traffic conflict analysis. An "Existing Hazard" is any potential interaction between vehicles. Whether it meets the definition of a *traffic conflict* depends on the subsequent events indicated in the figure. "Normal interaction between road users generally does not result in a traffic conflict. A precautionary measure to avoid a perceived dangerous situation is also not a conflict. A conflict only occurs when evasive action that is not part of normal driving is taken to avoid a *real* hazard" (Hamilton Associates, 1996, p. 5). Observers conducting a traffic conflict study must be able to say "yes" to each question along the left-hand side of Figure 6.7. In addition, the observer is expected to assess the severity of each observed traffic conflict, according to the criteria list below.

1. *Time to collision (TTC).*   Using estimated (or measured) approach speed and distance to the potential point of collision, an approximate TTC value can be computed. The severity score for each TTC value is given in Table 6.5.

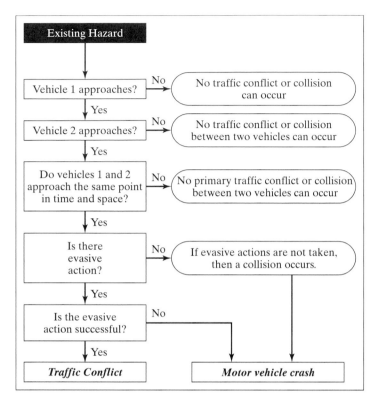

FIGURE 6.7
Traffic conflict concept. *Source*: Hamilton Associates, 1996, p. 5.

TABLE 6.5 Severity Scores for TTC and ROC Values

Severity Score	Time to Collision (seconds)	Risk of Collision
1	1.5–2.0	Low risk
2	1.0–1.5	Moderate risk
3	<1.0	High risk

Source: Hamilton Associates, 1996, p. 8.

2. *Risk of collision (ROC).* ROC is "a subjective measure of the collision potential, and is dependent on the perceived control that the road user appears to have over the traffic conflict event.... For example, a high ROC score would be associated with emergency braking or sudden abrupt swerving with limited maneuvering room, [while] light braking or mild lane changes on a wide and open roadway may result in a low ROC score" (Hamilton Associates, 1996, p. 7).

THINK ABOUT IT
The ROC value is based on a subjective decision by the observer. Not all observers would assign the same ROC value to a given conflict. How can the variation between observers be minimized?

The rear-end collision case mentioned earlier in this section is just one of seven traffic conflict types:

1. Left turn opposing
2. Left turn crossing
3. Crossing
4. Rear-end
5. Right turn
6. Weaving
7. Pedestrian

Just as a collision diagram (Figure 6.6) can be used to look for patterns in crash data for a site, noting the type of traffic conflict can assist in the development of corrective action at a location with too many traffic conflicts. Although traffic conflicts usually occur more frequently than collisions, they are still rare events. Observers use a form that looks much like a collision diagram to record the key elements of the observed conflict. When these sheets are summarized, an intersection conflict index (ICI) is computed. The ICI results from a series of simple calculations:

$$\text{Average hourly conflicts (AHC)} = \frac{\text{Total number of observed conflicts}}{\text{Number of observation hours}} \quad (6.7)$$

A typical value for AHC is 2.0–3.0 conflicts per hour.

$$\text{AHC per thousand entering vehicles, AHC/TEV} = \frac{\text{AHC} * 1000}{\text{AHEV}} \quad (6.8)$$

where AHEV = average hourly entering volume. A typical value for AHC/TEV is about 1.5.
The next intermediate calculation is

$$\text{Total conflict severity (TCS)} = \text{TTC Severity} + \text{ROC Severity} \quad (6.9)$$

where the TTC Severity and ROC Severity scores come from Table 6.5. The TCS value applies to each observed conflict. When all conflicts at a site are combined, we get

$$\text{Overall average conflict severity (OACS)} = \frac{\Sigma(\text{TCS of each observed conflict})}{\text{Total number of observed conflicts}} \quad (6.10)$$

A typical value for OACS is just above 3.0.

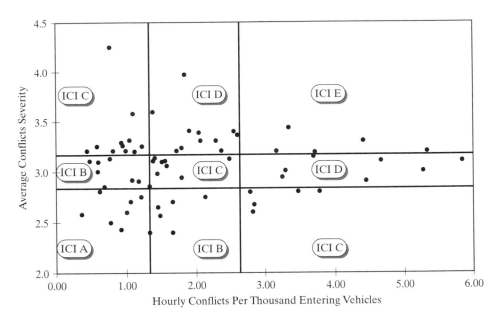

FIGURE 6.8
Intersection conflict index. *Source*: Hamilton Associates, 1996, p. 42.

As Figure 6.8 shows, the ICI value is much like the level of service rating in highway capacity analysis. A rating of A though E is assigned to an intersection, based on its location in the figure. For example, if an intersection has 2.5 hourly conflicts per thousand entering vehicles and an average conflict severity of 3.4, it falls in the "ICI D" region of Figure 6.8.

The boundaries in Figure 6.8 are based on a compilation of data gathered at numerous intersections. (Note the points plotted in Figure 6.8.) Cumulative plots of AHC and OACS values were studied, and the stratification shown in Table 6.6 was the result. Although this modern approach to traffic conflict analysis is still being refined, it represents a less subjective and less primitive procedure than simply counting the number of times brake lights are observed. The results of this procedure can form the basis for a Highway Safety Improvement Program before the

TABLE 6.6 Intersection Conflict Index Summary

Intersection Conflict Index	Conflict Risk (Frequency and Severity Combined)
A	Minor
B	Low
C	Moderate
D	High
E	Extreme

Source: Hamilton Associates, 1996, p. 42.

number of collisions at a site reaches the level that triggers the consideration of countermeasures.

Example 6.5

The intersection of Wyckliffe Boulevard and Kolfax Avenue may not have qualified as a hazardous intersection in Example 6.3, but many drivers perceive it as unsafe. A team of observers spent 40 hours at the intersection and collected the following information:

- 94 total conflicts, with 54 being of the rear-end conflict type
- Average hourly approach volume = 1205 vehicles
- Total TTC severity = 190 for the 94 conflicts, using the TTC scores in Table 6.5
- Total ROC severity = 201 for the 94 conflicts, using the ROC scores in Table 6.5

What intersection conflict index (ICI) value applies to this intersection?

Solution to Example 6.5   According to Equation 6.7, AHC = (94 observed conflicts/40 hours observed) = 2.35 conflicts per hour. Using Equation 6.8, AHC/TEV = (2.35 * 1000)/1205 = 1.905 conflicts per thousand entering vehicles. From Equation 6.9, TCS = Total TTC + Total ROC = 190 + 201 = 391. The average observed conflict severity comes from Equation 6.10: OACS = (Σ(TCS of each observed conflict))/Total number of observed conflicts = 391/94 = 4.16. When the values AHC/TEV = 1.905 and OACS = 4.16 are plotted in Figure 6.8, the resulting point lies in the region labeled "ICI D."

THINK ABOUT IT

Given the results of Example 6.5, what measures would you recommend be considered?

6.2   HUMAN FACTORS AND TRANSPORTATION ENGINEERING

A pavement resurfacing project on I-25 causes the two northbound (NB) lanes to be closed. NB traffic must cross the median and use one of the two SB lanes until the 3-month project is completed. The contractor follows the procedures for workzone signs and marking given in the *Manual of Uniform Traffic Control Devices*, but a fatal crash and several other collisions occur on the SB approach to the median crossover in the first few weeks of the project. The county highway engineer takes his video camera to an overpass with a clear view of the NB approach during the Sunday afternoon peak period (Figure 6.9). In the first 10 minutes, he records several dangerous maneuvers on videotape. What can be done to make the workzone safe?

6.2.1   Human Factors Concepts for Design

Human factors, also called *ergonomics* or *engineering psychology* (Wickens, 1999), is the study of how human beings function in their natural or constructed surroundings

FIGURE 6.9
Northbound approach to median crossover. More than 2 miles after being told by a series of signs to merge left, drivers of five vehicles in the right-hand lane (four of which are next to the semitrailer) slow down, looking for gaps in traffic in the left-hand lane. Traffic seeking to use the off-ramp just ahead must either wait behind them in the right-hand lane or use the shoulder, as two drivers are doing here. Photo: Jon D. Fricker, 20 August 2000.

(Kantowitz and Sorkin, 1983). There are many examples in everyday life. The design of some devices may have significant consequences with respect to safety:

- A punch press.
- The unfamiliar position of the various controls in a rented or borrowed car, especially in the dark, while many designs (or the lack of a standard design) may cause inconvenience or inefficiency.
- On which side of *this* car is the gasoline filler cap?
- Where on a particular TV remote control is the "previous channel" button?
- Which way does this door swing? In or out? Are the hinges on the left or right?

In the case of the punch press, an inefficient design may be the best design. By requiring that the operator use both hands to activate the machine's functions, neither hand is in danger.

THINK ABOUT IT

Think of at least one example of a design (good or bad) that has safety consequences and then provide at least two examples of design that causes (or avoids) inefficiency or inconvenience. Your examples do not have to be related to transportation activity.

A classic transportation example is the aircraft pilot who is surrounded by cockpit instruments. The location of the instruments, how readings are displayed, and what physical actions the pilot must take to achieve the desired results are all elements of the design of the cockpit. A more familiar situation is the driving task. Each driver is operating a vehicle on a roadway. The design of the vehicle and the design of the roadway will affect the performance of the driver. At the same time, the design of the vehicle and roadway must take into account the wide range of possible abilities, attitudes, backgrounds, and preferences of drivers using the roadway. Some drivers have slower reaction times than others. Some drivers' eyesight is not as keen as that of others. Some drivers are more aggressive than others. Some drivers are new to the area—or even new to the country. Some drivers prefer certain styles of driving that may not be compatible with other motorists' expectations. In this section, the challenge of designing for most (if not all) types of drivers and situations is presented. Other types of human factors applications will be used to illustrate the challenge of designing for a diverse set of users.

Of the approximately 40,000 highway deaths in the United States each year, more than 40 percent involve an intoxicated driver (NCSA, 1999). The remaining fatal crashes are due to highway design, weather, or "driver error." The road environment contributes to 17 to 34 percent of crashes and is the sole factor in 2 to 3 percent of the cases (O'Cinneide, 1995). Of crash causes, driver error is by far the most frequent. According to the Human Centered Systems Laboratories (TFHRC, 2001), inappropriate driver perceptions and behaviors are implicated in 80 to 90 percent of all highway crashes. Even if a roadway has been "adequately" designed to conventional standards, it may be possible for an enhanced roadway design to counteract some of the effects of weather or "driver error." If an enhanced design is possible, it would be helpful to know how to do it as cost-effectively as possible.

THINK ABOUT IT

Is driving while intoxicated a highway design issue? How about overly aggressive driving or road rage? If any or all of these are design issues, how would the study of human factors help the highway designer?

The Driving Task.   Operating a motor vehicle on a street or highway can be complex and demanding at times, but it can be boring at other times. This range of circumstances—coupled with the range of driver capabilities—presents a challenge to the highway designer. It is helpful to begin by considering the three essential elements of the driving task (Ogden, 1990; AASHTO, 2001):

- *Navigation (route selection).*   Because most trips are made repeatedly, or in familiar street networks, this is usually the least complex of the driving task elements. However, when a driver is looking for information to reach a destination in an unfamiliar network, that activity may detract from other driving task elements. Bad examples: street signs that are missing or hard to read. Good examples: signs to frequent destinations (downtown, university, stadium) within a city; notice of the next main cross street (Figure 6.10) before that intersection.
- *Guidance (vehicle tracking).*   Staying on the roadway and staying in the proper lane have obvious implications for safety. Examples: lane and edge markings

FIGURE 6.10
Overhead signs help motorists find desired path through intersection. Photo: Jon D. Fricker.

(Figure 6.11) on the pavement and delineators along the roadside. (These topics are covered in the section on traffic control devices later in this chapter.)

- *Control (object avoidance).* This activity involves proper application of steering and speed control skills. At the basic level, steering around clearly visible fixed objects and maintaining a safe distance from vehicles ahead and to the side constitutes the control element. However, unexpected maneuvers by other vehicles or objects that appear suddenly can require a high-level response (in reaction time, decision making, and action) that is quite complex.

FIGURE 6.11
White line marking roadway edge. Photo: Jon D. Fricker.

The three driving task elements are interrelated. For example, failure by one driver to accomplish the guidance element may cause another driver to exercise object avoidance to prevent a collision.

Although sounds and feel can provide useful information to a driver, most information comes in a visual form (Lay, 1986). A driver operates in a *zone of spatial commitment* that varies by driver and operating environment (ITE, 1982; Hulbert, 1972). In Figure 6.12, a vehicle is moving from left to right. The driver samples cues about what is ahead from a field of vision that is constantly changing. Examples of cues include other vehicles, pedestrians, traffic signs and markings, sharp curves, crests of hills, or any object or circumstance that could create an unsafe condition. At speeds around 30 kph, the driver's lateral (left-right) field of vision is about 100 degrees. This is the driver's effective field of vision—50 degrees to the left and 50 degrees to the right. Normally, the driver pays more attention to the objects and cues nearer the center of the visual field. Cues may also be detected in the peripheral vision of the driver, outside the normal effective visual field. At 100 kph, the driver's field of vision narrows to about 40 degrees (Cole, 1972). Within this visual field, the driver's eyes are moving from one object to another at a rate of four eye fixations per second, or less, depending on driver ability and attentiveness (Cole and Jenkins, 1982). The closer objects or visual cues require immediate decisions; the more distance cues provoke a *provisional commitment*. If the scene is cluttered with too many visual cues, the driver may miss important cues or get confused. The roadway designer's job is to reduce the number of negative cues, while providing just enough positive cues to assist the driver. Of course, many negative cues are beyond the control of the roadway designer, and driver responses to positive cues may vary. For the I-25 workzone described at the start of this section, how many warning signs are needed along the approach to the workzone, and where should they be placed?

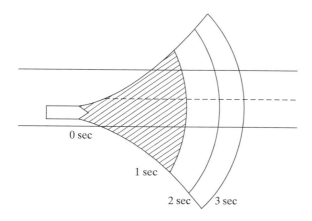

0 sec

1 sec

2 sec   3 sec

FIGURE 6.12
Driver's zone of spatial commitment. *Source:* ITE, 1982, based on Hulbert, 1972.

THINK ABOUT IT

Give examples of how sounds and feel can provide information to the driver that is useful in the driving task.

Perception-Reaction Times.   A driver sees most cues (signs, potential threats) soon enough to process them safely by routine driving actions—reduce speed, change vehicle path, or simply monitor the situation. On rare occasions, the cue appears suddenly and unexpectedly. It requires immediate action by the driver. The time needed for a driver to recognize and respond to the cue is called the *PIEV (perception/ identification/emotion/volition)* time (MUTCD, 2000, p. 2C-3). If the cue is a sign,

- Perception is the time it takes to see the sign. This is the time needed to locate the cue and classify the cue as a sign to be read. A commonly used sign will be classified quickly, if the driver is paying attention. Unusual cues may take up to two seconds to be perceived.

- Identification is the time to read and understand the sign. Section 2.6A of the Washington State DOT *Traffic Manual* (1996) states that "the average driver comprehends three words per second."

- Emotion is the time to consider the sign's meaning and make a decision. Sometimes, the decision is that no action is needed. In other cases, the type of action must be decided.

- Volition is the time to react or execute a maneuver. A typical driving maneuver is to apply the brakes or turn the steering wheel. Once the maneuver has begun, the volition time (and the PIEV time) has ended.

According to the MUTCD (2000, p. 2C-3), the "PIEV time can vary from several seconds for general warning signs to 6 seconds or more for warning signs requiring high road user judgment." Many sources prefer to use the term *perception-reaction time*, instead of PIEV time. Under the perception-reaction system, the perception, identification, and emotion phases of PIEV are replaced by *detection, identification,* and *recognition* phases of *perception* (Sanders and McCormick, 1993). "Volition" in PIEV is renamed the reaction phase of perception-reaction.

SOMETHING TO TRY

With a good Internet search engine, you can use the string "reaction time test" to find more than a dozen tests on the Web. Try several different reaction time tests. Describe the tests you tried, summarize your results, and comment on the validity of the tests.

If you tried a reaction time test as suggested in the box above, you were actually measuring your PIEV time, although under special circumstances. The cue was probably very well defined in terms of type of cue and location. The meaning of the cue and the proper response were also clear, at least after your first trial or so. Your PIEV or reaction times on the tests must be considered as your best-case performances. They will not transfer well to actual driving situations. Taoka (1989) looked at several studies of the brake reaction times for unalerted drivers. He found that the typical mean reaction time was about 1.2 seconds, with a standard deviation of about 0.7 seconds. The brake reaction times of drivers tend to be log-normally distributed. (See Figure 6.13.)

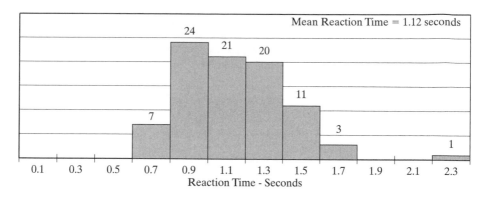

FIGURE 6.13
Distribution of unalerted driver brake reaction times. *Source*: Gazis et al. (1960).

AASHTO (2001) suggests using a driver perception-reaction time of 2.5 seconds for design purposes, but this value exceeds the 95th percentile reaction time found in most of the studies reviewed by Taoka. When designing roadways and placing traffic signs, clear sightlines and adequate decision sight distance must be provided, especially for the less capable driver.

The usual "braking-reaction-response time" for most persons is between 0.6 and 1 second. However, we must design public highways to accommodate a wide range of drivers, whose response characteristics are like those depicted in Figure 6.13. When people are surprised, their reaction time tends to be longer than reaction times that are measured under laboratory conditions. It has been determined that a response time of 2.5 seconds covers more than 90 percent of the drivers and should be used in making design decisions. By using standardized shapes, colors, and symbols, and locating the signs in consistent locations, the engineer can simplify the driving task. If traffic signs are easy to see and easy to read, the driver will have more time for the emotion and volition phases of PIEV.

THINK ABOUT IT

The AASHTO "Green Book" (2001) suggests using a driver perception-reaction time of 2.5 seconds for design purposes. Based on Taoka's findings, this is a very conservative design standard. Can a design ever be too conservative?

Example 6.6

Eastbound County Road 200 South ends at a T intersection with CR 300E. EB traffic on CR200S approaches this intersection on a crest vertical curve, requiring the placement of an advance warning sign, especially for nighttime traffic. If a typical EB driver is traveling at 50 mph when he sees the warning sign, how much distance will it take him to begin to brake?

Solution to Example 6.6   The "T Intersection Ahead" Advance Warning Sign is diamond-shaped, with a black "T" on a yellow background. This sign has a familiar shape and a symbolic (vs. verbal) message. According to Figure 6.13, the typical driver encountering the sign on an unexpected basis has a reaction time of about 1.12 seconds. At 50 mph (or 73.5 f/s), the PIEV distance is 1.12 * 73.5 = 82.32 feet. The braking distance calculation will be covered later in this chapter.

Example 6.7

A crash occurred in which the driver stated that she was driving at the 55 mph speed limit, when she came over the crest of a hill and spotted a deer crossing the road. However, the skid marks were found on the roadway for only the last 90 feet before the deer was struck. If the skid marks indicate the beginning of braking and the crest of the hill was about 250 feet from the point of impact, what was the driver's response time?

Solution to Example 6.7

$$T_{\text{Response}} = \frac{D_{\text{object}} - D_{\text{braking}}}{V} = \frac{250 \text{ ft} - 90 \text{ ft}}{55 \text{ mph} \times 1.47 \text{ fps/mph}} = 1.98 \text{ sec}$$

6.2.2   Human Factors Applications in Transportation

Subsection 6.2.1 introduced some basic ideas underlying the application of human factors to transportation problems. Elsewhere in this text, gap acceptance and the dilemma zone are topics that have a strong human factors component. In this subsection, several examples of how human factors can be used to analyze or improve certain situations are presented.

Changing the Status Quo.   Normally, *expectancy* is a design feature that helps motorists. A straight road will stay straight until a sign that warns of a curve ahead appears. Traffic signals are usually placed above the intersection, on cables or on masts. However, one kind of expectancy can be a problem—being too familiar with a location. Consider an intersection that, for many years, has been controlled by stop signs on two of its four approaches. Eventually, the traffic volumes or crash history at that intersection justifies the installation of stop signs on the previously "uncontrolled" approaches. At least some of the motorists who have driven on the uncontrolled approaches on a regular basis are not likely to notice the new stop signs, even if they are installed according to standards. What is the solution? Some localities install oversized stop signs on a temporary basis and then replace them with stop signs of standard size after a week or two. Another strategy is to place temporary stop signs on barrels at the previously "uncontrolled" approaches, to supplement the new stop signs. See Figure 6.14. This method overcomes the habits of drivers who are too familiar with the intersection.

Railroad Grade Crossings.   In 1998, there were 422 deaths at grade crossings, down one third from 5 years earlier (FRA, 1998). Of the 247 reported collisions, 183 were attributable to motor vehicle operator inattention or impatience (Farnham, 2000). Motorists failed to see the train, misjudged its speed, or simply lost a race to the tracks.

FIGURE 6.14
Temporary stop sign to supplement a new stop sign. Photo: Jon D. Fricker, 23 June
2001.

Of those fatal collisions, 114 occurred at crossings with active warning devices. There
were 254,017 grade-level railroad crossings in the country in 2000 (FRA, 2001), but
only about 62,000 are equipped with active warning systems, such as gates, lights or
bells. How many lives would installing more active warning devices save?

In some cases, limited sight distance would be addressed by installing active
warning devices. But, as Figure 6.15 illustrates, sound-only warnings may not be
enough. Leibowitz (1985) wrote about driver impatience and how poorly many drivers judge the speed of an approaching train. The size of the locomotive and the angle
at which the motorist views it deceives the motorist into thinking that the train is
much farther from the crossing than it really is and that it is moving much slower than
it actually is.

As is often the case in human factors, the time at which gates or other warning
devices are actuated with respect to the train's arrival is difficult to specify for all drivers. If the devices are not actuated early enough, some drivers may not have enough
time to clear the tracks comfortably. If the devices are actuated too early, they will be
too conservative for many drivers, especially the impatient ones.

Thirty-six percent of incidents at gated railroad grade crossings are caused by a
driver going around or through the gates (FRA, 1998). A segment on the NBC news-
magazine *Dateline* called "Blood on the Tracks," (first shown in October 1997)
showed a series of horrifying scenes in which motorists drove around functioning
gates, only to miss being hit by locomotives by mere seconds. A young man who was
interviewed for the program admitted to trying to beat a train to a grade crossing. He
saw the train coming, but he didn't quite clear the tracks and his passenger (his sister)
was killed.

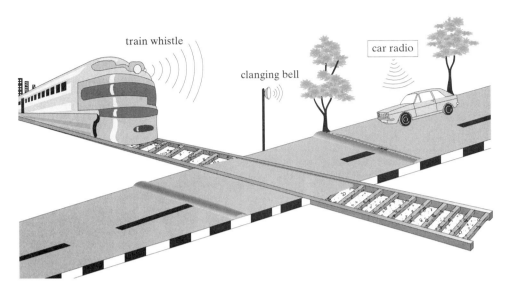

FIGURE 6.15
Sounds at a railroad grade crossing.

How should a transportation engineer respond to such driver (mis)behavior? The "easy" solution is to install barriers that cannot be circumnavigated by impatient or inattentive motorists. One such device is called a "four-quadrant gate." (See Figure 6.16.) The four-quadrant gate blocks vehicular access to the tracks on both sides of the roadway's centerline on both sides of the tracks. Another idea is a raised center median on the approach to the grade crossing, to keep motorists from driving around a lowered gate. Some railroads have begun to install remote cameras to document driver behavior and determine the need for the more extensive barriers. If the four-quadrant gate is such an easy solution, why aren't more of them being installed? The first problem is cost. (See Table 6.7 below.) Moreover, some people oppose the installation of the four-quadrant gate because it could trap motorists *on* the crossing.

The Federal Railroad Administration studied a variety *of supplemental safety measures (SSMs)* (FRA, 1999). The results are summarized in Table 6.7.

TABLE 6.7 Supplemental Safety Measures for Railroad Grade Crossings

Supplemental Safety Measure	Effectiveness	Costs
Temporary closure of grade crossing	100	
Four-quadrant gates	77–82	Gates and circuitry: $244,000–$318,000
		Annual maintenance: $3750
Mountable curb medians for 60 feet	75–80	$11,000
Photo enforcement	78	Capital: $55,000–$75,000
		Annual operations: $20,000–$30,000
Full grade separation	100	Bridge/road: $1,000,000 (added by authors)

Source: FRA, 1999.

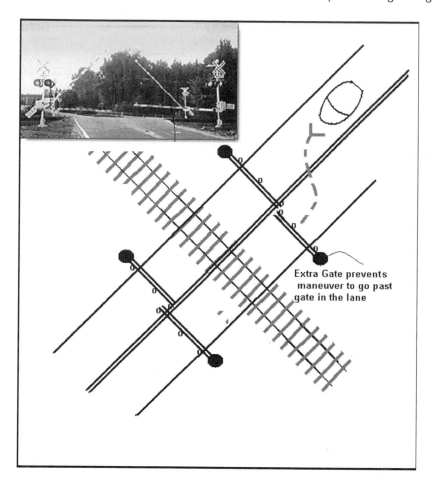

FIGURE 6.16
Four-quadrant gate. Insert from http://www.photocop.com/rail.htm.

Example 6.8

Using the information in Table 6.7, estimate the cost to install four-quadrant gates at all 62,000 grade crossings that now have active warning devices. If "effectiveness" in Table 6.7 means the percent reduction in fatalities, how many of the 114 fatalities per year at crossings with active warning devices would be prevented? Assuming a 25-year life for the four-quadrant gates and a discount rate of 4.0 percent per year, what would be the equivalent uniform annual cost of installing and maintaining four-quadrant gates at the 62,000 grade crossings? Is it possible to determine the apparent value of a human life, based on these calculations?

Solution to Example 6.8   Using the midpoint of $244,000 and $318,000 in Table 6.7, the average cost to install a four-quadrant gate system at a grade crossing is about $281,000. If all 62,000 grade crossings that now have active warning devices were to be upgraded in this way, the total installation cost would be 62,000 * $281,000 = $17.422 billion. Applying an 80 percent effectiveness (between 77 and 82 percent in Table 6.7) to the 114 fatalities means that 91 lives could be saved: $0.80*114 = 91.2$. The equivalent uniform annual cost of installing four-quadrant gates at the 62,000 grade crossings is found using the equation

$$A = P\left[\frac{i(i+i)^n}{(1+i)^n - 1}\right] = \$17.422B\left[\frac{0.04(1.04)^{25}}{(1.04)^{25} - 1}\right] = \$17.422B\left[\frac{0.106633}{1.665836}\right]$$

$$A = \$17.422B\,[0.064012] = \$1,115,212,000 \text{ per year}$$

Add to this value the \$3750 annual maintenance cost for the upgraded grade crossings: $62,000*\$3750 = \$232,500,000$

The equivalent uniform annual cost of installing and maintaining four-quadrant gates at the 62,000 grade crossings is \$1115.212 M + \$232.500 M = \$1347.712 M.

The question of the value of a human life is a sensitive one, but it must be confronted in some way. Given our calculations in this example, the cost to save each life is (\$1347.712 M/91 lives) = \$14.81 million word per life. Is a human life worth at least this much? It would have to be, for a rational analysis to support the installation of four-quadrant gates at the 62,000 grade crossings. Of course, it would not be feasible for all grade crossings to receive upgrades in 1 year. A 10-year program (Farnham, 2000) is analyzed later as an exercise. In this way, the economic burden of such a program may be spread out over time, but some of the safety benefits will be delayed.

License Plate Design and Law Enforcement. The license plate on a motor vehicle serves two principal functions: (1) to indicate that the vehicle is registered and (2) to uniquely identify the vehicle for law enforcement, data collection, or toll collection purposes. More recently, many states have made license plates a part of programs aimed at promoting a positive image of the state. How attractive a license plate is has become more important than how well it serves its two original functions. What's more, about 18 states now issue only one license plate, so that the identification function is further hampered.

A variety of human factors concepts can be applied to the design of an effective license plate (Fricker, 1986) The concepts can be used to address two key questions.

1. *Can the license plate be seen?* The size and form of the characters on the plate determine the legibility of the plate's "message." All U.S. license plates for cars are 6 in. by 12 in. Most states use characters that are 69 mm high. How far away can such letters and numbers be read? That depends on the eyesight of the observer. An observer's visual acuity can be measured in terms of the subtended visual angle shown in Figure 6.17.

   The degree to which visual acuity can vary from one person to another is illustrated in Smith's (1979) results from 2007 subjects. (See Figure 6.18.) Visual acuity also depends on the character being observed. If a character is easily confused with another character of similar appearance—(E vs. F or O vs. Q), an observer may need to be closer to the target to be sure of its true identity. For example, Townsend (1971) found that subjects identified the letter "Q" as an "O" in 28 percent of the cases he tested. (See Table 6.8.) Other factors that affect visual acuity are color contrast between character and background, lighting conditions, and the age of the observer. Older observers tend to have less visual acuity and need more illumination on the target than they did when younger. This is a major design consideration, especially for traffic control devices, as the average age of the driving public continues to increase.

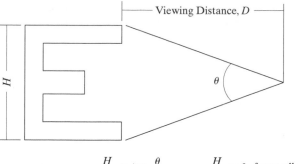

$$\frac{H}{2D} = \tan \frac{\theta}{2} \qquad \frac{H}{D} = \theta \quad \text{for small angles}$$

FIGURE 6.17
Subtended visual angle, θ. *Source*: Fricker, 1986.

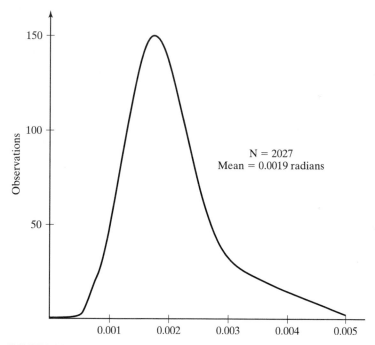

N = 2027
Mean = 0.0019 radians

FIGURE 6.18
Distribution of visual angle at the limit of legibility. Based on Smith, 1979.

TABLE 6.8 Excerpt from Confusion Matrix (Townsend, 1971)

Stimulus	Response	Percentage
Q	O	28
B	R	18
F	T	18
T	I	16
H	N	15
J	I	15

2. *Will the symbols be remembered?*   Even in an era when using videotape and optical character recognition are being explored as a way to automate the "reading" of license plates (Nelson, 1998), it is still important for eyewitnesses to remember the plate's message long enough to record it or report it. For this reason, the content of the license plate can be designed with human factors principles in mind. The license plate number needs to be long enough to uniquely identify a vehicle in a state with as many vehicles as California, but be as short as possible to assist an observer's *short-term memory (STM)*. The consensus (van der Heijden, 1981) is that individuals can process about seven "chunks" of information for retention in STM. The value of seven, however, can be affected by such things as

(a) the ability to "rehearse" the message content as the target is being viewed. This is much easier to do if an oncoming vehicle has a license plate on the front.

(b) the ability to combine individual characters into pronounceable chunks. "NID" (or even "NYD") is probably easier to remember than "PGW" or "H9X."

Example 6.9:   Visual Acuity

The county engineer was a front seat passenger in a car traveling on an interstate highway. When he saw that his car was closing the gap on the car ahead, he decided to try to read the license plate of the car ahead. As soon as he was sure of the number on the plate ahead, he started a stopwatch while noting (a) the location of the plate ahead with respect to a roadside object and (b) the speedometer reading (60 mph) for his car. Because they were traveling through Indiana, he was able to repeat this experiment several times for Indiana license plates. The average "time to target" for Indiana plates was 1.20 seconds. Later, the engineer determined that the numbers on Indiana license plates are 69 mm high. What was the engineer's visual angle θ for Indiana license plates under those conditions? Use metric units in the calculations.

Solution to Example 6.9   First, convert 60 mph to metric units, m/sec.

$$60 \frac{\text{mi}}{\text{hr}} * \frac{1609.3 \text{ m}}{\text{mi}} * \frac{\text{hr}}{3600 \text{ sec}} = 26.83 \text{ m/sec.}$$

Using the equation in Figure 6.17,

$$\theta = \frac{H}{D} = \tan \theta \tag{6.11}$$

where H = height of target, D = distance to target, and θ is small. So

$$\theta = \frac{0.069 \text{ m}}{26.83 \text{ m/sec} * 1.20 \text{ sec}} = 0.00214 \text{ radians}$$

Note that, in Figure 6.18, the mean θ in Smith's experiment was 0.0019, so the result here is reasonable.

Driving with Distractions.   How many things can humans do at once? That is a crucial question where the driving task is concerned. The National Highway Traffic Safety

Administration (Hendricks et al., 1999) estimates that 23 percent of the crashes reported by the nation's police each year are triggered by some form of distraction. When the Canadian Province of New Brunswick compiled a ranking of the most common causes of highway crashes, the order was as follows (New Brunswick, 2001):

1. Inattention.
2. Operating too fast for conditions.
3. Failure to grant right of way.
4. Alcohol.
5. Driver distraction.
6. View obstructed.
7. Following too closely.
8. Improper use of lanes

As cell phones gain in popularity, the issue of distracted drivers has become a topic for traffic engineers and legislators. Redelmeier (1998) found that the distraction caused by the use of a mobile phone, even a hands-free device, can delay an average driver's reaction time by 3 to 5 seconds, increasing a driver's risk of crashing fourfold.

On the other hand, a study of 32,303 vehicles involved in crashes in North Carolina from 1995 to 1999 (Stutts et al., 2001) seems to indicate that cell phone use is not a frequent cause of distraction. Table 6.9 shows the driver attention status in the North Carolina sample. Using or dialing a cell phone was the source of distraction in just 1.5 percent of the cases studied. Distractions that are more frequent than cell phones are listed in Table 6.10. Other links at the AAA Foundation for Traffic Safety Web site illustrate the debate over cell phone use by motorists (Figure 6.19).

TABLE 6.9 Driver Attention Status for All Crashes. (Listed next to the percentages are the 95% confidence intervals.)

Attentive (%)	48.6 ± 5.4
Distracted (%)	8.3 ± 1.2
Looked but did not see (%)	5.4 ± 1.4
Sleepy or fell asleep (%)	1.8 ± 0.8
Unknown/no driver (%)	35.9 ± 5.5

Source: Stutts et al., 2001.

Even as cell phones become more widely used, another potential distraction is beginning to emerge—in-vehicle devices such as Advanced Traveler Information Systems. These navigation aids can help a motorist to his/her destination in an unfamiliar city or alert a driver to congestion ahead and suggest a faster route. This information can be given in text form, as a map, in audio format, or in a combination of these (Yang et al., 1998). Designers of these devices must ensure that the navigation assistance they offer will not also interfere with the driving task (Collins, 1997).

Other studies point to drowsiness as a more frequent factor in highway crashes than previously thought. According to NHTSA (2001), "every year, falling asleep while driving is responsible for at least 100,000 automobile crashes, 40,000 injuries, and 1550

TABLE 6.10 Distribution of Distraction Activities. (Listed next to the percentages are the 95% confidence intervals.)

Outside person, object, or event (%)	29.4 ± 4.7
Adjusting radio/cassette/CD (%)	11.4 ± 7.2
Other occupant (%)	10.9 ± 3.3
Moving object in vehicle (%)	4.3 ± 3.2
Other device/object (%)	2.9 ± 1.6
Adjusting vehicle/climate controls (%)	2.8 ± 1.1
Eating and/or drinking (%)	1.7 ± 0.6
Using/dialing cell phone (%)	1.5 ± 0.9
Smoking related (%)	0.9 ± 0.4
Other distractions (%)	25.6 ± 6.0
Unknown distraction (%)	8.6 ± 5.3

Source: Stutts et al., 2001.

FIGURE 6.19
Is cell phone use a distraction from the driving task? Photo: AAA Foundation for Traffic Safety, 2001.

fatalities." Among the countermeasures proposed to counter drowsiness (NCSDR 2001), only rumble strips have a demonstrated effect on crashes. They reduce drive-off-the-road crashes by 30 to 50 percent.

Driver Habits. With the advent of cell phones, many drivers are acquiring a habit that, just a few years ago, was non-existent. If the use of cell phones by the operator of a motor vehicle is restricted or prohibited, some drivers will find it difficult to comply. Other habits and preferences have been the focus of traffic safety advocates in recent years. Driver behavior regarding the use of seat belts and motorcycle helmets has been the subject of debate between the advocates of private rights and public safety. Fricker and Larson (1989) looked at the relationship between a driver's use of seatbelts and his/her use of turn signals. Both seatbelt use and turn signal use were required by law but were seldom enforced. Their use was largely a matter of voluntary compliance by

> **THINK ABOUT IT**
>
> What concerns you more about Table 6.9: that so many of the crashes in-volved some form of inattention, or that such a large proportion of crashes (after "unknown" was excluded) involved *attentive* drivers? What implications are there in the table for highway designers?

the driver. The researchers found a positive correlation between seatbelts use and turn signal use. Apparently, driver behavior is made up of conscious decisions and accumulated habits. The traffic engineer must account for this when designing a roadway and its traffic control features.

Traffic Control at Workzones.   In 1999, 868 workers and motorists were killed in work zone-related crashes (Walls, 2001). Although the problems with the I-25 workzone described at the start of this section are not unusual, they still needed to be addressed. When the county engineer had a friend drive him through the work zone, he noted 12 signs over the 4-mile approach to the work zone, warning drivers of the potential hazard ahead. Still, as Figure 6.9 shows, some drivers do not merge until the last few yards.

A traffic engineer who wants to warn motorists of a work zone ahead faces several challenges. Temporary signs such as the 12 signs used along I-25 can be placed on the approach to the work zone. If the message is, for example, "Merge Right," most motorists will comply. The time between when the message becomes visible to a driver and when the desired action is taken will probably be widely distributed. Some drivers may not ever comply. Was the lack of compliance by these drivers a result of not having seen the sign or because of something related to driver attitude? Depending on the an-swer to this question, lack of compliance becomes a matter of better sign design and placement or a matter for law enforcement.

6.3   VEHICLE ATTRIBUTES THAT AFFECT SAFETY

In roadway situations that involve other cars, large trucks, motorcycles, bicycles, and pedestrians, the driver's ability to cause an automobile to stop, accelerate, or maneuver quickly may determine if a crash will occur. A key factor is the automobile's braking capability. The road surface and the tire tread also affect stopping distance. The capa-bility to steer the vehicle is the other major attribute that affects safety. However, with power steering, that limitation is less of a factor.

6.3.1   Forces Acting on Automobile

Consider the automobile traveling up an incline as shown in Figure 6.20.
The force that gives motion is derived from the engine acting through the wheels along the direction of travel:

$$F = F_f + F_r = \text{ma} = \frac{W}{g}\frac{dv}{dt} \text{ and } v = \frac{dx}{dt} \tag{6.12}$$

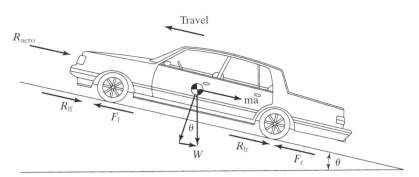

FIGURE 6.20
Forces acting on an automobile on an incline.

where F_f and F_r = the engine force applied at the front and rear wheels, respectively.

W = weight of the vehicle in pounds.

g = force of gravity.

Summing the forces acting on a car in motion, we get Equation 6.13:

$$F + R_r + R_{grade} + R_{aero} = \frac{W}{g}a + Wf_r \cos \Theta_g + W \sin \Theta_g + R_{aero} = 0 \quad (6.13)$$

where R_r = the sum of rolling resistance from each of the tires $R_r = R_{rlf} + R_{rrf} + R_{rlr} + R_{rrr} = f_r W \cos \Theta$

f_r = the coefficient of rolling resistance, usually $f_r = 0.01[1 + (V/1.47)]$, V in fps (Taborek, 1957).

R_{grade} is the component of gravity acting normal to the road.

In this text, we will ignore the aerodynamic force: $R_{aero} = 0.5 \, \rho C_D V^2$.

The same equation also applies to braking, where the rolling resistance is supplemented by with the force operating to stop the car through the friction applied to the highway. The braking acceleration (or deceleration) is usually assumed to be a constant, if the car does not go into a skid. If the coefficient of friction is f, the initial velocity is v_0, and the final velocity is v_f, the braking distance D_{br} from the time the brake is applied is given in Equation 6.14.

$$D_{br} = \frac{v_0^2 - v_f^2}{2a} = \frac{v_0^2 - v_f^2}{2f_r g} \quad (6.14)$$

On a level road, Equation 6.13 becomes

$$F + R_r = W\left(\frac{a}{g} + f_r\right) = 0; a = f_r g$$

This explains the rightmost form of Equation 6.14.

The time needed to go from v_o to v_f is given in Equation 6.15.

$$T_{v_o \text{ to } v_f} = \sqrt{\frac{D_{br}}{2gf}}$$ (6.15)

where f = the dimensionless coefficient of friction of the road

g = the force of gravity: 32.2 f/s² or 9.8 m/s². If braking takes place on a hill with a positive (uphill) grade G, the braking distance will be

$$D_{br} = \frac{v_o^2 - v_f^2}{2g(f \pm \tan \Theta_g)} = \frac{v_o^2 - v_f^2}{2g(f + G)}$$ (6.16)

where G is the grade in percent divided by 100.

Example 6.10

John is driving his 14-foot long automobile at 50 mph, when the traffic signal in front of him changes to yellow. He is 130 feet from the intersection when he applies the brakes after a 1-second reaction time.

A. If John's car can decelerate at a rate of 15 f/s², at what velocity will he be moving when he reaches the intersection?

B. If the yellow light is 4 seconds long, where will John be when the light turns red?

C. Based on the answers to part A and part B, will John be able to clear the intersection (width = 50 feet) before the light turns red? If not, let us assume he will continue through at the speed found in part B when the light changes to red. How long will the light have been red when he clears the intersection?

Solution to Example 6.10

A. Use Equation 6.14 to find John's speed upon reaching the intersection:

$$130 = \frac{(50 \times 1.47)^2 - v_f^2}{2 \times 15}; \quad v_f = 38.76 \text{ fps} = 26.4 \text{ mph}$$

B. When the light turns red, his speed will be

$$v_f = v_o - [(t_Y - t_r) * a] = (50 * 1.47) - [(4 - 1) * 15]$$
$$= 73.5 \text{ fps} - 45 \text{ fps} = 28.5 \text{ fps}$$

and the distance he will have traveled is

$$D_{br} = \frac{(50 \times 1.47)^2 - (28.5)^2}{2 \times 15} = 153.0 \text{ feet}$$

which is 23.0 feet after entering the 50-foot intersection.

C. The distance and time John would need to continue through the intersection at 28.5 fps (about 20 mph) will be:

$$\frac{(50 - 23.0) + 14}{28.5} = \frac{41 \text{ feet}}{28.5 \text{ fps}} = 1.44 \text{ sec.}$$

THINK ABOUT IT

Under the conditions described in the example, should John attempt to stop upon seeing the start of the yellow light, or should he proceed through the intersection?

6.3.2 Vehicle Braking

Vehicle dynamics play a crucial role in the design of the highway for safety. The design is based on the reaction time and the friction coefficient, which is related to the condition of the pavement on which the braking occurs. Although many individuals typically respond to stimuli in 1 second or less, the reaction time of 2.5 seconds is used in most design calculations.

Table 6.11 summarizes the results of using Equation 6.16 on level terrain ($G = 0$) with friction coefficient values that vary with speed. (See Figure 6.21.) The "computed *stopping sight distance*" *(SSD)* caption in the table means "the distance needed for a driver to detect an unexpected or otherwise difficult-to-perceive ... condition ..., select an appropriate speed and path, and initiate and complete the maneuver safely and efficiently?" [AASHTO, 2001, p. 115]. This distance is calculated from the following:

a. The distance covered during the natural delay or response time at the initial speed. That time may be as short as 0.5 seconds, if a person is very attentive

---

TABLE 6.11   Decision Sight Distances for Different Design Speeds

Design Speed (mph)	Reaction Time (sec)	Reaction Distance (ft)	Coefficient of Friction	Braking Distance on Level Terrain (ft)	Computed Stopping Sight Distance (ft)
20	2.5	73	0.40	33.3	107
25	2.5	92	0.38	55	147
30	2.5	110	0.35	86	196
35	2.5	128	0.34	120	248
40	2.5	147	0.32	167	313
45	2.5	165	0.31	218	383
50	2.5	183	0.30	278	461
55	2.5	202	0.30	336	538
60	2.5	220	0.29	414	634
65	2.5	238	0.29	486	724
70	2.5	257	0.28	583	840

and is capable of quick reactions, or it may extend to several seconds for elderly or drivers who are under the influence of alcohol or drugs, or are impaired or distracted.

b. The actual physical distance traveled while the car is being braked (decelerated) to a stop.

These two components combine to form Equation 6.17:

$$\text{SSD} = (t_{\text{response}} * v_{\text{o}}) + \frac{v_{\text{o}}^2}{2 * g * (f_{\text{braking}} \pm \text{Grade})} \quad (6.17)$$

The friction coefficient between the road and the tires can cover a wide range of values, depending on the pavement surface materials, the tread on the tires, and whether the road is dry or wet. Icy roads exhibit a coefficient of friction closer to 0.1, but because the ice can be detected, drivers are expected to use extreme caution and drive slowly. The coefficient of friction used in Table 6.11 has been assumed to be constant for a given design speed. However, as Figure 6.21 shows, the friction coefficient varies with speed. Because the calculations use a lower and constant value of friction

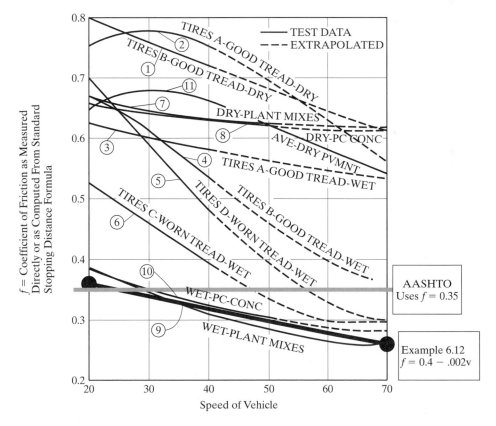

FIGURE 6.21
Skid resistance for various tire and pavement conditions. *Source:* AASHTO, 1990, Figure III-1.

coefficient for the table, the results should be conservative. The presumed road conditions chosen for design purposes are a wet concrete road.

It is assumed that the brakes are applied evenly and not "jammed-on," which would put the car into a skid. The friction coefficient is actually much less when skidding; hence, the stopping distance is greater. The table indicates stopping distance when on level terrain. If there is a hill, the stopping distance may be greater or smaller, depending on whether the car is traveling uphill (in which gravity will help the driver to stop) or downhill (where the effect is just the opposite). The computed distances in Table 6.11 can be used in highway alignment, traffic signal setting, passing (including road marking), and avoiding objects in the road when on a curve. The "design" values for stopping sight distance are given in Table 7.4.

Example 6.11

If f decreases according to the equation $f = 0.4 - 0.002$ v as a vehicle's speed decreases during braking (see the heavy solid line on Figure 6.21), show that the stopping distance and time to stop for an initial speed of 60 mph are more conservative than using a constant value of $f = 0.29$ for 60 mph from Table 6.11.

Solution to Example 6.11   By combining equation $f = 0.4 - 0.002$ v with gravity, the equation $a = g*f$ becomes $a = g*[0.4 - (0.002/1.47)v] = 32.2*(0.4 - .00136 v)$. The deceleration due to the friction coefficient can be approximated as $a = -12.9 + 0.044$ v, where v is in feet per second and a is in feet per second/second. Because a represents deceleration here, the signs change.

The braking time is governed by the equation $dv = a\ dt = (-12.9 + 0.044v)\ dt$. Solving this equation accounts for the changing value of f as the speed of the braking vehicle decreases:

$$\int_{v_0}^{0} \frac{dv}{-12.9 + .044v} = \int_{0}^{t_1} dt$$

$$t_1 = \frac{\ln 12.9}{0.044} - \frac{\ln(12.9 - 0.044*v_0)}{0.044}$$

If the initial speed $v_0 = 88$ fps, the time to brake from 88 fps is $t_{88} = 8.12$ seconds, when f is allowed to vary as the vehicle slows down. When the constant value $f = 0.29$ for $v = 60$ mph is used,

$$t_{88} = \frac{V}{32.2*f} = \frac{88}{32.2*0.29} = 9.42 \text{ sec}$$

Likewise, the stopping distance using Equation 6.14 is given by

$$x_{88} = \frac{88^2}{2*32.2*0.29} = 414 \text{ ft}$$

Using the constant friction factor yields a slightly more conservative result than the use of the more realistic friction factor that varies with speed. (See Table 6.12.) Using Table 6.11 is

TABLE 6.12   Answer to Example 6.11

	Constant f	Variable f
Time to stop (sec)	9.42	7.89
Distance to stop (ft)	414	367

conservative because it assumes a vehicle moving at 60 mph will need 414 feet to stop. Accounting for the increase in f as speed decreases during braking leads to a lower stopping distance of 367 feet.

6.3.3   Stopping Sight Distance

The SSD in Table 6.11 can be calculated for any given speed. For example, for $V = 55$ mph, the SSD includes the time to respond plus the time to brake. If 55 mph is the design speed, $f = 0.30$ in Table 6.11. Equation 6.17 gives us

$$\text{SSD} = (1.47 \times 55 \times 2.5) + \frac{(55 \times 1.47)^2}{2 \times 32.2 \times 0.30} = 202.1 + 338.3 = 540.5 \text{ ft}$$

This is close to the entry of 538 feet in Table 6.11.

When the design speed is 65 mph, the coefficient of friction is $f = 0.29$ and the stopping distance is longer:

$$\text{SSD} = (1.47 \times 65 \times 2.5) + \frac{(65 \times 1.47)^2}{2 \times 32.2 \times 0.29} = 238.9 + 488.9 = 727.8 \text{ ft}$$

In Table 6.11, the computed value is given as 724 feet. The design standard values of SSD in the far right-hand column of Table 7.4 are to be used for geometric design problems. If the minimum design standard cannot be met under the specified conditions of speed, grade, and/or radius, then at least one of the conditions must be altered in the design.

Example 6.12

A car is traveling down a 3 percent grade at 50 mph. How much longer will the stopping distance be than when it is traveling at 50 mph on a level surface?

Solution to Example 6.12   If the AASHTO Green Book's 2.5-second suggested perception-reaction time (see Section 6.2) is used in Equation 6.17, the stopping distance on level terrain is

$$D = 2.5*1.47*50 + \frac{(50 \times 1.47)^2}{2*32.2(0.3)} = 184 + 280 \text{ ft} = 464 \text{ ft}$$

On a downhill, however, gravity acts to increase the speed. Equation 6.17 is used.

$$D = 2.5*1.47*50 + \frac{(50 \times 1.47)^2}{2*32.2(0.30 - 0.03)} = 184 + 311 = 495 \text{ ft}$$

The difference when gravity is acting with you means that it takes 495 feet − 464 feet = 31 feet longer to stop. Table 6.13 shows how downhill grades up to 5 percent affect stopping distance.

TABLE 6.13   Braking Distance for Downhill Grades

V (mph)	f	No Grade	1%	2%	3%	4%	5%	Increase from 0% to 5%
40	0.32	315	320	326	332	339	346	9.87%
45	0.31	385	392	400	408	417	427	10.96%
50	0.30	464	473	483	495	506	519	12.07%
55	0.30	540	552	565	578	593	608	12.52%
60	0.29	637	652	668	685	704	724	13.62%

Example 6.13

The driving manual for the Department of Motor Vehicles in the State of Alaska states the braking distance for several speeds as indicated in Table 6.14. For the numbers given, what values have been assumed for driver response time and the coefficient of friction?

Solution to Example 6.13   The answers are given in the shaded portion of Table 6.14. For example, the calculations for v = 50 mph are

$$t_r = \frac{\text{dist}}{\text{speed}} = \frac{55}{74} = 0.74 \text{ sec}$$

$$f = \frac{v^2}{2*g*\text{dist}} = \frac{74^2}{2*32.2*160} = 0.53$$

TABLE 6.14   Braking Distances

Speed (mph)	Speed (fps)	Thinking Distance	Braking Distance	Total Distance	Response Time	Friction Coefficient
20	29	22	25	47	0.76	0.52
30	44	33	57	90	0.75	0.53
40	59	44	102	146	0.75	0.53
50	74	55	160	215	0.74	0.53
60	88	66	227	194	0.75	0.53
70	103	77	310	387	0.75	0.53

Source: Alaska DMV Manual, p. 44.

THINK ABOUT IT

Obviously, by using a reaction time of about 0.75 seconds, the Alaska DMV is not using the design value of 2.5 seconds. Why do you think they use such a fast reaction time? They also use a higher friction factor than the ones given in the table generated by the AASHTO. Why?

Example 6.14

You are asked to investigate a crash in which a teenage driver hit a barricade traveling at about 25 mph. Your observations at the scene indicate that the road is posted with a 55 mph speed limit. The road is straight and level. The first sign warning of the barricade was located 1000 feet before the barricade, and the second sign was 600 feet before the barricade. The skid marks from the car begin 300 feet before the barricade. You are asked to testify to a jury about your findings. What will you tell them about braking and response time? According to the Weather Service, the road was wet but visibility was good.

Solution to Example 6.14   The car traveled 700 feet from the first warning sign to the initiation of the skid marks. If the teenager was traveling at the speed limit, her response time would have been 700 ft/(55 * 1.47)fps = 8.66 sec.

If the velocity at impact was 25 mph, the initial velocity v_o before 300 feet of skidding was

$$v_o = \sqrt{v_f^2 + 2gfd_{skid}} = \sqrt{(25*1.47)^2 + 2(32.2)(0.30)(300)} = 84.5 \text{ fps} = 57.5 \text{ mph.}$$ From the

data, it would appear that the young driver did not respond to the first sign and reacted slowly to the second sign, with a response time of about $t_r = \dfrac{d_r}{v_o} = \dfrac{300 \text{ ft}}{57.5*1.47} = 3.55$ seconds. Moreover, she was driving above the speed limit.

THINK ABOUT IT

If you were on the jury, what would you decide? Was the road marked properly?

6.4   TRAFFIC CONTROL DEVICES

Among his many duties, the Mythaca County Highway Engineer must ensure that roadway signs and markings in the County are properly installed and maintained (Figure 6.22). If a crash occurs where someone thinks that a sign should have been installed, the county may be sued. If the engineer installs signs wherever there is even the slightest justification for them, the County Highway Department will probably not have enough left in its annual budget to maintain them. If any sign is stolen, vandalized, or allowed to become unreadable, and a crash occurs, the county may be sued. In theory, the rules for installing *traffic control devices (TCDs)* is quite simple. In practice, placing and maintaining TCDs requires diligence and good management practices—or else the public safety may be compromised—and the county may get sued.

FIGURE 6.22
Traffic signs.

6.4.1   TCDs Needed for Safety

Traffic control devices (TCDs)—otherwise known as roadway signs and markings—are used to regulate, warn, and guide drivers as they operate their vehicles. Although roadway signs and markings are familiar to everyone who uses the roads, there are well-established procedures to determine where certain TCDs are needed and how they should be installed. This section introduces those procedures and the references on which they are based. In doing so, numerous examples (good and bad) are presented. Some of the material in this section is based on a slide show used in a Training Course on Placement of Traffic Signs and Markings, produced by the Institute of Transportation Engineers (1974). Other major sources are the *Manual of Uniform Traffic Control Devices* (FHWA, 2003) and Richard C. Moeur's excellent Web site (1998).

6.4.2   Rules Governing Traffic Control Devices

A traffic control device (TCD) is a sign or pavement marking that is used to regulate, warn, or guide drivers as they operate their vehicles. An effective TCD meets five basic requirements. The sign must:

1. *Be needed.*   The transportation engineer must identify the need and select the most appropriate TCD. This section covers *warrant analysis*, which helps an analyst decide

when a *regulatory* TCD is justified. Too often, the demands of citizens groups or politicians are so great, that TCDs are installed where they are not *warranted*.

THINK ABOUT IT:

Have you ever seen a stop sign that is clearly unnecessary? What made you think it was unnecessary?

---

2. *Command attention.*   To be effective, the sign or marking must be seen, and seen without distracting the driver from his/her driving task. A sign blocked by other vehicles or foliage (Figure 6.23) will not get the desired response.
3. *Convey a clear, simple message.*   Using a standard combination of shape, color, and other graphic design elements, the TCD should be immediately recognizable to the driver, and its intended message should be unambiguous.
4. *Command respect.*   Signs and markings that are poorly designed and fabricated (Figure 6.24) will not have the credibility of properly installed TCDs. Likewise, unwarranted TCDs can instill a greater degree of disregard for similar TCDs among some drivers. A common example is stop signs used in residential neighborhoods in an attempt to control speeds.
5. *Be placed to get the proper response from the driver.*   Sign location is important. So is the placement of pavement markings. The advance warning sign must be placed far enough ahead of the potential hazard to allow the driver time to respond. The solid line that denotes a no-passing zone must not start too early or too late with respect to the actual section of highway that has inadequate passing sight distance.

FIGURE 6.23
Speed limit sign obscured by branches.
Source: ITE, 1974.

FIGURE 6.24
A poorly fabricated set of traffic signs.
Source: Jon D. Fricker.

6.4.3   Signs as TCDs

Traffic control devices can be placed in three major categories, namely, regulatory, warning, and guide signs. A sample of signs from each category, organized by the chapters in which they appear in the MUTCD, are shown in Figure 6.25.

FIGURE 6.25
Traffic sign classifications (FHWA, 2003).

Although there are several exceptions, in general, traffic signs can also be categorized by shape and color:

- Regulatory: rectangular, black on white
- Warning: diamond, black on yellow
- School zone: Schoolhouse shape, black on yellow
- Work zone: Diamond, black on orange
- Recreational and cultural interest: rectangular, white on brown

THINK ABOUT IT

Besides the signs shown above, can you think of any other common signs that do not follow the standard shape or color for their category?

6.4.4 Roadway Markings as TCDs

In addition to the signs categorized above, the other kind of TCDs are roadway markings. They consist of pavement markings, delineators, and object markings.

Pavement Markings. Pavement markings are used like roadway signs to warn, regulate, and inform motorists. Knowing what various pavement markings mean is important to the motorist, because they have the same force of law as signs. Dashed lines usually indicate that the driver has permission to pass, if it is safe to do so. Solid lines always indicate that maneuvering across them may bring risk to the driver. Figure 6.26 shows several common pavement markings for two-lane and multilane roadways.

Yellow markings

- Yellow markings, such as centerlines, separate traffic flow going in opposite directions. They always occur in pairs.
- Dashed yellow lines on the motorist's side indicate where passing is permitted on two-lane, two-way roads.

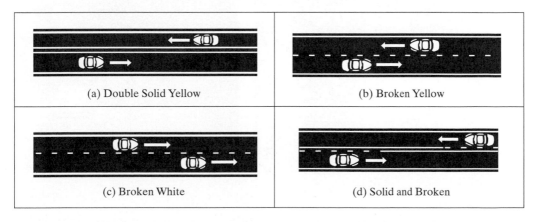

(a) Double Solid Yellow

(b) Broken Yellow

(c) Broken White

(d) Solid and Broken

FIGURE 6.26
Examples of centerline markings on highways. *Source*: Florida Department of Highway Safety and Motor Vehicles, 2000.

- Solid yellow lines indicate where passing is not permitted, although turning into a driveway across them is allowed where not prohibited.
- A single yellow line indicates the left edge of a divided roadway.

White lines

- White markings, such as lane lines, separate traffic going in the same direction on multilane or one-way roads.
- Dashed lines separate lanes of travel where changing lanes is not restricted and where the lane use is not restricted.
- Solid white lines are restrictive.

  1. They tell the driver to remain within a lane and do not move from it until it is safe to do so.
  2. They indicate the edges of lanes specified for certain uses where changing lanes is to be discouraged.
  3. They also mark the outside edge of the pavement or to indicate the edge of a shoulder. (See Figure 6.11.)

Transverse pavement markings

- Include crosswalks, stop lines, turn movement restrictions, and parking spaces.
- Whenever possible, turn movement restrictions marked on the pavement should be supplemented with signs over the respective lanes. (See Figure 6.10.)

Delineators.   Delineators are used to guide drivers through turns, especially at night or at times of poor visibility. The reflecting head of a delineator [Figure 6.27(a)] should be 4 feet above the roadway, between 2 and 6 feet from the outer edge of the shoulder. In Figure 6.27(b), chevrons are used effectively to guide drivers through a right-hand curve on a one-way urban arterial.

(a)   (b)

FIGURE 6.27
Delineators. (a) New delineators along dangerous curve and (b) Good use of chevrons as delineators. Photos: Jon D. Fricker.

Object markers are used to mark obstructions within or adjacent to the roadway. The three types of object markers are designated as OM-1, OM-2, and OM-3, as illustrated in Figure 6.28. The type 3 object marker features diagonal yellow or white stripes on a black background. The stripes should slope downward and toward the vehicle being warned. Thus, as shown in Figure 6.28(c), an OM-3 on the left side of the roadway should have its stripes sloping from "northwest" to "southeast." The stripes on an OM-3R (R = right) would slope from "northeast" to "southwest," as shown in Figure 6.28(c).

(a)

(b)

(c)

FIGURE 6.28
Object markers. (a) OM-1, (b) OM-2, and (c) OM-3L and OM-3R. *Source*: Photos by Jon D. Fricker.

6.4.5 Installation of Signs

The placement of traffic control devices is not an advanced engineering topic, but there can be serious consequences for incorrect installation. The Manual for Uniform Traffic Control Devices (MUTCD) (FHWA, 2003) is a valuable source of guidance, but some judgment may still be required. Figure 6.29 shows some examples. Figure 6.29(a) shows a shopping center exit that was "controlled" by a stop sign that not only was on a very short post but also it was placed behind the corner of a fence. In rural areas, the bottom of the sign should be at least 5 feet above the level of the pavement. In areas where cars are parked, the standard height becomes 7 feet. Although this sign was located on private property, for safety and liability reasons, the MUTCD requirements should be followed. After several years, the sign was replaced and the fence was removed, as shown in Figure 6.29(b).

(a)   (b)

FIGURE 6.29
Examples of obstructed sign. (a) Stop sign on short post is obstructed by fence and shrubs. (b) Taller post and removal of obstructions make stop sign visible at same site. *Source*: Photos: Jon D. Fricker.

Signs that are "cute" as in Figure 6.30(a) should be used sparingly, if at all, because of their tendency to distract rather than inform. On the other hand, just because a sign is not in the MUTCD, does not mean that a minor variation on a standard sign cannot be used effectively as shown in Figure 6.30(b).

THINK ABOUT IT
Based on the "message" conveyed by the sign in Figure 6.30(b), what would you expect to see as you continue to drive along the road ahead?

Because of the many languages that may be involved in traveling even modest distances in Europe, road signs there have relied more heavily on graphic representations than written text. Some of the European designs are being adopted by the

(a)   (b)   (c)

FIGURE 6.30

Other examples of bad and good TCD use. (a) Caution: Peacock Crossing. (b) Nonstandard sign with clear message. (c) Signs with international origins. *Sources*: (a) and (b): Jon D. Fricker; (c) FHWA 2003, Section 2B.30.

MUTCD. Examples of sign designs that originated in Europe but are now used throughout the United States are the DO NOT ENTER signs in Figure 6.30(c). Instead of words, a "slash" is used to indicate prohibited behavior, as shown by the second sign in Figure 6.30(c).

6.4.6   Stop Signs for Speed Control

When citizens get concerned about what they consider excessive speeds in their neighborhoods, they often call on their city engineer or elected officials to install stop signs at intersection approaches where they do not already exist. However, numerous studies have found that not only is such a strategy ineffective in reducing speeds, it also lessens the respect that motorists have for other stop signs in the vicinity. When Mythaca's City Engineer was faced with neighbors requesting stop signs for speed control, he mentioned it to the county engineer. Example 6.16 describes the subsequent study. (*Note*: Some students may not have been exposed in previous courses to the hypothesis-testing procedure that is essential to this example's solution. In that case, the student will have to trust the method shown and concentrate on the interpretation of the results.)

Example 6.15

When the county engineer's son Darren was in sixth grade, he had to choose a science project. Even before his father mentioned it, Darren had heard his friends' parents complain about the speed of cars in their neighborhood, known as Archer. (See Figures 6.31 and 6.32.) Those parents wanted the city traffic engineer to install stop signs as a speed control measure. In response, the city traffic engineer said that it would be an improper use of stop signs and,

FIGURE 6.31
Dodge Street at Allen Street, looking east. Note stop signs facing Allen Street traffic only.
One block ahead, traffic on Evergreen must stop at Garfield. Photo: Jon D. Fricker.

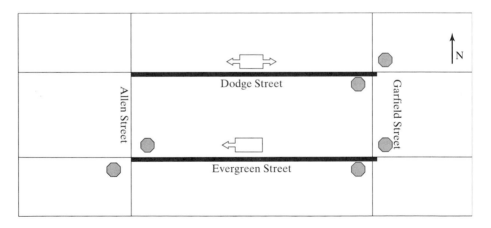

FIGURE 6.32
The Archer neighborhood. Only stop signs that faced EB or WB traffic on Dodge and Evergreen are shown.
Block arrows show location and direction(s) of radar gun.

furthermore, she said that it would not slow vehicles down. Darren wanted to verify or challenge the city traffic engineer's last statement. He borrowed the county's old radar detector—
it was heavy, but it still was accurate—and asked his father help him design the data collection
activity. After discussing it with his father, Darren decided that getting speeds for EB traffic on

Evergreen and both directions on Dodge would give him a good basis for checking out the city traffic engineer's theories. Then, Darren and his father went to those midblock locations in the Archer neighborhood, hooked the radar gun onto a side window of the family station wagon, and recorded vehicle speeds.

THINK ABOUT IT

Why did Darren choose the locations and directions he did? Would you have done anything differently?

A. Because there isn't much traffic on the Archer neighborhood streets, it may take a long time to get an adequate sample. Darren wanted to determine the minimum sample size to (a) avoid spending any more time than necessary collecting data while (b) avoiding the need to return to any sites in the study area. How many vehicle speeds should Darren record at each site? *Hint:* Recall how the sample size for the speed study in Example 2.8 was determined. From previous speed studies on urban two-lane streets in Mythaca, standard deviation $S = 4.8$ mph.

B. The speeds Darren recorded are summarized in Table 6.15. Did Darren collect a large enough sample?

C. Darren's dad explained hypothesis testing to him. Darren decided that the generic hypothesis should be: Vehicle speeds at midblock are the same, whether the block ends with a stop sign or not. After all, this seemed to be the city traffic engineer's position. Did the data that Darren collected support this hypothesis?

Solution to Example 6.15

A. If the mean midblock speed is desired, Equation 2.4 applies. The standard $z = 1.96$ value is chosen from Table 2.6. The choice of E is less obvious. Knowing that the radar gun has a digital readout that only registers integer speeds, and guarding against the case in which mean speeds at different sites are not significantly different, a small value of E was chosen: $E = 1.0$ mph. Equation 2.4 becomes

$$N = \frac{S^2 z^2}{E^2} = \left(4.8 \frac{1.96}{1.0}\right)^2 = (9.408)^2 = 89 \text{ speeds}$$

B. Table 6.15 shows that Darren was able to record only 39 midblock speeds on EB Dodge between Garfield and Allen before he and his dad had to go home. This was far fewer than the 89 speeds Darren wanted to collect. However, when Darren calculated the standard deviation for his sample, he found that $S = 3.837$. If he had used $S = 3.837$ in part A, he would have calculated $N = 57$. What if he relaxed the $E = 1.0$ requirement to, say, $E = 1.25$ mph? Then $N = 37$. The good things about $E = 1.25$ are that it is only 0.25 mph higher than the $E = 1.0$ value originally chosen and it makes another trip to the neighborhood unnecessary. A problem, however, is that the mean speeds shown in Table 6.15 are about 0.9 mph apart. It is probably unwise to let your "precision parameter" be larger than the differences in the variables you are trying to compare. For a paid consultant's job, more data might have to be collected. For a sixth grade science project, let us concentrate on the methodology and learn from the experience.

TABLE 6.15 Darren's speed data for Archer neighborhood

Fri. 10 Apr 98, 3:50–4:30 PM Dodge, bet. Garfield and Allen			Sat. 4 Apr 98 and Mon. 20 Apr 98 Evergreen, bet. Allen and Garfield	
Eastbound (U-S)		WB (S-U)	EB (S-S)	
Intersection	Midblock	Midblock	Midblock	
31	20	21	23	
19	18	24	21	Darren
22	21	23	21	Ms. Anderson
28	27	27	25	4/20/98
34	**33**	23	21	
24	18	24	19	S = stop-controlled intersection
14	22	21	24	U = uncontrolled intersection
24	22	24	24	1st Ho: WB midblock speed = EB midblock
27	22	26	24	speed
26	26	22	24	Emb = Evergreen midblock speed
26	28	**27**	22	Dmb = Dodge midblock speed
32	24	25	**27**	2nd Ho: Emb = Dmb
21	**33**	20	**27**	3rd Ho: Presence of downstream stop sign
21	19	20	23	reduces midblock speed.
23	22	21	24	
23	23	**27**	26	
28	23	24	17	
31	27	23	17	
24	26	20	20	
23	21	20	19	
28	23	**18**	21	
16	22	20	22	
26	25	21	27	
19	20	**18**	**15**	
21	25	**27**	24	
19	21	21	21	
24	22	22	24	
30	**17**	19	20	
25	23	**27**	18	
25	31	19	23	
22	25	19	19	
25	18	26	19	
26	21	24	21	
21	24	24	18	
22	25		21	
22	22		23	
26	21		16	
	29		16	
	24		22	
			27	
			21	
24.324	23.410	22.559	21.610	← Mean
4.378	3.837	2.841	3.185	← Std Dev
1.96	1.96	1.96	1.96	95% confidence level
1	1	1	1	error permitted (mph)
73.6	56.6	31.0	39.0	sample size desired
37	39	34	41	# of cars observed

C. The first hypothesis Darren wants to test is: *WB and EB mean speeds on Dodge are equal.* This is of interest because, for the observed block, EB Dodge ends at a stop sign at Garfield, whereas WB Dodge at Allen has no stop sign. In Table 6.15, "S" = stop sign and "U" = unsigned. Thus, EB is the "U-S" direction and WB is the "S-U" direction. To test the hypothesis H_0: $\mu_{EB} = \mu_{WB}$, you may have to refer to the notes or the textbook from your probability and statistics course on "two-sided tests." [The reference used here is *Hypothesis Testing for Comparing Two Means* in Lapin (1990), Section 12-3.] If you haven't yet completed such a course, you will have to trust the method about to be demonstrated. The steps in the method are:

1. Because the number of observations in each direction is greater than 30, compute the test statistic z:

$$z = \frac{\overline{X}_{EB} - \overline{X}_{WB}}{\sqrt{\dfrac{s^2_{EB}}{n_{EB}} + \dfrac{s^2_{WB}}{n_{WB}}}} = \frac{23.410 - 22.559}{\sqrt{\dfrac{(3.837)^2}{39} + \dfrac{(2.841)^2}{34}}} = 1.086 \tag{6.18}$$

where $\overline{X}$ = the mean value of the speeds in the observed sample,
  s^2 = the variance of the sample speeds,
  n = the number of speeds observed in the sample.

2. For a two-sided test at a 95 percent confidence level, the *critical value* of z is ±1.96. If the test statistic falls outside the range $\{-1.96, +1.96\}$, the hypothesis must be rejected. Here, $z_{\text{test}} = 1.086$, which lies within the specified range, so the hypothesis H_0: $\mu_{EB} = \mu_{WB}$ cannot be rejected.

As a result of this analysis, Darren can say that there is no statistically significant difference between the 23.410 mph average speed on EB Dodge (heading toward a stop sign) and the 22.559 mph average speed on WB Dodge (with no end-of-block stop sign). In fact, what little difference that exists between the EB and WB speeds is the opposite of what was expected by the Archer neighborhood parents: EB traffic heading toward a stop sign is the (slightly) *faster* traffic! Perhaps not having to stop at the *previous* intersection is an important factor. That is a second hypothesis that we can test in the guise of a homework problem.

SUMMARY

Improving safety on a roadway depends on the performance of three components: the vehicle, the driver, and the roadway itself. Each component can contribute to the safety (or lack of it) on a roadway. By understanding each of the three components, the transportation engineer can design roadways for safer performance. A framework called the Highway Safety Improvement Program establishes the basis for diagnosing roadway hazards, proposing solutions, and evaluating their efficacy. An appreciation for how individuals operate their vehicles in the roadway environment—a field of study known as human factors—can contribute to a better roadway design. Especially important is the recognition that drivers have a wide range of capabilities when it comes to vision, reaction time, and decision making. Designing for the least capable driver may actually induce less safe behavior on the part of other drivers. Vehicles can have a wide range of capabilities, too. Size, weight, acceleration, and other characteristics can vary significantly from vehicle to vehicle. Something as simple and familiar as traffic signs can have an important impact on roadway safety. Following published standards and adapting them to particular circumstances can assist motorists in operating their vehicles safely on a roadway.

ABBREVIATIONS AND NOTATION

a_{acc}	acceleration capability of a vehicle (usually a constant in this class)
A	annual cost
AASHTO	American Association of State Highway and Transportation Officials
ADT	average daily traffic volume
AHEV	average hourly entering volume
AHC	average hourly traffic conflicts
C	crash rate, crashes per year
CAS	critical approach speed
C_D	coefficient of aerodynamic drag
CI	traffic conflict index
CP(k)	crashes prevented (year)
CRF	crash reduction factor or percent reduction in crashes
DUI	driving under the influence of alcohol
D_{br}	braking distance
D_{object}	distance to an object
EC	expected number of crashes over a specified time (usually a year) if the countermeasure is not implemented and the traffic volume remains the same
$F_f\ F_r$	engine force applied at the front and rear wheels, respectively
f	dimensionless coefficient of friction of the road
f_r	coefficient of rolling resistance at constant velocity (usually $f_r = .01(1 + V^2/147)$ V in fps
fps	feet per second
FRA	Federal Railroad Administration
g	force of gravity 32.2 feet per second per second or 9.8 meters/second/second
G	grade in percent divided by 100.
HSIP	Highway Safety Improvement Program
ICI	intersection conflict index
MEV	million entering vehicles
MUCTD	Manual for Uniform Traffic Control Devices
NHTSA	National Highway Traffic Safety Administration
OACS	overall average conflict severity
OM	object markers
P_o	initial cost or investment
PDO	property damage only crashes
PI	personal injury crashes
PIEV	time for perception/identification/emotion/volition
PWC	present worth of a series of costs
R_r	sum of rolling resistance from each of the tires $R_r = R_{rlf} + R_{rrf} + R_{rlr} + R_{rrr}$ $= f_r W$
RHMVM	crash rate per million vehicle miles
R_{grade}	component of gravity acting normal to the road
R_{aero}	component of aerodynamic resistance or drag
RMEV	crash rate per million entering vehicles
ROC	rate of collision

ρ	air density
SSD	stopping sight distance
SSM	supplemental safety measure
STM	short-term memory
TCD	traffic control devices
TCS	total conflict severity
TEV	thousand entering vehicles
T_r	reaction time when driving a car, usually in conjunction with braking
TTC	time to collision
W	weight of the vehicle in pounds
v_o	velocity at the beginning of vehicle changes in speed
v_f,	final velocity of the vehicle
v	instantaneous velocity
V	velocity
VKT	vehicle kilometers traveled
VMT	vehicle miles traveled

GLOSSARY

Collision diagram: A graphic summary of the collisions at an intersection for one year or another appropriate time period.

Countermeasure: A project that is intended to reduce the crash rate at a site, especially in response to causes identified as part of the HSIP.

Crash rate: Crashes per million entering vehicles at intersections; crashes per hundred million vehicle miles on road sections.

Crash Reduction Factor: An estimate of how effective a certain countermeasure will be, based on historical data on crash reductions after the countermeasure has been applied.

Critical rate analysis: A way to determine whether a particular site is dangerous, such as an intersection. Use a representative sample of intersections or roadways to establish a crash rate threshold, then determine whether the site's crash rate exceeds the threshold.

Delineators: Roadside markers used to guide drivers through turns, especially at night or at times of poor visibility.

Emotion: The time to consider the sign's meaning and make a decision.

Expectancy: A design feature that helps motorists by giving them consistent clues and guidance.

Exposure: A measure of the amount of travel against which crashes, injuries, and fatalities can be compared.

Highway Safety Improvement Program: A framework for planning, implementing, and evaluating safety programs and projects.

Human factors: The study of how human beings function in their natural or constructed surroundings.

Identification: The time to read and understand the sign.

Perception: The time it takes to see the sign.

Stopping sight distance: The distance needed by a driver to bring his/her vehicle to a safe stop, given roadway grades, surface conditions, and operating speeds.

Traffic conflict analysis: A procedure to assess the potential for actual collisions by collecting data on traffic conditions that are conducive to dangerous interactions between vehicles.

Traffic control device: A sign or pavement marking that is used to regulate, warn, or guide drivers as they operate their vehicles.

Volition: The time to execute a maneuver.

Zone of spatial commitment: The field of vision from which a driver samples cues about what is ahead.

REFERENCES

6.1   Safety on the Highway

[1] Blincoe, Lawrence J., "The Economic Cost of Motor Vehicle Crashes, 1994," NHTSA Technical Report, Plans and Policy, National Highway Traffic Safety Administration, U.S. Department of Transportation, Washington, DC 20590, www.nhtsa.dot.gov/people/economic/ecomvc1994.html#toc3.

[2] Box, Paul C. and Joseph C. Oppenlander, *Manual of Traffic Engineering Studies*, Institute of Transportation Engineers, 4th Edition, 1976.

[3] Creasey, Tom and Kenneth R. Agent, *Development of Accident Reduction Factors*, Kentucky Transportation Research Program, 1985.

[4] Ermer, Daniel J., Jon D. Fricker, and Kumares C. Sinha, *Accident Reduction Factors for Indiana*, Final Report, FHWA/IN/JHRP-91/11, April 1992.

[5] Federal Highway Administration, *Highway Safety Improvement Program*, Document FHWA-TS-81-218, US DOT, June 1981.

[6] Hamilton Associates, *Traffic Conflict Procedures Manual*, 2nd edition, Vancouver BC, November 1996.

[7] Miller, Ted R., J.G. Viner, S.B. Rosman, N.M. Pindus, G.W. Gellert, J.B. Douglass, A.E. Dillingham, and G.C. Plomquist, "The Costs of Highway Crashes," The Urban Institute, Washington DC, 1991, as cited in Blincoe (1994).

[8] National Highway Traffic Safety Administration, *Traffic Safety Facts*, US DOT, issued annually.

[9] United States Department of Transportation, *Nationwide Personal Transportation Survey of 1990*.

[10] Perkins, S.R. and J.L. Harris, *Criteria for Traffic Conflict Characteristics*, Report GMR 632, General Motors Corporation, Warren, MI, 1967, as cited in Hamilton Associates (1996).

[11] United States Department of Commerce, *Statistical Abstract of the United States, Various years*.

[12] United States Department of Transportation, *Highway Safety Engineering Studies Procedural Guide*, June 1981.

6.2   Human Factors

[13] AAA Foundation for Traffic Safety, www.aaafts.org/text/research/distraction_phase1.cfm, as viewed 1 July 2001.

[14] American Association of State Highway and Transportation Officials, *A Policy on Geometric Design of Highways and Streets*, 2001.

[15] Cole, B.L., "Visual Aspects of Road Engineering," *Proceedings*, 6th Australian Road Research Board Conference, 1972, p. 102–148. As cited in Ogden (1990).

[16] Cole, B.L. and S.E. Jenkins, "Conspicuity of Traffic Control Devices," *Australian Road Research* 12, No. 4, 1982, p. 223–238. As cited in Ogden (1990).

[17] Collins, Dennis James, *An Examination of Driver Performance Under Reduced Visibility Conditions When Using an In-Vehicle Signing Information System*, Thesis, Master of Science

Industrial and Systems Engineering, Virginia Polytechnic Institute and State University, Blacksburg VA, April 1997.

[18] Farnham, W.L., "An In-depth Analysis of the Most Effective Railroad Crossing Protection," October 18, 2000, www.lakesnet.net/mnnrscf/mnnrscf/farnham.htm.

[19] Federal Railroad Administration, *Crossing Statistics*, 1998. As cited in Farnham, 2000.

[20] Federal Railroad Administration, Office of Railroad Development, U.S. Department of Transportation, Draft Environmental Impact Statement, *Proposed Rule for the Use of Locomotive Horns at Highway-rail Grade Crossings*, Washington, D.C. 20590, December 1999: ES-1, 4-27. As cited in Farnham, 2000.

[21] Federal Railroad Administration, Office of Safety Analysis web site, http://safetydata.fra.dot.gov/officeofsafety/Default.asp, as updated June 29, 2001.

[22] Fricker, Jon D. and R.J. Larson, "Safety Belts and Turn Signals: Driver Disposition and the Law," *Transportation Research Record* 1210, 1989, p. 47–52.

[23] Fricker, Jon D., "Human Information Processing and License Plate Design," *Transportation Research Record* 1093, 1986, p. 22–28.

[24] Gazis, D., R. Herman, and A. Maradudin, "The Problem of the Amber Signal in Traffic Flow," *Operations Research*, vol. 8, March–April 1960, p. 112–132.

[25] Hendricks, D.L., J.C. Fell, and M. Freedman, *The Relative Frequency of Unsafe Driving Acts in Serious Traffic Crashes*, sponsored by National Highway Traffic Safety Administration, December 1999.

[26] Hulbert, Slade F., "Driver Information Systems," *Human Factors in Highway Traffic Safety*, T.W. Forbes, ed., Wiley, 1972, p. 111.

[27] Institute of Transportation Engineers *Transportation and Traffic Engineering Handbook*, Prentice Hall, 1982.

[28] Johannson, G. and K. Rumar, "Driver's Brake Reaction Time," *Human Factors*, Vol. 13, No. 1, 1971, p. 22–27. As cited in AASHTO 2001.

[29] Kantowitz, Barry H. and Robert D. Sorkin, *Human Factors: Understanding People-System Relationships*, John Wiley & Sons, 1983.

[30] Lay, M., *Handbook of Road Technology*, Gordan and Breach, London, 1986. As cited in Ogden (1990).

[31] Leibowitz, H.W., "Grade Crossing Accidents and Human Factors Engineering," *American Scientist*, Vol. 73, No. 6, 1985, pp. 558–562.

[32] *Manual on Uniform Traffic Control Devices*, Millennium Edition, Federal Highway Administration, U.S. Department of Transportation, December 2000, including Errata No. 1 dated June 14, 2001.

[33] National Center on Sleep Disorders Research and National Highway Traffic Safety Administration, Drowsy Driving and Automobile Crashes, NCSDR/NHTSA Expert Panel on Driver Fatigue and Sleepiness, www.nhtsa.dot.gov/people/perform/human/Drowsy.html, as viewed 1 July 2001.

[34] National Highway Traffic Safety Administration, "Mission of the Drowsy Driving Program," www.nhtsa.dot.gov/people/injury/drowsy_driving1/index.html, as viewed 1 July 2001.

[35] National Center for Statistics & Analysis, *Traffic Safety Facts 1999—Alcohol*, DOT HS 809 086, National Highway Traffic Safety Administration, U.S. Department of Transportation, 400 Seventh Street, S.W., Washington, DC 20590.

[36] Nelson, Lee J., "License Plate Recognition Systems," ETTM on the Web, www.ettm.com/news/lpr.html, March 1998.

[37] New Brunswick Province, *Driver's Handbook*, Part 5, Safe Driving, viewed June 2001 at www.gov.nb.ca/dot/mv/handbook/part5/driver.shtml.

[38] O'Cinneide, D., "The Relationship Between Geometric Design Standards and Safety," Conference Proceedings, International Symposium on Highway Geometric Design Practices, Boston, Massachusetts, August 30–September 1, 1995

[39] Ogden, K.W., "Human Factors in Traffic Engineering," *ITE Journal*, August 1990, p. 41–46.

[40] Redelmeier, Donald A., MD, "Talking Distractions," *Recovery*, Volume 9, Number 2, Summer 1998.

[41] Sanders, Mark S. and Ernest J. McCormack, *Human Factors in Engineering and Design*, Seventh Edition, McGraw-Hill, Inc., 1993.

[42] Sen, Atri, *A Methodology to Determine the Effective Deployment of Portable Dynamic Message Signs Using a Virtual Reality Based Driving Simulator*, unpublished Masters Thesis, School of Civil Engineering, Purdue University, West Lafayette, IN, July 2001.

[43] Sivak, Michael, Paul L. Olson, and Kenneth M. Farmer, "Radar-Measured Reaction Times of Unalerted Drivers to Brake Signals," *Perceptual and Motor Skills*, vol. 55, 1982, p. 594.

[44] Smith, S.L., "Letter Size and Legibility," *Human Factors*, vol. 21, no. 6, 1979, p. 661–670.

[45] Stutts, Jane C., Donald W. Reinfurt, Loren Staplin, and Eric A. Rodgman, *The Role of Driver Distraction in Traffic Crashes*, University of North Carolina Highway Safety Research Center, funded by the AAA Foundation for Traffic Safety, May 2001.

[46] Taoka, George T., "Brake and Reaction Times of Unalerted Drivers," *ITE Journal*, March 1989, p. 19–21.

[47] Townsend, J.T., "Theoretical Analysis of an Alphabetic Confusion Matrix," *Perception and Psychophysics*, vol. 9, no. 1A, 1971, p. 40–50.

[48] Turner-Fairbank Highway Research Center (TFHRC), Human Centered Systems Studies, Federal Highway Administration, U.S. Dept. of Transportation, www.tfhrc.gov/safety/safety.htm#Human, as viewed 22 June 2001.

[49] van der Heijden, A.H.C., *Short-Term Visual Information Forgetting*, Routledge & Kegan Paul, Ltd., Boston, MA, 1981

[50] Walls, Ann, "National Work-Zone Awareness Week Commemorated Across the Nation," *Public Roads*, May/June 2001, p. 40.

[51] Washington State DOT, *Traffic Manual*, Manual Number: M 51-02, 1 February 1996. Cited at www.wsdot.wa.gov/regions/northwest/signdgn/htm/dgnguide/Intrduction.htm.

[52] Wickens, C.D., *Engineering Psychology and Human Performance*, Third Edition, Harper Collins, Publishers, 1999.

[53] Yang, CYD, J.D. Fricker, and T. Kuczek, "Designing Advanced Traveler Information Systems from a Driver's Perspective: Results of a Driving Simulation Study," *Transportation Research Record* 1621, 1998, p. 20–26.

6.3   Vehicle Attributes That Affect Safety

[54] Taborek, J.J., "Mechanics of Vehicles," *Machine Design*, 1957.

6.4   TCDs

[55] Institute of Transportation Engineers (ITE), *Introduction to Signs and Markings*, Slides and Narrative Notes, 4 September 1974.

[56] Federal Highway Administration (FHWA), *Manual for Uniform Traffic Control Devices*, 2003. Latest version available online at http://muted.fhwa.dot.gov.

[57] Florida Department of Highway Safety and Motor Vehicles, *The 2000 Florida Driver's Handbook*, Chapter 4, www.hsmv.state.fl.us/handbooks/English/ch_204.html.

EXERCISES FOR CHAPTER 6: SAFETY ON THE HIGHWAY

Highway Safety—Data and Analysis

6.1 Crash Rate on Roadway Section.   As it passes through a fast-growing part of Mythaca, a 0.79-mile section of Gifford Street has begun to experience a much higher crash rate than in previous years. Over the past 3 years, there have been 14, 13, and 15 crashes on this section of Gifford. The two-way ADT on this section of Gifford is 16,474 vehicles per day. What is the annual crash rate for this section of Gifford Street?

6.2 Crash Rate at Intersection.   As Gifford approaches downtown Mythaca, it crosses Gregory Avenue at a signalized intersection. The four approach volumes per day at this intersection are 28,648, 23,856, 12,150, and 10,174. The crash totals for the last 3 years are 91, 96, and 87. What is the annual crash rate for this intersection?

6.3 Accident Rates.   The intersection of one-way (EB) South Street and one-way (NB) 4th Street in Lafayette had the fourth highest accident rate (3.229 per MEV) in Tippecanoe County in 1994. If there were 31 accidents and the ADT on South Street was 16,030 in 1994, what was the ADT on 4th Street?

6.4 Is an Intersection "Hazardous"?   Concerned about the crash rate at Gifford and Gregory, the Mythaca County staff has found 10 other intersections in the region that have the same characteristics as Gifford and Gregory. The crash rates at these "control sites" over the past 3 years are 2.83, 3.19, 2.37, 2.59, 1.61, 3.13, 3.68, 2.23, 3.88, and 2.80.

(a) What are the mean and standard deviation for the control sites?
(b) Use the critical rate method to determine whether Gifford and Gregory qualifies as a hazardous intersection. Use the 95 percent confidence level, which means $Z = 1.96$.

6.5 Traffic Crashes: A True Story.   At about 5 PM on a sunny Saturday (27 July 1996), the driver of a small vehicle traveling west on US52 (Sagamore Parkway in West Lafayette) is waiting to turn left onto Blackhawk Lane. For some reason, she begins her turn as an EB pickup truck is approaching the intersection in the left lane at high speed. The pickup truck's brakes were applied, but the small car was hit in the right front fender by the front of the pickup truck. The intersection has four approaches, and Sagamore Parkway has a raised median. There were no injuries.

(a) Draw the collision diagram (main box only) for this crash.
(b) To begin an analysis of this crash, propose what you think are the two most likely "probable causes" of this crash. Explain if necessary.

6.6 Highway Deaths.   Between 1982 and 1994, highway fatalities decreased from about 50,000 to about 40,000. One element was a contributing factor in 50 percent of fatal crashes in 1982, but this percentage went down to less than 40 percent in 1994. What was this "element" and why did its percentage decrease?

6.7 Crash Analysis.   Which of the three causes below is most often responsible for highway crashes?

- Mechanical failure of vehicle
- "Failure" of guideway or roadway
- Human (operator) error

Name two new technological developments that may help to reduce this cause of crashes.

6.8 Crash Rates.   The number of crashes at the intersection of US52 and SR26 increased from 39 in 1989 to 43 in 1990. The 1990 approach ADTs are estimated to be the following: NB 13,545; SB 12,535; EB 7200; and WB 9760. What was the 1990 crash rate at this intersection?

6.9 Crash Rates.   The intersection of South Street and Earl Avenue had the highest number of accidents (41) in Tippecanoe County in 1991. Its accident rate was 3.730/MEV. What was the total AADT of all four approaches to South and Earl in 1991?

6.10 Roundabouts.   The NB, SB, EB, and WB approach AADT at a roundabout in West Virginia are 5242, 854, 3877, and 944, respectively. The number of crashes at the roundabout for the years 1994, 1995, and 1996 were 9, 10, and 9, respectively. What is the crash rate at this roundabout?

Human Factors and Transportation Engineering

6.11 Human Factors in Daily Life.   Make a list of items or environments that you have experienced that serve as examples of good or bad design from the perspective of human factors. They do not have to be related to transportation, although transportation examples are preferred.

6.12 Reaction Time Tests.   With a good Internet search engine, use the words "reaction time test" to find two different reaction time tests on the Web. Try them. Describe the tests you tried, summarize your results, and comment on the validity of the tests.

6.13 Visual Acuity.   The state DOT wants to erect a sign warning drivers of a merge in the road ahead. If the average driver must be able to see the sign from a distance of 400 feet, how tall must the letters be? Use the visual acuity data from Figure 6.18.

6.14 Visual Acuity Test.   Print out some letters and numbers onto a sheet of paper, using Arial font, bold, point size 36. Attach the sheet to a wall. Ask someone else to start at the opposite side of the room and move forward until a character can be read. Note the character, its height, and the distance to the target. Have the subject continue moving forward until the subject has identified all characters on the target. Which characters were misidentified? Which characters were easiest to see? Compute the subtended visual angle for each case. What guidance does this visual acuity experiment offer for the design of traffic signs?

6.15 Safety Device Design.   What is the current status of the design and use of airbags in automobiles? Comment on air bags being one of the few safety devices that carry a warning label.

6.16 Human Factors.   Based on your study of human factors in this course, respond "True" or "False" to each statement below.

- An individual's ability to perform a task may vary over time and depend on working conditions.
- Drivers tend to overestimate the speed of very large vehicles, such as locomotives.

6.17 Human Factors at Railroad Grade Crossings.   What is it about trains at grade crossings that drivers often misjudge? Why is this a problem?

6.18 Stopping For a Train.   You are traveling at 70 mph on a slushy road (friction coefficient = 0.20) when you hear a train whistle. You then see the warning sign that is placed 1000 feet before the gate-protected railroad grade crossing. You know you must try to stop.

(a)   How close will you be to the gate when you come to a stop? Your reaction time is 1.5 seconds.

(b)   Where will the train be relative to the grade crossing when you come to a stop?

6.19 Racing the Train.   A friend is driving along a local road at 55 mph. This friend hates to wait for anything, even the train that he sees heading for the grade crossing ahead. There is no gate at this grade crossing—only a cross buck sign and a bell. Your friend makes the decision to try to beat the train to the crossing. Although he can only guess at these values, the train is 1000 feet from the crossing and moving at 40 mph when your friend first sees it. At that time, your friend is 800 feet from the crossing.

(a) Assume your friend has a reaction time of 0.6 seconds. How far from the crossing will he be when he begins to accelerate?

(b) Your friend's car can accelerate at the rate of 28 ft/sec/sec, but it has a maximum speed of 85 mph. How fast will it be going when it reaches the crossing?

(c) How much time did your friend take to reach the grade crossing? Did he beat the train?

6.20 Aging Society.   What must a transportation planner or engineer take into consideration in a society where an increasing number of people are over 65 years of age?

Vehicle Attributes That Affect Safety

6.21 Stopping on a Downhill Grade.   At one point on SR835, there is a 4.9 percent downhill grade. How long will it take to bring a car traveling at 48 mph to a stop on that downhill segment if the driver's reaction time is 2.0 seconds and $f = 0.29$?

6.22 Traffic Accident.   A transportation student is driving on a level road on a cold rainy night and sees a construction sign 520 feet ahead. The student strikes the sign at 35 mph. Furthermore, the student claims that he was not violating the 55 mph speed limit. You are investigating the accident and you will testify in court.

(a) What evidence will you seek?

(b) What will you tell the court? (Be specific about reaction times and possible initial speeds.)

Traffic Control Devices

6.23 Traffic Control Devices.   Recently, the county highway engineer observed a two-person crew about to install a traffic control devices on the campus of Mythaca State University. At the base of the signpost were two signs, both with the message "Two Way Traffic Ahead." One sign had a rectangular shape with black letters on white background; the other was diamond-shaped with black on yellow. If the crew was making the correct change, which sign would be the correct one to put up?

(a) Diamond-shaped with black on yellow

(b) Rectangular shape with black letters on white background

Briefly explain your answer.

6.24 Traffic Control Devices.   A driver on the northbound (NB) approach to a stop sign-controlled intersection sees the sign and supplemental plaque shown below. Drivers on which approaches to that intersection will have to stop? Circle the approach directions that comprise your answer.

(a) EB

(b) NB

(c) SB

(d) WB

6.25 Traffic Control Devices.   The three SB lanes on Northwestern Avenue (as it enters the "T" intersection with Cherry Lane) are marked as shown in the figure below. (SB traffic moves from right to left in the figure. Line "C-C" represents the curb; it flares out . Line "D-D" is the stop line at the intersection.)

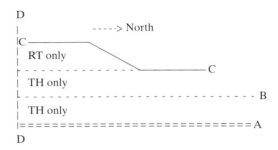

(a)   What color should the solid centerline "A" be?

(b)   What color should dashed line "B" be?

(c)   What change to dashed line "B" would probably be a good idea?

Highway Design for Safety

SCENARIO

As SR361 approaches Shoridan from the east, it crosses the Mythaca River by means of an old truss bridge (Figure 7.1). The bridge takes SR361 not only over the river but also over the road that runs alongside it—fittingly called River Road. Because of new trip patterns and increasing traffic volumes, Shoridan area officials have called for a direct connection between SR361 above and River Road below. At present, the connection must be made by driving to the outskirts of Shoridan and doubling back more than a mile to River Road. The other alternative is to leave SR361 several miles east of the river and use county roads with limited capacity and poor sight distances to reach River Road. The proposed direct connection would involve a significant change in elevation, from the bridge level, down a hillside, to the river valley level. The proposed connector road presents safety problems, in that grades, curves, and speeds will have to be controlled to maintain safe driving conditions. To make the design challenge even more difficult, the hillside to be used for the connector road has already been partially developed. Numerous homes have been built to take advantage of the views of the Mythaca River Valley and Murdoch Bay. Any connector road to the valley will have to remove several of the homes. Of course, it would be desirable to remove as few hillside homes as possible. Moreover, there may be environmentally sensitive areas in a preferred right of way that will have to be avoided. In many ways, highway design is a three-dimensional "puzzle" that must be solved by the engineer and supporting staff.

FIGURE 7.1
Hwy 361 bridge over River Road. Photo: Jon D. Fricker.

CHAPTER OBJECTIVES

By the end of this chapter, the student will be able to:

1. Determine the elevation of specified points along a vertical curve.
2. Calculate the key dimensions of a horizontal curve.
3. Design vertical and horizontal curves using safe stopping sight distance.
4. Calculate the appropriate distance for passing/no passing marking for a rural two-lane road.
5. Determine the appropriate bank angle (superelevation) for safe travel around a horizontal curve.

> **FYI:**   Many of the equations, tables, and design curves in this chapter are credited to "AASHTO." AASHTO stands for the American Association of State Highway and Transportation Officials, the organization that publishes A Policy on the Geometric Design of Highways and Streets. This publication, commonly called "The Green Book," contains the standards for geometric design that are followed by most jurisdictions in the United States.

7.1   THE CHALLENGE OF ROADWAY ALIGNMENT

7.1.1   Overview

Geometric Design is the term used to describe the way in which highway designers try to fit the highway to the terrain while maintaining design standards for safety and performance. Initially, a preliminary idea of the road orientation is laid out on the basis of a thorough survey. This is followed by geometric design that will:

A. Adhere to design standards pertaining to hills and curves. The ability of the driver to see (and avoid) obstacles ahead will be brought into the design specifications.

B. Provide a context in which to analyze and evaluate the construction and marking of a proposed highway section.

C. Determine the extent to which design standards will permit realignments to route a proposed roadway around specific locations of special value or to handle environmental situations such as wetland mitigation.

D. Produce "blueprints" for construction.

One construction question that usually arises is, how much material must be excavated to produce a highway section that meets design standards and at what cost? Highway engineers need to plan the excavation and the storage of fill material for the highway. Proximity to the construction site is important, because trucking soil over long distances to or from the "borrow pits" can be expensive. Figure 7.2 shows a simplified earthwork problem. The amount of material that is cut and filled as well as the transport cost to use excavated material for fill can be estimated. These costs must be included in the overall project cost structure. There are standard techniques and several computer software programs that enable the engineer to calculate the amount of

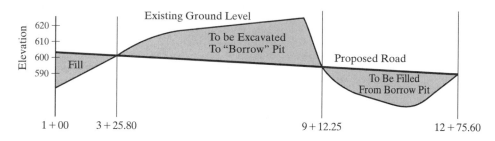

FIGURE 7.2
Seeking the optimal vertical alignment with respect to earthwork cost.

earthwork to be accomplished and how borrow pits are to be used. To the extent that material excavated in Figure 7.2 from station 3 + 25.80 to station 9 + 12.25 can be used to fill the segment from 9 + 12.25 to station 12 + 75.60, the amount of earth to be stored or borrowed can be reduced.

The results of geometric design are a set of coordinates through which the roadway will pass to connect specified starting and ending points. Figure 7.3 shows the "plan and profile" of an alignment that requires both horizontal and vertical changes in roadway direction and elevation. Note that the horizontal layout and the vertical layout appear on the same drawing. However, the engineer has to relate the station numbers on one diagram to the other. They do not necessarily appear directly above one another.

The challenge of geometric design is "fitting 'conventional curves' onto irregular terrain with specified benchmarks while maintaining smooth transitions between roadway segments" (AASHTO, 1994). To further illustrate the challenge of geometric design, consider Figure 7.4. The figure shows four alternative schemes for crossing the Charles River in Boston as part of the gigantic Central Artery and Third Harbor Tunnel project. These four alternatives were among a large number of alternatives that were evaluated. The criteria for evaluation included the homes and businesses that would have to be moved, the amount of open space that would be lost, the wetlands that would have to be protected or replaced, and adherence to modern highway safety standards. Furthermore, each element of the design should be aesthetically pleasing to drivers and to the road's neighbors.

The challenge in Boston is a vastly more complex version of the SR361-River Road project presented in this chapter's Scenario, because there were many more connections that had to be made or preserved. The photographs in Figure 7.5 show just how complex a modern urban alignment can be. Engineers must design curves, bridges, and other structures for the transportation facilities that will disrupt as little of the urban area as possible.

There are some important differences between vertical and horizontal curves:

- The vertical curve is represented by a parabola, and the length along the curve is simply the projected horizontal distance.
- The horizontal curve is circular, and its length is measured along the curve between its beginning and ending point.

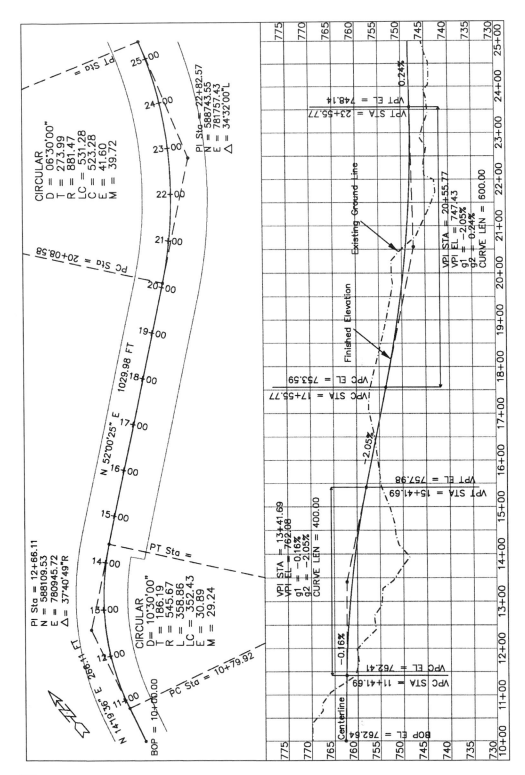

FIGURE 7.3
Typical output for a curve that requires both horizontal and vertical alignment. *Source:* Courtesy of D.C.A. Engineering Software, Inc.

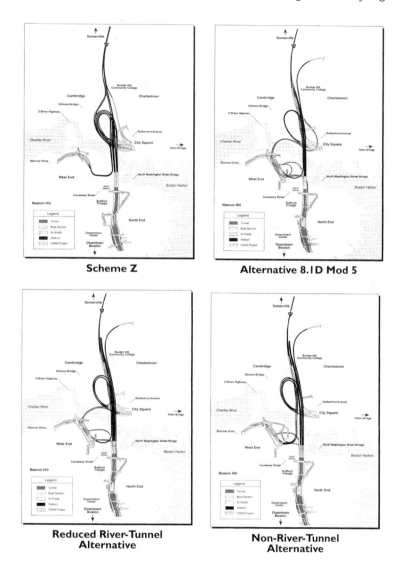

Scheme Z

Alternative 8.1D Mod 5

**Reduced River-Tunnel
Alternative**

**Non-River-Tunnel
Alternative**

FIGURE 7.4
Four alternatives for crossing the Charles River. *Source*: Holly Sutherland, Central
Artery/Tunnel Project.

7.1.2   Geometry of Vertical Curves

Vertical alignment of highways consists of grade tangents connected by parabolic vertical curves. The beginning of the vertical curve is designated as *vertical point of curvature (VPC)* and the end is the *vertical point of tangency (VPT)*. (See Figure 7.6.) The length of vertical curve is measured *horizontally* between the VPC and the VPT, not along the roadway surface. Likewise, any distance *x* from the VPC is measured *horizontally*.

The *vertical point of intersection (VPI)* is where the two tangent sections, when extended, intersect. "A parabolic curve ... centered on the VPI is usually used in roadway

FIGURE 7.5
Aerial photos of Boston's Central Artery Project. Left: Bridge construction over the Charles River. At Right: The elevated Central Artery (I-93) winds its way through downtown Boston. Photos used by permission of Dennis Rahilly, Central Artery/Tunnel Project, Boston MA, www.bigdig.com.

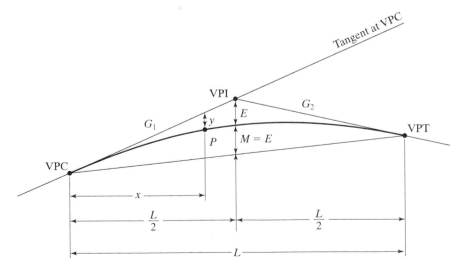

FIGURE 7.6
Layout of an equal tangent crest vertical curve.

design" [AASHTO, 2001, p. 270] This means that the VPI is at the halfway point (L/2) of the vertical curve, even when the approaching grade is different from the departing grade. (See Figure 7.7.) Such curves are called *equal tangent vertical curves* or *symmetric vertical curves*. The usual maximum grade for urban streets is about 6 percent. Higher grades may exist in areas where the topography requires it.

When the tangent grades are equal, the VPI will occur over (or below) the highest (or the lowest) point of the curve. The percent grade entering the vertical curve is denoted as G_1, and the percent grade of the tangent leaving the curve is G_2. Vertical curves in which the initial grade G_1 is greater than the final grade G_2 are crest curves. Sag curves (see Figure 7.7) have $G_1 < G_2$. Grades can be expressed in terms of percent, or they can be expressed as change in elevation (feet) in relation to change in horizontal distance (feet). For example, a 4 percent grade would have the equivalent value 0.04 feet/foot (ft/ft), which is actually dimensionless. The equations introduced in the remainder of this section use the ft/ft units for grade and feet for distances and elevations.

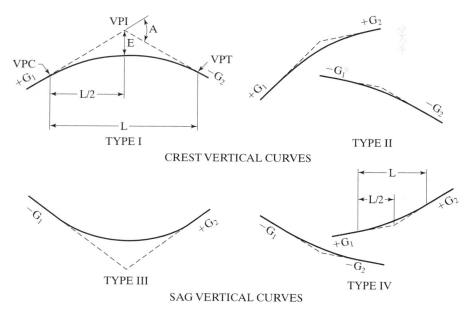

G_1 and G_2 = Tangent grades in percent
A = Algebraic difference in grade
L = Length of vertical curve

FIGURE 7.7
Types of vertical curves. *Source:* AASHTO, 2001, p. 269.

The total change in grade A between the VPC and the VPT on a vertical curve is

$$A = |G_2 - G_1| \tag{7.1}$$

The grades G_1 and G_2 can be extended as shown in Figure 7.7. These extensions of the entering and departing grades are called the VPC tangent and VPT tangent, respectively.

The elevation Y of the VPC tangent at any distance x from the VPC is

$$Y_{tan} = Y(VPC) + (G_1 * x) \tag{7.2}$$

Elevations Y along the VPT tangent can be computed by applying G_2 from the VPT or the VPI:

$$Y_{tan} = Y(VPT) - (G_2 * (L - x)) = Y(VPI) + ((x - L/2) * G_2)$$

The Vertical Curve as a Parabola.   The general equation of the parabola that describes a symmetric vertical curve is

$$Y = a + bx + cx^2 \tag{7.3}$$

where Y = the elevation of the vertical curve at a distance x feet from the VPC

   a = the elevation of VPC (feet)

   b = the slope of the entering tangent (ft/ft) = G_1

   c = the rate of change in grade, as derived below.

The first two terms in Equation 7.3 give the elevation of the extended tangent at a distance x from the VPC. This distance between the curve and the extended tangent is called the *offset*. The last term in Equation 7.3 gives the offset distance *down* to a point on the crest vertical curve, or the offset distance *up* to a point on the sag vertical curve. By taking the first and second derivative of Equation 7.3, two things can be observed. First, $dY/dx = b + 2cx$. At the origin $x = 0$, the slope is G_1. Therefore, $b = G_1$. The second derivative of Equation 7.3 is 2c. It is also the rate of change of the slope, which is $(G_2 - G_1)/L$. Therefore, $2c = (G_2 - G_1)/L$ and

$$c = \frac{G_2 - G_1}{2L} \tag{7.4}$$

The first derivative is zero at the high (or low) point

$$\frac{dY}{dx} = G_1 + \frac{G_2 - G_1}{L}x = 0.$$

Therefore, the high (or low) point of the parabola occurs at

$$x_{hi/lo} = \frac{-G_1}{G_2 - G_1} * L = L * \frac{G_1}{G_1 - G_2} \tag{7.5}$$

   Although the parabola is more convenient to use for most vertical curve analyses, sometimes more traditional equations are useful. One such equation is for the offset y between the tangent extended and the vertical curve.

$$y = \frac{|G_2 - G_1| * L}{2}\left(\frac{x}{L}\right)^2 = \frac{A * L}{2}\left(\frac{x}{L}\right)^2 = \frac{Ax^2}{2L} \tag{7.6}$$

Because $A = |G_2 - G_1|$, Equation 7.6 always produces a positive value of y. The value of y indicates the offset distance from the tangent extended *down* to the crest vertical curve and *up* to the sag vertical curve.

THINK ABOUT IT

Consider Equation 7.3. If $G_1 > 0$ and $G_2 < 0$ (as in type I in Figure 7.7), does the sign of x (positive or negative) make sense? If the vertical curve is type II in Figure 7.7, will the sign of x make sense?

Example 7.1

A crest vertical curve on SR562 begins with a 4.4 percent upgrade and ends with a 4.9 percent downgrade. The VPC is at Station 88 + 19.18 with an elevation of 711.10 feet. The VPT is at Station 94 + 44.95. A station is 100 feet of distance. Find the location (in stations), the offset with respect to the VPC tangent, and the elevation of the high point of the vertical curve.

Solution to Example 7.1   The length of the vertical curve is the difference of the VPC and VPT stations: 9444.95 − 8819.18 = 625.77 feet. Use Equation 7.5 to find the location of the high point of the crest vertical curve.

$$x_{high} = L*\frac{G_1}{G_1 - G_2} = 625.77*\frac{0.044}{0.044 - (-0.049)} = 625.77*0.473 = 296.06 \text{ feet from the VPC}$$

This location in stations is 8819.18 + 296.06 = 9115.24 feet = 91 + 15.24 stations. The offset of x_{high} is the difference in elevation between the tangent extended and the vertical curve. Using Equation 7.2 to find the elevation of the tangent extended,

$$Y = Y(\text{VPC}) + (G_1*x) = 711.10 \text{ feet} + (0.044 \text{ ft/ft}*296.06 \text{ ft})$$
$$= 711.10 \text{ ft} + 13.03 \text{ ft} = 724.13 \text{ ft}$$

The elevation of the vertical curve at x = 296.06 feet is found by using Equation 7.3 with

$$a = 711.10 \text{ ft}, b = G_1 = 0.044, c = \frac{G_2 - G_1}{2L} = \frac{-0.049 - 0.044}{2*625.77} = -0.000074$$

Equation 7.3 leads to $Y = a + bx + cx^2 = 711.10 + (0.044*296.06) + (-0.000074*(296.06)^2)$ = 711.10 + 13.03 − 6.49 = 717.64 feet. This is the elevation of the high point on the vertical curve. Note that the first two terms in Equation 7.3 equal the elevation of the tangent extended at x = 296.06. This leaves the third term in Equation 7.3 equal to the offset, y = −6.49 feet. The negative sign indicates that the vertical curve lies *below* the tangent.

Example 7.2

A 600-foot vertical curve connects a +4 percent grade to a −2 percent grade with VPI at Station 45 + 60.55. The VPI has elevation 348.64 feet. Calculate the location and elevation of the VPC, VPT, and high point of the curve.

Solutions to Example 7.2   The VPC is L/2 = 600/2 = 300 feet before the VPI, making its location at Station 45 + 60.55 − 3 + 00 = 42 + 60.55. (See Figure 7.8.) To find the elevation of VPC, work back from the VPI along the 4 percent tangent that is extended from the incoming grade, until the VPC is reached. Reducing the VPI's elevation at a rate of 4 percent for 300 feet, the VPC's elevation is found to be Y = 348.64 − (300*0.04) = 336.64 feet.

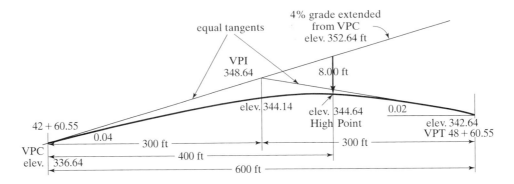

FIGURE 7.8
Sketch of the relationships from Example 7.2.

The VPT is 600 feet from the VPC, or at Station 48 + 60.55. Its elevation is found by using the 2 percent grade of the outgoing tangent's extension from the VPI: $Y = 348.64 + (-0.02 * 300) = 342.64$.

Using Equation 7.5, the high point of the vertical curve is at

$$x_{hi} = \frac{L * G_1}{G_1 - G_2} = \frac{600 * 0.04}{0.04 - (-0.02)} = \frac{24.00}{0.06} = 400 \text{ ft from the VPC}$$

The elevation of the high point is found using Equation 7.3 with $c = (G_2 - G_1)/(2L) = (-0.02 - 0.04)/(2 * 600) = -0.000050$:

$$Y = a + bx + cx^2 = 336.64 + (0.04 * 400) + (-0.000050 * (400)^2)$$
$$= 336.64 + 16.00 - 8.00 = 344.64 \text{ ft}$$

Similar calculations can be carried out for any point on the vertical curve. For example, the point at Station 47 + 00.00 is $x = 47 + 00 - (42 + 60.55) = 439.45$ feet from the VPC. Because this point is on the same side of the VPI as the high point, the offset and elevation calculations will be quite similar. The results are shown in Table 7.1.

TABLE 7.1 Solutions to Example 7.2

Station Point on Curve	Distance from PVC	Tangent Elevation	Offset y	Curve Elevation	
42 + 60.55	000.00	336.64	0.00	336.64	← VPC
44 + 00.00	139.45	342.22	−0.97	341.25	
45 + 60.55	300.00	348.64	−4.50	344.14	
46 + 60.55	400.00	352.64	−8.00	344.64	← High Point
47 + 00.00	439.45	354.22	−9.66	344.56	
48 + 60.55	600.00	360.64	−18.00	342.64	← VPT

Example 7.3

In a rugged portion of Mythaca County, a road must be built over a hill that begins with a 4.4 percent uphill grade and ends with a steep 9.4 percent downhill grade. Preliminary drawings indicate that the distance between the VPC and the VPT can be 720 feet. The VPC would have elevation 204.00 feet. Find (a) the location and elevation of the high point of the crest vertical curve and (b) the elevation of the VPI.

Solution to Example 7.3

(a) Equation 7.5 provides

$$x = \frac{L*G_1}{G_1 - G_2} = \frac{720*0.044}{0.044 - (-0.094)} = \frac{31.68}{0.138} = 229.56 \text{ feet.}$$

If $G_1 = 4.4$ percent and $G_2 = -9.4$ percent had been used instead of the decimal form of the grade, the answer would be the same. To find the elevation of the high point, use Equation 7.3 with $c = (G_2 - G_1)/(2L) = (-0.094 - 0.044)/(2*720) = -0.000096$:

$$Y = a + bx + cx^2 = 204.00 + (0.044*229.56) + (-0.000096*(229.56)^2)$$
$$= 204.00 + 10.10 - 5.05 = 209.05 \text{ feet}$$

(b) The elevation of the VPI is made easy by the fact that the VPI's location is $x = L/2$. Therefore,

$$Y(\text{VPI}) = 204.00 + (0.044*(720/2)) = 219.84.$$

TRY THIS

Find the elevation of the VPT in Example 7.3. Once that is found, repeat Example 7.3 by using the 9.4 percent grade as G1 and the previous VPT point as the VPC. Does the direction of the analysis make any difference?

Example 7.4

A sag vertical curve will connect two tangents, a -3 percent grade and a $+5$ percent grade, 800 feet apart. The ravine in which the sag curve will be situated needs to be built up. The elevation of the curve at its deepest point must be at elevation 168.00 feet or higher to clear a major pipe installation. If the VPC is at Station $00 + 00.00$ with an elevation of 176.20 feet, determine the location of the deepest point of the curve and the height of fill required above the culvert.

Solution to Example 7.4   The location of the lowest point is (using Equation 7.5) at $x = 800*(-0.03)/(-0.03 - 0.05) = 300$ feet or Station $3 + 00.00$. The vertical curve at $x = 300$ feet, according to Equation 7.3 with

$$c = (G_2 - G_1)/(2L) = (0.05 - (-0.03))/(2*800) = 0.000050,$$

will be at elevation

$$Y = a + bx + cx^2 = 176.20 + (-0.03*300.00) + (0.000050*(300.00)^2)$$
$$= 176.20 - 9.00 + 4.50 = 171.70 \text{ ft}$$

This is 3.7 feet above the minimum elevation required.

7.1.3   Geometry of Horizontal Curves

The horizontal curves are, by definition, circular curves of radius R. The elements of a horizontal curve are shown in Figure 7.9 and summarized (with units) in Table 7.2.

Equations 7.8 through 7.13, which apply to the analysis of the horizontal curve, are given below.

$$D = \frac{36{,}000}{2\pi R} = \frac{5729.6}{R} \tag{7.8}$$

$$L = \frac{100\Delta}{D} \tag{7.9}$$

$$T = R\tan\frac{1}{2}\Delta \tag{7.10}$$

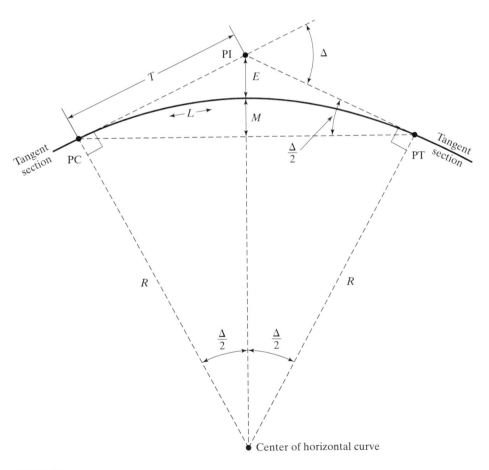

FIGURE 7.9
The elements of a horizontal curve.

TABLE 7.2 A Summary of Horizontal Curve Elements

Symbol	Name	Units
PC	Point of curvature, start of horizontal curve	
PT	Point of tangency, end of horizontal curve	
PI	Point of tangent intersection	
D	Degree of curvature	degrees per 100 feet of center-line
R	Radius of curve (measured from center of curve to centerline)	feet
L	Length of curve (measured along centerline)	feet
Δ	Central (subtended) angle of curve, PC to PT	degrees
T	Tangent length	feet
M	Middle ordinate	feet
LC	Length of long chord, from PC to PT	feet
E	External distance	feet

$$M = R\left(1 - \cos\frac{1}{2}\Delta\right) \tag{7.11}$$

$$LC = 2R\sin\frac{1}{2}\Delta \tag{7.12}$$

$$E = R\left(\frac{1}{\cos\frac{1}{2}\Delta} - 1\right) \tag{7.13}$$

Example 7.5

A 7-degree horizontal curve covers an angle of 63°15′34″. Determine the radius, the length of the curve, and the distance M from the circle to the chord.

Solution to Example 7.5 Rearranging Equation 7.8, with $D = 7$ degrees, the curve's radius R can be computed. Equation 7.9 allows calculation of the curve's length, L, once the curve's central angle is converted from 63°15′34″ to 63.2594 degrees. The middle ordinate calculation uses Equation 7.11. These computations are shown below.

$$R = \frac{5729.6}{7} = 818.5 \text{ ft}$$

$$L = \frac{100 \times 63.2594°}{7} = 903.7 \text{ ft}$$

$$M = 818.5 * (1 - \cos 31.6297°) = 121.6 \text{ ft}$$

If metric units are used, the definition of the degree of the curve must be carefully examined. Because the definition of the degree of curvature D is the central angle subtended by a 100-foot arc, then a "metric D" would be the angle subtended by a 30.5-meter arc. The subtended angle Δ does not change, but the metric values of R, L, and M (with 3.28 ft per meter) become

$$R = \frac{5729.6}{7 * 3.28} = 249.55 \text{ m}$$

$$\Delta = 63.2594°; \frac{1}{2}\Delta = 31.6297$$

$$L = \frac{100 * 63.2594°}{7 * 3.28} = 275.52 \text{ m}$$

$$M = 249.55 * (1 - \cos 31.6297°) = 37.07 \text{ m}$$

7.1.4   Pavement Widening at Curves

At an intersection in urban settings, extra space for right-angle turns must be provided. Even along a highway, horizontal curves often need to be made wider than the tangent sections of the road. The principal reasons for lane widening are large vehicles, whose bumpers have large overhangs and whose rear wheels track inside the paths taken by the lead wheels. This consideration is especially important on roadways where the lanes are less than 12 feet wide. The amount of widening in a curve depends on the vehicles that use the curve. If the traffic is principally automobiles, then an extra 2 feet of lane width in the curve or turn is satisfactory. If a large number of trucks or buses use the road, widening the curves must account for the turning track of the *design vehicle*. Figure 7.10 shows the types of conditions that can arise in a horizontal curve.

Table 7.3 offers general widening values on roads with few large vehicles. If there is 24 feet of pavement, most widening is not necessary. With a 20-foot pavement width, an extra 2 to 3 feet is usually enough. These curves illustrate the design parameters

TABLE 7.3   Calculated and Design Values for Pavement Widening on Open Highway Curves (Two-Lane Pavements, One-Way or Two-Way)

Degree of Curve	24 Foot Design Speed					20 Foot Design Speed			
	30	40	50	60	70	30	40	50	60
1	0.0	0.0	0.0	0.0	0.0	1.5	1.5	1.5	2.0
2	0.0	0.0	0.0	0.5	0.5	2.0	2.0	2.0	2.5
3	0.0	0.0	0.5	0.5	1.0	2.0	2.0	2.5	2.5
4	0.0	0.5	0.5	1.0	1.0	2.0	2.5	2.5	3.0
5	0.5	0.5	1.0	1.0		2.5	2.5	2.5	3.0
6	0.5	1.0	1.0	1.5		2.5	3.0	3.0	3.5
7	0.5	1.0	1.5			2.5	3.0	3.5	
8	1.0	1.0	1.5			3.0	3.0	3.5	
9–11	1.0	1.5	2.0			3.0	3.5		
12–14.5	1.5	2.0				3.5	4.0		
Over 15	2 to 3.5					4 to 5.5			

For four-lane roads, double the numbers in the table. Values less than 2 feet may be disregarded. For very large trucks, see Figure 7.10.
Adapted from AASHTO, 2001, Exhibit 3-51, p. 216.

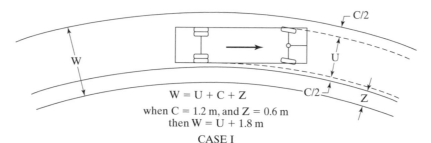

$$W = U + C + Z$$
when C = 1.2 m, and Z = 0.6 m
then W = U + 1.8 m

CASE I

ONE-LANE ONE-WAY OPERATION-NO PASSING

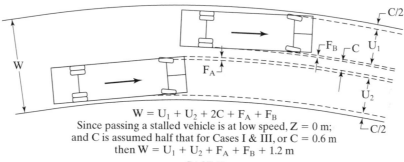

$$W = U_1 + U_2 + 2C + F_A + F_B$$
Since passing a stalled vehicle is at low speed, Z = 0 m;
and C is assumed half that for Cases I & III, or C = 0.6 m
then W = $U_1 + U_2 + F_A + F_B$ + 1.2 m

CASE II

ONE-LANE ONE-WAY OPERATION PROVISION FOR PASSING STALLED VEHICLE

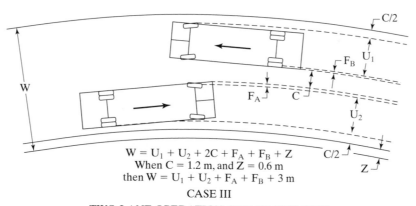

$$W = U_1 + U_2 + 2C + F_A + F_B + Z$$
When C = 1.2 m, and Z = 0.6 m
then W = $U_1 + U_2 + F_A + F_B$ + 3 m

CASE III

TWO-LANE OPERATION-ONE OR TWO WAY

U = Track width of vehicle (out-to-out tires), m
F_A = Width of front overhang, m
F_B = Width of rear overhang, m

C = Total lateral clearance per vehicle, m
Z = Extra width allowance due to difficulty
    of driving on curves, m

FIGURE 7.10

Tracking for automobiles. *Source:* AASHTO, 2001, p. 219.

necessary for one-way and two-lane roads. The two-lane road is the usual design situation. This is case III in Figure 7.10. For this case, the minimum width is 19 feet, unless the lateral clearance is carried to the edge of the pavement, in which case it is 23 feet, as shown below.

$$W = \frac{C}{2} + U_1 + U_2 + C + F_A + F_B + \frac{C}{2} + Z$$

$$= 2 + 6 + 6 + 4 + 0.75 + 0.25 + 2 + 2 = 23 \text{ ft} \tag{7.14}$$

The one-lane situation in which there is a stalled car in the road offers an interesting example. Although a one-lane road and normal passing of the stalled vehicle could occur on a tangent section 12 feet wide with a 2-foot shoulder, the curve calls for an increase of 3 feet due to vehicle tracking:

$$W = U_1 + U_2 + 2C + F_A + F_B = 6 + 6 + 2 \times 2 + 0.75 + 0.25 = 17 \text{ ft} \tag{7.15}$$

The overhang and tracking problem of large vehicles turning corners or tracking around curves is illustrated in Figures 7.11 and 7.12. Long vehicles, where there is no articulation between parts of the vehicle, such as a bus, generally exhibit the largest amount of overhang, as shown in Figure 7.11. On the other hand, a long vehicle must be carefully driven, because its track will be close to the curb and have overhang as well, as seen in Figure 7.12.

Actually, there is in the AASHTO policy a family of curves for different lengths of trucks and vehicles. In Equations 7.14 and 7.15, the overhang of the vehicle, the speed, and the type of vehicle enter into the calculation. However, "widening is costly and very little is gained from a small amount of widening. It is suggested that a minimum widening of 2.0 ft be used and that lower values in [Table 7.3] be disregarded" (AASHTO, 2001, p. 214).

7.2   STOPPING SIGHT DISTANCE AND ALIGNMENT

The presence of curves, either horizontal or vertical, raises the important question of sight distance needed for safe stopping. Crest curves require seeing over the hill far enough to stop, should there be a stopped car or other object in the road ahead. Sag curves permit adequate sight distance during the day, but at night, the automobile's headlight beam needs to illuminate an object in the road ahead. On horizontal curves, driver must be able to see across the arc of the curve far enough to detect objects in the road. Even on a level tangent portion of highway, there must be adequate passing sight distance when crossing the centerline to overtake a slower vehicle.

7.2.1   Sight Distance on a Hill

There are geometric standards that must be met if roads on hills are to be safe. The length of a vertical curve is critical in determining if there is ample stopping distance. At anytime during the day, the crest curve must be designed so that, as one comes over the crest of the hill, there is ample stopping distance when an object of height h_2 appears in the road ahead. Figure 7.13 shows the geometry.

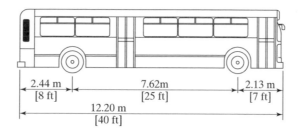

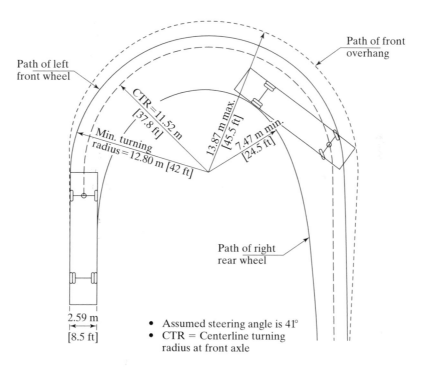

FIGURE 7.11
Turning radii of a typical city transit bus. *Source:* AASHTO, 2001, p. 25.

For the line of sight, the height of the driver's eyes (h_1) is important. Table 7.4 is based on the current standard driver eye level of 3.5 feet above the surface of the road and an object in the road ahead of height 2 feet, with friction coefficient $f = 0.35$.

Equations 7.16 below cover the possibilities that the curve length is greater than or less than the *stopping sight distance* (*SSD*) prescribed in Table 7.4 (AASHTO, 2001, p. 271).

$$L_{min\,1} = \frac{|A|S^2}{100\left(\sqrt{2h_1} + \sqrt{2h_2}\right)^2} \text{ for } S \leq L$$

$$L_{min\,2} = 2S - \frac{200\left(\sqrt{h_1} + \sqrt{h_2}\right)^2}{|A|} \text{ for } S \geq L \qquad (7.16)$$

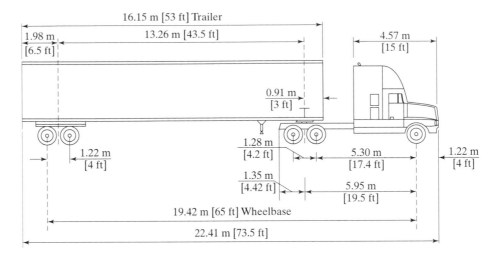

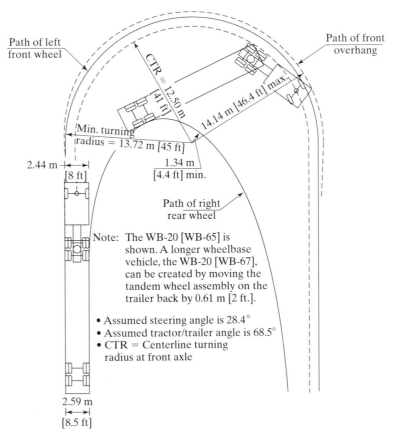

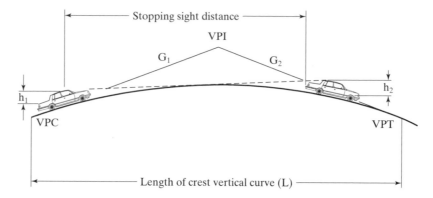

FIGURE 7.13
Stopping sight distance on a crest vertical curve or hill. *Source:* AASHTO, 2001, p. 271.

TABLE 7.4   Braking and Stopping Sight Distance

Design Speed (mph)	Brake Reaction Distance (feet)	Braking Distance on Level Terrain (feet)	Stopping Site Distance	
			Calculated (feet)	Design (feet)
15	55.1	21.6	76.7	80
20	73.5	38.4	111.9	115
25	91.9	60.0	151.9	155
30	110.3	86.4	196.7	200
35	128.6	117.6	246.2	250
40	147.0	153.6	300.6	305
45	165.4	194.4	359.8	360
50	183.8	240.0	423.8	425
55	202.1	290.3	492.4	495
60	220.5	345.5	566.0	570
65	238.9	405.5	644.4	645
70	257.3	470.3	727.6	730
75	275.6	539.9	815.5	820
80	294.0	614.3	908.3	910

Source: AASHTO, 2001, p. 112.

where

L = length of the curve
S = stopping sight distance
$|A| = |G_2 - G_1|\%$
h_1 = height of the driver's eyes (ft)
h_2 = height of the object (ft)

Substituting $h_1 = 3.5$ feet and $h_2 = 2$ feet into Equations 7.16 results in

$$L_{\min 1} = \frac{A \times SSD^2}{2158} \quad \text{for } SSD < L \quad L_{\min 2} = 2SSD - \frac{2158}{A} \quad \text{for} \quad SSD > L \quad (7.17)$$

Example 7.6

A highway is being designed to AASHTO standards with a 70 mph design speed. For one section, an equal tangent curve is designed to connect grades of +2.5 percent and −1 percent. Determine the minimum length of the curve to provide adequate stopping sight distance.

Solution to Example 7.6   The stopping sight distance at 70 mph from Table 7.4 is 730 feet and $A = 3.5$.

$$L_{\min 1} = \frac{3.5 \times 730^2}{2158} = 864 \text{ ft; this result holds if } SSD < L.$$

$$L_{\min 2} = 2(730) - \frac{2158}{3.5} = 843 \text{ ft; this result holds if } SSD > L.$$

Because 730 feet $<$ 864 feet, $L_{\min 1}$ applies. Note that the $L_{\min}$ values are fairly close to each other.

Example 7.7

A crest vertical curve with $A = 4$ percent is being designed on a roadway where falling rocks are a potential hazard. To be safe, it has been decided that the highway should be designed for 60 mph speeds and the object to be seen is a 4-inch rock. How long should the vertical curve be?

Solution to Example 7.7   Using the stopping sight distance for 60 mph as 570 feet from Table 7.4 and applying Equations 7.16 with $h_2 = 4$ in. $= 1/3$ ft leads to

$$L_{\min 1} = \frac{4*570^2}{100(\sqrt{2*3.5} + \sqrt{2/3})^2} = 1084 \text{ ft}$$

$$L_{\min 2} = 2(570) - \frac{200(\sqrt{3.5} + \sqrt{1/3})^2}{4} = 840 \text{ ft}$$

In both cases, $L_{\min} > SSD$, so the equation for $L_{\min 1}$ applies (See Table 7.5.). The correct answer is 1084 ft.

TABLE 7.5   Steps in Solving for Stopping Sight Distance

---

1. Using Table 7.4, convert the design speed (proposed or existing) into a stopping sight distance (SSD) value for use as S in Equations 7.16.
2. Try both Equations 7.16 to find the minimum length of the vertical curve, given input values of G_1 and G_2. Call the results of the first equation $L_{\min 1}$ and the results of the second equation $L_{\min 2}$.
3. If SSD $< L_{\min 1}$, choose the $L_{\min 1}$ value. If SSD $> L_{\min 2}$, choose the $L_{\min 2}$ value.
4. If the vertical curve's L (existing or proposed) is at least as long as the L value chosen in step 3, the vertical curve design (existing or proposed) is OK with respect to stopping sight distance.

In Example 7.8, these steps are rearranged to permit an analyst to find the maximum speed at which an adequate SSD can be maintained on a vertical curve with given length L.

Example 7.8

A 600-foot vertical crest curve has a 4 percent tangent and -2 percent tangent. What speed is necessary if there is to be ample stopping sight distance?

Solution to Example 7.8   Assuming standard values for h_1 and h_2, Equations 7.17 apply. For this curve, $A = |-2-4| = 6$ and $L = 600$, and

$$\text{If } S \le L,\ 600 = \frac{6S^2}{2158}\quad S = 464.5 \text{ ft}$$

$$\text{If } S \ge L,\ 600 = 2S - \frac{2158}{6};\quad S = 480 \text{ ft}$$

Because the calculated S is less than the specified L, the first equation is used to determine the stopping sight distance as 464.5 feet. Going to Table 7.4, the maximum speed should be 50 mph to meet the design SSD. If the crest curve is to be on a road with a general higher speed limit, then the designer of the highway will need to flatten out the curve or make it longer.

THINK ABOUT IT

The last example ends with the statement "the designer of the highway will need to flatten out the curve or make it longer." Explain how to make a crest vertical curve "longer."

7.2.2   Stopping Sight Distance in a Valley

The stopping sight distance for a sag vertical curve is treated differently from a crest vertical curve. During the daytime, there is no problem with sight distance (Figure 7.14). However, at night, the headlight beams must throw light on an object that is not self-illuminated soon enough for the car to be braked. Figure 7.15 shows the concept.

The critical dimensions are on the automobile—the height of the headlamp and the angle of the headlamp. The headlight forms an angle β with the horizontal. The equations governing ample sight distance are given in Equation 7.18.

$$L_{\min 1} = \frac{|A|S^2}{200(h + S \tan \beta)}\quad \text{for } S < L \quad \text{and}$$

$$L_{\min 2} = 2S - \frac{200(h + S \tan \beta)}{|A|}\quad \text{for } S > L \tag{7.18}$$

FIGURE 7.14
Sight distance on a sag vertical curve. Photo courtesy of Alaska Department of Transportation and Public Facilities (Vision, 2020).

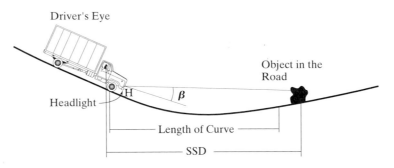

FIGURE 7.15
Stopping sight distance geometry for a sag vertical curve.

where

h = height of the headlamp, usually 2 ft.

β = beam angle, usually 1° from the horizontal

Example 7.9

Fred bought a new sports car that was very sleek and low to the ground. When checking the headlamp that was only 12 inches from the road, he thought they needed to be adjusted. What should the angle of the beam be so that he can drive safely at night at 55 mph?

Solution to Example 7.9 Fred wants to make sure that his headlights can illuminate the highway a distance SSD ahead, as shown in Figure 7.15. However, for his sports car, $h = 1$ foot, not the standard $h = 2$ feet value. For stopping distances at 55 mph, the desired SSD (from

Table 7.4) is $S = 495$ feet. Rearranging Equations 7.18 to isolate the two variables that are changing (h and β) produces

$$\frac{|A|S^2}{200L} = h + S \tan \beta$$

and

$$\frac{-|A|(L - 2S)}{200} = h + S \tan \beta.$$

In the right-hand side of either equation, $h = 2$ feet becomes $h = 1$ foot. To preserve the value of $h + S \tan \beta$ when $h = 1$ instead of $h = 2$, $\beta = 1$ degree must be adjusted in accordance with the calculations below.

$$2 + 550 \tan 1° = 1 + 550 \tan \beta; \quad 2 + (550*0.017455) = 1 + 550 \tan \beta$$

$$\tan \beta = \frac{2 + 9.5988 - 1}{550} = 0.01927; \quad \beta = \tan^{-1}(0.01927) = 1.104 \text{ degrees}$$

Fred needs to adjust his headlights by only 0.1 degree.

7.2.3   Using Rate of Curvature in Design of Vertical Curves with Adequate SSD

Another relationship that is useful in design is called the *K factor*, or the rate of curvature. It is defined as the length of the curve divided by the overall change in grade A. Using Equations 7.17 and 7.18 in combination with the stopping distances in Table 7.4, Table 7.6 can be constructed with the standard values of driver height, object height,

TABLE 7.6   K Factors for Stopping Sight Distance on Vertical Curves

Design Speed (mph)	Stopping Sight Distance feet (from Table 7.4)	Rate of Vertical Curvature K	
		Crest Curves	Sag Curves
15	80	3	10
20	115	7	17
25	155	12	26
30	200	19	37
35	250	29	49
40	305	44	64
45	360	61	79
50	425	84	96
55	495	114	115
60	570	151	136
65	645	193	157
70	730	247	181
75	820	312	206
80	910	384	231

Rate of vertical curvature K is the length of the curve per percent algebraic difference intersecting grades (A), $K = L/A$.

Source: AASHTO, 2001, Exhibits 3-76 and 3-79.

headlight beam angle and headlight height above the road. The maximum rate of curvature for safe stopping sight distances on both crest and sag vertical curves is given Table 7.6.

Example 7.10

An 1800-foot section of the scenic coastal highway just north of Mazurka was subject to heavy winds and waves of a major storm. A section was washed away. The south end of the washed out section is on a 3 percent grade at an elevation 52 feet. The north end (1800 feet away) has a 2 percent grade at an elevation of 22 feet. Figure 7.16 shows the situation. Design a vertical alignment that maximizes safe stopping sight distances over the section to be rebuilt for a speed limit of 45 mph.

Solution to Example 7.10 There are three segments over the 1800-foot section that need to be designed: (1) the crest vertical curve at the south end, (2) the sag curve at the north end, and (3) a middle segment of constant slope G_m that will be needed to connect the VPT of the crest curve to the VPC of the sag curve. We can use two different methods of expressing the difference in elevation between Point A above the VPT of the crest curve and Point B below the VPC of the sag curve in Figure 7.16. In the expressions below, the subscript **c** stands for the crest curve and the subscript **s** refers to the sag curve.

- Method 1: Point A lies along the tangent extended from the crest curve's VPC at grade G_{c1}. With L_c being the length of the crest curve, Point A's elevation is $G_{c1} * L_c$ feet higher than the crest curve's VPC: $\Delta y_c = G_{c1} * L_c$. Point B lies along the tangent extended from the sag curve's VPT at grade G_{s2}. Point B's elevation is $G_{s2} * L_s$ feet lower than the sag curve's VPT: $\Delta y_s = G_{s2} * L_s$. The elevation difference between Points A and B is $\Delta y_{AB} = G_{c1} * L_c + 30 + G_{s2} * L_s$.
- Method 2: The offset Y_c from Point A to VPT_c can be calculated using Equation 7.6:

$$Y_c = \frac{Ax^2}{2L} = \frac{(G_m - G_{c1})}{2L_c}L_c^2.$$

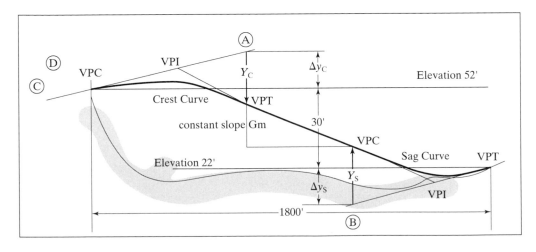

FIGURE 7.16
Desired vertical alignment for Example 7.10.

Likewise, the offset $Y_s = \dfrac{(G_{s2} - G_m)}{2L_s} L_s^2$. The horizontal distance between VPT_c and $VPC_s = 1800 - L_c - L_s$, so the elevation difference between VPT_c and $VPC_s = G_m * (1800 - L_c - L_s)$. Using Method 2,

$$\Delta y_{AB} = (G_m - G_{c1})\frac{L_c}{2} + [G_m * (1800 - L_c - L_s)] + (G_{s2} - G_m)\frac{L_s}{2}$$

or

$$\Delta y_{AB} = \frac{A_c L_c}{2} + [G_m * (1800 - L_c - L_s)] + \frac{A_s L_s}{2}.$$

After setting the Method 1 and Method 2 expressions for Δy_{AB} equal, using $L = K * A$ (see Table 7.6), and substituting the known values for $G_{c1} = 0.03$ and $G_{s2} = 0.02$, we get

$$(0.03 * K_c * A_c) + 30 + (0.02 * K_s * A_s) = (G_m - 0.03)\frac{K_c A_c}{2}$$

$$+ [G_m * (1800 - K_c * A_c - K_s * A_s)] + (0.02 - G_m)\frac{K_s A_s}{2}.$$

Using $A = |G_2 - G_1|$, we have

$$A_c = |G_m - G_{c1}| = |G_m - 0.03| = |G_m| + 0.03 \text{ if } G_m < 0$$

and

$$A_s = |G_{s1} - G_m| = |0.02 - G_m| = 0.02 + |G_m| \text{ if } G_m < 0.$$

Substituting these terms into the Method 1 side of the equation gives us

$$(0.03 * K_c * (G_m + 0.03)) + 30 + (0.02 * K_s * (0.02 + G_m)).$$

The Method 2 side becomes

$$(G_m - 0.03)^2\frac{K_c}{2} + [G_m * (1800 - K_c * (G_m + 0.03) - K_s * (0.02 + G_m))]$$

$$+ (0.02 + G_m)^2\frac{K_s}{2}.$$

At this point, we have three unknowns: G_m, K_c, and K_s. For a design speed of 45 mph, Table 7.6 gives us maximum values of $K_c = 61$ and $K_s = 79$. Because the K values in Table 7.6 are based on A in percent, and we have been using ft/ft values of G and A, the K values are converted to 6100 and 7900. The only unknown that remains in our equation is G_m. The Method 1 side is now

$$(0.03 * 6100 * (G_m + 0.03)) + 30 + (0.02 * 7900 * (0.02 + G_m)) = 341G_m + 38.65.$$

The Method 2 side is now

$$(G_m + 0.03)^2\frac{6100}{2} + [G_m * (1800 - 6100 * (G_m + 0.03) - 7900 * (0.02 + G_m))]$$

$$+ (0.02 + G_m)^2\frac{7900}{2} = -7000\,G_m^2 + 1800\,G_m + 4.325.$$

After combining the two sides, we have $7000\,G_m^2 - 1459\,G_m - 34.325 = 0$. The solution to this equation is $G_m = 0.027$. Because the slope of the middle segment is negative, $G_m = -2.7$ percent. The horizontal length of the middle segment will be

$$1800 - 6100*(0.03 + 0.027) - 7900*(0.02 + 0.027) = 1800 - 347.7 - 371.3 = 1081\text{ ft}.$$

The lengths of the crest and sag curves can be increased without affecting the required SSD, which is based on speed. This shortens the length of the middle segment and increases its grade. Example 7.11 will demonstrate.

Example 7.11

Determine the slope of the middle segment in Example 7.10 if the crest and sag curves are each 500 feet long.

Solution to Example 7.11   With longer curves, the K values for the maximum length of curve for SSD no longer apply. The vertical distance from Point A to Point B in Figure 7.16 is $G_c L_c + G_c L_c + 30 = (0.03*500) + (0.02*500) + 30 = 55$ ft.

Applying the offset equation,

$$\frac{A_c L_c}{2} + \frac{A_s L_s}{2} + |G_m|[1800 - 500 - 500] = 55\text{ ft}.$$

$$\frac{(0.03 + |G_m|)*500}{2} + \frac{(0.02 + |G_m|)*500}{2} + 800*|G_m| = 55\text{ ft}$$

$$250*(0.03 + G_m) + 250*(0.02 + G_m) + 800\,G_m = 55\text{ ft};$$

$$G_m = \frac{55 - 7.5 - 5.0}{250 + 250 + 800} = \frac{42.5}{1300} = 0.037.$$

Because the slope of the middle segment is negative, $G_m = -3.27$ percent.

7.2.4   Stopping Sight Distance for Horizontal Curves

Horizontal curves occur frequently as the alignment of roads and highways move through hilly country, along winding rivers, or around obstacles that cannot be moved. Figure 7.17 shows a simple circular horizontal curve. The stopping sight distance on a horizontal curve depends on the ability of the driver to see an object in the road across an arc of the curve in time to perceive the object and then to stop. The stopping sight distances from Table 7.4 are measured along the curve itself.

Equation 6.17 is used to define the stopping sight distance (SSD in Figure 7.17). For design purposes, the *effective radius* of the curve R_v is measured to the *middle of the innermost lane*. The other critical dimension in this analysis is the middle ordinate M_s. As shown in Figure 7.17, M_s is measured from the sight obstruction to the middle of the innermost lane.

The sight distance will govern the setback of any building or other sight obstruction, such as hedges and trees (Figure 7.18). The distance that is needed is governed by the arc of the curve that comprises the stopping sight distance. The distance from the

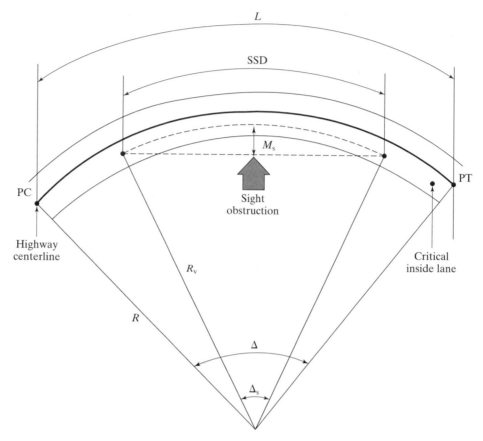

FIGURE 7.17
SSD analysis on horizontal curve.

middle of the innermost lane to the line drawn between the vehicle and the potential object in the road is the middle ordinate, which will govern the setback.

> **THINK ABOUT IT**
>
> (a) Why is the innermost lane used to govern the design with respect to SSD on a horizontal curve? (b) Why are the effective radius and the middle ordinate of the curve measured to the *middle* of the innermost lane? After all, the steering wheel is not in the middle of the dashboard.

The equation for the stopping sight distance (SSD) can be derived by first finding the central angle Δ_{SSD} for an arc equal to the required SSD from Table 7.4. Assuming that the arc of the curve is longer than the SSD, we have

$$ SSD = \frac{\pi}{180} R_v * \Delta_{SSD} \ \text{ or } \ \Delta_{SSD} = \frac{SSD * 180}{\pi * R_v} $$

FIGURE 7.18
Trees on inside of this horizontal curve reduce sight distance.
Photo by Robert K. Whitford

Substituting this into the equation for the middle ordinate M_s, we get

$$M_s = R_v\left(1 - \cos\frac{\Delta}{2}\right) = R_v\left(1 - \cos\frac{90 \times \text{SSD}}{\pi \times R_v}\right) \tag{7.19}$$

The argument of the cosine in Equation 7.19 has the units *degrees*. Solving for SSD, Equation 7.19 becomes Equation 7.20.

$$\text{SSD} = \frac{\pi \times R_v}{90}\left(\cos^{-1}\frac{R_v - M_{\text{SSD}}}{R_v}\right) \tag{7.20}$$

Note: Velocity is an important part of the SSD determination from Table 7.4.

Example 7.12

The mayor of Shoridan wants to erect a large sign welcoming visitors as they enter the city on SR361. (The mayor's name will also be on the sign.) As SR361 enters Shoridan, it has four 12-foot lanes with 4-foot shoulders. Unfortunately, it is also a horizontal curve with a radius of 1200 feet. The speed limit is 45 mph. How far in from the inside shoulder of the highway must the sign be placed to avoid potential stopping sight distance problems?

Solution to Example 7.12   From Table 7.4, the desired design SSD at 45 mph is 360 feet. Correct the radius from the centerline of the highway to the driver on the innermost of the four lanes. If the typical driver in the innermost lane is 6 feet from the pavement edge,

$$R_{\text{vehicle}} = 1200 - 24 + 6 = 1182 \text{ ft}$$

$$M_{\text{SSD}} = 1182\left(1 - \cos\frac{90 \times 360}{\pi \times 1182}\right) = 13.7 \text{ ft}$$

The welcome sign should be placed at least 13.7 feet inside the centerline of the inner lane or 3.7 feet inside the shoulder.

7.2.5   Passing Sight Distance

The sight distance required when pulling out to pass another vehicle on a two-lane road is critical both for the driver and for those who determine where no-passing zones should exist. AASHTO has established some guidelines for passing sight distance that depend on the speed of the car passing and the presumed speed of the car coming from the other direction.

The maneuver is well known to any driver. When driving behind a slow car or truck on a two-lane road, to pass, we must wait until the dashed line is on our side of the double striped centerline and there is no traffic in the opposing direction to interfere with the desired passing maneuver. When we pull out (the first phase in Figure 7.19), we accelerate and begin to pass. We proceed to pass (the second phase) and, as soon as we are two or three car lengths in front of the vehicle we are passing, we pull back into our lane. All that time, any vehicle coming toward us in the opposing lane grows nearer. The distance needed by the passing car to ensure safe passing is the sum of $d_1 + d_2 + d_3 + d_4$ in Figure 7.19. These distances are defined as:

d_1 The distance traveled while the driver observes an opportunity to pass and decides to begin the passing maneuver. The corresponding duration is t_1.

d_2^a The distance traveled while the vehicle is accelerating from "following speed" V_0 to "passing speed" V_1. The corresponding duration is t_2^a.

d_2^p The distance traveled while the vehicle is passing other vehicle(s) at speed V_1, until the passing maneuver is completed. The corresponding duration is t_2^p.

d_3 The margin of safety between the passing and opposing vehicles. The corresponding duration is t_3.

d_4 The distance traveled by the opposing vehicle at known speed V_2 during the time $t_4 = t_1 + t_2^a + t_2^p$.

The curve in Figure 7.20 gives the AASHTO guidelines (*AASHTO, A Policy of Geometric Design*, p. 123) based on the highway design speed and the average speed of the passing car, when the passing vehicle passes at a speed 10 mph faster than the vehicle being overtaken and the opposing vehicle's speed is the same as the passing vehicle's passing speed.

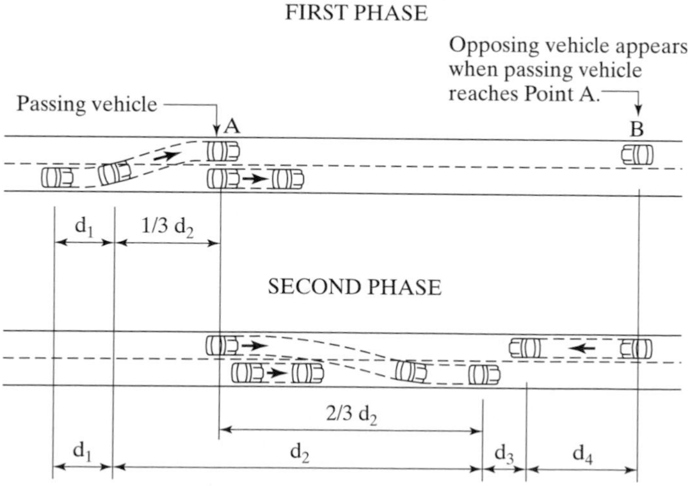

FIGURE 7.19
The passing maneuver. *Source:* AASHTO, 2001, p. 119.

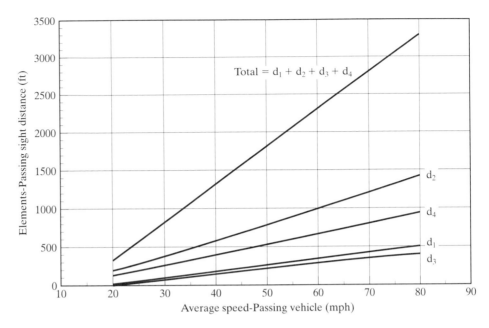

FIGURE 7.20
Passing sight distance on two-lane highways. *Source:* AASHTO, 2001, p. 123.

THINK ABOUT IT

For practical purposes, what maximum "passing speed" would you recommend using in a passing sight distance analysis?

Example 7.13

You are driving behind two loaded farm trucks that are traveling 45 miles per hour. After following the second truck by one second (66 feet) for a long time, the road ahead finally looks clear and you have a dashed line on your side. After using 1.5 seconds to make your decision to pass the trucks, you pull out to pass. You accelerate at a constant rate of 10 ft/s^2, until you reach a passing speed of 60 mph. You pass the trucks (a total length of 130 feet, including the space between them) and return to your lane once you have passed them by 40 feet. Shortly after you pull into your lane, the driver of the car coming from the other direction at 60 mph goes by, honking his horn. The oncoming driver had been flashing his headlights and is obviously agitated by your maneuver. How much time and distance were consumed during your passing maneuver?

Solution to Example 7.13 Let us solve this problem step-by-step, using the distance definitions in Figure 7.18 (and in the table above it) to organize the analysis.

- (d_1) Your observation-decision time is $t_1 = 1.5$ seconds and, at 45 mph, your observation decision distance is $d_1 = 1.5*45*1.47 = 99.22$ feet.

- (d_2^a) You increased your speed from 45 mph (converted to 66.15 fps) to 60 mph (converted to 88.20 fps). This took

$$t_2^a = \frac{88.2 - 66.15 \text{ fps}}{10 \text{ fps/sec}} = 2.205 \text{ seconds.}$$

Using a basic physics equation, the distance traveled during this acceleration was

$$d_2^a = v_0 t + \frac{1}{2}at^2 = (66.15*2.205) + \frac{1}{2}(10)(2.205)^2 = 145.86 + 24.31 = 170.17 \text{ feet.}$$

To accomplish your passing maneuver as described, you needed to gain a net of 66 + 130 + 40 = 236 feet on the lead truck. While you were accelerating to 60 mph from 45 mph, the trucks covered a distance of 2.205*45*1.47 = 145.86 feet. At this point, you still needed to gain 236 − (170.17 − 145.86) = 211.69 feet on the lead truck.

- (d_2^P) As your passing maneuver continues at 60 mph for t_2^P seconds, the trucks continue moving at 45 mph. Let us relate the distance you need to cover to reach the point at which you could pull in 40 feet ahead of the lead truck to the distance covered by the farm trucks during the same amount of time: $88.20*t_2^P = (66.15*t_2^P) + 211.69$. This can be solved as

$$t_2^P = \frac{211.69}{(88.20 - 66.15)} = 9.60 \text{ seconds.}$$

At 60 mph, this additional distance was d_2^P = 9.60 sec*88.20 fps = 846.76 feet.

- (d_4) During the time it took you to pass the two trucks, $t_4 = t_1 + t_2^a + t_2^P = 1.5 + 2.205 + 9.60 = 13.30$ seconds, a vehicle in the opposing lane at 60 mph would cover 13.30*60* 1.47 = 1173.5 feet.

You needed $d_1 + d_2^a + d_2^P$ = 99.22 + 170.17 + 846.76 = 1116.15 feet to execute the passing maneuver as described in the problem statement. Assuming for the moment that $d_3 = 0$, if the oncoming vehicle had been just coming into view when you made your decision to pass the trucks, the minimum required passing sight distance is 1116.15 + 1173.5 = 2289.65 feet.

Because Figure 7.20 assumes a speed differential of 10 mph between passing and passed vehicles (AASHTO, 2001, p. 122) and in Example 7.13 the differential is 15 mph, the figure offers only an approximate guide for *passing sight distance (PSD)* standards. Nevertheless, the $d_3 = 300$ feet value in Figure 7.20 for an average passing speed just under 60 mph is a good number to use in setting a desirable PSD value. It is reasonable to conclude that an opposing vehicle closer than 2289.65 feet to the passing vehicle at the start of time t_1 will have to slow down to avoid risking a head-on collision. Adding $d_3 = 300$ feet provides an adequate "cushion" for the opposing vehicle and increases the PSD standard to 2589.65 feet. Note that the AASHTO total is less than 2300 feet for the 10 mph differential only when the passing speed is below 60 mph.

THINK ABOUT IT

Why do you think the slower passing difference is lower than the calculation made for Example 7.13?

To complete the analysis, the actual sight distance at the point the decision to pass was made would have to be established. Some shortcuts are available to carry out the calculations done above. For example, equation (6.14) allows you to calculate d_2^a as

$$d_2^a = \frac{v_2^2 - v_1^2}{2a} = \frac{88.20^2 - 66.15^2}{2 \times 10} = 170.17 \text{ ft}$$

According to the AASHTO Policy of Geometric Design, other shortcuts are:

$d_1 = 1.47t_1(v - m) + \dfrac{1}{2}at^2$ = perception and reaction distance

$d_2 = 1.47vt_2$ = distance traveled while the passing vehicle occupies the passing lane

$d_4 = \dfrac{2}{3}d_2$ = distance traveled by oncoming vehicle for 2/3 the time spent by passing vehicle in passing lane

where

t_1 = time for initial perception and reaction
t_2 = time passing vehicle occupies the left lane
v = average speed of passing vehicle (mph)
m = difference in speed of passing vehicle and passed vehicle (mph)
a = acceleration rate to passing speed (ft/s^2)

THINK ABOUT IT

To what extent do the AASHTO shortcuts replicate the step-by-step calculations in Example 7.13? If there are any differences, can you explain them?

THINK ABOUT IT

In Figure 7.19, d_2^a and d_2^p as defined in Example 7.13 are $\frac{1}{3}d_2$ and $\frac{2}{3}d_2$, respectively. Are these relationships approximate or can they be verified mathematically as being exact?

7.2.6   Marking Passing Lanes

Passing lanes can only occur on fairly level roads, where passing sight distance can be maintained. Horizontal curves with any significant degree of curvature are marked as no-passing zones.

A good number for passing sight distance for a highway with a design speed of 60 mph is 2135 feet (AASHTO 2001, p. 124). AASHTO has specified that the design height is $h_2 = 4.25$ feet. In that case, Equations 7.16 must be satisfied with sight distance sufficient for passing.

Example 7.14

Determine the minimum length of a crest vertical curve that has a design speed of 60 mph, an entering grade of 2 percent, and a departing grade of 3 percent, so that sight distance sufficient for passing exists.

Solution to Example 7.14

$$|A| = |G_1 - G_2| = |+2 - (-3)| = 5 \text{ percent. Equations 7.16 become}$$

$$L_{\min 1} = \frac{5 \times 2135^2}{100(\sqrt{2*3.5} + \sqrt{2*4.25})^2} = 7369 \text{ ft for } S \leq L$$

$$L_{\min 2} = 2(2135) - \frac{200*(\sqrt{3.5} + \sqrt{4.25})^2}{5} = 3651 \text{ ft for } S \geq L$$

Because $S < L$, the minimum curve length is 7369 feet. If that distance is not available, then the road must be marked as a no-passing zone for vehicles entering the hill from either direction. Otherwise, the hill needs to be cut down.

7.3 BANKING CURVES (HORIZONTAL ALIGNMENT)

The AASHTO manual states that "For balance in highway design all geometric elements should, as far as economically feasible, be determined to provide safe, continuous operation at a speed likely under the general condition for that highway or street. This is done through the use of design speed as the overall control" (AASHTO 2001, p. 123). To adhere to this policy, horizontal curves must often be banked. The amount of banking is called superelevation.

Because the superelevation is a sloping curve, there is a practical limit to the superelevation; it cannot be greater than that which would let a standing car on snow or ice slide down. When traveling slowly around a curve of high superelevation, the lateral forces on the car are negative, causing the driver to do unnatural steering by trying to steer up the slope. Heavy trucks (and cars with loose steering when traveling below the design speed) place a larger than normal stress on the inside tires. Thus, there are realistic limits to superelevation, due primarily to the fact that except on race tracks, the speed can be slower than the design speed.

7.3.1 Forces When Cornering

If the radius of a flat horizontal curve is sufficiently large, an automobile may be driven through the curve safely at 70 mph. However, as the curve's radius is shortened, the centripetal forces from turning will increase. Eventually, they will reach a level where the car cannot negotiate a flat curve without overturning. At this point, the cross section of the curve must be banked. The banking of a curve is called the *superelevation rate (e)*. It is usually expressed in inches (or meters) of cross section rise per inch (or meter) of cross section. With the wide variation of speed on curves whether banked or not, there is usually some unbalanced force.

Consider the vehicle shown in Figure 7.21 traveling along a circular curve on the verge of sliding, held by the *coefficient of side friction* (f_{side}). This friction coefficient is different from the coefficient of friction used in the stopping distance calculations in Section 6.3.

In the diagram, the weight W of the vehicle in pounds (newtons) has both a component W_n normal to the road and a component W_p along the road surface. F_f is the frictional side force. The centripetal force F_c (lateral acceleration x mass) has its two components normal to and parallel to the road surface. R_v is the radius of the curve, measured from the center of the horizontal curve to center of the car. Summing the forces yields Equation 7.21.

$$W \sin \alpha + f_{side}\left(W \cos \alpha + \frac{W}{g} a_{cp} \sin \alpha\right) = \frac{W}{g} \times a_{cp} \times \cos \alpha$$

$$\text{where } a_{cp} = \frac{V^2}{R_v} \tag{7.21}$$

Dividing by W cos α

$$\tan \alpha + f_{side} = \frac{V^2}{g \times R_v}(1 - f_{side} \tan \alpha) \tag{7.22}$$

Noting from Figures 7.22 and 7.23 that tan α is usually between zero and 0.12 superelevation and the coefficient of friction is less than 0.17, a good approximation is that tan $\alpha \times f_{side} \ll 1$. Thus, Equation 7.22 becomes

$$e + f_{side} = \frac{V^2}{gR_v} \tag{7.23}$$

Equation 7.23 can be rearranged to express the radius of the curve as,

$$R_v = \frac{V^2}{g(e + f_{side})} \tag{7.24}$$

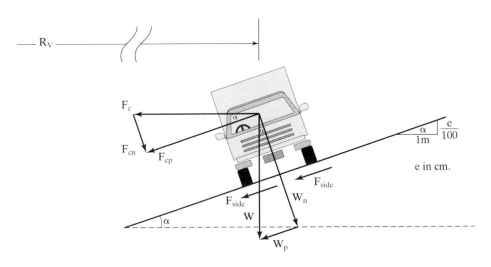

FIGURE 7.21
Cornering forces on a vehicle.

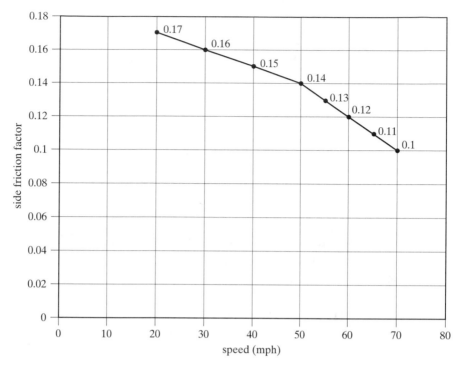

FIGURE 7.22
Maximum side friction factor in highway design. Plotted from data in AASHTO, 2001.

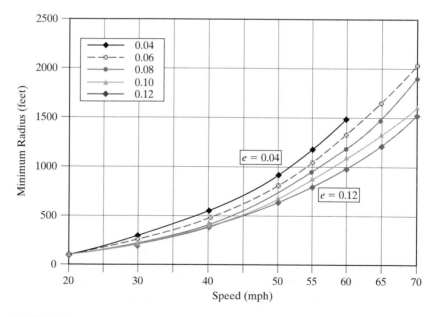

FIGURE 7.23
Minimum safe radius as a function of superelevation and speed. *Source*: The curve constructed
from data in AASHTO Policy on Geometric Design, 2001, Table III-6.

7.3.2 Design Parameters for Horizontal Curve

The coefficient f_{side} is called the side friction factor. The condition when the tire is about to enter into a skid, with a suitable factor of safety to account for differences in tires and vehicles, sets the upper limit on the value of f_{side}. Tests have indicated that with new tires and wet pavement the factor is about 0.35 at 60 mph. With bald tires and wet pavement the value is also 0.35, but at 45 mph. Note that f_{side} decreases as speed increases. The values used for f_{side} in design are shown in Figure 7.22 and range from 0.17 at 20 mph to 0.1 at 70 mph.

The choice of f_{side} then depends solely on the design speed. The choice of bank angle or superelevation will be based on the speed, radius, and other design factors. (See the curves in Figure 7.23.) In Figure 7.24(a), the cross slope on a race track is obvious. On a county road (Figure 7.24(b)), it is much subtler. Superelevation on urban roadways is usually 0.04 or below.

(a)   (b)

FIGURE 7.24
Illustrations of a banked curve (a) on an Indy Car racetrack, and (b) on a recently-constructed county road. (a) Photo by Greg S. McGehee; (b) Photo by Jon D. Fricker.

Example 7.15

A roadway is being designed capable of allowing 70 mph vehicle speeds. The superelevation around one curve is 0.05 inches per inch and the coefficient of side friction is 0.09. Determine the minimum radius of the curve to the centerline of the vehicle to provide safe travel.

Solution to Example 7.15

$$\text{From Equation 7.24, } R_v = \frac{(70 * 1.47)^2}{32.2 * (0.05 + 0.09)} = 2349 \text{ ft}$$

Example 7.16

Suppose the radius of the curve in Example 7.15 to the center of the vehicle is only 1500 feet. What speed limit will need to be posted for safe travel through the curve?

Solution to Example 7.16   Rearranging Equation 7.23,

$$V = \sqrt{32.2 * (0.05 + 0.09) * 1500} = 82.2 \text{ fps} = 55.9 \text{ mph}$$

The curve should be posted for 55 mph.

Example 7.17

A vehicle is traveling along a horizontal curve of 800-foot radius at the legal speed limit of 55 mph. The car is on the innermost lane of a four-lane highway (12-foot lanes). Given that the coefficient of friction is 0.13, determine the superelevation and the bank angle of the curve.

Solution to Example 7.17   Rearranging Equation 7.23 leads to

$$e = \frac{V^2}{g \times R_v} - f_{side} = \frac{(55 \times 1.47)^2}{32.2 \times (800 - 12 - 6)} - 0.13 = 0.2596 - 0.13 = 0.1296 \text{ in/in};$$

$$\alpha = \tan^{-1}(0.1296) = 7.38 \text{ degrees, which is a large } \alpha.$$

Example 7.18

A given section of 2-lane highway is constrained to have a maximum bank angle of 3 degrees and a radius of curvature of 630 feet. If the usual coefficient of friction is 0.16, what speed limit should be placed on the curve?

Solution to Example 7.18   Rearranging Equation 7.23 leads to

$$V = \sqrt{R_v \times g \times (e + f_{side})} = \sqrt{(630 - 6) \times 32.2 \times (0.16 + \tan 3°)} = 65.3 \text{ fps} = 44.4 \text{ mph}$$

Use a speed limit of 45 mph on the curve. Note that, for a speed of 45 mph, the coefficient of side friction in Figure 7.22 is 0.145. Solving the equation for $f = 0.145$, instead of $f = 0.16$, lowers the speed to 42.8 mph. Because many drivers push the posted speed around turns, posting a speed limit of 40 mph should be considered for added safety.

THINK ABOUT IT

When you see an advisory speed plate under a diamond-shaped sign warning you of a curve or turn in the road ahead, do you take the suggested speed seriously? If not, what assumptions do you make and how fast (relative to the suggested speed) do you enter the curve/turn?

7.3.3   Spiral Transition from Horizontal Curves to Tangent

A natural way of designing the change from one horizontal curve to another curve or to a tangent section is to use a spiral. The spiral can be used whether the curve is banked or not. It has a smoothly changing radius from a given radius to an infinite radius or tangent section of highway. The minimum length of the spiral is given by Equation 7.25, often called the Shortt formula.

$$L_s \text{ (feet)} = \frac{V_{fps}^3}{RC} = \left[\frac{3.15 V_{mph}^3}{RC} \right] \quad \text{or} \quad L_s \text{ (meters)} = \frac{V_{kph}^3}{45.7\,RC} \qquad (7.25)$$

where

L = the minimum length of the spiral in ft [ft] or (meters)
V = the speed in fps [mph] or (kph)

R = the radius of the curve in feet [feet] or (meters)

C = the rate of increase of centripetal acceleration [ft/s^2] or m/sec^2

It is more aesthetically pleasing and probably easier to drive along a spiral than to go directly from a circular curve to a tangent section (see Figure 7.25). Still, many highway departments use the simple runoff of a tangent to PC or PT of the circular curve.

For highway design, C in Equation 7.25 is approximately 2 ft/sec^2 or 0.6 m/sec^2. The equation applies directly for flat curves. However, when the curve is banked, the correct equation includes a term to account for the fact that the highway is actually going from a superelevation of "e" to 0 over the length of the spiral in Equation 7.26.

$$L_s = \frac{3.15V}{C}\left(\frac{V^2}{R} - 15e\right) \tag{7.26}$$

FIGURE 7.25
Visual impact of spiral transition. *Source*: AASHTO, *A Policy of Geometric Design*, 2001, p. 178.

Example 7.19

Compute the difference of spiral length for a 1200-foot radius curve from using the Shortt formula (Equation 7.26) above and the modified Shortt formula with no e correction (Equation 7.25). If the speed is 60 mph, $C = 2$ and $e = 1.2$ inches per foot.

Solution to Example 7.19

From Equation 7.26,
$$L_s = \frac{3.15(60)}{2}\left(\frac{60^2}{1200} - 15(0.1)\right) = 141.5 \text{ ft}$$

From Equation 7.25,
$$L_s = \frac{3.15(60)^3}{2*1200} = 283.5 \text{ ft}$$

If there is sufficient distance, the formula without the "e" correction is used for most design because it is conservative. Because it often gives a high value, overdesign can take place.

Another design need is joining two curves of differing radii. The minimum distance of a spiral to join the two curves will be the difference between the length of the spiral to tangent for the curve with the smaller radius and the length of the spiral with the larger radius. The AASHTO policy manual indicates that the ratio of the radii when connecting two circular curves be not more than 1.5, except in urban areas, where that ratio can be increased to 2.

Example 7.20

Two curves of differing radii are to be joined. The first curve has a radius of 1500 feet and the second curve a radius of 1000 feet. What is the length of the spiral curve connecting the two? The design speed for the roadway is 70 mph.

Solution to Example 7.20

Using Equation 7.25,
$$L_s = \frac{3.15(70)^3}{2*1000} - \frac{3.15(70)^3}{2*1500} = 180.7 \text{ ft}$$

7.3.4   Superelevation Runoff Transition

The change from a superelevated roadway to a level or normal tangent roadway is called *superelevation runoff*. Where the transition usually begins is illustrated in Figure 7.26 by examining the shift from the normal crown at point A to the fully banked curve at point E. Note that the roadway from point A to point D is part of the tangent or spiral, whichever is the case.

Like the superelevation runoff, the spiral has about one third of the length as part of the fully superelevated portion of the curve before getting to the tangent section. The transition design will typically find 65 to 75 percent of the transition actually occurring on the tangent section, not on the curve. Two thirds to three fourths of the transition occurs between point B on the tangent section and point D on the curve. The full bank angle is achieved at point E. The full transition is called the *superelevation runoff*.

Equation 7.25 gives the necessary length in feet for the runoff from fully superelevated to tangent curve as a function of the elevation rate and the design velocity in miles

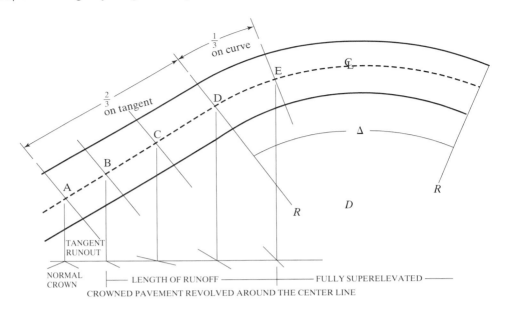

FIGURE 7.26
Typical superelevation runoff for normally crowned pavement.

per hour for the spiral. A table in the AASHTO "Green Book", governing runoff distances can be reduced to Equations 7.27 and 7.28.

$$L_{\text{runoff}} \approx 30e(V + 32) \text{ for two-lane roads with 12-foot lanes} \qquad (7.27)$$

$$L_{\text{runoff}} \approx 25e(V + 32) \text{ for two-lane roads with 10-foot lanes} \qquad (7.28)$$

Equations 7.27 and 7.28 are good approximations for speeds 20 mph or more. Equations 7.27 and 7.28 are for a two-lane road where only one lane is rotated (see Figure 7.27).

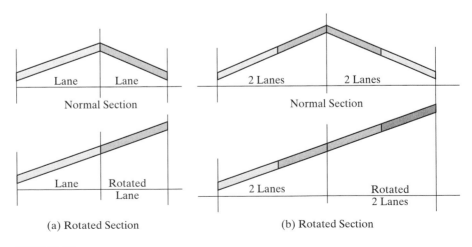

(a) Rotated Section   (b) Rotated Section

FIGURE 7.27
Rotation of lanes for superelevation. *Source:* AASHTO, 2001, p. 172.

If two lanes are rotated, as on a four-lane highway (see Figure 7.27(b)), the length of runoff will be 1.5 times as long as indicated by Equations 7.27 and 7.28.

Figure 7.28 illustrates two common methods of developing the transition to full superelevation. The first shows the case where the rotation is about the centerline of a crowned pavement. The points showing the pavement cross section refer to points A, B, C, D, and E in Figure 7.26. At point E, the pavement is fully superelevated. Two other common methods (not shown) are rotation about the inside pavement edge or about the outside pavement edge, respectively (AASHTO, 2001, p. 172). The lower of the two diagrams depicts pavements that are revolved about the outside edge. This type of pavement is found on the opposite sides of the median in a divided multilane highway.

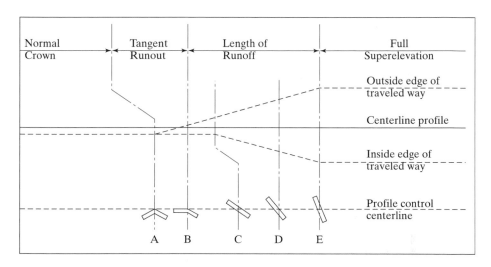

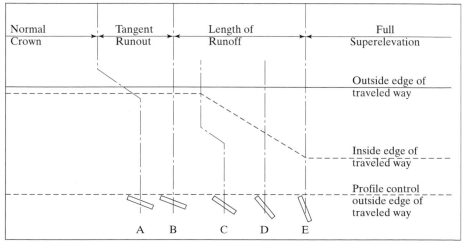

FIGURE 7.28
Two methods of attaining superelevation. *Source:* AASHTO, 2001, p. 185.

SUMMARY

The challenge of the geometric design of a highway is to fit a continuous section of roadway to its environment in such a way that it (a) maintains safe operating conditions, (b) minimizes costs (construction and operating), and (c) disturbs the environment (rural or urban) as little as possible. Design uses standard shapes—parabolic for vertical alignments and circular for horizontal curves—with dimensions that permit a driver to avoid hitting an obstacle in the curve. In hilly terrain, the trade-off is between (a) a level roadway achieved with cut and fill costs that may become prohibitive and (b) a hilly roadway that is cheaper to build but may be less safe and more costly to use. An enormous urban project, such as Boston's new Central Artery (Figure 7.5), must weave its way through important and historic buildings. Likewise, a rural project may have to wind its way around environmentally sensitive areas.

This chapter has introduced basic concepts needed to meet the challenge of geometric design. The components of vertical and horizontal curves were defined. The relationships between these components were demonstrated. Important applications of these relationships—especially safe stopping and passing sight distances—were used in sample design problems. Variations involving spiral transitions and superelevations were also introduced.

ABBREVIATIONS AND NOTATION

a_{acc}	acceleration capability of a vehicle (usually a constant in this class)
$\lvert A \rvert$	$\lvert \text{Grade}_2 - \text{Grade}_1 \rvert$ for vertical curves
a_{cp}	lateral or centripetal acceleration
C	total lateral clearance per vehicle
d_1	distance traveled while the driver observes an opportunity to pass and decides to begin the passing maneuver; the corresponding duration is t_1.
d_2^a	distance traveled while the vehicle is accelerating from "following speed" V_0 to "passing speed" V_1
d_2^p	distance traveled while the vehicle is passing other vehicle(s) at speed V_1, until the passing maneuver is completed
d_3	margin of safety between the passing and opposing vehicles
d_4	distance traveled by the opposing vehicle at known speed V_2 during the time $t_4 = t_1$
e	superelevation rate
F_f, F_r	engine force applied at the front and rear wheels, respectively
F_A	width of front overhang
F_B	width of rear overhang
f	dimensionless coefficient of friction of the road
f_{side}	coefficient of side friction
fps	feet per second
g	force of gravity 32.2 feet per second per second or 9.8 meters/second/second
G	grade in percent divided by 100.
h_1	height of the driver's eyes (ft) usually 3.5 ft
h_2	height of the object (ft) usually 2 ft
H	height of the headlights above the road
L	length of the curve
$L_{min\,1}$	minimum length of curve for site distance when SSD $>$ L
$L_{min\,2}$	minimum length of curve for site distance when SSD $>$ L
L_S	length of the spiral

M	middle ordinate in the arc of a circle
m	mass of the vehicle
R	radius of a horizontal curve
R_v	radius of the curve to driver of vehicle traveling on the innermost lane
R_{grade}	component of gravity acting normal to the road
S	sight distance
SSD	stopping sight distance
U	track width of the vehicle
W	weight of the vehicle in pounds
VMT	vehicle miles traveled
VKT	vehicle kilometers traveled in metric units
VPC	vertical point of curvature; beginning point of the curve from the tangent
VPI	intersection of the tangents from VPC and to VPT
VPT	vertical point of tangency; terminal point of the curve
Z	extra width allowance due to difficulty of driving on curves
α	angle of bank
β	angle the headlight beam makes with the horizontal
θ_g	grade of the road in degrees from the horizontal

GLOSSARY

Crest vertical curve: Two grade tangents connected by a parabolic vertical curve, in which the incoming grade is positive (uphill) and the outgoing grade is negative (downhill) or less than the incoming grade.

Degree of curvature: The number of degrees subtended by an angle defined by 100 feet of centerline along a horizontal curve.

Design vehicle: A vehicle chosen to represent the size and operating characteristics (such as turning radius) of vehicles expected to use the roadway facility being designed.

Geometric design: The way in which highway designers try to fit the highway to the terrain while maintaining design standards for safety and performance.

Green book: The common name for *A Policy on the Geometric Design of Highways and Streets*, which contains standards for geometric design that are followed by most jurisdictions in the United States.

Middle ordinate: Distance from the chord connecting the VPC and VPC to the centerline of the horizontal curve.

Overhang: The lateral distance between the wheels of a vehicle and the outside of its frame or body.

Passing sight distance: The sight distance required when pulling out to pass another vehicle on a two-lane road.

Radius of curve: Distance from the center of a circular horizontal curve to the centerline of the curve.

Rate of curvature: The length of the curve divided by the overall change in grade.

Runoff: The shift from a superelevated roadway to a level or normal tangent roadway.

Sag vertical curve: Two grade tangents connected by a parabolic vertical curve, in which the incoming grade is negative (downhill) and the outgoing grade is positive (uphill).

Spiral: Transition from one horizontal curve to another curve or to a tangent section.

Station: 100 ft of distance.

Superelevation: The banking of a curve.

REFERENCES

[1] American Association of State Highway and Transportation Officials (AASHTO), *A Policy on the Geometric Design of Highways and Streets*, 1994 and 2001.

EXERCISES FOR CHAPTER 7: HIGHWAY DESIGN FOR SAFETY

Geometric Design

7.1 Vertical Curves. If a vertical curve has its VPC at Station 22 + 66 (elev 325 ft), its VPT at Station 44 + 34 (elev 325 ft), with $g_1 = 4$ percent and $g_2 = -4$ percent. What is the location (in stations) and elevation (in feet) of the highest point of the curve?

7.2 Vertical Alignment. The following data describes a crest vertical curve. The VPC is at Station 22 + 60.55 and elevation 648.64 feet, $L = 800$ ft, $g_1 = +5$ percent, $g_2 = -3$ percent. At what location and elevation is the peak? Find the curve's elevation at Station 25 + 00.00, at Station 28 + 00.00, and at the VPT.

	Elevation	**Station**
PVC	950.00'	109 + 00
PVI	947.11'	110 + 92.5
Low point		110 + 65

7.3 Vertical Curve. An "equal tangent" sag curve has the following properties.

Determine the length of the vertical curve, the initial slope, G_1 and the elevation of the sag curve at its low point.

7.4 Horizontal Alignment. Given a horizontal curve of radius 1400 feet, an intersection angle = 46°21′ located at PI 500 + 00.00 on the centerline of a two-lane road with 12-foot wide lanes, find: a) degree of curve; b) station of PC; c) station of PT; d) length of the curve; e) middle ordinate; f) length of the chord connecting PC to PT.

7.5 Horizontal Alignment. A circular curve, planned to be located in Garden Township, connects two tangents with an external angle of 48 degrees. The PI is located at Station 948 + 67.3. The middle ordinate distance is 100 feet. To the nearest 0.1 foot, determine the locations (stations) of the PC and PT for the curve and the length of the curve.

Stopping Sight Distance

7.6 Stopping Sight Distance (SSD) on Vertical Curves. A particular crest vertical curve (VC) on SR361 between Middletown and Shoridan has long been criticized as being unsafe. Its length is 400 feet. The prevailing speed on the crest VC is 50 mph. In its present condition, its tangent grades are $G_1 = +2$ percent and $G_2 = -3$ percent. Does the crest VC meet minimum SSD standards? Because of a grant from the Governor's Highway Safety Initiative, it will be possible to enforce a speed limit of 45 mph on the crest VC. Will the VC meet minimum SSD standards if speeds can be limited to 45 mph?

7.7 Speed Limit in School Zone. The section of Gore Avenue between Sunrise Lane and Steak Street is used by school children as they walk to Mythaca's Sleepy Gulch Elementary School.

Unfortunately, it is a crest vertical curve. The vertical curve's characteristic data are

G_1 = +7.6 percent
G_2 = −5.4 percent
L = 190 feet

What is the maximum speed (to the nearest 1 mph) at which the "SSD for design" standard is maintained?

7.8 Vertical Alignment.   A 600-foot crest vertical curve is to be developed from a plus 1.70 percent grade intersecting with a minus 3.3 percent grade at Station 15 + 60.00 at an elevation of 312.88. a) Calculate the centerline elevation at each 100-foot station. b) Find the station and elevation of the peak of the curve. (*Hint:* Write the equation of the curve with its correction terms and set the derivative equal to zero; the peak always shifts toward the grade with the lowest absolute value.) c) Is the curve long enough for a design speed of 60 mph? Why or why not?

7.9 Vertical Alignment.   A sag vertical curve has a minus 3.3 percent grade that intersects with a plus 4.7 percent grade at Station 50 + 00.00 and an elevation of PVI of 532.20. To provide proper cover for a culvert, it is desired that the elevation at 50 + 75.00 be 538.50 or higher. a) What minimum length of vertical curve is necessary to provide the necessary cover? (*Hint:* It can be determined directly by determining the change in elevation due to the grade and solving the quadratic equation resulting from setting the remainder of the needed correction equal to the parabolic error term at Station 50 + 75.00); b) If a 900-foot curve is used, determine the station and the elevation of the low point of the curve; c) Find the station of PC and the elevation of PC for the 900-foot curve. d) Find the minimum length of curve for nighttime driving that will provide a stopping sight distance of 600 feet ($S < L$).

7.10 Vertical Alignment.   PVT of an equal tangent vertical curve is at Station 100 + 00 at an elevation of 320 feet with a grade of minus 4 percent. The curve provides for an overpass across railroad tracks from Station 85 + 70 to 86 + 30. (The maximum of the curve will occur at 86 + 00). The bridge and road are being reconstructed to provide for an increase of track clearance to allow for electrification of a high-speed train. The grade in the area of PVC is 2.5 percent. Find the station and elevation of PVC and the elevation of the bridge at 86 + 00 relative to the elevation of PVT.

(a) Find the length of the curve.
(b) the station of PVC.
(c) the elevation at PVC.
(d) the height of the curve at the railroad bridge above the elevation of PVT.
(e) assuming SSD < L, determine the SSD this vertical curve design accommodates, and determine if this SSD suggests placing speed controls on the curve. What speed?.

7.11 Vertical Curves and SSD.   A 3.3 percent downhill grade enters a vertical curve at Station 12 + 09 and emerges 702.6 feet later as a 2.6 percent uphill grade. What is the maximum design speed (to the nearest 5 mph) that will allow this vertical curve to meet computed desirable SSD standards?

7.12 Vertical Curve.   A road for 65 mph is being designed over a hill. The grade on one side is 3 percent while the grade on the other side is 1 percent.

(a) Determine the minimum length of the curve for adequate stopping sight distance. Is the length curve longer or shorter than the SSD?
(b) Assuming because of the terrain the curve is 1000 feet long, where will the high point be relative to PVC?

7.13 Horizontal Sight Distance.   A 1650-foot flat horizontal curve is to be used in a new segment of two-lane (each 12-foot wide) road. The surveyor has indicated that there is a very large clump of trees, the edge of which will be about 35 feet from the edge of the pavement at Station 32 + 50.00. If PC is at Station 28 + 00.00, how much, if any, will the edge of the trees be needed to be cut back if a desirable safe sight distance of 820 feet is to be observed?

7.14 Horizontal Curve.   A freeway exit ramp has a single 12-foot lane and consists entirely of a curve with a central angle of 90 degrees and L = 628 feet. If the distance cleared of obstacles is 19.4 feet from the centerline, what design speed was used?

7.15 Stopping Sight Distance on Horizontal Curves.   As SR361 enters Middletown from the NW, it must curve to the south between two historic buildings. At this point, SR361 consists of two 11-foot lanes. The building to the right of traffic is only 10 feet from the edge of the 5-foot-wide shoulder. Local policy is that the superelevation rates not exceed 6 percent. Will the minimum SSD be available for vehicles moving at 60 mph through this curve?

7.16 Combined Curves.   A bridge is to be built across the Mythaca River. For adequate clearance, the elevation of the bridge deck will be 344 feet. The road on the east side of the river is level at an elevation of 318 feet, and the road on the west side of the river about 1700 feet from the end of the bridge has a grade of plus 1.5 percent at an elevation of 328 feet. Determine the curves to join the bridge to the highway, ensuring adequate SSD at the design speed of 65 mph and no grade greater than 4 percent.

7.17 Scenic Coastal Highway.   Assume that a SSD is required at a design speed of 65 mph. What will the alignment be for Example 7.10?

7.18 Combined Curves.   A crest curve with a grade at VPC (elevation 126.5 feet) of 2 percent is to be joined with sag curve with a VPT (elevation 111.5 feet) of +1 percent. Determine the distance between the curves VPC and VPT if the common section between them is to be at a grade of minus 2 percent Adequate sight distance is to apply at 55 mph.

7.19 Combined Curves.   Repeat Exercise 7.18 with the angle of the VPT of the sag curve = −1 percent rather than +1 percent.

Banking the Curves and Runoff

7.20 Superelevation.   Find the minimum safe radius R and the maximum safe degree of curvature D for the following situations: a) design speed of 40 mph, e_{max} = 0.10 ft/ft; b) a design speed of 55 mph, e_{max} = 0.08 ft/ft; c) a design speed of 70 mph, e_{max} = 0.05 ft/ft.

7.21 Horizontal Alignment.   What is the maximum allowable degree of curve for a two-lane highway if e_{max} = 0.08, f_{max} = 0.12, and the design speed is 50 mph?

7.22 Superelevation.   A rural road (750 South) in Mythaca County is in poor condition and is experiencing increasing traffic. The county engineer wants to rehabilitate the road, to include superelevating a 600-foot curve. What value of e_{max} (rate of superelevation) do you recommend for this curve? If the design speed for this curve is 50 mph, what f-value (side friction factor) do you recommend? If e_{max} = 0.1, what speed limit should be placed on the curve?

7.23 Geometric Design.   You are asked to fit a horizontal (circular) curve between two tangent sections of highway. The "incoming" section runs in a northeast direction (compass reading 45 degrees) and the "outgoing" section runs directly east (90 degrees on the compass). The PC and PT are 361 feet apart. What will be the length of the horizontal curve (to nearest 0.1 feet)? Regardless of your calculations in part a, assume the curve has a radius of 566 feet. If the design speed is proposed to be 60 mph, demonstrate whether an appropriate superelevation value can be found to ensure a safe curve.

7.24 Spiral Transition Curves.   A transition curve is needed to connect two circular sections—one with a radius of 1424 feet and the other with an 841-foot radius. If the design speed of the highway is 50 mph, how long should the transition curve be?

7.25 Superelevation and Transition Curves.   SR 361 is a four-lane open highway with 12-foot lanes and 70 mph design speed in a region not susceptible to snow and ice. In one section of the highway, a transition must be made between a horizontal curve with radius 1828.2 feet and a tangent section. (a) What is the minimum required length of the spiral transition curve? (b) Using the equation, how long should the superelevation runoff be? (c) Which of the two values will you adopt as your transition curve length? Why? (d) If the curve is banked at a 3° angle, how will the spiral length change?

7.26 Spiral Transition Curves.   On NB U.S. Highway 12, just before it reaches CR21, the road turns NE toward I-25. The turn actually consists of two circular horizontal curves, connected by a spiral curve. The radius of the first HC is 689 feet and the radius of the second HC is 529 feet. If the design speed for this part of U.S. 12 is 65 mph, what should be the minimum length of the spiral? If the superelevation rate is 0.06, does the distance between the two curves change?

7.27 Superelevation Runoff.   For each of the horizontal curves on NB U.S. 12 described in the previous problem, how long should the superelevation runoff be, if a superelevation rate of 0.06 is used? Is there enough distance between the two circular curves for the required runoff(s)?

7.28 Spiral Curves.   A Mythaca County highway designer wants to connect two circular horizontal curves, with radii $R_1 = 322$ ft and $R_2 = 525$ ft using a spiral transition curve. If 955 feet are available for the spiral, what is the maximum design speed permitted? (Express your answer to the nearest 0.1 mph.)

7.29 Spiral Curves.   The transition curve on a section of highway between two circular curves (having radii $R_1 = 4583.79$ feet and $R_2 = 2864.79$ feet, respectively) runs between Stations 43 + 95 and 44 + 91. The highway's stated design speed is 60 mph. Is the spiral transition curve's design adequate for the conditions stated above?

7.30 Passing Sight Distance.   Repeat Example 7.13 with a passing speed of 65 mph and the car in the opposing direction moving at 55 mph.

Design of Intersections for Safety and Efficiency

SCENARIO

What used to be a rural intersection of two county roads has become a busy intersection. County Road 75E has long been the easiest way to get from Mythaca to the northern parts of the county. CR600N was just one of many east-west roads that crossed CR75E and were controlled by stop signs. In recent years, several large residential developments have caused rapid increase in enrollment at nearby Old Hickory High School (See Figure 8.1.). Consequently, traffic at the intersection of CR75E and CR600N has also increased dramatically. A recent rash of collisions at this intersection has residents in the area calling for the county highway engineer to "do something." What alternatives are available to the engineer, and what methods can he use to analyze the situation?

CHAPTER OBJECTIVES

By the end of this chapter, the student will be able to:

1. Find the critical gap for an unsignalized intersection approach.
2. Analyze the sight distances at an intersection for possible installation of stop or yield signs.
3. Determine when an intersection warrants the installation of a traffic signal.
4. Design signal settings for intersections so that drivers will not face a *dilemma zone*.
5. Estimate the *stopped delay* for an approach to an intersection.

FIGURE 8.1
Rural intersection near a high school. Photo: Jon D. Fricker.

CONTEXT

Much of our driving involves stopping and starting at yield signs, stop signs, and traffic signals. Many times we stop on a side street and must wait for traffic to have a sufficient break to allow us to go through the intersection, to turn right, or to turn left. Every intersection where traffic from several different directions meet is a potential safety hazard; someone has the right of way and others do not. This chapter deals with

1. Traffic situations at non-signalized intersections
2. Critical Gap criteria
3. Criteria for stop and yield signs where sight distances can be critical
4. Criteria for warrants to install a signal light
5. Determining whether a "dilemma zone" exists
6. A simple way to estimate average delay at a signalized intersection.

8.1   ANALYSIS OF NON-SIGNALIZED INTERSECTIONS

8.1.1   Gap Acceptance

A driver on a side street reaches the intersection with a busy arterial street. (See Figure 8.2.) She immediately looks to her left, hoping to find a gap in the traffic large enough to permit her to make a right turn. Several vehicles pass in front of her car, spaced too closely together for her to feel comfortable enough to attempt the right turn. Eventually, she finds a gap in the arterial street traffic that is large enough, and she makes the turn. By the time she makes the turn, another driver has arrived, also wanting to turn right. The second driver begins to check the traffic coming from his left, looking for an acceptable gap.

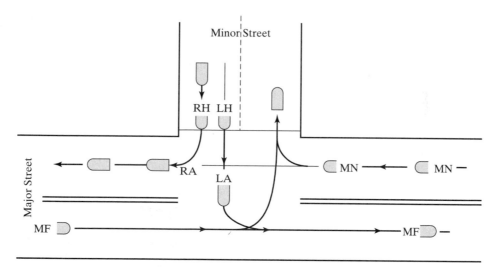

FIGURE 8.2
Events in gap acceptance analysis. *Source:* Fricker et al. 1991.

THINK ABOUT IT

How long a gap in the cross traffic does a driver need before he or she will attempt to make the right turn? Will the length of the gap vary from driver to driver? How would you collect data to answer these questions?

Gap acceptance analysis makes use of the following definitions (Robertson, 1994):

- *Headway.*   time elapsed between the front bumper of one vehicle and the front bumper of the following vehicle passing a given point.
- *Gap.*   time elapsed between the rear bumper of one vehicle and the front bumper of the following vehicle passing a given point. Gap is thus always smaller than headway.
- *Lag.*   time elapsed between the arrival of a minor-street vehicle ready to move into the major street and the arrival of the front bumper of the next vehicle in the major traffic stream. It may be that drivers treat lags differently than they treat gaps.
- *Accepted gap or accepted lag.*   gap or lag that the driver of a minor street vehicle uses to move into the major street.
- *Rejected gap or rejected lag.*   gap or lag that the driver of a minor street vehicle waiting to enter the major street does not accept.
- *Untested gap.*   no minor street vehicle was present.
- *Critical gap.*   the minimum size gap that a particular driver will accept.

THINK ABOUT IT

How many gaps (including the lag) will each driver accept? How many gaps (including the lag) will each driver reject? Express your answer as a range of values, if necessary.

Data for gap acceptance studies have been collected by having an observer hit a key on an electronic counting board or laptop computer each time a major street vehicle arrives. An alternative is to videotape the intersection and either record "event times" using a "time stamp" on the videotape or using a stopwatch to measure gap lengths on the videotape.

The "events" referred to in the paragraph above and shown in Figure 8.2 are (Fricker et al., 1991):

LH.   **L**eft-turning vehicle on minor street reaches **H**ead of minor street queue.

RH.   **R**ight-turning vehicle on minor street reaches **H**ead of minor street queue.

MN.   A **M**ajor street vehicle on the **N**ear side of the street passes in front of the minor street vehicle.

MF.   A **M**ajor street vehicle on the **F**ar side of the street passes in front of the minor street vehicle.

LA.   The **L**H vehicle **A**ccepts the gap.

RA.   The **R**H vehicle **A**ccepts the gap.

These definitions are used in the next example.

Example 8.1

Using videotape of a "T" intersection, an analyst creates the database of gap acceptance events shown in columns 1 and 2 of the table below. Fill in column 3 with the length of gaps, and note in column 3 whether the gap was accepted (A) or rejected (R).

(1) Event	(2) Time (sec)	(3) Gap (sec)
RH	0.0	
MN	3.4	
MN	7.5	
RA	11.4	
MN	15.3	
RH	17.6	
RA	17.6	
RH	23.4	
RA	23.4	
MN	32.7	

Solution to Example 8.1   In the 32.7 seconds of data shown in the table above, three right-turning vehicles "RH" are observed. The first gap (really a lag) of 3.4 seconds is rejected, as is the next 4.1-second gap. These gaps are marked "R1" in the column below, because the driver of vehicle 1 rejected them. The first RH vehicle accepts a gap that begins at 7.5 seconds and ends at 15.3 seconds, which is a 7.8-second gap. This gap is marked "A1."

The second vehicle moves into the "ready to turn" position at 17.6 seconds, at which time its lag begins. The next MN vehicle arrives at 32.7 seconds, by which time the second and third vehicles have made right turns. Vehicle 2 has a long gap to accept: $32.7 - 17.6 = 15.1$. The accepted gap for vehicle 3 is "closed" by the same MN vehicle: $A3 = 32.7 - 23.4 = 9.3$ seconds. Note in the table above that the analyst apparently decided to assign an RA event time equal to the RH event time if the vehicle turned without having to wait for an acceptable gap. Note also that the last major street gap shown in the data was actually $32.7 - 15.3 = 17.4$ seconds long, which is shown in the table below without an "A" or "R" mark. This gap was accepted by both

Gap (sec)
R1 3.4
R1 4.1
A1 7.8
A2 15.1
A3 9.3
17.4

vehicle 2 and vehicle 3, but their drivers saw the gap as a lag that began with their respective RH events.

THINK ABOUT IT

In the example above, a major street gap is calculated as MN(i+1) − MN(i). Is the resulting value really a gap, or is it a headway? For major street vehicles that are 15 feet long and are traveling at 35 mph, how much difference would it make?

THINK ABOUT IT

Is it possible for one driver to have rejected a gap that is longer than a gap accepted by another driver? Is it possible for the same driver to have rejected a gap that is longer than a subsequent gap he/she accepted?

Miller (1971) identified nine methods for converting observations such as those made in Example 8.1 into a value for critical gap. In the paragraphs below, we present another method. Table 8.1 shows a summary of gap acceptance data collected at the intersection of a county road and a state highway on the outskirts of a medium-sized city. In column 2, it indicates that 2 of 115 accepted gaps were less than 1 second long and four of the 115 accepted gaps were less than 2 seconds long. (Because these numbers are cumulative, the four accepted gaps that were less than 2 seconds in duration include the first two that were less than 1 second long.) In column 3, all 217 rejected gaps were greater than 0.0 seconds, but only five of 217 rejected gaps were longer than 8.0 seconds.

Because the number of rejected and accepted gaps are not equal, using cumulative *percentages* instead of cumulative *counts* allows the analyst to normalize the data and treat accepted gaps on the same basis as rejected gaps. The cumulative totals in columns 4 and 5 of Table 8.1 are plotted in Figure 8.3. The two cumulative curves appear to cross at a point between 5 and 6 seconds. This point is defined as the *critical gap* (t_c). Interpolating between the four points in Table 8.1 produces an estimate of t_c.

(1) Length of Gap (t sec)	(4) Percent Accepted Gaps (less than t sec)	(5) Percent Rejected Gaps (greater than t sec)
$5.0 = t_1$	$12.2 = A(t_1)$	$16.6 = R(t_1)$
$6.0 = t_2$	$16.5 = A(t_2)$	$8.8 = R(t_2)$

TABLE 8.1   Data for Computation of Critical Gap

(1) Length of Gap (t sec)	(2) Number of Accepted Gaps (less than t sec)	(3) Number of Rejected Gaps (greater than t sec)	(4) Percent Accepted Gaps $A(t)$ (less than t sec)	(5) Percent Rejected Gaps $R(t)$ (greater than t sec)
0.0	0	217	0.0	100.0
1.0	2	204	1.7	94.0
2.0	4	138	3.5	63.6
3.0	6	96	5.2	44.2
4.0	8	60	7.0	27.6
5.0	14	36	12.2	16.6
6.0	19	19	16.5	8.8
7.0	26	7	22.6	3.2
8.0	36	5	31.3	2.3
9.0	45		39.1	0.0
10.0	52		45.2	0.0
11.0	64		55.7	0.0
12.0	71		61.7	0.0
13.0	74		64.3	0.0
14.0	78		67.8	0.0
15.0	85		73.9	0.0
17.0	92		80.0	0.0
20.0	100		87.0	0.0
25.0	106		92.2	0.0
30.0	112		97.4	0.0
35.0	115		100.0	0.0

An equation that accomplishes a linear interpolation is

$$t_c = t_1 + \frac{(t_2 - t_1)[R(t_1) - A(t_1)]}{[A(t_2) - R(t_2)] + [R(t_1) - A(t_1)]} \tag{8.1}$$

Using Equation 8.1 with the four points that surround the point at which the two cumulative curves cross in Figure 8.3 leads to

$$t_c = 5.0 + \frac{(6.0 - 5.0)[16.6 - 12.2]}{[16.5 - 8.8] + [16.6 - 12.2]} = 5.0 + \frac{4.4}{7.7 + 4.4} = 5.36 \text{ sec.}$$

The critical gap for the data in Table 8.1 is 5.36 seconds. A "typical" driver at the intersection studied is just as likely to accept a gap of that duration as he/she is to reject such a gap. Like the nine gap acceptance analysis methods Miller (1971) evaluated, this graphic method has its flaws, but it illustrates driver behavior at work in the gap acceptance process.

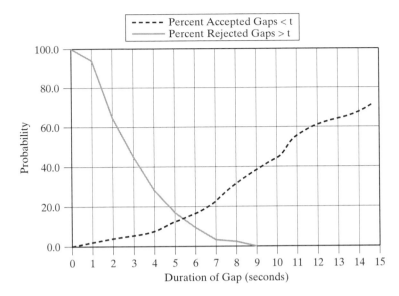

FIGURE 8.3
Cumulative distribution curves for accepted and rejected gaps.

THINK ABOUT IT

Do you think that the critical gap can vary significantly from one intersection to another? If so, what factors may affect the length of the critical gap?

Gap acceptance has several important applications. Once the gap acceptance behavior of drivers at an intersection is established, the need to add stop signs on the major street or install traffic signals at the intersection can be considered. Another action to consider is adding a second approach lane, so that minor street traffic turning left and right will not have to share the same approach lane. In the next section, we look at the critical approach speed as a way to decide when Stop or Yield signs may be called for at an intersection. At the end of that section, we return to critical gap as a design parameter.

THINK ABOUT IT

If a thorough gap acceptance study at a given intersection were done separately for right-turning vehicles and left-turning vehicles, which vehicles do you think would have the larger critical gap? Explain your reasoning.

8.1.2   Stop, Yield, or No Control at Urban Intersections (Based on Sight Distance)

In many parts of Europe, most intersections in residential neighborhoods have no traffic control at all. Drivers are expected to yield to traffic on the right. In the United States, most intersections have some kind of control. Because Yield signs at intersections are rare in the United States, the decision usually is between two-way and four-way stop-controlled intersections In Figure 8.4, through and left-turning traffic must stop, but traffic turning right must yield.

THINK ABOUT IT
Many traffic engineers do not like to install four-way stop signs at intersections, unless absolutely necessary. Why do you think that might be the case?

The *critical approach speed (CAS)* method outlined below illustrates many of the elements that affect the decision whether to install a traffic control device, and what kind. The most important elements are the distances that a driver can see in each direction along the cross street. When combined with assumptions regarding a driver's vantage point, reaction time, and deceleration rate, an acceptable approach speed can be computed. According to the CAS method, if the computed speed is between 10 and 15 mph, a Yield sign is appropriate. If the computed speed is less than 10 mph, a Stop

FIGURE 8.4
Stop and Yield signs. *Source*: Jon D. Fricker, 20 October 2001.

a', b', c', and d' are the distances from the drivers to the curb line.

a'', b'', c'', and d'' are the distances from view obstructions to the curb line.

a, b, c, and d are the distances from the drivers to the view obstructions.

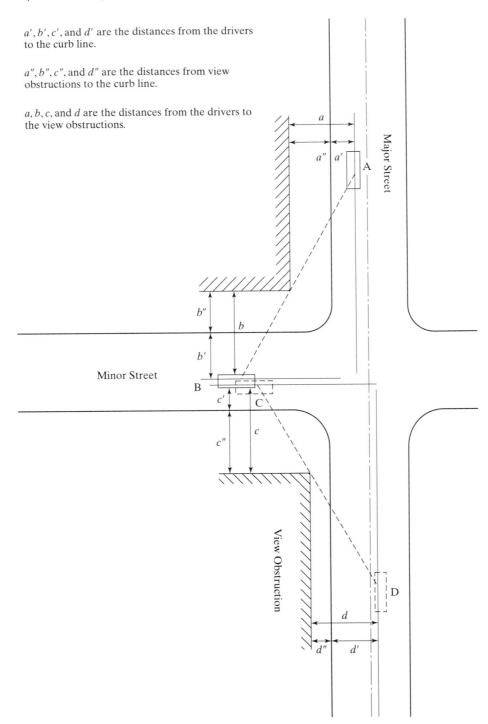

FIGURE 8.5
Typical critical approach speed problem. *Source*: Federal Highway Administration, 1983, p. 2-18.

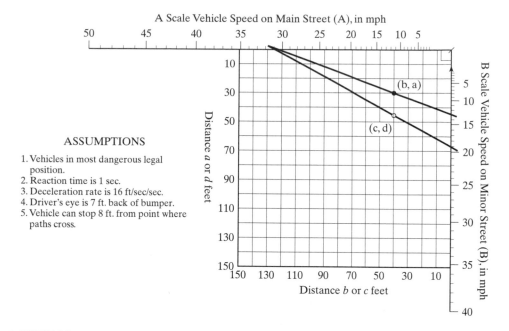

A Scale Vehicle Speed on Main Street (A), in mph

FIGURE 8.6
Critical approach speed chart. *Source*: Federal Highway Administration, 1983, p. 2-17.

sign must be installed. At computed speeds higher than 15 mph, the CAS does not by itself justify a Stop or Yield sign. Further analyses may be needed.

Example 8.2 and Figures 8.5 and 8.6 are taken from the *Traffic Control Devices Handbook* (FHWA, 1983). Take note in Figure 8.6 of the assumptions on which the design chart is based.

Example 8.2: Critical Approach Speed at Minor and Major Streets

The Minor Street approach to Major Street is affected by buildings built fairly close to the corner. See Figure 8.5. The posted speed on Major Street is 30 mph, but its 85th percentile speed is 31 mph. Parking is prohibited on Major Street but is allowed on Minor Street. Major Street is 60 feet wide and Minor Street is 36 feet curb-to-curb. The measured x'' distances (obstruction to curb) needed by the CAS method are:

$$a'' = 24 \text{ ft}, \, b'' = 19 \text{ ft}, \, c'' = 28 \text{ ft}, \, d'' = 12 \text{ ft}$$

Solution to Example 8.2 Step 1 in the CAS method is to determine the x′ distances (driver to curb) in Figure 8.5.

- $a' = 12$ feet with parking, $a' = 6$ feet without parking. Because a' pertains to Major Street, which has no parking, choose $a' = 6$ feet.
- $b' = \min\{\frac{1}{2}W + 3 \text{ ft}; W - 12 \text{ ft}\} = \min\{\frac{1}{2}(36 \text{ ft}) + 3 \text{ ft}; 36 \text{ ft} - 12 \text{ ft}\} = \min\{21 \text{ ft}; 24 \text{ ft}\} = 21$ feet, where $W = $ the width of Minor Street
- $c' = 12$ feet with parking, $c' = 6$ feet without parking. Because c' pertains to Minor Street, which has parking, choose $c' = 12$ feet.
- $d' = \frac{1}{2}W + 3 \text{ ft} = \frac{1}{2}(60 \text{ ft}) + 3 \text{ ft} = 33 \text{ ft}$, where $W = $ the width of Major Street

In step 2, calculate the overall dimensions $x = x' + x''$

- $a = a' + a'' = 6 + 24 = 30$ ft
- $b = b' + b'' = 21 + 19 = 40$ ft
- $c = c' + c'' = 12 + 28 = 40$ ft
- $d = d' + d'' = 33 + 12 = 45$ ft

CAS step 3 involves use of the design chart in Figure 8.6. Plot two points on the chart: (b, a) and (c, d). In this example, the point $(b = 40, a = 30)$ is the solid dot on the design chart and the point $(c = 40, d = 45)$ is the hollow dot.

In step 4, draw a line from the Major Street speed (31 mph) on the A scale, through each of the dots plotted in step 3, and read the speed on the B scale through which each line passes. The solid dot leads to a Minor Street speed of 13 mph on the B scale. The line through the hollow dot crosses the B scale at 19 mph.

Step 5. Choose the smaller of the two speeds found in step 4. In this example, the smaller speed is 13 mph. This is in the 10 to 15 mph range that calls for a Yield sign.

In Figure 8.5, the distance along the Major Street from Driver D to a point on the Major Street directly in front of Driver C is called the *intersection sight distance (ISD)* (AASHTO, 2001, Exhibit 9-50). Harwood et al. (1998) recommend using the critical gap at an intersection to establish the intersection's sight distance:

$$ISD = (1.47 * V) * t_c \qquad (8.2)$$

where ISD = intersection sight distance (ft), V = major street design speed (mph), and t_c = critical gap for a vehicle on the minor street to turn onto the major street (sec). For two-lane streets and highways, the values of t_c are 7.5 seconds for passenger cars, 9.5 seconds for single-unit trucks, and 11.5 seconds for combination trucks (AASHTO, 2001, p. 664). If adequate ISD cannot be provided, stop signs on the major street may be called for.

8.2   SIGNAL WARRANTS AND STOPPING DISTANCE AT SIGNALIZED INTERSECTIONS

Introduction

This section describes and discusses two important issues in the control of intersections: (1) When should traffic signals be installed? and (2) When do signal timings create a hazardous situation for drivers?

8.2.1   Warrant Analysis

In this section, procedures to determine when to install a traffic signal are introduced (see Figure 8.7.). This topic is of great interest to many citizens, as letters to the editor like the one in Figure 8.8 indicate. Fortunately, there doesn't have to be much subjectivity involved in the decisions regarding whether to install intersection control, and which type. *Warrants* use threshold values to help a traffic engineer make those determinations. The threshold values may be expressed in terms of traffic volumes or

FIGURE 8.7
Looking north at SB Approach to intersection in Example 8.3. Photo: Jon D. Fricker.

approach speeds (based on sight distance calculations). Some procedures are introduced below.

The *Manual on Uniform Traffic Control Devices for Streets and Highways* (or MUTCD) contains 11 possible conditions (warrants) to be examined in determining whether a traffic signal is justified at an intersection. They are (FHWA, 2003, Section 4C.01):

1. Eight-Hour vehicular volume
2. Four-hour vehicular volume
3. Peak hour
4. Pedestrian volume

Traffic light needed at SR 361/900E.

We need a stoplight at the intersection of Highway 361 and County Road 900 E. I have wondered for years why one was not put there a long time ago. This intersection is a fatality waiting to happen.

The opening of the Walmart more than three years ago, the strip mall on the opposite corner that followed, and new housing nearby have had a major impact on the increase in traffic in that area.

This intersection is very busy with left and right turns from semi trucks and dump trucks, not to mention automobiles. There are stop signs for traffic on the county road. If someone hasn't been killed or seriously injured, then it's just a matter of time.

This is a horrible intersection. It doesn't take a brain surgeon to figure out a traffic light is needed with arrows for the people to turn onto SR 361 safely.

Traffic planners, why not go to this intersection and see just how dangerous it is. Let's get a light and turn signals at this busy intersection.

Chris Doe, Mythaca

FIGURE 8.8
Letter to the editor demanding traffic signals.

5. School crossing
6. Coordinated signal system
7. Crash experience
8. Roadway network

While a "traffic" signal should not be installed unless one or more of the factors ... are met," the "satisfaction of a traffic signal warrant ... shall not in itself require installation of a traffic control signal" (FHWA 2003, p. 4C-1).

Each of the warrants has a procedure associated with it to determine if the warrant justifies a traffic signal.

The procedure for applying warrant 1 follows the steps given below.

1. Determine the number of lanes of moving traffic on each approach to the intersection. Doing this allows the analyst to choose the appropriate row to use in Table 8.2.

2. Collect hourly traffic volume data for each approach to the intersection, preferably for 24 hours on a "typical" day. If doing counts for 24 hours is not practical, collect data for all hours in which it is possible that the threshold values in Table 8.2 may be exceeded on any approach.

3. For each hour of data collected, add the two approach volumes on the major street. Once this has been done, count the number of hours in which the major street threshold was exceeded. If the number is less than eight, the warrant has not been met and analysis can be terminated. If the number is at least eight, continue to step 4.

4. For each of the eight or more hours in which the major street threshold has been exceeded, determine the higher volume on the minor approaches. Count the number of hours in which the minor street threshold is also exceeded. If that number is eight or greater, then warrant 1 has been met and a traffic signal is justified.

TABLE 8.2   Minimum Vehicular Volumes for Traffic Signal Warrant

Number of Lanes for Moving Traffic on Each Approach		Vehicles per Hour on Major Street (Total of Both Approaches)	Vehicles per Hour on Higher-Volume Minor-Street Approach (One Direction Only)
Major Street	Minor Street		
1	1	500	150
2 or more	1	600	150
2 or more	2 or more	600	200
1	2 or more	500	200

Source: FHWA 2003.

Example 8.3 illustrates the use of the four-step procedure for warrant 1.

Example 8.3:   Volume Warrant of Traffic Signal

An intersection on the Mythaca State University campus is currently controlled by three stop signs. Stadium Avenue is a four-lane arterial that runs east-west through campus. Speeds on Stadium do not exceed 40 mph. N. Russell Street is a one-way street. It approaches Stadium Avenue from the north with one wide lane that splits into three 12-foot lanes just before the intersection—one for left turns, one for through movements, and one for right turns. (See Figure 8.7.) The intersection is near a large residence hall and some fraternities and sororities. Recent increases in pedestrian and vehicle traffic have caused some students to ask the city that a traffic signal be installed at that intersection. Counts of automobiles and trucks were taken from 7 AM to 7 PM on several "typical" days and averaged. Based on the count data summarized in Table 8.3, is a traffic signal at Stadium and Russell justified under warrant 1?

TABLE 8.3   Approach Volumes, Stadium Avenue at N. Russell Street

| | Mean volume counts, each hour | | | | | |
| | Automobiles | | | Trucks and Buses | | |
Hour Beginning	SB Approach	EB Approach	WB Approach	SB Approach	EB Approach	WB Approach
7 AM	321	403	168	2	13	6
8 AM	289	366	189	6	21	17
9 AM	161	284	137	***	***	***
10 AM	139	304	167	2	18	12
11 AM	167	354	223	2	15	15
12 PM	281	413	301	4	18	14
1 PM	256	397	284	1	14	11
2 PM	226	433	306	2	18	15
3 PM	204	340	258	15	31	34
4 PM	239	440	322	1	12	14
5 PM	243	517	393	0	11	6
6 PM	227	423	355	3	4	3

*** These are actual data collected by students. At least one group of students at 9 AM forgot to identify trucks and buses in their data, so the mean "automobile" counts for that hour include trucks and buses.

Solution to Example 8.3   Following the four-step procedure:

1. There are two approach lanes in each direction on Stadium Avenue. The number of lanes on SB Russell is not so obvious. (See Figure 8.7.) Is it one lane or three?

THINK ABOUT IT

How many approach lanes on SB Russell are there for purposes of the warrant analysis? Give the reason(s) for your decision.

Let us decide (for now) that there is one approach lane on SB Russell. (Do not let this decision influence your thinking about the question posed in the box above. It is possible that three approach lanes is more correct. A good class discussion may help decide the issue.) Because the major street (Stadium) has "2 or more" approach lanes and the minor street (Russell) is assumed to have one approach lane, the second row in Table 8.2 is the one to use.

2. The mean approach volumes for each of 12 separate hours on a typical day are summarized in Table 8.3.

3. For the major street (Stadium), add the EB and WB auto, truck, and bus traffic together for each hour. For example, for the hour beginning at 7 AM, the total major street approach volume is 403 + 168 + 13 + 6 = 590. This value fails to reach the threshold value of 600. Likewise, the hours beginning 8 AM through 10 AM do not reach the total major street approach volume threshold of 600 vehicles. At 11 AM, however, the total is 354 + 223 + 15 + 15 = 607. This value exceeds the threshold, as does the volume for each of the next 7 hours. Because 8 hours have met the major street threshold, we continue on to step 4.

4. For each of the 8 hours identified in step 3, determine the approach volume on Russell and compare it against the threshold of 150 vehicles. (*Note*: Because Russell is a one-way street, there is no other approach for the minor street.) For each of the 8 hours identified in step 3, the minor street approach exceeds the threshold volume of 150 vehicles. Warrant 1 has been satisfied. A traffic signal can be justified at the intersection of Stadium and Russell.

8.2.2   Reacting to a Changing Traffic Signal

There is an important decision problem that a driver faces each time a traffic signal changes from green to yellow as he/she approaches an intersection. The driver must decide: Go through the intersection or try to stop? (Years ago, the official meaning of the amber phase in a signal cycle was "clear the intersection." The practical meaning of this definition was "stop if you can, or else continue through the intersection." In subsequent years, the term "amber" has been changed to "yellow," and the yellow phase simply means "the red phase is coming.") The preferred reaction is to try to stop. However, it is possible that there is not sufficient distance for the car to stop, and the car would be in the intersection when the cross traffic gets the green light. The driver could be accused of "running the red light."

Examine the situation in Figure 8.9. When traveling at a speed v_o, stopping before the reaching the "stopping line" means that the braking maneuver must occur within the distance D. Clearing the intersection requires going a distance of $D + W + L$ before the light turns red.

Assuming a reaction time of t_r for the driver and t_Y for the length of the caution or yellow light, the inequalities given in Equations 8.3 must be satisfied.

$$\text{to stop:}\quad D \geq v_o t_r + \left(v_o t_s - \frac{1}{2} d_{\text{brake}}\, t_s^2 \right) \quad \text{or} \quad D \geq v_o t_r + \frac{v_o^2}{2 d_{\text{brake}}} \quad (8.3a)$$

$$\text{to clear:}\quad D + W + L \leq v_o t_r + v_o(t_Y - t_r) + \frac{1}{2} a_{\text{acc}}(t_Y - t_r)^2 \quad (8.3b)$$

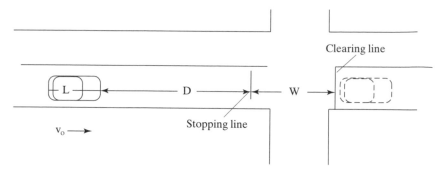

FIGURE 8.9
Vehicle approaching a signalized intersection.

where d_{brake} = the deceleration rate for braking to a stop

t_s = time to stop = $\dfrac{v_o}{d_{brake}}$

a_{acc} = the acceleration rate to increase speed when going through the intersection

THINK ABOUT IT

What limits would you put on the values of a_{acc} and the speed after acceleration, and why?

    The ability to stop depends on the driver/vehicle braking capability. The ability to continue through the intersection depends not only on the speed but also on the width of the intersection, the length of the car, and the duration of the yellow caution light. So, for some very wide intersections, either the approach speed must be changed or the duration of the yellow light would have to be very long. However, if the duration of every yellow light in an area were different, then drivers could become confused.

THINK ABOUT IT

Why would drivers become confused if the duration of yellow lights varied from intersection to intersection?

    The most common length of the yellow phase is about 4 seconds. If longer clearance time is needed, the lights for all approaches remain red for the necessary added time. This is known as the *all-red clearance interval*, which is covered in the next section. Consider Figure 8.10, which has three parts. In the top part, a driver in any part of the approach is able to make a correct decision and, in the non-overlapping section, either decision is correct. The middle part of the figure shows the limiting case, in which stopping is the correct decision until the driver is distance X_c from the intersection and then continuing is correct. The lower part illustrates the situation in which a dilemma zone exists. In other words, a driver approaching the intersection at the legal speed can execute neither of the two maneuvers safely, comfortably, or legally, if the signal changes to yellow while the vehicle is between X_c and X_o, i.e., in the dilemma zone.

THINK ABOUT IT

If you were a traffic engineer, what would you do to eliminate the dilemma zone?

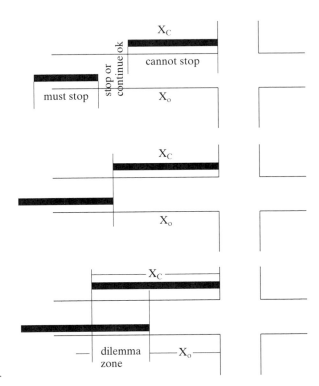

FIGURE 8.10
The dilemma zone situation.

Example 8.4

A driver traveling in his 16-foot SUV at the speed limit of 30 mph was arrested for running a red light at 15th and Main in Shoridan, an intersection that is 60 feet wide. The driver claimed innocence, on the grounds that the traffic signals were not set properly. The yellow light was on for the standard 4 seconds. The SUV driver's reaction time is assumed to be 1.5 seconds. Comfortable deceleration is at a rate of 10 ft/sec/sec. Did a dilemma zone exist on this intersection approach? If so, how long was it?

Solution to Example 8.4 The minimum distance needed to stop X_s is, from Equation 8.3a, computed as

$$t_s = \text{time to stop} = \frac{v_o}{d_{brake}} = \frac{30*1.47}{10} = 4.41 \text{ sec}$$

$$X_s = v_o t_r + v_o t_s - \frac{1}{2} d_{brake} \, t_s^2 = (30*1.47*1.5) + (30*1.47*4.41) - \frac{1}{2}(10)(4.41)^2$$

$$= 66.15 + 194.48 - 97.24 = 163.39 \text{ ft}$$

If the driver was any closer to th.e intersection than the X_s distance just calculated, he must be able to drive $W + L = 60 + 16$ feet farther than X_s to clear the intersection, or $X_c = 163.39 + 76 = 239.39$ feet. Using Equation 8.3b with $a_{acc} = 0$, the distance the driver can travel in t_Y seconds at speed v_o is:

$$X_Y = v_o t_r + v_o(t_Y - t_r) + \frac{1}{2} a_{acc}(t_Y - t_r)^2 = (30*1.47*1.5) + (30*1.47)(4.0 - 1.5) + 0$$

$$= 66.15 + 110.25 + 0 = 176.40 \text{ ft}$$

Because $X_Y < X_c$, a dilemma zone exists. Its length is $X_c - X_Y = 239.39 - 176.40 = 62.99$ ft. The driver's claim cannot be dismissed, but now the issue becomes whether the vehicle was in the dilemma zone when the light changed to yellow.

THINK ABOUT IT

In Example 8.4, the deceleration rate used was a "comfortable" 10 ft/sec/sec. Given the potential danger implied by the circumstances in the example, shouldn't a more "urgent" deceleration rate be used to analyze the possible existence of a dilemma zone?

Just before Example 8.4, the reader was asked to think about solutions to the dilemma zone. Now that Example 8.4 has been presented, some further analysis can be attempted. To facilitate such an analysis, a spreadsheet has been constructed to repeat the calculations of Example 8.4. (See column A of Figure 8.11.) The spreadsheet permits a series of proposed solutions to be tested. For example, what if the approach speed is reduced from 30 to 20 mph? The dilemma zone length shrinks to 45.69 feet (cell C21), but a dilemma zone still exists. Why not permit—even encourage—a higher approach speed, such as 40 mph? "Calc. 3" in column D shows that the dilemma zone grows to 101.42 feet. If we could somehow reduce reaction time to 1.0 seconds (column E), the dilemma zone shrinks to 40.80 feet. Although these trial solutions are somewhat interesting, they are of little practical use and do not seem to solve the problem at this intersection. What is left is to find the length of the yellow phase that will eliminate the dilemma zone. Using a spreadsheet's solver feature to adjust cell F7 so that cell F21 = 0, we find that t_Y must be at least 5.06 seconds. If it is preferred that a standard

	A	B	C	D	E	F
				Calc. 2	Calc. 3	Calc. 4	Calc. 5
1		Dilemma Zone calculations				
2			Calc. 2	Calc. 3	Calc. 4	Calc. 5
3	Input:					
4	30	v(0) = approach speed (mph)	20	40	30	20
5	1.5	t(r) = reaction time (sec)	1.5	1.5	1.0	1
6	−10	a < 0 = vehicle declaration rate (ft/sec/sec/)	−10	−10	−10	−10
7	4	t(Y) = duration of yellow interval (sec)	4	4	4	5.06
8	60	W = width of intersection (feet)	60	60	60	60
9						
10	Intermediate calcs:					
11	44.10	v(0) = approach speed (ft/sec)	29.33	58.67	44.00	29.33
12	66.15	x(r) = reaction distance (feet)	44.00	88.00	44.00	29.33
13	4.41	t(s) = time to stop (sec)	2.93	5.87	4.40	2.93
14	97.24	x(b) = braking distance (feet)	43.02	172.09	96.80	43.02
15	163.39	x(s) = stopping distance (feet)	87.02	260.09	140.80	72.36
16	176.40	x(Y) = dist. traveled during yellow (feet)	117.33	234.67	176.00	148.36
17	16.00	L = length of vehicle (feet), if desired	16	16	16	16
18	239.39	x(c) = distance to clear intersection (feet)	163.02	336.09	216.80	148.36
19						
20	Results:					
21	62.99	x(DZ) = [x(s) + W + L] − x(Y)	45.69	101.42	40.80	0.00
22		If x(DZ) > 0, x(DZ) = length of Dilemma zone (feet)				

FIGURE 8.11

Spreadsheet for dilemma zone sensitivity analysis. *Note:* Changes from the original input values are shown in **larger, bold font**.

t_Y = 4 seconds be maintained, it would be acceptable to introduce an all-red clearance interval of about 1 second.

Example 8.5

Using the vehicle and driver described in Example 8.4:

A. Determine the length of an all-red clearance interval needed for the vehicle to clear the intersection.

B. If the all-red clearance interval is 2 seconds long, determine the range of approach speeds at which a vehicle can clear the intersection.

Solutions to Example 8.5

A. In Example 8.4, the stopping distance with v_0 = 30 mph was 163.39 feet.

$$t_{allRed} = \text{time to clear intersection } t_C - \text{length of yellow phase } t_Y$$

From the second of Equations 8.3 and the data in Column A of Figure 8.11, the distance to clear the intersection $X_C = D + L + W = 163.39 + 16 + 60 = 239.39$ feet.

$$\text{Time to clear the intersection } t_C = \frac{239.39 \text{ ft}}{30 \text{ mph} * 1.47 \text{ fps/mph}} = 5.43 \text{ seconds.}$$

$$t_{allRed} = t_C - t_Y = 5.43 - 4.0 = 1.43 \text{ seconds}$$

B. Using time to stop $t_s = \dfrac{v_0}{d_{brake}}$ in $X_s = v_0 t_r + v_0 t_s - \dfrac{1}{2} d_{brake} t_s^2$ produces

$$X_s = v_0 t_r + v_0 \frac{v_0}{d_{brake}} - \frac{1}{2} d_{brake} \frac{v_0^2}{d_{brake}^2} = v_0 t_r + \frac{v_0^2}{d_{brake}} - \frac{v_0^2}{2 * d_{brake}} = v_0 t_r + \frac{v_0^2}{2 * d_{brake}}.$$

$$X_s = (1.5 * v_0) + \frac{v_0^2}{2 * 10}.$$

The distance to clear the intersection equals the last chance to stop when

$$X_C = X_s + L + W = (1.5 * v_0) + \frac{v_0^2}{2 * 10} + 16 + 60 = 0.05 v_0^2 + 1.5 v_0 + 76.$$

When $t_{allRed} = 2.0$ seconds, this distance is

$$X_C = (t_Y + t_{allRed}) * v_0 = (4 + 2) * v_0 = 6 v_0.$$

Equating the two expressions for X_C and combining terms, we have

$$0.05 \, v_0^2 - 4.5 \, v_0 + 76 = 0.$$

The solution is

$$v_0 = 67.5 \text{ fps} = 45.9 \text{ mph}.$$

At this speed or higher, the vehicle can clear the intersection.

THINK ABOUT IT

If John driving at an approach speed less than the v_0 value found in Part B of Example 8.5, will a dilemma zone exist?

8.3 ANALYSIS OF SIGNALIZED INTERSECTIONS

Introduction

It is a little-known fact that motorists can drive through downtown Mythaca on eastbound Acorn Avenue and never have to stop for a red light (see Figure 8.12.). The reason that this fact is such a secret is that it only works if eastbound drivers maintain a constant speed of a little less than 20 mph. This section introduces ways to summarize a signal-timing plan and analyze how well a series of signals are coordinated. In addition, actuated traffic signals are explained, and a method to estimate the delay caused by traffic signals is presented.

8.3.1 Traffic Signal Timing

The standard traffic signal cycle consists of green, yellow, and red phases, but how long should each phase last? As the green phase on one street is lengthened, less time out of

FIGURE 8.12
The light is green at 3rd Street, 4th Street, 5th Street, and 9th Street. Photo: Jon D. Fricker.

the cycle is available for the green phase on the cross street. The respective green phase durations depend in large part on the traffic flows on each intersection approach and the capacity of the intersection to allow certain traffic movements to take place. If a large number of drivers want to turn left, a separate left-turn-only phase may be necessary. At this point, some basic definitions are in order.

- *Phase.* The indication (green, yellow, red, arrow, etc.) that a driver sees on the traffic signal as (s)he approaches the intersection.
- *Cycle.* The length of time between consecutive occurrences of the same point in the green-yellow-red sequence of phases. A common point to use is the start of the green phase on the major street approach. If the time from one "start of green" to the next "start of green" is constant, the signal timing is "pretimed" or "fixed timed." If the signal timing varies in response to traffic flows detected by loops in the pavement, the signal is "actuated."
- *Interval.* At any time during a signal's cycle, the phase in effect for each approach (and turning movement, if present) can be noted. This combination of phases is known as the *interval*. The simplest example is when all approaches show a red indication. This is called the *all-red clearance interval.* When any of the phases that define the current interval change, a new interval has begun.

A simple version of a traffic signal-timing form is shown in Table 8.4. (The timing plans used by traffic engineers use a much more detailed and complex format.) Note that there are nine intervals at the intersection being studied. The first interval consists of a green phase seen by both left-turning (LT) and through (TH) drivers on Wildcat Avenue's NB (northbound) approach. Drivers on SB Wildcat must wait for the northbound LT green phase to end. The SB drivers on Wildcat, and all drivers on Coliseum Avenue (the cross street), see a red phase during interval 1. Interval 2 begins when the LT green on NB Wildcat changes to yellow. When this LT yellow goes to red, interval 3 begins. The form in Table 8.4 has one row for each interval. After the duration of each *phase* has been entered into the form from observations, the length of each *interval* can be deduced. (It is a good idea to measure the duration of each phase at least three

TABLE 8.4   Traffic Signal-Timing Form

Observer(s): RKW and JDF

Day & Date: Thurs. 13 Nov. 2002   Start Time: 7:30 AM   End Time: 8:30 AM

	Wildcat Ave.			Coliseum Avenue	
Interval	NB LT	NB TH/RT	SB Vehs	EB Vehs	WB Vehs
1	G = 14	G = 38	R = 19	R = 44	R = 44
2	Y = 4				
3	R = 57				
4			G = 19		
5		Y = 4	Y = 4		
6		R = 33	R = 33		
7				G = 26	G = 26
8				Y = 4	Y = 4
9				R = 1	R = 1

Cycle length: 75 seconds.

times. One method is to use a stopwatch with a sufficient number of memories, entering a "split time" each time a new phase begins. Before computing the mean duration of each phase, verify that the cycle length is constant.) Example 8.6 illustrates.

Example 8.6:   Traffic Signal Timing Form

Each of the phase durations in Table 8.4 is the mean of at least three observations made at the intersection, rounded off to the nearest second. It has been verified that the signal is pretimed, with a cycle length of 75 seconds. What is the length of each interval?

Solution to Example 8.6   Interval 1 is 14 seconds long, during which time NB LT and NB TH movements are *protected*. All other movements during interval 1 face a red phase. A *protected movement* is one in which no conflicting movement is legally possible. For example, if SB TH and NB LT drivers both see a green phase, the SB TH drivers have the priority, but NB LT drivers are *permitted* to complete their movement when it is safe to do so. Interval 2 simply ends the protected NB LT by showing a yellow phase 4 seconds long. NB TH continues as in interval 1. Interval 3 ends the red phase seen by SB TH drivers. All drivers except NB TH are facing a red phase. The duration of interval 3 is $19 - (14 + 4) = 1$ second. This calculation is possible because each interval is defined by the end of each one phase. Interval 3 ended 19 seconds into the signal cycle. Interval 2 ended $14 + 4 = 18$ seconds into the cycle. Interval 4 is 19 seconds long, after which both NB and SB traffic will see the start of yellow. Interval 5 is distinguished by that yellow phase, lasting 4 seconds. Interval 6 is the last interval in which EB and WB traffic will see red for a while. In fact, interval 6 is an *all-red clearance interval*—all drivers face a red phase. Intervals 7, 8, and 9 are 26, 4, and 1 second long, respectively. Interval 9 is another *all-red clearance interval*. In the case described by Table 8.4, the interval after interval 9 is interval 1, and a new signal cycle begins.

8.3.2   Time-Space Diagrams

If a traffic engineer observes long queues on one or more approaches to a signalized intersection, the traffic signal-timing forms presented in the previous section can help the engineer to look for ways to reduce the backups. The solution may involve taking green time away from one or more approaches and giving it to the traffic movements experiencing the longest delays. In fact, software exists to find a new signal-timing plan that minimizes total delay at an intersection, given the volumes for each traffic movement. A separate optimal timing plan can be generated for each distinct time of day. For example, an arterial may carry heavier traffic toward employment centers in the morning, but the directions of heavy flows are reversed in the afternoon. The amount of time dedicated to protected left turns may vary by direction and time of day. In the hours between the AM and PM peak periods, more green time may be allocated to the cross street. However, this optimization strategy applies only to one intersection. On major arterial streets, attention must be given to coordinating the signal timing plans at each intersection along the arterial. When signals are coordinated, a platoon of vehicles discharged from an upstream green light will not be stopped by a downstream red light. Ideally, platoons traveling along major arterials at reasonable speeds should be able to pass through a series of green lights. This would minimize delay and the emissions caused by stop-and-go driving. Traffic simulation software packages are available to help the traffic engineer establish a coordinated system of traffic signals. *Time-space diagrams (TSD)* illustrate the factors and the complications involved in trying to coordinate traffic signals.

Figure 8.13 tracks a vehicle's journey along a street in terms of time (horizontal axis) and space (vertical axis). During the time period marked "(a)" in Figure 8.13, the vehicle is waiting at a red light. The "clock" for our time-space analysis is running, but the vehicle is not covering any distance. The solid line marked "(b)" indicates that the vehicle covers a certain distance (the vertical dimension in the figure) during a certain amount of time (the horizontal dimension). The slope of this line is the speed of the vehicle. When the vehicle encounters a red light at "(c)," no distance is being covered, but time continues. The solid "vehicle trajectory" line "(c)" has zero slope. Once the light turns green, the vehicle can continue along a line "(d)," perhaps at a different speed. In Figure 8.13, a dashed line represents the time-space trajectory of a second vehicle. It passes through the first intersection (distance = 0) well behind the first vehicle, but catches it at the red light at the second intersection. The dashed-line vehicle waits behind the solid-line vehicle until the light turns green and then (according to the figure) follows behind the solid-line vehicle at the same speed.

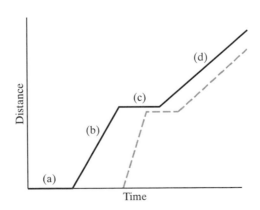

FIGURE 8.13
Vehicle trajectories.

It would have been nice for the drivers of both vehicles if the signal at point (c) in Figure 8.13 had been green. A time-space diagram allows a traffic engineer to plot existing or proposed signal-timing plans for a corridor, superimpose expected vehicle trajectories over the signal timings, and evaluate and adjust the timings to remove obvious delays and improve traffic flow. Here is how to draw a TSD:

1. On the left margin of the TSD, draw a plan of the arterial street section *to scale.* (See Figure 8.14) To be precise, the cross street widths should be defined by the location of the stop lines on the arterial.

2. Establish a "time frame" for the TSD that is sufficient for a vehicle to move through the corridor being studied. (*Hints*: A. Do a rough sketch of the TSD, see how the data you collect fit into the TSD, and be prepared to draw a new one. B. Collect your signal-timing data by using real clock times, not stopwatch times. This is explained in an example problem below.)

3. Check your data to verify that all the signals on the arterial are pretimed (i.e., all phases have constant duration). If any have variable phase durations, a TSD is of almost no value in signal coordination. In addition, the cycle lengths at each signalized intersection must be the same. If they are not, the solid bars that represent red phases "float" through the diagram, making it of no use in coordinating signals.

4. Determine which signal in the corridor is the *master intersection*. Normally, this is the intersection at which the master signal controller for the corridor is (or will be) located.

5. For the master signal:

   a. Find its location along the corridor and then draw a hollow bar that is as wide (vertically in the TSD) as the distance between the stop lines in opposing directions and as long (horizontally in the TSD) as the duration of the green + yellow phases for through traffic at that signal. (*Notes*: (i) Instead of

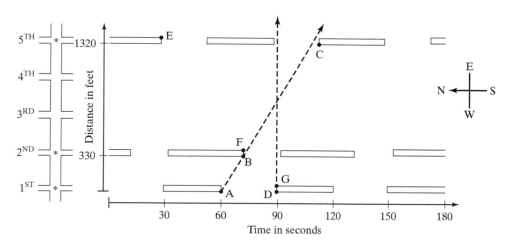

FIGURE 8.14
Time-space diagram for traffic signal timing.

drawing a hollow band, leaving the green + yellow phases as blank space between solid red phase bars is acceptable. In fact, it keeps the TSD less cluttered. (ii) For simplicity, treat yellow as part of the green phase. Besides, most
drivers do anyway. (iii) It is presumed that the objective is to reduce delays
for traffic moving <u>through</u> the corridor. Variations for other dominant traffic
paths are possible.)

b. At the end of the green + yellow phases at the master intersection, draw a
solid bar that is as long (horizontally in the TSD) as the duration of the red
light at that signal.

c. Repeat drawing the hollow and solid bars in the TSD for the master intersection until the end of the time frame (see step 2) is reached.

6. For each other signalized intersection in the corridor, find its location along the
corridor and:

a. Determine its *offset* with respect to the master signal. The signal *offset* is the
amount of time between the start of green at the master signal and the *next*
start of green at the particular signal. This is known as the *green offset*.

b. Beginning at the green offset value, draw a hollow bar as wide as the distance
between stop lines and as long as the green + yellow phases. Then draw a
solid bar that is as long as the duration of the red light at that signal.

c. Repeat drawing the hollow and solid bars in the TSD for each signal until the
end of the time frame (see step 2) is reached.

7. For each signal, fill in any phases (or parts of phases) that take place *before* the
green offset. This will allow the impact on traffic traveling in either direction
through the corridor to be analyzed.

Your completed time-space diagram should look much like Figure 8.14.

A common error in using a TSD is how the vehicle trajectories are drawn with respect to the red light bar. The key is the location of the stop line for the given direction.
In Figure 8.15, the SB dashed car has actually entered the intersection on red. The signal turns green as the SB dashed car *leaves* the intersection. The SB solid car enters the
intersection at the start of green. The NB dashed car is entering the intersection during

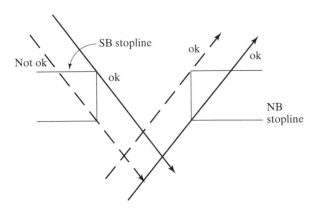

FIGURE 8.15
Important TSD detail.

green and is leaving the intersection as the signal turns red. The NB solid vehicle is entering the intersection as the signal turns red. The lesson: Establish rules that represent driver behavior and then draw vehicle trajectories according to those rules. In the example that follows, the SB dashed vehicle is illegal, but the other three vehicle movements are permitted.

Example 8.7:   Drawing a Time-Space Diagram

After numerous complaints about the signal progression (or lack of it) on Acorn Avenue from 1st to 5th Streets have gone unheeded, some citizens decide to study the situation themselves. They collect data on current signal timings and put them into a data file with the format shown in Table 8.5. (The citizens used their wristwatches and recorded events to the nearest second. They could have asked the city traffic engineer for the signal-timing plans, but (a) the plans might be too complex for untrained citizens to read and (b) they wanted to record what the *actual* timings are.)

On this section of Acorn Avenue, the cross streets are 50 feet wide and 330 feet apart, but only 1st, 2nd, and 5th Streets are signalized. The westernmost signal is at 1st Street. The speed limit on Acorn is 35 mph. The "04" appearing before "First Street" in Table 8.5 means that four cycles were observed at that signal. The first three fields contain the clock times (Hr Min Sec) at which the signal turned green. The last three fields in each data row contain the clock times for the start of red. Thus, the signal on Acorn at 1st Street turned green at 9:57:51 AM and turned red at 9:58:20 AM.

A. Determine the length of the G + Y and Red phases at each signal. Are the signals pretimed? What is the cycle length for each signal?

B. Is it possible for a vehicle to maintain a constant speed and proceed through the three signals on Acorn? If so, in which direction(s) and at what speed(s)?

Solutions to Example 8.7

A. During the four cycles observed at 1st Street, the G + Y phases lasted approximately 29, 29, 29, and 30 seconds. The slight variation is probably because the citizens used wristwatches, not stopwatches. Use G + Y = 29 seconds. The first red phase started

---

TABLE 8.5   Signal Timing Data for Acorn Avenue

04 First Street
09 57 51 09 58 20
09 58 51 09 59 20
09 59 50 10 00 19
10 00 50 10 01 20
04 Second Street
10 03 02 10 03 23
10 04 02 10 04 23
10 05 02 10 05 23
10 06 02 10 06 23
03 Fifth Street
10 09 18 10 09 44
10 10 18 10 10 43
10 11 18 10 11 43

at 9:58:20 AM and ended at 9:58:51 AM—a duration of 31 seconds. The next two red phases were measured as 30 and 31 seconds. Use 31 seconds. The cycle length is 31 + 29 = 60 seconds. Enter these data into Table 8.6 for 1st Street and carry out similar calculations for 2nd and 5th Streets. Because the phases are constant (within measurement error), the signals appear to be pretimed. All the cycles are of the same length, so signal coordination is possible.

B. A TSD allows us to look at the quality of the signal progression in either direction. One necessary ingredient is the offset of the signalized intersections with respect to the master signal. Let us adopt 1st Street as the master signal. The master signal has been observed to turn green about 50 or 51 seconds into each minute. At 2nd Street, green starts about 2 seconds into each minute. Thus, for any given minute, you can expect the signal at 2nd Street to turn green about 11 or 12 seconds after the green starts at 1st Street. A green offset of 11.5 seconds is entered into Table 8.6 for 2nd Street. Likewise, the 5th Street signal turns green either 27 or 28 seconds after the 1st Street green. Use 27.5 seconds as the green offset. We are now ready to draw the TSD for this section of Acorn Avenue. A good way to organize the drawing of a TSD is to determine the coordinates of each event to be included in the TSD. For example, because Acorn at 1st is the master intersection at the west end of the corridor being studied, it becomes the point of reference or "origin" for the time-distance coordinates used to draw a TSD. In this example:

- The stopline for EB traffic on Acorn at 1st Street is the point where distance = 0.
- The start of green for EB traffic on Acorn at 1st Street is the point where time = 0.

The time and distance coordinates for start-of-green and start-of-red events at each intersection for a time frame 3 minutes long have been computed and entered into Table 8.7. Plotting these points and using them as the corners of the bars to represent red phases produces the TSD shown in Figure 8.14. The effort to draw the TSD is worth it. It is clear that an EB vehicle could avoid red lights at all three signalized intersections, if it could maintain the right speed. One example is the vehicle represented by the dashed line starting at Point A (coordinates

TABLE 8.6   Summary of Data for Acorn Avenue

Cross Street	G + Y (sec)	Red (sec)	Cycle (sec)	Green Offset (sec)
1st (master)	29	31	60	0
2nd	21	39	60	11.5
5th	25	35	60	27.5

TABLE 8.7   Event Coordinates for Acorn Avenue TSD

Approach	Distance	G1	R1	G2	R2	G3	R3
EB at 1st	0	0	29	60	89	120	149
WB at 1st	50	0	29	60	89	120	149
EB at 2nd	330	11.5	32.5	71.5	92.5	131.5	152.5
WB at 2nd	380	11.5	32.5	71.5	92.5	131.5	152.5
EB at 5th	1320	27.5	52.5	87.5	112.5	147.5	172.5
WB at 5th	1370	27.5	52.5	87.5	112.5	147.5	172.5

time = 60, distance = 0) in Figure 8.14. The dashed line is defined by the time-distance coordinates (60,0) at 1st Street and Point B (71.5,330) at 2nd Street. The corresponding speed for the dashed line is

$$\frac{(330 - 0)\text{ft}}{(71.5 - 60)\text{sec}} = 28.7 \text{ fps} = 19.5 \text{ mph}.$$

Part B asked for a *range* of speeds at which a vehicle could avoid all red lights. What this means is finding (1) the "flattest" vehicle trajectory that can fit through all the G + Y phases in the TSD and (2) the "steepest" vehicle trajectory that does not exceed the speed limit. In the Acorn Avenue TSD, the flattest non-red trajectory is from the start of green at 1st Street until the end of G + Y (or start of red) at 5th Street (Point C in Figure 8.14). The corresponding speed is

$$\frac{(1320 - 0)\text{ft}}{(112.5 - 60)\text{sec}} = 21.1 \text{ fps} = 14.3 \text{ mph}.$$

The steepest no-red vehicle trajectory starts at the end of G + Y at 1st Street (Point D). A vertical line (infinite speed) will not encounter a red signal. Therefore, the acceptable "green wave" speed range is 14.3 to 35 mph (the speed limit) for EB traffic.

In the WB direction, a straight line can be drawn through Points E and F that will reach 1st Street during a green phase. Therefore, WB traffic on Acorn Avenue will also enjoy a "green wave" if the corresponding speeds are legal.

The method for drawing a TSD used in this section involved several simplifications. One was the rule that all drivers would enter an intersection on green or yellow but would not enter an intersection when the signal was showing red. A greater simplification is that vehicles assume a constant speed S instantaneously—no acceleration from zero to S is accounted for. Likewise, the deceleration from speed S to zero is assumed to take place immediately. Once the fundamentals of drawing a TSD are learned, an analyst can incorporate a non-linear trajectory that reflects vehicle acceleration and deceleration.

THINK ABOUT IT

Using a facsimile of Figure 8.14, show what the vehicle trajectories would look like for starting and stopping vehicles with realistic acceleration and deceleration characteristics. In what way would these details improve the analysis in Example 8.7? Besides the extra effort to draw non-linear trajectories, in what way do these details make the analysis more complicated?

8.3.3   Actuated Traffic Signals

Pretimed signals are easy to understand and analyze, but they can be an inefficient way to use the available intersection capacity. Each intersection approach has a *saturation flow rate*, which is the number of vehicles per hour that could enter the intersection if the signal were always green. Of course, the various approach directions must share the amount of time out of each cycle that is devoted to green phases. As more green time is allocated to, say, the EB and WB traffic, there is less for the other approaches to use. If the EB and WB traffic clears the intersection well before its pretimed green phase is finished, and traffic on other approaches are waiting, the result is unnecessary delay. Likewise, pretimed intervals for pedestrian use that are not fully used are also wasteful. The alternative is *demand-actuated signals*.

Actuated signals adjust signal timings in response to information provided to the traffic signal controller by detectors. Normally, the detectors are inductive loops placed in the pavement on the approaches to an intersection. However, video technologies are becoming more common. (See section 2.1.4.)

THINK ABOUT IT

Without measuring the length of phases and signal cycles at a signalized intersection, what is a good (but not perfectly reliable) way for an observer to determine whether the signals are actuated?

If only the minor street approaches are equipped with detectors, the major street will be given a green indication until a vehicle is detected on the minor street. This is known as *semiactuated* control (See Figure 8.16.). *Fully actuated signals* rely on detectors on all approaches. The typical data gathered from detectors are:

- The presence of a vehicle in the detector's "field of view"
- The time at which that presence was detected.

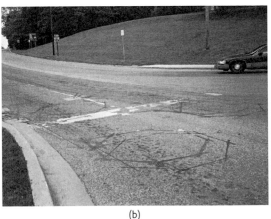

(a)   (b)

FIGURE 8.16
(a) Semiactuated intersection. (b) At same intersection, the loop beyond the stop bar detects each right turn on red. Photos: Jon D. Fricker.

The traffic signal controller translates these data into information needed to apply some rules for traffic signal control. Examples of the information needed are:

- Presence time. How long has a vehicle been *continuously* detected by a detector? The detector for this purpose is probably located at the stop line for the approach.

THINK ABOUT IT

How does the signal controller convert detector data into presence time information?

- Passage time: How long has it been since the most recent vehicle has been detected? The detectors for this purpose are probably located 100 feet or more "upstream" of the approach's stop line.

THINK ABOUT IT

How does the signal controller convert detector data into passage time information?

To illustrate how such information can be combined with some rules to produce more efficient signal timing for an intersection, consider the case of the intersection shown in Figure 8.17. Stadium Avenue is the major street. University Street (one-way

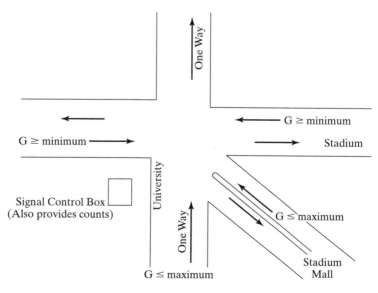

FIGURE 8.17
Intersection with actuated signal control.

NB) is considered the minor street, although it can carry heavy traffic at times. Stadium Mall is a dead end street with few vehicles, except at peak periods.

In the box below, rules convert the information generated by detectors so that (a) Stadium Avenue keeps the green whenever possible, (b) University Street gets green until there is too great a gap between consecutive vehicles or it reaches its limit, and (c) Stadium Mall gets a little green only when necessary.

TRAFFIC SIGNAL LOGIC AT STADIUM AND UNIVERSITY STREETS

0. Initial status = G on Stadium.

1. Wait until Stadium has had its min G, then go to step 2.

2. If no vehicle is present on University approach, go to step 7.

3. Give G to University for "min G_U" seconds.

4. (Optional) If passage time gap on University is too large, go to step 7.

5. If current G on University has exceeded its "max G_U," go to step 7.

6. Go to step 4.

7. If no vehicle is present on Stadium Mall approach, go to step 11.

8. Give G to Stadium Mall for "min G_M" seconds (optional).

9. If passage time gap on Stadium Mall is too large, go to step 11. If current G on Stadium Mall has exceeded its "min G_M," go to step 11.

10. Go to Step 9.

11. Give G to Stadium. If LT loop is "active" and actuated, the first G_{LA} seconds of G_S include a leading protected LT phase. Go to step 1.

Notes:

A. G = duration of green time, S = Stadium St, U = University St, M = Stadium Mall

B. LA = left turn arrow for WB Stadium (which may not always be active)

C. "Give G" implies standard process of Y, all-red, etc., as appropriate.

D. min G_{LA} = 5 with passage time (PT) = 1; min G_S = 20 with PT = 5

E. min G_U = 10 and min G_M = 7 with PT = 2; all-red = 1.5

8.3.4 Estimating Delays at Intersections

Of the several commonly used definitions of intersection delay, *stopped delay* has been used most often. It is easy to measure and has been used by the *Highway Capacity Manual* as a basis for intersection *level of service*. A manual procedure for measuring stopped delay at an intersection (HCM, 2000) is based on a set of observations made at regular intervals. As the field data sheet in Figure 8.18 indicates, a count V_s is made of the number of

INTERSECTION CONTROL DELAY WORKSHEET

General Information				Site Information					
Analyst(s): MG, AN				Approach street: Coliseum Avenue					
Day and Date: Thurs. 14 Nov 2002				Cross street: Wildcat Avenue					
Time: 5:30 PM to 5:40 PM				Approach direction: WB					
Weather: clear, 45°				Lane(s): Left, Through, Right					
Jurisdiction: City of Mythaca				Area Type: ___CBD _X_ Other					

$\dfrac{\text{sec}}{\text{min}}$	Number of stopped vehicles, V_s								
	0	15	30	45					
0	0	4	7	8					
1	8	2	0	2					
2	2	4	5	6					
3	7	4	0	1					
4	2	3	5	7					
5	8	0	0	1					
6	4	6	5	2					
7	0	1	1	3					
8	4	6	8	0					
9	1	2	3	4					
Totals:	36	32	34	34					

Computations	
Total vehicles in queue, ΣV_s = 136	
Total approach volume, V = 41	Average delay $= \dfrac{\Sigma V_s * I}{V} = $ 49.8 sec/veh
Interval, I = 15 seconds	

FIGURE 8.18

Form for recording the number of stopped vehicles and total approach volume. *Source:* Based on HCM 2000, Appendix 16A.

vehicles *stopped* on a given approach each 15 seconds. The assumption is that each vehicle counted in this way will, on the average, wait 15 seconds. If a particular vehicle is counted in consecutive time intervals, it contributes more to the total time spent waiting by vehicles on the approach. The only other quantity is the total approach volume V during the time the stopped vehicles are counted. This quantity counts each approaching vehicle once, whereas the V_s value may include some vehicles zero, one, or more times, depending on whether they were stopped on the approach when an interval count was taken. In Figure 8.18, the total number of stopped vehicles observed during the 10-minute study period was 136. The total approach volume during those 10 minutes was 41. This is evidence that many vehicles were counted as stopped more than once. Because each stopped vehicle contributes 15 seconds to the total intersection stopped delay, the total delay is

$$136 \text{ veh} * 15 \text{ sec/veh} = 2040 \text{ sec.}$$

Therefore, the average stopped delay is

$$\frac{2040 \text{ sec}}{41 \text{ veh}} = 49.8 \text{ sec/veh}$$

Variations and refinements of this basic technique are given in Appendix 16A of the Highway Capacity Manual (HCM, 2000). A critique of the method is given in Mousa (2002).

THINK ABOUT IT

Does the length of the time interval between V_s counts always have to be 15 seconds? If another interval length can be chosen,

A. How would the data collection and calculations need to be modified?
B. How would you decide which is the best interval length to use?

SUMMARY

The chapter began with the description of a rural intersection of two county roads. A recent rash of collisions indicates that the intersection, with stop signs controlling two of the four approaches, needs to be reexamined. New residential developments and the corresponding growth in enrollment at the nearby high school have caused increased traffic at the intersection. Sight distances on the approaches may vary by time of year, if tall crops are grown near the intersection. The critical approach speed analysis described in this chapter can be applied to such a case. The increased traffic flows—especially at peak periods—may have reduced the number of acceptable gaps on the main county road, causing impatience and risky behavior on the part of drivers on the minor road. The gap acceptance analysis covered in this chapter may help diagnose such a problem. It may be time for the first signalized intersection in this part of the county, if the conditions at the intersection pass the traffic signal warrant analysis demonstrated in this chapter. If a signal is installed, the signal timings should be set so a driver can make a decision whether to stop or continue upon seeing the yellow signal phase. Creating a dilemma zone should be avoided. Finally, a simple method to estimate average stop delay at the intersection can be carried out. The methods contained in this chapter are examples of techniques traffic engineers can use to create intersections that are safe and efficient.

ABBREVIATIONS AND NOTATION

a_{acc}	acceleration capability of a vehicle (usually a constant in this class)
A, a, a', a''	dimensions on major street lane closest to minor street to determine CAS
B, b, b', b''	dimensions on left side of minor street to determine CAS
C, c, c', c''	dimensions on right side of minor street to determine CAS
CAS	critical approach speed
D, d, d', d''	dimensions on major street lane nearest centerline on lane farthest from minor street to determine CAS
D_{br}	braking distance
EB approach	traffic approaching going east
LA	LH vehicle accepts the gap
LH	left-turning vehicle at the stop line or head of the minor street queue
MN	vehicle on major street in lane nearest the minor street intersection
MF	vehicle on major street in far lane from the minor street intersection
MUCTD	the Manual for Uniform Traffic Control Devices
NB approach	traffic approaching going north
RA	RH vehicle accepts the gap
RH	right-turning vehicle at the stop line or head of the minor street queue
SB approach	traffic approaching going south
TSD	time-space diagram
t_s	time to stop, in conjunction with braking
t_r	reaction time when driving a car, usually in conjunction with braking
t_Y	duration of the caution or amber or yellow signal light
t_{red}	time under the red signal light
WB approach	traffic approaching going west
v_o	velocity at the beginning of vehicle changes in speed
$v_f,$	final velocity of the vehicle
W	width of the intersection from stop line to stop line
X_s	distance to stop

GLOSSARY

Accepted gap or accepted lag: Gap or lag that the driver of a minor-street vehicle uses to move into the major street.

Actuated traffic signal: A signal timing whose phase durations are influenced by the presence or absence of vehicles on some or all of the intersection approaches.

All-red clearance interval: A time when drivers on all intersection approaches see a red signal.

Critical gap: The minimum size gap that a particular driver will accept.

Cycle: The length of time between consecutive occurrences of the same point in the green-yellow-red sequence of phases.

Fixed time signal: See pretimed traffic signal.

Gap: Time elapsed between the rear bumper of one vehicle and the front bumper of the following vehicle passing a given point. Gap is thus always smaller than headway.

Green wave: The situation when signal offsets are such that a vehicle traveling at a reasonable speed in a specified direction can encounter only green lights at a sequence of intersections.

Headway: Time elapsed between the front bumper of one vehicle and the front bumper of the following vehicle passing a given point.

Interval (traffic signal):   A period when none of the signal phases seen at a signalized intersection changes.

Lag: Time elapsed between the arrival of a minor-street vehicle ready to move into the major street and the arrival of the front bumper of the next vehicle in the major traffic stream. It may be that drivers treat lags differently than they treat gaps.

Master intersection: The intersection whose signal timings are used as the point of reference for such measurements as offsets.

Offset: The time of an event in a signal cycle with respect to that same event occurring at the master intersection (e.g., the start of the green phase).

Permitted phase: A green phase during which a given movement (e.g., a left turn) can be made by a driver only when gaps in conflicting traffic are acceptable to the driver.

Phase (traffic signal):   The indication (red, green, etc.) given to an intersection approach.

Platoons: Groups of vehicles approaching an intersection.

Pretimed traffic signal: A signal whose intervals are fixed in duration.

Protected phase: A time during the signal cycle in which a movement can be made when all conflicting movements are prohibited by red phases.

Rejected gap or rejected lag: Gap or lag that the driver of a minor-street vehicle waiting to enter the major street does not accept.

Saturation flow rate: The maximum number of vehicles that can pass through an intersection approach in an hour, assuming that the signal is always green.

Untested gap: No minor street vehicle was present.

REFERENCES

[1] American Association of State Highway and Transportation Officials, *A Policy on the Geometric Design of Highways and Streets*, Washington, DC, 2001.

[2] Federal Highway Administration, *Traffic Control Devices Handbook*, U.S. Department of Transportation, 1983.

[3] Federal Highway Administration, *Manual on Uniform Traffic Control Devices for Streets and Highways*, U.S. Department of Transportation, 2003. Latest version available online at http://mutcd.fhwa.dot.gov.

[4] Fricker, Jon D., Marlene Gutierrez, and David Moffett, "Gap Acceptance and Wait Time at Unsignalized Intersections," Proceedings of an International Workshop, *Intersections without Traffic Signals II*, 18–19 July 1991, Bochum, Germany, p. 297–307.

[5] Harwood, D.W., J.M. Mason, and R.E. Brydia, "Design policies for sight distance at stop-controlled intersections based on gap acceptance," *Transportation Research Part C*, December 1998, p. 199–216.

[6] Miller, Alan J., "Estimators of Gap-Acceptance Parameters," *Proceedings of the Fifth International Symposium on the Theory of Traffic Flow and Transportation*, Gordon Newell, ed., Berkeley CA, 16–18 June 1971, p. 215–235.

[7] Mousa, Ragab M., "Accuracy of Stopped Delay Method Measured by Stopped-Vehicle Counts Method," *Journal of Transportation Engineering*, American Society of Civil Engineers, September/October 2002, p. 439–446.

[8] Robertson, H. Douglas, editor, *Manual of Transportation Engineering Studies*, Prentice-Hall, 1994.

EXERCISES FOR CHAPTER 8: DESIGN OF INTERSECTIONS FOR SAFETY AND EFFICIENCY

Analysis of Non-Signalized Intersections

8.1 Estimating Critical Gap. The Mythaca County Highway Engineer's office has prepared a spreadsheet that contains gap acceptance data. The spreadsheet will be emailed to you, be available by FTP, or on a CD-ROM provided. The first 12 lines of data are shown below.

Arrive	Dir	TTW	TIQ	DAG	Nrej	DRG	TSR
41	R	5	0	11.7	1	4.6	0
48.9	R	0	0	8.4	0	0	0
62.4	R	12	0	6.6	4	1.4	0
0	R	0	0	0	0	4.1	1.4
0	R	0	0	0	0	3.9	5.5
0	R	0	0	0	0	2.2	9.4
65	R	16.5	9.4	12.1	1	6.2	0
67.3	R	15.7	15.7	9.7	0	0	0
68.4	R	28.5	18.4	4.3	2	5.9	0
0	R	0	0	0	0	3	5.9
70.2	R	30.1	26.7	14	1	3.1	0

The spreadsheet's column headings have the following meanings:

(1) Arrive = time vehicle entered minor street queue

(2) Dir = turn direction (left or right) of minor street vehicle

(3) TTW = total time waited on minor street

(4) TIQ = time spent in queue

(5) DAG = duration of gap accepted by the minor street vehicle

(6) Nrej = number of gaps rejected by the minor street vehicle

(7) DRG = duration of i-th rejected gap (DRG$(i), i = 0$, Nrej)

(8) TSR = wait time from *H until start of i-th rejected gap.

Using the DAG and DRG columns in the spreadsheet, estimate the critical gap for the vehicles on the minor street. Use the percentage version of the method illustrated in Figure 8.3. Explain the steps you took to organize the data. A clear plot of the data—either by hand or using spreadsheet software—is expected.

8.2 Critical Approach Speed. Example 8.2 just happens to describe the situation at Eighth Ave. and Edgemont Drive in Mythaca. Eighth Avenue is the major street.

(a) How high must the approach speed on 8th Ave. be, so that a stop sign on Edgemont is required? Explain or show clearly how you found that speed.

(b) As shown in Example 8.2, the critical approach speed method indicates that a Yield Sign on Edgemont is the appropriate traffic control device (TCD). What if parking is no longer permitted on Edgemont? Will that change the situation as far as the best TCD is concerned?

(c) If the 85th percentile speed on 8th Avenue is 39 mph, rather than 31 mph, what TCD is appropriate?

(d) If a'' and c'' are both 16 ft in the example, what TCD is needed with the speed on 8th Avenue at 31 mph?

8.3 Critical Approach Speed. Somehow, a TCD has never been installed at the corner of Crowell and Ft. Jekk Streets in a residential neighborhood. The longtime residents have gotten used to it, but new families are wondering how safe an uncontrolled intersection can be. They contact the city. The city sends an intern to the intersection. The intern thinks the EB approach on Crowell is the one most likely to need a TCD, so he takes the following measurements:

- Ft. Jekk is 31 feet wide; Crowell is 23 feet wide.
- 34 feet from building on NW corner to curb on SB Ft. Jekk
- 20 feet from building on SW corner to curb on NB Ft. Jekk
- 21 feet from building on NW corner to curb on EB Crowell
- 18 feet from building on SW corner to curb on EB Crowell
- Speed on Ft. Jekk = 31 mph
- Parking is allowed on both streets

According to the CAS method, should a TCD be installed on EB Crowell at Ft. Jekk? If so, should it be a Stop sign or a Yield sign? Show clearly how you carried out the steps in the CAS method.

8.4 Critical Approach Speed (CAS) Analysis for Stop/Yield/No Control. An uncontrolled intersection in a relatively new industrial park near Shoridan has been the site of several vehicle collisions that may have been attributable to limited sight distance. The 85th percentile speed on the major street is 48 mph. The major street is 76 feet wide; the minor street is 48 feet wide. Parking is permitted on both sides of both streets. The distances from curb to view obstruction, using the notation of Figure 8.4, are: $a'' = 40$ feet, $b'' = 73$ feet, $c'' = 94$ feet, $d'' = 85$ feet.

(a) Show clearly your calculations of a'', b'', c'', and d''.
(b) Submit a photocopy of Figure 8.6 with the (a, b) and (c, d) points and the "anchor point" speed clearly marked.
(c) Does your CAS analysis indicate that a TCD is needed at the intersection? If so, which type? What decision rule(s) did you use?

Signal Warrants and Stopping Distance at Signalized Intersections

8.5 Traffic Signal Warrants. Repeat the analysis in Example 8.3, this time using three approach lanes on SB Russell Street. Do the results change for whether a signal is warranted?

8.6 Other Traffic Signal Warrants. The Manual of Uniform Traffic Control Devices is online at http://mutcd.fhwa.dot.gov. Go to Section 4C.03 and apply Warrant 2 Four-Hour Vehicular Volume to the data used in Example 8.3. Is a signal justified?

8.7 Dilemma Zone. Drivers on Lincoln Street approaching Douglas Street do so at 25 mph. Douglas Street is 50 feet "deep," when measured from farside curb line to nearside stop bar. The yellow phase on Lincoln is 4 seconds long. Using a reaction time of 2 seconds, a deceleration rate of 10 ft/sec², and a vehicle length of 15 feet, answer the questions below.

(a) Is there a dilemma zone?
(b) Show your calculations for x_r, x_b, x_Y, etc.
(c) Also sketch the street section and show the key distance components you calculate.
(d) Regardless of your answer to part a, let us say there is a dilemma zone of length 45 feet.
(e) How much all-red or extra yellow time will be needed to eliminate it?

8.8 Dilemma Zone. If you want to reduce or eliminate the dilemma zone simply by changing the speed at which vehicles approach the intersection, would you want to increase or decrease that speed? On what do you base your answer?

8.9 Dilemma Zone. What other evidence might you be able to collect that would help decide whether the SUV driver in Example 8.4 was properly cited for running the red light? What questions would you ask, and of whom?

Analysis of Signalized Intersections

8.10 Traffic Signal Timing.

(a) In the diagram shown below how many EB phases are there?
(b) How many intervals are there during the time that NB TH/RT traffic sees G?
(c) What is the length of the all-red clearance interval?

	Wildcat Ave.			Coliseum Avenue	
	NB LT	NB TH/RT	SB Vehs	EB Vehs	WB Vehs
1	G = 9	G = 52	R = 13	R = 56	R = 56
2	Y = 3				
3	R = 85				
4			G = 39		
5		Y = 3	Y = 3		
6		R = 42	R = 42		
7				G = 37	G = 37
8				Y = 3	Y = 3
9				R = 1	

8.11 Phases and Intervals in Traffic Signal Timing.   Using the format shown in Table 8.4 and the data in Table 8.5 to 8.7, create a traffic signal-timing form for the intersection of First and Acorn. Assume the yellow phase is 4 seconds long and that a 1-second all-red clearance interval is used.

8.12 Time-Space Diagrams.   What lessons have the citizens learned from their study in Example 8.7? How did drawing a time-space diagram help them? What recommendations should they make to the city?

8.13 Red Offsets in Time-Space Diagrams.   It is also permissible to use *red offsets*, instead of green offsets. Repeat Example 8.7 using red offsets in a table like Table 8.6. Will the red offset values be the same as the green offsets? If not, will the TSD that is based on red offsets look any different? Is it easier to draw a TSD using green offsets or red offsets? If you have a preference, explain why.

8.14 Time-Space Diagrams.   Having seen a TSD being created in Example 8.7, do you think it is easy to coordinate signal timings along a typical arterial corridor? Is it possible that a green wave in both directions on Acorn can be provided? What factors would affect coordination of signals along an arterial?

8.15 Time-Space Diagrams.   Redraw the vehicle trajectories for Example 8.7 to account for the acceleration behavior of typical passenger vehicles. Use acceleration rate a = 4.5 ft/sec/sec and deceleration rate d = 6 ft/sec/sec. How do these details affect the analysis? Under what circumstances (if any) is the extra effort worth it?

8.16 Time-Space Diagrams.   Using Example 8.7 as a starting point, what change in green offset at what intersection(s) would permit a wider "green wave" in *both* WB and EB directions along Acorn Street? Are there any reasons *not* to make those changes?

8.17 Time-Space Diagrams.   Redraw the time-space diagram in Figure 8.14, such that WB traffic on Acorn Street can have a wider "green wave" at a legal speed limit. Do this by "sliding" the red bars for one or (only if necessary) two intersections along Acorn.

(a) What is the new offset for each intersection after your change(s)?
(b) What is the range of "green wave" speeds available to drivers in each direction (EB and WB) after your improvements?

8.18 Time-Space Diagrams. Sixth Street is 660 feet east of Fifth Street along Main Street in Mythaca. Along Main Street, the (pretimed) traffic signals all have the same cycle length, and the speed limit is 30 mph. At 12:00:17 PM, the green phase for EB and WB traffic on Main Street at Fifth is observed turning green. From 12:01:05 PM until 12:01:47 PM, the signal is seen to be red for EB and WB traffic. At 12:06:35 PM, the signal for EB and WB traffic on Main at Sixth Street turns green. The signal at Sixth turns red at 12:07:23 PM. What is the off-set of the signal at Sixth Street with respect to the signal at Fifth? What speed between the two signals would you recommend for EB and WB drivers to maintain on Main Street? Show your calculations.

8.19 Timing Plan. You have just received the traffic-timing data and distance data for Vine Street (below). Vine is a main north-south thoroughfare from 12th Street to Main Street that carries much traffic. Your office is getting many angry letters from motorists, and the news media call it the "Mile of Creeping Vine." Most complaints come from those who travel at the peak pe-riods. They say that too many cycle times are used to travel along this street, even when traffic is light. They say the speed limit is a myth. It should be posted as 15 mph.

- The timing data are in seconds.
- The cycle of each signal is fixed.
- The distance between the centers of adjacent intersections with signals is given in feet.
- The intersections are 60 feet wide.

Both 12th Street and Main Street are major thoroughfares and their 30-second street light timing at Vine is set by traffic along those routes, which flows quite well. Each light has a 4-second yellow that subtracts from the green. The speed limit along Vine is posted at 40 mph.

(a) Construct a time-space diagram for the present system.
(b) What speeds can the present light settings accommodate northbound and southbound?
(c) All of the lights from 1st Street to 9th Street can be reset with green as long as 45 seconds and red as short as 10 seconds. Adjusting your offset so that the speed limit is not exceed-ed, from the end of green when entering Vine southbound at Main to the beginning of

		G_{on}	R_{on}	G_{on}	R_{on}	G_{on}	R_{on}	G_{on}	R_{on}	G_{on}
	Main Street	0	30	60	90	120	150	180	210	240
880 ft										
	1st Street	0	30	40	70	80	110	120	150	160
550 ft										
	2nd Street	−20	−5	20	35	60	75	100	115	140
550 ft										
	3rd street	0	25	40	65	80	105	120	145	160
550 ft										
	4th Street	−15	0	25	40	65	80	105	120	145
1100 ft										
	6th Street	−5	25	35	65	75	105	115	145	155
1100 ft										
	8th Street	0	25	40	65	80	105	120	145	160
550 ft										
	9th Street	−15	0	25	40	65	80	105	120	145
1650 ft										
	12th Street	−10	20	50	80	110	140	170	200	230

green at 12th Street. Set the signals first to achieve flow in this direction. Set them first for the southbound trip (Main to 12th). Then check the flow for northbound traffic. Because the lights we presently have cannot be changed for time of day, what compromises will you make?

(d) What will your final solution permit you to achieve in speed in each direction?

8.20 Traffic Signal Study.   The traffic signals on Elm Street are observed to be operating on a 75-second cycle length. Two streets that cross Elm are First and Second Streets, which are 1/12 mile apart. The signal on Elm at First is observed "turning green" at 16:24:17 EST one afternoon. A few minutes later, we reach Second Street and see the beginning of the green phase on Elm at Second at 16:29:35.

(a) If the Elm Street signal at First is the "master signal," what is the offset of the signal on Elm at Second Street?

(b) The Green + Yellow phases on Elm at First total 47 seconds. These phases total 39 seconds at Second Street. What is the slowest speed any vehicle could travel through a green light at First Street and not be stopped by the next red light at Second Street?

8.21 Time-Space Diagrams.   The green phases on EB Main Street in Mythaca at First, Second, and Third Streets, spaced 1/12 mile apart, are observed to start at the times given in the list below.

Cross Street	Green Start
First St.	17:14:11
	17:15:26
	17:16:41
Second St.	17:22:03
	17:23:18
	17:24:33
Third St.	17:33:06
	17:34:21
	17:35:36

(a) What is the cycle length at these signals?
(b) If the master controller is at First Street, what are the offsets at Second and Third Streets?
(c) If a driver EB on Main Street crosses First Street at the start of that signal's green phase and is not impeded by other traffic, what range of speeds will permit her to proceed through the green phase at Second Street without delay? *Note*: The Main Street approach gets 40 seconds of effective green (G + Y).

8.22 Actuated Traffic Signals.   Go to a signalized intersection nearby that appears to have actuated signals. What physical evidence suggests that the signals are actuated? Time the durations of some phases and cycles. Do your measurements suggest that the signals are actuated?

8.23 Actuated Traffic Signals.   Is it possible for actuated signals to appear to an observer to be pre-timed? If so, explain how this could happen.

8.24 Intersection Analysis.

(a) Give the best reason for having a *pretimed* (vs. actuated) signal at an intersection on a major arterial street.
(b) What is the best reason to use an *actuated* signal at an intersection?

8.25 Data Collection for Traffic Signal Analysis.   Describe a strategy for determining the actual "min" and "max" values for the intersection pictured in Figure 8.17 that involves only observing the operation of the signals. (*Note*: There are some "min" and "max" values listed in the box below the figure that are not shown in the figure itself.) In other words, list the tasks that would have to be carried out. List the tasks in such a way that another person could implement them for you.

8.26 Intersection Delay.   Go to a nearby signalized intersection during a time in which long back-ups can be observed. Using a form that resembles Figure 8.18, estimate the average stopped delay at the intersection. For how many minutes did you collect data? Explain how you knew when to stop collecting data. Feel free to use a time interval between V_s counts other than 15 seconds. If you can arrange it, have another student estimate average stopped delay at the same intersection at the same time, using a time interval different from yours. Comment on the degree to which your results differ.

Highway Design for Rideability (Pavement Design)

SCENARIO

Each spring in Mythaca, as the snow and ice melts, water runs along gutters and seeps into cracks in the pavement. At night, the temperature again falls below freezing, and the "freeze-thaw cycle" continues. Soon the roads become rough and potholes begin to form. (See Figure 9.1.) Will the city have the resources to keep the potholes patched? What does it take to design and construct a pavement that will last more than a few years?

CHAPTER OBJECTIVES

By the end of this chapter the student should be able to:

1. Explain the effect that soil has on pavement strength.

2. Determine the loading that vehicles exert on the highway pavement system.

3. Design a section of flexible pavement.

4. Evaluate cost trade-offs between material attributes and the thickness of pavement layers.

5. Determine the thickness of a Portland cement concrete slab.

6. Discuss pavement management problems and the need for pavement repair.

FIGURE 9.1
Pothole and alligator cracking. Photo: Jon D. Fricker.

9.1   FACTORS IN PAVEMENT DESIGN

9.1.1   History

Before 1900, when there were very few cars, travel was either by train, water, or horse-drawn carriage. Most roads were dirt tracks cut into the land. Federal road officials have stated

> At the end of the century, approximately 300 years after the first settlement, the United States could claim little distinction because of the character of its roads. As in most parts of the world, the roads were largely plain earth surfaces almost impassable in wet weather. Neither state or federal officials had undertaken to provide funds on a scale that would permit general road improvement (Public Roads Administration, 1949).

Automobiles became more available to the general public and, by 1925, the United States was entering the Automobile Age. The initial pressure for better roads was from farmers who needed to get their products to market. The mass production innovations of Henry Ford reduced the price of the automobile, enabling many workers to purchase the Model-T automobiles. As the automobile became more prevalent, so did the demand for better roads. The Great Depression (1929–1936) saw some important road links built through federal projects created to provide work for the unemployed. These federal programs were done under the auspices of the Work Projects Administration and the Civilian Conservation Corps.

The end of World War II brought new land use patterns. Low-interest housing loans provided to World War II veterans resulted in houses being built farther and farther from the urban center, where land prices were low. This "urban sprawl" increased the demand for more and better roads. The Federal Aid Highway Act of 1956 authorized the National Defense and Interstate Highway System. The interstate road system provided new, high-speed links between the major metropolitan areas of the United States. Trucks began to replace the railroad as the premier carrier of goods. This was especially true for the higher value goods that required quick, reliable delivery.

The early roads were built mostly by "seat of the pants" engineering. Increased road use necessitated a more systematic approach to highway design. The American Association of State Highway Officials (AASHO; renamed AASHTO, with the "T" standing for Transportation) conducted a series of road tests near Ottawa, Illinois, from 1958 to 1960. (See section 9.3.1.) A design procedure, known as the "layered analysis for flexible pavement design," was subsequently developed and adopted by AASHO in 1961. That procedure was revised in 1972, in 1986, and again in 1994. The AASHO tests and subsequent revisions to design methods seemed to improve pavement longevity.

Since the 1950s, there has been significant improvement in the materials that are used in road building. At the same time, trucks have gotten larger and heavier. Certain climatic conditions, especially rain, have also been observed to have a greater effect on road life than was originally expected. With highways deteriorating faster than expected, the highway design process needed to undergo significant change. With the advent of complex computer programs, finite element analysis techniques explored some of the intricacies of road stress as it interacted with the materials. Some innovative testing machines, such as that shown in Figure 9.2, provided some added test data.

FIGURE 9.2
Accelerated pavement testing machine. Photo: Purdue University and Indiana
Department of Transportation.

A joint program between the Federal Highway Administration (FHWA) and the
Transportation Research Board (TRB), called the Strategic Highway Research Program (SHRP), began in 1985 to gather data on pavement performance nationwide. [It
has since been renamed the Long-Term Pavement Performance Program (LTPP)]. The
objective of the LTPP Program was to obtain data on pavements in different climates
and with differing characteristics regarding drainage, subsoil, pavement makeup, and
so forth. On hundreds of road sections, traffic was measured, weather recorded, and
pavement degradation measured. Over the past few years, enough data have been assembled to permit the development of a new design procedure.

NCHRP Project 1-37A, "Using Mechanistic Principles to Improve Pavement Design," is expected to be completed by the end of October 2003 (NCHRP, 2003). It may
take several years for states to test the new procedure before deciding whether to
adopt it. In the meantime, the engineer can use the present pavement design methods
successfully by taking careful note of traffic conditions, material attributes, and moisture and drainage potential.

9.1.2 Two Kinds of Pavements

Two general types of pavement form the basis for most paved road designs. Each is different in the materials used and the way the pavement and the soil under it react to applied loads. (See Figure 9.3.) One type is "flexible pavement," usually composed of
asphalt concrete, which flexes under loads applied by a vehicle's tires. Repeated applications of large loads often cause deformation in both the asphalt paving materials and
the soil subgrade underneath them. Figure 9.3 also shows rigid pavement, usually composed of Portland concrete cement. The action is that of a rigid "slab" that, under the

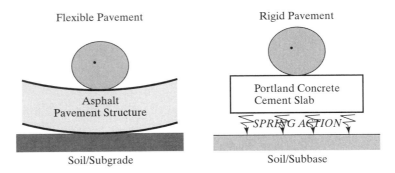

FIGURE 9.3
Wheel loading on flexible pavement and rigid pavement.

load, acts like a solid plate distributing the load evenly over the base material, which acts as a spring. Of special concern are the edges of the concrete slab. Under repeated loading, the edges can move out of place or fracture. In addition to the material atributes, primary design factors for both pavement types are the soil and the estimates of total loading from the wheels of heavy trucks.

9.1.3   Measuring Pavement Quality

This chapter is entitled "Highway Design for Rideability" because cracks, potholes, and other deformations, through repeated loads, arise in pavement and cause ride quality to be reduced. A pavement's rideability is evaluated by techniques that determine ride comfort at various speeds, up to the highway design speed. The pavement manager wants to know, for a given stretch of highway, when the ride becomes so bad that the highway is clearly in need of major maintenance or rehabilitation. The ability to evaluate an existing pavement is important, especially when it comes to deciding maintenance priorities with limited funds over many miles of roads.

Pavement failure is defined both by rideability (or comfort to the rider) and by visual observations of faults (such as cracking, rutting, and shoving) called pavement distresses. "Present Serviceability," the term used by the highway engineer, is developed on a somewhat arbitrary judgment of ride quality, on a scale from 0 to 5. The rating is called the *Present Serviceability Index (PSI)*. Flexible pavement design is typically based on, among other things, the integrated loading that occurs on the pavement until there is too much loss in serviceability or PSI.

Prior to the AASHO Road Test, there was little consensus as to the definition of pavement failure. Engineers saw pavement failure in terms of cracking, rutting, and the like. On the other hand, the motoring public thought that poor pavement was associated with a poor ride quality, especially washboarding and rutting. The Pavement Serviceability Performance Concept was developed in 1962 (Carey and Irick). The concept holds that a pavement begins its life in excellent condition. Then, as traffic loads are applied and the pavement is exposed to inclement weather, there will be deterioration of the pavement until it reaches an unacceptable level of service (primarily ride quality).

A panel of trained "raters" riding in cars assess the smoothness of ride and determine the *Present Serviceability Rating (PSR)*. A PSR value of 4.6 to 5 corresponds to a very smooth ride. The cumulative effects of weather and the loads inflicted by heavy trucks cause the pavement to eventually degrade in ride quality to an unacceptable level, usually called the *terminal serviceability index (TSI)*. Most agencies specify a TSI of 2.0 to 2.5, depending on roadway classification.

Historically, pavement ride quality has been manually measured with a California Profilograph, as shown in Figure 9.4a. Recently, instrumented vehicles called "profilometers" (Figure 9.4b) have put the PSI on a more standardized basis by removing some of the judgment of human raters and more accurately measuring ride quality using laser readings. The history of a typical pavement using PSI is presented in Figure 9.5a. Figure 9.5b shows the level of pavement serviceability when it needs to be repaired or rehabilitated. When the PSI reaches 2 to 2.5 (TSI), then maintenance or rehabilitation should be performed. Most new flexible pavements exhibit PSI of 4.2 to 4.5, so allowable PSI loss (ΔPSI) of about 2.0 is typical. After restoring the PSI to some

(a)   (b)

FIGURE 9.4
(a) California profilograph and (b) Lightweight profilometer. Photos courtesy of Adam Hand, Ph.D.

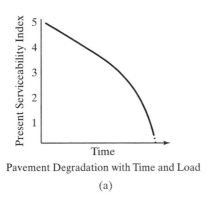
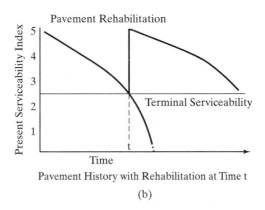

Pavement Degradation with Time and Load   Pavement History with Rehabilitation at Time t

(a)   (b)

FIGURE 9.5
Present serviceability index trends over life of pavement.

level near to that of a new pavement (through rehabilitation or maintenance), deterioration from loads and weather begin the process all over again. The time to major road repair for most properly designed highways is typically about 20 to 25 years after the road is constructed.

More recently, a more comprehensive measure for determining pavement rideability has been developed. It is called the *International Roughness Index (IRI)*. IRI is entirely ride-based, but is directly a function of pavement distress (cracking, patching, etc.). It is an equipment-based measure of pavement roughness reported in inches per mile (or in meters per kilometer), whereas the PSI includes variables such as rut depth, cracking and roughness. Low values of IRI mean that the pavement is relatively smooth and would give a good ride. Higher values bring in the various pavement distress factors that cause pavement failure. Although the relationship between PSI and IRI does not result in a one-to-one correspondence, comparisons are often made, such as those shown in Table 9.1. The table gives example performance measurements for a single highway. An IRI over 120 in/mi indicates time for rehabilitation and corresponds to a PSI of about 2.2.

Today, most state highway agency construction specifications include ride quality requirements. In fact, many include pay incentives to contractors for ride quality. This has prompted the highway contracting community to place more emphasis on building good riding (smooth) pavements. In fact, some contractors are using lightweight profilers, such as the one shown in Figure 9.4b, as a construction quality control tool to measure ride quality.

9.1.4   Soil Characteristics

The soil on which the pavement is built is an essential ingredient in determining the pavement's longevity and strength. Table 9.2 shows the characteristics of several soil types. For flexible pavement design, the elastic properties of the soil, expressed in the *modulus of resiliency (M_R)* (pounds of force per square inch, or Pascals), are essential.

If soil is gravelly or if gravel is placed down as a base or subbase, the pavement strength improves. However, when the soil is primarily clay, it has little resiliency and makes a very poor base. In those cases, the soil is often replaced with a much more resilient layer. If the soil is basically organic, it is totally unsuitable and must be replaced.

TABLE 9.1   Approximate Relationship of PSI to IRI

Present Serviceability Index	Pavement Condition	International Roughness Index (in/mi)
5.0	Excellent	0
4.3	Very good	45
4.0	Very good	59
3.5	Good	74
3.1	Fair	87
2.6	Poor to fair	105
2.1	Poor	133
1.8	Very poor	153

Source: Adam Hand, Ph.D.

TABLE 9.2 General Soil Characteristics

Major	Divisions	Letter	Name	Value as Base Directly Under Wearing Surface	Potential Frost Action	Compressibility and Expansion	Drainage Characteristics	Compaction Equipment	Unit Dry Weight (pcf)	Field CBR	Subgrade Modulus k (pci)
Coarse grained soils	Gravel and gravelly soils	GW	Gravel or sandy gravel Well graded	Good	None to very slight	Almost none	Excellent	1, 2, 3	125–140	60–80	>300
		GP	Gravel or sandy gravel Poorly graded	Poor to fair	None to very slight	Almost none	Excellent	1, 2, 3	120–130	35–60	>300
		GU	Gravel or sandy gravel Uniformly graded	Poor	None to very slight	Almost none	Excellent	1, 2	115–125	25–50	>300
		GM	Silty gravel or silty Sandy gravel	Fair to good	Slight to medium	Very slight	Fair to poor	2, 4*	130–145	40–80	>300
		GC	Clayey gravel or clayey Sandy gravel	Poor	Slight to medium	Slight	Poor to impervious	2, 4	120–140	20–40	200–300
	Sand and sandy soils	SW	Sand or gravelly sand Well graded	Poor	None to very slight	Almost none	Excellent	1, 2	110–130	20–40	200–300
		SP	Sand or gravelly sand Poorly graded	Poor to not suitable	None to very slight	Almost none	Excellent	1, 2	105–120	15–25	200–300
		SU	Sand or gravelly sand Uniformly graded	Not suitable	None to very slight	Almost none	Excellent	1, 2	100–115	10–20	200–300
		SM	Silty sand or silty gravelly sand	Poor	Slight to high	Very slight	Fair to poor	2, 4*	120–135	20–40	200–300

TABLE 9.2 Continued

Major	Divisions	Letter	Name	Value as Base Directly Under Wearing Surface	Potential Frost Action	Compressibility and Expansion	Drainage Characteristics	Compaction Equipment	Unit Dry Weight (pcf)	Field CBR	Subgrade Modulus k (pci)
		SC	Clayey sand or clayey Gravelly sand	Not suitable	Slight to high	Slight to medium	Poor to impervious	2, 4	105–130	10–20	200–300
Fine grained soils	Low compressibility LL < 50	ML	Silts, sandy silts gravelly silts or distomaceous	Not suitable	Medium to very high	Slight to medium	Fair to poor	2, 4*	100–125	5–15	100–200
		CL	Lean clays, sandy clays, or gravelly clays	Not suitable	Medium to high	Medium	Practically impervious	2, 4	100–125	5–15	100–200
		OL	Organic silts or lean organic clays	Not suitable	Medium to high	Medium to high	Poor	2, 4	90–105	4–8	100–200
	High compressibility LL > 50	MH	Micaceous clays or distomaceous soils	Not suitable	Medium to very high	High	Fair to poor	2, 5	80–100	4–8	100–200
		CH	Fat clays	Not suitable	Medium	High	Practically impervious	2, 4	90–110	3–6	50–100
		OH	Fat organic clays	Not suitable	Medium	High	Practically impervious	2, 4	80–105	3–6	50–100
Peat and fibrous organic soils		PT	Peat, Humus, etc.	Not suitable	Slight	Very high	Fair to poor	No			

*Requires close control of moisture.

Compaction equipment 1 = Crawler-type tractor; 2 = Rubber tired equipment; 3 = Steel #Wheeled roller; 4 = Rubber tired equipment; 4 = Sheepsfoot roller.

Source: FAA Advisory Circular 150/5335 Airport Pavement Design and Evaluation.

The soil chart shown in Table 9.2 relates the M_R to the California bearing ratio (CBR). When CBR <6, M_R is often approximated by $1500 \times$ CBR. The CBR is determined from a standard laboratory test, in which a specific size and weight piston is loaded and applied to the soil. The deflection of the soil sample is used to determine the CBR.

The M_R will vary during the year, depending on the moisture content and the low temperature/frost conditions of the soil. The same soil can have an M_R of 120 to 150 MPa (million pascals) (17,400 to 21,700 lb force/in^2) during winter, but only 25 to 40 MPa (3,600 to 5,800 psi) during spring thaw. Then, as the soil dries out, it stabilizes, becoming more resilient through the summer, with M_R increasing to 50 to 80 MPa (7,200 to 11,600 psi). An average M_R of 8,000 to 10,000 psi for the year is reasonably conservative (Raad, 1997). When we use the layered analysis approach to flexible pavement design, each layer will have its own modulus of resiliency. The soil properties will usually depend on the prevailing soil in the roadway corridor, unless the soil is treated or replaced. Preparation of the soil for the eventual placement of the pavement is very important.

9.2 DETERMINING LOADS FROM TRUCK TRAFFIC

The first step in the analysis of any pavement is to estimate the integrated load on the highway. The integrated load is determined by estimating the number of equivalent 18,000-pound single-axle loads [called equivalent single-axle loads (ESALs)] that are expected to occur on the pavement during its useful life. The objective is to estimate the ESALs that will cause a degradation of the PSI to 2 or 2.5 from its initial value of 4.5 to 5. The annual ESALs are calculated from the expected mix of daily truck traffic and summed for the year.

9.2.1 Determining Equivalent Single-Axle Load

The load that is imparted to the highway is primarily from heavy trucks. The ESALs can be approximated from actual truck axle weights using a fourth power formula as shown in Equation 9.1:

$$\text{ESAL}_{\text{Single}} = \left(\frac{W_{\text{single}}}{18,000 \text{ lb}}\right)^4 \tag{9.1}$$

Examine the five-axle heavy tractor-semitrailer-trailer or "double bottom" (2-S1-2) shown in Figure 9.6. If the trailers are full, bringing the truck weight up to the legal limit of 80,000 pounds, each single axle other than the steering axle will have a load of 17,000 pounds. That load on each single axle is less than the 18,000 pounds needed for one ESAL. Using the fourth power formula, the ESAL for one truck is 3.4. The calculation of the equivalent load on the highway is

$$\text{ESAL} = \left(\frac{12}{18}\right)^4 + 4 \times \left(\frac{17}{18}\right)^4 = 0.2 + 3.2 = 3.4$$

This means that the fully loaded double bottom applies a load to the pavement equivalent to 3.4 times a single-axle load of 18,000 pounds.

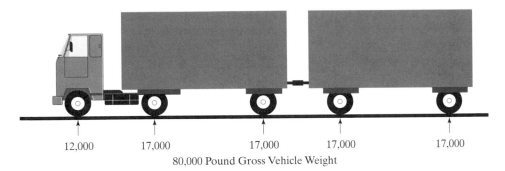

FIGURE 9.6
Twin tandem truck (2-S1-2).

By contrast, a 3000-pound automobile exerts about 0.0001 ESAL. It takes about 34,000 automobiles to create the same amount of road damage as a single tandem truck carrying its maximum legal load. Thus, automobile loads can be ignored when determining the vehicle load on the highway.

The heavy tractor-semitrailer truck or "18-wheeler" (3-S2) is shown in Figure 9.7 with its *tandem* axles. The tandem axles are counted differently. (See also Figure 2.31.) Because they ride on the highway close together, the pavement between them does not get the full effect of each single-axle weight. Thus, their effect is not double the load from a single axle, and they are treated in the calculations as one tandem axle. The overall effect on flexible (asphalt) reduces the load from 36,000 pounds (2 × 18,000 pounds) to about 33,200 pounds, as indicated in the fourth power Equation 9.2.

$$\text{ESAL}_{\text{Tandem}} = \left(\frac{W_{\text{Tandem}}}{33,200}\right)^4 \tag{9.2}$$

For example, the fully loaded 18-wheeler shown in Figure 9.7 will load the highway with 2.44 ESALs.

$$\text{ESAL}_{18-\text{Wheeler}} = \left(\frac{12}{18}\right)^4 + \left(\frac{36}{33.2}\right)^4 + \left(\frac{32}{33.2}\right)^4 = 0.2 + 1.38 + 0.86 = 2.44$$

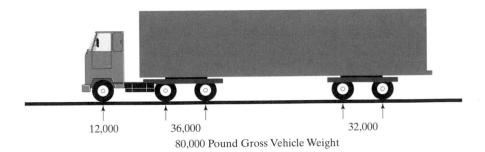

FIGURE 9.7
Semitrailer configuration 3-S2: the 18-wheeler.

For rigid (concrete) pavement, Equation 9.2 must be replaced by

$$ESAL_{Tandem} = \left(\frac{W_{Tandem}}{29,000}\right)^4 \tag{9.3}$$

By law in many states, the 3-S2 truck can have a *gross vehicle weight (GVW)* of no more than 80,000 pounds. The load on the steering axle is usually 10,000 to 12,000 pounds.

9.2.2   The Fourth Power Formula for Trucks

Highway design manuals often have *load equivalency factor (LEF)* tables indicating the number of 18K ESALs for each level of axle load. For example, 12,000 pounds acting through a single axle is equivalent to 0.189 of an 18,000-pound load on a single axle. These tables give different values, depending on the *structural number (SN)* of the pavement. (SN in Figure 9.8 indicates the structural capability of the pavement, as is discussed in Section 9.3.) However, for most axle weights, the weight is seldom higher than 30 kips for a single axle or 50 kips for a tandem axle. In this range, the fourth power formula is well within 10 or 15 percent of the table values for the usual structural number of the pavement, which is between 3 and 6. The curves in Figures 9.8a and b compare the load equivalents estimated by the fourth power Equations 9.1 and 9.2 to the values in the LEF tables for single and tandem axles on flexible pavement.

9.2.3   Identifying Differing Truck Configurations

There are many types of trucks on the road, and most state DOTs have a convention to identify various heavy trucks. The usual convention separates the individual elements of the truck, tractor and trailer(s) by a dash as shown in Table 9.3. Figure 9.9 shows several trucks with their designations.

9.2.4   Determining Lifetime ESALs for Pavement Design

To estimate the integrated ESALs that a highway will be subjected to, the present mix of truck traffic must be estimated and then projected over an appropriate planning

TABLE 9.3   Truck Identifiers

Designation	Truck Description	Common Name
2A	Single truck with two axles and four tires	Pick-up
2D	Heavy single-unit truck with two axles and 6 tires	Small delivery van
3SU	Heavy single-unit truck with three axles	Delivery van/dump truck
4SU	Heavy single-unit truck with 4 axles	Dump truck
2-S1	Heavy tractor-semitrailer truck with three axles	Semi
2-S2	Heavy tractor-semitrailer truck with four axles	Semi
3-S2	Heavy tractor-semitrailer truck with five axles	Semi (18-wheeler)
2-S1-2	Heavy tractor-semitrailer-trailer with five axles	Double bottom
3-S2-4	Heavy tractor-semitrailer-trailer with 9 axles	Turnpike double

Source: States of Wisconsin and New Jersey DOTs (website).

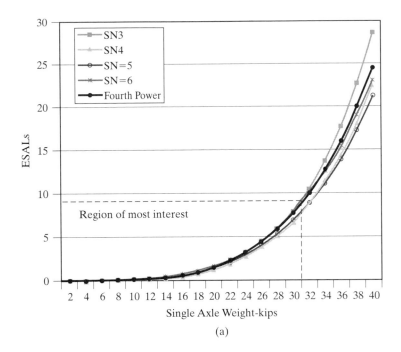

(a)

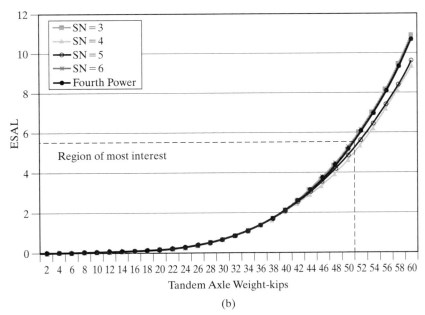

(b)

FIGURE 9.8
Comparison of Fourth power formula to LEF tables for (a) single axle and (b) tandem axles on flexible pavement.

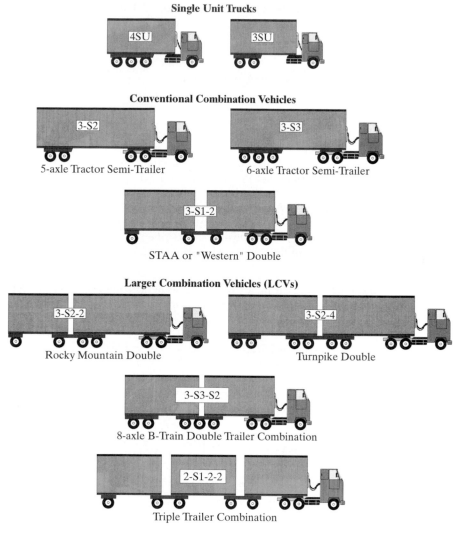

FIGURE 9.9
Axle Configurations for Trucks. *Source*: Public Roads 2001.

horizon, generally 20 or 25 years. Information from weigh stations is often used for the calculation of ESALs.

Although the percentage of trucks may be high on an interstate, a large number of the trucks will be empty. This is called an *empty backhaul*. Even though truckers try hard to get at least a partial load for their backhaul, in Indiana about 25 to 30 percent of heavy tractor-semitrailer trucks (3-S2) run empty (Lew et al., 1988). The effect of empty trucks on pavement life is minimal.

Because the pavement design equation requires that we consider integrated stress over the design pavement lifetime, the *expected growth* in truck traffic over the pavement life needs to be included. Equation 9.4 indicates the calculation of that growth when the annual growth is a constant percentage rate g and N is the expected lifetime for the pavement in years.

$$\text{ESAL} = \text{Annual ESAL} \times \frac{\lfloor (1 + g)^N - 1 \rfloor}{g} \tag{9.4}$$

Example 9.1: Multilane Freeway

There are 2000 trucks a day moving in one direction on a multilane freeway. For design purposes, we should assume that the trucks are operating on the outside lane (closest to the shoulder). Eight hundred of the trucks are empty, contributing at most about 0.25 ESALs per truck. For the truck mix shown in Table 9.4, determine the daily ESAL for the 2000 daily trucks if they have the gross vehicle weights indicated. (Assume that the weight remaining after subtracting the steering axle is evenly divided among the remaining axles.) Determine the annual ESALs and the ESALs for the pavement lifetime, if truck traffic is to grow at 3 percent per year for 20 years.

Solution to Example 9.1   In Table 9.4, the entries in the column entitled "ESAL per truck" are calculated with the fourth power formula. The last column extends that for the number of trucks per day. In computing the annual loads, 6-day weeks for 50 weeks per year are used:

$$\text{Annual ESAL} = 6 \times 50 \times 1950 = 585{,}000 \text{ per year}$$

If we look at history, it is clear that the truck traffic continues to grow as the economy grows. For this example, truck traffic is assumed to grow at a compound rate, $g = 3$ percent per year. Over the 20-year planning horizon, the integrated value of ESALs, using Equation 9.4 is

$$\text{ESAL}_{20\text{years}} = 585{,}000 \left(\frac{(1.03^{20} - 1)}{0.03} \right) = 15.72 \text{ million}$$

Thus the value of equivalent single-axle load to be used in highway design is 15.72 million ESALs.

TABLE 9.4   Daily ESALs for Example Traffic Mix

	Given			Calculated for solution		
Truck	GVW	Steering Axle	Remaining Axles	ESAL per Truck	Daily Traffic	Total ESAL per day
3-S2	80,000	12,000	34,000	2.39	100	239
3-S2	70,000	12,000	29,000	1.36	200	272
3-S2	60,000	12,000	24,000	0.75	300	225
2-S1-2	80,000	12,000	17,000	3.38	100	338
2-S1-2	70,000	12,000	14,500	1.88	200	376
2-S1-2	60,000	12,000	12,000	1.00	300	300
All others	30,000	12,000	4500	0.25	800	200
						1950

Example 9.2: Quarry Traffic

A road into a heavily used quarry (120 loaded trucks per day, 5 days per week) needs to be redesigned. The road is to last 10 years, and the traffic is expected to increase by about 2 percent per year. There are two types of trucks that are used equally each day, each of which will be fully loaded outbound and empty on return. One truck type is a 25-foot-long dump truck (4SU configuration) and the other type is a 3-S2 semitrailer. The weight of the full dump truck is 60,000 pounds and the full semitrailer is 80,000 pounds, with weight distributed as shown in Figure 9.7. During its empty backhaul, the semitrailer still has 12,000 pounds on the steering axle, 18,000 pounds on the front tandem, and 10,000 pounds on the back tandem.

$$\text{ESALs} = \left(\frac{12}{18}\right)^4 + \left(\frac{18}{33.2}\right)^4 + \left(\frac{10}{33.2}\right)^4 \cong 0.3$$

When full, the four-axle dump truck has 52,000 pounds on the rear three axles (called a tridem axle configuration), which gives an ESAL of 1.43, plus 0.04 for the 8,000 pounds on the steering axle. On its empty return trip, the dump truck's third axle is raised, so that it has a 3SU axle configuration. The weight on the front axle is 8,000 pounds and, with 16,000 pounds on the rear tandem, the truck's ESALs $= (8/18)^4 + (16/33.2)^4 \approx 0.1$.

Calculate the ESALs for the first year (assume, with holidays, a 50-week year) and for the next 10 years.

Solution to Example 9.2 In Table 9.5, Equations 9.1 and 9.2 are applied to the axle loads represented by each truck type, to estimate the ESALs for each. The total base year ESALs is 64,350. With 2 percent growth per year, Equation 9.4 is used,

$$\text{ESAL}_{10\text{years}} = \text{ESAL}_{1\text{st year}}\left(\frac{(1.02^{10} - 1)}{0.02}\right) = 10.95 * \text{ESAL}_{1\text{st year}}$$

To estimate the ESALs occurring over the next 10 years, multiply the current year ESALs by the 10-year growth factor (10.95). The resulting estimate of $64,350 * 10.95 = 705,000$ is used in Section 9.3 to determine the pavement thickness.

THINK ABOUT IT

In Example 9.2, a 52,000-pound load on the tridem axle of the four-axle dump truck has an ESAL of 1.43. Using the fourth power formula $\text{ESAL}_{\text{tridem}} = \left(52/X\right)^4$, what is X for a tridem axle?

TABLE 9.5 Truck ESALs for the First Year, Example 9.2

Truck Type	Number per Day	ESAL per Truck	ESAL per Day	ESAL per Year
Full semi	60	2.45	147	36,750
Empty semi	60	0.3	18	4500
Full dump truck	60	1.44	86.4	21,600
Empty dump truck	60	0.1	6	1500
Total	240		257.4	64,350

9.2.5   Obtaining Truck Weight Data

Truck weight, because it does so much damage to the highways, is controlled and trucks are heavily taxed. The maximum truck weight in most states is 80,000 pounds. However, in Michigan, there are special 11-axle semitrailer trucks (one called the "sled" and the other "Michigan train") that can carry loads up to a legal limit of 165,000 pounds. Figures 9.10a and b show the trucks. They are used to haul steel from northern Indiana into and throughout Michigan. The highway damage is not so high, because the weight is spread over the large number of axles.

Weight data are usually collected by two means: using static scales at roadside weigh stations and by devices embedded in the roadway. The in-roadway devices are called *weigh-in-motion (WIM)* sensors. Although not as accurate as static scales at the weigh stations, they have sufficient accuracy to obtain good axle weight data. The weighing is accomplished from sensitive devices generally placed in frames installed in the roadbed. The usual devices are:

- *Bending plates.*   These devices generate an electric current when subjected to pressure caused by an axle crossing the plate.
- *Capacitive strips.*   The degree of pressure on the strip as an axle crosses the strip allows calculation of the axle weight.
- *Piezoelectric sensors.*   These devices use a special material inside a tube. The material generates a varying electric current proportional to the weight of the axle crossing the sensor. (Pierre and Jacques Curie discovered the piezoelectric phenomenon.)

THINK ABOUT IT

Using the system introduced in Table 9.3, what would be the designations for the trucks shown in Figure 9.10?

(a)

(b)

FIGURE 9.10
Michigan steel hauling trucks. (a) 11-axle "sled." (b) 11-axle "Michigan Train." Photos: David Moffett.

One use of the WIM data is to convert the frequency of each axle loading into equivalent single-axle loads. Some WIM scales are portable, which allows them to be moved to various sites. However, one study (Lew et al., 1988) also found that the portable WIM scales underestimate the weights for steering axles (which are lighter) and overestimate the drive and trailer axle weights (which are heavier). The portable scales are also not easy to use properly.

WIM errors of 20 percent (Lew et al., 1988) demonstrate the challenge involved in acquiring dynamic truck weights. WIM scales are often used to screen trucks by weight as they are moving down the highway. Only those trucks whose weights may be near the legal limit are then sent to the static scales.

Several analyses have been done to estimate ESALs from vehicle classification data, which give only the axle configuration. For example, in Alaska, the equation for determining ESALs is (Walters and Whitford, 1998):

$$\text{Total Estimated ESAL} = (0.072 * \text{number of single axles})$$
$$+ (0.46 * \text{number of tandem axles}). \qquad (9.5)$$

where "single axles" is the number of single axles on trucks (class 5 or larger) crossing the classification counter in a given time period and "tandem axles" is the number of tandem (or tridem) axles on trucks (class 5 or larger) crossing the classification counter in the same time period. Note that this equation only applies for the truck mix seen on most Alaska highways. The equations will vary for other states, depending on their truck traffic mix. It is expected that similar equations will also be valid with corrected numbers for other states as well.

Example 9.3

Consider a classification site with daily traffic level of 300 trucks: 100 2-S1 and 200 3-S2 trucks. Estimate the ESALs using Equation 9.5.

Solution to Example 9.3   Each 2-S1 truck has three single axles. (See Table 9.3 and Figure 9.9.) Each 3-S2 truck has one single and two tandem axles. Collectively, the trucks at the site have $(100 * 3) + (200 * 1) = 500$ single axles and $(200 * 2) = 400$ tandem axles. Using Equation 9.5, the daily ESALs for the classification site would be estimated as ESALs = $(0.072 * 500) + (0.46 * 400) = 36.00 + 184.00 = 220.00$.

9.3   FLEXIBLE PAVEMENT DESIGN

9.3.1   The AASHO Road Test

The modern era of flexible pavement design began with a series of road tests near Ottawa, Illinois. (See Figure 9.11.) Those tests produced results that were used by the various state Departments of Highways (now Departments of Transportation) beginning in 1962. As roads were built and experience was gained, there have been a series of changes to the design equation. Oglesby and Hicks say

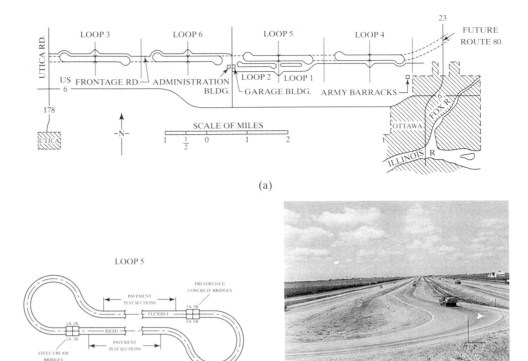

(a)

(b)   (c)

FIGURE 9.11
(a) The six test loops (Fisher and Viest, 1962). Reproduced with permission of the Transportation Research Board. In Special Report 73: *The AASHO Road Test*, Highway Research Board, National Research Council, Washington, D.C., 1962, Figure 1, p. 19. (b) Test loop 5 (Fisher and Viest, 1962). Reproduced with permission of the Transportation Research Board. In Special Report 73: *The AASHO Road Test*, Highway Research Board, National Research Council, Washington, D.C., 1962, Figure 2, p. 20. (c) AASHO road test track, near Ottawa, Illinois. The test track. FHWA by Day. *Source*: FHWA By Day, www.fhwa.dot.gov/byday/fhbd11xx.htm.

The AASHO Road Test was a $27 million cooperative effort sponsored by the then American Association of State Highway Officials. It involved both bituminous and Portland-cement-concrete pavements and a group of 50 ft single-span bridges of steel with composite and non composite design, prestressed concrete, both pretensioned and posttensioned, and reinforced concrete T beams.

Planning the project began about 1950. The site near Ottawa, Ill., was selected in 1954; construction was carried on in the years 1956–1958; testing began in October 1958 and ended in 1960; data analysis and final reports were completed in 1962. In all, the road contained six loops, each with two lanes. Single-axle loading ranged from 2000 to 30,000 lb; tandems from 24,000 to 48,000 lb. Field-testing and measurement, laboratory work and analysis made use of the most modern equipment and statistical methods. The final reports totaled more than 1600 pages (Oglesby and Hicks, 1982, p. 667–668).

9.3.2   Materials for Asphalt Pavements

Asphalt materials used in building flexible pavements are most frequently obtained as a product of the distillation of crude petroleum, although they are sometimes found naturally in places, such as the LaBrea tar pits in Los Angeles, California, where seeping has occurred. In the distillation process, they are the heaviest (residue) materials after the removal of better grades of gasoline, kerosene, and diesel fuel. There are several grades of asphalt retrieved from the distillation process. Depending on their material characteristics, there are grades used in road building for:

- Asphalt concrete in roads, parking areas, driveways, etc.
- Road mix, base course, patching, etc.
- Surface treatment, tack coat, and sealing
- Embankments and reservoir linings

The asphalt concrete used in most road construction is a well-mixed combination of asphalt cement, coarse aggregate, fine aggregate, and other materials, depending on whether it is cold or hot mix, cold or hot laid. When used in highway pavements, it must resist deformation from loads, be skid resistant (even when wet), and be impervious to most weather and deicing chemicals. Figure 9.12 shows the laying of hot mix asphalt. Recently, machines were developed to permit the reuse of the asphalt. Old pavement is taken up and processed to recover the asphalt, then the reprocessed asphalt is relaid.

Figures 9.13 and 9.14 show that most flexible pavements are constructed with several layers of material, usually placed on top of compacted subgrade/soil. The flexible pavement deflects under load, so that the stress (deflection) will be transmitted through the pavement structure all the way to the subgrade. In this way, the stress is spread over an area much larger than the contact area of the tires. This is called the *cone effect*, as shown in Figure 9.14. Thus, the dynamic characteristics of the subgrade (along with the materials in the other layers) play an important part in the overall pavement structural strength. The more resilient the subgrade, the less thick the paving materials in the layers above the subgrade must be to achieve a given overall pavement strength.

FIGURE 9.12
Constructing flexible pavement. Photo courtesy of Professor Adam Hand.

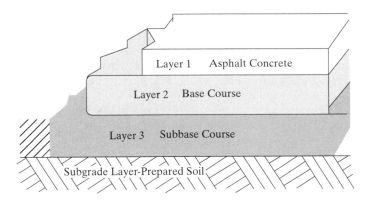

FIGURE 9.13
Load distribution on a flexible pavement.

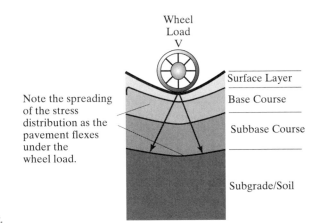

FIGURE 9.14
Layers of a flexible asphalt pavement.

9.3.3 Pavement Layers and the Structural Number

There have been efforts underway for many years to find a pavement design process that will allow the engineer to build a pavement that would last for at least 20 years. The tests performed by AASHO at Ottawa, Illinois, provided the first real pavement performance data and led to one of the early approaches for designing flexible pavement. At present, each state is free to develop its own design procedures, but many of the states have adopted the AASHTO (1993) method. The equations developed by AASHTO led to the layer design approach, with the pavement being composed of from one to three layers of material. Figures 9.13 and 9.14 illustrate the concept. The features of each layer are:

- *Surface layer of hot mix asphalt.* Sometimes this layer is composed of a coarse bituminous layer, with a very thin glaze or smoothing surface on top of it.
- *Base course.* Usually 6 to 12 inches of an unbound, untreated course aggregate. Sometimes the base course is stabilized with some additive such as emulsified asphalt.

- *Subbase course.*   Frequently composed of small gravel or aggregate. Where frost susceptibility is a problem, control of the materials comprising this course can be extremely important. In Alaska, the subbase is often as thick as 34 inches, with heavy control of the fines in the subbase material. This is to reduce capillary action of moisture under the pavement.

- *Subgrade is the soil under the pavement.*   If the soil is heavily organic (muck), so that it has little resiliency after the load is removed, pavement deterioration occurs quickly. Such soil is usually replaced with a more suitable soil, such as compacted sand. The subgrade is, in effect, the bottom layer and is always included when examining the integrity of the pavement. The subgrade is usually represented by its modulus of resiliency, which can vary by season, moisture content, or material. (See Table 9.2.)

The design concern is whether the resultant pavement composed of these layers will have sufficient structural strength to withstand the load (number of ESALs) it will experience over its lifetime before it must undergo replacement or rehabilitation. We begin with consideration of the structural number for the pavement. The structural number reflects the total pavement thickness, including the capability of the soil or subgrade to provide sufficient resiliency during the repeated loadings to maintain the wanted serviceability (PSI). As can be seen in Equation 9.6, coefficients a_1, a_2, and a_3 reflect the structural strength of the material used for each specific layer. The variables d_i represent the thickness of each layer in inches and m_i are modifiers for moisture.

$$SN = a_1d_1 + a_2d_2m_2 + a_3d_3m_3$$

$$SN = a_{hma}d_{hma} + (a*d*m)_{base} + (a*d*m)_{subbase} \qquad (9.6)$$

where

a_1, a_2, a_3 = coefficients of relative strength

d_1, d_2, d_3 = thickness in inches of the hot mixed asphalt or bituminous course, base course, and subbase course, respectively

m_i = modifiers for more than normal amounts of moisture;

TABLE 9.6   Recommended m_i Values for Modifying Structural Layer Coefficients of Untreated Base and Subbase Materials in Flexible Pavements

Quality of Drainage	Percent of Time Pavement Structure Is Exposed to Moisture Levels Approaching Saturation			
	Less Than 1%	1–5%	5–25%	Greater Than 25%
Excellent	1.40–1.35	1.35–1.30	1.30–1.20	1.20
Good	1.35–1.25	1.25–1.15	1.15–1.00	1.00
Fair	1.25–1.15	1.15–1.05	1.00–0.80	0.80
Poor	1.15–1.05	1.05–0.80	0.80–0.60	0.60
Very poor	1.05–0.95	0.95–0.75	0.75–0.40	0.40

Source: AASHTO, 1993, p. II-25.

$m_i = 1.0$ unless otherwise specified.

hma = hot mix asphalt

The drainage coefficients used in flexible pavement design are given in Table 9.6. Good drainage design is a necessity, especially in areas where substantial rainfall is common. Moisture has long been the nemesis of pavement engineers. Failure to account for it has been one reason why some pavements to have had a lower actual structural capacity than the design value.

9.3.4   Fitting the Design Variables Together

After analyzing the data from their road test, the AASHO engineers developed a complex equation. The dependent variable is the structural number (SN) of the pavement. The structural number needs to be computed so that the pavement will have the structural capacity to carry the anticipated load and will experience no more than the specified loss in serviceability. The properties (i.e., layer coefficients in Equation 9.6) of each layer play a major role in determining the thickness of each layer (SHRP-LTPP, 1994).

Once the required structural number of a pavement section is determined, the layers are computed by using the AASHTO layered analysis method. In this method, it is assumed that the structural capacity of the pavement is the sum of the structural capacity of each of its layers. Equation 9.6 reflects this. The structural capacity of a pavement is the product of its layer coefficient, drainage coefficient, and thickness. That is, increasing the thickness of a relatively weak pavement will improve its structural capacity. Equation 9.7 is the result of the AASHO Road Test (and subsequent updates) and has been proved to be fairly representative of pavements, especially those not subjected to freeze-thaw cycles (AASHTO, 1986).

$$\log_{10} W_{18} = Z_R \times S_o + 9.36 \times \log_{10}(SN + 1) - .20$$

$$+ \frac{\log_{10} \dfrac{\Delta PSI}{2.7}}{0.4 + \dfrac{1094}{(SN + 1)^{5.19}}} + 2.32 \log_{10} M_R - 8.07 \qquad (9.7)$$

where

W_{18} = the expected cumulative 18K ESALs over the anticipated life of the pavement

Z_R = standard normal deviate for reliability level R

S_o = overall standard deviation

ΔPSI = allowable loss in serviceability

SN = required structural number

M_R = resilient modulus of the soil/subgrade

The equation has been reduced to the nomograph (or design chart) presented in Figure 9.15. This nomograph is first used to identify the overall structural number of the pavement. Then the design chart is used to identify the structural contribution of each layer, in its structural coefficient and thickness. We must first develop independently the overall SN for the pavement in its reliability, standard deviation,

NOMOGRAPH SOLVES:

$$\log_{10} W_{18} = Z_R * S_o + 9.36 * \log_{10}(SN+1) - 0.20 + \frac{\log_{10}\left[\dfrac{\Delta PSI}{4.2-1.5}\right]}{0.40 + \dfrac{1094}{(SN+1)^{5.19}}} + 2.32 * \log_{10} M_R - 8.07$$

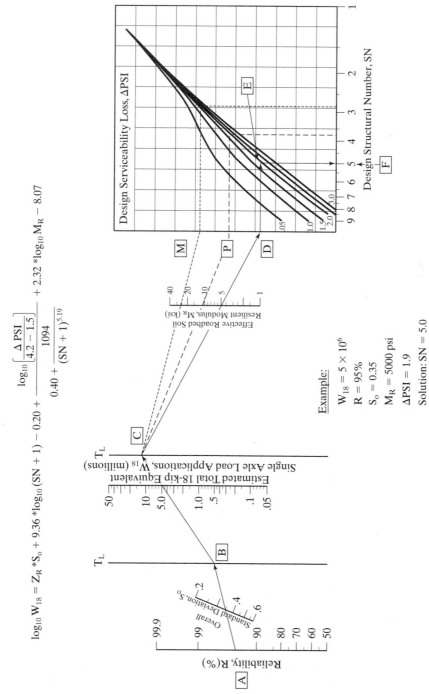

Example:

$W_{18} = 5 \times 10^6$
$R = 95\%$
$S_o = 0.35$
$M_R = 5000$ psi
$\Delta PSI = 1.9$
Solution: SN = 5.0

FIGURE 9.15

The AASHTO equation in graphical form. *Source:* AASHTO, 1993, p. II-32.

lifetime ESALs, resilient modulus of the soil, and the acceptable ΔPSI. That is done on the nomograph according to the following steps:

Step 1: Select the reliability desired, based on an assumption of the errors probable in the lifetime ESAL determination. "Reliability is the probability that the load applications a pavement can withstand in reaching a specified minimum serviceability level is not exceeded by the number of load applications that are actually applied to the pavement" (Kher and Darter, 1973). The values in practice are usually 0.95 to 0.99 for interstates, 0.9 to 0.8 for major arterials, and 0.5 to 0.8 for local roads. In Equation 9.7, the R values are represented by Z_R values, according to the table shown below. (The example in Figure 9.15 uses $R = 0.95$. See point A on the nonlinear reliability scale.)

Reliability (%)	Standard Normal Deviate, Z_R
80	−0.841
90	−1.282
95	−1.645
99	−2.327

Step 2: Draw a line through the standard deviation or standard error scale ($S_o = 0.35$) to the first turning line "T_L" (point B). The overall standard deviation S_o is a measure of the spread of the probability distribution for ESALs vs. PSI, considering all the parameters used to design a pavement. In many states, $S_o = 0.35$ to 0.45 is used for the design of flexible pavements and $S_o = 0.25$ to 0.35 is used for rigid pavement design.

Step 3: Draw a line through the estimated ESALs (5 million) to the next transfer line (point C).

Step 4: Draw a line through *the resilient modulus of the soil* (5000 psi) to the edge of the SN grid (point D).

Step 5: Draw a *horizontal* line to the selected ΔPSI of 1.9 (point E).

Step 6: Drop that point *vertically* to the bottom of the SN grid to find the structural number of the *overall* pavement, which is 5.0 (point F).

9.3.5 Performing the Layered Analysis

Once the overall structural number for the pavement has been found, the layered analysis to find the thickness of each pavement layer can begin. The nomograph in Figure 9.15 was used to find a structural number of 5.0 for the three-layer pavement. The same design chart will be used to find the one-layer and two-layer components of the three-layer pavement. The material characteristics of the layers to be used in this flexible pavement are:

Hot mix asphalt elastic modulus = 380,000 lb/in^2 → a_1 = 0.41 (from Figure 9.16)

Untreated granular base course, CBR = 55, M_r = 25,000 lb/in^2 → a_2
  = 0.12 (from Figure 9.17)

If a cement-treated base material were to be used, Figure 9.18 would apply.

Untreated granular subbase course, CBR = 20,
  M_r = 13,000 lb/in^2 → a_3 = 0.098 (from Figure 9.19)

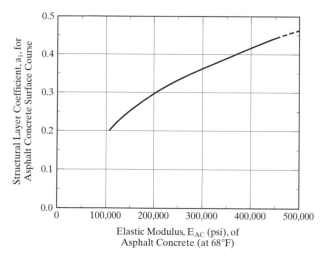

FIGURE 9.16

Chart for estimating structural layer coefficient a_1 of dense graded asphalt concrete based on the elastic resilient modulus. *Source:* AASHTO, 1993, p. II-18.

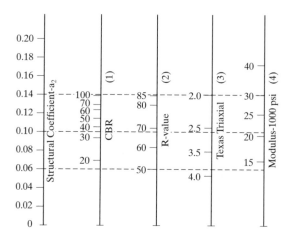

(1) Scale derived by averaging correlations obtained from Illinois.
(2) Scale derived by averaging correlations obtained from California, New Mexico, and Wyoming.
(3) Scale derived by averaging correlations obtained from Texas.
(4) Scale derived on NCHRP project (3).

FIGURE 9.17

Variation in granular base layer coefficient a_2 with various base strength parameters for untreated base. *Source:* AASHTO, 1993, p. II-19.

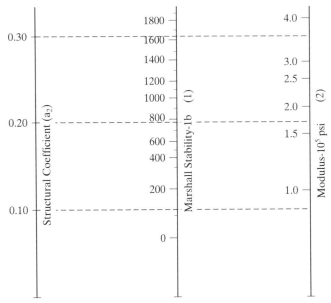

(1) Scale derived by correlation obtained from Illinois.
(2) Scale derived on NCHRP project (3).

FIGURE 9.18
Variation in granular base layer coefficient a_2 with various base strength
parameters for cement-treated base. *Source:* AASHTO, 1993, p. II-21.

The steps in the layered analysis proceed as follows.

Step 1: Find the structural number SN_1 for a one-layer pavement in the same design chart in which the three-layer structural number $SN_3 = 5.0$ was found. The one-layer pavement will rest on the second layer (the base), so the M_R value for the base layer (25,000 psi) will be used in the design chart. No other value to be used in the design chart will change. From point C on the second transfer line in Figure 9.15, draw a line through the resilient modulus of the base course (25,000 lb/in^2) to intersect the SN grid (point M in Figure 9.15). From the intersection of the horizontal line with the ΔPSI, drop a vertical line down to determine the SN for the top layer (SN = 3.0).

Step 2: Repeat step 1 to find the structural number SN_2 for a two-layer pavement that rests on the third layer (the subbase), which has $M_R = 13,000$ psi. From the transfer line in Figure 9.15, draw a line through the resilient modulus of the subbase (13,000 lb/in^2) to the SN diagram (point P) and, dropping a vertical line down from the proper ΔPSI, the SN for the top two layers is found to be about 3.9.

Step 3: Use the values $SN_1 = 3.0$, $SN_2 = 3.9$, and $SN_3 = 5.0$, with $a_1 = 0.41$, $a_2 = 0.12$, and $a_3 = 0.098$ in a modified form of Equation 9.6 to find layer thicknesses d_1, d_2, and d_3. The value of d_1 is found as follows: $SN_1 = a_1 * d_1$; $3.0 = 0.41 * d_1$; $d_1 = (3.0/0.41) = 7.32$. Check Table 9.7 to make sure that the recommended minimum thickness of the layer has been satisfied. Because the pavement is being designed for 5 million ESALs, the surface layer should be at least 3.5 inches thick. The computed value of d_1 exceeds 3.5, so $d_1 = 7.32$ inches is rounded up to $d_1^* = 7.5$ inches.

TABLE 9.7 Minimum Thickness Standards

Traffic (ESALs)	HMA (inches)	Aggregate Base*
<50,000	1	4
50,000–150,000	2	4
150,001–500,000	2.5	4
500,001–2 million	3	6
2 million to 7 million	3.5	6
7 million	4	6

*Includes cement, lime, and asphalt.

Source: AASHTO, 1993, p. II-35.

Step 4: The two-layer pavement has $SN_2 = 3.9$. The surface layer of the two-layer pavement contributes

$$(a_1 * d_1^*) = 0.41 * 7.5 = 3.075 \text{ to the two-layer } SN_2:$$
$$SN_2 = (a_1 * d_1^*) + (a_2 * d_2); 3.9 = (0.41 * 7.5) + (0.12 * d_2);$$
$$d_2 = \frac{3.9 - 3.075}{0.12} = 6.875 \text{ inches.}$$

Table 9.7 suggests a minimum base thickness of 6 inches for 5 million ESALs. The computed d_2 satisfies that minimum and is rounded up to $d_2^* = 7.0$ inches.

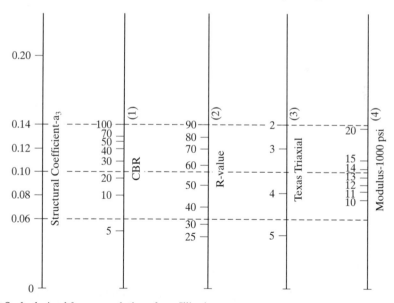

(1) Scale derived from correlations from Illinois.

(2) Scale derived from correlations obtained from The Asphalt Institute, California, New Mexico, and Wyoming.

(3) Scale derived from correlations obtained from Texas.

(4) Scale derived on NCHRP project (3).

FIGURE 9.19

Variation of granular subbase coefficient a_3 with various subbase strength parameters. *Source:* AASHTO, 1993, p. II-21.

Step 5:   Equation 9.6 for the three-layer pavement has become $SN_3 = (a_1 * d_1^*) + (a_2 * d_2^*) + (a_3 * d_3)$. In this equation, only d_3 is unknown. The subbase thickness will be

$$d_3 = \frac{5.0 - 3.075 - (0.12 * 7.0)}{0.098} = \frac{1.085}{0.098} = 11.1 \text{ inches, rounded up to } 11.5 \text{ inches.}$$

According to this design, the pavement will be $7.5 + 7.0 + 11.5 = 26$ inches thick and have an SN of $(7.5 * 0.41) + (7.5 * 0.12) + (11.5 * 0.098) = 5.04$. This SN exceeds the requirement of 5.0 for the entire three-layer pavement.

Example 9.4

Determine the least cost pavement for the design begun above, without going to higher-grade materials. (*Hint*: One approach is to specify the thickness of the surface layer of hot mix asphalt and then determine the thickness of the base and subbase courses in the steps above. In this solution, 4−, 6−, 8−, and 10-inch surface thicknesses will be tried. Be aware that additional excavation will be required for thicker pavement. The cost of added grading and hauling to a borrow pit will also be necessary. The extra excavation and haul costs may offset the saving from using thicker layers of less expensive material. For the cost estimates in this problem, use the values indicated in Table 9.8a.

Solution to Example 9.4   First, to provide an upper bound for the analysis, "design" what is called a *full-depth asphalt* pavement. This means that only hot mix asphalt (HMA) will be used between the pavement's surface and the subgrade. Because HMA is much more expensive than base and subbase materials, a full depth asphalt pavement is expected to be a very costly design. The thickness of the full-depth asphalt pavement would be

$$d_{\text{fulldepth}} = \frac{SN}{a_1} = \frac{5}{0.41} = 12.2 \text{ inches.}$$

Standard practice is to round up the computed thickness to the next one-half inch, making the full-depth asphalt design 12.5 inches thick, as shown at the bottom of the fourth column in Table 9.8b.

For a standard lane 12-feet wide and 1-mile long, the material cost *for each inch* thickness of HMA would be estimated as follows:

TABLE 9.8a   Cost of Materials for Example 9.4

Material	Layer	Layer Coefficient	Specific Gravity	Cost/ton Delivered	Cost per Lane-Mile per Inch Depth
Asphalt concrete	Top/wearing	0.36–0.44	2.65	$90.00	$39,352.50
Asphatic stabilized base	Alternate base	0.24–0.26	2.40	$40.00	$15,840.00
Base-compacted dense aggregate, well graded	Base	0.12–0.16	2.70	$13.00	$5,791.50
Course aggregate	Subbase	0.10–0.11	2.30	$7.00	$2,656.50

TABLE 9.8b Pavement Cost plus Added Excavation Costs in Example 9.4

Asphalt Concrete ($a = 0.41$)	Crushed Stone Base (well graded) ($a = 0.12$)	Subbase Crushed Stone ($a = 0.10$)	Overall Depth of Pavement	SN	Paving Cost	Added Grading @ $3/CY	Estimated Cost/mi to Lay Pavement	Added Grading @ $15/CY	Estimated Cost/mi to Lay Pavement
4"	14.0"	17"	35"	5.02	$283,652	13,200	$296,852	66,000	$349,652
6"	10"	13.5"	29.5"	5.01	$329,893	9,973	$339,866	49,867	$379,759
8"	7"	9"	24"	5.02	$379,269	6,747	$386,016	33,733	$413,002
10"	7.5"	0"	17.5"	5.00	$436,961	2,933	$439,894	14,667	$451,628
12.2"	0"	0"	12.5"	5.125	$491,906	0	$491,906	0	$491,906

$$\text{HMA volume: } 5280 \text{ ft/mi} * 12 \text{ ft} * \frac{1}{12} \text{ ft} = 5280 \text{ ft}^3 \text{ per mile}$$

$$\text{HMA weight: } 2.65 * 62.5 \text{ lb/ft}^3 * 5280 \text{ ft}^3 = 874,500 \text{ lb} = 437.25 \text{ tons}$$

$$\text{HMA cost: } 437.25 \text{ tons} * \$90.00/\text{ton} = \$39,352.50$$

This cost per lane-mile per inch depth appears in Table 9.8a for asphalt concrete. The "costs per lane-mile per inch" for the other materials in Table 9.8a can be computed and verified in a similar fashion. The cost of 12.5 inches of HMA would be $39,352.50 * 12.5 inches = $491,906.25, as shown in the last line of the "Paving Cost" column of Table 9.8b.

In Table 9.8b, the full-depth asphalt design serves as the frame of reference for excavation costs. Any other pavement design of the same strength will require more excavation, because the base and subbase layers will use greater thicknesses of less expensive, less strong materials. The hope is that the additional excavation and hauling costs will be more than offset by the cost savings from the use of less expensive materials. In an attempt to save money, use only 10 inches of HMA and enough base material to maintain a structural number of 5.0. From Equation 9.6,

$$d_2 = \frac{SN - (a_1 * d_1)}{a_2} = \frac{5.0 - (0.41 * 10.0)}{0.12} = \frac{5.0 - 4.1}{0.12} = \frac{0.9}{0.12} = 7.5 \text{ in}$$

There is no need to round up this value, because it is an exact multiple of one-half inch. This layer combination is shown in the next-to-last line of Table 9.8b.

The *extra* depth of excavation needed for the 7.5" + 10" design is $(7.5 + 10) - 12.5 = 5.0$ inches. This translates into $(5.0/12)$ ft thick * 12 ft wide * 5280 ft long $= 26,400$ ft$^3 = 977.8$ cubic yards. The extra cost to excavate and haul soil will differ from site to site. However, for this problem, the cost of extra excavation is set at a low value of $3.00 per cubic yard. At $3/CY, the extra cost to excavate and haul the extra soil in the new two-layer design is $2933. The cost to place 10.0 inches of HMA at $39,352.50/inch is $10.0 * \$39,352.50 = \$393,525$. The cost saved by reducing the HMA thickness from 12.5 inches to 10.0 inches is $(12.5 - 10.0) * \$39,352.50 = \$98,381$. The cost to place 7.5 inches of base material at $5791.50/inch (see Table 9.8a) is $7.5 * \$5791.50 = \$43,436$. The total paving cost becomes $\$393,525 + \$43,436 = \$436,961$. When the added excavation and hauling costs are included, the overall cost becomes $\$436,961 + \$2933 = \$439,894$. Trading in a unit thickness of expensive surface material for a greater thickness of less expensive base material, even with the added excavation costs, saved money while maintaining the strength of the pavement.

The alternative in Table 9.8b that uses 8″ of HMA introduces a subbase layer. The equation for finding d_2 used for the 10″ HMA alternative can be repeated here for 8″ of HMA:

$$d_2 = \frac{SN - (a_1 * d_1)}{a_2} = \frac{5.0 - (0.41 * 8.0)}{0.12} = \frac{5.0 - 3.28}{0.12} = \frac{1.72}{0.12} = 14.3″ \text{ of base}$$

Of the many possible values of d_2 that could be chosen, let us choose 7 inches. This means that the subbase layer will have to be thick enough to ensure that the overall pavement structural number is met.

$$d_3 = \frac{SN - (a_1 * d_1) - (a_2 * d_2)}{a_3} = \frac{5.0 - (0.41 * 8.0) - (0.12 * 7.0)}{0.10}$$

$$= \frac{5.0 - 3.28 - 0.84}{0.10} = \frac{0.88}{0.10} = 8.8″ \text{ of subbase}$$

In Table 9.8b, this is rounded to 9.0 inches. Again the trade-off is economical—the cost continues to decline.

Using different combinations of layer thickness that preserve the overall SN = 5.0 value, a variety of layer designs have been created in Table 9.8b. The total cost for each layer design is found in the right-hand column of Table 9.8b. It indicates that, for the costs given, the 4″ HMA + 14″ base + 17″ subbase will be the least expensive overall. In practice, the pavement design depends on the availability (and therefore the price) of the materials. Even if, because the borrow pits are some distance away, the grading/excavation costs were $15 (see shaded columns in Table 9.8b), the first pavement listed in Table 9.8b still has the lowest cost. However, operational considerations, like the availability of paving equipment, might swing the construction recommendation in favor of some other design.

THINK ABOUT IT

Would the cheapest pavement to put into place always be the one that uses the minimum thicknesses of the more expensive materials?

Example 9.5

In Example 9.2, we determined that a road to a quarry would carry 705,000 ESALs. The road from the interstate to the quarry road rests on a mucky soil (CBR = 5; modulus of resiliency = 7500). Because rock will be the cheapest material to get, it has been determined that the pavement will maximize the use of crushed rock ($a_2 = 0.12$; $a_3 = 0.10$). The state standard for the minimum thickness of a hot mix asphalt concrete surface layer (4 inches) will be observed. How should the pavement be designed? Reliability is to be 80 percent and the standard deviation 0.5, with ΔPSI = 2.5. The asphalt coefficient $a_1 = 0.32$, reflecting a lower grade asphalt.

Solution to Example 9.5   The pavement structural number is determined from the nomograph in Figure 9.20. The structural number for the pavement is 3.0.

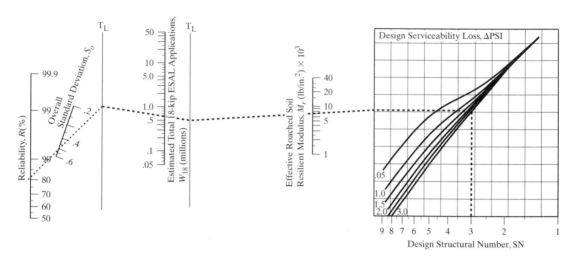

FIGURE 9.20
Nomograph for Example 9.5.

Four inches of HMA will give a structural number of 1.28, leaving 1.72 of SN to be made up from crushed stone. Six inches of base at $a_2 = 0.12$ will add 0.72 to the SN, leaving 1.0 of the SN to be made up from 10 inches of the subbase.

9.4 RIGID PAVEMENT DESIGN

9.4.1 Rigid Pavement Materials

Rigid pavements are normally constructed of *Portland cement concrete (PCC)*, and they may have a base course between the bottom of the concrete slab and the subgrade. The base course is also referred to as the subbase. The thickness of PCC slabs used for roads runs between 6 and 13 inches. The action of a rigid slab on the subbase or soil is that of a distributed load as shown in Figure 9.21. The materials used in PCC

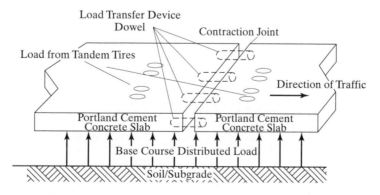

FIGURE 9.21
Rigid pavement components.

are Portland cement, aggregates (both coarse and fine), and water. Steel (called temperature steel) is often used as a mesh to control cracking due to temperature variation. Steel bars 1 to 2 inches in diameter are inserted every foot or so along the transverse direction to transfer loads as the vehicle passes from one section of slab to the next. (See Figures 9.21 and 9.22.) These are called dowel bars and act to transfer the load as the tires move across the joint.

As Figure 9.23 shows, 3/4-inch bars with hooks or irregular shapes are fashioned to enhance the bonding of two sections.

FIGURE 9.22
Rigid pavement being placed. Photo courtesy of Adam Hand, Ph.D.

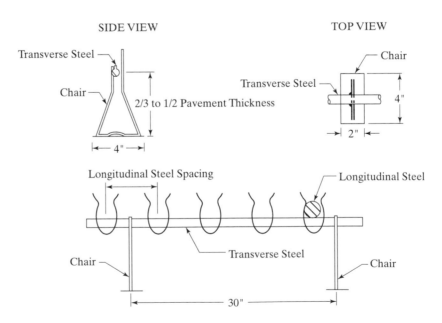

FIGURE 9.23
Transverse steel detail for rigid pavement. *Source*: FHWA, 1990.

Portland cement is composed of very fine calcium silicate compounds mixed with some gypsum. The mixtures generally used for roads have been blended for high durability with a high resistance to deicing salt compounds. The coarse aggregate is well graded and usually remains in a No. 4 sieve. There are special tests to determine the aggregate's ability to resist abrasion and withstand freezing and thawing without breaking up. The fine aggregate is usually well-controlled sand. Both aggregates must be thoroughly clean. The water to be used is to be at least as pure as drinking water.

Portland cement concrete, once placed, will gain strength over time as it is allowed to cure. The usual time for measuring the structural capability of the PCC is 28 days after it is placed. PCC often shows early cracking due to the many stresses that are set up in the slab as it is curing. Most of these cracks do not appreciably reduce the structural capability of the pavement.

9.4.2   ESALs for Rigid Pavement

Because of the way rigid pavement behaves, the ESAL determination is different from flexible pavement. Tables (AASHTO, 1993, Appendix D) have been prepared for the precise determination of ESALs for various thicknesses of the PCC slab. However, recall that the fourth power formula for tandem axles on rigid pavement (Equation 9.3) has a reference weight of 29,000 pounds, instead of the 33,200 pounds used for flexible pavement. (The single-axle reference weight is still 18,000 pounds.) Figure 9.24 shows how well the fourth-power approximation works. In fact, the fourth power curve in Figure 9.24 is almost identical to the plot of values in the AASHTO ESAL tables for a slab that is 8 inches thick.

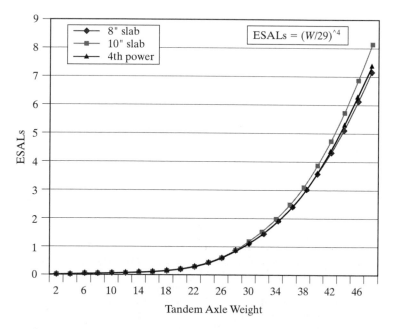

FIGURE 9.24
Comparison of tandem wheel loads fourth power formula vs. ESAL tables for 8″ and 10″ PCC.

Example 9.6

Determine the ESALs on rigid pavement for an 18-wheeler with 12 kips on the steering axle, 36 kips on the tractor's tandem axle, and 32 kips on the semitrailer's tandem axle.

Solution to Example 9.6

Equation 9.3 gives us $\text{ESAL}_{\text{rigid}} = \left(\dfrac{12}{18}\right)^4 + \left(\dfrac{36}{29}\right)^4 + \left(\dfrac{32}{29}\right)^4 = 0.2 + 2.37 + 1.48 = 4.05$

In the AASHTO tables, a 9-, 10-, or 11-inch slab would have yielded an ESAL for this truck of 4.0, 4.15, or 4.2, respectively.

9.4.3   Rigid Pavement Subgrade and Subbase

Figure 9.21 shows how the load is distributed on a rigid pavement section. The slab of concrete distributes the load evenly across the whole slab. Hence, the subbase (if any) and soil underneath the slab act as a large spring. Its effective spring constant is important and must be determined first. There are usually two layers involved below the PCC slab that affect modulus. First is the soil with its modulus. Second is the subbase material, which may be any number of materials, ranging from cement-treated aggregate to lime-stabilized soil. The *elastic modulus (E)* for each common subbase material is shown in Table 9.9. For the equivalent pavement strength with a given slab thickness, the poorer the soil, the stronger the subgrade should be. (The Loss of Support values to use in homework exercises in this text book are given at the right end of each row in Table 9.9.)

The resilient modulus of the subgrade (discussed briefly in section 9.1) is combined with the elastic modulus of the subbase and the depth of the subbase in inches to form a *composite spring constant modulus* for the subgrade, with units pounds per cubic inch. (The units for the composite modulus are actually pounds per square inch per inch of deformation or simply pounds per cubic inch).

Figure 9.25 shows how combining the subbase and subgrade moduli forms the composite modulus k that supports the PCC slab. For example, consider a 6-inch PCC slab on a lime-stabilized subbase material with $E = 50,000$ psi over a roadbed soil with a modulus of resiliency of 7000 psi, the composite modulus of subgrade reaction is

TABLE 9.9   Typical Ranges of Loss of Support Factors for Various Subbase Materials

Type of Material	Loss of Support
Cement-treated granular base ($E = 1,000,000–2,000,000$ lb/in^2)	0.0–1.0 (use 0.0)
Cement aggregate mixtures ($E = 500,000–1,000,000$ lb/in^2)	0.0–1.0 (use 0.0)
Asphalt-treated base ($E = 350,000–1,000,000$ lb/in^2)	0.0–1.0 (use 1.0)
Bituminous stabilized mixtures ($E = 40,000–300,000$ lb/in^2)	0.0–1.0 (use 1.0)
Lime-stabilized mixtures ($E = 20,000–70,000$ lb/in^2)	1.0–3.0 (use 1.5)
Unbound granular materials ($E = 15,000–45,000$ lb/in^2)	1.0–3.0 (use 2.0)
Fine grained or natural subgrade materials ($E = 3,000–40,000$ lb/in^2)	2.0–3.0 (use 3.0)

Source: AASHTO, 1993, Table 2.7, p. II-27.

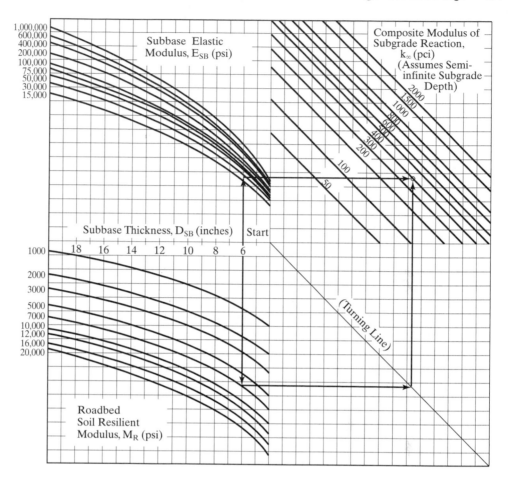

FIGURE 9.25
Chart for estimating composite modulus of subgrade reaction. *Source:* AASHTO, 1993, Figure 3.3, p. II-39.

400 pci. Start at the $D_{SB} = 6$ inches point shown in Figure 9.25. Draw a line up to the $E_{SB} = 50,000$ curve. From that curve, draw a horizontal line through the k_∞ quadrant of the figure. From the same $D_{SB} = 6$ inches starting point, draw a line down to the MR $= 7000$ curve. From the curve, draw a horizontal line to the turning line. From the turning line, draw a vertical line up through the k_∞ quadrant. Where the two lines in the k_∞ quadrant intersect, read the value of k_∞. This is the *uncorrected composite modulus.*

When the composite modulus k is corrected for the type of material, the correction is called the *loss of support* LS. The correction is based on the materials used for the subbase or base. As indicated in Table 9.9, when the elastic modulus of the base is more than 500,000, there is little loss of support. Accounting for the loss of support produces an *effective modulus* K_{LS}. Equation 9.8 uses the factors a and b given in Table 9.10.

$$k_{LS} = a*(k_\infty)^b \qquad (9.8)$$

For a lime-stabilized mixture with $E = 30,000$ lb/in^2 and a composite modulus of subgrade reaction $K_\infty = 400$ pci, Table 9.9 indicates that LS $= 1.5$ should be used. Using

TABLE 9.10 Loss of Support Factors for Example 9.7

Loss of Support	a	b	Example 9.7	
0	1.00	1.00	500	1000
1	0.93	0.82	151	268
1.5	0.87	0.7	67	110
2	0.80	0.635	41	64
3	0.56	0.56	18	27

Equation 9.8 and Table 9.10, $K_{LS} = 0.88(400)^{0.7} \approx 58$. The curves used in Figure 9.25 are for a subgrade that is "semi-infinite." Thus, another correction might be necessary if there is a solid or rigid foundation within 10 feet of the surface.

Example 9.7

Determine the loss of support when the composite modulus obtained from Figure 9.25 is 500 lb/in^3.

Solution to Example 9.7 The solution uses Equation 9.8 and is shown in the shaded portion of Table 9.10.

Example 9.8

Determine the subbase thickness required to have a composite modulus of 200, when the soil has a modulus of 5000 lb/in^2 and the asphalt treated base has an elastic modulus of 400,000 lb/in^2.

Solution to Example 9.8 According to Table 9.9, the loss of support is 1.0. Using Figure 9.25, Table 9.10, and Equation 9.8, various thicknesses of subbase are tried, resulting in the answers shown in Table 9.11. With the loss of support (LS) of 1.0, the required modulus of 200 results in a slab depth of 12 inches.

TABLE 9.11 Solution for Example 9.8

Subbase Thickness (in)	Composite Modulus, k_∞	K_{LS} when LS = 1
14	800	223
12	700	200
10	600	176
8	500	152

Example 9.9

The soil on which you plan to place a PCC road has an *average* M_R of 5000 lb/in^2. (Note that the M_R changes by month or season as a function of temperature and moisture.) You plan to use a lime treatment on the top 8 inches of the soil. The resulting lime-treated soil (loss of support = 2) will have an elastic modulus of 50,000 lb/in^2. What is the composite modulus of the subgrade and subbase?

Solution to Example 9.9   Enter Figure 9.25 at 8 inches and draw the line up to 50,000 and down to 5,000 and then find the uncorrected composite reaction modulus, k_{uc} of 450 lb/in^3. The loss of support will be about 2. The final value for the corrected composite modulus from Equation 9.8 will be $k_{ls} = 0.8 k_{uc}^{.635} = 0.8 \times (450)^{.635} \approx 40$

9.4.4   Equation to Determine Rigid Pavement Thickness

Equation 9.9 below has been developed for the designer to determine the thickness of the rigid pavement. In a manner similar to the flexible pavement, a nomograph has been developed to help the pavement designer find an acceptable pavement thickness. The rigid pavement nomograph has two segments. The first segment (Figure 9.26) has to do with the *materials* being used in the pavement. The second segment (Figure 9.27) deals with the impact of the *highway loads* on the pavement. The AASHTO rigid pavement design considering all the variables is shown in Equation 9.9:

$$\log_{10} W_{18} = Z_R S_o + 7.35 \log(D + 1) - 0.06 + \frac{\log\left[\dfrac{\Delta PSI}{4.5 - 1.5}\right]}{1 + \dfrac{1.624 * 10^7}{(D + 1)^{8.46}}}$$

$$+ (4.22 - 0.32\, p_t) \log\left[\frac{S_c'\, C_d(D^{0.75} - 1.132)}{215.63J\left(D^{0.75} - \dfrac{18.42}{(E_c/k)^{0.25}}\right)}\right] \qquad (9.9)$$

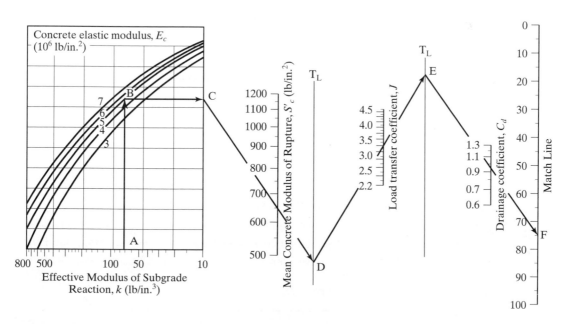

FIGURE 9.26
Materials portion of rigid pavement nomograph. *Source:* AASHTO, 1993, Figure 3.7, p. II-45.

where

$$D = \text{thickness of PCC pavement (inches)}$$
$$S'_c = \text{modulus of rupture of the concrete (psi)}$$
$$E_c = \text{elastic modulus of the concrete (psi)}$$
$$k = \text{modulus of subgrade reaction (lb/in}^3)$$
$$J = \text{joint load transfer coefficient}$$
$$C_d = \text{drainage coefficient}$$
$$\Delta\text{PSI} = \text{loss of serviceability } (p_i - p_t)$$
$$p_i = \text{initial serviceability}$$
$$p_t = \text{terminal serviceability}$$
$$W_{18} = \text{cumulative 18K ESALs over time } t$$
$$Z_R = \text{standard normal deviate for reliability R}$$
$$S_o = \text{overall standard deviation}$$

The last six quantities are the same as those discussed in Section 9.3. They are used in the second part of the rigid pavement nomograph—Figure 9.27. The five properties listed after D are specific to rigid pavement. The thickness **D** of PCC pavement in inches is what we want to determine. This thickness must be able to withstand the loads over a given period of time, with a loss in serviceability ΔPSI no greater than that specified.

9.4.5   Materials Portion of the Design Nomograph

The nomograph of Figure 9.26 is used to introduce the material characteristics to the pavement design. The right edge of Figure 9.26 interfaces with the left edge of Figure 9.27, which contains the highway input, to determine the pavement thickness.

Example 9.10

The composite elastic modulus is 72 lb/in³ and the elastic modulus of the concrete is 5×10^6 lb/in². The modulus of rupture is 650 lb/in². The load transfer is 3.2 and the moisture damage coefficient is 1.0. Determine the intersection of the match line on the materials portion of the nomograph.

Solution to Example 9.10

Step 1:   Determine the effective modulus of subgrade reaction k (72 lb/in²). Point A

Step 2:   Locate the intersection with the concrete elastic modulus E_c (5×10^6 lb/in²). Point B

Step 3:   Draw a horizontal line to chart edge. Point C

Step 4:   Draw line through modulus of rupture S'_c (650 lb/in²) to the first turning line Point D

Step 5:   Draw line through the load transfer coefficient J (3.2) to the second turning line. Point E

Step 6:   Draw line through drainage coefficient C_j (1.0) to the match line for the materials portion. Point F

Step 7:   Record the match line value of 75.

9.4.6 Field Determination of PCC Characteristics

To maintain quality in the field when concrete is placed, a test sample is prepared. The elastic modulus of the concrete in pounds per square inch (E_c) and the modulus of rupture of the concrete in pounds per square inch (S'_c) are developed from tests on core samples of the PCC to be used. The 6-inch core samples are tested to failure, from which both quantities may be determined. The formulas are given in Equations 9.9 and 9.10. (Source: Professor Julio Ramirez, Purdue University)

$$E_c = 33\omega^{1.5} \times f_c^{0.5} \tag{9.10}$$

$$S'_c = 7.5 \times f_c^{0.5} \tag{9.11}$$

where

ω = the density of the sample (pcf)
f_c = pressure at which the sample fails (psi)
S'_c = modulus of rupture. (psi)

Example 9.11

A 6-inch diameter test specimen is being used to determine the modulus of rupture and the elastic modulus of the PCC. The specimen buckles at an applied force of 325,000 pounds. The density of the specimen is 141 pounds per cubic foot. Determine the modulus of rupture and the elastic modulus of the PCC.

Solution to Example 9.11 The force applied to the cross-sectional area is equivalent to:

$$f_c = \frac{325,000 \text{ lb}}{\pi * \left(\dfrac{6}{2}\right)^2} = 11,495 \text{ lb/in}^2$$

Using Equation 9.10, the elastic modulus of the PCC is computed to be

$$E_c = 33 * \omega^{1.5} * f_c^{0.5} = 33 * (141)^{1.5} * (11,495)^{0.5} = 33 * 1674.28 * 107.21 = 5.923 * 10^6 \text{ lb/in}^2.$$

The modulus of rupture is found by using Equation 9.11:

$$S_c = 7.5 * f_c^{0.5} = 7.5 * (11,495)^{0.5} = 7.5 * 107.21 = 804 \text{ lb/in}^2.$$

9.4.7 Highway Portion of the Design Nomograph

The steps below describe how to incorporate highway performance criteria (e.g., load, reliability, and standard deviation) and serviceability loss in determining concrete slab thickness (see Figure 9.27).

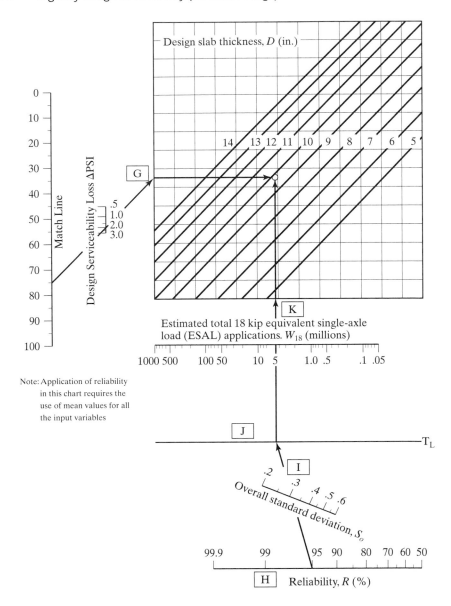

FIGURE 9.27
Highway performance portion of rigid pavement nomograph. *Source:* AASHTO, 1993, Figure 3.7, p. II-46.

Example 9.12

Complete the pavement thickness determination of Example 9.10 by using the steps shown below.

Solution to Example 9.12

Step 8:   Beginning with the match line point from step 7 in Example 9.10 (Figure 9.26) on the second part of the nomograph (Figure 9.27), draw a line through the design serviceability loss (ΔPSI = 2.0) to the edge of the design slab thickness grid. Point G

Step 9:   Find the desired reliability on the reliability scale (95 percent). Point H

Step 10:   Draw a line through the standard deviation S_o (0.3) at Point I to the intermediate turning line. Point J

Step 11:   Draw a straight line from there through the W_{18} load application (5 million ESAL) to the edge of the design slab thickness grid. Point K. Do not bend the line being drawn at the W_{18} scale line!

Step 12:   At the intersection of the *horizontal* line from point G and the *vertical* line from point K, read the slab thickness (10 inches).

Example 9.13

An adequate depth of rigid pavement for a rural stretch of highway is sought, where the first year ESALs are 200,000 and the expected annual growth rate is 4 percent. The pavement should last 20 years, with a serviceability loss of no more than 2.0. The soil and subbase have an effective reaction modulus k of 170 lb/in^3. The concrete has an elastic modulus of 5.5×10^6, a modulus of rupture of 700 psi, a load transfer of 3.2, and drainage 1.0. Use 95 percent reliability and $S_o = 0.28$.

Solution to Example 9.13   The dashed lines in Figure 9.28 hit the match line at a value of 66. The ESAL growth has a 20-year compound growth of 29.78. (See Equation 9.4 in Section 9.2.) The design value of ESAL is 6 million. In Figure 9.29, the dashed lines again apply. The

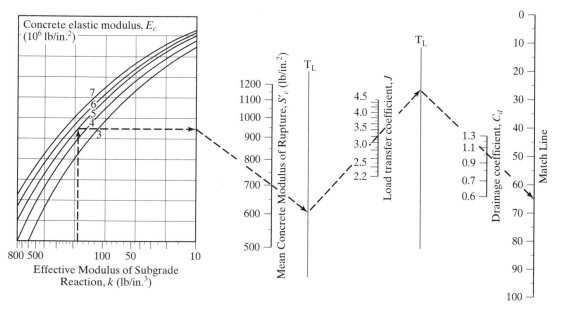

FIGURE 9.28
Materials portion of design nomograph solution for Example 9.13.

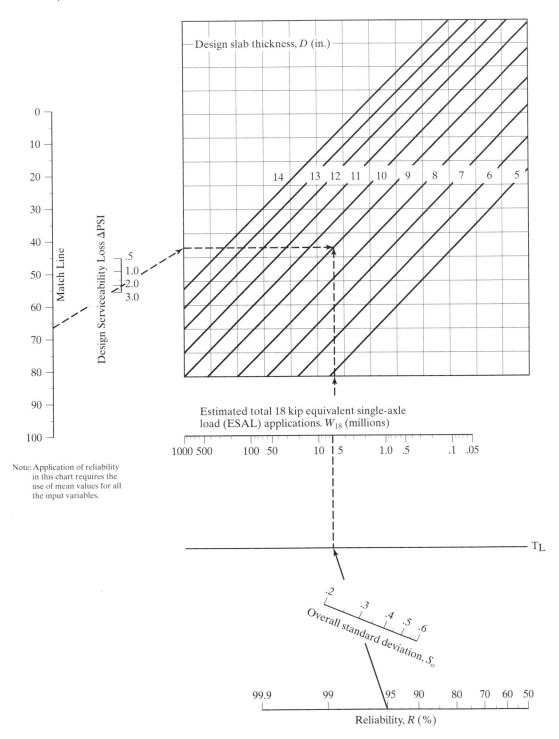

FIGURE 9.29
Highway portion of design nomograph solution for Example 9.13.

horizontal and vertical dashed lines within the "Design slab thickness" grid intersect at $D = 9$ inches. This is the design thickness for the pavement.

___

Example 9.14

Determine the thickness of a rigid pavement section of freeway (95% reliability) to accommodate 20 million ESALs (standard deviation of 0.4) sustaining a serviceability loss from PSI of 4.7 to 2.2. The soil has been stabilized with lime and compacted, giving a composite spring modulus of 40 lb/in³ (obtained from Example 9.10). Six-inch diameter specimens of Portland cement concrete with a density of 134 lb per cubic foot have been tested in the lab after 28 days of curing. The test specimens fail under a load of 416,000 lb. How thick should the PCC be if the effective load transfer across the joints is 3.8?

Solution to Example 9.14 From standard materials data, we learn that the elastic modulus E_c and modulus of rupture S_c' can be found from the test. The load applied to the 6-inch diameter test specimen means that the 416,000 lb is equivalent to f_c of 14,712 lb/in². With a density of 134 lb per cubic foot, Equation 9.10 produces

$$E_c = 33\omega^{1.5} \times f_c^{0.5} = 6.21 \times 10^6 \text{ lb/in}^2,$$

and Equation 9.11 leads to

$$S_c' = 7.5 \times f_c^{0.5} = 900 \text{ lb/in}^2.$$

In Figure 9.30, the match line value = 68. In Figure 9.31, after entering 95 percent reliability, 0.4 standard deviation, and 20 million ESALs, the required slab thickness is 11.5 inches.

___

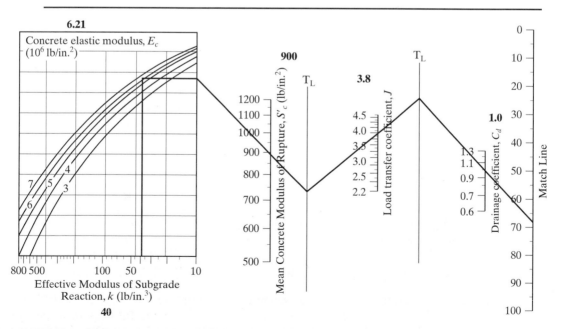

FIGURE 9.30
Materials portion of design nomograph solution for Example 9.14.

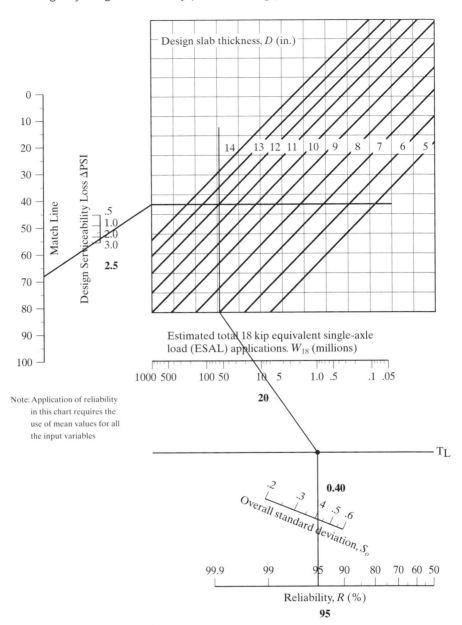

FIGURE 9.31
Highway portion of design nomograph solution for Example 9.14.

9.5   PAVEMENT MANAGEMENT SYSTEMS

For many governments, the road system is among the largest capital assets for which they are responsible. As pavements near the end of their useful lives, or as pavement deficiencies develop, decisions must be made about where, when, and how to apply limited budgets to address those problems (see Figure 9.32). That is pavement management.

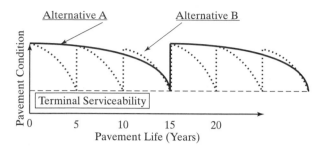

FIGURE 9.32
Performance curves for two rehabilitation or maintenance strategies.

Pavement Management, in its broadest sense, includes all the activities involved in the planning and programming, design, construction, maintenance, and rehabilitation of the pavement portion of a public works program. A *pavement management system (PMS)* is a set of tools or methods that assist decision makers in finding optimum strategies for providing and maintaining pavements in a serviceable condition over a given period of time. (Haas et al., 1994).

9.5.1   Road Inventory Data

An effective pavement management system requires good information about the road system. Important data items are as follows:

- *Pavement type.*   Flexible or rigid surface, details about underlying layers
- *Pavement performance data.*   Roughness, surface distress, deflection, friction, layer material properties
- *Road type.*   Function of road segment in road network, AADT, peak hour directional volumes
- *Historic data.*   Past maintenance and construction, traffic accidents, cost data, weather records
- *Geometric data.*   lane widths, curvature, slope, grade, shoulder/curb

Table 9.12 shows data from a section of major state road in one county in Indiana. It can be helpful for analysis and presentation if the road inventory can be tied into a geographic information system. One of the critical tasks is to keep the database up to date. Changing surface conditions, changing traffic conditions, and maintenance activities performed on roadway segments must be entered into the database on a regular basis, so that PMS decisions can be based on current roadway characteristics.

9.5.2   Forms of Pavement Distress

In section 9.1, the way in which pavement condition is measured was discussed. Over the years, measurement techniques have become more sophisticated. Being able to categorize the highway performance through its deficiencies, such as roughness, washboarding, cracking, patch deficiency, rutting, spalling, rigid pavement blowup, transverse joint deterioration, lane-to-shoulder dropoff, water bleeding and

TABLE 9.12 Typical Road Inventory for 5.26 Mile Stretch of US 27

County	Description	Route	Length—ft.	Urban or Rural	Jurisdiction (Fed, St., Co., etc.)	Functional Class	Number of Lanes	Lane Width—ft. Undivided	Divided	Width Left Shoulder	Width Right Shoulder	Road Width	International Roughness Index (date measured)	Capacity—vph	Service Flow vph	PM PkHr Volume Vehicles	AM PkHr Volume Vehicles	Average Annual Daily Traffic (AADT)	Percent Heavy Trucks
1	US 27 Jay Co LINE	US27	0.86	Rural	St	4	2	12	U	4	4	24	62(01)	2800	2164	302	259	6197	18
1	Detail Item Change	US27	0.06	Rural	St	4	2	12	U	4	4	24	62(01)	2800	2164	302	259	6197	18
1	Geneva Corp Line	US27	0.1	Urban	St	4	2	12	U	6	6	24	56(01)	2800	2373	302	259	6197	18
1	North St. (IR174 RT and IR 6 LT)	US27	0.36	Urban	St	4	2	12	U	6	6	24	56(01)	2800	2373	443	381	9095	18
1	6th Street RT	US27	0.07	Urban	St	4	2	12	U	6	6	24	56(01)	2800	2373	443	381	9095	18
1	Shackley St.	US27	0.07	Urban	St	4	2	12	U	6	0	32	95(02)	2800	1746	443	381	9095	18
1	SR 116 (Line St)	US27	0.48	Urban	St	4	2	12	U	4	0	24	95(02)	2800	1746	375	322	7695	18
1	Detail Item Change	US27	0.61	Urban	St	4	2	12	U	6	6	24	83(00)	2800	2373	375	322	7695	18
1	Geneva Corp Line	US27	0.61	Urban	St	4	2	12	U	6	6	24	83(00)	2800	2373	375	322	7695	18
1	Detail Item Change	US27	0.2	Rural	St	4	2	12	U	9	6	24	83(00)	2800	2373	375	322	7695	18
1	Detail Item Change	US27	1.73	Rural	St	4	2	12	U	7	8	24	83(00)	2800	2373	375	322	7695	18
1	IR 20 (700S)	US27	0.11	Rural	St	4	2	12	U	7	8	24	83(00)	2800	2373	547	470	11,222	18

Source: Adapted from Indiana Department of Transportation Road Inventory.

pumping, is essential to the beginning of the pavement management system. The other basic need is a full inventory of the highway system by section and type of pavement. Figures 9.33 and 9.34 show several of the more common pavement distress conditions. Figure 9.33 indicates problems with flexible pavement, except for Figure 9.33c, which shows the problem when asphalt is used as an overlay for Portland cement concrete pavement. Figure 9.34 shows distress in PCC pavement. Machines called profilometers, profilographs, or simply profilers make measurements from which the pavement strength can be calculated. Two versions of profilers are pictured in Figure 9.4.

9.5.3 Pavement Management Strategies

An effective pavement management system (PMS) requires that engineers and budget analysts work together in developing trade-offs between sections of pavement that need attention. Some of the trade-offs will be the economic evaluation of alternative strategies, such as deciding

- Which materials and thickness of layers are to be used.
- Between pavement overlay and complete rehabilitation.
- Between patching and section replacement.
- Between various road geometric alterations and signage to improve safety.

The final selection of a design strategy for implementation is partially subjective. Although the economic analysis must form a key basis for the decision, no firm decision rules exist that can be followed exactly (Haas et al., 1994).

A typical structure for a pavement management system is shown in Figure 9.35. A review of the flow diagram will show extensive use of the pavement design methods discussed earlier in this chapter and of the evaluation techniques discussed in Chapter 5. The standard inputs for a PMS are (Haas et al., 1994):

- Material properties.
- Environmental and serviceability parameters.
- Load and traffic variables.
- Maintenance variables.
- Program control variables such as (1) minimum time to first overlay, (2) minimum time between overlays, (3) maximum funds available for construction, (4) maximum total thickness of construction, (5) minimum thickness of a single overlay and (6) maximum accumulated total thickness of all overlays.
- Traffic delay variables. Traffic control in work zones will include concern for waiting times and delays that may well determine whether work will be scheduled during the day or night.

The implementation of a PMS brings in the economic trade-offs indicated above. The result will be a set of projects ranked by need and, preferably, scheduled by year for the long-term good of the road network.

FIGURE 9.33
Pavement distress in asphalt concrete pavements. (a) Alligator pattern cracking/spalled interconnected cracks occurs in areas subjected to repeated traffic loads. Shows fatigue due to repeated wheel loads. (b) High severity longitudinal cracking occurs predominately parallel to pavement centerline, may be wide gap with adjacent random cracking. (c) Cracks in asphalt concrete overlay surfaces that occur over joints in concrete pavements. High severity reflection cracking at joints. (d) High severity pothole: Close-up view, showing the loss of material as well as the associated fatigue cracking and spalling. (e) Rutting: longitudinal surface depression in wheel path. (f) Water bleeding and pumping: seeping or ejection of water from beneath the pavement through cracks, sometimes leaves deposits of fine material on surface. *Source*: Strategic Highway Research Program, 1993.

FIGURE 9.34

Pavement distress in jointed portland cement concrete pavements. (a) Corner breaks: portion of the slab separated by a crack at a 45 degree angle with direction of traffic. Medium severity crack shown. (b) Durability cracking: closely spaced, crescent-shaped hairline cracks. Some small pieces displaced. Moderate severity shown. (c) Longitudinal cracking: cracking that occurs predominately parallel to the pavement centerline. Spalling occurs as the severity increases. (d) Joint seal failure: enables a significant amount of water to infiltrate joint from surface. Results from bonding failure, splitting or loss of sealant. (e) Scaling is the deterioration of the upper concrete slab surface. May occur anywhere over the surface. (f) Blowups: localized upward movement of the pavement surface at transverse joints or cracks, often accompanied by shattering the concrete in that area and exposing reinforcement steel. *Source*: Strategic Highway Research Program, 1993.

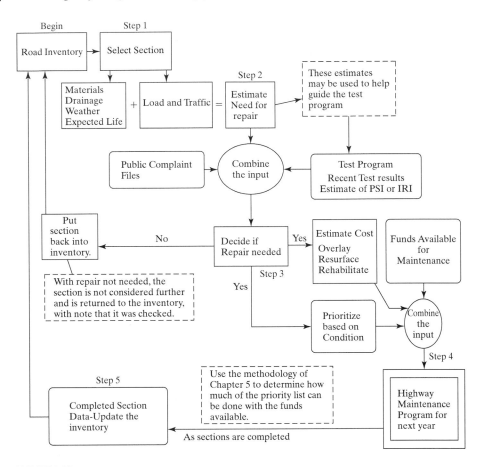

FIGURE 9.35
Flow diagram of a pavement management system.

SUMMARY

"Highway Design for Rideability" is an activity that involves a variety of considerations, including:

- What is the amount and nature of the traffic using (and expected to use) the roadway?
- For how many years should the roadway's surface be expected to be "rideable"?
- How much maintenance effort will be devoted to this roadway during its design life?
- What materials are available for the construction of the roadway?
- What are the soil and drainage conditions that affect the roadway?
- How much money is available for the construction and maintenance of the roadway?

Pavement design decisions often involve trade-offs. For example, should an expensive pavement be designed, hoping for a long life, or should a less expensive pavement be designed, with the knowledge that it will need to be resurfaced or reconstructed years earlier?

In this chapter, the role of vehicle loadings, soil conditions, material characteristics, and other factors that affect pavement design have been discussed. Standard methods to design flexible and rigid pavements have been presented, based on the two principles of load distribution and the results of the AASHO road test. For the typical local government, the road system is among its largest capital assets. Pavement conditions are apparent and important to the traveling public. A system for managing these assets is described in the section on pavement management systems.

ABBREVIATIONS AND NOTATION

a_i	coefficients of flexible pavement layer strength
AASHTO	The American Association of State Highway and Transportation Officials
base course	layer under the surface layer
blowup	rigid pavement distress condition
C	soil classification clay
CBR	California bearing ratio
CH	soil classification fat clays
CL	soil classification lean clays
C_d	drainage coefficient
cracking	flexible pavement distress condition,
d_i	thickness in inches of the pavement layer
D	thickness of rigid pavement (inches)
E	elastic modulus
E_{sb}	subbase elastic modulus
E_c	elastic modulus of the concrete (psi)
ESALs	equivalent single-axle (18,000 pound) loads
FHWA	Federal Highway Administration
f_c	load under which test specimen fails
flexible pavement	asphalt concrete pavement flexes and deforms under application of large loads
g	compound growth rate in percent per year
G	soil classification gravel
GIS	geographic information system
GVW	gross vehicle weight
hma	hot mix asphalt
IRI	International Roughness Index
J	joint load transfer coefficient
k	subgrade modulus (pci)
k_∞	composite modulus of subgrade reaction semi-infinite subgrade depth
k_{LS}	composite modulus accounts for loss of support
kips	thousands of pounds per square inch
LTPP	Long-Term Pavement Performance Program
m_i	modifiers for more or less than normal amounts of moisture
M	soil classification silty or micaceous
MH	soil classification clays or distomaceous soils
Mpa	million pascals
M_R	modulus of resiliency
O	soil classification organic
OH	soil classification fat organic clays

OL	soil classification lean organic clays
P	soil classification poorly graded
PCC	Portland cement concrete
PMS	pavement management system
PSI	Present Serviceability Index
p_i	initial serviceability
p_t	terminal serviceability
profilometers	measures pavement profile
profilographs	pavement profiles
psi	pounds per square inch
PT	soil classification peat
rigid pavement	composed of portland concrete cement
roughness	pavement distress condition
rutting	flexible pavement distress condition
S	soil classification sandy
SHRP	Strategic Highway Research Program
spalling	flexible pavement distress condition
transverse joint deterioration	rigid pavement distress condition
TRB	Transportation Research Board
SAMP	systems analysis method for pavements
Single axle	one axle apart from all other axles
S_o	overall standard deviation of reliability of ESAL calculation
SN	required structural number of layer or pavement
S'_c	modulus of rupture of the concrete (psi)
subbase course	layer between base and subgrade/soil
subgrade	soil under the pavement
surface layer	usually hot mix asphalt layer is often coarse bituminous layer with a very thin glaze
tandem axle	two truck axles close together
tridem axle	three truck axles close together
TSI	Terminal Serviceability Index
U	soil classification uniformly graded
W	soil classification well graded
water bleeding and pumping	rigid pavement distress condition
washboarding	flexible pavement distress condition
WIM	weigh-in-motion
Z_R	standard normal deviate
2A	single truck with two axles and four tires or pick-up
2D	heavy single-unit truck with two axles and six tires
3SU	heavy single-unit truck with three axles
4SU	heavy single-unit truck with four axles
2-S1	heavy tractor-semitrailer truck with three axles
2-S2	heavy tractor-semitrailer truck with four axles
3-S2	heavy tractor-semitrailer truck with five axles or 18-wheeler
2-S1-2	heavy tractor-semitrailer-trailer with five axles twin-tandem double
3-S2-4	heavy tractor-semitrailer-trailer with nine axles
ω	density of concrete (pounds per cubic foot)

REFERENCES

[1] Alaska Department of Transportation and Public Facilities, *Alaska Pavement Design Manual*, Alaska DOT, 1982.

[2] American Association of State Highway and Transportation Officials, Guide for the Design of Pavements, AASHTO, Washington, 1986.

[3] Asphalt Institute, *Asphalt Pavements for Highways and Streets*, Manual #1 (*Thickness Design*), Asphalt Institute, 1996.

[4] Carey, W.N. and P.E. Irick, "The Pavement Serviceability-Performance Concept," HRB Bulletin 250, National Research Council, Washington, DC. 1960.

[5] Federal Aviation Administration Advisory Circular 150/5345, *Pavement Design for Aircraft Runways* (Soils Chart).

[6] FAA, *Civil Airfield Pavements for Seasonal Frost and Perma Frost Conditions*, from Report FAA-RD-74-30, 1974.

[7] Garber, N.J. and L.A. Hoel, *Traffic and Highway Engineering*, Second Edition, PWS Publishing, Boston, MA, 1996.

[8] Gopinath, D, M. Ben-Akiva, and R. Ramaswamy, "Modeling Performance of Highway Pavements," Transportation Research Report 1449, *Design and Rehabilitation of Pavements*, TRB, Washington DC, 1994, pp. 1–7.

[9] Haas, R., W.R. Hudson, and J. Zaniewski, *Modern Pavement Management*, Krieger Publishing, FL, 1994.

[10] Hall, K.D. and M.R. Thompson, "Soil Property-Based Subgrade Resilient Modulus Estimation for Flexible Pavement Design," Transportation Research Report 1449, *Design and Rehabilitation of Pavements*, TRB, Washington DC, 1994, pp. 30–38.

[11] Kher, R. K. and M. I. Darter, "Probabilistic Concepts and Their Applications to AASHTO Interim Guide for Design of Rigid Pavement," Highway Research Record No. 466. Highway Research Board. Washington, DC, 1973.

[12] Lew, J.J., G.D. Hooker, and K.J. Kercher, *Pilot Study of a Portable High Speed Weigh-in-Motion System*, FHWA/IN/RD-88/1. Report by Indiana DOT on HPR Contract 2378 (025), August 1988.

[13] McCall, B., *WIM Handbook*, Center for Iowa Transportation, Iowa State University, Ames, IA, August 1997.

[14] Moore, R.K., G.N. Clark, and G.N. Plumb, "Present Serviceability-Roughness Correlation's Using Rating Panel Data," Transportation Research Record 1117, *Pavement Evaluation and Rehabilitation*, TRB, Washington D.C., 1987, pp. 152–158.

[15] National Research Council, *Distress Identification Manual for the Long-term Pavement Performance Project*. Strategic Highway Research Program, (SHRP), Washington DC, 1993 (Distress Pictures only).

[16] National Cooperative Highway Research Program, NCHRP Project 1-37A, "Development of the 2002 Guide for the Design of New and Rehabilitated Pavement Structures," http://www.2002designguide.com/, retrieved 19 April 2003.

[17] Oglesby, C.H. and R.G. Hicks, *Highway Engineering*, Fourth Edition, John Wiley & Sons, 1982, pp. 667–668.

[18] Office of Research, Development, and Technology, Federal Highway Administration, U.S. Department of Transportation, *Public Roads*, Washington, DC, March/April 2001, p. 4.

[19] Public Roads Administration, *Highway Practice in the United States*, 1949.

[20] Raad, L., *Parks Highway Load Restriction Study Field Data Analysis*, Transportation Research Center, Institute for Northern Engineering, University of Fairbanks, January 1997.

[21] Paterson, W.D.O., "International Roughness Index: Relationship to Other Measures of Roughness and Ride Quality," Transportation Research Report 1084, *Pavement Roughness and Skid Resistance*, TRB, Washington DC, 1986, pp. 49–59.

[22] Siddharthan, R., P.E. Sebaaly, and M. Javaregowda, "Influence of Statistical Variation in Falling Weight Deflectometers on Pavement Analysis," Transportation Research Report 1377, *Nondestructive Deflection Testing and Back-Calculations for Pavements*, TRB, Washington, DC, 1992, pp. 57–66.

[23] Strategic Highway Research Program, *Early Analysis of LTPP General Pavement Studies Data—Executive Summary*, SHRP-P-392, National Research Council, Washington, DC, April 1994.

[24] Strategic Highway Research Program, *SHRP-LTPP Overview, Five year report, Strategic Highway Research Program*, SHRP-P416, National Research Council, 1994, pp. 236ff.

[25] Stubstad, R.N., S.D. Tayabji, and E.O. Lukanen, *LTPP Data Analysis: Variations in Pavement Design Inputs*, NHCRP Project 20-50 Web Document #48, TRB, Washington, DC, 2002.

[26] Wu, S-S., "Procedure to Estimate Loading from Weigh-in-Motion Data," Transportation Research Report 1536, *Traffic and Pavement Surface Monitoring Issues*, TRB, Washington D.C., 1996, pp. 19–24.

[27] Walters, B.L. and R.K. Whitford, *Strategy for Handling the Statistics of Truck Weight Data in Alaska*, Proceedings, North American Travel Monitoring Exhibition and Conference Proceedings, Charlotte, NC, May 11–15, 1998.

EXERCISES FOR CHAPTER 9: HIGHWAY DESIGN FOR RIDEABILITY (PAVEMENT DESIGN)

Determining Loads from Truck Traffic

9.1 Axle Configurations.   Go to a location that affords a good view of a highway with significant semi-trailer truck traffic. Do all 3-S2 trucks have the same axle positions? Support your answer with digital photos, if possible.

9.2 ESAL Calculations—SUV vs. 3S-2.   A sport utility vehicle weighs 5000 lbs. How many of these SUVs would it take to have the same total ESALs as the fully-loaded 3S-2 truck shown in Figure 9.7?

9.3 ESAL Calculations—Trucks to Railhead.   The Ajax Metal and Electrical Products Company uses a Western Double Truck (see Figure 9.9) to transport goods from their plant to the nearest railhead some 60 miles away. The firm has obtained an overload permit for the road to the railhead. The firm loads 80 trucks (160 trailers) daily with 30,000 lbs of product in the front trailer and 60,000 lbs in the rear trailer, 300 days/year. The state highway and trucking company have struck a deal that the company will maintain the road between the plant and railhead. The tractor weighs 25,000 lbs, with a steering axle weight of 10,000 lbs. Each trailer has a tare weight of 10,000 lbs. Determine the annual ESAL from the company trucks.

9.4 ESAL Calculations—Port Access.   The Port Authority ramp at Mazurka handles 615 buses per day, 365 days per year. The weight on the front (single) axle is 20,000 pounds, whereas the weight on the rear (tandem) axle is 36,000 pounds. The access road is to last for 20 years and there is 2 percent per year growth in bus traffic expected. For how many ESAL should the access road be designed?

9.5 ESAL Calculations—Truck Data.   Compute the annual ESAL for the traffic summarized in the table below, using the fourth power formula and assuming a 320-day year. (Using 320 days per year compensates for reduced traffic on weekends.)

Weekday Traffic Data

Truck GVW(lbs)	Average Weight (k = 1000 lbs)	Number of trucks		
		3 Axles, All Single	5-Axle 3-S2 (Single Trailer 40-53')	5-Axle 2-S1-2 Twin Tandem truck
20,000–40,000	32K	50	260	240
40,100–66,000	58K	0	450	150
66,100–80,000	76K	0	250	250
80,100–110,000	108K	0	5	20
Load on steering axle		8,000 lbs	12,000 lbs	14,000 lbs
Load distribution. front/rear		equal	60/40	equal

9.6 ESAL Calculations—WIM Data.   Data from one day of the weigh-in-motion device on design lane of SR535 is given below. Using the fourth power formula and a 320-day year, complete the entries in the table below as you calculate a total 25-year ESAL value for this road for flexible pavement.

Weigh-In Motion Data

i	Kips/Axle	Axle Type	Design Lane Freq/day N (i)	ESALs	Growth Rate r	Compounded G	ESAL (i)
1	1.5	Single	19,699		0.018		
2	1.5	Single	2,500		0.062		
3	3.5	Single	2,500		0.062		
4	3.5	Single	1,344		0.022		
5	10	Single	2,550		0.045		
6	12	Tandem	1,344		0.022		
7	20	Tandem	2,550		0.045		
8	34	Tandem	2,550		0.045		
				Total Design Life ESAL =			

Flexible Pavement Design

9.7 Subbase Thickness.   A flexible pavement is to be designed. The moisture factor is 1.0. The subgrade has been prepared to a CBR of 5. The structural coefficient for the 2-inch hot mix asphalt is 0.4, and there should be 6 inches of asphalt base with a structural coefficient of 0.3. The structural coefficient of the subbase = 0.10. The reliability is 90 percent, standard deviation is 0.4, and 20,000,000 ESAL are expected. The PSI when built was 4.8 and the PSI before rehabilitation is to be no less than 2.3. How much subbase aggregate is required? Express your answer in inches of subbase thickness.

9.8 Subbase Thickness—Drainage Problems.   For the previous problem, if there is a high amount of water and difficult drainage and m is 0.65 for the subbase, how much subbase aggregate is required? (Hint: $m_1 = m_2 = 1$.)

9.9 Layer Thicknesses. With the expected growth in business by Ajax Metal and Electrical Products, 10,000,000 ESAL are expected over the next 15 years. At present, the road has become in need of major rehabilitation. The moisture coefficient is 1.0. The state requires 90 percent reliability with a standard deviation of 0.4 and a change in Serviceability Index no greater than 1.5. Assuming a clay soil subgrade with a CBR of 6, determine the thickness of hot mix asphalt concrete ($a_1 = 0.40$) that will be needed for the combined surface and base course, if the subbase is 12 inches of crushed packed stone with a structural layer coefficient of 0.10.

9.10 Pavement Life. Give three reasons why a pavement with a valid 20-year design might "fail" before 20 years.

9.11 Structural Number. What is the Structural Number for the data presented below?

If the coefficients for a_1, a_2, and a_3 are 0.35, 0.14, and 0.11, respectively, determine the layer thicknesses of pavement using the design process given in section 9.3.5.

Reliability	95 percent	Standard deviation	0.46	Δ PSI	2.0
18K ESAL	2.7×10^6	Resilient modulus	7800 psi		

9.12 Flexible Pavement Design. Using the data below, determine the design structural numbers of layers SN (3), SN (2), and SN (1) for the design lane of SR535. Use an enlarged copy of the design chart in Figure 9.15 and submit it with your assignment. Complete the design by determining the layer thickness if the hot mixed asphalt structural coefficient is 0.37, the structural coefficient of the cement treated base is 0.21 and the subbase structural coefficient is 0.10.

Subgrade CBR = 4	Subbase M_r = 10,800 psi	ESAL = 39×10^6	PSI(t) = 2.5
R = 95%	Base M_r = 22,000 psi	S(0) = 0.35	PSI(i) = 4.2

9.13 Layer Design and Cost Analysis. Determine the least cost flexible pavement design per lane mile in the design lane on SR535 in the previous exercise. The lane is 12 feet wide. The subgrade soil extends 10 feet below the surface, so that drainage is not a factor in the design [i.e., $m(i) = 1.0$]. The table below indicates the materials that you must use. This problem is best done by setting up a spreadsheet and determining the costs per lane mile for various layer designs. Your design alternatives (and their associated costs) should include:

(a) A full-depth hot mix asphalt concrete pavement that uses a design structural number of 4.5, regardless of your answer to the previous problem.
(b) A layer design that follows the guidelines for layer thickness given in section 9.3.5.
(c) Use SN(1) = 2.8, SN(2) = 3.8, and SN(3) = 4.5 as your design structural numbers.
(d) An easy-to-follow search for the values of D_1^*, D_2^*, and D_3^* that are consistent with having the asphalt pavement surface layer be no less than 4 inches thick. (A spreadsheet would greatly facilitate your search and your client's review of your work. Write out sample calculations for key cells.)
(e) Which of your layer designs is the most economical? Show your analyses and state your conclusions clearly.

Data for Layer Design and Cost Analysis

Material	Layer	a(i)	M(R)	Specific Gravity	Unit Cost (delivered)
Hot mix asphalt concrete	Top/wearing	0.36	—	2.65	$59.75/ton
Compacted dense aggregate	Base	0.15	32,000	2.75	$10.45/ton
Coarse aggregate	Subbase	0.11	24,000	2.55	$9.50/ton

Note: The subgrade soil has a CBR of 15.
Note: Excavation costs are $4.87 per cubic yard of soil displaced.
Note: The subgrade soil has an M(R) value of 7750 psi.

9.14 Cost of Pavement Designed.   Using the ESAL from the design lane on SR535, determine the thickness of flexible pavement using the AASHTO layered analysis. Assume a CBR of 3 for the clay subgrade. The reliability is to be 95, standard deviation is 0.5, and the moisture content is such that m is 1. Use HMAC, emulsified aggregate, and course aggregate ($a = 0.44, 0.3,$ and 0.11, respectively). For the change in PSI, use a 2.5 value. Using the costs and specific gravity in the table for the previous problem, what is the cost per lane-mile of laying this pavement.

Rigid Pavement Design

9.15 Moduli of Pavement Components.   A 6-inch diameter test specimen is being used to determine the modulus of rupture and the elastic modulus of the PCC. The specimen buckles at an applied force of 412,000 pounds. The density of the specimen is 136 pounds per cubic foot. Determine the modulus of rupture and the elastic modulus of the PCC.

9.16 Slab Thickness.   Using the data from the previous problem, how thick should a pavement be using the PCC of the previous problem if the subgrade modulus is 7000 psi and the subbase is an asphalt-treated base with E of 400,000 psi? Let $J = 3.0$, reliability = 85 percent, $C_d = 0.92$, and standard deviation = 0.35. The net change in PSI allowed is 2.5 (from 4.7 to 2.2) and 15 million ESALS are projected.

9.17 ESAL Calculations.   An interstate highway is to be built by using PCC with a 12-inch slab. During construction, the 6-inch test core buckled at 378,000 pounds of pressure and tested very dense at 146 pounds per cubic foot. The soil is clay with a CBR of 5. The subbase is a lime-treated soil 10 inches thick with an elastic modulus of 50,000 psi. Interstate reliability is 0.98 with a standard deviation of 0.35. Assuming the usual PSI degradation before rehabilitation is 2.5 and $J = 3.7$, determine how many ESALs the pavement can withstand over 20 years. Assume a 300-day year and a 1.5 percent per year constant compound growth rate.

9.18 Slab Thickness.   Determine the slab thickness for the design lane on SR535 with 18 Million ESALS using the AASHTO rigid pavement design method. The rigid pavement characteristics are given in the table below. Make a copy of Figure 9.26 for "Segment 1" of the rigid pavement design chart and a copy of Figure 9.27 for "Segment 2". Submit these copies with your assignment. Use a = 0.80 and b = 0.635 in Equation 9.8 to convert K into the effective modulus K_{LS}.

Data for Rigid Pavement Design

$C(d) = 1.0$	$k = 800$ pci	$R = 95\%$
$E(c) = 5 \times 10^6$	$p(i)$ or PSI(o) $= 4.2$	$S'(c) = 600$ psi
$J = 3.2$	$p(t)$ or TSI $= 2.5$	$S(o) = 0.35$

Public Mass Transportation

SCENARIO

Vehicle miles traveled is increasing at a rate faster than population growth in the Mythaca area. Most of this new travel is by private car, but there is little room—and even less money—to build enough extra roadways to accommodate this extra travel. What's more, the percentage of trucks in the vehicle mix is increasing fastest of all. Add to all this the growing number of drivers who are in their 70s, 80s, and even older. At what age will they have to stop driving? How do individuals who are physically unable to drive make the trips necessary for a satisfactory lifestyle?

As these problems of congestion and personal mobility intensify, the Mythaca Bus Company (MBC) is trying to make its contribution to the solutions. However, the MBC also operates with limited resources. The MBC would like to provide frequent service that covers the entire Mythaca area, but it is already heavily subsidized by local, state, and federal programs that are funded by taxpayers. Despite MBC's best efforts, citizens of Mythaca often complain that the bus service "never" takes them where they want to go, when they want to go there. How can MBC provide the best level of service that its limited budget will allow?

FIGURE 10.1
A Flushing-bound IRT 7 train enters the upper level of the Queensboro Plaza station in New York City. Photo © Hank Eisenstein.

CHAPTER OBJECTIVES

By the end of this chapter, the student will be able to:

1. Design rail service (Figure 10.1) in a corridor with respect to station spacing and vehicle capabilities.

2. Calculate changes in transit ridership in response to changes in fare or service.

3. Measure and compare the performance of public transportation operations.

4. Evaluate equipment replacement alternatives by applying the life cycle cost method.
5. Discuss public transportation's role in addressing certain public issues.

10.1 TRANSIT MODES

The term *public mass transportation* can have several meanings. The particular meaning used in this chapter is based on several distinguishing characteristics:

1. *A common carrier.* The service is available to the general public. Special cases, such as service required by the Americans with Disabilities Act (ADA), are covered separately in this chapter.
2. *Fixed route and fixed schedule.* Some transit service is *demand responsive*, meaning that a vehicle is sent to a rider's location as close to the desired pickup time as possible. However, most transit service is provided along a fixed route and according to a fixed schedule.
3. *The area served is limited to an urban area or a rural area.* Service *between* cities by bus, rail, air, and so forth is treated separately as intercity mass transportation.

In this chapter, we focus on the *urban mass transit* form of public mass transportation. Service is provided to the general public on fixed routes using a fixed schedule within an urban area. This service is usually provided by one or more of the following transit technologies:

1. *Bus.* The standard 35-foot bus has 35 to 45 seats and can carry about 70 passengers, including standees (HCM, 1985).
2. *Rail.* For areas with higher densities of ridership, some form of rail transit provides greater capacity than does bus. Most often, rail transit vehicles have steel wheels that run on steel rails (Figure 10.1), but rubber-tired vehicles operating in their own guideways (Figure 10.2) could also be placed in this category. The two principal subdivisions of rail transit are ...

   A. *Light-rail transit.* LRT (Figure 10.3b) is the modern form of what used to be called *streetcars* (Figure 10.3a). Light-rail vehicles can operate on dedicated right-of-way (ROW) or on ROW shared with cars, trucks, and buses. A typical LRT car can carry 100–200 passengers, including standing passengers. (TCRP13, 1996)
   B. *Heavy rail transit.* HRT, also called Rail Rapid Transit (RRT), is the mode associated with subways in major cities. HRT service features exclusive ROW, trains with as many as 10 cars, and speeds up to 60 mph (96 kph). Most HRT rail cars can carry 125 to 180 passengers, including standees (TCRP13, 1996). The exclusive ROW could be in tunnels, on elevated structures (see Figure 10.1), or along expressway medians.

Although most U.S. cities do not have waterborne ferry service, ferries can be important in locations where bridges that cross bodies of water have insufficient capacity or are non-existent. U.S. cities, such as Seattle (Figure 10.4a), San Francisco, and New York City have ferries that operate as part of the area's transit system. Ferry service is also important in non-urban areas (Figure 10.4b) and in developing countries.

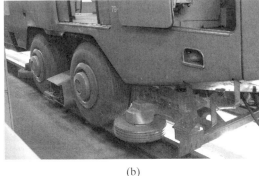

(a)   (b)

FIGURE 10.2
Rubber-tired transit vehicles in the Montréal Métro. (a) MR-73 (Bombardier 1975) waiting to be serviced. (b) Closeup of the "truck" or "bogie". Note the horizontal direction wheels. Photos © 2002, Marc Dufour.

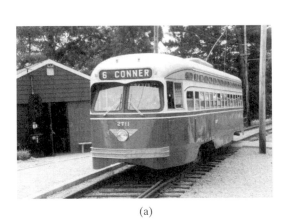

(a)   (b)

FIGURE 10.3
Light-rail transit—Old and new. (a) PCC streetcar. (b) LRT in Salt Lake City. *Sources*: (a) Photo by Jon D. Fricker, (b) Photo by Darcy Bullock.

Other forms of urban mass transit exist, but it may be possible to put them into one of the technology categories above. For example, a monorail (Figure 10.5) seems to be a unique transit mode, but it has the same basic characteristics as heavy rail—separate ROW, multicar trains, and high speeds.

Another variation of mass transportation is *taxicab*, a common carrier whose service area is usually limited to a certain urban area, but whose service is demand responsive. Rather than putting taxi in the "bus" category, we use another category called *paratransit*, which can also include special handicapped (ADA) service, vanpools, and carpools. It may already be apparent that classifying transit services is not always easy. Vuchic (1981) proposed several methods, including one based on ROW category and vehicle technology. Table 10.1 is based on the Vuchic method.

(a)

(b)

FIGURE 10.4
Ferry boats. (a) Washington State Ferry viewed from the dock of another ferry. (b) Passengers leaving a ferry in Southeast Alaska. *Sources*: (a) http://www.wsdot.wa.gov/ferries/your_wsf/photo_gallery/, (b) Photo by Peter Metcalfe, used with permission of the Alaska Marine Highway System.

(a)

(b)

FIGURE 10.5
Monorails. (a) At left: Monorail in elevated downtown Seattle station. Photo Jon D. Fricker (b) Above: Monorail at Walt Disney World. Photo: Drew Sommer.

TABLE 10.1 Classification of Urban Public Transportation Modes

| | Technology | | | |
ROW Category	Highway, Driver-Steered	Rubber-Tired, Guided or Partially Guided	Rail	Special
Surface streets with mixed traffic	Regular bus Paratransit	Trolleybus	Streetcar Cable car	Ferryboat Hydrofoil Helicopter
Separated from traffic but with grade crossings		Dual-mode bus	Light-rail transit	
Fully controlled	Bus on buswayd	Rubber-tired monorails Automated guideway transit	Rail rapid transit Commuter rail	Funicular Aerial tramway

Source: Vuchic, 1981.

Automated guideway transit (AGT) in Table 10.1 is a system that does not have an on-board operator (driver). An AGT is essentially a horizontal elevator. It may sound exotic or even dangerous, but they are becoming rather common. The place you are most likely to find one is in the form of *automated people movers (APMs)* at a large airport, such as Dallas-Fort Worth and Orlando (Figure 10.6). Driverless *rail rapid transit (RRT)* has been possible for several decades, but passengers do not like to ride in an RRT train that doesn't have an operator. Instead, some transit systems use an automated RRT operation, but put a driver on board to manually override the automated system, if necessary.

FIGURE 10.6
People mover at Orlando Airport. Photo: Jon D. Fricker.

10.2   DESIGNING A RAIL TRANSIT LINE

Some of the worst traffic congestion in Mythaca County occurs along the Mythaca River Valley between the City of Mythaca and Shoridan. Residential and commercial development in this corridor has increased recently and is expected to continue. Because high concentrations of tripmakers arranged in a linear pattern are conducive to service by high-capacity rail, the county has begun to study the feasibility of building a rail transit line along the valley.

10.2.1   Transit Vehicle Travel Analysis

One of the most important attributes of a transit vehicle is the time it takes to carry passengers from their origins to their destinations. The ideal situation might be for the vehicle to carry its passengers from a single origin to a single destination, without any intermediate stops to slow it down and increase the trip time. This is not ideal, however, if there are other persons desiring service along the non-stop route. The trade-off becomes:

> **A.** *Increase the number of stops* along a route, to improve access to the transit service, and perhaps increase ridership.
> **B.** *Reduce the number of stops*, to increase average operating speed and reduce travel time along the route.

Strategy A is analyzed in Section 10.3. In this section, strategy B can be revised and refined as follows:

> **B.1** Determine the best distance between transit stops on a route to make the best use of the performance characteristics of the transit vehicles assigned to that route.
> **B.2** Determine the best performance characteristics for transit vehicles assigned to a particular route, given a specified spacing between transit stops on that route.

In conducting an analysis for either version of strategy B, the key vehicle performance characteristics are acceleration, deceleration, and maximum speed. Acceleration rates are limited by three principal considerations (Lang and Soberman, 1964): (1) the physical limit of the frictional force that can be developed between wheel and track, (2) passenger comfort, and (3) power costs. Normally, passenger comfort is the governing factor. The most commonly used acceleration rates are between 3 and 4 mph per second.

THINK ABOUT IT

Explain in practical terms what is meant by *passenger comfort* when the acceleration of a transit vehicle is the issue.

Deceleration, like acceleration, is governed by passenger comfort. Although a transit vehicle can decelerate at a rate greater than 3 to 4 mph per second, a deceleration rate in that range is normally used. Actually, there is a factor more important to passenger comfort than acceleration or deceleration rate; it is called *jerk*, the rate of change of acceleration or deceleration per unit time. A constant rate of acceleration is not as uncomfortable as a situation in which the acceleration rate is changing. According to Vuchic (1981), designers and operators of transit vehicles try to limit the jerk value to 0.6 m/sec³. Lang and Soberman (1964) cited a limit of 7 ft/sec³, which is within the range of vehicle characteristics listed in CUTS (1992)—1.12–2.68 m/sec³. (1 meter = 3.28 feet.)

10.2.2 Transit Vehicle Travel Regimes

In either strategy B.1 or B.2, the objective is to maximize the average operating speed along the route. A transit vehicle with a high maximum speed seems desirable, but (a) such a vehicle will cost more to purchase, (b) it will probably consume more energy, and (c) it may spend little or no time at its maximum speed, depending on the distance between stations. To analyze this situation, five regimes of motion must be defined and examined. See Figure 10.7.

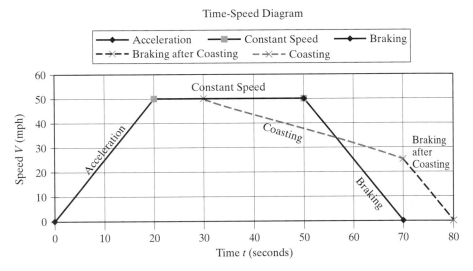

FIGURE 10.7
Diagram of five transit travel regimes.

1. *Acceleration regime.*   The rail vehicle leaves the station (or transit stop) and begins to increase speed, until the desired top speed V_{top} is reached. If an average acceleration rate $\bar{a}$ is used, the time t_a needed to accelerate to V_{top} is $t_a = V_{top}/\bar{a}$. The distance s_a needed to reach V_{top} is found by substituting $t_a = V_{top}/\bar{a}$ into $s_a = \frac{1}{2}(\bar{a}\, t_a^2)$ to get $s_a = \frac{1}{2}(V_{top}^2/\bar{a})$.

2. *Constant speed regime.*   The speed V_{top} may be the maximum possible speed V_{max} for a particular vehicle, the maximum speed allowed by company policy for safety or other reasons, or the speed reached before deceleration must begin. The

(a)

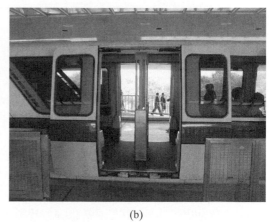
(b)

FIGURE 10.8
Dwell time policies vary with operating environment. (a) New York City subway: Only as long as needed to board passengers. (b) Walt Disney World: Long enough to keep vacationers in a relaxed mood. Photos: (a) Koh Ee Huei, (b) Kurt Sommer.

time spent in this regime depends on the time spent in all other regimes between stations or stops.

3. *Coasting regime.* An advantage to steel wheels on steel rail is the very low rolling resistance. A rail vehicle's motors could be shut off, with little loss in speed but with significant energy savings. Coasting is simply deceleration at a constant rate c, primarily because of friction and air resistance. During the coasting regime, the transit vehicle's speed is reduced from V_{top} to V_{ec}. The coasting regime lasts t_c time units. Therefore, $V_{ec} = V_{top} - c\,t_c$ and $t_c = (V_{top} - V_{ec})/c$.

4. *Braking regime.* The brakes are applied with an average deceleration rate $\bar{b}$, and the vehicle is stopped from an initial speed V_i, where $V_i = V_{top}$ or $V_i = V_{ec}$. The time needed to stop is $t_b = V_i/\bar{b}$. The corresponding braking distance is $s_b = \frac{1}{2}\left(V_i^2/\bar{b}\right)$.

5. *Station standing time, or dwell time.* Time is needed to allow passengers to board and leave the transit vehicle (Figure 10.8). The greater the number of passengers using a given stop, the longer must be the time the vehicle remains motionless at that stop. At busy stations in large-city RRT systems, the mean dwell time may be as long as 60 seconds (TCRP13, 1996). An equation that estimates dwell time t_d is

$$t_d = P_{on} t_{on} + P_{off} t_{off} + t_{oc}$$

where P_{on} and P_{off} are the number of passengers boarding and alighting through the busiest doors, respectively; t_{on} and t_{off} are the average boarding and alighting times (seconds per passenger); and t_{oc} is the combined time (seconds) to open and close the vehicle doors (Danaher, 1999).

THINK ABOUT IT

Another familiar example of *dwell time* is the time the doors of an elevator stay open at a floor. Have you seen instances in which the doors seem to stay open too long? Have you seen instances in which the doors start to close too soon? How would you design the elevator door system to improve this situation? Are there any lessons from the elevator problem that apply to the transit vehicle dwell time problem? Are there any important differences?

Figure 10.7 is a *travel regime diagram* that shows how the five transit travel regimes fit together during a transit vehicle's movement between stops or stations. At time $t = 0$, the vehicle leaves a station and accelerates with average rate $\bar{a}$ until a speed V_{top} or V_{max} is reached. The acceleration regime lasts for t_a units of time. The vehicle could maintain the constant speed V_{top} for t_V units of time and then begin braking at an average rate $\bar{b}$, t_b units of time before reaching the next station. If, instead, coasting is used, the vehicle could be allowed to decelerate at a rate c, until braking at average rate $\bar{b}$ is initiated $t_{b,c}$ units of time before reaching the next station.

For any time spent in a given travel regime, there is the corresponding distance spent in that same regime. Table 10.2 summarizes the equations involved.

The equations in Table 10.2 that involve coasting require some explanation. During the coasting regime, the transit vehicle's speed is reduced from V_{top} to V_{ec}. The coasting regime lasts t_c time units with an average speed

$$\bar{V_c} = \frac{1}{2}(V_{top} + V_{ec}) \tag{10.11}$$

The total distance between two consecutive stations, S, is the sum of the acceleration, coasting, and deceleration regime distances:

$$S = s_a + s_c + s_{b,c} \tag{10.12}$$

The length of the constant speed regime, if there is one, can be considered later. The distance s_a needed to reach V_{top} is

$$s_a = \frac{1}{2}\frac{V_{top}^2}{\bar{a}} \tag{10.2a}$$

The distance spent coasting uses the relationship $s = vt$, with v from Equation (10.11) and t from Equation (10.5):

$$s_c = \bar{V_c} t_c = \frac{1}{2}(V_{top} + V_{ec})t_c \tag{10.6b}$$

The distance $s_{b,c}$ needed to brake to a stop from the end-of-coasting speed V_{ec} is

$$s_{b,c} = \frac{1}{2}\frac{V_{ec}^2}{\bar{b}} \tag{10.8b}$$

TABLE 10.2　Equations for Transit Vehicle Travel Regimes

Regime	Time in regime	Distance in regime
1. Acceleration regime	$(10.1\text{a})\ t_a = \dfrac{V_{top}}{a}$	$(10.2\text{a})\ s_a = \dfrac{1}{2}\dfrac{V_{top}^2}{a}$
	$(10.1\text{b})\ t_{a,c} = \dfrac{V_{top} - V_{ec}}{a}$	$(10.2\text{b})\ s_{a,c} = \dfrac{1}{2}(V_{top} + V_{ec})t_{a,c}$
2. Constant speed regime	$(10.3\text{a})\ t_V = \dfrac{s_V}{V_{top}}$	$(10.4\text{a})\ s_V = S - (s_a + s_b)$
	$(10.3\text{b})\ t_{V,c} = \dfrac{s_{V,c}}{V_{top}}$	$(10.4\text{b})\ s_{V,c} = S - (s_a + s_c + s_{b,c})$
3. Coasting regime	$(10.5)\ t_c = \dfrac{V_{top} - V_{ec}}{c}$	$(10.6\text{a})\ s_c = S - (s_a + s_{b,c})$
		$(10.6\text{b})\ s_c = \dfrac{1}{2}(V_{top} + V_{ec})t_c$
4. Braking regime	$(10.7\text{a})\ t_b = \dfrac{V_{top}}{b}$	$(10.8\text{a})\ s_b = \dfrac{1}{2}\dfrac{V_{top}^2}{b}$
	$(10.7\text{b})\ t_{b,c} = \dfrac{V_{ec}}{b}$	$(10.8\text{b})\ s_{b,c} = \dfrac{1}{2}\dfrac{V_{ec}^2}{b}$
5. Station standing time, or dwell time	$(10.9)\ t_d = P_{on}t_{on} + P_{off}t_{off} + t_{oc}$	$s_d = 0$
6. Speed at end of coasting regime	$(10.10)\ V_{ec} = \sqrt{\dfrac{2\bar{a}bcS - V_{top}^2(c + \bar{a})\bar{b}}{\bar{a}(c - \bar{b})}}$	

S = the distance between stations.

Substituting Equations (10.2a), (10.6b), (10.5), and (10.8b) into Equation (10.12) produces

$$S = \frac{1}{2}\frac{V_{top}^2}{\bar{a}} + \frac{1}{2}\frac{V_{top}^2 - V_{ec}^2}{c} + \frac{1}{2}\frac{V_{ec}^2}{\bar{b}} \tag{10.13}$$

The only unknown in Equation (10.13) is V_{ec}. Rearranging the terms can isolate V_{ec} in the total distance equation:

$$2S - \frac{V_{top}^2}{\bar{a}} - \frac{V_{top}^2}{c} = \frac{V_{ec}^2}{\bar{b}} - \frac{V_{ec}^2}{c}$$

$$\frac{2\bar{a}bcS - V_{top}^2(c + \bar{a})\bar{b}}{\bar{a}(c - \bar{b})} = V_{ec}^2$$

The resulting solution for V_{ec} is

$$V_{ec} = \sqrt{\frac{2\bar{a}\bar{b}cS - V_{top}^2(c + \bar{a})\bar{b}}{\bar{a}(c - \bar{b})}}$$

(10.10)

Equation (10.10) only applies to the case in which there is no constant speed regime. Equation (10.10) is an important calculation to do early in an analysis, if a coasting regime is being considered. If the calculated V_{ec} is unreasonably low, either the coasting time should be reduced or it should be eliminated. Equations (10.1b) and (10.2b) will be explained in an example to follow.

THINK ABOUT IT

If coasting saves energy, why should a low value of V_{ec} be the reason to reject coasting? What other factor(s) would become more important?

Example 10.1

"Proposal A" for rail service between Mythaca and Shoridan is for a light-rail transit line. The LRT line would be 24.4 miles long, with a stop every 0.66 mile. The LRT vehicle would have an average acceleration rate of 3.0 mph/second and an average deceleration rate of 2.85 mph/second. The vehicle would be able to achieve the desired top operating speed of 44 mph. Coasting is not part of the proposal.

Draw the time-speed diagram for one interstation spacing using proposal A. Label the key points of the diagram, including the corresponding values. How long (in time and distance) would the LRT vehicle be able to maintain the desired 44 mph constant speed? How far apart (or close together) would two stations have to be, so that the LRT would have to begin braking as soon as it reached 44 mph?

Solution to Example 10.1   By using Equation (10.1), the time spent in the acceleration regime is

$$t_a = \frac{44 \text{ mph}}{3.0 \text{ mph/sec}} = 14.7 \text{ sec.}$$

According to Equation (10.7a), the deceleration regime lasts

$$t_b = \frac{44 \text{ mph}}{2.85 \text{ mph/sec}} = 15.4 \text{ sec.}$$

To determine the length of the constant speed regime, first t_a and t_b have to be converted into their corresponding distances:

$$s_a = \frac{1}{2} \frac{V_{top}^2}{\bar{a}} = \frac{1}{2} \frac{(1.47 * 44)^2}{(1.47 * 3.0)} = \frac{4183.5 \text{ ft}^2/\text{sec}^2}{8.82 \text{ ft/sec}^2} = 474.3 \text{ ft}$$

(10.2a)

$$s_b = \frac{1}{2} \frac{V_{top}^2}{\bar{b}} = \frac{1}{2} \frac{(1.47 * 44)^2}{(1.47 * 2.85)} = \frac{4183.5 \text{ ft}^2/\text{sec}^2}{8.38 \text{ ft/sec}^2} = 499.2 \text{ ft}$$

(10.8a)

$s_a = 474.1$ feet. How many additional feet are needed to coast from 64.68 ft/sec down to 30 mph (or 44.1 ft/sec)?

$$t_c = \frac{64.68 - 44.1}{0.294} = 70.0 \text{ sec} \tag{10.5}$$

$$s_c = \frac{1}{2} * (64.68 + 44.1) * 70.0 = 3807.3 \text{ ft} \tag{10.6b}$$

To brake to a stop from 44.1 ft/sec takes

$$s_{b,c} = \frac{1}{2} * \frac{(44.1)^2}{4.19} = 232.1 \text{ ft} \tag{10.8b}$$

After these calculations, which assume $V_{ec} = 44.1$ ft/sec, $s_a + s_c + s_{b,c} = 474.1 + 3807.3 + 232.1 = 4513.5$ feet. Because this is less than the available S of 12,408 feet, there is actually too much distance between stations for just one coasting regime. Some "re-acceleration" back to 64.68 ft/sec is needed. It appears that another cycle of "accelerate to 44 mph, coast to 30 mph," followed by a cycle of a accelerating to a speed lower than 44 mph and then coasting and/or braking to a stop would work. However, the problem calls for a revised V_{ec} value that will allow an integer number of equivalent cycles—probably three cycles. The equation to set up and try is

$$S = s_a + s_c + s_{a,c} + s_c + s_{a,c} + s_c + s_{b,c} = 12,408 \text{ ft}$$

With $V_{ec} = 44.1$ ft/sec, $S = 12,636$ feet using this equation. Apparently, V_{ec} does not need to be changed very much. Except for s_a, each term in the equation can be written as a function of V_{ec}. To simplify these calculations and set up a framework for subsequent calculations, a spreadsheet can be used. The spreadsheet in Table 10.3 shows the results of using the "Tools Solver" feature, in which the V_{ec} value in cell C8 was changed so that the sum in cell C24 would equal the given S value in cell C3. The V_{ec} value didn't have to change much—from 44.1 ft/sec to 44.59 ft/sec. Figure 10.10 shows the three *cycles* used—acceleration-coast, reacceleration-coast, and reacceleration-coast—until the braking regime was begun 12170.7 feet and 228.7 seconds after leaving the previous station.

TABLE 10.3   Calculations for Example 10.3

	A	B	C
1		With Coasting & Reaccel cycles	
2		(mph units)	(fps units)
3	S given	2.35	12408
4	a bar	3	4.41
5	b bar	2.85	4.19
6	c rate	0.2	0.294
7	V top	44	64.68
8	V ec	30	44.59
9			
10	sa		474.3
11	ta		14.7
12	sc		3732.878
13	tc		68.32182
14	sac		248.8586
15	tac		4.55
16	sbc		237.3

(*Continued*)

TABLE 10.3 (*Continued*)

	A	B	C
1		With Coasting & Reaccel cycles	
2		(mph units)	(fps units)
17	tbc		15.4
18			
19	sa		474.3
20	3 sc		11198.64
21	2 sac		497.7171
22	sbc		237.3
23			
24	S sum		12408.0

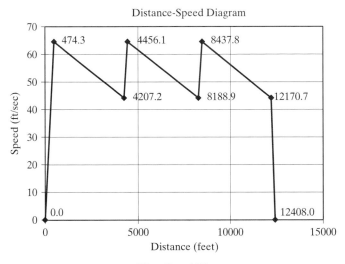

Distance-Speed Diagram

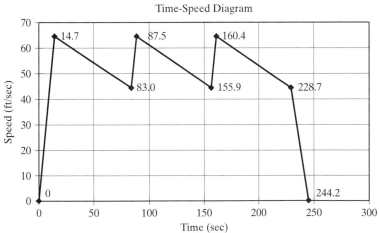

Time-Speed Diagram

FIGURE 10.10
Travel regime diagrams for Example 10.3.

10.3   PREDICTING TRANSIT RIDERSHIP CHANGES

Introduction

For 10 years, the Mythaca Bus Company (MBC) has kept its base adult fare at 50 cents. Over the past 7 years, however, operating costs have risen steadily (see Figure 10.14), while ridership levels have been stagnant (Figure 10.11). The result has been an increasing operating deficit that must be subsidized by income and property taxes. Some local government officials want a portion of that rising deficit shifted back onto the users of the bus service, which means raising the bus fares. The MBC is worried that most bus users have low incomes, and they can least afford a fare increase. It seems reasonable to expect that a fare increase will cause some MBC patrons to cease using its bus service. How many riders will be "driven away" by a fare increase to 75 cents?

10.3.1   Transit Elasticity

Let us say that bus route #8 in Mythaca carried an average of 1000 passengers per day in the past year. The base fare during the past year was 50 cents. We can represent these two bits of information about the "base case" as $Q_o = 1000$ and $P_o = 0.50$, where

- Q stands for "quantity of service purchased," or ridership
- P stands for price of service, or fare.

If a fare increase to 75 cents was implemented, and the resulting ridership was 805 per day, we could calculate something called the *shrinkage ratio*. This is the percent change in ridership for each percent change in fare. The value of the shrinkage ratio is one way of measuring the *demand elasticity of transit ridership with respect to fare*. The

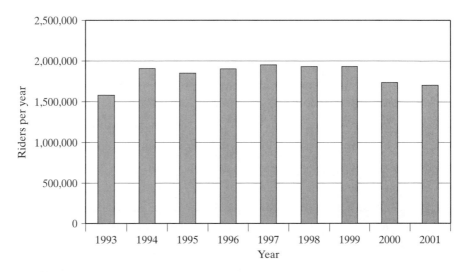

FIGURE 10.11
Transit ridership trends.

shrinkage ratio elasticity is

$$\varepsilon_{shr} = \frac{\%\,\Delta Q}{\%\,\Delta P} = \frac{\left(\dfrac{Q_1 - Q_0}{Q_0}\right)}{\left(\dfrac{P_1 - P_0}{P_0}\right)} = \frac{(Q_1 - Q_0)P_0}{(P_1 - P_0)Q_0} \tag{10.14}$$

In the case at hand, $P_1 = 0.75$ and $Q_1 = 805$. The shrinkage ratio elasticity would be

$$\varepsilon_{shr} = \frac{\%\,\Delta Q}{\%\,\Delta P} = \frac{\left(\dfrac{805 - 1000}{1000}\right)}{\left(\dfrac{0.75 - 0.50}{0.50}\right)} = \frac{-0.195}{0.50} = -0.39$$

Based on the fare and ridership data given, a shrinkage ratio elasticity of -0.39 has been calculated. Note that the ε_{shr} value is negative, because ridership *decreased* as fare *increased*.

If enough transit operators keep track of their ridership changes after they make fare changes, and the resulting shrinkage ratios are calculated, it may be possible for transit operators to learn from the experience of others. In the absence of reliable local data, a common value for ε_{shr} is -0.33, which is based on a study done by Curtin (1968). Despite numerous more recent studies, $\varepsilon_{shr} = -0.33$ remains the most commonly used transit fare elasticity value. If the Mythaca Bus Company (MBC) were to adopt $\varepsilon_{shr} = -0.33$ for its estimation of the impact of a fare increase from 50 cents to 75 cents, rearranging Equation 10.14 produces

$$\varepsilon_{shr} = \frac{(Q_1 - Q_0)P_0}{(P_1 - P_0)Q_0} = \frac{(Q_1 - 1000)0.50}{(0.75 - 0.50)1000} = -0.33 \tag{10.15}$$

Solving for Q_1...

$$Q_1 - 1000 = \frac{(0.25)(1000)(-0.33)}{0.50} = -165; \quad Q_1 = 835.$$

Using the Curtin value for ε_{shr} provides a reasonable estimate for ridership loss due to a fare increase. In the MBC example, further study would need to be carried out to determine if a certain category of rider would be more adversely affected than other categories, and whether this is compatible with public policy.

THINK ABOUT IT

What categories of riders might be deserving of special protection from fare increases? Why? How might "special protection" be afforded?

If MBC would lose 165 riders out of 1000, how would this help reduce the rising operating deficit? Calculate the farebox (operating) revenue before and after the fare change:

Before: $0.50 * 1000 = $500
After: $0.75 * 835 = $626.25

Because ridership is expected to decrease only 0.33 percent for every 1 percent increase in fare, operating revenue would actually increase. In other words,

- If revenues will increase, despite a fare increase, the demand is "inelastic." Riders may have been enjoying "consumer surplus" (Chapter 5) or they have few alternatives to transit.
- If revenues will decrease as the fare increases, the demand is "elastic."

At this point, MBC has two points on a demand curve for its transit service—one based on current ridership ($Q_0 = 1000$ with $P_0 = 0.50$) and one based on an estimate ($Q_0 = 835$ if $P_0 = 0.75$). These points on the demand curve are shown in Figure 10.12. If MBC had enough data over a wide enough range of fares, it could draw a complete demand curve and not have to use elasticity equations. The general shape of the transit demand curve in Figure 10.13 is based on research by Fricker (1972), which confirmed findings made by Meyer et al. (1959).

Unfortunately, it is rare to have reliable information on ridership for the fare ranges shown by the dashed lines in Figure 10.13. The solid curve between 40 cents and 75 cents can be based on recent experience in charging fares within that range. The common recourse is to assume that ridership responses outside the known fare range can be estimated by using elasticity calculations such as those demonstrated earlier in this section.

So far, the discussion in this section has been about "fare elasticity," which is shorthand for *elasticity of transit ridership with respect to fare*. There is also such a thing as "service

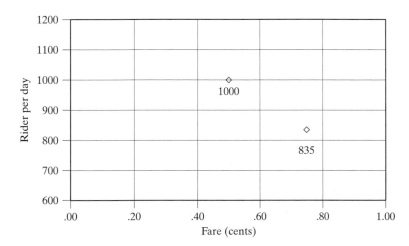

FIGURE 10.12
Only two points on a transit demand curve.

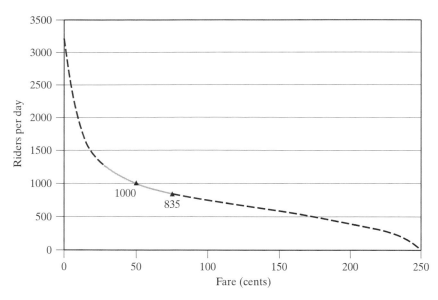

FIGURE 10.13
A transit demand curve.

elasticity," or *elasticity of transit ridership with respect to service*. Suppose the MBC, instead of raising fares to try to increase operating revenue, were to reduce service in an attempt to reduce operating costs. Where headway (time between buses) had been 30 minutes, buses would run only every 60 minutes. The MBC would be interested in any data collected by other transit operators after they had reduced service in this way. The input variable could be *headway* (time between buses) or *frequency* (buses per hour). Representative bus *headway* elasticities are −0.37 during peak hours and −0.46 in the off-peak hours (Mayworm et al., 1980). A shrinkage ratio can be developed for this or for any other measurable service variable. The service elasticity form of Equation 10.14 is simply

$$\varepsilon_{\text{shr}} = \frac{\% \Delta Q}{\% \Delta S} = \frac{\left(\dfrac{Q_1 - Q_0}{Q_0}\right)}{\left(\dfrac{S_1 - S_0}{S_0}\right)} = \frac{(Q_1 - Q_0)S_0}{(S_1 - S_0)Q_0} \tag{10.16}$$

where S stands for some measure of service, such as time between scheduled vehicles (headway) or the frequency of scheduled service (vehicles per hour).

Example 10.4

MBC's bus route #8 carried an average of 1000 passengers per day in the past year. Half of the route #8 passengers on a typical day rode during the AM and PM peak periods. The peak and off-peak headways were 30 minutes. What would ridership on route #8 be if the headways were increased to 60 minutes between buses? Use the headway elasticities reported by Mayworm et al. (1980).

Solution to Example 10.4   Mayworm et al. reported a peak headway elasticity of -0.37. The service form of demand elasticity Equation 10.15 is

$$\varepsilon_{shr} = \frac{(Q_1 - Q_0)S_0}{(S_1 - S_0)Q_0} = \frac{(Q_1 - 500)*30}{(60 - 30)*500} = -0.37$$

where S = service = headway, expressed in minutes between buses. Solving for the new peak Q_1:

$$Q_1 - 500 = \frac{(30)(500)(-0.37)}{30} = -185; \text{ peak period } Q_1 = 315.$$

In the off-peak period,

$$Q_1 - 500 = \frac{(30)(500)(-0.46)}{30} = -230; \text{ off-peak } Q_1 = 270.$$

So total Q on route #8 would drop from 1000 passengers per day to $315 + 270 = 585$. Of course, the new Q values are just estimates, but this analysis may help the MBC staff and the MBC Board of Directors formulate their deficit-cutting strategies.

THINK ABOUT IT

Imagine that you are a member of the MBC Board of Directors. Having seen the fare and service elasticity calculations above, which deficit-cutting strategy would you prefer? Why? What other information, if any, would you request from the MBC staff?

Example 10.5

An alternative to *headway elasticity* to quantify ridership response to service changes is *service frequency elasticity*. Use the service changes proposed in Example 10.4 and the ridership responses found in Example 10.4 to calculate the peak and off-peak service *frequency* elasticities.

Solution to Example 10.5   The initial peak period ridership in Example 10.4 was 500 when peak headway was 30 minutes. That service was equivalent to two buses per hour. When peak service was reduced to one bus per hour (60-minute headway), peak ridership fell to 315. Using Equation 10.16 with S = service frequency:

$$\varepsilon_{shr} = \frac{(Q_1 - Q_0)S_0}{(S_1 - S_0)Q_0} = \frac{(315 - 500)*2}{(1 - 2)*500} = +0.74 \text{ in the peak period.}$$

In the off-peak periods,

$$\varepsilon_{shr} = \frac{(Q_1 - Q_0)S_0}{(S_1 - S_0)Q_0} = \frac{(270 - 500)*2}{(1 - 2)*500} = +0.92.$$

Examples 10.4 and 10.5 used the same service changes and ridership changes, but different ways to quantify the relationship between service and ridership. In this case, the distinction between *headway elasticity* and *service frequency elasticity* is important, even though the same situation is being analyzed.

THINK ABOUT IT

Curtin's demand elasticity is negative. So were the service elasticities from Mayworm et al. Why was the elasticity calculated in Example 10.5 positive?

Besides the usual problem in using travel behavior data collected in another city, there is another impediment to using "borrowed" data. Quite often, a transit agency needing to make a change because of financial difficulty will not just raise fares or reduce service—it will do both at the same time. This makes a separate computation of fare elasticity and service elasticity impossible. Some other definitions and equations for elasticity exist, but most of them also need the input variables—fare and service—to change one at a time.

10.4   PERFORMANCE MEASURES IN PUBLIC TRANSPORTATION

Introduction

Ellen McCullough is a new member of the Mythaca Bus Company Board of Directors. She sees the Mythaca County Highway Engineer before an elementary school musical and says to him, quietly, "After only a few months on the board, I am already tired of people who criticize our bus company. Maybe I shouldn't take it so personally, but I think we have a pretty good bus service." Then she asks him, "Is there any way I can satisfy myself that our service is good? Or, if it turns out that our service is deficient in some way, how can I determine what is wrong and what will make it better?" The county highway engineer responds with two words, "Performance measures." (See Figure 10.14.)

10.4.1   Transit Performance Measures

What if an acquaintance asked you to identify the "best" transit agency in the state? First, you would need to determine what the acquaintance meant by "best." Did he mean the most effective, the most efficient, the most popular, or did he have something else in mind? (In measuring transit performance, there is a difference between *effectiveness* and *efficiency*, as we shall see.) Once his meaning is clarified, appropriate criteria can be considered for adoption—preferably criteria that can be easily and reliably measured. We are seeking *performance measures* (Fielding et al., 1978) to help us compare two or more transit operations.

What if an acquaintance asked you if a particular transit agency's performance is "getting better"? Again, we would need to clarify the basis for answering such a question. We would seek performance measures and appropriate time frames to carry out a *longitudinal analysis* of the transit agency's operations.

These questions, if coming from an acquaintance, may be asked out of genuine curiosity. If asked by a transit board member or a member of the transit agency's staff, the answers may trigger some important management decisions. In either case, the answers ought to be based on good data that are properly processed and analyzed.

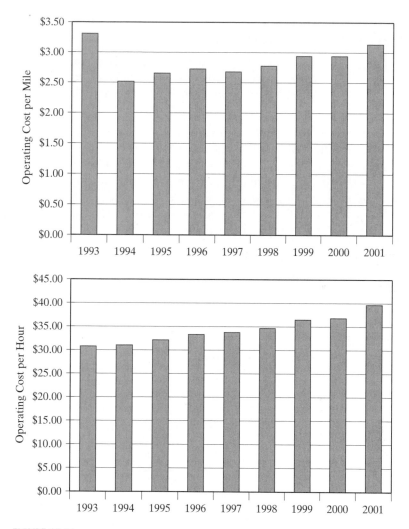

FIGURE 10.14

Examples of transit performance measures. *Source:* Greater Lafayette Public Transportation Corporation.

Table 10.4 shows selected performance measures for the larger bus transit systems in Indiana for the year 1999. The systems have been placed into groups. These groups are called *peer groups*, because the members of each group are intended to have more in common with each other than with transit operators in other peer groups. (Groups 3 and 4 in Indiana are for Urban Demand Response and Rural Demand Response Systems, respectively, and are not presented here.) The peer groups in Table 10.4 were set up to allocate about $20 million in state-operating subsidies. If a transit agency wanted to perform its own peer group analysis, it is free to define whatever membership criteria it thinks are appropriate.

THINK ABOUT IT

Given only the information in Table 10.4, would you consider reassigning one or more systems to a group different from that shown in the table? Which system(s) should be reassigned, and why?

In Table 10.4, *TVM (total vehicle miles)* accounts for all miles covered by the buses in the fleet during the year. A subset of TVM is *revenue vehicle miles (RVM)*, which counts only those miles covered by buses while in revenue service. The difference is the distance traveled between the route and the bus garage, or other nonrevenue mileage. In Table 10.4, "Recovery" is short for *farebox recovery ratio*, which is the fraction of the system's total operating (not capital) costs that are funded from farebox revenues and charter income.

THINK ABOUT IT

Do you think that a transit system whose operating revenues are about 25 percent of its operating expenses is doing a good job? Explain your answer.

TABLE 10.4   Performance Measures by System—1999

	Peak Vehs	TVM	Ridership	Cost/mi	Cost/trip	Recovery	% Labor
Group 1 System							
Evansville	36	1,316,693	1,315,275	$2.81	$2.82	0.19	80
Fort Wayne	30	1,314,267	1,282,639	4.15	4.25	0.12	82
Gary	33	1,755,503	2,472,305	4.66	3.31	0.18	67
Indianapolis	129	9,730,537	11,239,155	3.16	2.74	0.24	54
Lafayette	44	1,368,050	2,135,333	3.13	2.01	0.23	74
Muncie	35	1,145,288	1,308,846	3.55	3.10	0.06	71
South Bend	49	1,891,181	2,486,602	3.23	2.45	0.21	78
Group Avg.		18,521,519	22,240,155	$3.38	$2.81	0.20	64
Group 2 System							
Anderson	10	499,642	279,413	$3.57	$6.38	0.07	83
Bloomington	17	780,075	1,044,344	3.15	2.35	0.21	57
Columbus	6	277,254	168,479	2.58	4.24	0.06	81
East Chicago	3	197,101	238,841	4.90	4.04	0.00	70
Hammond	9	457,289	346,617	4.11	5.42	0.15	6
Marion	4	152,568	129,924	3.19	3.75	0.06	79
Michigan City	6	239,717	196,713	3.02	3.68	0.11	84
Richmond	16	338,256	325,871	2.05	2.13	0.23	81
Southern Indiana	10	481,517	229,659	2.62	5.50	0.22	68
Terre Haute	8	288,578	188,321	3.11	4.76	0.14	80
Group Avg.		3,711,997	3,148,182	$3.20	$3.77	0.14	62

Peak vehs = numbers of transit vehicles in service during peak period; TVM = total vehicle miles; Cost = operating expenses; Recovery = ratio of farebox revenues to operating expenses; % Labor = percent of operating costs spent on salaries, wages, and fringe benefits.

Source: INDOT, 2000.

Example 10.6

The Mythaca Bus Company has a fleet of 46 active vehicles, 37 of which operate during the peak period and 20 of which are lift-equipped. The average seating capacity of an MBC bus is 43, and its average vehicle age is 12 years. In 1999, MBC carried 1.672 million riders, with total operating expenses of $5.808 million; $4.285 million went to salaries, wages, and fringe benefits. MBC buses covered a total of 1.247 million vehicle-miles that year and collected $1.402 million in fares and charter income.

A. If you wanted to compare MBC's 1999 performance measures against its peers in the Indiana list, which systems would you choose from Table 10.4 to serve as MBC's peer group? Explain.

B. Compute for MBC the value of its performance measures in 1999, using those performance measures shown in Table 10.4.

C. Compare MBC's performance measure values against its peer group members. In what areas (if any) does MBC seem to be doing well? In what areas (if any) does MBC need to improve?

Solution to Example 10.6

A. MBC's total peak period fleet of 37 is bigger than all systems in group 2 and is near the median of the 7 bus systems in group 1. A reasonable peer group to use is group 1 in Table 10.4. The largest group 1 system (Indianapolis) has 129 peak buses. If discrepancies arise during our peer group analysis, we can take fleet size into account, if it seems appropriate.

B. Computing MBC's performance measures for 1999 proceeds as follows:

$$\text{Cost/TVM} = \frac{\$5.808 \text{ M}}{1.247 \, TVM} = \$4.65$$

$$\text{Cost/Trip} = \frac{\$5.808 \text{ M}}{1.672 \text{ M trips}} = \$3.47$$

$$\text{Fare recovery} = \frac{\$1.402 \text{ M}}{\$5.808 \text{ M}} = 0.24$$

$$\% \text{ Labor} = \frac{\$4.285 \text{ M}}{\$5.808 \text{ M}} = 0.74$$

C. Mythaca Bus Company's rank as one of eight members of group 1 for each performance measure is:

- *Cost/TVM.*   Second highest (second worst), just $0.01 below Gary
- *Cost/Trip.*   Second highest (second worst)
- *Fare recovery.*   Tied with Indianapolis as highest (best)
- *% Labor.*   Tied for third lowest (third best) with Lafayette

MBC ranks first in farebox recovery ratio. This could mean that its fares are relatively high or that its operating expenses are fairly low. The cost/trip and cost/TVM values for MBC indicate that the latter theory is false—MBC's operating expenses are among the worst in its peer group. MBC's average fare of 84 cents is quite high. This might be discouraging otherwise-willing customers from using the service. It is possible that reducing the MBC fare will attract more riders, thereby reducing cost/trip but also probably hurting farebox recovery. Cost/TVM will not be affected, unless the increase in ridership will require that more capacity (more TVM) be provided. If the service area and/or service frequency cannot be reduced because of political reasons, the operating expenses are being spread over relatively few passenger trips and TVM. Because "% Labor" is not high, MBC can't blame that cost category for its cost problems. The group 1 breakdown of operating expenditures by category is 16 percent services, 10 percent materials and supplies, 2 percent utilities, 5 percent purchased transport, and 1 percent other. MBC ought to check its breakdown against these group 1 averages to find a category where savings are possible.

THINK ABOUT IT

Can the average fare value be computed from the information given in Example 10.6? Explain how to do it.

Without a peer group to provide a basis for comparison, performance measure values can be difficult to interpret, especially for someone who is not an experienced transit analyst. Example 10.6 demonstrated how peer group analysis could be used to evaluate a transit system's operation and help diagnose any problems it may have. Example 10.7 shows how *longitudinal analysis* can be helpful, if applied with care.

Example 10.7

The Mythaca Bus Company may charge high fares on its regular city routes, but in August 1999, it began an experimental *free* service on the campus of Mythaca State University. At the February 2000 MBC Board of Directors meeting, the campus ridership figures for January 1999 and for January 2000 were presented. These ridership figures were 30,392 and 114,423, respectively.

A. By how much did ridership increase (in percent) between January 1999 and January 2000?

B. Why is the January 2000 ridership figure being compared to January 1999, instead of to either December 1999 (the previous month) or to July 1999 (the last full month before the free fare was introduced)?

C. Are there any other pitfalls to avoid in comparing one month's ridership to that of another month's?

Solution to Example 10.7

A. The percent increase in ridership since the previous January is

$$\frac{114{,}423 - 30{,}392}{30{,}392} \times 100 = 276.5\%.$$

B. The month most like January 2000 (except for the fare change) is January 1999. December of any year has characteristics that makes a ridership comparison with the next month (January) almost meaningless. On a campus, the principal reason is usually that the number of days that class is in session may be significantly different. Comparing January with July is even more foolish. July may have no classes in session and the weather in a northern climate is certainly different. Weather can affect ridership from season to season, or even day to day.

C. Even a January-to-January comparison must be done carefully. The spring semester may have started several days later in one of those Januarys, denying the bus company its campus riders for those "lost" days. Good ridership days may be gained as weather deteriorates, but days may also be lost if extremely bad weather causes cancellation of classes. Before any ridership change is used as the basis for a management decision, any hidden reason for such changes must be sought out.

Performance measures can be classified as measures of *effectiveness* or measures of *efficiency*. In simplest terms, effectiveness is "doing the right thing" and efficiency is "doing something well." A transit operator may have certain objectives, such as "providing fixed route bus service that is within one-quarter mile of 55 percent of the city's residents." To the extent that the operator reaches the 55 percent target, the service can be considered effective. If, in doing so, the operator spends $3.25 per revenue vehicle mile, the efficiency of the service provided can be compared against the cost per revenue vehicle mile of peer group operations.

As revealed in the "Recovery" column of Table 10.4, many transit agencies recover less than 20 percent of their operating costs from revenue sources such as fares, charters, and advertising. The deficit must be covered with subsidies from local, state, and federal governments (i.e., taxpayers). Normally, state and federal subsidies are allocated to individual transit agencies according to formulas. These formulas tend to be made up of performance measures.

10.5   LIFE CYCLE COSTS IN PUBLIC TRANSPORTATION

Introduction

The MBC has just received notification from the Federal Transit Administration that its application for a $1.2 million capital improvement grant has been approved. This means that MBC can order four new buses and use the remainder of the grant to upgrade its maintenance facilities. After sending out specifications and requests for bids from qualified bus manufacturers, the choice comes down to two brands of buses. The key attributes of these buses are summarized in Table 10.5. Which bus should MBC order?

TABLE 10.5   Data for Bus Purchase Decision

Bus Model, Manufacturer	Purchase Price	Fuel Efficiency	Maintenance Cost
Tornado, American Bus Company	$258,370	4.7 mi/gal	$9797/year
Typhoon, National Transit Vehicle Corp.	$233,400	4.1 mi/gal	$7372/year

10.5.1   Life Cycle Cost Calculations

The simplest way to make a purchase decision would be to choose the option with the lowest purchase price. In Table 10.5, that would mean that MBC should buy as many of the Typhoon buses (each about $25,000 cheaper than the Tornado) as its $1.2 million federal grant would allow. However, such a decision would ignore the costs that would result from such a purchase. According to the data in Table 10.5, the Typhoon also has lower estimated annual maintenance costs, but the Tornado is more fuel-efficient.

THINK ABOUT IT

Do the data in Table 10.5 provide a sufficient basis for you to make a purchase decision? If not, what additional information do you need?

The Typhoon bus "wins" two of the three categories in Table 10.5. Is that enough to justify a purchase decision involving more than $1 million of taxpayer money? Because the maintenance costs are expressed in $/year units, it may be helpful to know how long these buses can be expected to be in service for MBC. It turns out that one of the requirements attached to federal grants for bus purchases is that the bus be kept for at least 12 years. During the years the bus is used, it will be consuming fuel for each mile it travels. The average MBC bus travels 46,000 miles each year. Other data may also be important to a *life cycle cost (LCC)* analysis:

A. Fuel costs $0.80 per gallon.

B. The average payroll cost for a driver at MBC is $19.10 per hour. *Payroll cost* includes wages and fringe benefits, such as contributions to health insurance premiums and retirement programs. The fringe benefit costs to the employer are often not seen by the employee.

C. While in service, the average bus speed is 12 mph.

D. The *discount rate*—the time value of money used to establish an equivalence between $1 now (in Year 0) and X dollars in Year n – is 4.5 percent.

Table 10.6 is a spreadsheet that summarizes the LCC analysis for the Tornado and Typhoon buses. For each of the 12 years in the analysis, the value of a cost component is expressed in terms of its *present worth (PW)* at year zero. For example, fuel costs for the Tornado bus in year 12 are computed as follows:

$$\frac{\$0.80/gal}{4.7 \ mi/gal} = \$0.170/mi$$

$$46,000 \ mi/yr * \$0.170/mi = \$7830/yr$$

The present worth of $7830 in year 12, brought back to year 0 at 4.5 percent per year is

$$\$7830(1.045)^{-12} = \$4617.$$

This is the value that appears in the "Fuel 1" column and year "12" row of the Table 10.6 spreadsheet. In a similar way, the present worth of the Typhoon fuel costs, and maintenance costs for both buses are computed. Driver costs for both buses are the same, but they are computed here for the sake of completeness. Because the purchase price and all the cost components have been brought back to present worth values at year zero, all that remains is to add up the four LCC components for each bus: purchase + fuel + maintenance + driver. The Typhoon's LCC total is less than the

TABLE 10.6   Twelve-Year Life Cycle Cost Analysis of Bus Purchase.

Tornado (Bus 1)		Typhoon (Bus 2)	
fuel $/gal	$0.80	fuel $/gal	$0.80
mpg	**4.7**	mpg	**4.1**
fuel cost/mi	$0.170	fuel cost/mi	$0.195
bus-mi/yr	46,000	bus-mi/yr	46,000
fuel cost/yr	$7830	fuel cost/yr	$8976
Driver $/hr	$19.10	Driver $/hr	$19.10
bus mph	12.0	bus mph	12.0
Driver $/yr	$73,217	Driver $/yr	$73,217
Maint $/yr	**$9797**	Maint $/yr	**$7372**
discount rate	4.5%	discount rate	4.5%
"inflation"	0%	"inflation"	0%
LCC(1)	$1,086,733	LCC(2)	$1,050,099

Yr	Purch 1	Fuel 1	Maint 1	Driver	Purch 2	Fuel 2	Maint 2
PW	**$258,370**	$71,397	$89,335	$667,632	**$233,400**	$81,845	$67,222
1		$7493	$9375	$70,064		$8589	$7055
2		$7170	$8971	$67,047		$8219	$6751
3		$6861	$8585	$64,160		$7865	$6460
4		$6566	$8215	$61,397		$7527	$6182
5		$6283	$7862	$58,753		$7202	$5916
6		$6012	$7523	$56,223		$6892	$5661
7		$5754	$7199	$53,802		$6596	$5417
8		$5506	$6889	$51,485		$6312	$5184
9		$5269	$6592	$49,268		$6040	$4961
10		$5042	$6309	$47,146		$5780	$4747
11		$4825	$6037	$45,116		$5531	$4543
12		$4617	$5777	$43,173		$5293	$4347

Bold entries = bus-specific input values.

Tornado's, so it is the preferred purchase option, but the Tornado's superior fuel efficiency made the decision a lot closer than it might have appeared at first glance.

10.5.2   Equipment Replacement

In practice, buses not only *can* last longer than 12 years, but they often *must* last longer than the federally mandated life span. This is because transit agencies must rely on federal funding mechanisms that do not guarantee capital replacement funds whenever a transit system wants to buy new buses. Furthermore, a transit agency must plan to get new buses before they need them. As this is written, it takes 82 weeks from the time an order for new buses is placed to the time delivery is made. For these reasons, it is important that a transit agency have a rational procedure for anticipating the need for new buses. This procedure could also be used to justify an application for federal funds, when so many agencies are competing for federal grants.

Table 10.7 summarizes the key cost components for the bus preferred by the MBC. Note that the cost data extend to year 20 of the bus life. As it ages, the bus is placed in service for fewer miles per year, but maintenance costs increase.

TABLE 10.7   Bus Replacement Analysis by Life-Cycle Cost

$260,000 Initial purchase price
12 expected life (years)

Age	Miles/Yr	Deprec Cost	Fuel Cost	Maint Cost	Tot Ann Cost	Ann LCC/Mi
1	46000	$43,333.33	$9660.00	$4420.00	$57,413.33	$1.25
2	46000	36,111.11	9660.00	4906.20	50,677.31	1.10
3	46000	30,092.59	9660.00	5445.88	45,198.47	0.98
4	46000	25,077.16	9660.00	6044.93	40,782.09	0.89
5	46000	20,897.63	9660.00	6709.87	37,267.50	0.81
6	41000	17,414.69	9020.00	7447.96	33,882.65	0.83
7	41000	14,512.25	9020.00	8966.09	32,498.34	0.79
8	41000	12,093.54	9020.00	9952.36	31,065.90	0.76
9	41000	10,077.95	9020.00	11,047.12	30,145.07	0.74
10	41000	8398.29	9020.00	12,262.30	29,680.59	0.72
11	38000	6998.58	9500.00	11,089.56	27,588.14	0.73
12	38000	5832.15	9500.00	12,309.41	27,641.56	0.73
13	38000	—	9500.00	17,056.79	26,556.79	0.70
14	36000	—	9000.00	15,456.89	24,456.89	0.68
15	36000	—	9000.00	17,157.15	26,157.15	0.73
16	34000	—	8160.00	19,044.43	27,204.43	0.80
17	34000	—	8160.00	21,139.32	29,299.32	0.86
18	32000	—	7680.00	23,464.65	31,144.65	0.97
19	32000	—	7680.00	26,045.76	33,725.76	1.05
20	30000	—	7200.00	28,910.79	36,110.79	1.20

Source: Lynch, 1995.

The initial bus purchase price of $260,000 is depreciated according to the *double declining balance method*:

- Assume a 12-year standard bus life.
- At the end of each year, subtract from the previous year's balance (remaining bus value) two-twelfths of that balance. For example, at the end of year 1:

$$\$260,000 - \frac{2}{12}(\$260,000) = \$260,000 - \$43,333.33 = \$216,666.67$$

- The $43,333.33 is the depreciation of the bus's value during the first year. This appears as "Deprec Cost" in Table 10.7.
- At the end of year 2, the depreciation calculation is:

$$\$216,666.67 - \frac{2}{12}(\$216,666.67) = \$216,666.67 - \$36,111.11 = \$180,555.56$$

- Because a 12-year life was used, no depreciation cost occurs beyond year 12.

The "fuel cost" column in Table 10.7 is based on a fuel cost of $0.21/mile for the first 5 years, $0.22/mi for years 6 to 10, $0.25/mi for years 11 to 15, and $0.24/mi for years 16 to 20. The "Tot Ann Cost" column is the sum of the three columns to its left for each of the 20 years. To its right, the "Ann LCC/Mi" column contains the year-by-year quotient

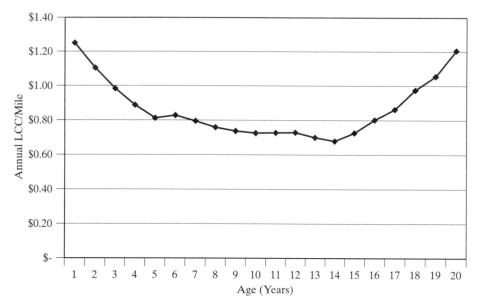

FIGURE 10.15
LCC Curve for Standard 40' Bus.

of "Tot Ann Cost" and "Miles/Yr." Note how the "Ann LCC/Mi" continues to decrease for the first 14 years and then begins to climb. This behavior is shown in Figure 10.15, which is a plot of "Annual LCC/Mi" vs. "Age."

THINK ABOUT IT

At what age should the bus described in Table 10.7 and Figure 10.13 be replaced by a new version of the same bus?

SUMMARY

The provision of public mass transportation service takes place in an unusual environment. Transit agencies are expected to serve affluent commuters from the suburbs and low-income riders in the inner city. Transit stops should be far enough apart to permit good in-vehicle travel speeds, but stops should be close enough together to keep access distance (especially on foot) to reasonable distance. Transit agencies are expected to provide expensive curb-to-curb service required by the Americans with Disabilities Act but are given no additional funding to do so. No public transit agency covers its operating costs. Raising fares may increase revenues, but such a strategy hurts those passengers who need it the most—those with few transportation alternatives. The government units that provide the necessary subsidies often attach strings to their contributions that may not be conducive to efficient service. Public transit is often viewed as a public utility—necessary for the public welfare.

This chapter introduced topics that illustrate various aspects of transit operations. What transit modes are available and how do their characteristics compare? What alternatives are available in operating a rail system in a corridor? How will riders react to a change in fare and/or service? When should a bus be replaced with a new one? The answers to these and many other questions help define the nature of public transit service in a community.

ABBREVIATIONS AND NOTATION

$\bar{a}$	average acceleration rate, mph/sec or m/sec^2
ADA	Americans with Disabilities Act
AGT	automated guideway transit
APM	automated people mover
$\bar{b}$	average deceleration rate, mph/sec or m/sec^2
BRT	bus rapid transit
c	deceleration rate due to coasting, mph/sec or m/sec^2
ε_{shr}	demand elasticity, shrinkage ratio definition
HRT	heavy-rail transit
LRT	light-rail transit
P_{off}	number of passengers leaving the vehicle through the busiest door
P_{on}	number of passengers boarding the vehicle through the busiest door
ROW	right-of-way
RRT	rail rapid transit
S	distance between transit stops or stations
s_a	distance spent by transit vehicle in acceleration regime
$s_{a,c}$	distance spent by transit vehicle reaccelerating after a coasting regime
s_b	distance spent by transit vehicle in deceleration (braking) regime
$s_{b,c}$	distance spent by transit vehicle in deceleration (braking) regime, after a coasting regime is used
s_c	distance spent by transit vehicle in coasting regime
s_d	distance spent by transit vehicle in station-standing regime; always zero
s_V	distance spent by transit vehicle in constant speed regime
$s_{V,c}$	distance spent by transit vehicle in constant speed regime, if coasting is to follow
t_a	time spent by transit vehicle in acceleration regime
$t_{a,c}$	time spent by transit vehicle reaccelerating after a coasting regime
t_b	time spent by transit vehicle in deceleration (braking) regime
$t_{b,c}$	time spent by transit vehicle in deceleration (braking) regime, after a coasting regime is used
t_c	time spent by transit vehicle in coasting regime
t_d	time spent by transit vehicle in station-standing regime; dwell time
t_{oc}	door opening and closing time (combined), sec.
t_{on}	passenger boarding time, seconds per passenger
t_{off}	passenger alighting time, seconds per passenger
t_V	time spent by transit vehicle in constant speed regime
$t_{V,c}$	time spent by transit vehicle in constant speed regime, if coasting is to follow
V_{ec}	speed of vehicle at end of coasting regime
V_{max}	maximum speed of which vehicle is capable
V_{top}	top operating speed for a vehicle, normally set by policy rather than vehicle capability

GLOSSARY

This glossary contains only those terms used in this chapter. For an extensive glossary of transit terms, see the website of the American Public Transit Association at www.apta.com/info/online/glossary.htm.

Acceleration regime: The first transit travel regime, in which the transit vehicle leaves the station or stop and begins to increase speed, until the desired top speed V_{top} is reached.

Americans with Disabilities Act: The 1990 federal legislation that states that public transportation systems cannot deny to people with disabilities services that are available to people without disabilities. This means that, for example, that public transit buses must be "accessible" to individuals with disabilities.

Automated guideway transit: Vehicles whose speed, spacing, and other functions are automatically controlled by a computer system rather than an in-vehicle driver. A requirement for AGT is exclusive right-of-way.

Automated people mover: See *automated guideway transit*.

Braking regime: The last moving transit travel regime, in which the transit vehicle ends its constant-speed or coasting regime and brakes to a stop to pick up or discharge passengers.

Coasting regime: A stage of rail vehicle operation, usually after the desired top speed has been reached, in which no power is applied and the vehicle coasts to save energy and take advantage of the low rolling resistance of steel wheels on steel rails.

Constant-speed regime: The desired top speed that a vehicle should maintain after acceleration and before either coasting or braking.

Deceleration regime: See *braking regime*.

Demand-responsive: Transit service that is provided in response to requests from eligible individuals. Eligibility may be determined on the basis of geographic location with respect to a service area, status as "disabled" under the ADA, age, or other factors.

Dwell time: The time spent by a transit vehicle at a station or stop taking on and discharging passengers. In Equation 10.9, dwell time includes the time needed to open and close the vehicle's doors.

Elasticity: The percent change in quantity demanded for every 1 percent change in price or service.

Frequency of service: Number of bus/train arrivals per hour.

Headway: The time between bus/train arrivals at a point.

Heavy-rail transit: The form of transit characterized by fully grade-separated rights-of-way, high speeds, greater station spacing, and electric multiple-unit cars.

Jerk: The rate of change in acceleration (or, less often, deceleration) rate, ft/sec^3 or m/sec^3.

Light-rail transit: A modern form of the streetcar, more likely to have its own ROW than its streetcar predecessors did.

Paratransit: Variations on public mass transportation that have some of its characteristics. Taxi cabs, carpools, and vanpools are examples.

Rail rapid transit: See *heavy rail transit*.

Transit travel regime: Any one of five possible states of motion (acceleration, constant-speed, coasting, braking, and dwell) that a transit vehicle can attain during its movement from one station or stop to the next.

Travel regime diagram: A graphical representation of a transit vehicle's "trajectory" through time and space, as it moves from one station or stop to the next.

REFERENCES

[1] Black, Alan, *Urban Mass Transportation Planning*, McGraw-Hill, Inc., 1995.

[2] Curtin, John F., "Effect of Fares on Transit Riding," *Highway Research Record* 213, Highway Research Board (now Transportation Research Board), 1968.

[3] CUTS 1992. "Characteristics of Urban Transportation Systems," prepared for Federal Transit Administration under a grant to The Urban Institute, September 1992, Tables 2-23, 2-25, 3-16, 3-18. www.fta.dot.gov/library/reference/CUTS/frchap3.htm.

[4] Danaher, A.R., "Development of Transit Capacity and Quality of Service: Principles, Practices and Procedures," TCRP Web Report 6, Transit Cooperative Research Program, Project A-15, January 1999, Part 2, Chapter 1. www.nas.edu/trb/publications/tcrp/tcrp_webdoc_6-a.pdf.

[5] Dufour, Marc, website (http://emdx.org).

[6] Eisenstein, Hank, "Queens-bound IRT 7 train," www.quuxuum.org/~nixon/transport/page2.html, retrieved 26 November 2002.

[7] Fielding, G.J., R.E. Glauthier, and C.A. Lave, "Performance Indicators for Transit Management," *Transportation*, Vol. 7, 1978, p. 365–379.

[8] Fricker, Jon. D., *Economic Efficiency, Government Intervention, and Measurable Social Benefits in Urban Mass Transit Operations*, Report No. CMUTRI-TP-72-089, Transportation Research Institute, Carnegie-Mellon University, Pittsburgh PA, May 1972.

[9] Fuhs, Charles A., "High-Occupancy Vehicle Facilities: A Planning, Design, and Operation Manual", Parsons, Brinckerhoff, Quade, & Douglas, 1990, as cited in CUTS 1992.

[10] HCM 1985. *Highway Capacity Manual*, Special Report 209, Transportation Research Board, National Academy of Sciences, 1985, as cited in CUTS 1992.

[11] Indiana Department of Transportation (INDOT), Public Transit Section, *1999 Annual Report, Indiana Public Transit*, August 2000. Annual updates at http://www.in.gov/dot/modetrans/bus/tran_5.html.

[12] Lang, A. Scheffer and Richard M. Soberman, *Urban Rail Transit: Its Economics and Technology*, The M.I.T. Press, 1964, Chapter 4.

[13] Lynch, John P., *Elements of the Indiana Public Transportation System*, Master's Research Report, School of Civil Engineering, Purdue University, West Lafayette, IN, December 1995.

[14] MacIsaac, James W., *Analysis of the PSRC Draft 2001 MTP Update*, www.gt-wa.com/PITF, May 2001.

[15] Mayworm, P., A. Lago, and J.M. McEnroe, *Patronage Impacts of Changes in Transit Fares and Services*, Report DOT-UT 90014, prepared by Ecosometrics, Inc. for the U.S. Dept. of Transportation, September 3, 1980, as quoted in Meyer, Michael and Eric J. Miller, *Urban Transportation Planning: A Decision-Making Approach*, McGraw-Hill, 1984.

[16] Meyer, John R., Merton J. Peck, John Stenason, and Charles Zwick, *The Economics of Competition in the Transportation Industries*, Harvard University Press, January 1959.

[17] TCRP 13. *Rail Transit Capacity*, Transit Cooperative Research Program Report 13, Transportation Research Board, National Academy of Sciences, 1996. Tables 4.16, 5.8, 5.9.

[18] Vuchic, Vukan R., *Urban Public Transportation: Systems and Technology*, Prentice-Hall, Inc., 1981, Chapter 3.

EXERCISES FOR CHAPTER 10: PUBLIC MASS TRANSPORTATION

Transit Modes

10.1 Classification of Urban Public Transportation Modes.

(a) Where in Table 10.1 does the monorail at Walt Disney World belong? Explain.

(b) What is a "funicular" in Table 10.1? Where is the nearest funicular to where you live?

(c) Have you ever ridden an "aerial tramway"? If so, would the aerial tramway you rode qualify as an urban transit mode?

(d) What is a dual-mode bus? How does it operate?

10.2 Bus Rapid Transit.   A new form of urban mass transit is *bus rapid transit* (*BRT*), in which buses operate on their own exclusive right-of-way. See the Federal Transit Administration website on BRT at http://www.fta.dot.gov/brt/. Where in Table 10.1 do you think "BRT" should be placed?

10.3 Transit Prerequisites.   What conditions are conducive to service by rail transit?

10.4 Be Careful What You Wish for.   What are possible problems associated with encouraging travelers to make greater use of public transit? What are possible problems associated with encouraging travelers to make greater use of ridesharing?

Designing a Rail Transit Line

10.5 Keeping Top Speed.   Repeat Example 10.3, but use only one coasting regime before the braking regime. After initial acceleration to V_{top}, sustain that constant speed until coasting should begin. How long (in time and distance) should the constant speed regime last?

10.6 Constant Speed.   Repeat Example 10.3, but maintain the top speed until the braking regime needs to begin. Draw the resulting time-speed diagram and label the key points with time and distance values.

10.7 Average Speeds.   Compare the total travel time and average speeds of the three operational alternatives represented in Example 10.3 and the "Keeping Top Speed" and "Constant Speed" problems above. Be sure to state all operating parameters, including average dwell time.

10.8 Coasting.   In Example 10.3, the first use of Equation 10.10 produced an irrational number. To explore this further, calculate the distance the transit vehicle would coast until it came to a complete stop (i.e., $V_{ec} = 0$).

10.9 Transit Time-Speed Diagram.   As in Example 10.3, the longest distance between stations in the proposed Mythaca-Shoridan LRT corridor is expected to be 2.35 miles. Use the same $\bar{a}, \bar{b}$, c, and V_{top} values as in Example 10.3. If V_{ec} was found in Example 10.3 to be slower than 30 mph, a new acceleration regime must begin when $V = 30$ mph. Determine the points—in time and distance units—at which the transit vehicle (a) coasts to 30 mph, (b) accelerates to V_{top}, and (c) must cease accelerating to begin the braking regime. Draw the resulting time-speed diagram and label the key points with time and distance values.

10.10 German LRT Operating Regimes.   The K4000 is a German-made low-floor LRT vehicle. It has jerk-free acceleration (1.3 m/sec^2) and deceleration (1.4 m/sec^2), with a maximum speed of 80 kph. What minimum station spacing will allow the K4000 to reach its maximum speed before it has to begin braking for the next station?

10.11 Non-Zero Jerk.   In drawing the time-speed diagram in Figure 10.8, the value of jerk was ignored. Redraw the diagram with a reasonable value of jerk, so that the rate of acceleration is not constant. Which of the values of jerk mentioned in section 10.2.1 did you use?

10.12 Station Spacing and Rider Response.   The designer of a rail transit line is trying to decide how close together to put the stations. What are the advantages of closer station spacing? What are the disadvantages?

10.13 Busways.   Buses can have their own rights-of-way, too. (See the lower left cell in Table 10.1.) *Busways* have been built in Pittsburgh and other cities. The average bus, however, does not

Typical Transit Bus Performance Characteristics

Speeds (mph)	mph/sec	Speeds (kph)	kph/sec
Acceleration			
0-10	3.33	0-16	5.36
10-30	2.22	16-48	3.57
30-50	0.95	48-80	1.53
Normal deceleration			
2–3 mph/sec		3.2-4.8 m/sec^2	
Top speed			
65 mph		105 kph	

Source: Fuhs 1990 as cited in CUTS 1992, Table 3-15.

exhibit jerk-free acceleration. See the table above for acceleration rates in various speed ranges. Using either U.S. or metric units, draw the acceleration portion of the time-speed diagram. Label the key points in the diagram with both distance and time values.

Predicting Transit Ridership Changes

10.14 Elasticity.   The Mythaca Bus Company is planning a fare-free day on the first Tuesday in January. This means that the usual 75-cent cash fare will not be charged that day. Using a fare elasticity shrinkage ratio of -0.30 and a typical January weekday "cash" (non-pass) ridership of 1400 when the university is not in session, how many non-pass riders can MBC expect on that Tuesday in January?

10.15 Transit Ridership Forecast.   The Mythaca Bus Company offers special service to Mythaca State University students in a service area near campus called the Wildcat Zone. MBC plans to increase the Wildcat Zone fare from 25 cents to 35 cents on next January 1st. Last spring, 227,000 riders paid the Wildcat Zone fare. How many Wildcat Zone riders can MBC expect next spring? Use the shrinkage ratio formula and the standard elasticity value.

10.16 Transit Demand Elasticities.

(a) The MBC is considering reducing its peak period headway on Route 6 from 30 to 15 minutes. Which is the more likely value of transit demand elasticity with respect to service, if service is measured in terms of frequency, -0.70 or $+0.70$? Explain in one or two sentences.

(b) Last year, when MBC raised its average fare from 40 cents to 50 cents, daily ridership fell from 6604 to 5943. Compute MBC's transit demand elasticity with respect to fare, based on last year's experience.

10.17 Transit Pricing Practices.   In the first paragraph of section 10.3, the public transit operator reacts to a bad financial situation by considering a fare increase. Would an airline in the same situation raise its fares? "Compare and contrast" the two transportation providers—transit and airline—and their economic strategies.

10.18 Elasticity by Rider Class.   Is it sufficient to apply one service or fare elasticity to all riders? Explain by identifying possible ridership categories and guessing at how they might react to a change in fare or service.

10.19 Shrinkage Ratio.   Two years ago, the Mythaca Bus Company carried 5964 cash-paying riders when the cash fare was $0.50. Last year, after a cash fare increase to 75 cents, the cash-paying ridership fell to 4711.

(a) What is the demand elasticity of cash-paying MBC riders with respect to fare?

(b) Is the transit demand elastic or inelastic with respect to fare? On what do you base your conclusion?

(c) Some members of the MBC Board of Directors say they want to restore the old 50 cents cash fare next year. Estimate cash-paying ridership next year at the lower fare.

Performance Measures in Public Transportation

10.20 Transit System Size. In 1997, an operator of one of the monorail trains at Walt Disney World (WDW) told the Mythaca County Highway Engineer that the WDW transportation system was the fourth largest in the United States. How would you measure the "size" of a transit system? If you propose more than one possible size measure, rank the measures in your order of preference, and explain your preferences.

10.21 Subsidy per Transit Trip. Use the MBC data presented in Example 10.6 to compute the subsidy needed to pay for the cost of providing an average transit trip.

10.22 Public Transportation Performance Measures. One measure of transit performance is the average load factor (LF), which is the average occupancy of transit vehicles in service. The units for LF are "passengers per seat." For a bus system with the following data, what is the LF?

- 1 million passengers per year
- Average transit trip length = 3.0 miles
- In-service VMT for all vehicles in the fleet = 420,000
- Average seats per bus = 45

10.23 Mass Transit Performance Indicators. Inbound morning peak hour service on rail rapid transit Line 18 in Major City involves five trains per hour. Each train consists of two 80-seat cars. The inbound AM peak hour ridership on this line is 1880 passengers. Line 18 is 8.8 miles long, but the average rider travels only 4.7 miles. What is the load factor on Major City Transit Line 18 for inbound peak hour service only?

Life Cycle Costs in Public Transportation

10.24 Bus Purchase Decision. The American Bus Company has just introduced its newest model, the Whirlwind. Its purchase price is $285,000. Because it features a new design, ABC is offering MBC a special offer—if average maintenance costs during a 12-year warranty period go above $6200 per year, ABC will pay the excess. Using life-cycle cost analysis and the data for the Tornado and Typhoon in Table 10.6, should MBC buy the Whirlwind? Why or why not?

10.25 Breakeven Fuel Price. Based on the data in Table 10.6, at what fuel price ($/gal) does the life cycle cost of the Tornado bus become lower than the LCC of the Typhoon bus?

10.26 Bus Replacement Policy. Some MBC board members have ideas that might help them use the data in Table 10.7 to decide on the year to plan to replace the bus. Help them implement their ideas.

(a) Add a column to the right of the "Ann LCC/mi" column called "Cumul Cost/mi." For each year y, this column will contain the ratio of cumulative cost (years 1 to y) to cumulative miles. Using the entries in this column, when should the bus be replaced? Explain as necessary.

(b) Using a discount rate of 0.045, add a column labeled "PWC" that shows the PW of costs for each year. Finally, add a column "EUAC" that shows the equivalent uniform annual costs of the cumulative costs incurred through each year. Using the entries in these columns, when should the bus be replaced? Explain as necessary.

(c) Which replacement decision method should be used by MBC: Ann LCC/mi, Cumul Cost/mi, PWC, or EUAC? Explain.

Air Transportation and Airports

SCENARIO

Mythaca Airport is like many other airports across the country. Year after year, more air travelers have been using its air services to reach other cities and other countries. The airport's management has been struggling to keep up with the growth in commercial air passenger, air freight, and general aviation (private) traffic. Delayed flight departures are becoming more frequent, as it becomes obvious that Mythaca must invest in airport expansion or risk becoming a "second-class city." The expansion plans must take into account the number, spacing, and orientation of runways. Also on the "airside" of the airport, an adequate number of gates and apron space must be provided. (See Figure 11.1.) In the terminal, a variety of customer amenities (food, shops, etc.) must compete for space with such necessities as ticketing and baggage handling. As the airport increases in size, internal transportation becomes an issue. Is a people mover needed? On the "landside," access and automobile parking head the list of design and operational challenges. And, of course, security has become an even more important consideration. How will Mythaca be able to pay for the expansion? Can the city and the region afford not to?

CHAPTER OBJECTIVES

By the end of this chapter, the student will be able to:

1. Describe and discuss some of the issues that face air transportation in the 21st century.
2. Explain how airport improvements can be financed and carry out sample calculations.
3. Develop a forecast of future traffic for a medium-sized airport.
4. Calculate the capacity of an airport.
5. Calculate an estimate of delay for specified operating conditions.
6. Explain the way in which winds determine runway orientation.

FIGURE 11.1
Activity on the apron of an airport. Photo: Jon D. Fricker.

7. Determine runway length for a specific aircraft.
8. Use airport and aircraft characteristics to determine the allowable takeoff weight.
9. Determine the size of important landside components, including the terminal, parking, and curbside for a medium-sized airport.
10. Describe the role played by air in the transport of freight.

11.1 OVERVIEW OF THE AIR TRANSPORTATION SYSTEM

The air system is a system of transportation that is different from the other forms of passenger and freight movement. Because it possesses a special vocabulary, the reader is directed to the Glossary at the end of the chapter whenever a new term is encountered.

11.1.1 Aviation Is Growing Fast

Over the last 50 years, air transportation has been one of the fastest growing segments of the U.S. economy. According to the Bureau of Transportation Statistics (BTS), U.S. commercial airlines flew 5.3 billion *revenue passenger miles (RPM)* in 1945, growing to 104.1 billion RPM in 1970 and reaching 668 billion RPM in 1999. The RPM in 1999 was 5.1 percent higher than in 1998. Commercial and commuter air carriers increased their boardings from 312 million persons in 1982 to 612.6 million in 1998. The events of September 11, 2001 had a major impact on the growth of aviation in the U.S. and the economic vitality of domestic airlines. The amount of air passenger travel is gradually returning to previous levels.

There are more than 4000 aircraft operating under the guidance of the Federal Aviation Administration's (FAA) Air Traffic Control (ATC) system (mostly commercial aircraft) that are airborne over the United States at any one time. Figure 11.2 shows the national aircraft traffic situation that is displayed at enroute traffic control centers. The paths of the aircraft are evident (e.g., look at aircraft around Dallas–Ft. Worth) on the large display board, updated with data from every air traffic control center in the country every 2 minutes. It is possible to know where any commercial air carrier is at any time it is in the air. Each aircraft has an electronic transponder that transmits flight number, altitude, aircraft, heading, speed, and origin-destination to the air traffic controller.

The principal mission of the FAA is to maintain safety. Even though major aircraft crashes make the headlines, air transport is still the safest way to travel. Over the 10 years from 1992 to 2001, there were only 1.99 fatalities per billion passenger miles traveled (PMT) by air vs. 10.19 per billion PMT by automobile (BTS, 2003).

THINK ABOUT IT

Convert the 40,000 deaths on U.S. highways per year into number of highway deaths per day and compare that with the capacity of a typical commercial airliner. Which would get more publicity: the crash of the airliner or the collective deaths on U.S. highways on a typical day? Why?

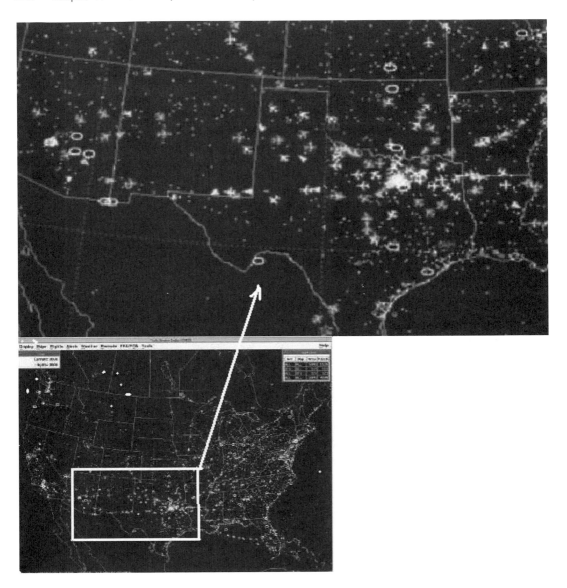

FIGURE 11.2
Aircraft Traffic Situation Display. This picture depicts the Traffic Situation Display (TSD) of all aircraft in flight at one time over the entire United States. Highlighted is the region around Dallas Forth Worth Airport (DFW). Arriving flights and departing flights can be distinguished by the direction of flight of the icons. The white circles denote aircraft in holding patterns, circling waiting for clearance to land from Air Traffic Control (ATC). The operator can click on an icon and find the name of the flight (e.g., AA35), destination, altitude, ground speed and type of aircraft. The display is in color so that departing and arriving aircraft are easily distinguished from each other. The display was developed at Volpe National Transportation Systems Center. Data are fed in from all radar sites. The display is available at the FAA's Air Traffic Control System Command Center and at all enroute centers and many terminal ATC sites. *Source*: Picture courtesy of Volpe National Transportation Systems Center.

Aviation's present impact on the economy has been no less spectacular than its growth in usage. It provided the basis for about 5.6 percent of the gross domestic product in 1989, with more than 2 million persons directly employed in aviation (Smith, 1990). The air transportation system, particularly airports, is considered by many to be "an engine for economic growth." Chicago's O'Hare Airport is the nation's busiest airport, serving more than 66 million passengers in 1994. O'Hare is credited with adding an estimated $13.3 billion in wages alone to the Chicago economy (al Chalabi, 1993). In 1989, all aviation in the New York metro area was estimated to contribute about $30 billion to that economy (Smith, 1990). This significant contribution to the economy is predicted to grow, as air travel increases about 4 times as fast as the population.

11.1.2   Airports in the United States

As can be seen in Figure 11.3, airports in the United States are mostly privately owned, mostly closed to public use, and mostly unpaved, unlighted airstrips. However, all the large, most medium, and many small airports are open to public use. It is those airports that are lighted, with runways that are usually paved. Most importantly, airports are classified under an FAA program called the National Plan of Integrated Airport Systems (NPIAS). Under NPIAS, the airports are categorized on the basis of the activity at the airport. This plan is used to guide the federal funding for airports. Table 11.1 shows the number of airports in each classification and how that classification is determined. Figure 11.4 shows how the public use airports support the travel needs of most citizens in the United States.

11.1.3   Aviation Issues: Safety, Security, and System Efficiency

In the fall of 1997, the FAA completed work on developing goals that would carry the agency and aerospace into the next century. The 1998 FAA Strategic Plan lays out goals, areas for strategic focus, projects to implement them, and outcome measures. The

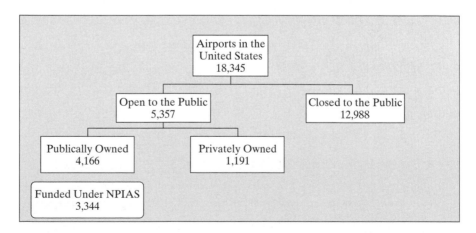

FIGURE 11.3
Airport ownership in the United States. *Source*: FAA, <u>National Plan of Integrated Airport Systems (1998 to 2002)</u>.

TABLE 11.1   Categorization and Distribution of Activity for NPIAS Airports

Number of Airports	Airport Type	Airport National Enplanements (E)	All Enplanements (%)	Active GA Aircraft (%)
29	Large hub primary	E > 1%	67.3	1.3
42	Medium hub primary	$0.25\% \leq E \leq 1\%$	22.2	3.8
70	Small hub primary	$0.05\% \leq E \leq 0.25\%$	7.1	4.7
272	Nonhub primary	$10{,}000 \leq E \leq 0.05\%$	3.3	11.4
125	Other commercial	$2{,}500 \leq E \leq 10{,}000$	0.1	2.1
334	Relievers		0.0	31.5
2,472	General aviation		0.0	37.3
3,344	Total existing NPIAS		100.0	92.1
15,000	Low activity landing sites (Non-NPIAS)		0.0	7.9

Source: FAA, *National Plan of Integrated Airport Systems (1998 to 2002)*, Table 1.

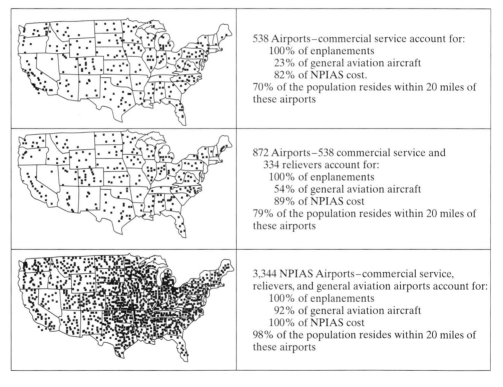

FIGURE 11.4
Geographic Coverage of Airports in the United States. *Source*: FAA, National Plan of Integrated Airport Systems (1998 to 2002), Table 1.

FAA Strategic Plan is focused around three mission goals: safety, security, and system efficiency, as shown in Table 11.2.

As the demand for air travel increases, concern about the delay inherent in a system that operates close to saturation has caused the FAA to embark on a program to

TABLE 11.2   FAA's Strategic Plan

In the area of safety

Overall safety	By 2007, reduce U.S. aviation fatal accident rates by 80 percent from 1996 levels.
Fatal aircraft accident rate	By 2007, reduce the U.S. aviation fatal accident rate per aircraft departure, as measured by a year moving average, by 80 percent from the year average for 1994 to 1996.
Overall aircraft accident rate	Reduce the rate per aircraft departure.
Fatalities and losses by type of accident	Reduce the number and type of fatalities and losses from accidents that occur for each major type of accident.
Occupant risk	Reduce the risk of mortality to a passenger or flight crew member on a typical flight.

In the area of security

Explosive device and weapons detection	Increase ability to detect improvised explosive devices (through use of simulants) and weapons in checked and carry-on baggage and on the person with no significant increase in operational impact by 2003.
Compliance with security requirements	Increase as measured by compliance audits.
Risk/vulnerability at airports and airway facilities	Reduce by year 2005, as measured by risk assessments.

In the area of system efficiency

Overall efficiency	Provide an aerospace transportation system that meets the needs of users and is efficient in the application of FAA and aerospace resources.
System flexibility	Reduce total number of published air traffic control (ATC) preferential routes by 7 percent from the 1994 baseline by 1998.
User access	Reduce the average call waiting times for Automated Flight Service Stations (AFSS) by 20 percent from the 1994 baseline by 1999.
System delays	Reduce the rates of volume- and equipment-related delays by 20 percent from the 1994 baseline by the year 2000.

Source: FAA website.

identify the needs for expanded capacity in the next 10 to 20 years (FAA, 1991). Additional capacity is expected to be provided through a number of changes to the system. At many airports, more runways are being provided. At others, high-speed exits have been added by orienting the connection from the runway to the taxiway at 30 to 45 degrees (rather than at 90 degrees). This allows the aircraft to get off the runway without stopping, thereby saving valuable runway occupancy time. In addition, an increased number of reliever airports are planned, along with improved instrument approach procedures, changes on limitations on runway spacing, on-site weather stations and a more efficient air traffic control (ATC) system. The U.S.DOT now measures the delay at airports and indicates the performance of the major airline carriers to the public. In fact, the FAA maintains a website (www.fly.faa.gov/flyFAA/index.html) that provides airport status reports in near-real time. Table 11.3 lists (a) carriers in on-time performance over the 10 years from 1985 to 1996 and (b) the airports where the greatest delay during the year 1997 was observed. The capacity and delay problem is a function of both the

TABLE 11.3   Delay by Carriers and at Airports

Percentage of Operations to Arrive on Time from September 1987 to December 1995 by Major Carrier			Airports with Greater than 5 minutes' Delay per Operation in 1997
Southwest	85.5	(1)	Newark International
America West	83.8	(2)	Atlanta Hartsfield
Northwest	82.4	(3)	LaGuardia New York
Alaska	81.1	(4)	Philadelphia
American	80.9	(5)	Dallas-Fort Worth
USAir	79.5	(6)	Detroit
Continental	78.9	(7)	St. Louis
Delta	78.5	(8)	Minneapolis-St. Paul
TWA	78.3	(9)	J.F. Kennedy
United	77.9	(10)	Boston Logan
			Cincinnati
			Cleveland Hopkins
			San Francisco

Sources: Air Traveler Consumer Report, U.S. DOT and FAA (National Plan for Integrated Airport systems, 1998 to 2002).

THINK ABOUT IT

Is delay more important to the air traveler than to the driver on a congested freeway? Why or why not?

carriers and the airport. The carriers' operational difficulties may lead to excessive delays, and airports can physically handle only so many aircraft per hour. In light of intolerable delays, much has to be done to improve the situation, especially in the face of increasing travel by Americans, as well as increases in international flights and visitors.

11.1.4   The Airport System

The airport functions as a harbor for aircraft, as a dock for unloading, and as a point of modal connection for the passenger. At the ends of a trip, the airport facilitates changing from ground mode to air mode, or vice versa. As a result, the airport is often analyzed by using the schematic portrayed in Figure 11.5, with the following definitions:

- *Airside* consists of approach airspace, landing aids, runway, taxiway, and apron that serve to bring the passenger to the gate.
- *Landside* consists of those areas where the passenger (or cargo) finds the facilities for further movement on land, such as the arrival and departure concourses, baggage handling, curbside, access to parking lots, to roads, and to other forms of transit.

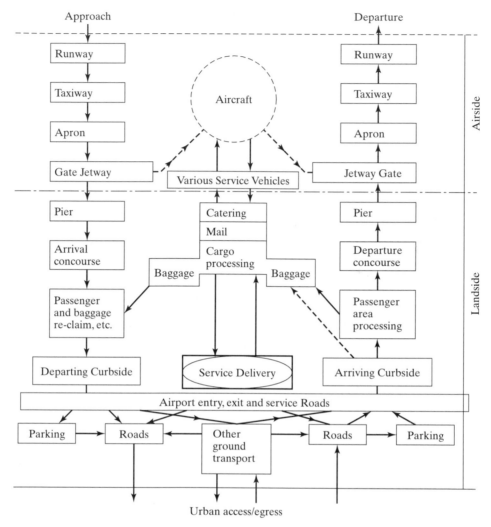

FIGURE 11.5
The airport system. Based on Ashford, Stanton, and Moore, 1991.

FYI: Note that there is no "tarmac" shown in Figure 11.5. Tarmac is a bituminous material used in paving, and some airport surfaces are paved with this substance. However, when a news reporter uses the term "tarmac," he/she almost always means the *apron*, which is the hard-surfaced or paved area around a hangar or terminal. Source: AEROFILES AvSpeak, A Glossary of Aviation Terms and Abbreviations, www.aerofiles.com/glossary.html

An airport consumes considerable land area. Most of that land will be devoted to runways, taxiways, apron, and restricted areas. Large open areas are needed between runways and at the ends of the runway to facilitate simultaneous operations, not only for safety but also to compensate for the severe noise levels of today's jet engine fleet. Because of the desire to have relatively flat runways, grading and drainage play an important part in airport site design. The terminal buildings and other airport service buildings, such as fire stations, fixed base operators, hangers, and general aviation areas, although important, consume much less land area than do the airside activities. Next to the airfield itself, automobile parking is often the largest user of land. The space relationship can be observed in Figure 11.6. The diagonal high-speed exits from Runway 8L-26R to adjacent taxiways are one way to increase capacity.

11.1.5   Ownership and Management

As was discussed in section 11.1.2, most airports used by the public are owned and operated by the municipal government(s) of the political jurisdictions in which they are located. The oversight and management is often under the purview of an airport board or authority whose management is usually tuned to the needs of the local community. Sometimes, as has happened in Indianapolis and Pittsburgh, the local board hires a private sector company to manage the airport. One major exception is the State of Alaska.

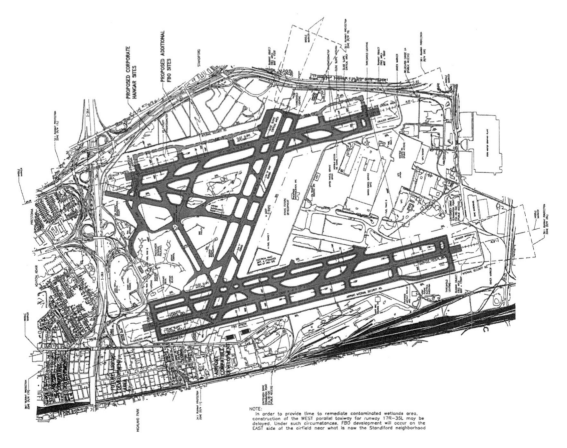

FIGURE 11.6
Conceptual layout of an airport (Standiford Field, Louisville, KY). *Source*: Airport Layout Plan, Standiford Field, 1996, Louisville, KY.

Because of the many isolated communities that can only be served by air, the State of Alaska owns and operates more than 260 airports. These airports range from Anchorage International, with its three 10,000-foot long by 150-foot wide runways, to the 3300 by 60-foot unpaved airstrip at Kwethluk, an isolated (no roads) village of about 500 residents in western Alaska.

Where multiple jurisdictions are near airport boundaries or have significant use of the airport, an authority or board is established with representatives either selected or appointed from the jurisdictions, usually with some joint operating and funding arrangement. For example, the major airports around New York City—LaGuardia, John F. Kennedy International, and Newark International—are managed by the Port Authority of New York and New Jersey. The Port Authority also manages a general aviation airport at Teterboro and two heliports in the area within about a 25-mile radius of the Statue of Liberty (Port Authority of NY and NJ, 1992). The airports of Chicago—O'Hare and Midway—are managed by the Airport Authority of the City of Chicago.

Many of the large and medium U.S. hubs have negotiated long-term agreements with their major airlines under some form of *residual cost* management. Residual cost means that the airlines assume responsibility for paying any residual (uncovered) expenses the airport incurs in the year. The airlines wield a considerable amount of power in the management and investment decisions affecting the operation of these airports.

Other airports operate with the more usual *compensatory cost* approach. Compensatory cost management gives the airport management the responsibility for all airport cost accounts. The agreements with the airport are shorter term, but the local airport board has considerably more flexibility in making changes and improvements to the airport. The local airport authority or board has the latitude to make plans more reflective of the community needs. Many airports raise the cash they need for terminal improvements or to provide the local match for federal and state grants through increases in the community's tax base.

11.1.6 Investment and Financing of Airports

Airports fund improvements using grant money from the FAA for airside (and related) improvements. Figure 11.7 shows NPIAS costs by the type of development proposed in the airport. The 35 percent for "standards" in Figure 11.7 is for the maintenance and upgrading of the air traffic control system. The federal government provides funding for airside investments through the Airport and Airways Trust Fund. The trust fund is largely supported by the 8 percent tax on each airline ticket. In 1988, the federal outlays for airports and airways were about $6 billion, with $2.9 billion from the trust fund matched by $3 billion from general revenue (CBO, 1988).

Almost all of the federal funding has been for airside and navigation improvements, with virtually none devoted to terminal or landside improvement. Recently, however, more federal money is going to landside projects, especially when an airport's capacity limitations and delay are related to some landside constraint. Overall, the larger share of funding for airports nationally comes from the state and local governments through taxes and revenue bonds (CBO, 1984). The federal share for these airside (navigation, landing aids, or runway/taxiway) improvements ranges from 75 percent for large and medium

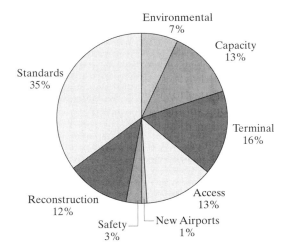

FIGURE 11.7
NPIAS cost by type of airport development. *Source*: National Plan for Integrated Airports [1998–2002].

hubs to 92.5 percent for smaller airports (Ashford and Wright, 1992). Figure 11.8 shows how NPIAS funds are spent as a function of airport size.

The Airport Safety and Capacity Expansion Act of 1990 allows airports to charge each passenger enplaning a *passenger facility charge (PFC)* (FAA, 1994a). The PFC allows airports to charge all users a fee not to exceed $4.50 for boarding at the airport. These funds are used for airside and terminal improvements but not for improvements related to concessions or parking. PFC projects must be approved by the FAA.

Money is raised locally to fund the airport's share (7.5 to 50 percent) of the cost of improvements, usually through bond issues. With the continuing growth of the air system and the significant economic development around airports, airports have very good

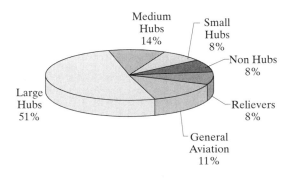

FIGURE 11.8
NPIAS Expenditure by Type of Airport. *Source*: National Plan for Integrated Airports [1998–2002].

bond ratings. The debt is serviced from either local taxes or from airport revenue (e.g., landing fees or airport concessions) or a combination of both.

Example 11.1:   A Simple Airport Investment Problem

The Shoridan airport (airport ID = SHD) has reached a high enough level of traffic (1,000,000 enplanements per year) that its governing board has decided that it must expand its runway and terminal. The cost estimate for the local share of the expansion is $16 million. The cumulative weight of the aircraft landing at the airport is almost 1,200,000,000 pounds per year. The airport is operated by a municipal board that has the authority to raise bonds that can be paid from local tax money and/or from revenues from landing fees, although no single local project can exceed $10 million. The general obligation bonding authority is up to a maximum of 1 percent of the community's assessed valuation of $2235 million.

What increase will occur in the property tax base (in cents per $100 of assessed valuation) and what increase will occur to the present 92 cents per thousand pounds landing fee to fund the improvements? The board requires that a *sinking fund* (like a savings account) be set up so that the return of principal at the end of the bond's life will be available in the bank. The bank is paying 7 percent on its long-term deposits.

Local property tax funds will repay the *general obligation (GO)* bond, which is due in 12 years at 10 percent interest. The landing fees will provide the annual payment for the revenue bond, over 15 years at 9 percent interest.

Solution for Example 11.1   The entries in Table 11.4 were made by following the steps shown below the table.

TABLE 11.4   Bond Data for SHO Expansion Project for Example 11.1

Type	Type	Principal	Annual Interest	Annual Sinking Fund Deposit	Total
General obligation bond	12 yr, 10%	$10,000,000	$1,000,000	$559,000	$1,559,000
Revenue bond	15 yr, 9%	$6,000,000	$540,000	$238,800	$778,000

Step 1:   Determine the bond annual payments from the percent interest times the principal. For the GO bond, annual interest is $0.10 * \$10,000,000 = \$1,000,000$. For the revenue bond, annual interest is $0.09 * \$6,000,000 = \$540,000$.

Step 2:   Determine the annual payment into the sinking fund account for the bonds. For the GO bond, the annual sinking fund payment is

$$SF = F * [A|F, i = 0.07, n = 12] = F * \left[\frac{i}{(1 + i)^n - 1} \right] = 10,000,000 * \left[\frac{0.07}{(1.07)^{12} - 1} \right]$$

$$SF = 10,000,000 * [0.0559] = 559,000.$$

For the revenue bond, the annual sinking fund payment to the bank is

$$SF = 6,000,000 * \left[\frac{0.07}{(1.07)^{15} - 1} \right] = 6,000,000 * [0.0398] = \$238,800$$

Step 3:   Determine the impact on the taxes and landing fee.
The increase in taxes is $1,559,000/2,235,000,000 = 7 cents per $100 valuation.
The increase in landing fees is $778,000/1,200,000,000 = 65 cents per thousand pounds.

Note the increase of 70 percent in landing fees over the current 92 cents. This may be too high. The airport management may have to look for other sources of funding to keep their landing fees competitive. A passenger facility charge is a real possibility.

11.2   AIRPORT PLANNING AND FORECASTING

11.2.1   FAA Sets the Planning Framework

Each state collects and presents its requests for airport improvement grants to the FAA. To aid in its review, the FAA creates, each year, a national forecast of enplanements, operations, and revenue passenger miles for both domestic and international travel. Its forecast covers a period of 11 years and is a key ingredient in evaluating grant requests. After sustaining a high growth rate through the last half of the 1990s, domestic aviation activity is experiencing a lull because of the terrorist attacks of September 11, 2001.

Only about 15 percent of the FAA workforce is in Washington, DC. The remaining personnel are spread around the United States. The map in Figure 11.9 shows the FAA regions and regional offices. In addition, there are FAA personnel who staff the air traffic control towers and centers around the country, plus a number who enforce regulation in various aspects of the industry. Having the funding and the regulatory power in the same agency has its problems and potential conflicts, but the FAA has separated those two functions and provides an environment in which the country's aviation system appears to be safe and functioning well.

THINK ABOUT IT
Why would having the funding and the regulatory power in the same agency cause problems and potential conflicts?

11.2.2   The FAA Forecast

In Table 11.5, the FAA forecasts for 1998 through 2009 and for 2000 through 2011 are shown. The national enplanement forecasts are used directly in planning for terminal and parking at individual airports. The operations at airports are a major determinant for future air traffic control needs and airside capacity. Forecasts of revenue passenger miles are available to the air carriers for their use in planning services.

THINK ABOUT IT
The enplanements increase by 5.5 percent per year, but the operations grow by only about 1.9 percent. Why are they not the same?

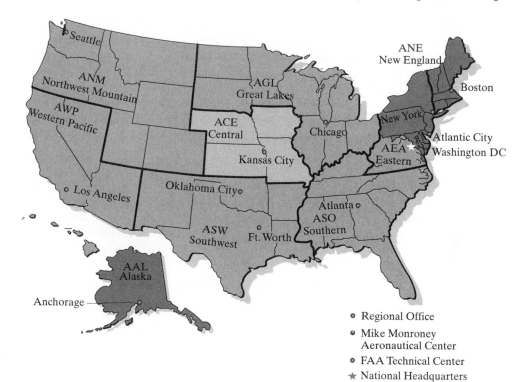

FIGURE 11.9
Regional headquarters and centers of FAA. *Source*: FAA website.

TABLE 11.5 FAA Forecasts (1998 and 2000)

	1998 FAA Forecast (11 Years)			2000 FAA Forecast (11 Years)		
Aviation Activity	1997 Actual	2009 Forecast	Annual Growth (%)	1999 Actual	2011 Forecast	Annual Growth (%)
Enplanements (millions)						
+ Domestic	542.3	821.5	3.5	576.1	880.1	3.6
+ International	52.5	102.8	5.8	53.3	101.7	5.5
+ Atlantic	16.5	27.8	4.4	19.1	32.2	4.4
+ Latin America	20.2	41.5	6.2	21.9	44.4	6.1
+ Pacific	15.8	33.5	6.5	12.3	25.1	6.1
+ System (domestic and international)	594.7	924.3	3.7	629.4	981.8	3.8
+ Commuter/regional	61.9	117.0	5.5	72.4	137.5	5.5
Aircraft operations (millions)						
+ Air carrier	14.2	18.6	2.3	14.4	20.1	2.8
+ Commuter/regional	10.0	12.8	2.1	9.3	12.4	2.4
+ General aviation	36.6	41.5	1.0	29.1	34.0	1.3
+ Military	2.5	2.5	−0.2	2.2	2.2	0.0
+ Total	63.4	75.4	1.5	55.1	68.8	1.9

Source: FAA Aviation Forecasts are updated March of each year. See http://api.hq.faa.gov/.

11.2.3 The Master Plan

The individual airport master plan is the most important document for an airport. It details long-range needs and implementation plans for each airport. It defines to the airport's governing board or authority, to the state, and to the FAA the airport's future funding requirements. A small, *general aviation (GA)* airport with 20,000 operations per year may require a plan of a few pages accompanied by the site description of the airport, called the *airport layout plan (ALP)*. Large, sophisticated airports will usually have ongoing studies, often involving several consultants and documented in several volumes. For example, the Master Plan for Chicago's O'Hare Airport consists of about 15 volumes containing more than 6000 pages. Airside improvements, such as added or extended runways and taxiways, or apron expansion and landside (terminal and parking) needs, are included in the master plan. Like most transportation plans, coordination with various levels of government is required and public participation (often through public hearings) is an important part of the process.

Airports receiving federal funds are required to keep their ALP current, to reflect current airport modifications and off-airport development. However, the entire Master Plan only needs to be updated on a 5- to 10-year cycle, unless substantive changes in the community or in the airport's function within the air system have occurred or are expected.

11.2.4 Forecasting Airport Traffic at Hubs

Planning for an airport requires that future traffic be forecasted in a thorough, sensible manner. An overly optimistic forecast could cause premature expenditure of funds, resulting in higher-than-needed operating costs. On the other hand, an overly conservative forecast could delay necessary investments, so that congestion, lost revenues, and high levels of delay result. Forecasting provides information about two important areas of need for the design engineer or architect.

1. Prediction of enplanements (people who will use the airport terminal and parking) to aid in planning for terminal facilities
2. Prediction of the number of operations (aircraft that will use the airport) needed for an appraisal of the adequacy of runways, taxiways, apron, and air traffic control to handle the traffic without significant delay.

Of the several methods used by airport analysts to forecast demand, the simplest uses a travel forecast based on the recent history of the airport and the *stature* of that airport within the U.S. domestic system. This method is often called the *market share approach*. It is based on the concept that the U.S. Commercial Aviation system is such that most airports maintain a given share of the total U.S. enplanements based on their historical share. This concept works well, unless the community around the airport is growing and/or changing much faster than the overall nation.

Table 11.6 presents a spreadsheet in which a simple market share forecast has been developed for Mythaca Airport (MYH). For this case, the historical data and their sources are:

- Enplanements from the airlines (column C)
- Interline or transferring passengers from the airlines (column D)
- History and forecast of National Enplanements (column E). See Table 11.5.
- Commercial departures from the FAA air traffic control (column H)
- Average departing seats per aircraft from the airlines (column I)
- Total operations from the FAA air traffic control (column K)

The remaining columns in Table 11.6 are calculated. The calculations are divided into two tiers. The first tier of calculations are based on historical data to determine appropriate planning factors for the future. These data are combined with the actual enplanements nationally to determine the stature of the airport in the National Air System—its market share. The second-tier calculations use the planning factors to develop a future estimate of enplanements and operations.

The historical market share is usually adequate to determine the demand, at an airport unless unusual conditions exist in an area. Unusual conditions may occur in large urban areas like Chicago, in a city with fast-changing demographics, in a smaller area where a large company relocates, or in a resort region experiencing rapid growth. The demand can be simply stated as the percentage demand that an airport has, compared to the national air system as a total. We call it the "stature" of the airport in the U.S. air system. In more complex areas, the demand forecast would be refined by using more detail about local economic conditions, other transportation facilities, about the airline operations, aircraft to be used as demand changes, etc. The data in Table 11.7 have been developed by the FAA to aid in longer-range planning than the official 11-year plan.

Table 11.6 indicates that the example airport has experienced a fairly constant share of the market, namely, 0.25 percent. The first step in the market share forecast method would be to say the airport would continue to have the existing market share (column F). Likewise, unless some airline carrier has major plans to use the airport for more or less hubbing than at present, those passengers who *interline* will continue at about 27.5 percent of the enplanements (column G).

The analyst must carefully distinguish between passengers served (a number frequently used by many airport managers) and enplanements (a number of particular importance for airlines and terminal design). Equations 11.1 and 11.2 are used to convert passenger data into enplanements. *Origin-destination (O-D)* passengers are those whose trips either begin or end at the airport. They either live in the area served by the airport or fly into the area for business or pleasure. Transferring passengers are those who move from one airplane to another and don't leave the airport.

$$ENP = (0.5 \times PAX_{O-D}) + PAX_{TRANSFER} \qquad (11.1)$$

$$PAX_{TOTAL} = PAX_{O-D} + PAX_{TRANSFER} \qquad (11.2)$$

TABLE 11.6 Forecast for the MYH Airport

Historical data for forecasting

A	B	C	D	E	F	G	H	I	J	K	L	M
Source:	Calc.	Airlines	Airlines	FAA	Calc.	Calc.	FAA/ATC	Airlines	Calc.	FAA/ATC	Calc.	Calc.
Year	Total O-D Comml Pax	Commercial Enplanements	Interline Pax	National Enplanements (10⁶)	Stature of Airport (%)	Transfer (%)	Commercial Departures	Departing Seats	Load Factor (%)	Total Annual Operations	General Aviation Operations	General Aviation (%)
—	2*(C−D)	Given	Given	Given	C/E	D/C	Given	Given	C/(I*H)	Given	K−(2*H)	L/K
1990	1,860,000	1,276,000	346,000	495.0	0.258	27.1	34,800	61.5	59.6	82,500	12,900	15.6
1991	1,903,000	1,276,500	325,000	487.0	0.262	25.5	33,890	63.0	59.8	80,780	13,000	16.1
1992	1,896,000	1,300,000	352,000	510.7	0.255	27.1	33,560	65.2	59.4	80,620	13,500	16.7
1993	1,926,400	1,308,200	345,000	525.7	0.249	26.4	33,800	68.0	56.9	81,600	14,000	17.2
1994	2,034,000	1,379,000	362,000	557.6	0.247	26.3	33,900	69.4	58.6	80,400	12,600	15.7
1995	2,203,450	1,476,725	375,000	597.7	0.247	25.4	34,300	71.9	59.9	80,600	12,000	14.9
1996	2,256,600	1,528,300	400,000	608.1	0.251	26.2	36,800	72.8	57.0	86,600	13,000	15.0
1997	2,293,500	1,586,750	440,000	630.6	0.252	27.7	37,200	74.5	57.3	88,900	14,500	16.3
1998	2,366,240	1,638,120	455,000	642.4	0.255	27.8	37,800	76.3	56.8	90,050	14,450	16.0
1999	2,455,660	1,687,830	460,000	664.5	0.254	27.3	38,200	77.8	56.8	91,023	14,623	16.1
2000	2,522,930	1,733,185	471,720	688.1	0.252	27.2	37,914	80.4	56.9	91,590	15,762	17.2
2001	2,565,826	1,777,088	494,175	710.5	0.250	27.8	38,283	81.8	56.7	91,886	15,320	16.7
2002	2,643,524	1,829,953	508,191	733.2	0.250	27.8	38,651	83.2	56.9	92,570	15,268	16.5

The forecast

Step:	3	1	2	National Forecast(10⁶)	Planning Factor	Planning Factor	4	Plg Factor Change/Year	Planning Factor	5	6	Planning Factor
	Calc	Calc	Calc	Given	0.25%	28%	Calc	1.4	57%	Calc	Calc	16.5%
	2*(C−D)	F*E	G*C		0.25	28	C/(I*J)		57	(2*H)/(1−M)	M*K	
2003	2,726,280	1,893,250	530,110	757.3	0.25	28	39,261	84.6	57	94,039	15,516	16.5
2004	2,837,880	1,970,750	551,810	788.3	0.25	28	40,203	86.0	57	96,295	15,889	16.5
2005	2,961,720	2,056,750	575,890	822.7	0.25	28	41,285	87.4	57	98,887	16,316	16.5
2010	3,625,200	2,517,500	704,900	1007.0	0.25	28	46,787	94.4	57	112,064	18,491	16.5
2015	4,384,800	3,045,000	852,600	1218.0	0.25	28	55,531	96.2#	57	133,009	21,946	16.5
2020	5,205,600	3,615,000	1,012,200	1446.0	0.25	28	65,047	97.5#	57	155,802	25,707	16.5
2025	6,264,000	4,350,000	1,218,000	1740.0	0.25	28	77,087	99.0#	57	184,639	30,465	16.5

Terminal Sizing

Runway Capacity

#Growth slows after 2010

TABLE 11.7   FAA Forecast Data for Enplanements

Aviation Activity	Average Annual Percent Changes	
Passenger enplanements	1999–2011	2011–2025
U.S. domestic	3.6	3.2
Regional commuters	5.5	4.9
International (to and from U.S.)	5.1	4.9

Source: FAA, <u>Long-Range Aerospace Forecasts</u>, Fiscal Years 2015, 2020, and 2025, Office of Policy and Plans, June 2000.

where
$\quad$ ENP = enplaning passengers

$\text{PAX}_{\text{O–D}}$ = passengers who live/work in the airport market area or are visiting that area

$\text{PAX}_{\text{TRANSFER}}$ = passengers who do not live/work in the airport market area

$\text{PAX}_{\text{TOTAL}}$ = total passengers; a number often quoted by airports

From these equations, the number of origin-destination passengers can be determined. The total number of passengers served is the sum of columns B and D in Table 11.6.

The link between operations and enplanements is the capacity of the average aircraft (departing seats) coupled with the average passenger load factor as shown in Equation 11.3.

$$\text{OPS}_{\text{departures}} = \frac{\text{ENP}}{\text{SEATS}_{\text{departing}} \times \text{LF}} \tag{11.3}$$

where $\quad\text{OPS}_{\text{departures}}$ = commercial aircraft departures

LF = load factor

ENP = enplaning passengers

$\text{SEATS}_{\text{departing}}$ = equals the average number of seats on each departing aircraft.

The next calculation is to determine the planning factor for "Load Factor" in column J using Equation 11.3. In forecasting, it is reasonable to use the recent history of load factors to determine the planning factor. The trend in load factors is not likely to change much over time. On the other hand, as the number of passengers increase, the airlines may begin to substitute larger planes for the previous smaller ones, causing the number of average seats per aircraft to increase. In this case, the trend is about 1.8 more seats per aircraft per year. In forecasting load factor and seats, Table 11.8 can be used to add to the planner's knowledge. Conversations with air carriers about their plans will also help the planner.

The infusion of larger planes into the fleet is often done in response to an increase in demand. What often happens is that the number of enplanements increases, but because the planes are larger, the load factor will remain about constant, or even decline. Table 11.6 shows that this has happened at MYH. However, the forecast by the FAA suggests that both the number of seats per aircraft and the load factor will grow from 2015 through 2025. The trend in the average number of departing seats is

TABLE 11.8   FAA Long-Range Forecast of Seats and Load Factor

Year	Average Seats per Aircraft		Load Factor in Percent	
	U.S. Domestic Carriers	Regional/Commuter Carriers	U.S. Domestic Carriers	Regional/Commuter Carriers
1994	146.6	29.1	64.2	50.4
1995	143.4	30.3	65.2	49.3
1996	141.8	30.5	67.4	52.3
1997	142.5	31.4	68.9	53.6
1998	142.1	33.3	70.1	56.5
1999	141.3	36.0	69.8	57.6
2000	142.4	37.3	69.3	58.2
2001	142.8	38.3	68.8	58.7
2002	143.0	39.2	68.3	59.1
2003	143.1	39.9	68.3	59.4
2004	143.2	40.5	69.0	59.8
2005	143.3	41.1	70.0	60.1
2010	147.7	43.8	70.0	61.4
2015	152.7	47.1	70.0	63.0
2020	158.0	51.0	70.0	64.5
2025	161.0	54.5	70.0	66.0

Source: FAA, <u>Long-Range Aerospace Forecasts</u>, Fiscal Years 2015, 2020, and 2025, Office of Policy and Plans, June 2000.

airport-specific. It depends on the aircraft used and the markets served. An airport with many commuter aircraft may see some increase in load factor, but larger and medium airports are not likely to.

THINK ABOUT IT

What data will you want from the airlines to develop an airportwide database from which you can determine seats and load factors? Because one can be calculated from the other, why get both?

When given the commercial departures and total operations, the number of GA operations can be determined by noting that, for each commercial departure, there was also an arrival. Thus, subtracting twice the commercial departures from the total operations gives a number of operations left over for freight, military, and other aircraft. For most airports, these will be largely GA operations. This computation leads to establishing the last planning factor, namely, the percent of GA operations out of total operations. One assumption in Table 11.6 is that GA operations will continue at about 16 percent of the total. Another possibility that is often true at busy airports is that the *number* of GA operations will continue at about the same level. With these planning factors, the calculations can begin. The order of the calculations is indicated on the spreadsheet. In 2025, the terminal will need to accommodate about 7.5 million passengers—6.3 million origin-destination passengers plus 1.2 million interline or transfer passengers. The airport would see about 153,000 annual operations. The first result will be used to project the physical needs in the terminal and apron area. The second value will determine the number of runways needed. (This topic is covered in section 11.4.)

It may be necessary to alter the FAA forecast if changes predicted in the locality for variables such as disposable income, jobs, and population, are vastly different from the national assumptions used to generate the national forecast. Those persons charged with forecasting traffic at an airport are wise to compare several different forecast approaches. If the market share approach is close to other approaches, it should be used, primarily because the FAA data are used to construct the forecast and the FAA also will need to approve the forecast as part of the master plan. Two other approaches are often used:

1. *Trends.* The past number of enplanements is often used to develop a trend.
2. *Regression of important variables.* Variables that tend to determine the future use of the air system (local wages, employment, and population) are used to develop a regression equation from historical data. That equation is assumed to apply to the future and is used to forecast future airport usage.

Example 11.2: Compare the trend analysis and a regression analysis for the Shoridan Airport.

The Shoridan Airport is a fast growing airport, as is the community of Shoridan. There are only regional commuter aircraft, with a few 737s and MD80s, now stopping in Shoridan. There are no interlines (transferring passengers). Past demographic and enplanement data are presented in Table 11.9. The mayor's office provided the forecast values for population and average income indicated in Table 11.10.

Solution to Example 11.2

Step 1: Determine the market share forecast. The airport has a stature in the National Air System of 0.17 percent. Using those data in conjunction with the FAA data in Table 11.6, forecast the enplanements.

TABLE 11.9 Basic Demographic Data for Example 11.2

	Population	Average Income ($)	Enplanements
1994	370,000	16,233	935,778
1995	373,235	16,343	976,345
1996	376,321	16,479	1,018,443
1997	379,452	16,521	1,076,987
1998	381,890	16,721	1,100,322
1999	383,677	16,790	1,123,500

TABLE 11.10 Forecast of Demographic Data for Example 11.2

	Population Forecast	Average Income Forecast ($)	Enplanements
2000	386,500	16,831	??
2005	400,000	17,700	??
2010	413,000	18,600	??
2015	428,000	19,400	??
2020	439,000	20,000	??
2025	450,000	20,600	??

Step 2: Use the historic population data to develop a regression for population. Using a spreadsheet, the best-fit equation for Enplanements/Population = 2.4575 + 0.0835(x − 1993). The regression line has an $R^2 = 0.9732$.

Step 3: Use the average income as the basis for enplanements. Using a spreadsheet, the equation for Enplanements per Dollar Average Income = 56.068 + 1.941(x − 1993). $R^2 = 0.9612$.

Steps 4 and 5: Use the trend analysis to develop the straight line forecast and the "average annual percent growth rate." The equations are:

$$\text{Trend straightline enplanements} = 901{,}654 + 39117(x − 1993) \text{ with } R^2 = 0.9792$$
$$\text{Trend exponential enplanements} = 907403\ e^{0.038(x−1993)} \text{ with } R^2 = 0.9749$$

Step 6: Develop the multiple regression equation using both population and income. Because the coefficient of determination is so high for the simple regressions in Steps 4 and 5, multiple regression was not deemed necessary.

Step 7: Going back and entering the proper years in the equation gives the answers shown in Table 11.11. The last column is the accepted forecast used for planning the airport. Look at 2025 and note the wide divergence between forecasts. (See also Figure 11.10.) The choice of 2.75 million

TABLE 11.11   Forecast Results for Five Different Approaches

Year	Market Share	Enp/Pop	Enp/Inc	Trend-SL	Trend-Exp.	Forecast
2000	1,169,770	1,175,733	1,172,195	1,175,473	1,183,921	**1,170,000**
2005	1,398,590	1,383,800	1,404,495	1,371,058	1,431,655	**1,400,000**
2010	1,711,900	1,601,201	1,656,423	1,566,643	1,731,229	**1,700,000**
2015	2,070,600	1,838,046	1,915,944	1,762,228	2,093,488	**2,000,000**
2020	2,492,200	2,068,568	2,169,300	1,957,813	2,531,549	**2,300,000**
2025	2,958,000	2,308,275	2,434,302	2,153,398	3,061,275	**2,750,000**

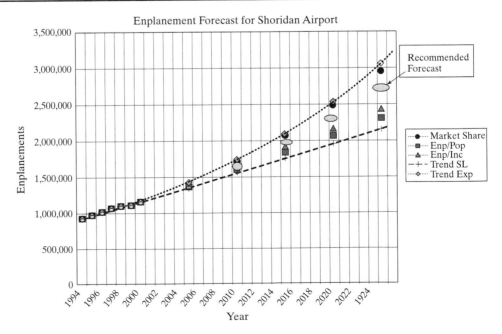

FIGURE 11.10
Plot of various forecasts of enplanements for Shoridan Airport.

here is more conservative than the market share approach. Plotting each of the results on the same graph indicates how the forecasts relate to each other. Picking a central value is often a good compromise. (Recall that the FAA must approve the forecast as part of the airport's master plan.)

THINK ABOUT IT

Why didn't the consultant use the FAA market share forecast?

11.3   CAPACITY AND DELAY

11.3.1   Delay Is Painful and Expensive

The four most-delayed U.S. airports in 1998 were Newark (6,907.9 delays per 100,000 operations), San Francisco (6,599.2), Chicago O'Hare (3,207.3), and Atlanta (3,306.0). An aircraft delay is registered when an aircraft departs the gate or arrives at the gate more than 15 minutes after the scheduled time for that operation. A passenger can encounter delay at any of several points in an airport—most often at check-in (Figure 11.11), security, or in the boarding area. Statistics compiled by the FAA include delays resulting from National Airspace System equipment problems, traffic volume, weather, and runway closures. Denver International Airport (DIA) registered the fewest delays among the nation's 20 busiest airports, with an average of 174.2 delays per 100,000 operations.

As indicated in Table 11.12, the costs associated with delay are enormous. The costs approached 10 billion dollars in 1994, and they have grown since. This is an

FIGURE 11.11
Air passengers wait in line. Photo: Jon D. Fricker.

TABLE 11.12 Total Delay Costs All U.S. Air Carriers 1987 to 1994 (millions of dollars)

	1987	1988	1989	1990	1991	1992	1993	1994
Total operating cost	1895	2159	2365	2829	2485	2478	2509	2514
Total passenger delay cost	4649	4959	5363	5861	5695	6272	6495	7014
Total delay cost	6544	7118	7728	8690	8180	8750	9004	9528

Source: U.S. DOT/FAA, *Total Cost for Air Carrier Delay*, 1995.

important national problem—one that has received much attention, particularly with the continued growth rate of aviation at 3 to 4 percent each year.

Congestion occurs at airports as well as on highways. The goal in both cases is to have sufficient capacity to meet the demand without subjecting the users to significant delay. As we learned in Chapter 3, capacity is defined as the processing capability of a facility to serve its user demands over some period of time. In aviation, the equivalent to the roadway traffic jam occurs when a large number of airplanes are circling the airport waiting to land and/or there is a long line of aircraft on the ground waiting for a gate or waiting to take off (as shown in Figure 11.12). Aircraft waiting to land always have the priority over aircraft waiting to take off. Departing aircraft will be permitted to "roll out" when there is ample time between successive landing aircraft for a safe takeoff. When the demand becomes very great, large delays will occur, indicating that the facility is in need of added capacity. The difficulty is that added runways, just like added roads, require a sizable investment and take a long time to construct.

FIGURE 11.12
An aircraft queue at an international airport—waiting to take off. *Source*: Getty Images, Inc. Liaison

A little average delay is tolerable—an average of about 0.5 minutes for all aircraft is the usual planning value. Maximum delays of 5 to 15 times the average delay can often occur at peak times and are still considered reasonable. On the other hand, there will always be some, albeit small, average arrival delay, because aircraft arrivals are a Poisson process.

11.3.2 Operations

The *geometry of operations* is important when analyzing the capacity of an airport. For example, consider an *instrument landing system (ILS)*-equipped runway found at most large and medium airports. The glide slope for ILS is three degrees above the horizontal. As illustrated in Figure 11.13, the pilot flies along an invisible wire from the beginning to the end of the glide slope. If the pilot gets off course, his instruments tell him what steering corrections are needed to return to the glide slope.

The glide slope of the aircraft begins at a point in space called the *entry gate*, which is a distance γ (usually 5 to 7 miles) from the threshold of the runway. Figure 11.14 shows that geometry.

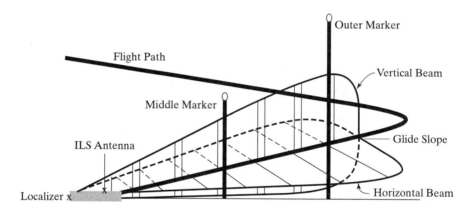

FIGURE 11.13
Schematic of ILS glide slope.

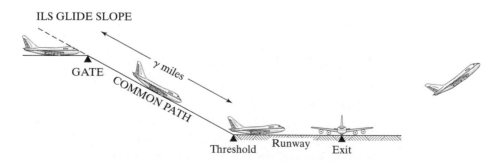

FIGURE 11.14
Geometry of runway operations. *Source*: Mitre, 1972.

Also important to an airport's capacity is the location of exits from the runway to a taxiway, so that an aircraft (when it has slowed sufficiently after landing) can exit the runway. Because only one aircraft—either landing or taking off—is allowed to occupy the runway at a time, minimizing the time at which a landing aircraft spends taxiing on the runway before exiting will help maximize the capacity.

The separation δ between consecutive landing aircraft must be maintained or exceeded along the entire common glide slope. Maintaining the required separation is fairly easy when two consecutive aircraft have the same approach speed. This is case 1 in the analyses to follow. When the trailing plane is faster than the lead aircraft, this is the *overtaking case* (case 2), in which the separation at the entry gate must set so that the required minimum separation is never violated on the glide path to the threshold. See Figure 11.15(a). The *opening case* (case 3) occurs when following plane is slower. See Figure 11.15(b). Here, the minimum δ occurs at the entry gate, and the separation between the aircraft increases along the common path.

11.3.3 Capacity Computation

As with the highway mode, the determination of delay begins by first determining capacity (or supply) and then comparing that with the demand. Four elements control the airborne queues:

1. Unlike the extensive highway system, diversion routes are usually not possible and certainly not desirable.

2. Landing or arriving aircraft always have priority over aircraft taking off or departing. This is because of the higher level of risk for safe landing and the fuel expenditure for aircraft in the air.

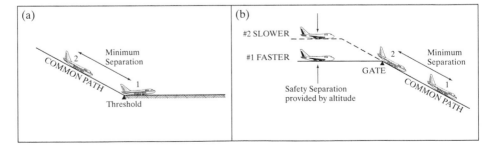

FIGURE 11.15
Closest point of approach on the common path. (a) Cases 1 and 2. The following plane is of equal speed or is faster. In the "closing" or "overtaking" case, $V_2 \geq V_1$. (b) Case 3. Slower plane following a faster aircraft. In the "opening" case, $V_2 < V_1$. *Source*: Mitre, 1972.

3. Weather [Instrument Flight Rules (IFR) vs. Visual Flight Rules (VFR)] and the potential of harmful wake vortices constrain the headway between successive aircraft.
4. Landing speed and aircraft size are very important.

THINK ABOUT IT

Why does the landing (or arriving) aircraft always have priority, even if a Cessna 182 is landing and a fully loaded commercial airliner is waiting to take off?

As in the case of ground transportation, the airport engineer uses distance-time diagrams to help define the process. The capacity will be determined by the length of the time between successive operations, either landings or takeoffs, summed over an hour. Important distances are the distance γ from the entry gate to the threshold (Figure 11.14) and the separation standard δ_{ij} between two successive landing aircraft required for safe landing as determined by the air traffic controller. A distance-time diagram is shown in Figure 11.16. Note that the figure contains equations to determine the time ΔT_{ij} between successive landings of aircraft. At no point between the entry gate and the runway threshold may the separation standard δ_{ij} be violated.

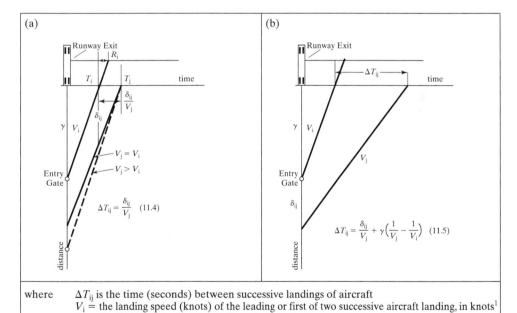

where ΔT_{ij} is the time (seconds) between successive landings of aircraft
   V_i = the landing speed (knots) of the leading or first of two successive aircraft landing, in knots[1]
   V_j = the landing speed (knots) of the trailing or second of two successive aircraft landing
   δ_{ij} = the separation distance
   γ = the distance of the entry gate from the threshold

FIGURE 11.16
Space-time diagram for aircraft landing. (a) Same speed (case 1, $V_i = V_j$) or closing case (case 2, $V_j > V_i$).
(b) Opening case (case 3), $V_i > V_j$.

For case 1 or 2 with a separation distance $\delta = 4$ nautical miles and trailing aircraft speed of 150 n.mi. per hour,

$$\Delta T_{ij} = \delta_{ij}/V_j = (4 \text{ n.mi.}/150 \text{ kn}) * 3600 \text{ sec/hr} = 96 \text{ sec}$$

The "opening case" is different from the other cases, in that the closest point of approach, where the separation standards apply, occurs at the entry gate, not at the threshold of the runway. Because the slower aircraft is trailing, there is an additional term in Equation 11.15 to account for the extra time the slower aircraft takes to fly from the entry gate to the threshold.

11.3.4 Aircraft-Generated Vortices

An aircraft, when generating lift, creates a pair of counterrotating cylindrical air masses trailing each wing, called *wake vortices*. This rotating air mass can be thought of as a horizontal tornado, which can impinge on the following aircraft with a large overturning moment. Wake vortices, particularly those generated by large aircraft, are a potential hazard to following aircraft near the ground, where the vortices move laterally and do not dissipate as rapidly as they do at higher altitudes. In the absence of operational means to locate and track vortices, safety is currently maintained by imposing longer arrival and departure separations between aircraft in the terminal area, so that the vortex will have mainly dissipated. Figure 11.17 shows the effects of a large jet aircraft that has just flown past a smoke tower. The existence of circular air masses or wake turbulence is evident. As you examine the smoke tower picture, note that one vortex is still very active, whereas the other is practically dissipated.

As the aircraft gets close to the ground while landing, the vortices have a tendency to move laterally away, perpendicular to the aircraft's path, at a speed of about 6 knots. A 6-knot crosswind would tend to hold one vortex over the runway, whereas the other vortex would dissipate by moving away at 12 knots. This is illustrated in Figure 11.18. Winds

FIGURE 11.17
Smoke tower showing wake vortices. Photo courtesy of Volpe
National Transportation Systems Center.

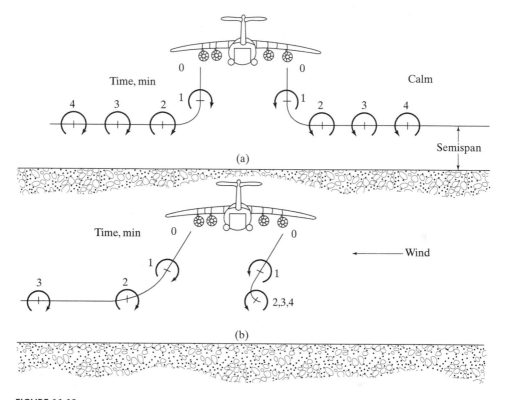

FIGURE 11.18
Vortex movement near the ground showing the effect of wind. *Source*: National Aeronautics and Space Administration.

with crosswind components greater than 10 knots diminish the vortex for the following aircraft. The requirement to maintain the vortex-imposed separation standards constrains the capacity of many airports.

11.3.5 Separation Standards

Air traffic controllers maintain the separation between successive aircraft, depending on weather conditions and the potential for wake vortex hazard. When there is no likelihood of a wake vortex hazard and when the weather is VFR, the separation between successive landings is about 2 nautical miles, regardless of aircraft type. Under IFR, without wake vortex, the separation is maintained at about 3 nautical miles.

THINK ABOUT IT
Under what wind conditions do you think the hazard of wake vortex would not be present? What about winds along the runway?

A trailing aircraft can crash when landing too close behind the aircraft in front of it. Because the magnitude of the force produced by rotating air mass is roughly proportional to the area of the wing and increases with the aircraft weight, separation standards for small light planes landing behind heavy planes must be increased. The safe distance is 6 nautical miles when a Cessna 208 lands behind a Boeing 747. Table 11.13 shows the separation standards that apply when the danger of wake vortex exists.

TABLE 11.13 Spacing δ_{ij} Required for Landing in IFR Weather with Wake Vortex*

Trailing aircraft (j)	Lead Aircraft (i)			Type	Weight (lb)
	Heavy (n.mi.)	Large (n.mi.)	Small (n.mi.)		
Heavy	4	3	3	D	>255,000
Large	5	3	3	C	41,000–300,000
Small	6	4	3	A&B	<41,500

*To convert for VFR, replace 3, 4, 5 and 6 n.mi. with 1.9, 2.7, 3.6, and 4.5 n.mi., respectively.
Source: Horonjeff and McKelvey, 1994, p. 158.

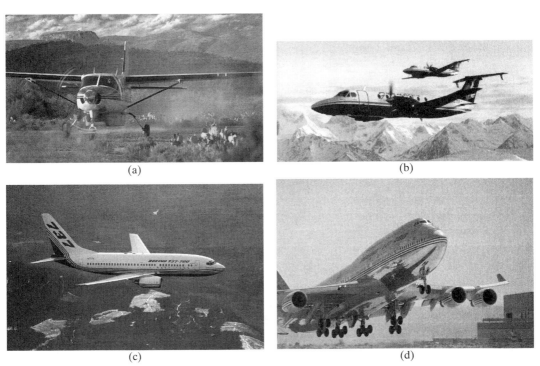

(a)   (b)

(c)   (d)

FIGURE 11.19
Examples of aircraft categories for capacity calculations. MATOW = maximum allowable takeoff weight. (a) Small aircraft <12,500 pounds. Example: Cessna 208; 8,750 pounds MATOW, seats 9; Courtesy of Peninsula Airways. (b) Small aircraft <41,000 pounds. Example: Raytheon Beechcraft 1900; 17,600 pounds MATOW, seats 19; Photo courtesy of Frontier Flying Service. (c) Large aircraft 41,000 to 255,000 pounds. Example Boeing 737-600; 154,500 pounds MATOW, seats 128 to 144;. (d) Heavy aircraft >300,000 pounds. Example: Boeing 777-300; 660,000 pounds MATOW, seats up to 550.

The calculation of capacity for aircraft arrivals alone is presented in Example 11.3 below. Heavy aircraft include such aircraft as the Boeing 747, MD-11, and DC-10, whereas examples of large aircraft include the smaller workhorses in the U.S. commercial market, such as the 727, 737, MD-80s, and DC-9s. The small aircraft category, according to the latest FAA standards (1996), has an upper bound of 41,000 pounds. Small aircraft include general aviation aircraft, air taxies, and many commuter aircraft. Figure 11.19 shows several examples of aircraft categories for capacity calculations. In most airports, it is common to have operations involving small aircraft and large aircraft at the same time.

Example 11.3

Determine the capacity in IFR weather for a runway with only arrivals. The landing speeds and the separation standards δ_{ij} for this example are as indicated in Table 11.14.

The entry gate is 5 nautical miles from the threshold. The aircraft mix and landing speeds are:

10 percent heavy aircraft (H) landing at 150 knots (kn)

70 percent large aircraft (L) landing at 120 kn

20 percent small general aviation and business aircraft (S) landing at 100 kn

Solution for Example 11.3   Case 1 and 2 calculations shown in Table 11.15 are based on Table 11.14. The results are identical, because the trailing aircraft is at least as fast as the lead aircraft. Case 3 has the added problem of a slower aircraft trailing. The air controller will provide safe separation at the entry gate. This means that, in the case of the small aircraft following a heavy aircraft, its 6 nautical mile minimum separation at the entry gate will grow to 7.7 miles when landing occurs. The minimum distance/time associated with the 6 miles separation at either the entry gate or the threshold that governs. The total separation time for case 3 using Equation 11.5 is

$$T_{ij} = 3600 \text{ sec/hr } (6 \text{ n.mi./100 kn} + 5 \text{ n.mi. } [1/100 \text{ kn} - 1/150 \text{ kn}]) = 216 + 60 \text{ sec} = 276 \text{ sec}$$

The "weighted time" between successive landings of aircraft in categories *i* and *j* is the probability two successive aircraft will land in an *i, j* sequence, multiplied by the separation time required for that *i, j* sequence. For example, the probability of a heavy aircraft leading (10%) and a heavy aircraft trailing (10%) is $0.10*0.10 = 0.01$. [See the "H, H" entry in Table 11.15(b).] The required separation time for "heavy aircraft follows heavy aircraft" is 96 seconds. [See the "H, H" entry in Table 11.15(a).] The contribution of the heavy follows heavy sequence to the total weighted time between landings is $0.10*0.10*96 = 0.96$ seconds. Similar calculations for all other sequences of the three aircraft categories are shown in the two rightmost columns of Table 11.16.

TABLE 11.14   Separation Distances and Landing Speeds for Example 11.3

Trailing Aircraft (j)		Lead Aircraft (i)		
		Heavy $V_i = 150$ kn	Large $V_i = 120$ kn	Small $V_i = 100$ kn
Heavy	$V_j = 150$ kn	Case 1 (4 n.mi.)	Case 2 (3 n.mi.)	Case 2 (3 n.mi.)
Large	$V_j = 120$ kn	Case 3 (5 n.mi.)	Case 1 (3 n.mi.)	Case 2 (3 n.mi.)
Small	$V_j = 100$ kn	Case 3 (6 n.mi.)	Case 3 (4 n.mi.)	Case 1 (3 n.mi.)

TABLE 11.15   Calculated Data for Specific Landing Orders of Aircraft

Trailing Aircraft	Lead Aircraft H	L	S		Lead Aircraft H	L	S
H	96	72	72	H	0.01	0.07	0.02
L	180	90	90	L	0.07	0.49	0.14
S	276	210	108	S	0.02	0.14	0.04

(a) Separation times in seconds.   (b) Probability of successive landings.

TABLE 11.16   Spreadsheet for Computing Average Time Between Landings

Lead Aircraft Type	Trail Aircraft Type	Case	Sep. Std. (n.mi.)	V_i (kn)	V_j (kn)	T_{ij} (sec)	Time Added (sec)	Total Time (sec)	Pr i,j	Weighted Time sec
H	H	1	4	150	150	96	0	96	0.01	0.96
L	L	1	3	120	120	90	0	90	0.49	44.10
S	S	1	3	100	100	108	0	108	0.04	4.32
L	H	2	3	120	150	72	0	72	0.07	5.04
S	H	2	3	100	150	72	0	72	0.02	1.44
S	L	2	3	100	120	90	0	90	0.14	12.60
H	L	3	5	150	120	150	30	180	0.07	12.60
H	S	3	6	150	100	216	60	276	0.02	5.52
L	S	3	4	120	100	144	30	174	0.14	24.36
								Weighted average:		**110.94**

The weighted sum of the average time between landings is 110.94 seconds for the mixture of aircraft arriving in a random sequence. The capacity for landings is 3600 seconds per hour/110.94 seconds per landing, or 32.45 aircraft landings per hour.

11.3.6   One Runway—Mixed Landings and Takeoffs

When an airport has a single runway in operation, takeoffs must be mixed in with landings. Figure 11.20 shows schematically the runway occupancy rule, where the landing interval must be greater than the runway occupancy time. Figure 11.21 gives a pictorial representation of the departure rule. Takeoffs must be at least 60 seconds apart. To intersperse aircraft, the controller assumes that it takes an average of 45 seconds for a landing aircraft, once it touches down, to exit the runway if there are high-speed exits. Because only one aircraft can occupy a runway at one time, the next landing aircraft must be at least 2 nautical miles away at the time a waiting aircraft is released for takeoff. Take the case of an aircraft with an approach speed of 120 knots. From 2 nautical miles out, it takes 60 seconds to reach the threshold. Table 11.17 summarizes rules for interspersed landing and takeoff operations.

Landing interval ≥ occupancy time

Threshold   Exit

FIGURE 11.20
Runway occupancy rule. *Source:* Mitre, 1972.

D_1/D_2 interval ≥ 60 seconds

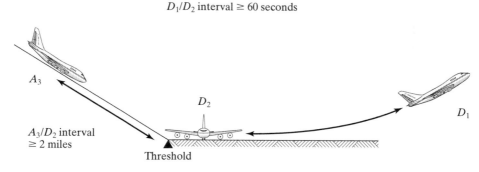

A_3

A_3/D_2 interval
≥ 2 miles

D_2

D_1

Threshold

FIGURE 11.21
Departure Rules. *Source:* Mitre, 1972.

TABLE 11.17   Minimum Runway Operations Times for Mixed Landings and Takeoffs

---

1. For consecutive arriving aircraft, calculate time between landings using Equations 11.4 and 11.5.
2. 45 sec. for an arriving (landing) aircraft to leave runway after touching down.
3. 5 sec. for a departing aircraft to enter runway after previous aircraft (arriving or departing) clears runway.
4. 45 sec. for a departing aircraft to "roll out and rotate", that is, to take off after being cleared by air traffic control.
5. Landing aircraft must be at least 2 n.mi. from runway when a departing aircraft on runway begins rollout. Calculate time for the landing aircraft to reach runway using T = 2 n.mi./landing speed.
6. Because the time T calculated in Item 5 will take place during the departing aircraft's 45-second takeoff (Item 4), do not double count the time during which both events are taking place.
7. Regardless of the rules above, takeoffs must be at least 60 seconds apart.

Example 11.4:   Airport capacity with a known sequence of mixed operations.

Because of scheduled flights and filed general aviation flight plans, the expected sequence of operations at Georgetown (GTN) Airport for the time period beginning 5:30 PM today can be given below. Subscript "a" means the aircraft (H, L, or S) is arriving (landing) and "d" means departing (taking off).

$$L_d, L_d, L_a, L_a, L_d, H_d, L_a, L_a, L_d, L_d, H_a, L_a, L_a, L_a, L_d, S_d, L_a, S_a, L_a, L_a$$

A. Using the rules in Table 11.17, create a table that summarizes the time between operations for each pair of consecutive aircraft. What is the total elapsed time between the first and last operations listed above?

B. What is the capacity (ops/hr) of GTN during the period studied?

Solution to Example 11.4

A. In the summary Table 11.18, the first operation is the departure of a large aircraft. It takes 5 seconds for the aircraft to enter the runway and 45 seconds to take off. Operation 2 is the departure of another large aircraft. It also takes 5 seconds to enter the runway (Item 3 in Table 11.17), but it is held 10 seconds to ensure that its takeoff does not occur within 60 seconds of the takeoff in Operation 1 (Item 7). The 45 seconds Operation 2 needs to roll out and rotate after being released for takeoff (Item 6) are part of the 60 seconds used by the arriving large aircraft (Operation 3) to fly the last 2 n.mi. of its glide path, so Operation 2 is "charged" with only the first 15 seconds.

Including the time for the first aircraft to take off, the total elapsed time is 1542 seconds or 25.70 minutes.

B. To calculate the capacity (ops/hr) of GTN during the period studied, first compute 1542 sec/20 ops = 77.1 sec/op. Then (3600 sec/hr)/(77.1 sec/op) = 46.7 ops/hr. Note that GTN's runway capacity depends on the mix of operations and aircraft sizes that use it, just as a highway's capacity depends on the percentage of trucks and other factors.

TABLE 11.18   Summary of Operations Times for Example 11.4

Op Nr	Op	Approach Speed (kt)	Item in Table 11.17 or Equation Used	Time for op (sec.)	Comments
1	Ld		3,4	50	to take off
2	Ld		3,6,7	15	until rollout
3	La	120	5	60	after Ld rollout
4	La	120	(11.4)	120	after previous La landing
5	Ld		7	60	after La landing
6	Hd		3,6,7	15	until rollout
7	La	120	5	60	after rollout
8	La	120	(11.4)	120	after previous La landing
9	Ld		7	60	after La landing
10	Ld		3,6	5	until rollout
11	Ha	150	5	48	after Ld rollout
12	La	120	(11.5)	180	after Ha landing
13	La	120	(11.4)	120	after previous La landing
14	La	120	(11.4)	120	after previous La landing
15	Ld		3,4	50	after La landing
16	Sd		3,6,7	15	until rollout
17	La	120	5	60	after Sd rollout
18	Sa	100	(11.5)	174	after La landing
19	La	120	(11.4)	90	after Sa landing
20	La	120	(11.4)	120	after previous La landing
				1542	25.70 minutes total
				77.1	average delay
				46.7	ops/hr

11.3.7   Delay

The effect of operations exceeding capacity at an airport [because (due to separation) each operation takes a specific time] is delay. It costs the airlines an average of about $1600 for each hour of operational delay (DOT/FAA, 1995). In addition, it has been estimated to cost (in passenger time, visitor time, and airport services) an average of about $44.00 per passenger for each hour of delay. So delay costs money. More importantly, delay is also a factor in how the person flying perceives the airline. Delay is especially aggravating when changing airplanes in the middle of a trip and the first outbound leg of a trip is delayed to the point where a traveler misses the connecting flight.

Consider the arrival diagram pictured in Figure 11.22. Beginning at 1600 hours, the number of operations that must be accommodated exceeds the capacity. Arriving aircraft will begin to "stack up" and circle the airport until the air traffic controller can give them clearance to land. The queues of aircraft in the air and those on the ground waiting to take off build up, and the delay for all the aircraft accumulates until 1900 hours, when the air traffic controller can begin to "work off" the queue.

Because aircraft arrive in a Poisson manner, when a single runway serves arriving aircraft only, the mean delay is given by Equation 11.6. (See Chapter 3 for a discussion of Poisson equations.)

$$W_a = \frac{\lambda_a \left(\sigma_a^2 + \dfrac{1}{\mu_a^2} \right)}{2 \left(1 - \dfrac{\lambda_a}{\mu_a} \right)} \tag{11.6}$$

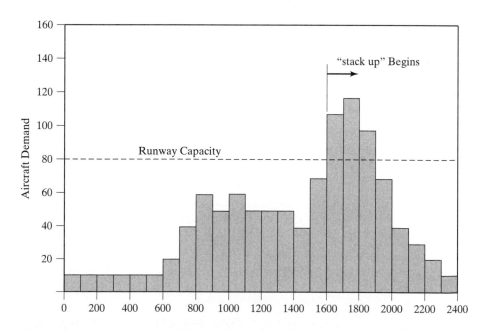

FIGURE 11.22
Typical pattern of hourly demand for an airport operating at overcapacity. *Source*: FAA: AC150/5060A
Airport Capacity and Delay.

where W_a = mean delay to arriving aircraft

λ_a = mean arrival rate, aircraft per unit time

μ_a = mean service rate, or reciprocal of the mean service time (time to land and depart the runway)

σ_a = standard deviation of mean service time of arriving aircraft

$\lambda_a < \mu_a$

When all aircraft are departing, Equation 11.6 is used by replacing the subscript "a" with "d." When aircraft share the same runway for landing and takeoff, arriving aircraft always have priority, so the delay for arriving aircraft differs from that of the arriving aircraft. Equation 11.6 expresses the delay for arriving aircraft, whereas Equation 11.7 can be used to evaluate the delay for departing aircraft.

$$W_d = \frac{\lambda_d\left(\sigma_j^2 + j^2\right)}{2\left(1 - \lambda_d j\right)} + \frac{g\left(\sigma_f^2 + f^2\right)}{2\left(1 - \lambda_a f\right)} \tag{11.7}$$

where W_d = mean delay to departing aircraft on a shared runway

λ_a = mean arrival rate, aircraft per unit time

λ_d = mean departure rate, aircraft per unit time

j = mean interval of time between two successive departures

σ_j = standard deviation of mean interval of time between two successive departures

g = mean rate at which gaps between successive aircraft occur

f = mean interval of time in which no departure can be released, and

σ_f = standard deviation of mean interval of time in which no departure can be released.

Example 11.5

An airport has a capacity of 32 aircraft arriving per hour. Assume the standard deviation of the mean service time for arriving aircraft is 20 seconds.

A. Determine the average delay when the demand for arrivals is 24 aircraft per hour.

B. Determine the maximum arrival rate with a capacity of 32 aircraft arriving per hour such that average delay does not exceed 0.5 minutes.

C. What is the capacity required to maintain an average delay of 0.5 minutes or less with an arrival rate of 24 aircraft per hour?

Solution for Example 11.5

A. The values to use in Equation 11.6 are

λ_a = mean arrival rate = 24 aircraft per hour = 0.4 aircraft per minute

μ_a = mean service rate = 32 aircraft per hour = 0.533 aircraft per minute

σ_a = standard deviation of mean service time for landing aircraft = 0.33 minutes

Now that the units have been made consistent, Equation 11.6 yields an average delay time of

$$W_a = [0.4 * (0.109 + 3.52)]/[2 * (1 - 0.755]) = 2.901 \text{ min}$$

B. Equation 11.6 is

$$W_a = \frac{\lambda_a \left(\sigma_a^2 + \dfrac{1}{\mu_a^2} \right)}{2 \left(1 - \dfrac{\lambda_a}{\mu_a} \right)}.$$

Find λ_a such that $W_a = 0.5$ with all the other variable values as in part A. Using a spreadsheet solver, an arrival rate of $\lambda_a = 10.91$ landings per hour or less is needed to maintain $W_a = 0.5$.

C. In Equation 11.6, find μ_a such that $W_a = 0.5$ with all the other variable values as in part A. Using a spreadsheet solver, a capacity of $\mu_a = 53.36$ landings per hour or more is needed to maintain $W_a = 0.5$.

Besides being affected by separation on landing, the capacity and delay are functions of the configuration of runways, runway exit geometry, landing speed and braking ability. Air traffic control measures for noise abatement, heavy wind conditions, arriving and departing flight paths, and the type of navigational aids also add complexity to the determination of capacity/delay. Safe spacing must always exist between successive aircraft.

11.3.8   Using Annual Service Volume for Estimating Delay

To aid planning, capacity and delay may be estimated by using the annual service volume (ASV) for the airport in combination with the annual demand. Eight of the many possible runway configurations are depicted in Figure 11.23. Crosswind runways may also be present, but they do not add significantly to the airport's ASV. The other simplifying assumptions to introduce ASV computations in this textbook are:

1. Percent arrivals equal percent departures.
2. Full-length parallel taxiway with ample entrances and no taxiway crossing problems.
3. No airspace limitations exist that would adversely impact flight operations.
4. At least one runway is equipped with an ILS and the ATC facilities to operate in a radar environment.
5. IFR weather conditions occur 10 percent of the time.
6. Roughly 80 percent of the time the runway configuration that produces the greatest hourly capacity is used.

Any deviations from these assumed conditions would, of course, affect the actual ASV value.

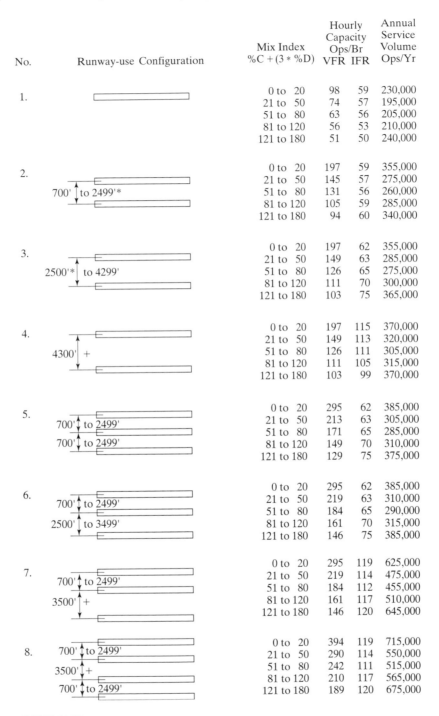

No.	Runway-use Configuration	Mix Index %C + (3 * %D)	Hourly Capacity Ops/Br VFR	Hourly Capacity Ops/Br IFR	Annual Service Volume Ops/Yr
1.		0 to 20	98	59	230,000
		21 to 50	74	57	195,000
		51 to 80	63	56	205,000
		81 to 120	56	53	210,000
		121 to 180	51	50	240,000
2.	700' to 2499'*	0 to 20	197	59	355,000
		21 to 50	145	57	275,000
		51 to 80	131	56	260,000
		81 to 120	105	59	285,000
		121 to 180	94	60	340,000
3.	2500'* to 4299'	0 to 20	197	62	355,000
		21 to 50	149	63	285,000
		51 to 80	126	65	275,000
		81 to 120	111	70	300,000
		121 to 180	103	75	365,000
4.	4300' +	0 to 20	197	115	370,000
		21 to 50	149	113	320,000
		51 to 80	126	111	305,000
		81 to 120	111	105	315,000
		121 to 180	103	99	370,000
5.	700' to 2499' / 700' to 2499'	0 to 20	295	62	385,000
		21 to 50	213	63	305,000
		51 to 80	171	65	285,000
		81 to 120	149	70	310,000
		121 to 180	129	75	375,000
6.	700' to 2499' / 2500' to 3499'	0 to 20	295	62	385,000
		21 to 50	219	63	310,000
		51 to 80	184	65	290,000
		81 to 120	161	70	315,000
		121 to 180	146	75	385,000
7.	700' to 2499' / 3500' +	0 to 20	295	119	625,000
		21 to 50	219	114	475,000
		51 to 80	184	112	455,000
		81 to 120	161	117	510,000
		121 to 180	146	120	645,000
8.	700' to 2499' / 3500' + / 700' to 2499'	0 to 20	394	119	715,000
		21 to 50	290	114	550,000
		51 to 80	242	111	515,000
		81 to 120	210	117	565,000
		121 to 180	189	120	675,000

FIGURE 11.23
Runway configurations—Capacity and ASV for long-range planning. *Source*: FAA, 1995, <u>Airport Capacity and Delay</u>, Figure 2-1.

Example 11.6

Assume an airport has a present annual demand of 160,000 operations, with peak hour traffic of 56 aircraft per hour. It has two parallel runways separated by 1000 feet. In a forecast similar to that done in section 11.2.3, it was learned that in the year 2020, the airport would have about 245,000 operations and have the type of aircraft traffic estimated in Table 11.19.

TABLE 11.19 Aircraft mix for peak hour operation at airport

Aircraft Class	Typical Aircraft	Peak Hour Ops 1992		Peak Hour Ops 2020	
		VFR	IFR	VFR	IFR
A	Cessna, Piper, etc	4 (10%)	0	6 (9%)	0
B	Lear Jet, Shorts	8 (22%)	4 (14%)	18 (27%)	10 (18%)
C	DC-9, B-727, MD-80	22 (61%)	22 (79%)	35 (51%)	35 (65%)
D	DC-10, B-747, B-757	2 (6%)	2 (7%)	9 (13%)	9 (17%)
Mix index		79	100	90	116

A. What ASV must the airport have in year 2020, if the current average delay is to be maintained?

B. If a new runway configuration will be needed by year 2020, suggest one using Figure 11.23.

Solution to Example 11.6 The *mix index (MI)* calculated in Equation 11.8 is an important parameter that is used in the delay calculations. Note that it only uses the percentage of aircraft that are class C or D in its computations.

$$MI = (\%C) + 3*(\%D) \quad \text{(For peak hour)} \tag{11.8}$$

A. The calculated mix indices for 1992 are $MI_{VFR} = 61 + (3*6) = 79$ and $MI_{IFR} = 79 + (3*7) = 100$. Two parallel runways 1000 feet apart match runway configuration 2 in Figure 11.23. The $MI_{VFR} = 79$ value for that configuration has an ASV = 260,000 operations/year. With 160,000 operations in 1992, the *delay factor (DF)* can be calculated in Equation 11.9:

$$DF = \text{Annual Demand/ASV} = 160,000/260,000 = 0.615 \tag{11.9}$$

Figure 11.24 can be used to convert DF into an average delay. The dotted line in Figure 11.24 crosses DF = 0.615 at about 0.5 minutes. To maintain this average delay with 245,000 operations in year 2020, find ASV so that DF = 0.615. ASV = Annual Demand/DF = 245,000/0.615 = 398,000.

B. In 2020, the mix indices will grow to $MI_{VFR} = 51 + (3*13) = 90$ and $MI_{IFR} = 65 + (3*17) = 116$. For runway configuration 2 at the predicted MI_{VFR} value, ASV = 285,000 is possible, but not ASV = 398,000. Scanning down Figure 11.23, the first runway configuration that provides ASV > 398,000 with $MI_{VFR} = 90$ is configuration 7. A third runway must be constructed at least 3500 feet from the other two runways. The

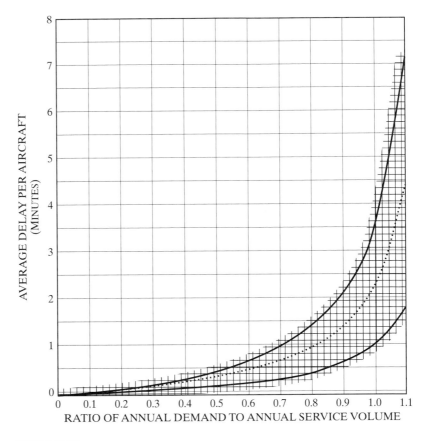

FIGURE 11.24
Average aircraft delay for long-range planning. *Note:* For airports with a high level of commercial traffic use the upper portion of the range. For smaller airports use the whole range. *Source:* FAA, 1995, Capacity and Delay.

hourly capacity for VFR would be 161 operations/hour and 117 for IFR. This is well below the estimated hourly operations from Table 11.23 of 68 and 54, respectively. The average delay in 2020 with the added runway would be less than 0.5 minutes.

It must be pointed out that an *average* delay on the order of 2 minutes can mean that much higher delays can occur regularly on specific flights, especially during peak period operations, when *scheduled* arrivals may actually exceed the short-term capacity of the airport.

11.4 AIR TRAFFIC MANAGEMENT

Air traffic controllers play a vital role in the exercise of air traffic control standards, such as separation standards. Controllers communicate with pilots by radio, taking their cues from radar beacons on each commercial aircraft. Even though each pilot has ultimate responsibility for his aircraft, paying attention to the advice and requests of the air traffic controller is a must. Pilots and controllers work closely together to maintain the safety of the air transportation system. Of particular interest here are

1. Air traffic control procedures for a particular volume of terminal airspace
2. Determination of what constitutes an obstruction to air navigation
3. Provision of electronic and visual approach aids related to the landing, ground control and takeoff at the airport.

11.4.1   Landing

Each airport provides certain aids to landing that serve to help the pilot flying in both good (VFR) and poor (IFR) weather. VFR flying is based on a pilot beware, "see and be seen," strategy, except in airspace under the positive control of air traffic controllers. This usually occurs in the proximity of major airports. Commercial aircraft and many business aircraft are equipped with beacons, radar, and other equipment that permit them to fly in instrument weather and/or in controlled airspace. General aviation pilots flying in VFR need only a functioning radio and altimeter. Capability for landing on a given runway (both directions) and the use of navigation aids varies from airport to airport.

In Figure 11.25, the typical pattern of flight for landing, the approach commences at the Initial Approach Fix (IAF). The courses to be flown by all aircraft in the intermediate segment from *intermediate fix (IF)* to *final approach fix (FAF)* and during the final approach segment (FAF or inner marker to touchdown) are identical, except in special cases. The intermediate fix point is usually 5 to 9 miles from the threshold of the runway. The distance from IAF to threshold is the distance of the entry gate to threshold γ.

11.4.2   Terminal Airspace

The class of aircraft and amount of traffic play a significant role in determining the extent of controlled airspace around the airport. The five approach speeds and typical aircraft types are listed in the Glossary at the end of this chapter. The controlled airspace, with its safety or buffer zones, is configured with these speeds in mind. The volume of sky called *controlled airspace* shown in Figure 11.26 gives the appearance of "an upside-down wedding cake." Starting at the bottom, altitudes are stated in terms of feet *above ground level (AGL)*. The federal airways go from 1200 feet AGL to 18,000 feet above *mean sea level (MSL)*. The jet routes are controlled above that to a *flight level (FL)* of 45,000 feet.

11.5   AIRPORT SITE DETERMINATION AND RUNWAY ORIENTATION

Figure 11.27 shows some of the flight tracks generated by landing aircraft over time, whereas Figure 11.28 shows the runway configuration of an airport, with its *runway protection zones (RPZ)*. Notice the dashed circle with radius 10 miles from the center of the airport in Figure 11.27. This marks the border between controlled and uncontrolled airspace. (See also Figure 11.26.) The airspace and associated ground tracks

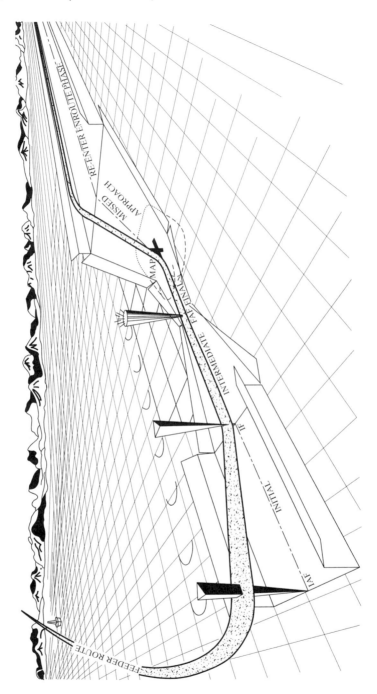

FIGURE 11.25
Typical landing flight path. *Source*: FAA, 1975, <u>Terminal Procedures</u> (TERPS).

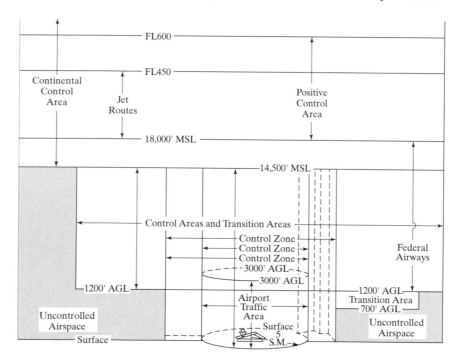

FIGURE 11.26
Controlled airspace. *Source:* FAA, 1985, <u>Master Plans</u>.

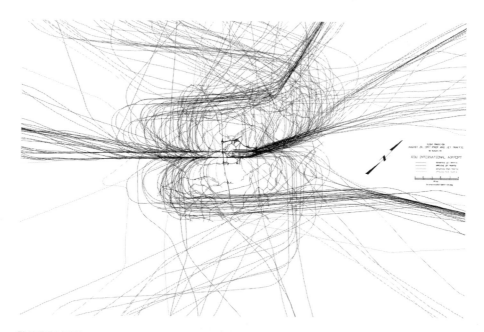

FIGURE 11.27
Typical flight tracks for runway 5 at Raleigh-Durham Airport. *Source:* Raleigh-Durham International Airport Noise Information, www.rduaircraftnoise.com/noise-info/runway5.html.

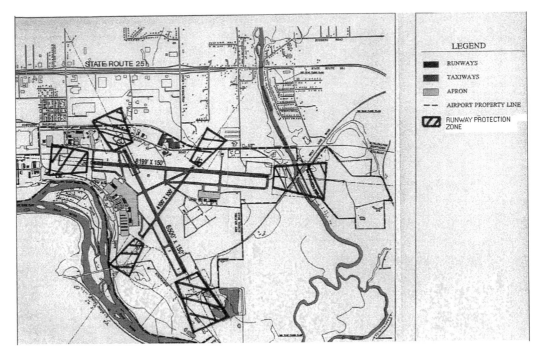

FIGURE 11.28
Typical airport layout showing runways and RPZs. *Source*: Greater Rockford Airport Master Plan.

along the takeoff and landing corridors are critical, not only to site location but also for runway orientation, because they define:

- Where safe landing of aircraft in more than 95 percent of the wind conditions must occur
- Where obstacles projecting into the flight path must be eliminated
- Where homes and other buildings would be subjected to unacceptably high levels of aircraft noise.

Siting of runways must accommodate all three of these constraints. In addition, runways must avoid landing and takeoff paths that are over landfills and other areas that are prime bird habitats. In recognition of the severity of aircraft crashes when they occur in the vicinity of public assembly buildings, particularly schools, communities are encouraged to control the land use within 3 miles from the airport center (FAA, 1983). Other site considerations include soil condition, required grading and earthwork, wetlands, and suitable access connecting the airport with the community being served and with major businesses nearby.

11.5.1 Property Constraints

The land from the outer edge of the RPZ in Figure 11.29 to the runway threshold is the minimum amount of land beyond that associated with the runways and the terminal

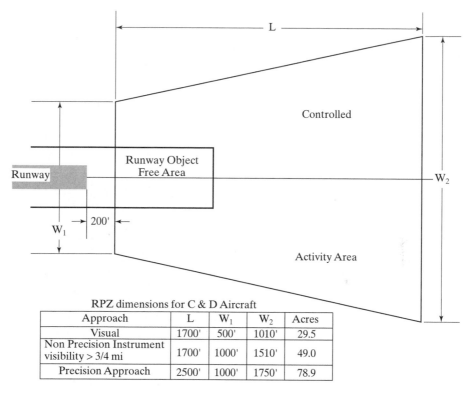

RPZ dimensions for C & D Aircraft				
Approach	L	W_1	W_2	Acres
Visual	1700'	500'	1010'	29.5
Non Precision Instrument visibility > 3/4 mi	1700'	1000'	1510'	49.0
Precision Approach	2500'	1000'	1750'	78.9

FIGURE 11.29
Runway protection zone layout. *Source*: FAA, 1998, <u>Airport Design</u>.

that should be under the direct control of the airport management. Sometimes special easements or other legal instruments are needed to ensure this control.

The more land that the airport owns, especially land extending from the ends of major runways, the better the airport will be able to expand to meet growing demand for air travel. Good land use planning will help maintain acceptable relationships with the community surrounding the airport. Superimposing the flight tracks in Figure 11.27 on a city map (see also Figure 11.33) shows that some flight paths will create more disturbance and require control of high obstacles such as smoke stacks and radio towers near the flight path.

11.5.2   Obstacles

For the pilot on final approach, the runway is an extension of the glide path. The length and slope of the glide path depend on the airport's traffic and the approach capability of the runway landing system—visual or instrument, non-precision or precision. The glide path for landing and taking off aircraft must be under the control of the airport to the extent that obstacles are avoided and the safest possible navigation and landing is facilitated. Obstacles along the glide path pose a most severe situation. On a 50:1 slope, the distance from the end of the runway to clear a 200-foot (60 m) obstacle by 250 feet

(75 m) is 22,500 feet (6850 m or 4.3 miles). Imaginary surface models are used to define the limits on a potential obstacle on or near the glide slope. For takeoff, these criteria are also critical, because it is required that a transport aircraft be able to successfully take off even if one engine is out. In terms of safety, the FAA has established object height requirements in the vicinity of the airport, as shown in Figure 11.30.

11.5.3 Runway Orientation for Winds

Airplanes operate best when they are flown heading into the wind, so the runway choice, if there is one, is always to land (or to take off) heading directly into the wind. Landing into the wind has also resulted in the convention for numbering runways where the runway number consists of the azimuth of the runway rotated by 180 degrees to account for the direction from which the wind would be coming, with the last digit removed. For example, an aircraft landing directly into a wind from the east would use runway 9. The same runway could be used in the opposite direction, but it would then be called runway 27. In two parallel runways, they would be further designated as "left" and "right": 27L and 27R. Because the azimuth of the wind varies and the runway is fixed, landing directly into the wind is usually not possible. Figure 11.31 shows an aircraft landing on runway 24 in a 25-knot wind blowing from 280 degrees azimuth. Thus, the pilot landing on runway 24 will have the headwind component of 19.2 knots. The crosswind component of wind is 16 knots. The polar plot displaying the history of wind conditions at an airport is called a *wind rose*.

Until a few years ago, because of possible structural limitations on some aircraft and pilot control difficulties, aircraft were allowed to land only if the crosswind was less than a specified velocity. The new rules relate the crosswind restriction only to the

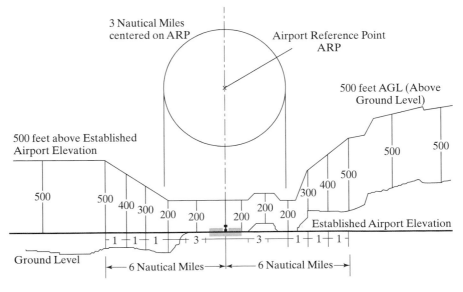

FIGURE 11.30
Obstacle clearance requirements. *Source:* FAR 77.23(a)(2) 1975.

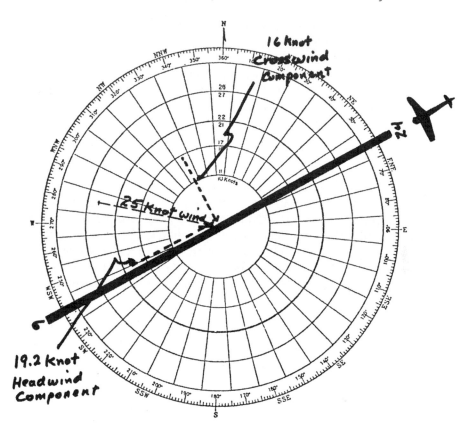

FIGURE 11.31
Runway orientation and wind direction. *Source:* CRC *Civil Engineering Handbook,* 1995.

width of the runway, as indicated in Table 11.20. The crosswind restriction, for example, has been changed for basic transport aircraft to 20 knots (Ashford and Wright, 1992). However, there is a trade-off between allowable crosswind and runway width for lighter planes that are difficult to control in heavy crosswinds. For example, a 200-foot-wide runway gives the pilot of a small aircraft much more latitude for maintaining control in a heavy (20-knot) crosswind (provided the structural integrity of the aircraft is not exceeded) than for landing on a 75-foot-wide runway.

TABLE 11.20   Runway Width and Allowable Crosswind for Landing

Runway Width W	Allowable Crosswind Component (knots)
$W < 75$ feet	10.5
75 feet $\leq W <$ 100 feet	13
100 feet $\leq W <$ 150 feet	16
$W \geq 150$ feet	20

Source: FAA, 1998, <u>Airport Design.</u>

The FAA standards, given in the U.S. Code (CFR Title 14, Part 25), require that the airport must be able to accommodate landings on its runway(s) 95 percent of the time, subject to the crosswinds indicated in Table 11.20. When this cannot be accomplished with one runway, then the airport must have a "crosswind" runway. The two runways together should reduce the frequency of unacceptable crosswinds to less than 5 percent of the time. If possible, a 10-year sample of wind soundings taken hourly is used to establish a model of the wind velocity and direction. The wind data, a sample of which is shown in Table 11.21, are then analyzed and placed in the appropriate cell as shown in the wind rose of Figure 11.32. Each cell shows the percentage for a particular combination of wind amplitude and direction. The proposed runways are placed on the wind rose to be analyzed for minimum crosswinds in excess of the crosswind criterion. For each orientation, the cells outside the runway template in Figure 11.32 are summed to determine if the 95 percent criterion has been met. If the template covers only part of a cell, the percentage in the cell is prorated according to the fraction of the cell that is (un)covered. The acceptable practice for most airports has been to ensure that the runway configuration provides for a minimum of 95 percent against a 13-knot crosswind. In the case shown in

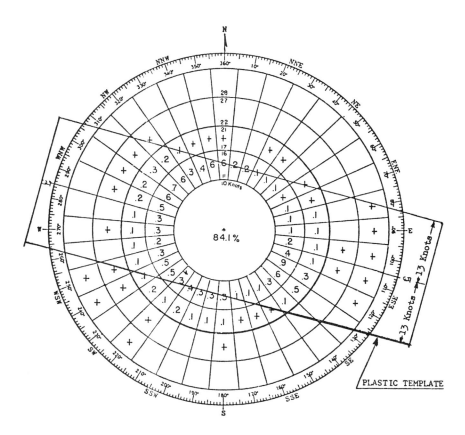

FIGURE 11.32
Wind rose analysis. *Source:* FAA, 1998, Airport Design.

TABLE 11.21   Example of Wind Data for Airport Layout

	Wind Direction Versus Wind Speed		
Station: <u>Anywhere, USA</u>	Hours: <u>24 Observations/Day</u>		Period of Record: <u>1964–1973</u>

	Hourly Observations of Wind Speed										Average Speed	
						Knots			41			
	0–3	4–6	7–10	11–16	17–21	22–27	28–33	34–40	Over			
						MPH			47			
Direction	0–3	4–7	8–12	13–18	19–24	25–31	32–38	39–46	Over	Total	Knots	MPH
01	469	842	568	212						2091	6.2	7.1
02	568	1263	820	169						2820	6.0	6.9
03	294	775	519	73	9					1670	5.7	6.6
04	317	872	509	62	11					1771	5.7	6.6
05	260	861	437	106						1672	5.6	6.4
06	357	534	151	42	8					1092	4.9	5.6
07	369	403	273	64	36	10				1175	6.6	7.6
08	158	261	138	69	73	52	41	22		814	7.6	8.8
09	167	352	176	128	68	59	21			971	7.5	8.6
10	119	303	127	100	98	41	9			877	9.3	10.7
11	323	586	268	312	111	23	28			1651	7.9	9.1
12	610	1397	624	779	271	69	21			3779	8.3	9.6
13	472	1375	674	501	452	67				3571	8.4	9.7
14	647	1077	574	261	129					3008	6.2	7.1
15	338	1093	348	105	27					1941	5.6	6.4
16	560	1399	523	121	19					2622	5.5	6.3
17	507	883	469	120	12					2079	5.4	6.2
18	1046	1904	1068	297	03	10				4496	5.8	6.7
19	499	793	586	241	92					2211	6.2	7.1
20	371	946	615	243	64					2239	6.6	7.6
21	340	732	528	323	147	0				2070	7.6	8.8
22	479	768	603	231	115	38	19			2253	7.7	8.9
23	107	1000	915	413	192					2715	7.9	9.1
24	450	943	800	453	96	11	18			2779	7.2	8.2
25	351	890	752	297	102	21	9			2431	7.2	8.2
26	368	731	379	206	53					1739	6.3	7.2
27	411	748	469	232	110	19				1997	6.7	7.7
28	191	554	276	287	118					1426	7.3	8.4
29	271	642	548	479	143	17				2100	8.0	9.3
30	379	673	526	543	200	34				2563	8.0	9.3
31	299	643	597	618	222	19				2398	8.5	9.8
32	397	852	521	559	150	23				2510	7.9	9.1
33	236	721	324	238	48					1567	6.7	7.7
34	280	916	845	307	24					2372	6.9	7.9
35	252	931	918	487	23					2611	6.9	7.9
36	501	1568	1381	560	27					4046	7.0	8.0
00	7729									7720	0.0	0.0
Total	21,676	31,826	19,849	10,437	3357	529	166	22		87,864	6.9	7.9

Source: FAA, 1998 <u>Airport Design.</u>

Figure 11.32, 97.22 percent of the winds would be contained if the runway was oriented at 105 to 285 true azimuth (runway 11/29). The likely landing would be from the northwest on runway 11. Once the possible best directions of runways are established, then other factors that impinge on direction, such as obstacles and noise, become critical.

11.5.4   Noise

Airport noise produced from aircraft engines and from the flow of air over the aerodynamic surfaces is a major concern. The advent of jet turbine-driven aircraft greatly increased the noise levels above their piston engine predecessors. Noise from airports is an unwanted sound that has evoked individual clamor, numerous law suits, and extensive media attention, much to the frustration of airport officials who neither own nor operate the aircraft. Noise has restrained development, constrained operations, and restricted the expansion of many airports in the United States.

The critical factors in considering noise impacts are:

- Length or duration of the sound
- Repetition of the sound
- Predominate frequencies generated
- Time of day when the noise occurs.

Loudness is the subjective magnitude of noise, which is considered to double with each increase of 10 decibels. The human ear is not sensitive to all noise in the aircraft-generated frequency range of 20 to 20,000 Hz. Usually it perceives noise in the middle of the range 50 to 2000 Hz, called the *A range*. Sound measuring devices generally measure noise in the *A range in decibels (dBA)*. However, with aircraft noise, the simple dBA or sound intensity lacks any correlation to the perceived noise disturbance heard by the human ear. Because the irritation from noise comes not from a single fly-by but from the integrated or cumulative effect of many fly-bys, an integrated metric called L_{DN} (DN stands for day/night) has been developed for modeling the impact from aircraft noise in the vicinity of an airport.

There is a standard computer module for determining the impact of noise around airport is called the *Integrated Noise Model (INM)*. It gives the contours of equal noise exposure and has the facility to overlay the noise contours over a map of the community so that the specific noise impacts at places like schools and churches can be determined. Figure 11.33 shows the noise contours for 65, 70, and 75 L_{DN} around Standiford Field in Louisville for one set of aircraft mix and flight tracks. Each aircraft has its own noise signature, which are combined over each 24-hour period of expected fights. Flights occurring between 10 PM and 7 AM are weighted 10 times as heavily as those occurring during the day. The nighttime operations at the UPS hub in Louisville exacerbate the noise concerns at Standiford Field. Incorporating new quiet engines in existing aircraft has reduced the noise impact considerably. However, the number of flights is increasing, causing the noise situation to become worse.

Having computed the noise level generated by each specific aircraft using the schedule of flights, it is then necessary to determine the effect the noise will have on the

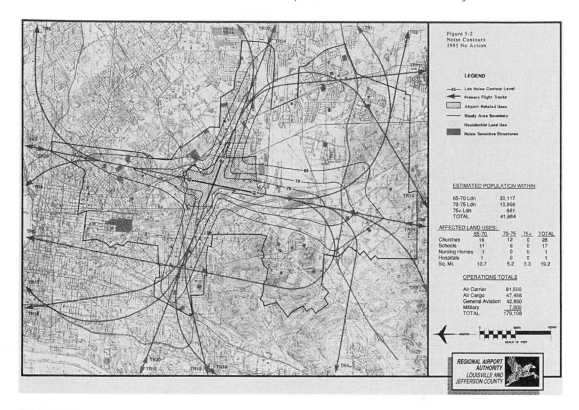

FIGURE 11.33
Noise contours and primary flight tracks. *Source*: FAA, 1990 Final Environmental Impact Statement: Standiford Field Airport: Louisville, KY.

community. How much noise is too much? In what situations? Table 11.22 describes how noise impacts relate to land use-planning recommendations (or regulations) in the community. Airport designers pay special attention to areas impacted or potentially impacted by noise levels of L_{DN} 60 to 65. Experience indicates that owners of property in this range frequently file noise-related lawsuits. Even so, the FAA has set L_{DN} limits at 65. Practitioners consider residential land uses, regardless of density, as unacceptable when the noise level is above 60 dB. This is particularly true under the glide paths or tracks for landing or takeoff. EPA (1974) suggests that the safe L_{DN} criterion for "health and welfare" is 55 dB.

In Figure 11.33, both major flight tracks are shown, as are the zones for various decibel levels. From the noise contours, it can be observed that there are nine churches and four schools at the 65 L_{DN} level or above with the proposed runway orientation. Those schools and churches would have to be relocated, usually at the airport's expense, with some help from the FAA. The number of affected residences can also be counted, although they are not listed on the chart in Figure 11.33.

TABLE 11.22   Land Use for Various Levels of Airport Noise

Land Use Zones	Noise Exposure Class	L_{DN} (db)	HUD Noise Guidelines	Suggested Land Use Controls	Recommended Land Use
A	Minimal	0–55	Acceptable	Normally requires no special consideration	Residential, cultural, public assembly, schools, resorts, mobile homes, parks, service
B	Low moderate	55–60	Normally acceptable	Some sound-reducing controls may be useful	Residential hotels, apartments, business services, office complexes, light industry
BC	High moderate	60–65	Sometimes acceptable	Some sound-reducing controls may be useful	Residential land use unacceptable above 60 dB.
C	Significant	65–75	Normally unacceptable	Noise easements required with strict land use controls	Manufacturing, retail trade, construction services, refining, paper/pulp
D	Severe	>75	Clearly unacceptable	Should be within the airport boundary; use of positive compatibility controls required	Highway right-of-way, motor vehicle transportation, rail transit, undeveloped area, heavy industry, farming

Source: Adapted from FAA, 1983, <u>Airport Land Use Compatibility Planning</u> Advisory Circular AC150/5050-6.

11.6   AIRFIELD DESIGN

At most major airports, one sees a continuous wave of construction and reconstruction. The opportunities for planning, engineering, and construction have been and will continue to be significant. From 1993 to 2002, the federal government has spent about $65 billion to upgrade and improve more than 2000 airports. Local projects include building and rebuilding runways, constructing new airports, adding capacity, making safety improvements, and mitigating environmental impacts, especially noise. In addition to the federal expenditures, local airport boards and authorities approve bonds ($4.1 billion in 1996) and use passenger facility charges ($1.85 billion collected in 2002) to build and upgrade terminal facilities (FAA, 2003). The new midfield terminal at Detroit's airport (Figure 11.34) cost $1.2 billion to construct. Airport design must take into account expected trends in demand for air travel and the aviation technology necessary to satisfy that demand.

11.6.1   Runway Length

The layout and design of an airport involves numerous trade-offs, including the effects of altitude, temperature, runway gradient, winds, and aircraft performance. Because the airlines will assign aircraft to meet the range/payload requirements of the markets they want to serve, it behooves the airport planner to make sure that the runway is long enough to serve the most distant markets that will attract airlines. In this section, emphasis is on the determination of runway length and how runway length affects the operational capabilities of the aircraft.

Airport Elevation and Air Temperature.   The lift generated by any aircraft in motion is proportional to air density, a dimensionless lift coefficient, the wing area, and the

FIGURE 11.34
Midfield Terminal Construction at Detroit Airport. *Source*: Don Coles, used with permission.

forward speed of the aircraft. Because the density of the air is determined by elevation of the airport and by the air temperature, an airport high in the mountains will need a longer runway than one at sea level for the same aircraft to take off. Likewise, on a hot summer day, an aircraft will need more runway (or be more limited in takeoff weight) than the same aircraft on a cool fall day.

Figure 11.35 shows the altitude and temperature effects for a Boeing 727-200 with JT8D-15 engines, the flaps set at 20 degrees, and 150,000 pounds of takeoff weight. It is clear that a longer runway is needed as the altitude of the site above MSL increases. At a temperature of 65 degrees Fahrenheit, the increase in required runway length is from 4900 feet at sea level to 8860 feet at an altitude of 8000 feet, or about 370 feet of added runway for each 1000-foot increase in altitude. The increase in runway length due to temperature, especially when the temperature is high, is equally dramatic. Going from 65 to 80 degrees Fahrenheit for an airport at 4000-foot elevation requires an increase in runway length of about 24 ft/°F. While going from 95 to 105°F for an airport at the same elevation, the rate of increase in runway length is 58 ft/°F. For a given aircraft operating on a specific runway at a given airport and temperature, the *maximum allowable takeoff weight (MATOW)* can be determined.

Runway Constraints. In addition to incorporating aircraft characteristics (takeoff weights, engine capabilities, landing and braking ability, flap settings, and required safety factors), the determination of runway length also takes into account the average high

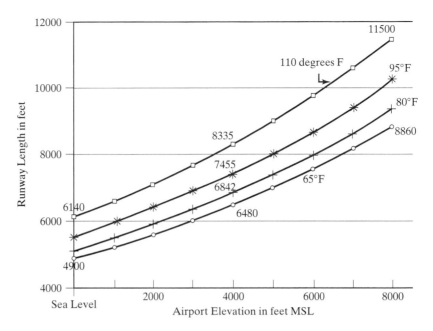

FIGURE 11.35
Runway length for a Boeing 727 as a function of elevation and temperature. Calculated from
FAA Advisory Circular AC150 5325-4A, 1990.

temperature (average temperature for the hottest month) and the elevation of the airport. The runway must be long enough to allow an obstacle-free takeoff by an aircraft with one engine out. There must be ample paved stopping distance in case the pilot chooses to abort just before rotating the aircraft to become airborne. This extra portion of the runway is called the *stopway*. On the other hand, the runway length for landing must be capable of permitting safe braking if touchdown occurs one third of the length of the runway past the threshold.

The advisory circular on runway length (FAA, 1990) gives the takeoff weight for several different flap angle settings. Taking off with a low flap angle permits a higher MATOW, but it takes a longer runway to attain the speed to become airborne. Figure 11.36 plots the MATOW for various flap angles for a temperature of 90°F at a given airport. The curve starting at the lower left is constrained by the length of runway (in this case, 9500 feet), whereas the curve starting at the upper left is constrained by the engine thrust capability of the JT8D-15 engines and is independent of runway length. A third curve not shown is at the structural limit of the aircraft, which for this aircraft is 181,500 pounds. A flap angle setting of 17 to 18 degrees will provide the MATOW of 167,300 pounds for the specified temperature (90°F) and altitude (4000 feet) on a 9500-foot runway. Under these conditions, no greater takeoff weight is possible.

Calculating Range from Runway Length and Aircraft Performance. If a takeoff weight limitation exists, the major operational constraint is the range that can be

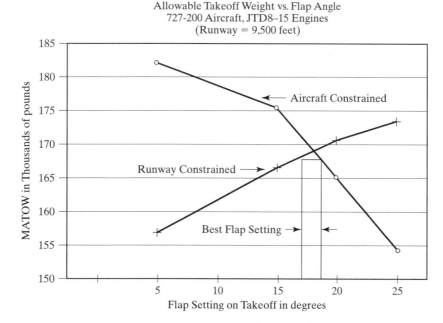

FIGURE 11.36
Takeoff weight as a function of flap angle setting. Calculated from FAA Circular AC150 5325-4A.

achieved with the payload that can be carried. The 727-200 with JT8D-15 engines has an empty weight of 109,211 pounds and consumes 22 pounds of fuel per mile of flight. Entries (A) and (B) in Table 11.23 are for an airport at 4000-foot elevation with a 90 degree Fahrenheit *design temperature* (the mean daily maximum temperature during the hottest month of the year).

The MATOW determined from Table 11.23 at point A is 175,400 pounds. Using the reference factor of 86.9, the necessary runway length for that MATOW is found by interpolation to be 10,680 feet at point C in the bottom portion of Table 11.23. This is longer than the 9500 feet available for the runway introduced in Figure 11.36. As found to the far left of point D in Table 11.23, the allowable takeoff weight is 166,500 pounds on the 9500-foot runway. The results of this preliminary step will apply to all three cases that will be studied below.

Case 1. In Table 11.24, for the column marked "Max Payload (Case 1)," the calculation begins with 166,500 pounds. The aircraft empty weight of 109,211 pounds is subtracted from the maximum payload. We can fill the aircraft with passengers and cargo to the maximum structural limit of 40,339 pounds. Because we want to carry all the passengers we can, we include 162 passengers at 200 pounds, or 32,400 pounds. The airplane can carry the remaining weight as fuel or cargo. After the weight of the passengers (200 pounds each, including their luggage) is subtracted, 7939 pounds is left for freight. The remaining takeoff weight is 16,950 pounds, which will be fuel. Because fuel is consumed at a rate of 22 pounds per mile, in this example, the aircraft can serve a market 770 miles away.

TABLE 11.23   Aircraft performance, takeoff Boeing 727-200 JT8D-15 Engine, 15° Flaps

Maximum Allowable Takeoff Weight (1000 lb)

Airport Elevation (ft)

TEMP (°F)	0	1000	2000	3000	4000	5000	6000	7000
50	197.0	197.0	197.0	193.4	186.9	180.2	173.5	166.9
60	197.0	197.0	197.0	193.4	186.9	180.2	173.5	166.9
70	197.0	197.0	197.0	193.4	186.9	180.2	173.5	166.9
80	197.0	197.0	196.2	189.0	182.2	175.8	169.5	163.4
90	197.0	195.8	188.7	181.9	**175.4 (A)**	169.0	162.9	156.8
100	194.5	187.7	181.0	174.5	168.2	162.0	156.0	150.2
110	186.5	179.7	173.1	166.8	160.7	154.8	149.0	143.4

Reference Factor

Airport Elevation (ft)

TEMP (°F)	0	1000	2000	3000	4000	5000	6000	7000
50	58.6	61.6	65.7	70.7	76.5	83.0	90.2	97.9
60	58.4	62.0	66.3	71.3	77.0	83.4	90.5	98.4
70	59.4	63.5	68.1	73.2	78.9	85.4	92.6	100.6
80	61.4	65.9	70.8	76.2	82.2	88.9	96.4	104.9
90	64.5	69.4	74.7	80.5	**86.9 (B)**	94.0	102.0	111.0
100	68.7	73.9	79.6	85.9	92.9	100.7	109.4	119.0
110	74.0	79.4	85.5	92.5	100.3	109.0	118.5	129.0

Effective Runway Length (1000 ft)

Reference Factor

Weight (1000 lb)	58	68	78	86.9 (Interpolate)	88	98	108	118	128
130	3.96	4.55	5.20		5.90	6.61	7.28	7.88	8.39
135	4.23	4.89	5.59		6.30	7.00	7.66	8.28	8.82
140	4.51	5.25	5.99		6.73	7.44	8.13	8.78	9.39
145	4.81	5.63	6.42		7.19	7.94	8.67	9.38	10.08
150	5.13	6.02	6.87		7.69	8.49	9.29	10.09	10.90
155	5.46	6.44	7.35		8.23	9.10	9.98	10.89	11.85
160	5.82	6.86	7.85		8.81	9.77	10.76	11.80	12.93
165	6.18	7.31	8.38	9.31	9.42	10.49	11.61	12.81	14.13
166.5				**9.50 (D)**					
170	6.57	7.77	8.92	9.95	10.08	11.26	12.53	13.92	
175	6.97	8.25	9.49	10.62	10.76	12.10	13.54		
175.4				**10.68 (C)**					
180	7.37	8.74	10.09	11.34	11.49	12.98			
185	7.83	9.25	10.71		12.25	13.92			
190	8.28	9.78	11.35		13.05				

Source: FAA, 1990, <u>Runway Length Requirements Airport Design.</u>

TABLE 11.24   Range/Payload Calculation for 727-200 with JT8D-15 Engines and 15 Degree Flaps

Characteristic	Units/Notes	Max Payload (Case 1)	1500-Mile Range (Case 2)	50% Load Factor (Case 3)
Maximum allowable takeoff weight for aircraft	Table 11.23 (point A) gives 175,400 lb as the maximum takeoff weight. However, that much weight requires a runway length of 10,680 feet. See (C) in Table 11.23.			
Takeoff weight, if runway constrained	Calculated using a reference factor of 86.9 found in Table 11.23 (point D)	166,500 lb (76,050 kg)	166,500 lb (76,050 kg)	166,500 lb (76,050 kg)
Typical operating empty weight plus fuel reserve	Given (1.25 hr of fuel reserve required for domestic flight)	109,211 lb (49,650 kg)	109,211 lb (49,650 kg)	109,211 lb (49,650 kg)
Remaining for payload and fuel		57,289 lb (26,040 kg)	57,289 lb (26,040 kg)	57,289 lb (26,040 kg)
Number of passengers	Maximum = 162	162 (100% load factor)	141 (87% load factor)	81 (50% load factor)
Weight of passengers and luggage	200 lb (90 kg) per passenger	32,400 lb* (14,730 kg)	24,289 lb (11,040 kg)	16,200 lb (7365 kg)
Max belly air cargo (does not include fuel)	Structural payload limit of 40,339 lb (18,335 kg)	7,939 lb (3605 kg)	0 lb	0 lb
Weight of non-reserve fuel	Capacity 44,187 lb	16,950 lb (7,700 kg)	33,000 lb (15,000 kg)	41,089 lb (18,675 kg)
Distance to market served (range)	Fuel rate given as 22 lb/mi(6.2 kg/km)	770 miles (1240 km)	1500 miles (2415 km)	1867 miles (3005 km)

*Source: Calculated from data in FAA, 1990, Runway Length Requirements for Airport Design.

Case 2.   Determine the payload (number of passengers) possible for the 727 to serve a market 1500 miles away. In this case, the fixed components of weight are the *operating empty weight (OEM)* and the fuel for 1500-mile range. Without any belly cargo, this leaves 24,289 pounds for passengers, which, at 200 pounds per passenger, gives a maximum passenger load of 141. The maximum load factor is about 87 percent of the 162-passenger plane.

Case 3.   Given a 50 percent load factor, determine the maximum range of the aircraft. The load factor of 50 percent is equivalent to 81 passengers. At 200 pounds per passenger (16,200 pounds) and an OEM of 109,211 pounds, 41,089 pounds is available for fuel. At 22 pounds of fuel per mile, the range is 1867 miles.

Example 11.7

An airport is at an altitude of 5000 feet. It is a hot day, with the temperature in the shade hovering at 100 degrees Fahrenheit. The end of the 10,000-foot runway is being repaired, limiting the runway for takeoffs to 7500 feet. We have 70 passengers who want to go to Albuquerque, which is 1000 miles away. Can we accommodate all 70 of them on the Boeing 727-200 with JT8D-15 engines?

Solution to Example 11.7   In the top portion of Table 11.23 for elevation 5000 feet and Temp 100°F, the MATOW is "162.0," or 162,000 pounds. In the middle portion of Table 11.23, the reference point **R** for Elevation 5000 feet and Temp 100°F is 100.7.

Step 1: The entries in the bottom portion of Table 11.23 contain runway lengths for various combinations of reference factor and weight (MATOW). In this example, $R = 100.7$, which is between column headings $R = 98$ and $R = 108$. We need to find the maximum allowable takeoff weight for a runway length of 7500 feet and $R = 100.7$. Interpolation is required to do this. An interpolation table (below) is constructed with rows labeled MATOW $= 135$ and MATOW $= 140$, because the four bold runway length entries "bracket" the desired value of 7.5.

	Reference Point		
MATOW lbs	98	100.7	108
135,000	**7.00**		**7.66**
		7.50	
140,000	**7.44**		**8.13**

Step 2: Determine runway lengths for 135,000 and 140,000 pounds with $R = 100.7$ by interpolation.

	Reference Point		
MATOW lbs	98	100.7	108
135,000	7.00	*7.18*	7.66
		7.50	
140,000	7.44	*7.63*	8.13

Step 3: Determine the takeoff weight that corresponds to the 7500-foot runway length with $R = 100.7$ by interpolation. The result is 3555 pounds more than 135,000 pounds.

	Reference Point		
MATOW lbs	98	100.7	108
135,000	7.00	*7.18*	7.66
138,555		7.50	
140,000	7.44	*7.63*	8.13

Step 4: Determine how much payload weight is available

Maximum takeoff weight	138,555 lb
Operating empty weight	109,211 lb
Amount left for payload + fuel	29,344 lb
Fuel 1000 miles $\times$ 22 lb/mi	22,000 lb
Amount left for passengers	7,344 lb

Result: At 200 pounds per passenger, we can accommodate only 36 of the 70 passengers who want to fly to Albuquerque.

11.6.2   Maximum Landing Weight

Critical items are temperature, elevation, and landing flap settings. At our sample airport—4000-feet elevation with 90 degrees Fahrenheit temperature—the *maximum allowable landing weight* (*MALW*) is 154,500 pounds with 30 degree flaps. This requires 5720 feet of runway. MALW is 142,500 pounds with 40 degree flaps, requiring 5490 feet of runway. There is considerable margin for landing. It is generally true that the takeoff conditions for most aircraft are more critical in determining runway length than landing conditions.

11.6.3   Runway Geometry, Markings, and Lights

The FAA has developed a set of standard dimensions that cover runway width, separations between runways and taxiways, safety areas around runways and taxiways, shoulder width, pads to deflect jet blast, object-free areas, and the like. These standards are the function of approach speed and aircraft size. Many large airports opt for 200-foot-wide runways and 150-foot-wide taxiways.

Longitudinal Gradient.   The desire at any airport site is to have the runways and taxiways as level as possible, allowing for drainage with the design of the transverse grade. In many locations, the grading for a perfectly level site would be too expensive, when most aircraft can easily accept a gradient of 1 percent. The usual penalty for gradients is to reduce the effective runway length by 10 feet per foot of the difference from the maximum and minimum elevation of the runway (FAA, 1995).

Example 11.8

How would the runway length calculations be impacted if the runway at an airport were 10,000 feet long, but there was a differential of 50 feet between the highest point and the lowest point?

Solution to Example 11.8   The effective runway length for MATOW calculations would be 10,000 ft − (50 ft × 10 ft/ft) or 9500 feet.

Line of Sight.   Any two points 5 feet above the runway centerline must be mutually visible for the entire runway or, if on a parallel runway or taxiway, for one half of the runway. Likewise, there needs to be a clear line of sight at the intersection of two runways, two taxiways, and taxiways that cross an active runway.

Drainage.   Transverse gradients are important to ensure adequate drainage from the runways and the taxiways. Because of their large contiguous area, aprons are critical and must have an adequate sewer system. Water treatment is required when there are fuel spills or, during the winter, when deicing chemicals are used.

Runway Markings.   The signing and lighting on the airfield consists of runway markings showing the threshold and various lengths along the runway. For runways with less than 0.75-mile visibility on their approaches, there will be light bars before the threshold that show the approach to the runway.

Airfield Markings. On the airfield itself, there are signs and lights that define the airfield and help the pilot navigate on the airport in the dark. Runways and taxiways have signs directing the pilot and defining the location of intersections. Blue lights, for example, indicate taxiways.

Airfield Traffic Lighting. Lighting and signing of the runway provide the pilot visual cues to ensure alignment with the runway, lateral displacement, and distance along the runway. Runway edge lights stand no more than 30 inches high and are located no more than 10 feet from the runway edge. They are 200 feet or less apart and are white except for the last 2000 feet of runway, when they show yellow. Centerline lights are white and are set 2 feet off the centerline of the runway, except for the last 3000 feet. In this area, they are alternating red and white for 2000 feet and they are red 1000 feet from the runway end. When an aircraft is approaching the runway to land, the pilot determines the threshold because a bar of green lights marks it. However, those lights appear red to the pilot of an aircraft approaching the end of the runway from the other direction.

11.6.4 Runway Pavement Design

Similar to highway design, airport pavement design methods are based on the accumulated gross weight of the aircraft over the pavement life. Because it is impracticable to develop design curves for each type of aircraft, composite aircraft are determined and loads are converted from the actual aircraft to the design aircraft. The design aircraft is usually the one that requires the greatest thickness of pavement. Furthermore, the traffic forecast, which includes the mix of aircraft anticipated, is converted to a traffic forecast of equivalent annual departures. This is equivalent to determining the number of equivalent single-axle loads for highway pavement design. Typically, the airport pavements are much thicker than highway pavements, where individual trucks may weigh no more than 80,000 pounds. Note that a 747 can weigh more than 1,000,000 pounds.

FYI: Pavements are thicker for airport taxiways and aprons than for runways. This is because the lift on the wing surfaces of aircraft at takeoff/landing speeds makes the loading less than from slow-moving or stationary aircraft (Sargious, 1975, p. 43).

11.7 AIRPORT LANDSIDE AND ACCESS

Shoridan's airport is old and small, especially compared to Denver International Airport (Figure 11.37), but air traffic has been growing rapidly in Shoridan. A complete renovation of the terminal is being proposed. The design committee is being inundated with demands. For example:

FIGURE 11.37
Main Terminal, Denver International Airport. Photo: Jon Fricker, July 2001.

1. Passengers should not have to walk too far from the curb to the boarding area.
2. Passengers should have access to all necessary services between the curb and the boarding area.

These two demands are the most obvious examples of the challenge that confronts the designer of an airport terminal. How can the design locate so many activities close to each other? The Denver airport lobby obviously is spacious, but at the cost of long walking distances.

11.7.1   Landside Functions

In section 11.1.4, we defined an airport's landside as those areas where the passenger (or cargo) finds the facilities for movement on land to or from the gate area. The aprons, arrival and departure concourses, baggage handling areas, and land transportation access facilities (curbside spaces, parking lots, access roads, and transit service) are all part of an airport's landside. Figure 11.38 shows the full range of functions that are involved in the terminal.

Important design issues include the type of terminal that is to be designed, the length of the curbside, the baggage facilities, the parking lots, and the manner in which the local roads feed into the airport. In many ways, terminals in large airports duplicate the functions of a small city. Terminals offer the same amenities, meeting the likely needs of four specific groups: (1) passengers, (2) visitors (greeters, friends, salespersons, etc.), (3) airline staff, and (4) airport/aircraft support persons and concessionaires. The functions of a terminal include the following:

1. *Passenger processing*, which encompasses all the activities associated with the airplane passenger's trip, such as baggage handling, ticket handling, seating, and security.

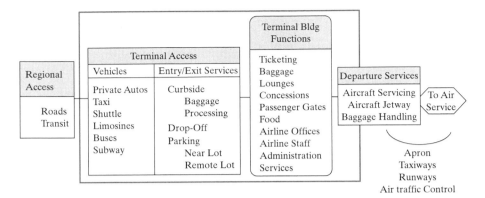

FIGURE 11.38
Functional view of airport landside.

2. *Support facilities* for passengers, employees, airline crew, air traffic controllers, and airport management.

3. *Change of transportation mode* for the local traveler who arrives by one mode and changes to the air mode.

4. *Change of aircraft*, which usually occurs in larger hubs. Passengers who change aircraft generally do not require as many of the ground facilities. Their baggage, however, must be transferred from one plane to another.

5. *Collection space for passengers* who are waiting to board an aircraft that may have 100 to 400 seats.

These functions for most airports are melded into an efficient flow of passengers, baggage, and visitors. Figure 11.39 illustrates the nature of the flow within the terminal. The collection of individual departing passengers in the departure lounge for enplanement on a single airplane is called the "many-to-one" process. Upon the arrival of an airplane, the functions are reversed as the passengers depart the airport in a "one-to-many" process.

11.7.2   Terminal Design Concepts

The terminal must be sized to meet the air passenger demand, which was computed in section 11.2.4. In addition, an airport's attractiveness is always an issue, because it is the introduction to a city for many people. Spaciousness and amenities for both greeters and for passengers who are waiting to change planes or to board a delayed flight are essential. Likewise, the terminal design ought to balance the needs of both departing and arriving passengers, as well as those transferring to another aircraft. In all respects, convenience and comfort will be the bywords of airport terminal design. The designer has several options to consider. These are discussed in the following sections.

11.7.3   Horizontal Distribution/Gate Layout Options

Air travelers may find themselves in one of the several terminal configurations illustrated in Figure 11.40. Airports have been growing fast. Some airport expansions have resulted in a mixture of design concepts due in part to the airlines' desires.

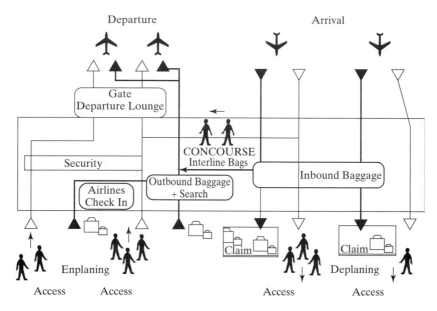

FIGURE 11.39
Passenger and baggage flow.

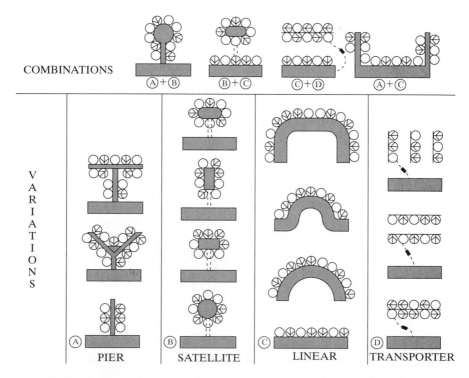

FIGURE 11.40
Terminal configurations. *Source*: FAA, 1988b.

The San Francisco Airport (SFO) layout shown in Figure 11.41 provides an example of one terminal that has grown over time and has added several different gate configurations. The circular access route provides options for parking and the gates for different airlines are adjacent to the internal access roads. Each little sector functions almost as a separate "subterminal." The different layout configurations—pier, linear, circular, and variations—are all present at SFO.

THINK ABOUT IT

For the San Francisco airport, label each gate type and discuss what might be the advantages or disadvantages for the airline occupying the particular gate type.

The dimensions along the piers will depend on the type of aircraft that will be using the gates. With jetways for the loading and unloading of passengers and with the various wingspans on the aircraft, gates are generally situated to accommodate a variety of aircraft. The mobility of the jetways allows for adjustments to be made for both

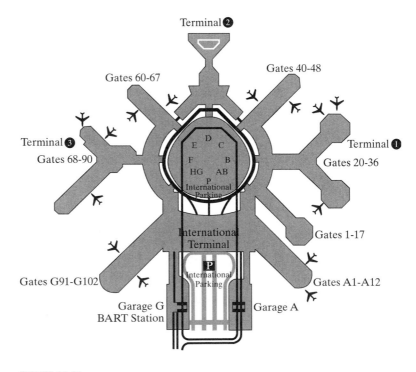

FIGURE 11.41
The layout of San Francisco Airport. *Source*: San Francisco Airport Authority, used with permission.

small and large aircraft. The loading of the commuter aircraft is often accomplished directly from the apron, requiring the passengers to walk down steps or take an elevator. Figure 11.42 indicates the dimensional differences between different class aircraft for the standard pushback configuration.

From Curb to Departure Gate—Vertical Distribution.   In many smaller airports, the passengers and baggage are handled on a single level. For others, the enplaning function is often separated from the deplaning function, especially where the curbside for departing passengers is on the upper level and the baggage claim and ground transportation for arriving passengers are reached on the lower level. Figure 11.43 shows examples of both. In the lower of the two schematics, enplaning and deplaning passengers are separated as they enter the airport from the aircraft.

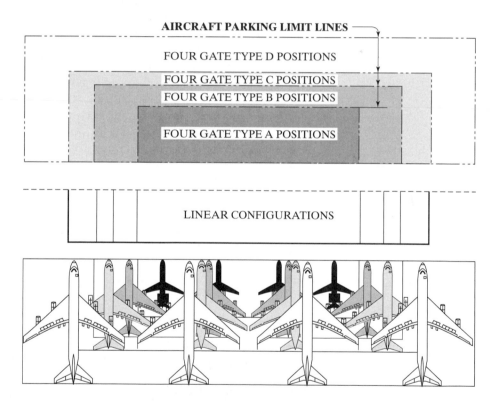

ALL WINGTIPS SHOW
20-FT CLEARANCE

SCALE IN FEET

FIGURE 11.42
Gate layout requirements for pushout. *Source:* FAA, 1988b.

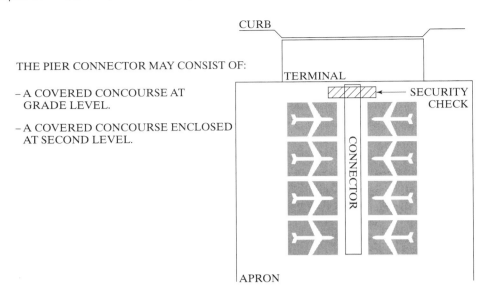

THE PIER CONNECTOR MAY CONSIST OF:

– A COVERED CONCOURSE AT
  GRADE LEVEL.

– A COVERED CONCOURSE ENCLOSED
  AT SECOND LEVEL.

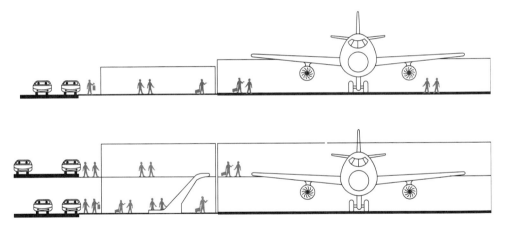

FIGURE 11.43
Vertical separation of passenger and baggage flows in pier concept. *Source*: FAA, 1988b.

11.7.4 Passenger Terminal Design

The sizing of the terminal requires knowledge of three overlapping requirements:

1. Passenger demand, including the anticipated requirements for transfer passengers.
2. Number of gates needed for boarding.
3. The anticipated aircraft size and mix.

Several concepts exist to assist the designer/planner in determining the gross terminal size. Here, we use a factor called the *typical peak hour passenger (TPHP)* estimate and apply design concepts related to the *equivalent aircraft gates*.

Typical Peak Hour Passengers.   Using the TPHP to determine the gross size of the terminal gives a good estimate that must later be refined. The TPHP does not represent the *maximum* passenger demand of the airport, but is well above the average demand and takes into consideration periods of high airport usage. The TPHP is an approximation based on annual enplanements using Equation 11.9 for larger airports and Equation 11.10 for smaller airports (Whitford, 1995).

For airports where annual enplanements exceed 500,000

$$\text{TPHP} = 0.004 \text{ ENP}^{0.9} \tag{11.9}$$

For airports where annual enplanements are less than 500,000

$$\text{TPHP} = 0.009 \text{ ENP}^{0.9} \tag{11.10}$$

where ENP equals annual enplanements.

The gross estimate for domestic airports is 120 to 150 square feet of terminal space for each TPHP. For an international airport, this planning number increases to about 250 square feet per TPHP (Ashford and Wright, 1992).

Example 11.9

The approximate size of the Mythaca Airport terminal is 700,000 square feet and Mythaca had 4,300,000 enplanements in Year 2002. At what level of enplanements will the size of the airport likely be exceeded?

Solution to Example 11.9   Using Equation 11.9, the TPHP is $0.004 * (4,300,000)^{0.9} = 3734$. Under the domestic planning number, we would expect the airport size to be in the range of $120 * 3734 = 450,000$ to $150 * 3734 = 560,000$ square feet. Mythaca has more space than the minimum needed, so airport terminal expansion based on enplanements is not justified. With 700,000 square feet, a level of TPHP $= 700,000/120 = 5830$ can be accommodated. According to Equation 11.9, this corresponds to

$$\text{ENP} = \left(\frac{5830}{0.004} \right)^{1/0.9} = 7,054,300 \text{ enplanements per year.}$$

This is 64 percent above the current value. In Table 11.6, a 64-percent increase over Year 2002 enplanements would be 733.2 million $* 1.64 = 1202$ million enplanements. This level is estimated to occur in about Year 2015. If traffic at Mythaca Airport grows at the same rate as national air travel, Mythaca should plan for terminal expansion in time to meet Year 2015 requirements.

Size Estimate Using the Equivalent Aircraft Factor.   FAA Advisory Circular 150/5360-13, entitled *Planning and Design Guidelines for Airport Terminal Facilities* (1988), is on the Web. The circular presents a set of design curves for the preliminary layout of airport functional areas such as ticket lobby, baggage claim, and concessions. Most of the space requirements in these areas depend on the number and types of aircraft (and available seats on them) and the number of gates that will be in use in the peak hour.

Two factors are involved in the development of many of the space requirements. One is called the equivalent aircraft factor (EQF) for each aircraft type. The other

TABLE 11.25   Equivalent Aircraft Factor and Gates

Type of A/C	No. A/C Peak Hr	Seat Range	Equiv. A/C Factor EQF	Gates Required	EQG	Departure Lounge (sq. feet)
B(a)	5	<80	0.6	2	1.2	1,200
C(b)	15	81–100	1.0	12	12.0	12,000
C(c)	11	111–160	1.4	11	15.4	15,400
D(d)	3	161–210	2.0	3	6.0	6,000
D(e)	1	211–280	2.5	2	5.0	5,000
D(f)	0	281–420	3.5	—	—	
D(g)	0	421–500	4.6	—	—	
	35		Total	30	39.6	39,600

Source: Computed from FAA, 1988b.

important factor is based on the number of gates required to service each aircraft type during the peak hour. The FAA advisory circular uses the term "Gates $\times$ EQA" for this factor; we will use the shorthand "EQG" here. Table 11.25 shows the peak hour aircraft at Mythaca Airport by type, the EQF for each aircraft type, the gates required to serve the aircraft, and the EQG for each aircraft and the airport. Note that only two gates are required to serve the commuter aircraft, because they are boarded from the apron. Also note that, for the smallest of medium jets (100 seats), only 12 gates are required for 15 aircraft in the peak hour, because the average time at the gate is about 48 minutes.

As can be seen from the table, the EQF is approximately 1/100 of the number of seats on each type of aircraft. Also shown in the table is the departure lounge space, which is directly related to EQG. The size of aircraft is given as a "seat range." "B" aircraft are small commuter jets, usually from 19 to 65 seats. "C" aircraft are the B737, B727, MD 80, DC 9s and similar aircraft that serve as the workhorses of the domestic fleet. "D" aircraft are jumbo jets like the DC 10, A320, B767, and B747.

Functional space allocation is described in Chapter 5 of FAA Advisory Circular 150/5360-13 (1988) through a series of design curves. The circular can be accessed at the FAA Web site, http://www1.faa.gov/arp/150acs.cfm?ARPnav=acs. For relatively large airports with a low percentage of transferring passengers (known as "origin-destination airports"), less accurate estimates are adequate for planning purposes. In the steps below, several of the design curves in AC 150/5360-13 have been replaced by equations that approximate space requirements for the specified airport function. The designations appear in Table 11.26.

I. Passenger/Baggage space

    a. The *arrival and departure curbside* serves as the entrance to the airport. The length of the curbside will be determined in Section 11.7.5. The adjoining building space is consumed as part of the airport structure that is covered in item III.e below.

    b. The *entry and ticket lobby* space requirement is non-linear with respect to EQG, as shown in Figure 11.44. Use the region above the dashed curve for O-D airports.

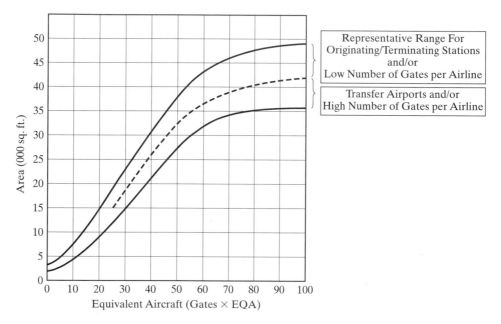

FIGURE 11.44
Space for ticketing lobby. *Source:* FAA 1988b, Figure 5-4.

   c. *Lobby Waiting Areas.* Airport travelers and visitors need places to sit down, wait, and rest. The space estimate is based on seating for 20 percent of TPHP (p. 55 of AC 150/5360-13) and area $= 1000 + 18*$seats) square feet (Figure 5-5 in AC 150/5360-13).

   d. The *departure lounge* area is the waiting or holding area for passengers immediately prior to boarding. The lounge normally includes space for one or more airline ticket agents, a waiting area, and a queueing area. Table 5-3 in AC 150/5360-13 includes a wide range of departure lounge space requirements. For an aircraft with 100 seats, about 1000 square feet is reasonable for each EQG.

   e. *Outbound baggage and transfer* space is about 500 square feet × EQG. (Figure 5-13 in AC 150/5360-13)

   f. *Outbound baggage inspection* in which every bag is searched was initiated as a security measure after 9/11/2001. The space for manual or machine search based on an estimate from international requirements (FAA, 1988b, p. 104) is about 30 square foot per bag and takes an estimated 5 minutes. The space is based on 1.3 bags*60 percent departures times the TPHP at a rate of 12 bags per hour times 30 square feet per inspection station.

   g. *Baggage claim* display space includes space taken up by the claim devices. The public space depends on the percentage of deplaning destination passengers who will retrieve bags, at an average of 1.3 bags per person. The space needed can be determined from Figure 11.45, based on the percentage of arrival aircraft

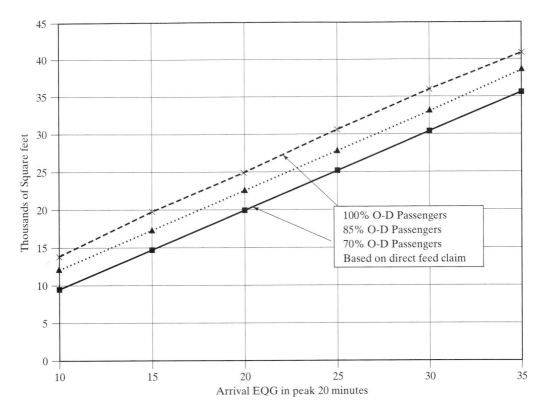

FIGURE 11.45
Baggage claim area as a function of 20-minute arrival EQG. *Source:* Adapted from FAA, 1988b, Figure 5-28.

multiplied by the EQG of aircraft arriving during the peak 20 minutes. The usual value for the peak 20 minutes is assumed to be 50 percent of the peak hourly traffic. During the typical peak hour, no more than 60 percent of the aircraft operations are landings. Baggage claim devices and their capacity and space needs are shown in Figure 11.46.

h. The *nonpublic baggage claim* space, usually hidden, is for the airlines to unload the bags from carts. For a direct feed T device add about 25 percent and for a flat bed U device add about 40 percent to the space determined in Figure 11.45.

i. The *baggage claim lobby* is that space required for passengers who are retrieving their bags and for their guests who have come to pick them up. Assuming that 80 percent of deplaning passengers will need to retrieve their bags, then the people waiting for approximately 20 minutes are 80 percent times 70 percent (those retrieving bags) time 60 percent (arrivals) times 50 percent (peak 20 minutes) times 3 (passenger plus greeters) [Ashford and Wright, p 245] times 21 square feet per person [IATA 1989; Whitford 1995, p. 2098)].

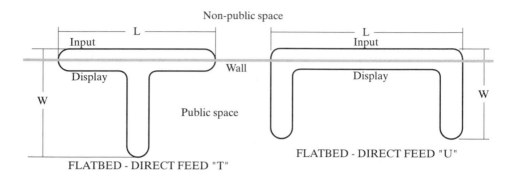

Shape	L & W (feet)	Claim Frontage	Bag Storage
⊥	85 × 45	180	216
⊥	85 × 65	220	264
⊔	50 × 45	190	228

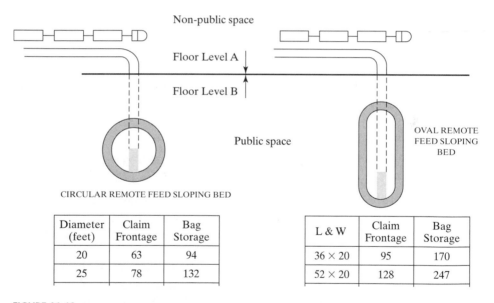

Diameter (feet)	Claim Frontage	Bag Storage
20	63	94
25	78	132

L & W	Claim Frontage	Bag Storage
36 × 20	95	170
52 × 20	128	247

FIGURE 11.46
Baggage claim devices and their capacities. *Source*: FAA 1988b, Figure 5.26.

II. Airline Space

a. *Airline ticket counter.*   A terminal with a high level of interconnection traffic (called hubbing) has a larger percentage of passengers who will be changing aircraft and who do not need ticketing. This affects the ticket counter area and, to some extent, the ticket lobby area. The airline ticket counter space is about 15 feet deep and (using Figure 5-10 in AC 150/5360-13) about 200 + 5 * EQG feet long.

b. *Airline office and support* space includes the airline crew lounge, employee space, office support, and clubs. It is approximately determined from Figure 5-10 in AC 150/5360-13 as 4500 + 75 * EQG sq.ft.

c. *Other airline operations.*   Airports need space to meet the administrative needs of the FAA in that terminal area. Twenty percent of the airline office and support is set aside for these functions. This does not include traffic controllers or tower operations, which are usually housed elsewhere on the airport.

d. *Departure gate* boarding areas are a part of or adjacent to each gate area. This is where the airlines check tickets, give seat assignments (at the gate), and take tickets. The estimate is about 200 square foot per gate.

III. Airport Space

a. *Airport administration offices* require about the same amount of space as calculated for the airline office and support space in Item II.b.

b. *Security.*   Before 9/11/01, Section 73 of AC 150/5360-13 suggested an area of 0.23 * TPHP square feet. Since then, security areas have been increased at all airports. In addition, the number of personnel involved as part of the Transportation Security Agency (TSA) has grown. If the screening rate has been reduced to 100 passengers per hour and about 50 percent of TPHP is the number of enplaning passengers needing to be screened, the required number of security stations can be calculated. Each screening area has increased about fourfold, to about 1000 sq. ft. The space estimate for security is now

$$\frac{0.5 * \text{TPHP}}{100} * 1000 \text{ sq. ft} = 5 * \text{TPHP sq. ft.}$$

c. Space for *food and beverage services* is influenced most by the percentage of long-haul flights using the airport. A reasonable fit to the curves in Figure 5-31 in AC 150/5360-13 is the equation 10,000 + (0.0034 * annual enplanements) for O-D airports.

d. Space for *concessions* (gift shops, auto rentals, etc.) *and building services* (rest rooms, post offices, etc.) varies with the number of origin-destination (non-interline) passengers. From Figure 5-32 in AC 150/5360-13, an average number of square feet for concessions is approximated by 10,000 + (0.007 * annual enplanements). Because we will need to consider restrooms separately in the next item, note that AC 150/5360-13 suggests 1333 sq. ft. per million annual enplanements as a restroom space estimate.

e. *Circulation.*   Walkways must be wide enough to accommodate the flow of passengers and permit the viewing of gate assignment monitors. The area needed for corridors is not explicitly given in FAA AC 150/5360-13, but Figure 5-3 in the Circular gives the space allocation shown below as being typical. Whitford (1995) uses 80 percent of the space taken up by all items except I.e through I.h and III.c through II.d as the basis for estimating corridor space.

f. *HVAC and Mechanical.*   The airport will have power, heating, ventilating, structural walls, elevators, stairs, overhang of curbside area and other functions. A reasonable estimate of space required is 25 percent of all space already calculated.

g. *Architectural.*   The airport is the first thing about a city that the first-time visitor sees close up. Therefore many airports have "welcome" space devoted to special displays, signage and art, or other aesthetically appealing architectural features beyond what is required functionally. This space typically amounts to about 5 percent of the total airport space.

Example 11.10

Twenty years from now, the Mythaca airport is expected to have 10 million enplanements, an EQG value of 70, 58 gates, and 75 percent O-D passengers. How large should the terminal be?

Solution to Example 11.10   Table 11.26 summarizes the estimation of the areas needed by the functional spaces in an airport. The table follows the outline of functional spaces presented in this section and, in its second column, indicates how the values in the two rightmost columns were computed. The space needs for each functional area of Mythaca Airport (MYH) were estimated using the planning (approximate) version of the functional analysis.

Summing the results for the functional areas, the current space needs of MYH would be about 565,000 sq. ft, or 151 sq. ft per TPHP. This value falls near the upper end of the 120–150 sq. ft per TPHP range recommended for domestic airports. The solution for 20 years from now is a total area of about 1,014,000 square feet. This is about 127 square feet per TPHP.

THINK ABOUT IT

Why would a large airport usually need fewer square feet per TPHP than a smaller airport, although both supply all the needed amenities?

A terminal with a high level of interconnection traffic (called hubbing) has a larger percentage of passengers who do not need ticketing, baggage claim, curb front access, and parking, and do not have visitors to meet them. Categorizing airports as hubbing airports or origin-destination airports will help guide estimates of space needs. When an airport is divided into several "mini-terminals," a calculation is made for each "mini-terminal." The results are then added together for the total.

11.7.5 Curb Frontage

The curbside dimensions will depend on the mode split of those coming to the airport. Each arriving vehicle will either be discharging passengers or picking up passengers.

TABLE 11.26   Functional Approach to Terminal Sizing

Characteristic	How developed	Base Year 2003	Horizon Year 2025
Enplanements	Forecast	4,300,000	10,000,000
Gates	Aircraft in peak hour	30	58
EQG	Aircraft-mix and gates	40	70

Functional Design Area	Basis	Base Year	Horizon Year
I. Pax/Baggage Space			
a. Curbside & Entry overhang	Section 11.7.5.		
b. Lobby and Ticketing	Fig. 11.44	28,000	44,000
c. Lobby waiting areas	$1000 + (18*0.20*TPHP*2)$ sq. ft.	28,000	58,600
d. Departure lounge	1000 sq. ft * EQG	40,000	70,000
e. Outbound Baggage room	500 sq. ft * EQG	20,000	35,000
f. Outbound Baggage Search	$75\% * 70\% * 60\% *50\% * TPHP * 30$ sq. ft/12 bags/hr	1,477	3,150
g. Public baggage display area	Fig. 11.45 Direct Feed	23,000	34,000
h. The non public baggage claim space	25 to 40% of public space	5,750	8,500
i. Baggage Claim Lobby	$75\% * 70\% * 60\% * 50\% * TPHP * 3 * 21$	37,209	79,380
II. Airline space			
a. Ticket operations	$15*(200 + 4$ EQG$)$	5,400	7,200
b. Office and support space	$4500 + (75*$EQG$)$	7,500	9,750
c. Other Airline Operations	Use 20% of II.b	1,500	1,950
d. Departure/Boarding areas	200 * EQG	8,000	14,000
III. Airport space			
a. Airport Administrative Offices	Same as II.b	7,500	9,750
b. Passenger Security Screening	300 * EQG	12,000	21,000
c. Food and beverage	$10,000 + (0.0034*$enpl$)$	24,620	44,000
d. Concessions	$10,000 + (0.007*$enpl$)$	40,100	80,000
e. Other circulation	0.80 * (all but I.e, I.f, I.g, I.h, III.c and III.d)	140,088	252,504
f. HVAC, mechanical, structure	0.25 * (all areas above)	107,536	193,196
g. Architectural Special	About 5% of total	26,884	48,299
Total sq. ft		564,563	1,014,279
Space per TPHP		151	127

Adapted and estimated from FAA 1988b.

For each type of vehicle and its function, there will be a certain dwell time at the curb. For example, a bus loading 20 passengers may take up to 15 or 20 minutes, whereas discharging a passenger from a taxi may take less than 2 minutes. Figure 11.47 shows how the curb front integrates with the airport for both arriving and departing passengers.

The average dwell times in Table 11.27 are from the FAA (1998b) design manual. The extension to feet-sec uses the midpoint dwell times. These dwell times are based on the number of each type vehicle that will be using the curb. Although theoretically, one lineal foot of curb front can provide 3600 foot-seconds of curb front in 1 hour, it has been suggested that the practical capacity is about 70 percent of this number (Cherwoney and Zabawski, 1983). For gross planning, 115 lineal feet per million originating passengers can be used (deNeufville, 1976).

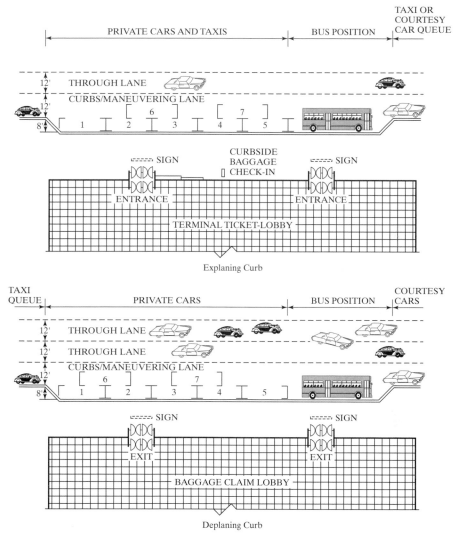

FIGURE 11.47
Curb configuration for enplaning and deplaning passengers. *Source*: FAA, 1988b.

TABLE 11.27   Dwell Time and Curb Front Needed by Vehicle Types at an Airport

Vehicle Type	Curb Dwell Time/in min. (sec.)		Vehicle Length (ft)	Curb Front in (feet-sec)	
	Enplane	Deplane		Enplane	Deplane
Private auto	1 to 3 (120)	2 to 4 (180)	25	3,000	4,500
Courtesy vehicles	1 to 3 (120)	2 to 4 (180)	25	3,000	4,500
Taxi	1 to 2 (90)	1 to 3 (120)	20	1,800	2,400
Limousine	2 to 4 (180)	2 to 5 (210)	35	6,300	9,450
Bus	2 to 5 (210)	5 to 10 (450)	50	13,500	22,500

Example 11.11

The mode split between cars, taxis, buses, and courtesy cars at an airport is as indicated in Table 11.28. From the expected curb traffic indicated in the table below, determine the appropriate enplaning and deplaning curb lengths for the airport.

TABLE 11.28 Mode Split Data for Curb Length Determination

Mode	Enplaning in Peak Hour			Deplaning in Peak Hour		
	Passengers	Vehicles	Peak ft-sec	Passengers	Vehicles	Peak ft-sec
Personal auto	250	200	600,000	320	300	1,350,000
Courtesy vehicle	160	40	120,000	140	60	270,000
Taxi	75	75	135,000	100	100	240,000
Limousine	80	15	94,500	80	12	113,400
Bus	200	15	202,500	130	10	225,000
Total	765	345	1,152,000	770	482	2,198,400

Solution to Example 11.11   The curb length should be

$$CL_{enplaning} = \frac{foot - sec_{peak}}{Capacity\ factor \times sec/hr} = \frac{1,152,000}{0.7 \times 3600} = 457\ ft \qquad (11.11)$$

$$CL_{deplaning} = \frac{2,198,400}{0.7 \times 3600} = 872\ ft$$

11.7.6 Parking

Parking requirements for airports vary widely, depending on the nature of an airport and the manner in which people come to the airport. Having a parking place for those coming to the airport is fundamental to the issue of access. Often there is not sufficient land near the terminals, so off-site parking with shuttle services is provided. Because an operating parking lot, including movement, consumes 350 to 400 square feet per vehicle (or 109 to 124 cars per acre), parking lots consume considerable land. The design of the airport parking is a bit more complicated than most parking lots, because there is continual coming and going as people come to the airport to board a plane and others deplane, get their baggage, and leave the parking lot.

Passenger/Public Parking.   Short-term parking (less than a day) accommodates well wishers and greeters, visitors to the airport itself, and salespersons, and is usually located next to the terminal. Medium-term parking (0.5 to 2 days) and long-term parking (3 or more days) are often combined as the long-term lot.

The demand for parking lots by passengers is usually controlled by price, with higher prices charged for spaces nearer the terminal building. Short-term parking rates much higher than long-term rates tend to discourage long-term parkers from clogging the short-term parking areas.

The coming and going of many cars means that sizing the long-term lot (and the medium-term lot, if there is one) requires a simulation. Designers often begin by obtaining data from an airport similar to the one being designed, noting the time and day a car arrives and the length of stay. Figures 11.48 and 11.49 indicate the number of cars by day in the Indianapolis airport parking lot over a 2-week period in June 1993. The arrival data are matched with airline flight schedules in a computer simulation. Historically, persons have arrived from 30 to 75 minutes before their flight. Since September 11, 2001 this time has grown to 90–150 minutes. It also takes from 15 to 60 minutes following the arrival of a flight for persons to retrieve their bags and get to their long-term parking space and exit.

The short-term lot can usually be sized on the basis of the long-term parking lot. One useful rule is that short-term parking should consume about 20 percent of

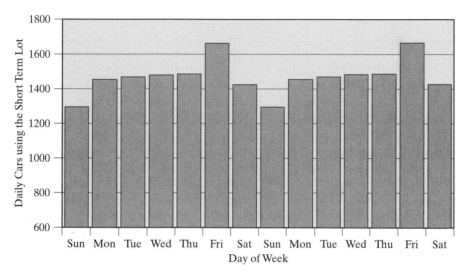

FIGURE 11.48
Daily number of vehicles in short-term parking lot—Indianapolis Airport. *Source*: R.K. Whitford.

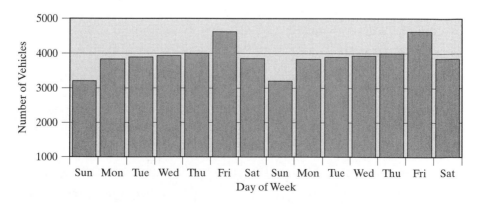

FIGURE 11.49
Daily average number of vehicles in long-term parking lot—Indianapolis Airport. *Source*: R.K. Whitford.

the total parking space. Approximately 70 to 85 percent of all cars coming to the airport will use the short-term lot. This means that the long-term lot, although consuming 80 percent of the space, will handle about 20 percent of the cars. Therefore, for preliminary planning data, for total passenger parking, designers will need to plan on 1.5 to 2.0 spaces for each TPHP *originating* passenger. For this calculation, variations in seasonal activity must be taken into account. The data for the Indianapolis airport are for June, a high month in which 1.05 times the average monthly passengers use the airport. The upper end of the range is for those airports where the transit connections to the airport are poor.

Example 11.12

The actual data for Indianapolis Airport for June 1993 are shown above. That year, there were about 3,000,000 originating passengers, with the parking needing to accommodate 6000 vehicles on Fridays. What was the factor related to TPHP? (See Section 11.7.4.)

Solution to Example 11.12   Using Equation 11.9, TPHP $= 0.004 * \text{ENP}^{0.9} = 2700$ passengers $*$ 1.05(June average) $= 2840$

Minimum parking spaces $= x * 2840 = 6000$

$x = 2.11$ spaces for each TPHP

Example 11.13

The medium-sized airport at Mazurka has 2,000,000 annual enplanements and 15 percent interline passengers. How many acres are needed for parking?

Solution to Example 11.13   Using Equation 11.9, TPHP $= 0.004 \, \text{ENP}^{0.9} = 1875$

Originating enplanements $= 0.85 * \text{Total enplanements} = 1600$

Parking lot size $= 3200$ spaces or approximately 28 acres of parking space, 6 acres for short term and 22 acres for long term.

Employee Parking.   With all the concessions, restaurants, airline crews, ground support personnel, and air traffic control, there is a sizeable workforce at an airport. Most airports have found it practical to set aside some relatively convenient but secured space to accommodate employees. Most of these persons are working a regular shift at the airport, although some airline crew will park for longer periods if this airport is their home base. From the data derived from 30 different airports in the United States, the following regression equation (Equation 11.12) gives a good estimate of the number of employees in terms of origin-destination, connecting, and international enplanements (TAMS in Chicago Airport Study). Airports that handle a lot of freight will have more employees, but many of them will park in other locations.

$$
\begin{aligned}
\text{Employment at Airports (Direct Jobs)} = \; & (0.00164 \times \text{O-D ENPL}) \\
& + (0.001207 \times \text{Connecting ENPL}) \\
& + (0.0054 \times \text{INTL ENPL}) \quad (11.12)
\end{aligned}
$$

The parking lot will need to accommodate some percentage of these workers, depending on the airline schedule, the times the concessions are busiest, and the level of service by transit. A rule of thumb is that about two thirds of the employees use the lot and, of those, about 30 percent carpool with an average occupancy of 2.5 persons per vehicle. The overlap between the early day and afternoon/evening shifts often causes the peak demand for parking spaces.

Example 11.14

From the data above, determine the employee parking requirements for the Indianapolis airport.

Solution to Example 11.14   Using Equation 11.12, Jobs $= 0.00164 * 3,000,000 = 4920$.

If persons coming in cars $= 3300$ and persons carpooling $= 1000$,

Cars $= (3300 - 1000) + 1000/2.5 = 2700$ cars

If the shifts overlap, the maximum need will be about 2700 employee parking spaces. This will require an additional 24 to 25 acres.

11.7.8   Access to the Airport

The best access routes are those that allow for channelization of traffic to each of the airlines within the terminals, or to the best parking areas. At some busy airports, the buses and courtesy vehicles are separated from the taxis and personal vehicles. Figure 11.50 shows two possible access routes. The access cannot be designed without full knowledge of the various routes that will feed into the airport entrance.

11.8   AIR FREIGHT

11.8.1   Overview of Air Freight

The movement of freight by air is very important to our economy—not so much for the amount of weight that it carries, but for the speed at which it moves. Air freight is an important part of the business community, next to e-mail and the fax. Overnight letters and parcels are an important part of business paperwork. Specialty companies, such as Federal Express and United Parcel Service, as well as the U.S. Postal Service, carry a significant amount of small packages and air mail. Much of the growth in air freight is found in overseas flights. Much of that cargo is on large super jets like the Boeing 747, which can lift more than 250,000 pounds. The amount of payload may be reduced if more fuel is needed for longer hops.

Table 11.29 summarizes the air freight shipments by U.S. air carriers and by parcel and postal courier services. Air carriers moved mainly electrical, electronic, computer equipment, pharmaceuticals, medical instruments and appliances, radios, and fresh foodstuffs. The courier services moved similar products, with the addition of textiles and printed matter. The table has three values that are indicative of the industry. The value per ton is very high—averaging for all movement $38,523 per ton or $19.26 per pound.

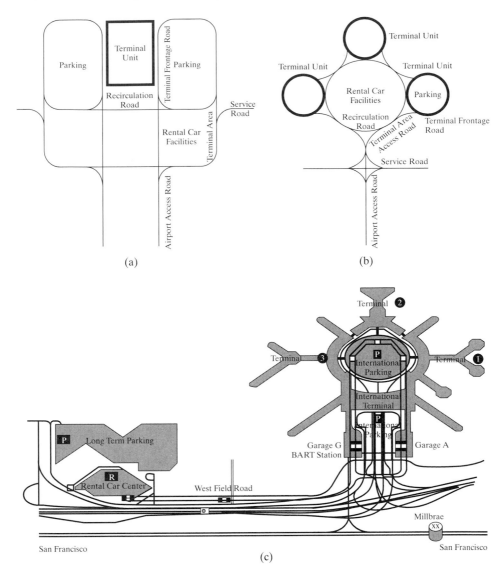

FIGURE 11.50
Access routes to an airport. (a) Single terminal, (b) Multiple terminals, and (c) San Francisco Airport showing road and transit and parking. *Source*: (a) & (b) FAA 1988; (c) San Francisco Airport Authority, used with permission.

TABLE 11.29 Value of Goods Moved by U.S. Air Carriers in 1997

Carrier	Tons Carried	Value Moved	Value per Ton	Miles per Ton
Air carriers	4,475,000	$229,062	$51,157	1393
Courier carriers	23,688,700	$855,897	$36,131	760

Source: USDOT, Transportation Statistics Annual Report, 2000.

FIGURE 11.51
Loading a pallet into the belly of a heavy jet aircraft. *Source:* Photo by Boeing, used by permission.

11.8.2 Preparing Air Freight

Cargo for air travel is often packaged in containers, the size of which depends on the aircraft moving the goods. Airline containers are called *unit load devices (ULD)* and are shaped to maximize the space within the aircraft. Each pallet or container is weighed so that when they are loaded on the aircraft, the weight balance (location of the CG) is best for the flight plan. Figure 11.52 shows four of the most common ULDs.

Aircraft are selected for different flights and different cargos.

- Different flights, because the destination airport's runway configuration may preclude certain flights from using the airport. If the destination is distant, fuel may take the place of some cargo to arrive at the destination without refueling.
- Different cargos, simply because the carrying capacity of aircraft differ. Table 11.30 shows the carrying capacity of several prominent aircraft. The Cessna is included because it has been the feeder aircraft for several carriers. The CASA 212 is a Spanish aircraft that has good range and relatively excellent payload. Its major advantage is that it can land and take off on very short runways.

11.8.3 Air Freight Terminals

Air freight also demands special facilities for the loading and unloading of freight. Departing freight generally arrives at a truck dock as shown in Figure 11.53. It will come in two forms: (1) already in a ULD or on a pallet ready to place on the aircraft or (2) as pieces needing to be assembled into a ULD or palletized for shipment. The freight that comes already packaged passes through what is called a by pass door leading directly to the aircraft loading dock. The remaining freight is packaged in a larger area that contains empty packing material and ULDs. The arriving freight is placed on the aircraft dock, and the containers or pallets are disassembled. This freight may be stored for a

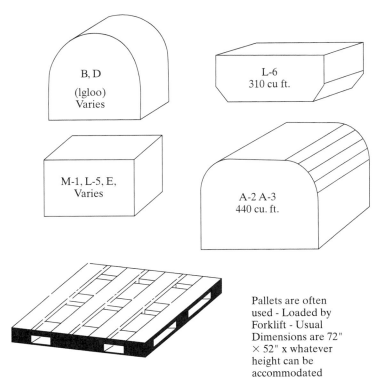

FIGURE 11.52
Several types of ULDs. *Source:* Ashford and Wright, 1992.

TABLE 11.30 Capacities of Selected Aircraft

Aircraft	Type	Average Payload Carrying Capacity in Lbs
B 747-200F	Jet	254,640
B 757-200PF	Jet	50,000
B 737-200C	Jet	34,996
B 727-100D	Jet	44,000
DC8-63F	Jet	155,700
DC 6	Prop	25,000
CASA 212	Turbo prop	10,000
Cessna Caravan	Turboprop	3430

short period of time until the truck to move it arrives. It may also be delayed if there is a customs inspection requirement or if it is hazardous material.

11.8.4 Air Cargo Markets Continue to Grow

Air freight is a special way to use aircraft. Because of the labor cost and shipping distance relative to the number of pounds being moved, the commodities shipped usually

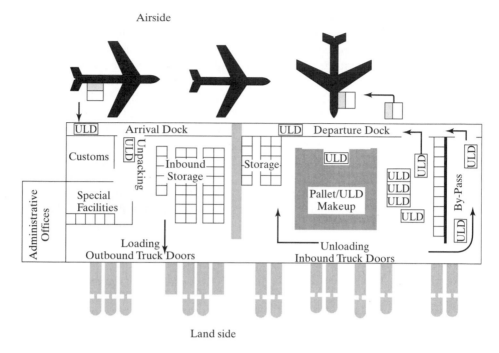

FIGURE 11.53
Schematic of an air freight terminal.

have very high value in terms of dollars per pound. Or as is often the case, freight must be delivered in a short time, especially high-technology items and fresh food. Another case in which the speed of air freight is needed and the expense is justified is when a tool or part is needed to repair an assembly line that has broken down. The cost of air freight can be small compared to the cost of lost production time.

Each year, Boeing issues a report forecasting the world cargo markets and estimating the need for freighter aircraft. The markets are expanding fast in all parts of the world. The Boeing forecast for 2002–2003 predicts a 20-year worldwide growth in air cargo in excess of 6 percent (Boeing, 2001). The Boeing study also states that the cargo markets are very vulnerable to both terrorist attacks and changes in the commodity movement decisions. The forecast indicates to Boeing where sales of new aircraft are needed to accommodate freight growth. The domestic markets for freight are largely those used for overnight or next day delivery, such as Federal Express and United Parcel Service and the U.S. Postal System. Much of the mail and many of the small freight items that forwarders handle are delivered to the destination city in the belly of existing regular passenger flights.

SUMMARY

Airport expansion and renovation are among the most active areas of transportation engineering. As the number of airline passengers continues to increase, airports must be prepared to serve more people and accommodate larger aircraft. Today's air passengers expect a minimum of delays and a wide variety of services within the airport. At the same time, airports are attempting to provide much more in the way of security, because of the threat of terrorist acts. All these activities require space, and the space and equipment involved require money. This chapter has introduced standard procedures for forecasting the number of passengers and number of flights that will be using a given airport in some specified future year. Methods of estimating delay for any existing or proposed situation at an airport—aircraft operations and airport capacity—have been demonstrated. Runways have been described both in how many there should be and how they should be oriented. Furthermore, runway lengths needed to accommodate certain aircraft and, conversely, aircraft takeoff weights permitted under specified runway conditions have been analyzed. Standards for designing the many components of the landside of an airport have been presented. Common ways in which airport improvements are funded are described, including sample calculations. The air transportation sector is an exciting one that can be expected to grow into the foreseeable future.

ABBREVIATIONS

AFSS	automated flight service stations
AGL	above ground level
ASV	annual service volume
DF	delay factor
EQA	equivalent aircraft
FAA	Federal Aviation Administration
FAF	final approach fix (outer marker to touchdown)
IAF	initial approach fix (IA)
FL	flight level
IF	intermediate fix (Middle Marker)
INM	integrated noise model
Kn.or Kt.	knot(s), nautical miles per hour
MALW	maximum allowable landing weight
MATOW	maximum allowable take-off weight
MSL	mean sea level
NPIAS	National Plan of Integrated Airport Systems
PFC	passenger facility charge
RPM	revenue passenger mile
RPZ	runway protection zone
ULD	unit load device
XXX	Three letter code that identifies most airports; e.g., ORD = O'Hare Airport and LGA = LaGuardia Airport in New York

GLOSSARY

The airport and air travel have several distinct terms and nomenclature with which the student must be aware. Most of the definitions below are taken from Whitford, R.K., "Airport Planning and Design," *CRC Civil Engineering Handbook*, Edited by W. Chen, 1995, pp. 2137–2139. Other good sources for aviation-related terms, definitions, acronyms and abbreviations are

- www.gps.tc.faa.gov/glossary.html
- nasdac.faa.gov/internet/fw_glossary.htm
- www.aopa.ch/xgloss.htm

- www.aviationnet.com/glossary.html
- www.aerofiles.com/glossary.html
- www.air-transport.org/public/publications/display1.asp?nid=971

Or just use the key words "aviation" and "glossary" with your favorite Internet search engine.

Air traffic control (ATC):   The positive control of aircraft movement by persons at the airport and between airports exercised by communication links between pilot and controller.

Aircraft approach category: A grouping of aircraft whose minimum landing speed is based on 1.3 times their stall speed (insufficient lift generated to fly) in their landing configuration at their maximum certificated landing weight. The categories are as follows:

A: Speed less than 91 knots

B: Speed 91 knots or more but less than 121 knots

C: Speed 121 knots or more but less than 141 knots

D: Speed 141 knots or more but less than 166 knots

E: Speed 166 knots or more.

Airplane design group (ADG):   A grouping of airplanes based on wingspan. The groups are as follows:

I: Up to but not including 49 feet (15 m)

II: 49 (15 m) up to but not including 79 feet (24 m)

III: 79 feet (24 m) up to but not including 118 feet (36 m)

IV: 118 feet (36 m) up to but not including 171 feet (52 m)

V: 171 feet (52 m) up to but not including 214 feet (65 m)

VI: 214 feet (65 m) up to but not including 262 feet (80 m).

Airport elevation: The highest point on an airport's usable runway, expressed in feet above mean sea level (MSL)

Airport layout plan (ALP):   The plan of an airport showing the layout of existing and proposed airport facilities

Airport reference point (ARP):   The latitude and longitude of the approximate center of the airport as defined by a combined geometric center of all the runways.

Airside: Reflects all the activities and facilities of the airport beginning at the gate in the terminal and includes the apron, taxiways, runways, navigation equipment, fire-fighting equipment, and service for the aircraft when parked at the gate.

Apron: The space adjacent to the terminal where the aircraft can park and be serviced.

Displaced threshold: The portion of pavement behind a displaced threshold may be available for takeoffs in either direction and for landings from the opposite direction.

Enplanement: A revenue passenger boarding an aircraft.

Entry gate: A point in space at which the glide path of the aircraft begins, which is a distance γ (usually 5 to 7 miles) from the threshold of the runway.

Fixed base operator: The contractor(s) at the airport that is/are responsible for aircraft services such as fueling, maintenance, unicom, and so forth.

General aviation (GA):   Private aircraft operation, including corporate aircraft.

Hazard to air navigation: An object which, as a result of an aeronautical study, the FAA determines will have a substantial adverse effect on the safe and efficient use of navigable airspace by aircraft, operation of air navigation facilities, or existing or potential airport capacity. Figure 11.26 shows controlled airspace in the vicinity of an airport. A more detailed diagram pertains to the airspace at the ends of a runway.

Heavy aircraft: An airplane of more than 255,000 pounds maximum certificated takeoff weight. B747, MD11, and L1011 are examples.

IFR (instrument flight rules):   Those rules required for flight when the weather is below the minimums for safe visual flight.

ILS (instrument landing system):   A navigation system that provides aircraft with horizontal and vertical guidance just before landing and, at certain fixed points, indicates the distance to the point of landing.

Interline passengers: Passengers who land in one airplane, deplane, and then enplane a different airplane and never leave the airport. Also called *transfer* passengers.

Landing interval: The time between landings of consecutive aircraft.

Landside: Begins at the departure lounge and extends through the airport terminal beyond the curbs to the parking and the interface with transit.

Large airplane: An airplane of more than 41,000 pounds maximum certificated takeoff weight, but less than 225,000 pounds. The DC-9, Boeing 727, Boeing 737, and Airbus 320 are examples.

Load factor: (a) The ratio of a specified load to the total weight of the aircraft. [www.vfa-13.com/preflight/glossary/basics.html] (b) The percentage of available seats that are filled with paying passengers, or the percent of freight capacity that is utilized. Technically, revenue passenger miles divided by available seat miles or cargo ton miles divided by available cargo ton miles. [www.air-transport.org/public/publications/display1.asp?nid=971]

MYH: The identifier for Mythaca International Airport.

Nautical mile: 6076 feet

NAVAIDS: Aids to navigation that are available at an airport or enroute to the airport.

Occupancy time: The length of time an aircraft is likely to spend on the runway during takeoff or landing.

OEM: operating empty weight

Operations: Any takeoff or landing is one operation. Touch-and-gos account for two operations on each pass.

Origin-destination passengers: Passengers from the local community or passengers who have entered the community for business, meetings, or recreation. Each origin-destination passenger is counted twice when calculating the number of passengers an airport serves. Once when they enter the airport to fly out and once when they enter the airport having returned by airplane.

Pilot: The recognized person responsible for the safe operation of an aircraft in all phases of flight and in landing and taxiing.

Reliever airport: An airport whose purpose is to relieve congestion at a nearby larger airport, normally by diverting general aviation traffic away from the larger airport.

Runway: A defined rectangular surface on an airport prepared or suitable for the landing or takeoff of airplanes.

Runway type: A runway use classification related to its associated aircraft approach procedure. The runway types are as follows:

- *Visual runway.*   A runway without an existing or planned straight-in instrument approach procedure.

- *Non-precision instrument runway.*   A runway with an approved or planned straight-in instrument approach procedure that has no existing precision instrument approach procedure.
- *Precision instrument runway.*   A runway with an existing or planned precision instrument approach procedure.

Shoulder: An area adjacent to the edge of paved runways, taxiways, or aprons providing a transition between the pavement and the adjacent surface: support for aircraft running off the pavement, enhanced drainage, and blast protection.

Small aircraft: An airplane of 41,000 pounds or less maximum certificated takeoff weight.

Stopway: A defined rectangular surface beyond the end of a runway prepared or suitable for use in lieu of runway to support an airplane, without causing structural damage to the airplane, during an aborted takeoff.

Taxiway: A defined path established for the taxing of aircraft from one part of an airport to another.

T-Hangers: An aircraft hangar in which aircraft are parked alternately tail to tail, each in the T-shaped space left by the other row of aircraft or aircraft compartments. [http://www.airport-stormwater.org/definitions_6.htm]

Threshold: The beginning of that portion of the runway available for landing. When the threshold is located a point other than at the beginning of the pavement, it is referred to as either a displaced or a relocated threshold, depending on how the pavement behind the threshold may be used.

Touch-and-go: Landing practice wherein an aircraft does not make a full stop after a landing but proceeds immediately to another takeoff. [www.aerofiles.com/glossary.html]

TPHP: Typical peak hour passenger; relates airport size with passengers.

VFR (visual flight rules):   Rules that apply where the weather is good, visibility is in excess of 1 mile, at altitudes usually higher than 1000 feet. "Under VFR, pilots are responsible for maintaining adequate separation from other aircraft, which is why these rules sometimes are called the "see and be seen" rules." [www.air-transport.org/public/publications/display1.asp?nid=968]

REFERENCES

BOOKS

[1] Ashford, N., and P.H. Wright, *Airport Engineering*, Third Edition, John Wiley & Sons, 1992.

[2] Ashford, N., H. Stanton, and P. Moore, *Airport Operations*, Pitman U.K., 1991.

[3] Horonjeff R. and F.X. McKelvey, *Planning and Design of Airports*, Fourth Edition, McGraw-Hill, 1994.

[4] Sargious, Michel, *Pavements and Surfacings for Highways and Airports*, Halstead Press, John Wiley & Sons, 1975.

[5] Whitford, R.K. Chapter 56, "Airport Planning and Design," *Civil Engineering Handbook*, Edited by W. Chen, CRC Publishing, 1995, p 2042–2162.

FAA ADVISORY CIRCULARS.   The circulars are available on the Web under www.FAA.gov

2000, <u>Advisory Circular Checklist</u>, Advisory Circular AC00-2.13.

1995, <u>Airport Capacity and Delay</u>, AC150/5060-5.

1998, <u>Airport Design</u>, AC150/5300-13.

1989, <u>Planning the State Aviation System</u>, AC 150/5050-3B.

1985, <u>Airport Master Plans</u>, AC150/5070-6A.

1994, <u>Passenger Facility Charge Application</u>, AC150/5000-12.

1990, <u>Runway Length Requirements for Airport Design</u>, Advisory Circular AC150/5325-4A.

1999, <u>Standards for Airport Markings</u>, AC150/5340-1H.

1983, <u>Noise Control and Compatibility Planning for Airports</u>, AC150/5020-1.

1983, <u>Airport Land Use Compatibility Planning</u>, AC 150/5050-6

1995, <u>Airport Pavement Design and Evaluation</u>, AC150/5320-6D.

1970, <u>Airport Drainage</u>, AC150/5320-5B.

1988b and 1994, <u>Planning and Design Guidelines for Airport Terminal Facilities</u>, AC150/5360-13.

OTHER SOURCES

[1] al Chalabi, M, *The Economic Impact of a Major Airport*, Urban Land Institute Research *Paper #622*, 1993.

[2] Boeing Aircraft Company, *World Air Cargo Forecasts, 2002–2003*, http://boeing.com/-commercial/cargo/pdf, published 2001.

[3] Bureau of Transportation Statistics (BTS), *National Transportation Statistics 2002*, Tables 2-1 and 1-34, http://www.bts.gov/publications/national_transportation_statistics/2002/index.html, retrieved 21 December 2003.

[4] Congressional Budget Office, *Financing U.S. Airports in the 1980's*, U.S. Congress, 1984.

[5] Congressional Budget Office, *The Status of the Airport and Airway Trust Fund*, U.S. Congress, 1988.

[6] Federal Aviation Administration website, www.fly.faa.gov/flyFAA/index.html. Accessed November 2002.

[7] Federal Aviation Administration, various years, *Aviation Forecasts*; FAA-APO.

[8] Federal Aviation Administration, *1991-92 Aviation System Capacity Plan*, U.S. Dept. of Transportation, DOT/FAA/ASC-91-1, 1991.

[9] Federal Aviation Administration, *Final Environmental Impact Statement*; Standiford Field Airport Louisville, KY, 1990b.

[10] Federal Aviation Administration, various years, *Terminal Area Forecasts*.

[11] Federal Aviation Administration, *1998 Strategic Plan* (Web).

[12] Federal Aviation Administration, *Long-Range Aerospace Forecasts*: FY 2015, 2020 and 2025, APO-00-5, 2000.

[13] Federal Aviation Administration, *United States Standard for Terminal Instrument Procedures*, Third Edition FAA Handbook 8260.3B July 1976).

[14] Federal Aviation Administration, *National Plan of Integrated Airport Systems 1998 to 2002.*, 1998.

[15] Harris, R, *Models for Runway Capacity Analysis*, The Mitre Corporation 1972 MTR 4102 Rev 2.

[16] International Air Transport Association, 1989. *Airport Terminals Reference Manual*, 7th edition, Montreal.

[17] Port Authority of New York and New Jersey, *Airport Highlights*, Annual Report, 1992.

[18] Transportation Research Board, *Aviation Forecasting Methodology*; A Special Workshop, TRB Circular 348, 1989.

[19] U.S. Department of Commerce (NOAA), Department of Defense, Federal Aviation Administration, 1999 *U.S. Government Flight Information Publication, U.S. Terminal Procedures and Supplement (various states).*

[20] Wilbur Smith Associates and Partnership for Improved Air Travel, *The Economic Impact of Civil Aviation on the U.S. Economy*, FAA, Washington, D.C. 1990.

EXERCISES FOR CHAPTER 11: AIR TRANSPORTATION AND AIRPORTS

11.1 Aviation Data. Search the Web to get more current values for the items listed below. For each item, provide the most recent value you can find, the year of that value, what search engine you used, and what key words you used to do the search.

(a) Total number of U.S. airports
(b) Number of commercial aircraft
(c) Number of general aviation aircraft
(d) Miles flown by commercial aircraft
(e) Miles flown by general aviation aircraft
(f) Passengers carried by commercial aircraft
(g) Persons carried by general aviation aircraft

11.2 Financing Airport Improvements. An $8.7 million improvement to MYH will allow the airport to serve about 10,000 more operations per year than it does now. It has been agreed that general obligation (GO) bonds and revenue bonds will each finance half of the project. The GO bonds will pay 10.5 percent over 20 years, and the revenue bonds will have a 30-year life at 10.0 percent. The airport authority's taxing jurisdiction encompasses property with an assessed valuation of $2000 million. Mythaca State Bank will pay 7.65 percent on funds deposited by the airport authority into a sinking fund established to pay off the principal on the bonds. There is currently no passenger facility charge (PFC) at MYH, but one is being proposed to pay off the revenue bond. Both the project and the bond issues to pay for it are planned to begin next year.

(a) By how much will the property tax rate (in cents per $100 assessed valuation) have to be increased to pay for the GO bond?
(b) Assuming no growth, how much will the new PFC (in cents per enplaning passenger) have to be to pay for the revenue bond?

11.3 National Plan. Find the latest version of NPIAS on the Web. Review the plan and indicate the manner in which the FAA treats the needs of small hubs and GA airports.

11.4 FAA "Share of National Enplanements Model." Construct a spreadsheet with exactly the format of Table 11.6 to forecast the total operations and total passenger traffic from 2005 to 2025. For national data and projections, refer to Table 11.6. For Middleville (MID) Airport data and projections, see the table below.

				Data and Projections for Middleville Airport	
Year	**Commercial Enplanements**	**No. of Transfers**	**Commercial Departures**	**Total Operations**	**Avg. Seats/ Comm Plane**
1996	19,024	734	1636	22,391	23.5
1997	22,101	862	2196	23,206	23.8
1998	19,615	918	1709	23,209	24.6
1999	22,982	1050	2082	23,611	26.4
2000	24,208	1053	1959	24,284	24.7
2001	23,485	991	1892	25,645	24.1
2002	26,708	1007	1998	26,871	23.6

11.5 Forecasting Air Travel Using an Equation.   Sometimes an airport's activity and expected changes over the foreseeable future are not compatible with the FAA share model forecasting method. Instead, equations based on local data can be developed.

(a) Using the data provided in the table below, repeat Example 11.2 and develop five different regression equations (enplanements/population, enplanements/income, trend straight-line, trend exponential and multiple regression on population and income) to forecast enplanements based on data from the year 1997 to 2002. Using the equations you developed, find the forecasted enplanements for the years 2003 to 2020.

Year	Georgetown Enplanements	Georgetown Population	Georgetown Employment	Avg. Income ($)	Yield (cents/pax)	%GA
1997	329,103	321,271	225,346	25,256	13.72	0.577
1998	341,695	325,411	231,131	26,298	13.80	0.589
1999	344,030	329,131	232,558	26,614	13.82	0.592
2000	352,366	332,461	235,756	27,026	14.36	0.520
2001	359,941	334,747	238,212	27,346	13.12	0.502
2002	369,031	337,103	242,273	28,192	13.15	0.502
			Forecast			
2003		339,504	244,118	28,455	13.30	0.502
2004		341,841	245,955	28,731	13.40	0.503
2005		344,326	247,821	28,455	13.50	0.526
2010		356,668	255,774	29,844	14.00	0.503
2015		370,030	267,190	31,297	14.50	0.504
2020		399,016	295,924	34,581	15.00	0.505

(b) The managers of Georgetown Airport prefer to use the EVV equation of the form:

$$CE = a \, RE^b \, YIELD^c \, GA^d$$

where CE = commercial enplanements
       RE = regional employment
    YIELD = average yield per passenger mile in cents
       GA = percentage of operations which are general aviation

Find the coefficients a, b, c, and d for the EVV equation based on the data from 1997 to 2002. Using the equation you have developed, find the forecasted values of enplanements for the year 2003 to 2020.

11.6 Capacity and Delay.   During the peak period for arrivals at MYH, medium (125 knots landing speed) and heavy (150 knots landing speed) aircraft arrive in an alternating sequence. They arrive at a rate of 10 of each size per hour. The standard deviation involved in handling these arrivals is 0.15 minutes. MYH entry gate is 6 nautical miles from the threshold.

(a) What is the capacity of MYH under this alternating pattern of aircraft arrivals? (Use the format of Table 11.16.)
(b) What is the average delay experienced by the aircraft?

(c) Concerned about the vorticity characteristics of the medium aircraft landing at MYH, FAA reclassifies them as heavy aircraft. Of course, the formerly medium aircraft still approach MYH at 125 knots. How does this reclassification affect MYH capacity and average delay time?

11.7 Air Capacity. There are 747s and 737s landing at your airport. They come in groups of four, first a 747 (a heavy aircraft) and then three 737s (medium aircraft) and then a 747 followed by three 737s, and so forth. Determine the capacity of the airport for landing only, if the separation required is that given in Table 11.13 for IFR condition. The ATC gate is 6 nautical miles out and the landing speeds are 150 knots for the 747 and 120 knots for the 737.

11.8 Airport Delay. The air traffic controller receives exactly 34 aircraft (Poisson arrival pattern) in her sector during the peak hour. One aircraft is a heavy (Boeing 747) and the next a large (Boeing 737) aircraft. Because of wake vortex turbulence created by the heavy, the separation for a large behind a heavy is 5 nautical miles and for a heavy behind a large is 3 nautical miles. Both aircraft land at a speed of 135 knots and always arrive into the sector alternately.

(a) If the standard deviation of handling the aircraft is 0.15 minutes, what average delay will the airplanes coming in, experience?
(b) If another runway is added to the airport so that simultaneous landings are permitted, what does the delay become?

11.9 Queueing at Airports. For each takeoff or landing, the runway is "dedicated" to the particular aircraft involved for an average of 75 seconds. This is the runway's mean service time. Flights arrive or depart every 2 minutes during this airport's peak hour. During this period:

(a) What are λ and μ?
(b) What is the average waiting time for an aircraft requesting use of the runway?
(c) On the average, how many flights are waiting to use the runway?
(d) What is the probability that two or more aircraft are "in the system" (i.e., either using the runway or waiting for it)?

11.10 Airport Runway Capacity. Two runway configurations are being considered for an airport that expects to serve no type A traffic but approximately equal percentages of type B, C, and D traffic. These configurations are the first two shown in Figure 11.23—Type 1 (single runway) and Type 2 (two parallel runways).

(a) What is the mix index for the airport?
(b) What are the approximate practical annual capacities for the two configurations under consideration?

11.11 Airport Terminal Design. An airport of modest size (eight gates at present) is being considered as a future hub by a regional airline. If the number of passengers is expected to triple in the next 10 years, what steps will the airport planners have to take to determine the number of gates at the end of that time?

11.12 Airport Runway Layout. At 10:10 PM on Tuesday 22 November 1994, TWA Flight 427 sheared off the top of a Cessna that had taxied into runway 30R at Lambert Field in St. Louis. According to an article in the 24 November 1994 issue of the St. Louis Post Dispatch, the Cessna "taxied south. It should have stopped at Runway 31, but instead taxied southward to Runway 30R, where TWA Flight 427 began its takeoff as assigned." Draw as much of Lambert Field's runway system as possible from this information.

11.13 Runway length and aircraft performance.   The Jamestown airport (JHW), whose runway configuration is a single runway with a length of 6700 ft, lies at an elevation 3724 ft above MSL. Boeing 727-200's are scheduled to use JHW on a regular basis.

(a) With a 15 degree flap setting and the temperature at 90° F, what is the MATOW for the Boeing 727-200 at JHW?

(b) While the runway is undergoing reconstruction that reduces its length to 5500 ft, what is the new MATOW for the Boeing 727-200 at JHW?

11.14 Runway Design.   You are the weight chief of Ajax Airlines operating from the Mythaca airport, which is at 2000-foot elevation. Mythaca is experiencing a heat wave, with the temperature reaching 100 degrees F. Your main runway is 7800 feet with an elevation differential of 50 feet. The crosswind runway is 6100 feet with virtually no elevation differential. Ajax flight 223 is a DC10-10 (CF6-6D engine), which, when full can carry 270 passengers. AX 223 is headed for Chicago, which is 1200 miles away.

(a) How many passengers can AX 223 carry when taking off from the main runway for each flap setting (0, 5, 10, 15, 20, and 25)?

(b) What flap setting will you use? (To determine this, plot the MATOW for each flap setting on the same curve as the MATOW for your runway length corrected for the gradient.) Refer to Section 11.6.3 for the gradient correction.

(c) Three days Flight AX 223 is ready to depart for Chicago, only now the main runway is closed because of heaving. The crosswind runway must be used. The temperature is 95 degrees. How many passengers can be carried with a 15-degree flap setting taking off from the crosswind runway? (Use 1 pax = 200 pounds.)

   Note: The runway/weight charts for the DC 10 in FAA AC 150/5325-4A can be found on the FAA Web site at www.dot.ca.gov/hq/planning/aeronaut/htmlfile/faa-acs.html. See Tables 64 to 71 in Part 6 of 7.

11.15 Airport Runway Length.   The weight chief of Icarus Airlines wants you to check his calculations. MYH can experience summer temperatures of 110 degrees F in the shade. The main runway at MYH is 8800 feet long with an elevation differential of 40 feet. The crosswind runway is 7300 feet with no differential. MYH elevation is 4000 feet above sea level. Icarus Flight IC354 is a 727-200 which, when full, can carry 162 passengers. The flight's destination is Benjamin, which is 850 miles away. For all OEM calculations, assume the domestic fuel reserve of 1.25 hours applies.

(a) How many passengers can IC354 carry when taking off from the main runway with the flap settings at 15 degrees?

(b) At this flap setting, how many passengers can the flight carry if operations restrict it to the crosswind runway?

11.16 Terminal Design at Mazurka Airport.   Because its present location is too close to the city and any expansion would require flying over a bird habitat, the City of Mazurka must relocate its airport. At present there are 1,200,000 enplanements per year, expected to grow in accordance with the FAA growth rates. There are now 33,000 commercial operations and 5000 GA operations, with the current load factor at 0.60. The planning horizon is 2025. The aircraft mix is 5 percent heavies (seating 200), 10 percent commuter jets (seating 38), 10 percent medium-sized aircrafts (seating 140), and 75 percent medium-sized aircrafts (seating 100). In 2025, the mix changes to 10 percent, 12 percent, 66 percent, and 12 percent, respectively. The average seating for all aircraft is 100, which will increase to 113 by the year 2025. Assume that peak day operations are 20 percent of the average weekly operations (using 50 weeks per year). Peak hour operations are expected to reduce from 25 percent of peak day operations to 20 percent in 2025.

(a) Determine the 2025 operations, enplanements, peak hour commercial departures, and TPHP, using a load factor of 0.70.
(b) Estimate the EQG and gates needed for the year 2000 and the year 2025.
(c) Using a function analysis, construct a table as shown in Table 11.26 to determine the number of square feet needed for the terminal in both 2000 and 2025.
(d) Determine the curb space needed if 25 percent of the passengers are dropoffs, 40 percent park and take shuttles, 10 percent come on courtesy vans, 10 percent come in taxis, and the rest arrive on buses. Deplaning passengers use landside modes in the same percentages as do enplaning passengers in both 2000 and 2025. Assume that enplaning passengers during peak hour is 50 percent of TPHP, while deplaning passengers are 60 percent of TPHP.
(e) Determine the parking requirements for the airport in years 2000 and 2025, assuming that the average occupancy for employee vehicles is 2.5 passengers per vehicle. Assume there is no overlap of shifts.

Moving Freight

SCENARIO

Long before Shoridan became a vacation city, it was important as a port on Lake Murdock. It still receives shipments of goods destined for Mythaca and points east. Like most ports, an increasing percentage of cargo is being moved in containers. (See Figure 12.1.) Yesterday, three major shipments were received at the Port of Shoridan: (1) 994 tons of coal for an inland power plant, (2) eight mainframe computers for the university, and (3) 247,000 gallons of heating oil for a regional distributor. In each case, the designee (receiver) is located in Mythaca. In each case, the shipper and receiver have agreed on the carrier (transport mode and company) to use on each leg of the freight shipment. What carrier is the best choice for each of the three shipments from Shoridan to Mythaca? What factors go into the decision as to the carrier to use?

CHAPTER OBJECTIVES

By the end of this chapter, the student will be able to:

1. Describe the role of freight movement in the economy.
2. Explain why certain commodities tend to move on specific transport modes.
3. Define the role of warehousing and distribution centers in the movement of goods.
4. Describe the role of containers in freight logistics.
5. Determine the size of power units to move a train, propel a ship, and send oil through a pipeline.

FIGURE 12.1
Container port. *Source*: Getty Images, Inc.–Photodisc.

12.1 FREIGHT—THE MOVEMENT BEHIND ECONOMIC WELL-BEING

Consider your furniture, your clothes, the food you eat, and everything else you use as part of your life. There is very little among those things that did not at some point undergo movement by at least one freight carrier.

Multiple movements using various modes of freight transport (rail, truck, barge, etc.) are involved in a product being processed from raw material to final product; final product to the store; and the store to your possession.

Anything that is not moving people is considered freight movement. As transportation engineers, we are concerned about the flow of goods to and from companies (called *logistics*) and how transportation affects corporate decisions about products, locations, and manufacturing and/or marketing practices. Logistics is much larger than just freight transportation. A comprehensive study of it would involve detailed review of not only manufacturing but also the way companies do business. Here, we focus on the movement of goods, which includes both physical movement and holding the goods (warehousing) between physical movements. The cost of freight movement in 2000 was about $575 billion. This is a little more than 6 percent of the total U.S. Gross Domestic Product. (Transportation Data Book 2002, Table 3.1) As indicated in Table 12.1, the bulk of freight movement expenditures are for trucking.

The last item in Table 12.1 accounts for the time goods spend being transported. While a good is in transit, on a loading dock, in a warehouse, or at a distribution center, it incurs costs because it must be insured and packaged. Unless the good is transferred to the carrier, that cost is borne by the shipper or the designee and is difficult to assess. The "Inventory under movement" line item is an estimate using an annual rate of 25 percent of the value of the goods transported for a typical transport/warehousing time of 3 days.

12.1.1 Logistics or Physical Distribution

The cost of transport (including inventory holding costs) can account for as much as 20 percent of the sale price of a given consumer product. The cost of transport relative to the value of goods is a key determinant in the movement of goods. Likewise, warehouses and *distribution centers (DCs)* play a vital role in the cost control of retail marketing. Companies like Wal-Mart and K-Mart set up large DCs to which they move the retail products very cheaply by the rail carload or large truck truckload. At a DC, the goods are repackaged into much smaller lots and placed on local trucks for delivery to stores

TABLE 12.1 The Economics of Freight Movement in the United States

Type of Movement	Revenue or Expenditures ($ Billion)
Truck movements/warehousing	456.8
Rail freight	35.9
Air freight	25.3
Water freight	24.5
Pipeline	9.1
Other	10.2
Inventory under movement	15.0
Total	576.8

Source: Transportation Data Book, 2002, Table 3.7.

within a radius of 150 to 200 miles. The cost of extra handling at the DC is more than off-set by the less expensive transportation of large stocks from the factory to the distribution center and the ease of transport to the local stores of several different goods. It also reduces in-store inventory, thereby minimizing floor space needed for each commodity. It is the traffic manager in a company who seeks the best strategies for:

- Moving raw material to his manufacturing plant
- Storing goods en route or maintaining an inventory
- Moving the finished product to the market.

THINK ABOUT IT

Consider the trade-off for a retail store that must choose between a U.S.-made product with a higher wholesale price and a less expensive foreign-made alternative. What role do transportation costs play in this trade-off?

Figure 12.2 presents schematically the physical distribution system for a consumer appliance manufacturer where suppliers transport parts and raw materials to the company. There they are combined, machined, and transported to the assembly line, which may be in another building or city. Once assembled, the appliances are transported to a distributor warehouse, where they are held until they are shipped to the retail store. The retail store may also deliver the appliance to the customer. Each of those steps involves both transportation and holding (warehousing) stock. A major logistic concept that affects freight transportation is discussed below.

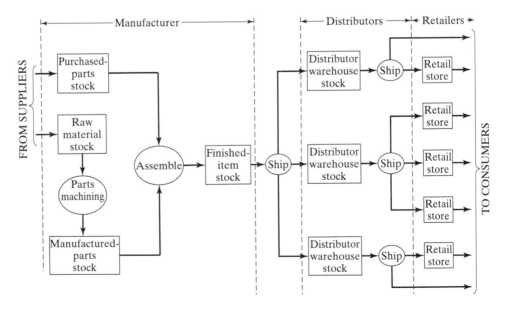

FIGURE 12.2
Physical distribution for a consumer appliance manufacturer. *Source*: Magee, 1961, Figure 1.5.

Just In Time.   In the early 1980s companies tended to keep a large quantity of parts at their plants to support their assembly lines. Keeping enough parts to cover at least 14 days of production guarded against disruptions from late parts deliveries, strikes at suppliers, and so on. With low inventory interest rates and storage costs, companies could justify this business strategy. As the interest rates rose and industrial land around the plants became more valuable, companies looked for a way to reduce the costs of assembly line inputs. They began to work with transportation firms and suppliers to ensure the arrival of parts only a few hours before they were needed on the assembly line. The saving of inventory costs was significant for many manufacturers.

12.1.2   Regional Trade Promotes Economic Growth

Often a particular country (or region of a country) has a specific capability for producing some products cheaply but cannot produce other products at competitive prices. Such a region will be better off selling some of its "homemade" products to another region, while importing the products that it cannot produce easily or cheaply. It is transportation that facilitates the trade between regions that makes regional specialization profitable.

Example 12.1   Trade Between Countries

Region A is the source of fine cloth and clothes that it makes quite cheaply, but it cannot produce dairy products (milk and cheese) at competitive prices. On the other hand, region B is blessed with many fine dairy farms for producing milk and cheese economically, but it does not have the resources to produce fine clothes. Region A, with 320,000 inhabitants, spends $32 million producing the 24 million pounds of clothes it needs and spends $80 million to produce the 32 million pounds of dairy products it consumes. Region B, with population 400,000, spends $60 million producing the 40 million pounds of dairy products it consumes and spends $60 million for the 30 million pounds of clothes it needs. There is no trade between the two regions. Determine the economic benefit that would come from trade between the two regions if the transportation cost between the two regions is $10/cwt for clothes movement and $15/cwt for dairy products.

FYI: The abbreviation *cwt* stands for *hundredweight*, or 100 pounds. It is a standard basis for setting freight transport rates, even where the good is normally measured in terms of volume (gallons of milk in Example 12.1) or counts (articles of clothing, computers, etc.).

Solution for Example 12.1   The steps in the analysis are:

1.   Determine the rate at which a particular product is consumed in each region. For example, in region A, 24 million pounds of clothes are used by its 320,000 residents. This is 75 pounds of clothes per person.
2.   Determine the cost to produce each product locally. In region A, it costs $32 million to produce 24 million pounds of clothes ($1.33/lb) and $80 million to produce 32 million pounds of dairy products ($2.50/lb). In region B, it costs $60 million to produce 30 million pounds of clothes ($2.00/lb) and $60 million to produce 40 million pounds of dairy products ($1.50/lb).
3.   Determine the cost to transport a cheaper good to the other region. Region A produces clothes more cheaply, at $1.33/lb, or $133/cwt. Region B consumes 30 million

pounds of clothes per year and produces clothes at $2.00/lb, or $200/cwt. Shipping 100 pounds of clothes from region A to region B would cost $10. The demand for clothes in region B is 30 million pounds. At region A prices, this is $40 million. The cost of shipping 30 million pounds from region A to region B @ $10/cwt is $3 million. Instead of spending $60 million for locally made clothes, region B consumers spend $40 million plus transport costs of $3 million for clothes from Region A.

The calculations associated with this analysis are summarized in Table 12.2. An assumption is made that each region will switch its resources to producing only the good that it can produce more efficiently. This would ensure adequate supply of that good for both regions.

TABLE 12.2   Solution to Example 12.1

	Region A	Region B
Population	320,000	400,000
Clothes consumption	75 lb/person	75 lb/person
Cost to produce clothes	$1.33/pound	$2.00/pound
Total cost clothes	$32 million	$60 million
Dairy products consumption	100 lb/person	100 lb/person
Cost to produce dairy products	$2.50/lb	$1.50/lb
Total cost dairy products	$80 million	$60 million
Cost without trade	$112 million	$120 million
Cost of buying other regions' cheaper good (w/o transport)	Region B dairy products: $48 million	Region A clothes: $40 million
Transport cost of other regions' cheaper good	Region B dairy products: $15/cwt = $4.8 million	Region A clothes: $10/cwt = $3 million
Total import costs, including transport, at current consumption rates	$52.8 million	$43 million
Money saved by importing cheaper good from the other region	$27.2 million	$17 million

If the transportation costs for clothes from region A to region B had been $67/cwt, the clothes produced in region A would have not been cheaper to buy in region B. If region A had instituted a tariff of ($200 − $133) − $10 = $57 on each 100 pounds of clothes, then region A clothes would no longer be cheaper than locally made clothes in region B.

THINK ABOUT IT

Why would a country whose residents can buy an imported product more cheaply than a locally made product put a tax on those imports?

12.1.3   Holding: A Cost Often Overlooked in Freight Movement

Once an item has been manufactured, it is shipped to some location. The cost of the shipment has two components:

1. The cost of carrying the product by whatever mode is used
2. The cost of the time of transport, which is called the holding cost.

The holding cost is that cost that must be paid for the storage of the product while it is being shipped. If the value of the goods being shipped is high, then high holding rates (e.g., 20 to 30 percent) will cause a faster mode of transport to be chosen. The holding rate often includes interest (or time value of money) on the value of the product, special packaging, and insurance.

Example 12.2

A logistics manager has to move 100 containers from the Far East to a plant in New Jersey. Each loaded container weighs 40,000 pounds. The container itself weighs 3600 pounds. It is loaded with an electrical product whose value is $4.25 per pound. There is about 15 percent dunnage (packing) by weight. His three choices can be seen in Figure 12.3. He can move the containers by an all-water route through the Panama Canal, which takes 24 days. He can sail to Seattle/Tacoma in 10 days and take the railroad (called *land bridge*) to New Jersey, a trip that take 6 days, or he can sail to Long Beach in 12 days and place the freight on a special double stack rail service that gives a guaranteed delivery of 72 hours. The prices for the three services are given in Table 12.3. Which route should the manager choose?

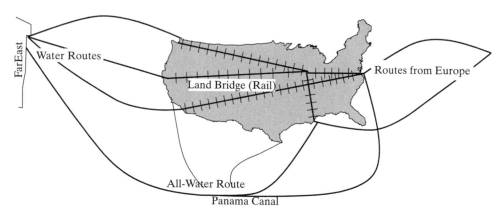

FIGURE 12.3
Land bridge and all water routes.

TABLE 12.3 Cost Data for Example 12.2

Transport	Rate
All-water route	$2.45 per cwt. through Panama Canal
Sail to West Coast port	$1.25 per cwt to Seattle or Long Beach
Handling at Port of Seattle	$300 per container
Rail from Seattle	$3.25 per cwt
Handling at Port of Long Beach	$250 per container
Drayage at Long Beach to LA	$150 per container
Double Stack Train from Los Angeles	$4.00 per cwt
Handling in New Jersey	$150 per container
Holding cost	35% of value per year

Solution to Example 12.2   Determine the value of the freight in each container. The container weighs 40,000 pounds, including the 3600 pounds tare (empty) weight of the container. Of the 36,400 pounds of contents, 15 percent is dunnage (packing), leaving 30,940 pounds of product in each container at $4.25 per pound. The value of each loaded container then is $131,495.

Determine the holding cost. The holding cost is computed at an annual rate of 35 percent per year. The daily rate is then $131,495 * 0.35/365 = $126.09 per container per day. Table 12.4

TABLE 12.4   Solution to Example 12.2

Options	Panama Canal	Sea-Tac + Rail	Long Beach + Rail
Days of shipping	24	16	15
Transport (water)	$98,000	$50,000	$50,000
Transport (rail)	$0	$130,000	$160,000
Holding @ 35%/yr	$302,619	$201,746	$189,137
Handling	$15,000	$30,000	$25,000
Drayage	$0	$0	$15,000
Total cost	$415,619	$411,746	$439,137
If holding is 25%/yr	$216,156	$144,104	$135,098
Total cost with 25%/yr	$329,156	$354,104	$385,098

extends the numbers for the 100-container shipment. The 35 percent holding cost and the extra days make the route through Seattle the least expensive. However, note that reducing the holding fee to 25 percent makes the all-water route less expensive.

12.1.4   Transportation Impacts Strategic Location Decisions

Manufacturing Plant Location.   Transportation and transportation costs often play a key role in the location decision of a manufacturing plant. Important in the cost of the product to the consumer is cost of moving raw materials to the plant and the cost of delivering the product to the market. Another important factor is the availability of a skilled workforce close to the location.

Example 12.3

A company's major markets are Atlanta, Chicago, Cleveland, and New York City (shown with stars in Figure 12.4). Its resources come from Duluth, MN, and Charleston, WV (the two "suns" in Figure 12.4). Given the volumes and transportation costs shown in the fifth and sixth columns of Table 12.5, where should the plant be located from a transportation point of view?

Solution to Example 12.3   The solution begins by plotting the sources and sinks (markets) on a map and adding the latitude and longitude to a table. (Any consistent grid system will do.) Determining the cost factor as a function of weight per unit distance for each source or sink is shown in Table 12.5.

The critical numbers are the X weights and Y weights. An X weight is obtained by multiplying the source's (or market's) longitude X, its volume to (or from) the plant (in thousands of pounds), and the transport cost rate. Longitude serves as a distance measure proxy for miles. For

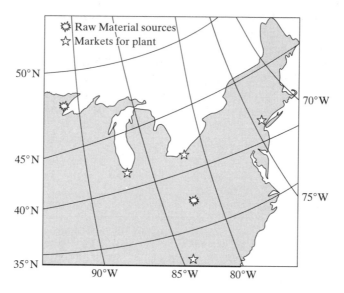

FIGURE 12.4
Location of raw material sources and markets for plant.

TABLE 12.5   Data for Example 12.3

Source or Market	Site	Longitude	Latitude	Annual Volume (tons)	Transport Rate ($/mi/cwt)	X wts: Long. * Vol* rate/1000	Y wts: Lat. * Vol* rate/1000
Raw materials	Duluth	92.2°	47.0°	10,200	$1.50 (rail)	28,213	14,382
Raw materials	Charleston, WV	82.0°	38.0°	6700	$2.40 (rail)	26,371	12,221
Market A	New York City	74.1°	40.8°	5800	$3.75 (truck)	32,234	17,748
Market B	Cleveland	81.7°	41.5°	2600	$3.75 (truck)	15,932	8,093
Market C	Chicago	87.6°	42.0°	4600	$3.75 (truck)	30,222	14,490
Market D	Atlanta	84.4°	33.8°	4400	$3.75 (truck)	27,852	11,154

Duluth as a source: $X = 92.2 * 10{,}200$ tons $* (2000 \text{ lb/ton})/(100 \text{ lb/cwt}) * \$1.50/\text{cwt} * (1/1000) = 28{,}213$. The Y weight is based on latitude. For Duluth: $Y = 47.0 * 10{,}200 * (2000/100) * \$1.50 * (1/1000) = 14{,}382$. By listing all the sources and markets in order from east to west (see top of Table 12.6), the best longitude can be determined using interpolation. By listing all the sources and markets in order from north to south (see bottom of Table 12.6), the best latitude can be determined by interpolation. The plant location that would minimize the weighted sum of transport costs has longitude that corresponds to the midpoint cumulative X weight, i.e., half the total cumulative X weight. The least-cost plant location has latitude that corresponds to the midpoint cumulative Y weight, i.e., half the total cumulative Y weight.

In this example, the plant location that would minimize transport costs would be near Mansfield, OH, between Columbus and Cleveland. Of course, considerations besides transport costs figure prominently in plant location decisions.

TABLE 12.6   Determining Plant Location to Minimize Transport Costs

(a) Longitude or X determination

Source or Market	Site	Longitude	Annual Volume (tons)	Transport Rate ($/mi/cwt)	Sorted X Weight	Cumulative X Weight
Market A	New York City	74.1°	5800	$3.75	32,234	32,234
Market B	Cleveland	81.7°	2600	$3.75	15,932	48,165
Raw materials	Charleston, WV	82.0°	6700	$2.40	26,371	74,536
(midpoint)	Plant longitude	**82.5°**				**80,412**
Market D	Atlanta	84.4°	4400	$3.75	27,852	102,388
Market C	Chicago	87.6°	4600	$3.75	30,222	132,610
Raw materials	Duluth	92.2°	10,200	$1.50	28,213	160,823

(b) Latitude or Y determination

Source or Market	Site	Latitude	Annual Volume (tons)	Transport Rate ($/mi/cwt)	Sorted Y Weight	Cumulative Y Weight
Market D	Atlanta	33.8°	4400	$3.75	11,154	11,154
Raw materials	Charleston, WV	38.0°	6700	$2.40	12,221	23,375
(midpoint)	Plant latitude	**40.5°**				**39,044**
Market A	New York City	40.8°	5800	$3.75	17,748	41,123
Market B	Cleveland	41.5°	2600	$3.75	8,093	49,215
Market C	Chicago	42.0°	4600	$3.75	14,490	63,705
Raw materials	Duluth	47.0°	10,200	$1.50	14,382	78,087

THINK ABOUT IT

Even if transport costs are the only consideration, in what way is Example 12.3 an oversimplification?

Warehouse or Distribution Center.   The purpose of warehouses and distribution centers is to provide an inexpensive place to store goods in the process of moving the goods from the manufacturer to the market. It is often a place where modes of transport are changed. Two examples show how this works.

Ajax Clothing Company has outlets in 85 eastern cities. Its main products come from several U.S. clothiers and from international ones in Hong Kong and Malaysia. It has established 10 centers along the eastern seaboard. Each center is located on a rail line and is near a major interstate. Clothes manufactured in three different plants in the Carolinas are brought north by rail to each of the DCs. The clothes are stored and distributed to the retail stores by a local delivery service. Likewise, clothes from overseas will come either by all-water or landbridge to the DCs, depending on the best cost and when the product is needed. By proper location of the DCs, Ajax found that they could reduce their delivery costs by 23 percent. In addition, there was a reduction in loss and damage of 8 percent.

MagicMart has stores all over the nation and sells many different products. Products come to one of five major centers by the least expensive transportation possible. The products are off-loaded at the major centers and stored or moved to complete full truckload shipments to one of their 85 regional distribution centers located along interstate highways. Each store deals only with its regional DC to replenish its inventory.

Example 12.4   Delivering Widgets

The Wright Corporation has a plant that produces widgets and appliances to improve homemaking. Two of Wright's major markets are in City A (200 miles southwest of the plant) and City B (200 miles to the southeast), but there are also many small stores in the region that carry Wright products. Cities A and B are 200 miles apart. The Wright Corporation is reviewing its delivery operations. The widgets can be delivered directly from the plant to stores. Direct delivery involves handling costs of $1.00 per widget at the plant and delivery costs of 6 cents per mile. An alternative is to make rail shipments to distribution centers in Cities A and B, and make delivery to stores from DCs. The cost to ship the average widget to City A by rail is $2.50, with warehousing and handling costs at City A equal to 40 cents per widget. Rail shipment to City B costs $3.05 per widget, plus a handling cost of 35 cents per widget. The warehouses at Cities A and B are both able to get contracts at about 4 cents per mile per widget to deliver widgets to stores.

(a)   Draw the triangular region defined by the plant location, City A, and City B. On that "map", draw cost contours that represent lines of constant total plant-to-store delivery costs throughout the region.

(b)   What boundaries would be appropriate for delivery between these markets? What is the likely highest distribution cost that would be charged?

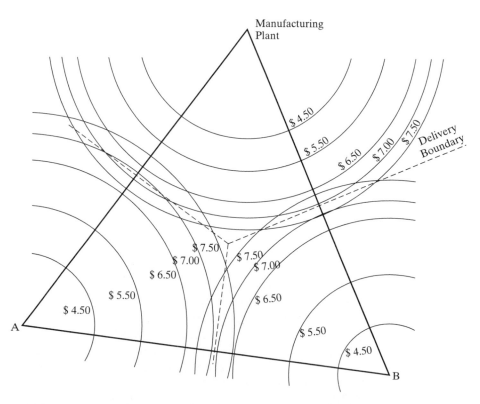

FIGURE 12.5
Curves of equal delivery costs in Example 12.4.

Solution to Example 12.4

A. Using the costs indicated, draw a map that shows contours of constant delivery cost centered on the plant, City A, and City B, as shown in Figure 12.5.

B. The points where the contours overlap with equal value are logical boundaries territories to be served by distribution centers at the plant, City A, and City B. In Figure 12.5, the highest distribution cost appears to be between $7.50 and $8.00, where the three delivery boundaries (dashed lines) meet.

Silos for Grain Storage. The secret for the farmer is to balance the cost of land and delivery. As the farmer moves farther from the market, his land is cheaper, but it costs more for his grain to be brought to market. Figure 12.6 displays the trade-off between land prices and truck delivery prices. As the farm gets farther out, it is less competitive because of larger truck transportation costs.

The secret to effective distribution is to use modes that reduce the long line-haul cost (e.g., rail) and then use a more expensive mode (e.g., truck) that is better suited for the urban areas or short delivery runs. Silos have been located strategically along existing rail lines to provide cheaper and wider access to markets for farmers. The location of a silo (e.g., a warehouse or distribution center) will be determined by transportation costs and the distance from market.

Figure 12.7 shows how the introduction of a grain elevator or silo 70 miles from market can reduce the cost of getting produce to market. The "Previous TC" curve

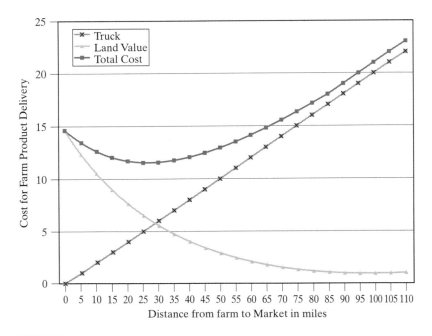

FIGURE 12.6
Effects of land cost and truck rates as a function of distance from farm to market.

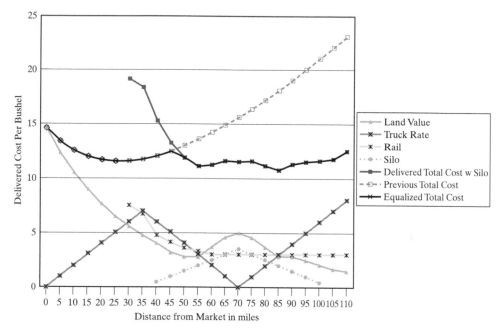

FIGURE 12.7
Effects of adding an elevator to a farming region.

matches the top curve ("Total Cost") in Figure 12.6. The "Land Value" curve is the same as in Figure 12.6 within 50 miles from market, but the presence of the silo will increase land values around the 70-mile mark—the "Silo Effect". An elevator offers a focal point for rail service, which is a less costly mode than truck. Trucks can now be used to reach the elevator or go directly to market, as is shown by the piecewise linear "Truck Cost" curve. The severe peak in the "Delivered Total Cost w Silo" curve demonstrates that using the silo can be much more costly for farmers located closer to market than is the silo. For those farmers, trucking produce directly to market is still less costly. In Figure 12.7, the "Min TC" curve refers to the least total cost option for any distance from market. For distances to market greater than 50 miles, using the elevator can offer significant savings in getting produce to market. A side benefit of the elevator is that it can hold grain, if desired, so farmers can wait for a better price than is possible with immediate truck delivery direct to market. Rail also provides access to markets over a larger geographic area.

12.1.5   Major Categories of Freight

A study of freight movement involves some understanding of the commodity being moved, including its value and the distance over which it must travel. Table 12.7 gives a general picture of which modes are used for which commodities.

TABLE 12.7   Cargo Types and the Modes Most Likely to Be Used

	Cargo Type	Dominant Mode(s)	Delivery
1	High value	Air, truck for short trips	Truck
2A	Medium value/individual package	Truck	Truck
2B	Medium value/containerized	Rail, marine	Truck or rail
3A	Low value/bulk commodities/dry	Rail, marine	Direct or truck
3B	Low value/bulk commodities/wet liquid oil	Pipeline, rail, marine	Usually pipe
4	Delivery vehicle is container with freight (ferry)	Marine, rail (roll on/roll off)	Delivery vehicle

In Table 12.7, "bulk" cargo is that which is not put into packages or containers. It is handled as a mass, such as coal, petroleum products, or natural gas. The two other categories of cargo are breakbulk and containerized. Breakbulk cargo is handled in units such as packages, crates, and bags. Cargo that is packed in containers can be easily transferred between modes.

THINK ABOUT IT

Which modes would you suggest be used to carry each of the three shipments described in this chapter's Scenario?

12.2   SURFACE FREIGHT MODES

12.2.1   Trucking

We have already seen in Chapter 9 that heavy trucks impose loads that inflict damage to highway pavement. Trucks pay a tax, depending on their weight and road use, that is intended to provide funds for repair of the road. We also learned in Chapters 2 and 3 that trucks reduce the capacity of our roads and can lead to congestion. Despite all the difficulties trucks seem to cause, they are an essential element in the movement of goods that sustain a national economy. In Table 12.1, trucking was the leading means of moving freight, at least over short distances. Goods are also often shipped long distances by truck, thanks to the extensive U.S. road system—especially the 42,500 miles of interstate highway.

Trucks move about 1 trillion ton-miles of freight in the United States each year. It costs the shipper more than other forms of freight movement (except air freight) and for that reason usually transports high-value goods. In most cases, trucks are limited to a total weight of 80,000 pounds on U.S. highways. The tractor of a tractor-semitrailer usually weighs 20,000 to 25,000 pounds and the tare weight of the semitrailer is 6,000 to 10,000 pounds. This means the useful freight haul is a maximum of 22 to 25 tons. Most trucks do not carry as much as they could. Because of both weight and dimensional restrictions, the load actually carried depends on the density of the goods. Some trucks "weigh out" (reach their gross weight limit) before they "cube-out" (reach their volume limit), whereas others "cube out" before they weigh out and must carry less than the maximum weight of goods allowed. Some states allow double or tandem trailers on their highways, similar to that shown in Figure 12.8. Table 12.8 lists several of the many type of trucks.

THINK ABOUT IT

Why would any unloading and reloading of packages be advantageous to the LTL carrier?

FIGURE 12.8
Line-haul over-the-road truck rig for LTL carrier. Photo Courtesy of Roadway Express.

TABLE 12.8   Types of Truck Carriers

General freight carriers	Move freight for companies that cannot afford, or who do not want, to own their own trucks. The types of goods being transported indicate the nature of the movement. Full truckload carriers and LTL carriers mentioned below are considered general freight carriers.
Full truckload carriers	Pick up a full load or full trailer at the dock of the shipper (origin) and carry it to the dock of the consignee or designee (destination). Because no large investment in terminal facilities is needed, large truckload quantities are moved at relatively low rates.
Less than truckload (LTL) carriers	Pick up packages or pieces of freight, bring them to a truck terminal, and then place them on trucks headed in the general direction of the consignee. The package may be unloaded and reloaded at one or more terminals enroute. It may ride on several different trucks before it is delivered to the recipient. The key to economy for the LTL carriers is having an adequate network of terminal facilities, so freight is unloaded and reloaded the minimum number of times.
Special freight carriers	Some goods require special trucks and/or special loading and unloading. Special freight carriers contract with individuals or companies for the movement of household goods (moving vans), liquids from milk to caustic chemicals and gases (tank trucks), automobiles (car carriers), steel and heavy metals (flatbed trucks), and grain, soil, and rocks (bottom- or rear-unloading trucks). Special freight carriage accounts for about 20 percent of truck ton-miles.
Private carriers	Many companies own and operate their own fleet of trucks. Private trucking constitutes about 50 percent of the trucks on the highway. The reason many companies choose to operate their own fleet of trucks is to maintain control of the goods between their own plants or distribution centers.
Drayage	Usually a very short movement. Draymen often move goods from a rail yard to a major truck terminal or port. For example, goods arriving at the Port of Long Beach must be moved about 2 miles overland to the rail terminal not located on the port facility.

Example 12.5

If a truck tractor weighs 22,000 pounds and its semi-trailer is 48-feet-long, 8-feet-wide, 9-feet-high, and weighs 7000 pounds, what would be the truck's weight if it were filled with a product that has a density of 15 pounds per cubic foot? Which would limit the truck's cargo— the 80,000-pound weight limit on the truck's gross weight or the size of its trailer?

Solution to Example 12.5   The truck trailer can carry, when totally loaded, $48 \times 8 \times 9$ or 3456 cubic feet of goods. At 15 pounds per cubic foot, the total weight of the load could be 51,840 pounds. But the trailer can only carry $80,000 - 22,000 - 7000 = 51,000$ pounds. Because the truck will reach its maximum over-the-road weight before it can be filled with the product, it will travel at the 80,000 pounds gross vehicle weight (GVW) limit.

Example 12.6

Suppose only a 40-foot semi-trailer with a 6000-pound tare weight capacity had been available for the load in Example 12.1. Would the truck "cube out" or "weigh out"?

Solution for Example 12.6   When the truck is totally loaded, it can carry $40 \times 8 \times 9 = 2880$ cubic feet. A semi-trailer full of cargo would weigh $2880 \times 15 = 43,200$ pounds. The total weight of the truck would be $22,000 + 6,000 + 43,200 = 71,200$ pounds. In this instance, the truck would have been filled before its maximum weight could be reached. The truck "cubed out" before it "weighed out."

Truck Operations.   There are more than 1.8 million large trucks in the United States. These trucks travel an average of 70,000 miles per year. Many travel more than 100,000 miles in a year. Truck tractors often have places as part of the cab for drivers to sleep on the road. Some trucks are kept going day and night with a team of drivers. The truck tractor couples to the trailer through what is called a *fifth wheel*. Thus, a tractor can drop off one trailer to be unloaded at a dock and then pick up another trailer without having to wait for the semi-trailer to be unloaded. Trucks have become more fuel efficient in the last 30 years, but at the same time, the growing U.S. economy has resulted in more shipments of goods by truck. Table 12.9 shows the growth in trucking since 1980.

Trucks are traveling farther per haul, consuming more gasoline, but at a lower rate. The seeming low fuel economy may be a bit misleading, because trucking has changed since 1980. For those shippers whose main products are low in density, trailers up to 53 feet in length may now be used. More and more states are permitting

TABLE 12.9   Truck Performance Since 1980

Year	1980	1985	1990	1995	1998
Trucks registered (thousands)	1417	1403	1709	1696	1831
Truck miles traveled (million miles)	68,700	78,100	99,500	115,500	128,160
Fuel consumed (million gallons)	13,037	14,005	16,133	19,777	21,100
Average miles traveled per vehicle per year	48,500	55,700	55,200	68,100	70,000
Average miles per gallon	5.3	5.6	5.8	5.8	6.1
Average length of haul (miles)	263	286	366	391	426
Average ton-miles (millions)	555,000	610,000	735,000	921,000	996,000
Average ton-miles/gallon	42.6	43.6	45.6	46.6	47.2

Source: U.S.DOT/Bureau of Transportation Statistics, 2000.

small 23- to 30-foot trailers to move in tandem on interstate highways. These tandems have helped the LTL carriers to simply unhook and reconnect the smaller trailer, instead of having to unload and then reload larger trailers. Truck productivity in terms of energy has improved, with the delivery of almost 50 ton-miles of goods per gallon of fuel consumed.

A trucking LTL or transfer terminal is schematically shown in Figure 12.9. The terminal provides a place where the freight and haul vehicles can be matched by destination. Freight is assembled (Figure 12.10a) from local delivery trucks for the long-haul highway tractor-trailers. When freight arrives on a long-haul truck, it is disassembled for local delivery or it is placed on another long haul truck (Figure 12.10b) to move closer to its intended destination. For many products, there are intermediate storage requirements. Another function of the truck terminal area is to provide facilities for dispatching and for the repair and maintenance of equipment.

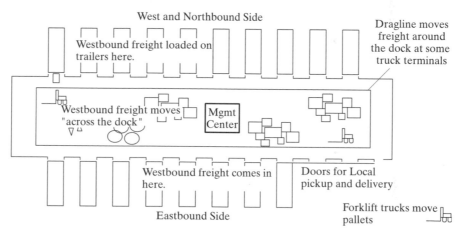

FIGURE 12.9
Schematic of a trucking terminal for LTL Carrier.

(a)

(b)

FIGURE 12.10
LTL terminal. (a) Moving freight across the dock of an LTL terminal and (b) Outside of LTL terminal. Photos courtesy of Roadway Incorporated.

12.2.2 Railroads

The railroads have been an important cornerstone in the history of the U.S. transportation system for both freight and passengers. In the early 1800s wagon trains and river rafts provided the major travel mode in the United States for both passenger and freight. Following the invention of the steam engine, the period from before the Civil War to the 1940s was the "heyday" of the railroad as the miles of track grew from 380 miles in 1833 to 30,000 miles at the time of the Civil War to more than 400,000 in 1950. In 1869, the first transcontinental railroad was completed. By 1960 the diesel electric locomotives had replaced the stream engines, improving the railroads' ability to be much more efficient.

In the late 1800s, railroads were the major source of movement on land. Because of their monopoly position, they could charge whatever they wanted. With many accusations of predatory pricing and discrimination from customers needing to move freight by rail, the U.S. Congress set up the Interstate Commerce Commission to regulate entry into markets and pricing. The railroads lived in a heavily regulated environment that lasted for almost 100 years, until 1980, when the Staggers Act was signed into law. The act gave considerable freedom to the railroads, which were no longer a dominant passenger carrier, and which had strong freight competition from truck, inland waterways, and pipeline. In 1940, the railroads moved 74 percent of all the ton-miles of freight movement in the United States. By 1999, their market share had been reduced to about 30 percent. Since deregulation, there has been a rash of mergers, along with shifts in management and operation. In the 1950s, there were 132 Class I or major railroads, with a total of 391,000 miles of track. Today, there are 7 Class I U.S. railroads, which own and maintain about 120,000 route miles.

America's freight railroads carry more than 40 percent of the nation's intercity freight, 70 percent of the motor vehicles made by domestic manufacturers, 64 percent of the nation's coal (which generates 36 percent of the nation's electricity), and 40 percent of the nation's grain. Class I railroads hauled 26 million carloads of freight in 1998, accounting for 1.38 trillion ton-miles. The other thing brought on by deregulation has been an increase in intermodalism, where the rail mode shares the origin-destination movement with other modes, primarily truck, barge, and ships from overseas. Intermodal rail freight has nearly tripled since 1980. The railroads moved about 9 million intermodal trailers and containers in 1998. In 1998, 20,261 locomotives pulled a fleet of 1,315,667 freight cars with an aggregate capacity of 127.8 million tons—an increase of 18 percent since 1990. According to the American Association of Railroads (AAR), it would take 3 million trucks to equal the capacity of the rail car fleet.

The effect of *deregulation* is clear when we examine the curves in Figure 12.11. Even the lower revenue has meant better profits, because the railroads' return on investment has roughly quadrupled, from a miniscule 2 percent to about 8 percent in 1999. Productivity is up more than 150 percent; shipping rates are down 53 percent.

The Way and Alignment.   Because the railroads are private companies, they own and maintain their own right-of-way (ROW). The track, like a highway, is in constant need of repair or replacement. Preparing the track roadbed involves grading and compacting the soil underneath. The roadbed provides the surface on which the ballast is placed. The intent of the ballast is to distribute the loads uniformly over the subgrade, to hold

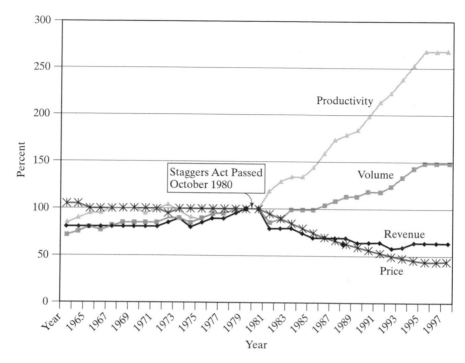

FIGURE 12.11
Class I railroad performance 1964 to 1998 (1981 = 100). *Source:* American Association of Railroads 2003.

the track structure firmly to its alignment, to provide for good drainage, to reduce dust, and to prevent the growth of weeds. Ballast is usually crushed rock, washed and screened gravel, slag and cinders. The interlocking effect of crushed rock aggregate provides a high degree of stability under the flexure caused by successive wheel loads of a moving train. This loading action causes displacement of the ballast at the ends of ties, requiring frequent tamping to avoid "center-bound" ties. Ballast, like pavement thickness, depends on the eventual loading expected before the track should be rehabilitated.

THINK ABOUT IT

What is meant by "center-bound" ties? Sketch what you think that situation would look like.

Figure 12.12 shows a typical cross section of a rail car, with its flanged wheels and track. The track is installed by spikes that attach to metal plates, called *tie plates*, that reduce the mechanical wear from loads on the ties and reduces some side sway of the trains. Rail is specified by the weight per yard. Rail usually runs from 50 to 160 pounds per yard. The railroad gage is a standardized 4 feet 8.5 inches

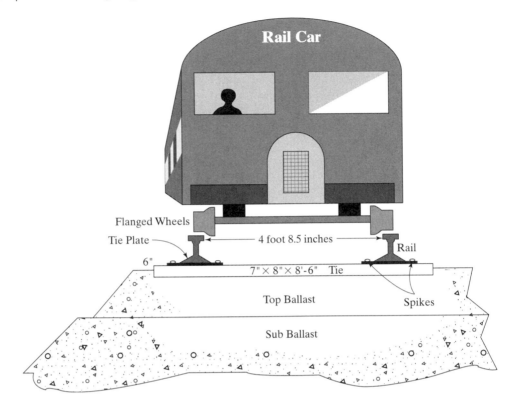

FIGURE 12.12
Track and ballast section.

throughout the country, permitting railcars from all railroads to ride on another company's track.

THINK ABOUT IT

How did the standard railroad gage come to be 4 feet 8.5 inches? Is this gage the same in every country?

Operations. There are a number of operating problems that confront the railroad carriers. Railcar ownership, a major problem when there were more than 100 Class I railroads, is easier now. A car owned by one railroad may go on another railroad's track to make a delivery. Once emptied, the car is put on a train headed back to the owning railroad. The AAR (American Association of Railroads) has a large database that keeps track of the location of all the cars in use. The railroads pay per diem rates when a car that they do not own is used in one of their hauls.

The car management problem has the interesting twist that there are many different kinds of cars, as Figure 12.13 indicates. Because of all the specialized cars, managing

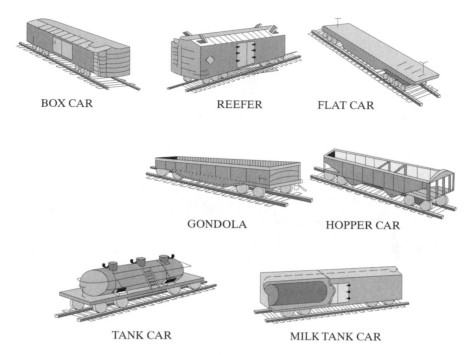

BOX CAR   REEFER   FLAT CAR

GONDOLA   HOPPER CAR

TANK CAR   MILK TANK CAR

FIGURE 12.13
Some types of freight cars. *Source*: Hazard, 1977.

the total fleet of cars can be difficult. Trains are "broken down" and "built up" in railroad "classification" yards. Figure 12.14 shows a classification yard where the cars are shuffled onto the correct train through a series of switches and *retarders* to slow them down. Individual cars are pushed by a work engine to the *hump* and, after the proper switches have been set, they are propelled by gravity to the correct track. There are also classification yards where all movement is performed by switch engines moving cars to the right train.

THINK ABOUT IT
Why is managing a fleet of specialized freight cars especially difficult?

Car utilization is a problem for railroads. Frequently, the cars are not where they are needed, or they are empty being returned for reloading. The typical freight car has a cycle from loading to the next loading of 25 to 30 days. The rail car spends more time being loaded, unloaded, in yards, or just empty than it does moving on the rail, which is a problem for improving rail productivity. Unit trains carrying coal, grain, or oil have the advantage of bypassing intermediate classification yards. The entire consist stays together as a unit to deliver its single commodity and then usually returns empty for the next load of that commodity.

Intermodal Rail/Truck.   One system technology that has brought vast improvements in rail/truck intermodal productivity is Roadrailer™. These combination rail/highway

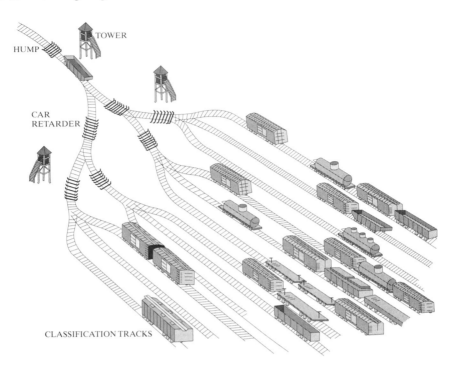

FIGURE 12.14
Hump classification yard. *Source:* Hazard 1977.

cars are used in low-density markets where costly intermodal yards are not possible. As shown in Figure 12.15, the cars of the train also have rubber tires lifted above the rail. The rail axles, with wheels called *bogies*, are placed between each trailer. The long-distance move is made on the more efficient rail line. When the rail portion of the move is over, the rail wheels are removed and the trailer is hooked up to a conventional truck

FIGURE 12.15
Roadrailer freight system. Photo: Jon D. Fricker.

tractor and continues using the rubber tires over the road like any other trailer. The trailer is loaded at the origin and remains loaded until it reaches its destination.

12.2.3 Barges and Inland Waterway

The inland waterway system consists of 26,000 miles of navigable rivers and canals within the United States. The major waterway routes are the Mississippi and Ohio Rivers. Most of the barge freight (65 percent) travels down the Mississippi to the Port of New Orleans. The Missouri River, the Ohio River Basin, the Illinois River (which connects Chicago and Lake Michigan to the Mississippi), and other navigable channels, such as the Monongahela, Tennessee, and Cumberland Rivers are also major waterway shipping routes. The inland waterway system opens the coalfields of West Virginia to the ports on the Gulf of Mexico. A map of the waterway system is shown in Figure 12.16. It is important to note that, because the flow proceeds along natural river channels, there is considerable circuity associated with barge traffic. Compared to the great circle distance from Minneapolis to New Orleans, a commodity carried by barge between those two cities travels almost 1.6 times the great circle route.

Vessels on the Inland Waterways. There are 33,500 nonpropelled barges carrying large quantities of freight—usually bulk commodities like grain (Figure 12.17a), coal (Figure 12.29a), steel, or oil—on the major U.S. rivers. Grain and coal are carried in several scows or barges, each 195 feet long and 35 feet wide (Figure 12.17b), that are lashed together to

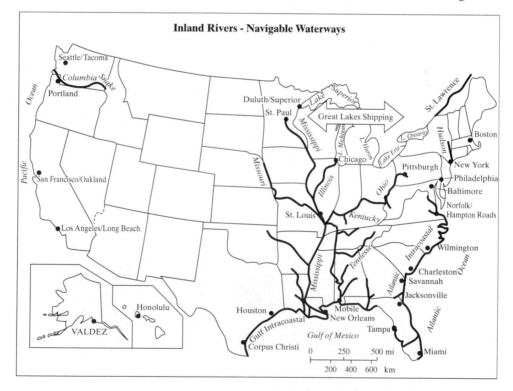

FIGURE 12.16
The U.S. domestic waterway system. *Source:* Army Corps of Engineers website.

form what is called a "tow." Tows and barges are propelled by river towboats (Figure 12.7c) that are 120 to 160 feet long, about 35 feet wide, and are powered by two to four engines of 1000 to 6000 horsepower. A towboat's pusher plates are lashed to one of the barges in a tow, so that the entire tow can be pushed along the river (American Waterways Operators, 1965). Tugboats tow barges along the Intracoastal Waterway and in open water (Figure 12.17d) along the Atlantic and Pacific coasts.

THINK ABOUT IT

Why is a towboat called by that name, when it actually *pushes* a collection of barges?

Locks and Dams.   A lock is for the purpose of stepping down (or up) from one elevation of water to the next. (See Figure 12.18.) The dams provide flood control to nearby property. In the late 19th and early 20th centuries, the Upper Mississippi River (from St. Paul, Minnesota to the river's confluence with the Missouri River) had a shallow but swift current, rock ledges, small waterfalls, uncharted shoals, and sandbars. It was the system of dams and locks that stabilized the Upper Mississippi and made navigation possible. The U.S. Army Corps of Engineers maintains the channels so they are navigable. The U.S. Coast Guard provides buoys and other safety measures used on the waterways.

FIGURE 12.17
(a) Covered grain tow. *Source*: Mike Lindgren, Riverway, Minneapolis MN. (b) Empty hopper barges or scows. *Source*: Towboat Joe, http://towboat20.tripod.com/hopperba.htm. (c) Towboat with its upright pusher plates. *Source*: Towboat Joe, http://www.dragg.net/jbrown/. (d) Tugboat tows barge in open water. *Source*: Photo courtesy of Lynden, Inc.

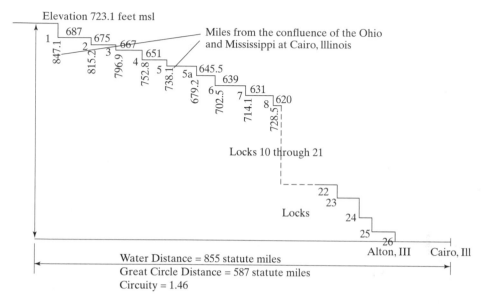

FIGURE 12.18
Elevation diagram of locks and dams on the Upper Mississippi. *Source:* U.S. Army Corps of Engineers website.

The operation of a lock is pictured in Figure 12.19. The lock has gates on each end. A vessel going downstream approaches the upstream gate at the higher water level. With the gates closed, the water is allowed to flow from the upper level into the lock. When the water in the lock reaches the upper level, the upstream gate is opened and the vessel enters. Then the gate is closed and the lower level valve is opened, allowing the water to lower. Once the water level reaches the downstream level, the bottom gate is opened and the vessel proceeds downstream.

The typical lock and dam along the Mississippi (Figure 12.18) can handle tows that are three barges wide and three barges long, so many tows have to be disassembled to proceed through the lock. (See Figure 12.20.) After all the barges have passed through the lock, the tow will be reassembled before proceeding along the river.

The Coastwise Shipping System/Intracoastal Waterway. There is a 3000-mile (4827 km) partly natural, partly artificial, waterway that provides mostly sheltered passage for commercial barges, low-draft vessels and leisure boats along the coast. As shown in Figure 12.16, it stretches along the Gulf of Mexico coast from Brownsville, Texas, on the Rio Grande River to the west coast of Florida, and along the Atlantic Seaboard from Key West, Florida, to Boston, Massachusetts. The waterway was authorized by Congress in 1919 and is, like most water routes, maintained by the Army Corps of Engineers (ACOE). The ACOE maintains a minimum depth of 12 feet (4 meters) for most of its length. Unfortunately, some portions of the shipping link have depths as shallow as 7 feet (2.1 meters). Barges on the Intracoastal waterway haul mainly petroleum and petroleum products. Other activity includes some regional movements of food, building materials, and manufactured goods.

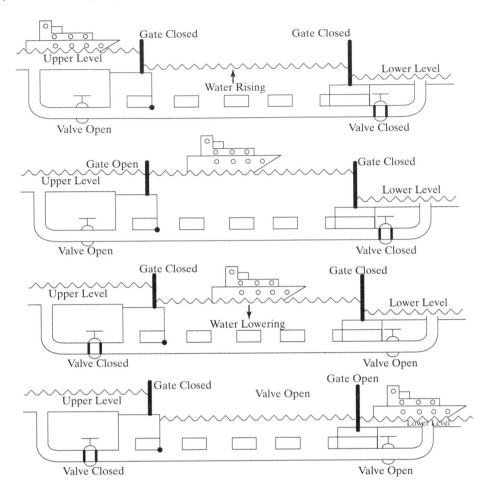

FIGURE 12.19
Operation of a lock. *Source*: U.S. Army Corps of Engineers website.

12.2.4 Shipping on the Great Lakes

Most Americans consider the United States to have three coastlines: the Atlantic, the Pacific, and the Gulf of Mexico. But when Congress designated the Great Lakes as our "Fourth Seacoast" in 1970, it recognized the Great Lakes as one of the country's important waterways. The map of Figure 12.16 shows Great Lakes Shipping and St. Lawrence Seaway in the upper portion.

The St. Lawrence Seaway has 19 locks and 6 short canals. It maintains a depth of 27 feet, and its locks can accommodate seagoing vessels of 740-foot length, 78-foot beam, and draft of 26.5 feet. It takes 8.5 sailing days to traverse the 2340 statute miles from the Atlantic Ocean to Duluth, Minnesota. Both overseas ships (called *Salties*) and Great Lakes vessels (called *Lakers*) navigate on the seaway.

FIGURE 12.20
Lock and Dam 8 on the Upper Mississippi. *Source:* The Carl Guell slide collection at the
Department of Geography, University of Wisconsin, Oshkosh.

In a typical shipping season, about 8 to 9 months long, the U.S.-Flag Lakes fleet will haul upward of 115 million tons. Each U.S.-Flag Laker is a *self-unloader*. This means that each ship is equipped so that its crew can unload the vessel quickly without the help of shoreside personnel or equipment. For example, any of the thirteen 1000-foot-long supercarriers can discharge as much as 70,000 tons of iron ore or coal in less than 10 hours. On the Great Lakes, dry-bulk commerce like the rock unloading shown in Figure 12.21 totaled 177,241,630 net tons during the 2000 navigation season. The vast majority of these cargos were carried by U.S.-Flag and Canadian-Flag Lakers. Third-Flag Salties participate almost entirely in the overseas export grain trade.

12.2.5   International Shipping and U.S. Ports

Ports and harbors are integral to the effective functioning of the U.S. transportation system and its access to world markets. In 1998 the United States imported approximately 750 billion tons of freight through its ports. The exports were only about 400 billion tons that same year. Almost two tons were imported for every ton exported. While the tonnage shipped has grown by about a factor of 3 since 1960, the imbalance between imports and exports has not changed appreciably. In economic terms, all freight modes in 2001 accounted for about $925 billion imports and $600 billion in exports. The freight mode shares 30 percent air, 30 percent land, and 40 percent water. While the land (to/from Canada and Mexico) and air modes carried about equal value of imports

FIGURE 12.21
Rock carrier unloading. Photo: Don Coles, used with permission.

and exports, the water mode's $600 billion share was split into $165B exports versus $435B imports.

THINK ABOUT IT

What is the importance to the U.S. economy that the United States has a negative balance of imports versus exports? What would it mean to the economy if that balance were to become positive? How might that be achieved?

THINK ABOUT IT

What do you think accounts for the large difference in economic value between the imports and the exports that travel through the ports?

Vessels.   At the many ports on any of the four U.S. seacoasts, a variety of vessels are used to effect the shipment of 1.2 billion tons of freight from water to land or vice versa in a year. The size and the draft of the vessels range from supertankers that are over 1500 feet long and carry 500,000 tons of petroleum to small general freight carriers with average length about 400 feet long and a draft of 20 feet carrying about 8000 tons. Containerships range in length from 400 feet to about 950 feet with a draft from 30 to 43 feet and a tonnage capability from 18,000 to 52,000 tons.

The ships that carry loose cargo in packages such as bundles, crates, barrels, and pallets, and the ports that handle such cargo, must be flexible. The equipment for loading and unloading may be part of the ship or be supplied by the port. Port-supplied loading equipment can be seen in Figure 12.22(a). One crane is in use, with three others standing

(a)

(b)

(c)

FIGURE 12.22

(a) Port of Charleston, SC, (b) Breakbulk cargo ship, and (c) Loading pipe on a general freight carrier. *Sources*: (a) Courtesy of the Port of Charleston, used by permission. (b) Port of Anchorage—State of Alaska, used with permission. (c) Port of New Orleans, used with permission.

by. Figure 12.22(b) shows one break-bulk cargo ship, which carries both general freight and containers serving the Port of Anchorage. Figure 12.22(c) shows another general dry goods freight carrier being loaded with pipe at the Port of New Orleans.

Some vessels allow trucks and railcars to roll on and roll off with their cargo. They serve as ferries for these vehicles. One example of a "RORO" vessel is the "ocean-going" barge shown in Figure 12.23. Note the railcars (and tracks) on its deck, which extend from one end of the vessel to the other. The route of this vessel forms the rail link between Seattle, Washington NY and Seward, Alaska. At the destination port, the train is reassembled to complete its journey. Other freight, including containers, can also seen being carried by this barge. Of special importance to international shipping is the time to load and unload as well as the cost of access to the port. Competition between ports to serve one market or another often centers around these factors.

Ports.   The top 10 U.S. ports and harbors are indicated in Table 12.10. Those 10 harbors handle about 50 percent of the freight. Note that 7 of the 10 are either in Texas or Louisiana. Most of these ports are coastal ports, handling overseas shipments.

Functions of a Harbor.   The function of the port is to provide safe and suitable accommodations for vessels that are involved in cargo transfer, refueling, in for repair

FIGURE 12.23
Barge carrying rail cars and container freight. Photo courtesy of Lynden, Inc.

TABLE 12.10   1999 Tonnage of Top 10 U.S. Water Ports, Ranked by Total Tons (Millions)

Rank	Location	Tons	Rank	Location	Tons
1	South Louisiana, LA	214.2	6	Beaumont, TX	69.5
2	Houston, TX	158.8	7	Baton Rouge, LA	63.7
3	New York, NY and NJ	133.7	8	Plaquemine, LA	62.4
4	New Orleans, LA	87.5	9	Long Beach, CA	60.9
5	Corpus Christi, TX	78.1	10	Valdez, AK	53.4

Source: BTS 2001, Table 1–50.

and maintenance, or simply seeking refuge from storm. The key functions of a harbor are to provide:

1. *Fast loading and unloading capability.*   One of the most important factors in the port design is the efficiency of handling unloaded goods. Port costs consume about 43 percent of all the costs of international transport from port of exit to port of entry. Thus, there is a large premium for loading and unloading vessels quickly to maximize the vessels' sailing time. Vessels often carry their own loading and unloading equipment to facilitate the process.

2. *Protection.*   This is done through breakwaters, jetties, levees, and so forth. These structures are designed to withstand the wave motion and other storms. See Figure 12.24.

3. *Waterside capacity.*   The harbor must be sized to accommodate several docking requirements.

   a. *Turning and holding areas.*   The area inside the breakwater must accommodate ship turning maneuvers and provide for anchorage when the port is at capacity.

   b. *Shoreside handling* area for containers and other cargo.

   c. *Depth.*   The Army Corps of Engineers maintains harbor depth to accommodate the ships that will be using the harbor.

   d. *Dry dockage* for ship repair.

4. *Vessel control and management.*   The movement and flow of vessels requires a moderate level of control. Ships are required to have communication on board when operating in a large harbor.

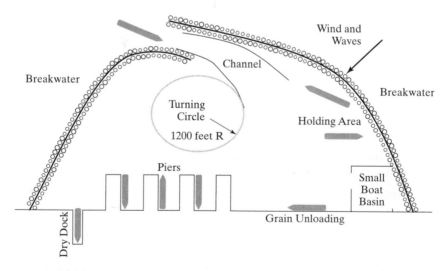

FIGURE 12.24
Simple schematic showing a large artificial harbor.

5. *Docks, berths, and space to process the goods delivered.*   Functions range from warehousing to loading land carriers.

   a. *Special purpose commodity handling.*   Frequently, a port will handle grain, coal, or other bulk commodities. The landside of the port then will need to have silos, coal repositories, oil tanks, and other equipment to support the offloading.

   b. *Wharf or pier usage.*   General breakbulk ships will require holding areas alongside their berths. Longshoremen are responsible for unloading general freight not in containers.

   c. *Transport in and out.*   Adequate space is needed for the transfer of freight from or to the particular mode of ground transport—truck or rail.

Like the airports, most ports are managed by local public or semi-public agencies. Port management is key to the success of the port, including marketing the port's capabilities. In addition to ease of water access and land access, the ability to store and to quickly handle large quantities of freight is paramount. In addition, ports are an important part of a community's economic development:

> Commercial port activities in 1996 provided employment for 1.4 million Americans. Port activity contributed $74.8 billion to the U.S. Gross Domestic Product (GDP), and personal income of $52.7 billion in 1996. Port activities in 1996 accounted for federal taxes of $14.7 billion, and state and local tax revenues amounting to $5.5 billion. (American Association of Port Authorities, 2000)

12.2.6   Oil and Gas Pipelines

One means of transport often overlooked is pipeline. The primary commodities moved over the pipeline are petroleum (both crude and refined products) and natural gas. There are *slurry pipelines*, where a solid like coal is broken up and suspended in a liquid. The result is called a *slurry*, which can be moved through a pipeline. In this section, the focus is on petroleum pipelines.

Background.   The first crude oil produced in the United States was collected in wooden barrels and transported to the refinery by boat, train, or horse-drawn wagon. The first successful pipelines were constructed in the middle 1860s and were mainly used to transport water from wells to users. Shortly after that, there was competition between pipelines and railroads for the transport of oil. By the 1930s, almost all pipeline movement in the U.S. was for crude oil.

A technological breakthrough for the pipeline came with the invention of "batching" in the 1930s. To minimize mixing of two products at the interface, the characteristics of oil being shipped are controlled to maximize the density difference at the interface. Occasionally, batches are separated by rubber spheres or other physical barriers called "pigs" that move easily within the pipe. Pigs allow shippers to deliver more than one product through a single pipeline.

There are two types of pipelines. *Gathering lines* of small diameter bring the oil from various wells and fields to a central dispatch center. *Line-haul pipelines* of sizeable diameter (e.g., 18 inches to 4 feet) move the oil long distances with the aid of strategically placed pumping stations along the line. One of the most recent domestic pipelines to be built was the Alaskan pipeline.

The Alaska Pipeline. The Alaska Pipeline was a magnificent engineering feat begun after oil was discovered on the North Slope of Alaska in 1968. It cost more than $8 billion to construct and can handle up to 2 million barrels of oil per day. The pipeline is about 800 miles long (the airline distance from New York City to Chicago)—crossing mountains, river valleys, muskeg, and the ice-infested soil of central Alaska. It was completed in 1977. Composed of 1/2-inch-thick steel, the 48-inch diameter pipe has about 4 inches of fiberglass insulation. The oil, when pumped from depths of several thousand feet, is about 160 to 180 degrees Fahrenheit. Heat exchangers cool the oil to about 120 degrees when it enters the pipeline. Even at that temperature, the pipe could not be buried in permafrost, because the heat of the oil would cause the icy soil to melt and cause the pipe to sink into the mushy soil and possibly leak. Therefore, where permafrost existed, some 400 miles of the pipeline was constructed above ground.

As can be seen in Figure 12.25, the pipeline is supported by "H" posts. There are 78,000 such posts placed about every 60 feet. The cross member of the "H" supports provide opportunity for the pipe to shift with the aid of Teflon-coated shoes as it expands and contracts with temperature fluctuations. The pattern also allows the pipeline to flex during earthquakes. A chemical liquid inside the posts cools down the base of the posts to well below freezing by taking away heat from the soil. As the liquid heats up, it changes state to become a gas that rises to the aluminum radiators, where heat exchange takes place, returning the chemical to a liquid that, by gravity, returns to the base to extract more heat. This passive control technology ensures that the ground stays frozen.

The temperature of the oil in the pipeline is nearly constant, even though the outside temperature can range from nearly 100°F to −80°F. The insulation keeps the oil warm enough to flow even on the coldest days. Powered by 10 pump stations, the oil flows at 5 to 7 mph, taking more than 5 days to traverse the 800 miles from the Arctic Ocean to

FIGURE 12.25
The Alaska pipeline. Photo courtesy of Alyeska Pipeline.

Valdez (the northernmost ice free port in the United States), where it is transferred to oil tankers. About 10 percent of all oil Americans use daily passes through the pipeline.

Oil Pipeline Regulation.   The Federal Energy Regulatory Commission (FERC), operating under congressional mandate (The Interstate Commerce Act), regulates the rates and practices of oil pipeline companies engaged in interstate transportation. It establishes "just and reasonable" rates to encourage maximum use of oil pipelines. The FERC also helps to prevent discrimination by ensuring shippers equal access to pipeline transportation, equal service conditions on a pipeline, and reasonable rates for moving petroleum and petroleum products by pipeline.

The pipeline companies act as common carriers and thus do not take possession of the oil being shipped, but rather provide the service and charge a fair rate. In 1998, there were about 180,000 statute miles of petroleum pipeline (including gathering lines) in the United States. The use of this pipeline system is about equally split between crude and refined product and is operated by about 2400 operators (BTS, 2000).

The Office of Pipeline Safety within the U.S. DOT establishes and monitors pipeline activities as they relate leakage from oil pipelines. Each carrier develops oil spill response plans for onshore oil pipelines.

12.3   COMMODITY FLOW

Every five years, the U.S. Department of Transportation and Department of Commerce jointly conduct a Commodity Flow Survey. The survey covers the major companies that ship commodities and a sample of transportation modal carriers. Origin-destination data are combined with a network model that covers truck, rail, water, and pipeline carriers.

Freight items are classified by using a five-digit Standard Classification of Transported Goods (SCTG) Code. Table 12.11 shows a few of the more prominent commodities at the two-digit level. Each successive level subdivides the commodity at a more refined level. The two-digit code is the basic code used by most analysts. Company marketing and logistics persons may be interested in third-, fourth-, or even fifth-level movements.

In Table 12.12, the length of shipments is summarized. Most shipments are below 500 miles, with almost 50 percent less than 50 miles.

TABLE 12.11   Sample of Standard Classification of Transported Goods (SCTG) at 2-Digit Level

SCTG	Title	SCTG	Title
02	Cereal grains	24	Rubber/plastic products
12	Gravel and crushed stone	26	Lumber and wood products
14	Metallic ores and concentrates	31	Non-metallic mineral products
15	Coal	35	Electronics
17	Gasoline/Other fuel	37	Transportation equipment
20	Basic chemical	40	Miscellaneous manufactured products

Source: Commodity Flow Survey Report, 1997.

TABLE 12.12   Distance All Commodities Travel

Distance	Tons (000)	Percent	Ton Miles (millions)	Percent	Cents per ton
<50 miles	225,203	46	5,445	3	1.19
50–499 miles	151,122	31	46,502	23	1.20
500–999 miles	77,389	16	84,306	42	1.27
>1000 miles	35,978	7	63,728	32	1.36

Source: Commodity Flow Survey Report, 1997.

12.3.1 The Increasing Role of Containers

From 1975 to 1986, the number of containers worldwide grew from 1.5 million TEU (20 foot equivalents) to 4.8 million TEUs. It is estimated that, in 2000, the number worldwide was about 52 million TEU. Standard ISO containers fit well on all the various modes of transportation and are easily lifted by common handling equipment from a ship, railcar, or truck. Containers can be stored several high and can be sealed with their contents unknown to discourage theft. Intermodal yards that handle the transfer of containers from rail to truck (or vice versa) have sprung up in most major cities. In many places, they have replaced rail classification yards.

The container movement began with railcars carrying truck trailers on flat cars, called *TOFC*. The trailers were moved from the railcar at a rail yard near the destination, hooked to a truck tractor, and hauled the remainder of the journey over the road. Figure 12.26 shows a TOFC movement. The advantage was that the rail movement for a long haul was more efficient than the truck and did not require a driver. The disadvantage to the railroad was the air spaces around the bottom of the container increased the aerodynamic drag, requiring considerable added power to attain the speeds necessary to be a competitive freight mode.

Gradually, railroads realized that containers on flat cars (COFC) did not have the aerodynamic drag disadvantages but that they required the added capital equipment of a full truck chassis at the terminal to go the remainder of the trip by truck. Containers going into a marine terminal by rail did not require the added chassis. The

FIGURE 12.26
A TOFC (trailer on flat car) train. *Source*: www.finnmoller.dk.

FIGURE 12.27
The evolution from TOFC service to double stack service.

saving in fuel by using double stack over a full TOFC consist is about 50 percent. See Figures 12.27 and 12.28.

12.3.2 Dry-Bulk Commodities

Because of their excellent price structure for hauling bulk freight, rail and barge share a major amount of the movement of dry-bulk commodities, mainly coal and grain. Truck is also a major player, but only for short distances, usually for delivery from a rail yard to a receiver that does not have rail access.

The movement of grain (SCTG 02)—in terms of both distance and modal movement—is shown in Tables 12.13 and 12.14. Truck carries about 46 percent of the tons of grain and earns more revenue than other carriers. Because of the

TABLE 12.13 Movement of Grain (SCTG02) by Single Modes

Mode	Value ($ million)	Percent	Tons (thousands)	Percent	Cents per ton	Ton Miles (millions)	Percent	Dollars per Ton-Mile
Truck	23,352	50	188,763	47	1.24	18,320	10	$1.275
Rail	15,826	34	142,257	35	1.11	116,247	61	$0.136
Barge	7,911	17	70,753	18	1.12	55,043	29	$0.144

Source: Commodity Flow Survey Report, 1997.

TABLE 12.14 Movement of Grain (SCTG02) by Distance

Percent	Tons (000)	Percent	Ton Miles Millions	Percent	Cents per Ton
<50 miles	225,203	46	5,445	3	1.19
50 to 499 miles	151,122	31	46,502	23	1.20
500 to 999 miles	77,389	16	84,306	42	1.27
>1000 miles	35,978	7	63,728	32	1.36

Source: Commodity Flow Survey Report, 1997.

(a)

(b)

(c)

FIGURE 12.28
Container movements: (a) Truck, (b) Rail double stack, and (c) Ship. *Sources*: (a) Photo by Jon Fricker. (b) Photo courtesy of www.finnmoller.dk. (c) Photo courtesy of Port of Charleston, SC.

distance-cost relationship, the truck mode accounts for only about 10 percent of the ton-miles.

Shown in Figure 12.29 are the two principal methods of moving coal. Railroad moves about 81 percent of the ton-miles of coal, largely because waterways do not have access to many of the nation's electric power plants. The movement of coal is critical to our economy. This fact is rediscovered whenever a railroad strike is threatened. Electric power companies maintain a limited (6- to 13-day) inventory of coal. During a railroad strike, electricity production is reduced, resulting in brownouts. In addition, many companies shut down or reduce production, and have to furlough employees.

(a)

(b)

FIGURE 12.29
Two ways to move coal: a 3 × 5 tow and a 100-car coal train. (a) A 3 × 5 tow on the Ohio River. Photo by Jon D. Fricker.
(b) A 130-car unit coal train. Photo by Bob Eisthen, used by permission.

Figure 12.30 illustrates the relative cargo capacity of three freight modes—barge, rail, and truck. A mode's capacity characteristic must be considered along with speed, reliability, cost, and other factors appropriate to the type of cargo to be transported.

12.3.3 Movement of Oil

More than 94 percent of the movement of oil is by water or pipeline. (See Table 12.15.) Crude oil for the refineries on the East Coast arrives mostly by supertanker at several deep-water ports, where it is often transferred to pipeline, truck, or rail for further

TABLE 12.15   Combined Crude and Petroleum Products—Ton-Miles (Billions)/Percent of Total

	1975	1980	1985	1990	1995	1999
Pipelines	507.0/59.9%	588.2/47.2%	564.3/47.2%	584.1/54.2%	601.1/57.5%	617.7/67.7%
Water carriers	298.0/35.2%	617.8/49.6%	590.4/49.4%	449.0/41.7%	400.9/38.4%	247.5/27.1%
Motor carriers	27.6/3.3%	26.8/2.2%	28.7/2.4%	29.7/2.8%	26.3/2.5%	29.0/3.2%
Railroads	14.1/1.7%	12.5/1.0%	12.1/1.0%	14.0/1.3%	16.6/1.6%	18.7/2.1%
Total	846.7	1245.3	1195.5	1076.8	1044.9	912.9

Source: BTS, 2001, Table 1-54.

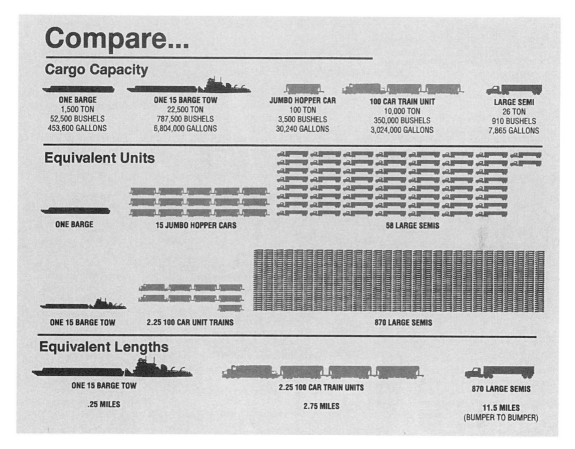

FIGURE 12.30
Comparison of bulk freight movement: Barge vs. truck vs. rail. *Source*: Iowa State Department of Transportation.

distribution within the United States. The maps of Figure 12.31 show that refined product is more evenly split between pipeline (60 percent) and water (30 percent). Most water movements of refined product are on the intercoastal waterway from Texas and Louisiana to the mid and lower Atlantic seaboard. Trucks are used more for distributing to gasoline stations and other retail outlets.

12.4   GEOMETRIC DESIGN CONSIDERATIONS FOR RAIL

Because a train rides on a fixed track, there is no freedom to compensate for side forces as an automobile, which is not constrained, can do on the road. The horizontal curvature, both superelevation and its runoff, must be very carefully designed. Banking (or superelevation) of curves is accomplished by elevating the outer rail above the inner rail by e inches, to the place where the wheels of the car or locomotive bear equally on the rails, producing no lateral force. Equilibrium occurs when the resultant of the weight of the vehicle W and the centrifugal force $(W/g)a$ is perpendicular to

Petroleum Products (Pipeline Shipments)

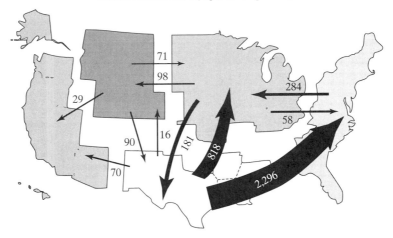

Petroleum Products (Tanker and Barge Shipments)

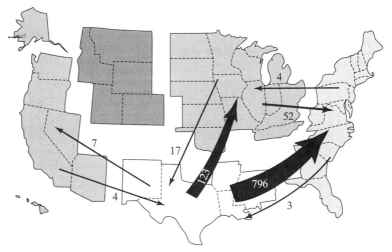

FIGURE 12.31
Interregional movement of petroleum products by pipeline (top) and water (below).
Source: Petroleum: An Energy Profile from the United States Department of Energy.

the roadway, and there is no tendency to slide or overturn. Figure 12.32 shows this relationship.

Because the outward acceleration a is a function of the vehicle speed V and the curve's radius R, the centrifugal force F is

$$F = \frac{W}{g}a = \frac{W}{g}\frac{(1.47\,V)^2}{R} = 2.15\frac{WV^2}{gR} \qquad (12.1)$$

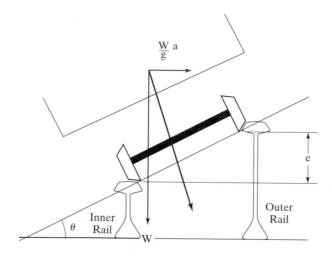

FIGURE 12.32
Forces and superelevation for rail vehicles.

where

W = vehicle weight (lbs)
g = acceleration of gravity (ft/sec^2)
V = speed of train (mph)
R = radius of the curve (ft)

The center-to-center distance between rails is called the *effective gage*, G. The typical value of G is 59 inches or 4.92 feet. In Figure 12.32, $F = W \tan \theta = W \sin \theta$ for small θ, so $F = W(e/G)$, or $e = (G/W)F$. To be consistent with the units in Equation 12.1, calculate the equilibrium elevation E, in feet, as

$$E_{\text{feet}} = \frac{G}{W} \times F = \frac{G}{W} \times 2.15 \frac{WV^2}{gR} = 2.15 \frac{GV^2}{gR} = 2.15 \frac{4.92}{32.2} \frac{V^2}{R} = 0.329 \times \frac{V^2}{R} \quad (12.2)$$

where

E = equilibrium elevation of the outer rail with respect to the inner rail

If e in inches is desired, Equation 7.8 applies.

$$e_{\text{inches}} = (12 \times 0.329) \times \frac{V^2}{R} = 3.95 \frac{V^2}{(5730/D)} = 0.0007V^2D \quad (12.3)$$

where

e = equilibrium elevation in inches
D = is the curvature of the rail

THINK ABOUT IT

In the equation above, G = 4.92 ft. In Figure 12.12, the gage is shown as 4 ft 8.5 inches. Can both be right?

Unlike the highway, where the driver has some latitude in steering, a train must travel on fixed track. The transition from curves to tangent track requires that a spiral be used. The length of the spiral in feet is usually 62 times E in inches

$$L_s = 62 \times 0.329 \times 12 \times \frac{V^2}{R} = 244.7 \frac{V^2}{R} \qquad (12.4)$$

where

V = speed of train (mph)

R = Radius of the curve (ft)

The *standard gage* is 56.5 inches or 4.708 feet. The standard gage is measured between the inside heads of the rail 5/8 inch below the top of the rail. The *effective gage* is the gage or distance between the actions of force exerted by the wheels on curves and is the standard gage + 2.5 to 3 inches. (See Figure 12.33.) Because of track wear, an effective gage of 59 inches or 4.92 feet is normally used.

The speed of the train is critical. When the speed is less than equilibrium, excessive forces exist toward the inside rail, causing a crushing force against the rail. However, when the speed exceeds the equilibrium value, there is a component of force that is closer to the outside rail, causing excessive lateral pressure and wear on that rail. One heavy rail transit property built its system with flat curves, causing an audible screeching as curves were traversed. In addition, the flanged steel wheels, intended for a life of 50,000 hours, had to be replaced after only 20,000 hours of use.

Much higher speed will cause an overturning of the train. Practice among railroad engineers is to limit the speed on curves to an unbalanced elevation of 3 inches. This speed is sometimes called the "comfortable speed." The overturning speed is between $170/\sqrt{D}$ and $185/\sqrt{D}$ mph, depending on the height of the center of gravity of the car—7 feet and 6 feet, respectively.

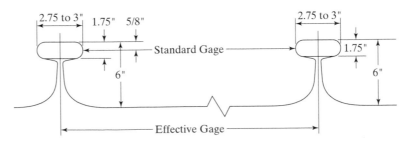

FIGURE 12.33
Relationship of track gage in the United States.

Example 12.7   Riverside Railroad

Because some railroad track follows a riverbed with oxbows, tight curves are required. The tightest curve has a radius of 700 feet.

(a) To what height must the outside rail be raised if an equilibrium speed of 40 mph is to be maintained?

(b) If the railroad's operating rules require that the elevation be no more than 5 inches, what is the equilibrium speed?

(c) What is the "comfortable speed" along the riverbed?

(d) What is the overturning speed?

(e) What is the length of the transition spiral with the 5-inch elevation if the equilibrium speed is used?

Solutions to Example 12.7

A. From Equation 12.2 with V in mph, $E = 0.329 \, (40^2/700) = 0.752$ ft or $e = 12 \times 0.752 = 9.02$ in.

B. Equation 12.2 for e in inches is

$$e_{\text{inches}} = 12 \times 0.329 \times \frac{V^2}{R} = 3.95 \frac{V^2}{R}.$$

Rearranging this equation to find V when $e = 5$ inches, $V = \sqrt{((700 \times 5)/3.95)} = 29.8$ mph.

C. At $e = 3$ inches for a comfortable speed, $V = \sqrt{\dfrac{700 \times 3}{3.95}} = 23.1$ mph.

D. The overturning speed with a center of gravity at 7 feet is

$$V_{\text{overturning}} = \frac{170}{\sqrt{D}} = \frac{170}{\sqrt{\dfrac{5730}{700}}} = \frac{170}{2.86} = 59.4 \text{ mph.}$$

With a center of gravity at 6 feet, the overturning speed is

$$V_{\text{overturning}} = \frac{185}{\sqrt{D}} = \frac{185}{2.86} = 64.7 \text{ mph.}$$

E. From Equation 12.4, $L_s = 244.7 \dfrac{29.8^2}{700} = 310.4$ ft.

Vertical curves are not a safety problem for railroads as much as they are a power problem. Railroads, because of long trains or *consists*, operate on the basis of having enough motive power to traverse the most severe vertical curve or grade on a given route. This grade is called the *ruling grade*. Sufficient motive power is needed to overcome resistive forces of about 20 pounds per ton of drawbar pull (the force applied by the locomotives on the remainder of the consist) for each percent of grade. This can be the largest resistance, even for a 1 percent grade, compared with every other resistive force the engine must overcome. Most rail lines limit any grade on the mainline high-speed tracks to 1 or 2 percent.

12.5   THE POWER NEEDED TO MOVE FREIGHT

Every freight mode requires power to move its cargo. Railroads have locomotives to pull the trains, barges have boats that push the tow, ships have internal marine engines that drive the shafts to move propellers, and pipelines have pumps. For each mode, there are resistances that have to be overcome. In this section, the power needed to move the cargo at a certain speed is determined.

12.5.1   Rail Resistance and Locomotive Power

Resistance.   For locomotives to pull a train, they must overcome resistance from four main causes:

  A. Resistance from the *weight* of the trailing train, which includes the small amount of frictional resistance from the wheels
  B. Resistance from the *aerodynamic drag* forces that occur (especially at higher velocities)
  C. Resistance from the *ruling grade* (R_{grade}),
  D. Resistance from the most severe *horizontal curve* (R_{curv}) encountered during the trip.

The effect of the first two causes of resistance are combined in Equation 12.5. (May 1982)

$$R_{tt} = 0.6 + \frac{20}{w} + 0.01V + K\frac{V^2}{wn} \tag{12.5}$$

where

  R_{tt} = resistance from wheels and air (lb/trailing ton)
  V = velocity (mph)
  K = aerodynamic coefficient
  n = number of axles per car
  w = weight of car per axle (ton)

The aerodynamic coefficient K can assume a variety of values, depending on the type of train. For example, $K = 0.07$ for most trains, 0.0935 for container trains (COFC or double stack), and 0.16 for truck trailers on flat car (TOFC). The third and fourth sources of resistance are handled separately in Equations 12.6 and 12.7. The resistance from the ruling grade increases by 20 pounds per ton for every additional one percent grade.

$$R_{grade} = 20\,G \tag{12.6}$$

where

G = Grade (percent)

As a train traverses horizontal curves, the resistance (R_{curv}) goes up in proportion to the degree of curvature.

$$R_{curv} = 0.8 \, \Delta \tag{12.7}$$

where

Δ = curvature of horizontal curve (degrees)

Note that the results of Equations 12.5 to 12.7 are resistance values with units lb/ton. The total resistance of a train having C cars, each with n axles, and an average weight per axle of w tons is

$$R_{total} = (R_{tt} + R_{grade} + R_{curv}) \text{ lb/ton} * (C * w * n) \text{ tons} = TE_{drawbar} \tag{12.8}$$

The result of using Equation 12.8 is *drawbar pull*, in pounds, needed to move the train specified at speed V.

Example 12.8   80-car Train on Level Terrain

What *tractive effort* ($TE_{drawbar}$) is required to pull a train of 80 flat cars (four axles each), each flat car carrying two 48,000 pound truck trailers (TOFC) at 70 mph on level tangent track? The average tare (unloaded) weight of a flat car is 64,000 pounds.

Solution to Example 12.8   To use Equation 12.5, first determine w to be

$$w = \frac{(2 * 48,000 \text{ lb}) + 64,000 \text{ lb}}{2000 \text{ lb/ton} * 4 \text{ axles}} = 20 \text{ tons per axle.}$$

Because the load is TOFC, K = 0.16 and

$$R_{tt} = 0.6 + \frac{20}{20} + (0.01 * 70) + \left(0.16 * \frac{70^2}{80}\right) = 1.6 \text{ lb/ton} + 0.7 \text{ lb/ton} + 9.8 \text{ lb/ton}$$

$$= 12.1 \text{ lb/ton.}$$

Now use Equation 12.8 to convert the resistance (lb/ton) of the trailing train into the *tractive effort* ($TE_{drawbar}$) needed to overcome the resistances.

$$TE_{drawbar} = 12.1 \text{ lb/ton} * 80 \text{ cars} * 80 \text{ tons/car} = 77,440 \text{ lb}$$

THINK ABOUT IT

What component caused the greatest contribution to the resistance R_{tt} value found in Example 12.8? Does it surprise you?

The drawbar pull is equal to the instantaneous *total tractive effort* minus the force required to move the weight of the locomotive(s). This implies that the locomotives must also generate power to overcome their own weight ($R_{\text{locomotive}}$). The resistance from the locomotives themselves is approximated by Equation 12.9.

$$R_{\text{locomotive}} \cong 1200N \text{ to } 1600N \tag{12.9}$$

where N = number of locomotives.

Here, the units for $R_{\text{locomotive}}$ are pounds, not lb/ton. Thus, the *total required tractive effort* TE $= R_{\text{total}} + R_{\text{locomotive}}$. Combining Equations 12.8 and 12.9,

$$\text{TE} = [(R_{\text{tt}} + R_{\text{grade}} + R_{\text{curv}}) * (C * w * n)] + R_{\text{locomotive}} \tag{12.10}$$

In Example 12.10, each locomotive used to pull the train would add 1200 to 1600 pounds of resistance to the train.

Propulsion.   The power that goes to the rails is generally delivered by series wound electric (torque) motors. The electricity for running the motors comes from either a diesel-powered generator or, where it is available, directly from an overhead power line. Locomotives provide the power and are usually 82 to 93 percent efficient.

Tractive effort is equal to the force necessary to move the train, which includes the locomotive(s) plus all of the railcars, at a given speed. Because horsepower is work per unit time and one horsepower is 550 ft-lb/sec, HP $= (F * V)/550$ with F in pounds and V given in fps. If V is given in mph,

$$\text{HP} = \frac{F * V * 1.47 \text{ fps/mph}}{550} = \frac{F * V}{375}.$$

The tractive effort (TE) in pounds of force imparted by the engine to overcome the resistance of the load it pulls is given by the formula

$$\text{TE} = \frac{375 \times \text{HP} \times e}{V} \tag{12.11}$$

where HP = rated locomotive horsepower
$\quad e$ = efficiency of locomotive (usually 0.82 to 0.93)
$\quad V$ = speed (mph)

The *available* tractive effort found in Equation 12.11 must be more than enough to overcome the resistance (or total *required* tractive effort) computed in Equation 12.10. Example 12.9 offers a simple illustration.

Example 12.9   Number of Locomotives Needed

How many 5000 horsepower locomotives at 85 percent efficiency will it take to provide the pull of 77,440 pounds from Example 12.8 if the resistance of each engine is 1200 pounds?

Solution to Example 12.9   The locomotives must pull the trailing train at 70 mph (previous example) and overcome their own resistance. On the other side of the equation is the horsepower and efficiency of the locomotives. By using Equation 12.11, we can solve for the number of engines N.

$$77{,}440 \text{ lb} + (1200 * N) = \frac{375 * 5000N * 0.85}{70}$$

$$N = 3.59 \rightarrow 4 \text{ engines}$$

Of course, not every track is straight and level. Example 12.10 introduces the influence of the ruling grade on train operations.

Example 12.10   Ruling Grade and Train Speed

Assuming the ruling grade in this series of examples is 0.75 percent, at what speed will a train with four 5000 horsepower locomotives (efficiency at 85%) be able to climb that grade? Use 1200 pounds resistance per locomotive.

Solution to Example 12.10   From Equation 12.6, the resistance due to the grade $G = 0.75$ is $R_{grade} = 20 * 0.75 = 15$ lb/ton. This R_{grade} value is added to R_{tt} expression used in Example 12.8, which must be adjusted for a speed V that is now unknown.

$$R = 0.6 + \frac{20}{20} + (0.01 * V) + \left(0.16 * \frac{V^2}{80}\right) + 15 = 16.6 + (0.01 * V) + (0.002 * V^2)$$

Recalling Equation 12.8, $R_{total} = [(R_{tt} + R_{grade}) * (C * w * n)] = TE_{drawbar}$.

$$TE_{drawbar} = (0.002 \, V^2 + 0.01 \, V + 16.6) \text{ lb/ton} * 80 \text{ cars} * 80 \text{ tons/car}$$

$$= (12.8 \, V^2 + 64 \, V + 106{,}240) \text{ lb}$$

Four 5000 HP locomotives provide power at 85 percent efficiency and speed V (mph), but they also add (1200*4) pounds of resistance. The resistance-propulsion relationship (with Equation 12.9 on the right-hand side) becomes

$$12.8 \, V^2 + 64 \, V + 106{,}240 + (1200 * 4) = \frac{375 * (5000 * 4) * 0.85}{V}$$

Rearranging terms produces $12.8 \, V^3 + 64 \, V^2 + 111{,}040 \, V = 6{,}375{,}000$. $V = 45.4$ mph.

Using the resistance-propulsion relationship, with Equation 12.8 on one side and Equation 12.9 on the other, many basic train analyses can be performed. Already, solutions for N (number of locomotives) and V (operating speed on the ruling grade) have been found. Example 12.11 shows how a train can be expected to perform on a track that is neither straight nor level.

Example 12.11   Coal Train on Hills and Curves

A 100-car unit coal train is moving from the mine to the coast through the West Virginia mountains. The most difficult section is a 1 percent grade on a curve with curvature of 3.8 degrees.

Each coal car weighs 60,000 pounds empty and carries 68,000 pounds of coal. Each loaded car weighs 64 tons. How many 6000 horsepower diesel-electric engines with 87 percent efficiency does it take to pull the train over that curve at 25 mph if each of the locomotives requires 1500 pounds of force?

Solution to Example 12.11   Applying Equations 12.3 though 12.5 with the default value of $K = 0.07$,

$$R_{tt} = 0.6 + \frac{20}{w} + 0.01V + K\frac{V^2}{wn} = 0.6 + \frac{20}{(64/4)} + 0.01V + 0.07\frac{25^2}{16*4} = 2.78 \text{ lb/ton}$$

$$R_{grade} = 20\,G = 20*1 = 20 \text{ lb/ton}$$

The resistance from the ruling grade increases by 20 pounds per ton for every additional 1 percent grade.

$$R_{curv} = 0.8\,\Delta = 0.8*3.8 = 3.04 \text{ lb/ton}$$

Converting to total drawbar pull resistance,

$$\text{TE}_{drawbar} = (2.78 + 20 + 3.04) \text{ lb/ton} * 100 \text{ cars} * 64 \text{ tons/car} = 165{,}248 \text{ lb}$$

Add the locomotive resistance to get total tractive effort required and then equate that sum with the propulsion side of the relationship:

$$165{,}248 + (1500*N) = \frac{375*6000*N*0.87}{25} = 78{,}300\,N$$

$$N = \frac{165{,}248}{78{,}300 - 1500} = 2.15 \rightarrow 3 \text{ engines}$$

12.5.2 Ship Resistance and Power

Buoyancy.   Figure 12.34 shows a block floating in water. Its submerged dimensions are $L \times B \times D$ feet. Recall that the density of freshwater is 62.4 pounds per cubic foot and, for saltwater, 64.0 pounds per cubic foot. Thus, in saltwater, the weight of the LBD block in tons is

$$\text{LBD wt (saltwater)} = L \times B \times D \times 64 \text{ lb/ft}^3 \times \frac{1 \text{ ton}}{2000 \text{ lb}} \qquad (12.12)$$

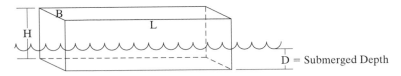

FIGURE 12.34
LBD block.

The volume of water displaced by an LBD block is

Saltwater	31.25 cubic ft/ton
Freshwater	32.05 cubic ft/ton

The hull of a ship, of course, is not a block, except in the case of barges. Ship hulls have a contoured shape for improved flow in water. Thus, each vessel has a *block coefficient* C_b, as defined in Equation 12.13.

$$C_b = \frac{\text{Actual underwater volume}}{\text{LBD block volume}} \qquad (12.13)$$

The *prismatic coefficient*, C_p, represents how well a vessel's hull moves through water. Barges look very much like a block, with a C_p of more than 0.9, but seagoing freighters have a C_p of about 0.65 and the most streamlined vessels have a C_p of about 0.5.

In the design of cargo vessels, three major sources of resistance must be considered.

1. Frictional resistance from movement through water.
2. Resistance from waves generated by the vessel.
3. Aerodynamic resistance.

Frictional Resistance. The frictional resistance from the motion of water against a smooth surface of a ship approximates the turbulent flow regime of liquid flowing through a pipe. This resistance accounts for 50 to 85 percent of the resistance for most freight-carrying vessels. Work by William Froude in the 1870s established Equation 12.14 to determine the skin friction resistance R_{sf} for ships, in pounds of force.

$$R_{sf} = f \times S_{wet} \times V^{1.825} \qquad (12.14)$$

where S_{wet} is the wetted (underwater) surface area in square feet, f is the friction coefficient defined by

$$f = 0.0106L^{-0.031} \qquad (12.15)$$

and L is the length of the waterline, in feet. The frictional coefficient varies from 0.01052 for a 20-foot vessel to 0.00852 for a 1200-foot vessel. Equation 12.15 comes from a regression fit of tabular data developed by Froude (Muckle 1981, Table 81). Most of the skin friction resistance comes from the friction between the wetted portion of the hull and still water. The calculation of the wetted surface S_{wet} is due to Taylor (Muckle 1981).

$$S_{wet} = C_s \sqrt{\Delta * L} \qquad (12.16)$$

where

> Δ = the weight of the displaced water, in tons
>
> C_s is a constant that depends on hull shape and usually varies from about 15.2 to 16.5.
>
> V = the velocity in knots

Resistance to Wavemaking Phenomena.   Ships in motion make waves. The resistance R_{WM} is called wavemaking. Marine engineers and naval architects often use a number developed by William Froude's son Robert Edmund Froude (circa 1885) as a means of making engineering judgments on ship design. R.E. Froude's work led to a simplified metric called the *speed to square root of length ratio* $V/\sqrt{L}$, where V is the velocity in knots and L is the wetted length in feet. Froude's number is

$$\text{Fr} = 0.298\,\frac{V}{\sqrt{L}}. \tag{12.17}$$

For Froude numbers greater than 0.2, wavemaking resistance is difficult to determine. As the speed of a given vessel increases, the waves produced by the bow interact with those produced in the stern. The result is a curve of resistance of the shape shown in Figure 12.35. Because hull shapes differ and small changes in Froude number can move a vessel from a state in which waves interfere with each other to a condition where they reinforce each other, it is an important effect. Because of the variability

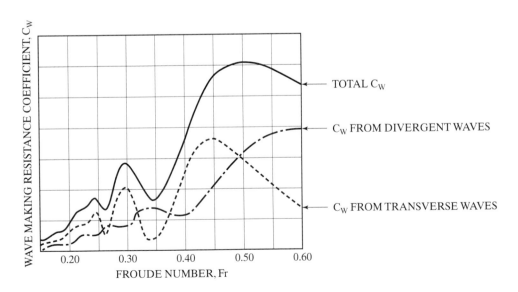

FIGURE 12.35
Contributions made by transverse and divergent wave systems to wave resistance. *Source*: Mandel, 1969, p. 4.16.

and specificity of the resistance, scale models of ships are built and tested in wave tanks. The wavemaking resistance is determined by subtracting out the skin friction resistance from the total resistance measured by the scale model after suitable scale factors are applied.

$$R_{WM} = R_{smt} - R_{sf} \tag{12.18}$$

Note that the resistance coefficient in Figure 12.35, which shows the general shape according to the Froude number, has no values shown on the vertical axis. If a vessel is being operated at a speed with a Froude number of 0.5, lowering the speed so that the Froude number is 0.35 reduces the wavemaking resistance to one-third its previous value, according to Figure 12.35. A small change in speed may make a large difference in wavemaking resistance.

THINK ABOUT IT

If a vessel is moving at a speed that has a Froude number of 0.30, what "nearby" Froude numbers offer significantly less wavemaking resistance? What percent change in speed will allow the vessel to operate with those better Froude numbers?

Example 12.12

The design for a large freighter (length 900 feet, beam 50 feet, and draft 28 feet) calls for travel at 30 knots when loaded. The current design has a block coefficient of 0.75 and $C_s = 15.6$. A scale model test, when the proper scaling is applied, indicates a total resistance of 546,445 pounds of force (lbf). From Figure 12.35, is there a small change in the Froude number that could significantly reduce the resistance from wavemaking? If so, determine that resistance.

Solution to Example 12.12   From Equation 12.15, $f = 0.0106 \, L^{-0.031} = 0.0106 * (900)^{-0.031} = 0.008585$. Equations 12.16, 12.14, 12.18, and 12.17 are used next, as follows.

$$S_{wet} = C_s \sqrt{\Delta * L} = 15.6 \sqrt{\frac{900 \times 50 \times 28 \times .75}{31.25}} \times 900 = 81,384 \text{ sq ft} \tag{12.16}$$

$$R_{sf} = f \times S_{wet} \times V^{1.825} = 0.008585 \times 81,834 \times 30^{1.825} = 346,743 \text{ lbf} \tag{12.14}$$

$$R_{WM} = R_{smt} - R_{sf} = 546,445 - 346,743 = 199,702 \text{ lbf} \tag{12.18}$$

$$Fr = 0.298 \frac{V}{\sqrt{L}} = 0.298 \times \frac{30}{\sqrt{900}} = 0.30 \tag{12.17}$$

Reducing the Froude number to 0.26 would reduce the wavemaking total resistance to about one-half its previous value. $R_{WM} = 0.5 \times 199,702 = 99,851$ lbf. By rearranging Equation 12.17, the corresponding velocity can be found.

$$V = \frac{Fr * \sqrt{L}}{0.298} = \frac{0.26 * \sqrt{900}}{0.298} = 26.2 \text{ knots}$$

The updated resistance calculations are:

$$R_{sf} = 0.008585 \times 81{,}384 \times 26.2^{1.825} = 270{,}808 \text{ lbf} \tag{12.14}$$

$$R_{total} = R_{sf} + R_{WM} = 270{,}808 + 99{,}851 = 370{,}659 \text{ lbf} \tag{12.18}$$

$$\text{Percent reduction in } R_{total} = \frac{546{,}445 - 370{,}659}{546{,}445} = 32.2 \text{ percent}$$

$$\text{Percent reduction in speed} = \frac{30 - 26.2}{30} = 12.7 \text{ percent}$$

Aerodynamic Resistance.   The third resistance is due to aerodynamic drag of the moving vessel. Because cargo vessels do not move very fast, the resistance is not very large, and is often ignored.

$$R_{aero} = \frac{\rho}{2} \times C_D \times A \times V^2 \cong 0.00408 \times A \times V^2 \tag{12.19}$$

where ρ = the density of the air (lb/cu ft)

C_D = the coefficient of drag

A = the effective projected area of the ship above the water (sq ft)

V = the speed (knots)

In the absence of specific data, use $\frac{\rho}{2} \times C_D = 0.00408$ (Muckle, 1981, p. 234).

Total Resistance.   The total resistance is the sum of the three resistive forces just covered.

$$R_{total} = R_{sf} + R_{WM} + R_{aero} \tag{12.20}$$

Example 12.13

An ocean vessel moving at 20 knots is 850 feet long, has a 135-foot beam and 40-foot draft. The vessel presents a cross-sectional area of 3000 square feet above the surface. A similar hull design (with $C_s = 15.6$) showed that the total resistance from scale model tests would be about 410,000 pounds of force at the Froude number associated with this situation. Determine the resistance to motion if the displacement is 100,000 tons.

Solution to Example 12.13   By Equation 12.17, the Froude number is $0.298 \times \dfrac{20}{\sqrt{850}} = 0.2043$.

The underwater volume is 100,000 tons $\times$ 31.25 cu ft/ton = 3.125 million cubic feet

From Equations 12.14, 12.15, and 12.16, $R_{sf} = 0.0106 \times 850^{-0.031} \times 15.6\sqrt{100{,}000 \times 850 \times 20^{1.83}}$
= 297,400 lb

Using Equation 12.18, $R_{WM} = 410{,}000 - 297{,}400 = 112{,}600$ lb

Equation 12.19 leads to $R_{aero} = \dfrac{\rho}{2} \times C_D \times A \times V^2 = 0.00408 \times A \times V^2 = 0.00408 \times 3000 \times 20^2 = 5769$ lb

From Equation 12.20, $R_{total} = 416{,}000$ pounds of force needed to be developed by the engine/propeller system.

Example 12.14

A 3 × 5 tow of coal barges consists of 15 identical barges that are each 195 feet long × 35 feet wide × 8.5 feet draft. What is the resistance of the tow moving along a river at a 6-knot speed?

Solution to Example 12.14   According to Equation 12.17, the Froude number is Fr = 0.298 * (6 knots/$\sqrt{5}$ * 195 ft) = 0.298 * 0.192 = 0.057. At a Froude number below 0.15, the wavemaking resistance is very small in comparison with the skin friction. (See Figure 12.35.) R_{WM} and the aerodynamic resistance at these velocities and low profile in the water can be reasonably ignored. (Comstock 1967, p. 312)

The tow has overall underwater dimensions 975 feet by 105 feet by 8.5 feet.

The wetted surface per tow assumes that the barges are so close, that the friction is essentially only on the outside and bottom areas. The bottom area is (35 * 195 * 15) = 102,375 square feet. The perimeter of the tow is 105 + 975 + 105 + 975 = 2160 feet. With a draft of 8.5 feet, the underwater outside area is 8.5 * 2160 = 18,360 square feet.

The wetted surface is 102,375 + 18,360 = 120,735 square feet.

Assuming a single body, the skin resistance (Equation 12.12) is

$$R_{sf} = f * S_{wet} * V^{1.825} = 0.0106 \times 975^{-0.031} \times 120,735 \times 6^{1.83}$$
$$= 0.008563 * 120,735 * 26.3 = 27,202 \text{ lbf}$$

Powering the Vessel.   Marine engines are coupled to propellers to provide the drive force needed to overcome the resistance determined above. The engine shafts are usually powered by diesel engines through a gear reduction to the propeller. Experience indicates that the overall power provided to the ship is usually between 0.6 and 0.8 of that delivered by the engines at the shaft. The equation used for railroad propulsion applies here, but with velocity in knots, not miles per hour:

$$R_{total} = \frac{550 \times HP \times e}{V \times 1.69} = 325 \times \frac{HP \times e}{V} \tag{12.21}$$

and 550 ft-lb/sec = 1 HP

e = the efficiency

V = the velocity in knots

Example 12.15

Determine the horsepower needed for Example 12.13 if the engine/propeller system is 68 percent efficient.

Solution to Example 12.15   After rearranging Equation 12.21,

$$HP = \frac{R_{total} \times V}{e \times 325} = \frac{416,000 \times 20}{0.68 \times 325} = 37,647 \text{ horsepower.}$$

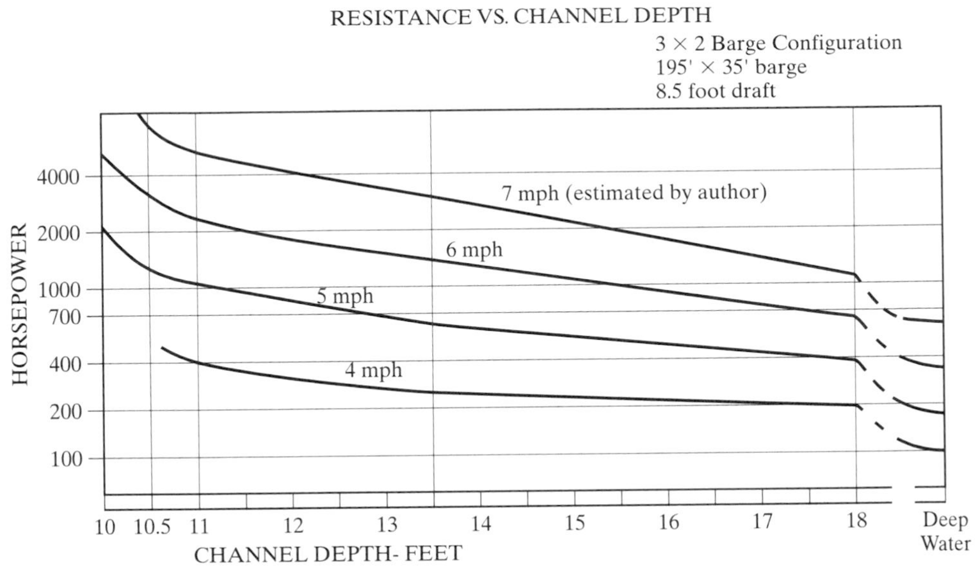

RESISTANCE VS. CHANNEL DEPTH

3 × 2 Barge Configuration
195' × 35' barge
8.5 foot draft

FIGURE 12.36
Horsepower needs to overcome resistance for shallow channels. *Source:* Plate 13, Resistance of Barge Tows, University of Michigan, 1960.

In sizing the engine, there are also other small resistances to be overcome, such as appendages, roughness of the hull, and weather. These can add as much as 40 percent to the needed horsepower (Muckle, 1981).

Example 12.16

How much horsepower must be delivered by a towboat to the tow described in Example 12.14 to achieve a 5-knot speed in a deep channel? Assume perfect application of power by the tugboat.

Solution to Example 12.16   Again, rearranging Equation 12.21 allows a calculation for HP:

$$HP = \frac{R_{total} \times V}{e \times 325} = \frac{27{,}202 * 5}{1.00 * 325} = 418 \text{ HP}$$

Barges, however, do not operate in deep water. When channels are dredged by the Army Corps of Engineers, the typical depth that is maintained is 13.5 feet. As might be expected, there is a significant increase in the action of the water between the bottom of the tow and the bottom of the channel. As the amount of water underneath the barges decreases, the resistance will increase, and more horsepower will be needed. Note that the curves of Figure 12.36 are in miles per hour (15 percent lower than knots). For instance, at 6 mph (5.2 knots), a channel of 18 feet would require about twice the horsepower that was needed in deep water (600 vs. 300). At the typical channel depth of 13.5 feet, the multiple is about 5 (1500 HP vs. 300 HP). Be sure to take into account the nonlinear vertical scale when carrying out interpolation and extrapolation in Figure 12.36.

Also note that Figure 12.36 applies to a 3 × 2 tow. For other barge configurations, an estimate of the HP needed can be based on the ratio of wetted surface of the tow being analyzed and S_{wet} for a 3 × 2 tow. Example 12.17 will demonstrate.

Example 12.17

If the tow in Example 12.16 is operating at 5 knots in a channel 12 feet deep, determine the horsepower that the tugboat needs to deliver to the tow.

Solution to Example 12.17   Five knots is equivalent to 5.75 mph. In Figure 12.36, the HP value at 12 feet is approximately 2000 for 5.75 mph. In deep water, the HP value for 5.75 mph is about 300. Apply the multiple 2000/300 = 6.7 to the 418-HP solution in Example 12.16. The tugboat would need to have 418 × 6.7, or about 2800 HP, if the tow were in a 3 × 2 barge configuration. In Example 12.14, this 3 × 5 tow was found to have S_{wet} = 120,735 ft². The perimeter of the 3 × 2 tow in Figure 12.36 is $(3*35*2) + (2*195*2) = 210 + 780 = 990$ ft. With a draft of 8.5 ft, the outside underwater surface is 8.5*990 = 8415 ft². The bottom surface is 6*35*195 = 40,950 ft², making the total wetted surface 49,365 ft². Because the ratio of the wetted surfaces is 120,735/49,365 = 2.45, the HP needed for the 3 × 5 tow is estimated to be 2.45*2800 ≈ 6850 HP.

Example 12.18

A small oceangoing freighter moves at 15 knots and is 450 feet long, with 62 foot beam and 28 foot draft. Its wetted surface area of 35,450 square feet has shown a total resistance of 69,547 lbf in scaled-up model tests. What are the wavemaking resistance and the rated horsepower if the efficiency is 0.66?

Solution to Example 12.18   Equation 12.15 provides $R_{sf} = f \times S_{wet} \times V^{1.825} = 0.0088 * 35{,}450 * 15^{1.825} = 43{,}555$ lbf.

From Equation 12.18, $R_{WM} = 69{,}547 - 43{,}555 = 25{,}992$ lbf

Equation 12.21 gives us HP $= \dfrac{R_{total} \times V}{e \times 325} = \dfrac{69{,}547 \times 15}{0.66 \times 325} = 4863$ HP.

12.5.3   Moving Oil in a Pipeline

Because pipes are not frictionless, pumps are often required along the route to overcome resistance to the flow of the material through the pipeline. As the fluid moves along the pipe, as shown in Figure 12.37, there is a pressure drop created by the internal viscosity of the fluid and the movement of the fluid along the walls of the pipe. The pipe may be smooth or (as in the case of concrete pipes) rough.

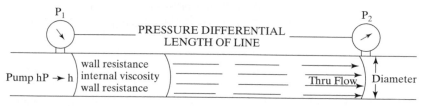

FIGURE 12.37
Pressure loss in a pipeline.

The friction in a pipe has been determined to be a function of the diameter of the pipe, the velocity of the fluid, the mass density of the fluid, and the viscosity of the fluid. The *Reynolds number*, used to determine the friction factor, is defined as

$$N = \frac{V d \gamma}{\mu g} \tag{12.22}$$

where

V = the velocity of flow (fps)
d = the pipe diameter (ft)
γ = the density of the fluid (lb/ft^3)
μ = the viscosity of the fluid (lb-sec/ft^2)
g = gravitational acceleration (32.2 ft/sec^2)

When the Reynolds number is below 2000, the flow is considered to be laminar and the resistance is due to the shear forces in the fluid. When the Reynolds number exceeds 3000, the flow is said to be turbulent. Between 2000 and 3000 is a transition zone. To be conservative, when N is between 2000 and 3000, the flow is usually considered to be turbulent. It is necessary to determine the required volume of flow Q, which has the units of cubic feet per second.

$$Q \text{ ft}^3/\text{sec} = V \text{ fps} \times A \text{ ft}^2 = \frac{\text{Vol gal/sec}}{7.48 \text{ gal/ft}^3} \tag{12.23}$$

where

V = the velocity of the fluid (fps)
A = the cross-sectional area of the pipe (sq. ft)
Vol = the volume of flow in gallons/second

and there are 7.48 gallons per cubic foot of fluid.

Example 12.19

Oil is to be transported in a smooth pipe at a temperature at which its viscosity is 0.004 lb-sec/ft^2 and its specific gravity is 0.8017. The anticipated flow rate is 100,000 barrels per day. What is the minimum size pipe needed so the oil will move at laminar flow?

Solution to Example 12.19   According to Equation 12.20, the flow Q is

$$Q = \frac{100{,}000 \text{ bbl/day} \times 42 \text{ gal/bbl}}{24*3600 \text{ sec/day} \times 7.48 \text{ gal/ft}^3} = 6.51 \text{ ft}^3/\text{sec}$$

With $A = \pi r^2 = \pi \left(\dfrac{d}{2}\right)^2$ in sq. ft,

$$V = \frac{Q}{A} = \frac{6.51}{\pi \left(\dfrac{d}{2}\right)^2} = \frac{8.289}{d^2} \text{ fps.}$$

To maintain laminar flow, the Reynolds number N must be less than or equal to 2000. Setting $N = 2000$ in Equation 12.22, we get

$$N = \frac{V \, d \gamma}{\mu g} = \frac{\dfrac{8.289}{d^2} * d * 0.8017 * 62.4}{0.004 * 32.2} = \frac{3219.4}{d} = 2000.$$

Solving for d, $d = \dfrac{3219.4}{2000} = 1.61$ ft $= 19.3$ inches. Choose the next larger standard pipe diameter (probably 20 or 24 inches) to maintain laminar flow.

The pressure loss along the pipe when the flow is laminar flow is given by

$$p = \frac{32 V l \mu}{d^2} \tag{12.24}$$

where

  $l =$ length of pipe (ft)
  $d =$ pipe diameter (in.)

Example 12.20

   Let us use a 24-inch pipe for the trunk line to carry the oil in Example 12.19. How far will it be between pumps if the pressure is to always stay above 100 psi with the maximum working pressure set at 750 psi?

Solution to Example 12.20   In Example 12.19, the velocity was determined with the pipe diameter given in feet. If the diameter d of the same pipe is given in inches,

$$V = \frac{Q}{A} = \frac{6.51}{\pi \left(\dfrac{d}{2 * 12} \right)^2} = \frac{1193}{d^2} = \frac{1193}{(24)^2} = 2.07 \text{ fps}$$

Substituting this V into the pressure loss Equation 12.24,

$$p = \frac{32 V / \mu}{d^2} = \frac{32 * 2.07 \text{ fps} * 5280 \text{ ft/mi} * 0.004 \text{ lb-sec/ft}^2}{(24 \text{ in.})^2} = 2.43 \text{ psi pressure loss per mile of pipe}$$

If the pipe is level and the pressure must stay between 100 psi and 750 psi, the maximum distance D between pumps can be given by

$$\Delta P = 750 \text{ psi} - 100 \text{ psi} = D \text{ miles} * 2.43 \text{ psi/mile}$$

$$D = \frac{650 \text{ psi}}{2.43 \text{ psi/mi}} = 267.5 \text{ miles}$$

When the Reynolds number N exceeds 3000, the flow is clearly turbulent. The flow is also usually treated as turbulent when N is in the transition zone between 2000 and 3000. The head loss is given by

$$h_L = f \frac{lV^2}{2gd} \tag{12.25}$$

where h_L = the head loss per unit length
f = the friction coefficient for turbulent flow from Figure 12.38.

Example 12.21

Heavy crude oil is to be transported over a 600-mile 20-inch trunk line at 200,000 barrels/day. The oil has a viscosity of 0.001 lb-sec/ft^2 and a density of 56 lb/ft^3 at the temperature being used. The maximum working pressure is 900 psi and should always be above 50 psi. Over the distance, the elevation decreases a total of 360 feet. How many pumps will it take to move the oil at the required flow rate?

Solution to Example 12.21   Recalling that there are 7.48 gallons in a cubic foot, the flow is

$$Q = \frac{200,000 \text{ bbl/day} \times 42 \text{ gal/bbl}}{24 \times 3600 \text{ sec/day} \times 7.48 \text{ gal/ft}^3} = 12.95 \text{ ft}^3/\text{sec}$$

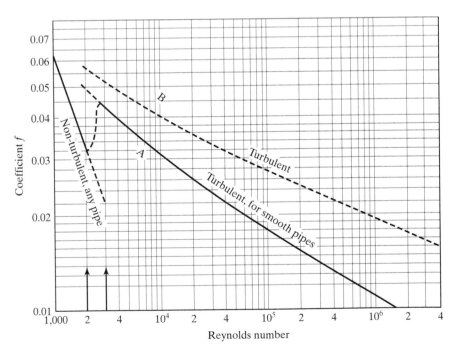

FIGURE 12.38
Friction factor f for partially turbulent flow of fluids. *Source:* Harris 1944, as cited in Haefner in 1986.

The velocity is

$$V = \frac{Q}{A} = \frac{12.95}{\frac{\pi}{4}\left(\frac{20}{12}\right)^2} = 5.94 \text{ fps.}$$

The Reynolds number is

$$N = \frac{Vd\gamma}{\mu g} = \frac{5.94 \times \frac{20}{12} \times 56}{0.001 \times 32.2} = 17{,}217.$$

From Figure 12.38, the friction factor is approximately 0.036, assuming a rough pipe. According to Equation 12.25, the head loss is

$$h_L = f\frac{lV^2}{2gd} = 0.036\frac{5280 \text{ ft/mi} \times (5.94)^2}{2 \times 32.2 \times \frac{20}{12}} = 62.48 \text{ ft/mi.}$$

The pressure loss per mile is the product of the head loss and the density of the fluid.

$$p = h_L * \gamma = 62.48 \text{ ft/mi} \times 56 \text{ lb/ft}^3 \times \frac{\text{ft}^2}{144 \text{ in}^2} = 24.3 \text{ psi/mi}$$

The number of pumps is equal to the total pressure drop (including the change due to difference in elevation) across the pipe divided by the allowable pressure differential.

$$\text{Pumps} = \frac{(600 \text{ mi} \times 24.3 \text{ psi/mi}) - \left(360 \text{ ft} \times 56 \text{ lb/ft}^3 \times \frac{\text{ft}^2}{144 \text{ in}^2}\right)}{900 \text{ psi} - 50 \text{ psi}} = 16.98 = 17 \text{ pumps}$$

Oil is forced through the pipe by a pumping pressure P, expressed in either feet of head or pounds per square inch. The total force to move the oil is the weight of the liquid in pounds of flow per second times the head in feet. Converting from foot pounds per second to horsepower and factoring in the pump efficiency, we get Equation 12.26.

$$HP = W_f \times \frac{h}{550} \times \frac{1}{\eta_{\text{pump}}} \tag{12.26}$$

where W_f = weight of fluid to flow in pounds per second: $W_f = \gamma$ (12.27)

h = the total head of the liquid in feet: $h = \frac{p}{\gamma}D$ (12.28)

η_{pump} = the efficiency of the pump.

Pumping pressure is provided initially and at various booster stations along the line of flow. Diesel engines and electric motors are the usual prime movers to provide the power to the pump. The pump itself is most likely a centrifugal pump. The quantity of discharge for a centrifugal pump varies directly with speed. Pumping pressure varies as the square of speed and horsepower with the cube of speed. The pumps are usually sized as delivering a given horsepower at a given speed. Engineering curves exist that relate the speed and head, capacity, horsepower, and efficiency. Two or more pumps may be placed in parallel to increase capacity or in series to raise the pressure.

Example 12.22

For the laminar flow case of Example 12.19, what pumping pressure in horsepower is required to go the distance of 267.5 miles and maintain the pressure in excess of 100 psi? Assume $\eta = 100$ percent.

Solution to Example 12.22   Equation 12.26 applies:

$$HP = W_f \times \frac{h}{550} \times \frac{1}{\eta_{pump}}.$$

From Equation 12.27,

$$W_f = 6.51 \text{ ft}^3/\text{sec} \times 49.9 \text{ lb/ft}^3 = 325.7 \text{ lb/sec}.$$

Using Equation 12.28,

$$h = \frac{p}{\gamma} \times D = \frac{2.43 \text{ psi/mi} \times 144 \text{ in}^2/\text{ft}^2}{49.9 \text{ lb/ft}^3} \times 267.5 \text{ mi} = 1872.1 \text{ ft}.$$

$$HP = (325.7 \text{ lb/sec} \times 1872.1 \text{ ft}) \frac{hp}{550 \text{ ft-lb/sec}} = 1109 \text{ hp}$$

Example 12.23

What will be the rating needed for each of the pumps in Example 12.21 if the efficiency is 85 percent?

Solution to Example 12.23   The head loss per mile is 62.48 feet and the pumps will be $600/17 = 35.3$ miles apart. The head loss per pump is 62.43 ft/mi $*$ 35.3 mi = 2205.5 feet. The horsepower of the pump must be (Equation 12.26)

$$HP = (12.95 \text{ ft}^3/\text{sec} \times 56 \text{ lb/ft}^3) \times 2205.5 \text{ ft} \times \frac{1 \text{ hp}}{550 \text{ ft-lb/sec}} \times \frac{1}{0.85} = 3421 \text{ hp}$$

12.6   ENERGY INTENSITY

The concept of energy intensity (EI) is an attempt to place all of transportation under the same performance measure, namely, British thermal units (BTU) per unit traveled or freight hauled. For freight, energy intensity identifies the amount of energy expended by each mode to move one ton of freight one mile.

Energy intensity is the reciprocal of energy efficiency. Army Corps of Engineers data state that the energy efficiency for water freight is 514 ton-miles on 1 gallon of fuel, rail moves 202 ton-miles on 1 gallon of fuel, and truck moves 59 ton-miles on 1 gallon of fuel. These energy efficiency numbers are very similar to the use of miles per gallon for automobiles.

Example 12.24

Convert the energy efficiencies given in the preceding paragraph to energy intensities. Assume that residual fuel for barges has a heating value of 150,000 BTU per gallon and diesel fuel for rail and trucks has 138,700 BTU per gallon.

Solution to Example 12.24

$$\text{For the water freight mode: } \frac{1 \text{ gal}}{514 \text{ ton-mi}} \times \frac{150,000 \text{ BTU}}{\text{gal}} = 291 \text{ BTU/ton-mi}$$

$$\text{For rail: } \frac{1 \text{ gal}}{202 \text{ ton-mi}} \times \frac{138,700 \text{ BTU}}{\text{gal}} = 687 \text{ BTU/ton-mi}$$

$$\text{For truck: } \frac{1 \text{ gal}}{59 \text{ ton-mi}} \times \frac{138,700 \text{ BTU}}{\text{gal}} = 2350 \text{ BTU/ton-mi}$$

From these data it would appear that the barge is the most efficient carrier of freight from an energy point of view. The problem is, of course, that the country is not crisscrossed with rivers and canals. If we are moving freight to places not served by water, we must rely on rail or trucking. If the railroad does not go where we want to go or if there is a short haul, then truck will generally move the commodity.

Circuity must be factored in. None of the modes can travel along a great circle from origin to destination. The barge generally has the longest route because it is more circuitous. Rivers meander more than roads or rail lines. If a barge (because of the meandering of the river) has to go 600 miles to deliver its freight, while the railroad goes 400 miles and a truck goes 300 miles, the barge's apparent superior energy efficiency in delivering that freight is diminished. The barge is still more efficient, but the difference is not as great. The other major determinant for barge is whether the tow is going upstream or downstream. The current is not great between one lock and dam and the next, because that is what the locks do—they allow a change in elevation. On the other hand, from Cairo, Illinois to New Orleans, there are no locks, and the barge moves at the speed of the current plus that added by the towboat. Furthermore, if the barge is moving upstream, then the towboat must overcome both the frictional resistance of the barge in the water and the flow of the water that would naturally push the tow downstream. Much less energy is required going down the Mississippi than returning against the current, especially with a load.

The numbers from the Army Corps of Engineers quoted in Example 12.23 appear to be for downstream energy efficiency. Table 12.16 gives the rail and water energy intensities based on an average number of BTUs expended across the entire system in a year divided by the ton-miles delivered in the year.

The reasons for the changes in the energy intensity over time are not all clear. Because mounting containers on truck chassis involves added weight, the carriage of containers will be less efficient than trucks and trailers built for long distance over the road. The rail carriers have been reducing their track mileage and there are fewer carriers, so their freight hauls may be consuming less energy for the ton-miles carried.

TABLE 12.16   Energy Intensity for Truck, Rail, and Water Modes

Mode	Truck	Rail	Water
1970	2241 BTU/ton-mile	691 BTU/ton-mile	545 BTU/ton-mile
1975	2245 BTU/ton-mile	687 BTU/ton-mile	549 BTU/ton-mile
1980	2293 BTU/ton-mile	597 BTU/ton-mile	358 BTU/ton-mile
1985	2964 BTU/ton-mile	497 BTU/ton-mile	446 BTU/ton-mile
1990	2642 BTU/ton-mile	420 BTU/ton-mile	387 BTU/ton-mile
1995	2978 BTU/ton-mile	372 BTU/ton-mile	374 BTU/ton-mile
2000	3200 BTU/ton-mile	352 BTU/ton-mile	508 BTU/ton-mile

Source: ORNL, 2002, and BTS, 2002.

Example 12.25   Energy Use and Circuity

Using a great circle distance of 500 miles with circuity of 1.1 for truck, 1.4 for rail, and 1.8 for barge, determine the energy used by each mode for the delivery of 1000 tons of cargo.

Solution to Example 12.25

1000 tons * 500 mi = 500,000 ton-miles. With a circuity of 1.1, truck ton-miles would actually be 550,000. With truck energy intensity of 3200 BTU/TM (see table below), 550,000 TM * 3200 BTU/TM = $1760 * 10^6$ BTU. The rail and water results are also shown in the table below.

Mode	Energy Intensity	Actual TM Moved with Circuity	Energy Used
Truck	3200 BTU/TM	550,000 TM	1760 MBTU
Rail	352 BTU/TM	700,000 TM	246.4 MBTU
Water	508 BTU/TM	900,000 TM	457.2 MBTU

To understand the real energy consumed, the unmodified energy intensity values must be applied in an operational context. A unit train is likely to have an empty backhaul, consuming energy on the return trip without earning any revenue. As a barge meanders with the river it uses as its route, any energy intensity advantage over competing modes becomes diminished.

Example 12.26   Energy Use in Operational Context

In the previous example, the truck can find a backhaul. The rail mode, because it was a unit train, will return empty. However, empty rail cars weigh about one half of their loaded weight and the energy expended on the empty return trip will be about 70% of that expended when the train is loaded. The barge will find a partial load of 700 tons to push upstream, requiring that the energy expended will be 50 percent more than was needed to push the 1000 tons downstream. In this case, what energy was expended for the round trip for each mode?

Solution to Example 12.26

Sample calculations for the water mode (barge). It will bring back 700 tons over its 1.8 * 500 = 900-mile return trip. 700 tons * 900 miles = 630,000 ton-miles. Because the trip is upstream, increase the energy consumed by 50 percent: 1.5 * 457.2 BTU = $685.8 * 10^6$ BTU.

Barge's round-trip energy used is $457.2 + 685.8 = 1143.0$ MBTU. Round-trip ton-miles $= 900,000 + 630,000 = 1,530,000$ TM. Without circuity, the ton-miles via the great circle route would have been $(1000 \text{ tons} * 500 \text{ mi}) + (700 \text{ tons} * 500 \text{ mi}) = 850,000$ TM. "Operational EI" is based on actual BTU and great circle TM: $1143.0/850,000 = 717$ BTU/TM.

Rail energy use on the return trip will be $0.7 * 246.4$ MBTU $= 172.48$ MBTU. The results of the remaining calculations are shown in the table below.

Mode	Return Load (tons)	Actual Return Distance	Actual TM on Return	Energy Intensity BTU/TM	Energy Used on Return MBTU	Energy Used on 1st Leg MBTU	Round Trip MBTU	Great Circle TM Delivered	Operational EI BTU/TM
Truck	1000	550 mi	550,000	3200	1760	1760	3520	1,000,000	3520
Rail	0	700 mi	0	352	172.5	246.4	418.9	500,000	838
Water	700	900 mi	630,000	508	685.8	457.2	1143.0	850,000	1345

SUMMARY

Moving freight is important to a society. Trucks on highways are a familiar sight, but most freight activity takes place out of the way or out of sight. Railroads have developed a separate network that they maintain for their own use. Aircraft and ships share airspace and waterways with passenger travel, but maintain separate terminal facilities.

Each freight mode has strengths and weaknesses. Depending on the nature of the cargo—its size, weight, density, value, perishability, need at the destination—one or more modes are likely to emerge as the best choice(s). Freight transportation providers (the carriers) try to design their services to suit their customers (the shippers and receivers). A railroad may offer unit train service to haul coal long distances, but barges may offer competing service, if the waterways connect the shipper and receiver. At the other extreme, trucks, airlines, or a combination of the two modes can offer overnight delivery for items of limited size and high value.

The movement of freight involves many decisions. Where should plants and distribution centers be located? Which shipping routes offer the best combination of cost and schedule reliability? How much horsepower is needed to move a shipment by rail or ship? This chapter introduces analytical techniques to support those decisions and provides an overview of how freight movement contributes to the vitality of a society.

ABBREVIATIONS AND NOTATION

A	the cross-sectional area of the pipe in square feet
ACOE	Army Corps of Engineers
A_w	wetted surface area,
bbl	barrel of oil equal to 42 gallons
BTU	British Thermal Unit
COFC	container on flat car
cwt	one-hundred weight $= 100$ pounds
DC	Distribution Centers
Δ	Curvature of horizontal Curve (degrees)
Δ	Displacement; amount of water displaced by a ship

E	equilibrium elevation of the outer rail
EI	energy intensity
FERC	Federal Energy Regulatory Commission
Fr	Froude number
G	effective gage (center to center of the rails or 4.92 feet)
GDP	Gross Domestic Product
GVW	gross vehicle weight
K	the aerodynamic coefficient
L	length of the vessel
LTL	less than truckload—freight loads of a volume and weight less than the truck can carry (LTL carriers carry a number of smaller freight packages)
N	Reynolds number
OPS	Office of Pipeline Safety
Q	flow of liquid in a pipeline, ft^3/sec
R	radius of the curve
R_{aero}	aerodynamic resistance
$R_{curvature}$	resistance from horizontal curve
R_{grade}	resistance of a grade
$R_{locomotive}$	the resistance of a locomotive
R_{sf}	the skin friction resistance of the hull moving through quiet water
R_{smt}	total resistance determined by scale model test
R_{total}	total resistance
R_{tt}	frictional resistance from the train wheels
R_{WM}	resistance from wavemaking
ROW	right of way
SCTG	Standard Classification of Transported Goods
S_{wet}	wetted surface area
TE	tractive effort
TEU	standard container measure, 20 foot Equivalent Units (e.g., a 40' $\times$ 8' $\times$ 8' container is 2 TEU's)
TOFC	trailer on freight car (often referred to as piggyback)
TT	trailing tons
V	speed of vehicle
Vol	volume of flow in gal/sec
W_f	weight of fluid to flow in lb/sec
g	gravity in ft/sec^2
d	the pipe diameter
f	skin friction factor for vessels
h	total head of the liquid in feet
h_L	head loss per unit length
v	velocity of the fluid in ft/sec
β	component indicating ratio of vessel volume to LBD block volume
γ	the density of the fluid in lb/ft^3
μ	the viscosity of the fluid in lb-sec/ft^2
η_{pump}	efficiency of the pump

GLOSSARY

Backhaul: The return trip to "home base" after a shipment has been delivered.

Ballast: Gravel used as the base below the track for rail.

Beam: Ship's breadth at its widest point.

Berth: Place where ship is anchored next to a pier or wharf.

British thermal unit: Heat to raise one pound of water one degree Fahrenheit.

Breakbulk cargo: Goods that are handled in units, packages, crates, bags, and are not containerized or handled as bulk commodities. Examples are palletized cargo, boxed cargo, large machinery, and trucks.

Breakwaters, jetties, levees, etc: Protect a harbor from storms and high seas.

Bulk cargo: Cargo of a homogeneous nature that is loaded directly into a ship's hold to be shipped in loose condition. Cargo that is shipped either dry (such as grain and ore) or liquid (such as petroleum products).

Car cycle: Time from when freight car loaded until it is reloaded for the next trip.

Circuity: Distance traveled divided by the great circle or shortest distance.

Classification yard: Place where railroad cars from one consist are reassembled into several new consists.

Coastwise: Water transportation along the eastern seaboard on protected waters and canals—Texas to New York.

Comfortable speed: Maximum speed in a curve that has an unbalanced rail elevation of 3 inches.

Consist: A group of vehicles coupled together to form a train.

Containerized cargo: Packed in vans or containers for easy handling to be transported as a unit.

Crude pipeline: A pipeline carrying crude oil.

Chemical tanker: Tanker carrying specific chemicals.

Density of water: 7.48 gal/ft^3.

Designee: Receiver of shipment.

Draft: The depth of water that a ship displaces.

Drayage: Short haul moving of goods between connecting major freight modes or between carriers.

Dry-bulk carrier: Carries grain, salt, and other bulk commodities.

Dam: Structure across a river used for flood control and to slow the stream speed.

Energy intensity: Number of BTUs needed to move one ton-mile of goods.

Equilibrium speed: Speed in a curve at which centrifugal force offsets the weight of the train.

Fifth wheel: The device on the truck that connects with the semi-trailer.

Fourth U.S. Seacoast: Expression denoting the St. Lawrence Seaway and Great Lakes.

General tanker: Carries liquid cargo such as oil or chemicals.

Gondola: An open freight railroad car.

Hopper car: Rail car that carries grain carrier and unloads from shoot below.

Jones Act: Law overseeing shipbuilding and shipping on U.S. waters.

Locks: Place where a vessel is lowered or raised to the level of the water on the other side of a dam.

Logistics: The utilization of resources and personal to accomplish a movement including the flow and storage of goods from origin to destination.

Nautical mile: 6076 feet

Overturning speed: Speed in a curve at which centrifugal force will cause the train to overturn.

Petroleum products: Products from the refining process (e.g., gasoline, diesel, kerosene, etc.).

Pier: A wharf.

Reefer: A refrigerated railroad car.

RORO (Roll-on/Roll-off):   A vessel (ferry) that allows wheeled vehicles such as trucks to drive on and drive off (ferry).

Ruling grade: The most severe vertical curve or grade on a given rail route.

Scow: A large flat-bottomed boat with square ends, used chiefly for transporting freight.

Staggers Act: 1980 law that deregulated much of rail traffic.

Tare weight: Weight of vehicle when empty.

Tow (3 × 2):   A set of barge scows lashed together to form a tow 3 across by 2 long.

Towboat: Boat used with barges—generally pushes rather than pulls.

Tugboat: All-purpose boat used on the river to move barges and to help larger ships maneuver.

Ton mile: One ton of freight moved one mile; the product of a shipment's weight and the distance it is carried.

Tractive effort: The force needed to move a train, which includes the locomotive(s) and all of the railcars, at a given speed.

Unit train: A train devoted to a single cargo, usually a bulk commodity such as coal.

Waterline length or wetted length: The length of a ship's hull at the waterline.

Wharf: A pier or place on land where cargo is placed in the process of loading or unloading.

REFERENCES

[1] American Association of Port Authorities, *U.S. Public Port Facts*, 2000, http://www.aapaports.org/industryinfo/portfact.htm, retrieved 30 May 2003.

[2] Association of American Railroads, The Impact of the Staggers Rail Act of 1980, January 2003, http://www.aar.org/GetFile.asp?File_ID = 151, retrieved 30 May 2003.

[3] American Waterways Operators, Inc., *Big Load Afloat*, Washington, DC, 1965.

[4] Bray, Larry G., Chrisman A. Dager, Ronald L. Henry, and M. Carolyn Koroa, "River Efficiency, Fuel Taxes, and Modal Shifts," *TR News*, Number 221, July–August 2002.

[5] Bruun, Per, *Port Engineering*, Third Edition, Gulf Publishing, Houston, TX, 1981.

[6] Bureau of Transportation Statistics, *National Transportation Statistics*, U.S. Department of Transportation, 2001.

[7] Comstock, John P., ed., *Principles of Naval Architecture*, Society of Naval Architects and Marine Engineers, New York, 1967.

[8] Federal Energy Regulation Commission website.

[9] Hay, W.C., *Railroad Engineering* 2nd Edition, John Wiley, New York, 1982.

[10] Hazard, J.L., *Transportation—Management, Economics, and Policy*, Cornell Maritime Press, 1977.

[11] Hennes, R.G and M.I. Eske, *Fundamentals of Transportation Engineering*, McGraw Hill, 1955.

[12] Kennedy, J., *Oil and Gas Pipeline Fundamentals*, 2nd edition, Pennwell Pub, 1993.

[13] Lake Carriers Association website.

[14] Magee, John F., *Industrial Logistics*, McGraw Hill, 1968.

[15] Mohitpour, M., H. Golshan, and A. Murray, *Pipeline Design and Construction*, ASME Press, New York, 2000.

[16] Moody, L. F., "Friction factors for pipe flow," *Transactions of the ASME*, Vol. 66, 1944.

[17] Muckle, W., *Naval Architecture for Marine Engineers*, Butterworth publishers, 3rd printing, 1981.

[18] National Research Council, *Inland Navigation System Planning: The Upper Mississippi River-Illinois Waterway*, National Academy Press; ISBN: 0309074053; 1st Edition, 2001.

[19] Quinn, A.D., *Design and Construction of Ports and Marine Structures*, 1961.

[20] Robeson, J.F., *Handbook of Logistics*, Free Press, 1994.

[21] Sampson, R.J., Ferris, M.T. and Shrock, D.L., *Domestic Transportation, Practice, Theory and Policy* 5th Edition, Houghton Mifflin Company, Boston, 1985.

[22] Taff, W.C., *Commercial Truck Transportation*, 6th Edition, Cornell Maritime Press, Centreville, MD, 1980.

[23] U.S. Army Corps of Engineers *Inland Waterway Fundamentals*.

[24] U.S. Army Corps of Engineers, *Resistance of Barge Tows, Model and Prototype Investigations, Civil Works Investigations 814 and 835*, U.S. Army Engineer Division, Ohio River, Cincinnati, OH, August 1960.

[25] U.S. Department of Energy, *Annual Energy Review*, 2001.

[26] Waterway and Port Association website.

EXERCISES FOR CHAPTER 12: MOVING FREIGHT

12.1 Plant Location.   In Example 12.3, two sources of raw materials and four major markets were considered.

(a) Suppose a new opportunity for a resource comes up with the identical level of resource needed, but at a landing site in Detroit for $1.00/mi/cwt rather than the previous transport cost from Duluth. Determine the longitude and latitude of the new plant location. How far is the new location from the location determined in the example?

(b) When the plant location decision was made in Example 12.3, only the four most important markets were considered. Now it appears that market opportunities also exist in Boston, Philadelphia, and St. Louis, all at the same level of demand and transport costs as Cleveland. What plant location would minimize the transport costs? Assume that the added resources needed to manufacture the added items for the three new markets come from the two original sources in their original proportion.

12.2 Landbridge.   It takes an average of 7 days (actually 6 to 9 days is promised) to ship containerized freight from the Far East to Long Beach Harbor. From there, the goods are put on a unit train for their final destination along the East Coast (3 to 5 added days is promised). Each container of 18 tons contains 12 tons of freight, dunnage of 20 percent, and has an unladen or tare weight of 3.6 tons. The usual train has 160 containers. The goods have a value of $4.25 per pound. The rate for a full train is $2.70/cwt and the rate for a full ship is $1.25/cwt. The holding value is 30 percent per year.

(a) What is the usual cost to ship from the Far East to the U.S. East Coast?

(b) The last ship had only a partial trainload of 80 containers. The next ship is due in 3 days (2 to 5 days). If the rail cost for the partial load is $3.50/cwt, can the shipper afford to wait for the next ship to bring the 80 containers to complete the load if the delivery times are average?

(c) How late would the ship have to be for the shipper to change his/her mind? Explain your answer.

12.3 Number of Distribution Centers.   You have a major plant in St. Louis and have been making all of your sales from that plant, mostly by taking telephone orders and hiring private parcel delivery services. This has allowed you to keep your inventory low, but late deliveries have caused some customer dissatisfaction. You want to evaluate what could happen if you develop a distribution center approach. If you decide your market is nationwide and is spread out like the population, devise an <u>approach</u> for determining the cost effectiveness of DCs. Explain how you would decide how many DCs to establish. What criteria would you use? What are the pros and cons of regional distribution centers?

12.4 Regional Distribution.   You are looking to expand your market coverage of electrical machinery and appliances (each of which weighs 250 lb) beyond the Chicago area by adding warehouses in Minneapolis, Kansas City, and Charleston, WV. The cost of rail to each of those cities $0.001 per cwt/mi. Handling costs at the warehouses you have purchased is indicated in the table below. A delivery cost in each of the markets is 5 cents per mile per item.

(a) Draw a map that indicates the locations of the proposed warehouses.

Distribution Center Location	Driving Distance from Chicago in Miles	Handling Cost (in cents) per item in DC
Minneapolis	480	30
Charleston, WV	520	20
Kansas City	520	25

(b) Indicate the approximate geographic locality where the delivery cost is the same from the factory and the new distribution centers and nearly the lowest.

(c) The centers in Kansas City and Minneapolis are fighting over a major delivery to Sioux City, IA. Which center should make the delivery? Why?

12.5 Piggyback.   Many LTL trucking companies, when they have large enough shipments to fill a trailer, will often send that trailer piggyback for part of its trip. Why do you think they would use a competing mode for some of their transportation?

12.6 Inventory and Transportation Costs.   You produce sophisticated industrial-use instruments in your plant at St. Louis. A typical instrument costs $1700.00, weighs 500 lb with packing, and occupies a special container with dimensions 8 ft × 4 ft × 6.5 ft. The markets are Chicago, Detroit, Cincinnati, and Indianapolis. The required weekly deliveries and driving times are:

Destination	Number of Instruments per Week	Driving Time (in Hours) from St. Louis
Chicago	50	7
Detroit	40	13
Cincinnati	30	8
Indianapolis	40	6

Your trucks' 48-feet trailers have inside dimensions 7.5 feet high by 8 feet wide with an empty weight of 5000 pounds. The tractor weighs 25,000 pounds and the roads have a weight limit of 80,000 lbs per truck.

(a) How many instruments fit into the trailer? Will the trailer weigh out before it cubes out?

(b) The transportation cost including loading and unloading is $150.00 + $0.50/mile from the plant. The average speed is 45 mph. What is the delivery cost per unit to each of the markets, including empty backhaul, which is 40 cents per mile?

(c) How many trucks are needed for delivery, assuming loading and unloading time is 30 minutes per instrument? Unloading can occur between 8 AM and 6 PM, and trucks can travel 8 hours per day.

12.7 Third-Party Transportation.   An alternative to the strategy analyzed in the previous problem, whereby the St. Louis-based instrument company delivers its products itself, is for the company to hire a separate firm to transport the product from factory to customer. TPL, a railroad company that also owns a trucking company based in Indianapolis, has offered to take over the delivery tasks. Their offer includes having a special monthly train move one month's supply of products from St. Louis to Indianapolis at a time, storing the instruments in their warehouse, and delivering them on a weekly basis, while insuring each instrument for $600. What are the arguments for and against accepting TPL's offer?

12.8 Trucking Widgets to Port.   Middletown Trucking Company (MTC) has as the standard vehicle in its fleet a 3-axle truck with the following characteristics: 2560 ft^3 of space for cargo; GVW limit of 51,400 lb, 16,200 lb of which is the empty vehicle weight. Mythaca Widget Works wants to hire MTC to take 85 tons of its product by truck to Shoridan's port. The product has a density of 15 lb/ft^3. Is this product more likely to "cube out" or "weigh out" when carried on trucks of the type owned by MTC? (*Note*: Widgets are small and hard to break; they can be packed in cartons of almost any size and shape.)

12.9 Rail Gage.   How did the standard U.S. railroad gage come to be 4 feet 8.5 inches? Is this gage the same in every country? Cite the sources you used to answer these questions.

12.10 High-Speed Rail Superelevation.   The high-speed (70 mph) train between Chicago and Los Angeles travels over horizontal curves of (a) $R = 600$ feet, (b) $R = 800$ feet, and (c) $R = 1000$ feet.

(a) What superelevation will be needed for each horizontal curve?
(b) If the rail level difference is limited to 6 inches in practical situations, will any of the curves above be limited in speed. If so, what will the comfortable and overturning speeds be?
(c) For each of the three horizontal curves, compute the length of spiral needed for transition to tangent track.

12.11 Graphing Rail Differential.   Graph the elevation of the outer rail versus train speed for horizontal curves with curvature of 1.5 degrees.

12.12 Mixed Freight Train.   A mixed freight train has 120 cars. Each car has an average tare weight of 70,000 pounds and a load of 60,000 lb. The train's route has a ruling grade of 0.25 percent. The desired speed for the train is 60 mph.

(a) If the train's aerodynamic coefficient is 0.09, what drawbar pull will be required?
(b) If the company has 5500-HP engines derated to 91 percent, how many will be needed? Use an engine resistance of 1500 lb per engine.
(c) If the ruling grade is 1 percent, how many 5500-HP engines will be needed to maintain a speed of 60 mph?
(d) How fast can the train, with the number of engines determined in Part B, pull the train up the 1-percent incline?

12.13 Powering a Supercontainer Ship.   A super container ship is 1200 feet long, with a 200-feet beam and an 80-feet draft. The speed desired is 35 knots. Scale tests indicate that the total water resistance is 2,850,000 lbf.

(a) Ignoring aerodynamics, how much horsepower does the engine have to deliver?
(b) Is there a better velocity for the ship, based on the curves of Figure 12.35?
(c) What will be the resistance under that velocity? Hint: The wavemaking resistance will scale according to the percentage change from Figure 12.35.
(d) If marine engines cost $10/horsepower and are paid for over three years and the ship's usual port time is 3 days and its usual distance is 5700 nautical miles, is the reduced velocity worth it?

12.14 Marine Engine Horsepower for Breakbulk Freighter. A breakbulk freighter is 423 feet long, 63 feet wide, with a 28 feet-draft and a displacement of 20,000 tons. The owners would like to increase its speed from 20 knots to 40 knots. At 20 knots, the wavemaking resistance is equal to the skin friction resistance. Aerodynamically, it presents a projected area of 2500 ft^2. What increase in horsepower is required if the efficiency through the drive train to the propeller is 0.7? Assume a C_s value of 15.8.

12.15 RORO Vessel Design. A ferry (RORO) ship is being designed. It is to be 394 feet long, 62 feet wide, with a 28-feet draft. The block coefficient is 0.64. The key design parameter is speed.

(a) What is the skin resistance at 21 knots, assuming that the wavemaking resistance was equal to the skin resistance?
(b) Should the speed be set at 19 knots, or 17.33 knots or 14.25 knots? Explain your decision.

12.16 Horsepower for 3 × 2 tow. A 3 × 2 tow (each scow is 35 × 195 feet) is carrying coal and has a draft of 8.5 feet. The desired speed in still water is 5 knots. If the water is at the depth of 13 feet, estimate the horsepower to be delivered to the tow.

12.17 Heavy Oil Pipeline. Heavy oil is to be transported in a smooth pipe at a temperature so that its viscosity is 0.006 lb-sec/ft^2 and its specific gravity is 0.88. The anticipated flow rate is 50,000 barrels per day. What is the minimum size pipe so that the oil will move at laminar flow?

12.18 Light Oil Pipeline. Very light oil is to be transported in a smooth pipe at a temperature so that its viscosity is 0.003 lb-sec/ft^2 and its specific gravity is 0.71. The anticipated flow rate is 250,000 barrels per day. What is the minimum size pipe so that the oil will move at laminar flow?

12.19 18-inch Pipeline. Let us use an 18-inch pipe for the trunk line to carry the very light oil in the previous problem. How far will it be between pumps on level ground if the pressure is to always stay above 100 psi with the maximum working pressure set at 750 psi?

12.20 Heavy Crude Pipeline. Heavy crude oil is to be transported over a 500-mile, 24-inch smooth pipe trunk line at 300,000 barrels/day. The oil has a viscosity of 0.0009 lb-sec/ft^2 and a density of 51 lb/feet^3 at the temperature being used. The maximum working pressure is 1200 psi and should always be above 150 psi. Over the distance, the elevation increases a total of 900 feet. How many pumps at 72-percent efficiency will it take to move the oil at the required flow rate?

12.21 Energy Intensity of Wheat Shipments by Rail. There is intense competition to move wheat from Minneapolis to New Orleans (great circle distance of 1051 miles) either by barge or by 100-car unit train. The jumbo hopper rail cars each carry 100 tons and each has a tare weight of 80 tons. The four locomotives weigh 250 tons each. The route has a circuity of 1.3. The BTU per car mile is 14,500.

(a) How many ton-miles does the loaded 100-car unit train plus four engines require for delivery to New Orleans?
(b) How many TM does the empty returning train consume?
(c) How many revenue TM were delivered?
(d) What percent efficiency exists, comparing delivered TM to expended TM?
(e) What is the efficiency with the great circle route distance TM delivered?
(f) How many BTU are utilized?
(g) What is the actual BTU per TM delivered with the great circle route?
(h) What is the BTU/Revenue TM?

12.22 Energy Intensity of Wheat Shipments by Barge. The barge route along the Mississippi River has a circuity of 1.7. There are no locks from Cairo, Illinois downstream to New Orleans. The river flows at about 3.5 knots. Therefore, coming upstream with a full load requires fighting against the current, consuming more energy. One major factor in the energy intensity of the competing rail mode is that rail cars weigh almost as much as the load they carry. The weight of each barge scow is determined by its displacement. The weight of grain in each 35-ft × 8.5-ft × 195-ft scow is 1500 tons. The typical scow has a prismatic coefficient of 0.9.

(a) Determine the weight of each scow loaded with 1500 tons of grain at an 8.5-foot draft.
(b) If the towboat weights 700 tons, determine the actual ton-miles going to New Orleans.
(c) Determine the actual revenue ton-miles and the great circle equivalent ton-miles.
(d) Compare the efficiency of GCRTM delivered versus TM expended.
(e) Determine the horsepower needed to pull the 3 × 5 tow at 5 knots. (Hint: Interpolate values from Figure 12.36.)
(f) Estimate the time in transit for 5-knot speed from Minneapolis to New Orleans. Assume that there are 30 lockings that average two hours each.
(g) Estimate the HP needed if the average channel depth is 13.5 foot depth.
(h) Using the conversion factor of 1 HP-sec = 0.7078 BTU, determine the anticipated expenditure of energy.
(i) Estimate the energy intensity of moving grain for this route from Minneapolis to New Orleans.

12.23 Comparing Energy Intensities by Mode.

(a) Using costs for actual ton-miles of 1.2 cents per TM by rail and 0.85 cents per TM by barge, compare the rail and barge transportation costs.
(b) Compare the actual great circle energy intensities for the rail and barge shipments in the previous two problems.
(c) Which mode appears superior for the data provided? Suggest other factors besides cost and energy intensity that could affect the choice of mode.

Toward a Sustainable Transportation System

SCENARIO

It is 7:15 AM. Because you overslept, you did not leave home at your usual 6:30 AM departure time. Because you got such a "late" start, you are stuck in traffic (see Figure 13.1) and your usual 30-minute commute to work will take more than an hour. There are already five lanes leading to downtown, and they are clogged with traffic. However, (a) finding the space for new lanes will be difficult and expensive, (b) the new lanes will probably fill up with vehicles soon after be constructed, and (c) more traffic will mean more air pollution and other environmental impacts. This urban area isn't as nice a place to live as it used to be. You would move, but your job is here. Can't something be done to make this city more livable?

CHAPTER OBJECTIVES

By the end of this chapter, the student will be able to:

1. Describe what sustainable transportation is and how it can be supported.
2. Explain the impacts of traffic congestion on a corridor or in a community.
3. Discuss transportation's energy problem and identify ways to alleviate its impacts.
4. Explain the environmental impacts of transportation activities and describe strategies being proposed to minimize those impacts.
5. Define Intelligent Transportation Systems (ITS), and explain how ITS can help make transportation activities more efficient and less harmful to the environment.

FIGURE 13.1
Traffic jam. *Source*: TTI 2002.

6. Describe some of the advances being made in transportation vehicles, and explain the extent to which these advances can contribute to a more sustainable transportation system.

13.1   SUSTAINABILITY

In Chapter 1, it was stated that a "New Transportation Engineer" is needed for the 21st century. It is toward sustainability that transportation must move, or else its users will face increasing hours spent in traffic jams, especially in our urban centers. The training for this "new" engineer must involve:

1. Standard highway design concepts, because autos and trucks will continue to account for more than 90 percent of personal travel.
2. Understanding other transportation modes, especially transit and air.
3. Conducting of thoughtful assessments (against a backdrop of planning and evaluation) of the problems faced by those who use the system today.

There are no easy answers. The growth of automotive travel in the United States has simply outpaced our ability to handle it efficiently. There is also a desire for more travel by the American public. Congestion on U.S. roadways wastes our time and strains our patience, our nation's high level of oil use has international political implications, automobiles use the energy in ways that create air pollution that can be dangerous to Americans' health, and travel costs are increasing. The key question becomes, How can we maintain for our traveling public a good quality of life in a pleasant place to live with an effective level of accessibility to desired destinations? As our communities and their economies continue to grow, that growth must be managed intelligently.

FYI: In recent years, transportation professionals have begun to make a distinction between *mobility* and *accessibility*. Mobility refers to amount of travel; accessibility means the ability to reach desired destinations. Thus, a community could have impressive values for total vehicle-miles traveled in a year but still have a large fraction of its citizens unable to reach certain neighborhoods, where jobs, recreation, or shopping opportunities may be located.

The transportation engineer of the 21st century will be expected to react to the important changes in public attitude. Our transportation system has provided unparalleled mobility for many, including access to places that were not thought possible several years ago.

The concept of *sustainable transportation* ties in with other community values, as persons become concerned about the nature of their lives and how transportation affects them. Much dialogue is necessary. A variety of creative ideas need to be analyzed and pursued. This chapter identifies several problems standing in the way of achieving sustainability. It lays some groundwork for the engineer to consider when undertaking the "thoughtful assessment" required of future transportation engineers.

13.1.1 The Pros and Cons of Transportation

In Chapter 12, we saw how the reliable, inexpensive transportation of goods gives consumers in various regions and countries a wider selection of products at reasonable prices. Not everyone would see this as a positive result, however. North Carolina's textile industry, Louisiana's sugar industry, Michigan's automobile industry, and Pennsylvania's steel industry are among those sectors that have been severely challenged by foreign competitors who can ship their products to the United States at rates that do not add significantly to their selling prices. As a result of transportation (and several other) factors, employment in these U.S. industries has suffered or has required protection by tariffs. Domestic industries, such as those mentioned above, would be much more secure if transportation were not so inexpensive. On the other hand, if freight transportation were more expensive and less reliable, domestic producers would have to pay more to get raw materials and send products to market. Prices to consumers would be appreciably higher, thereby reducing their purchasing power. Likewise, the way in which individuals travel can offer great convenience, even as negative impacts are created. The positive aspects of the U.S. transportation system include:

(+) An increasing number of vehicles per capita or per household, providing unprecedented individual mobility.

(+) Relatively inexpensive fuel prices and improved fuel efficiency, keeping the direct cost of automobile use from increasing.

At the same time as, and perhaps because of, these positive aspects, some serious problems need our attention:

(−) A significant fraction of the population that is without access to an automobile, because of age, physical limitation, or economic lack.

(−) Increased congestion and the accompanying pollution.

(−) The inefficient land use patterns that have emerged in response to the widespread use of personal vehicles.

(−) An increasing percentage of the fuel we use is imported.

(−) Public mass transportation is, except for a few large cities, a neglected travel option.

This chapter addresses the role transportation plays in the creation or intensification of some problems that affect quality of life. Ways in which transportation—and, therefore, the transportation engineer—can contribute to solutions are presented. These "ways" include careful analysis, new technologies, and a change in how travel choices are made.

Example 13.1

In the United States, there are about 100 million automobiles, each of which is driven an average of 11,800 miles per year. Thirty-one percent of that mileage is devoted to commuting (ORNL, 2002). Among the three possibilities listed below for reducing petroleum use, which one would have the greatest impact? Estimate the impact for each of the strategies as independent solutions on fuel consumption, driving time, and pollution.

1. Move from a 25-mpg automobile to a 50-mpg automobile (such as a Hybrid Electric Vehicle).
2. Continue to use the 25-mpg car, but increase the average occupancy from 1.1 to 2.5 persons per car.
3. Improve the flow on major arteries and freeways by using improved information technology, increasing average speed from 21 mph to 42 mph. This would will improve gas mileage by about 25 percent.

Solution to Example 3.1

- Alternative 1

100,000,000 cars $*$ 11,800 miles $*$ 31 percent $= 366 \times 10^9$ auto miles
Gas used today $= (366 \times 10^9)/25 = 14.6 \times 10^9$ gallons per year
Gas used at 50 mpg $= 7.3 \times 10^9$ gallons per year or
$$\frac{7.3 \times 10^9 \, \text{gal}}{42 \, \text{gal/bbl} \times 10^6} = 174 \; MBPY \; (million \; barrels \; per \; year)$$
Saving $= 174$ MBPY

- Alternative 2

Fuel use now $= 366 \times 10^9$ auto miles/25 mpg $= 14.64 \times 10^9$ gal/yr or 348 MBPY
Person miles today $= 366 \times 10^9$ auto miles $*$ 1.1 persons per car $= 402.6 \times 10^9$ person miles
402.6×10^9 person miles/2.5 persons per car $= 161 \times 10^9$ auto miles
At 25 mpg, fuel use would be 6.44×10^9 gallons or 153 MBPY
Saving $= (348$ MBPY $- 153$ MBPY$) = 195$ MBPY

- Alternative 3

Time on the road at 21 mph $= (366 \times 10^9$ auto miles$)/21$ mph $= 17.43 \times 10^9$ hours on the road, at 45 mph $= (366 \times 10^9$ auto miles$)/45$ mph $= 8.33 \times 10^9$ hours on the road. The reduced hours will reduce pollution, and the improved speed will improve gas mileage an estimated 25 percent, for a saving of $0.25 * 348$ MBPY $= 87$ MBPY

The following table summarizes the findings. After examining the table, which strategy do you recommend?

TABLE 13.1   Summary of Savings in Fuel Use, Driving Time, and Pollution for Example 13.1

Alternative	Fuel Savings	Driving Time	Effect on Environment
Fuel economy 25 to 50 mpg	174 MBPY	Little change	Significant improvement, because fuel consumption is cut in half
Vehicle occupancy 1.1 to 2.5	195 MBPY	Reduced, because of fewer vehicles on the road	Significant improvement, because of reduced fuel burned
Improved flow 21 mph to 45 mph	87 MBPY	Major reduction by 9.1 million hours	Significant improvement, because of more efficient travel speed and reduced time spent driving

All three steps would reduce fuel usage, reduce environmental impacts, and reduce driving time. The third one relies on improved technology, but the result of that technology is to improve productivity. The middle strategy has the most impact, but because it requires an attitudinal change rather than a technological change, it is the alternative least likely to be adopted, or the last one that people will accept.

13.1.2 Congestion Is Growing

The nation's longest-running study of traffic jams from 1982 to this year shows urban congestion growing in three increasingly visible ways.

1. The time penalty for peak period travelers has jumped from 16 hours per year in 1982 to 62 hours in 2000.
2. The period of time when travelers might experience congestion has increased from 4.5 hours in 1982 to 7 hours in 2000.
3. The volume of roadways where travel is congested has grown from 34 percent in 1982 to 58 percent in 2000 (TTI, 2002).

The good news is that there are solutions—more roadways and transit, ramp metering, *high-occupancy vehicle (HOV)* lanes, incentives to make trips at different times, and better incident management (clearing accident scenes more quickly, etc.)—that can reduce the magnitude of the problem. The bad news, researchers say, is that even if transportation officials do all the right things, the likely effect is that congestion will continue to grow, albeit more slowly. Even the best efficiency-boosting ideas can't entirely take the place of new roads. The researchers point out that although roadway expansion isn't a "wonder drug" to cure traffic jams, it does make a difference. "The few cities that have added roads at the same rate that they've added car travel have seen slower congestion growth. Even so, it is very difficult to sustain the funding, environmental approval and public confidence to pursue that as the only strategy" (TTI, 2002).

13.1.3 The Cost of Congestion

In many large U.S. cities, freeways during rush hour become "long parking lots." (See Figure 13.1.) The levels of congestion experienced across the United States cost a large amount of money. Not only is time lost but also fuel costs are higher. It is estimated that the overall cost of congestion in 1997 was $72 billion in added gasoline cost alone. As Table 13.2 indicates, the cost per driver differs by community. In Los Angeles, the

TABLE 13.2 Gasoline Wasted in Congestion, Six Highest Communities (Millions of Gallons)

Urban Area	1982	1986	1990	1992	1995	1996	1997
1 Los Angeles, CA	490	764	1044	1042	1066	1092	1108
2 New York, NY-Northeastern NJ	314	405	616	567	680	724	802
3 Chicago, IL-Northwestern IN	108	205	271	297	342	371	398
4 Washington, DC-MD-VA	101	164	225	248	300	326	327
5 Detroit, MI	98	131	211	241	245	266	288
6 San Francisco-Oakland, CA	149	241	288	276	286	291	280
Total fuel wasted in United States	**1991**	**3191**	**4586**	**4874**	**5786**	**6160**	**6570**
% of fuel used on U.S. highways	**1.95%**	**3.05%**	**4.19%**	**4.39%**	**4.94%**	**5.15%**	**5.43%**

Source: ORNL, 2001.

cost exceeded $1000 per driver in 1997. We wasted more than 5 percent of our annual gasoline consumption on congested highways. Table 13.2 shows that the six urban areas with the most congestion combine for almost 50 percent of the total gasoline wasted in the United States.

THINK ABOUT IT

Does 5.43 percent of the nation's fuel use seem like an excessive amount to waste sitting on a congested highway? Is "wasted fuel" a good measure of the impacts of congestion? What do the trends found by the TTI study say about the responses of Americans (and their governments) to congestion?

One must carefully assess the total transportation network when deciding to improve the system. For example, on some routes, adding lanes or somehow speeding up the flow of traffic does not do much to relieve the situation. Why? Because many drivers are already taking alternative routes to work. These drivers will shift back to the main route and its improved traffic flow, thereby increasing the demand on the improved highway and decreasing the gains in average speed.

Drivers have extremely good mobility with the extensive U.S. road system. When our automobile is parked in our garage, access to it is extremely easy—so motorists become used to driving alone. A very large fraction of the vehicles during any rush hour in most cities have only one occupant. This proclivity for being alone is, however, wasteful of one of the country's resources, namely, the unused passenger capacity in most vehicles.

Figure 13.2 shows how the costs of individual auto travel vary in terms of the size of city and to what extent travel takes place during the peak (vs. off-peak) periods. In

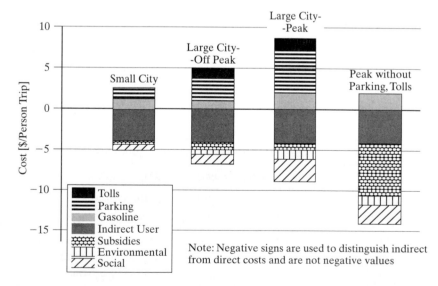

FIGURE 13.2
The total cost of automobile transportation. *Source:* Vuchic, 1999.

none of the four cases is gasoline a large fraction of the total cost of automobile own-ership. Costs other than gasoline are higher, but these are costs often not considered by the driver. The positive costs above the "0" line reflect *direct* or actual *out-of-pocket costs* that the driver is very well aware of. The costs below the "0" line reflect those items of expense that are not usually considered by the driver. However, they still rep-resent a social cost (e.g., taxes, employers, and productivity).

The indirect user costs shown among all indirect costs in Figure 13.2 do not vary as the amount of driving varies. The *indirect user costs* are borne by the vehicle owner to make use of the vehicle possible. They include insurance, depreciation, taxes, and license fees. Subsidies are the costs paid for by sources other than the car user. In cases where the operation and maintenance costs exceed any tolls and fuel tax revenues, other govern-ment revenues must be allocated as road subsidies. When an employer provides parking that is free or underpriced, that is another way driving is subsidized. Social costs of con-gestion include increased travel time and inconvenience. Environmental costs are diffi-cult to quantify exactly, but they reflect the effects of tailpipe emissions and the extra efforts to protect the environment, such as gasoline additives. There are many costs asso-ciated with auto travel. If the full cost of driving could be computed and charged to dri-vers, driving patterns might be different and transit might become more appealing.

THINK ABOUT IT

Is the passenger car the only transportation mode whose price is subsi-dized? If there are other subsidized modes, what do you think would hap-pen if their subsidies were eliminated?

13.1.4   Sustainability Defined

The notion of sustainability is to seek strategies that meet current community needs without denying necessary resources to future inhabitants. It can be thought of as a *systems* notion. To the extent that the components of a community (land use patterns, transportation, utilities, education, recreation, etc.) can be coordinated and made more efficient, the goals and desires of the present can be achieved without compromising the livability of the future.

For the United States, the concept of sustainable transportation means that transportation

- Is clean and quiet
- Is efficient
- Does not contribute to pollution and global warming
- Uses nonrenewable energy wisely (within the country's resources)
- Contributes positively to the country's economy

Stated this way, it becomes clear that the transportation engineer may need to expand the scope of even traditional analyses to develop an understanding of community de-sign and land use control. Consider what the U.S. Department of Energy says about sustainable transportation.

Sustainable Transportation

Many communities have reached a crossroads. If they build a new highway, traffic will stop backing up—at least that's the initial rationale. Citizens will stop calling to complain. Everyone presumably will be satisfied—for a while. This "solution," however, is short-lived.

When pavement is laid, more vehicles come. With more vehicles, comes more smog. Autos are a major contributor to global warming. Their pollution also causes severe health problems for many. Traffic congestion, already costing us an estimated $168 billion annually in lost productivity, is expected to triple in coming years, wasting more productivity and fuel and worsening our air quality.

Our auto habits have caused increasing dependency on oil imports, much of it coming from unstable parts of the world. In 1970, 23 percent of America's petroleum was imported. Today, we import more than 54 percent of our petroleum needs, and this number is estimated to reach more than 60 percent by 2010. The cost of oil imports to U.S. consumers totals some $50 billion annually. And in addition to the cost of oil imports, the cost of productivity loss, and the cost of congestion, we must add other social costs of transportation, such as traffic deaths and injuries, and pollution" (U.S. Department of Energy, 2002).

Some communities have found a promising new approach to handle growth and its transportation consequences. Planners hope that the new approach will produce livable or sustainable communities. These plans focus on moving people, rather than moving cars. The intent of this chapter is to help the engineer to think along the path to sustainability. The emphasis has to be on broad systems design with the meaningful involvement of all stakeholders, based on a clear statement of the values and goals of a community.

13.1.5 Some Ways to Consider Sustainability

Planning for sustainable transportation is based on the examination of several possible avenues to achieve community goals. The goals are community-specific, accounting for its needs and those areas that are expected to impact the community mobility in the best way. The Surface Transportation Policy Project (STPP) in its report "Ten Years of Progress" describes more than 100 projects that have been tried since 1990, with an indication of their successes, failures, and lessons learned. U.S. Department of Energy's Smart Growth program indicates areas where they expect planning to have the most value (STPP, 2002). Some of the experiments are described below.

Employee Incentives for Carpooling.   As we saw in Example 13.1, increased ridesharing and other multiple-occupancy vehicle travel can be a large factor in reducing congestion, fuel consumption, and pollution. Higher parking costs and tolls for highway use during peak hours could be a part of the eventual answer.

Integrated Land Use Planning.   Transportation planning for sustainability will integrate land use planning with transportation by focusing on moving people (rather than cars) on their feet, on bikes, on transit. This type of planning recognizes that many destinations are (or could be made) close enough that, given a safe facility, more people would use nonmotorized transportation instead of a motor vehicle.

Transit-Oriented Design (TOD).   TOD integrates residential and commercial land uses with accessible transit service. Persons living in a TOD area will have not only convenient shops and restaurants but also transit access to activities outside the community. The convenience of transit for others can attract customers to businesses in the area without requiring large amounts of land devoted to parking lots.

Foster Centralized Growth, Stop Urban Sprawl. One of the results of the TTI study is the notion that urban sprawl is contributing to congestion more than other factors. A number of ideas being tried are as follows: (1) zoning to keep urban growth contained, (2) impact fees to encourage developers to control development, (3) incentives offered by property tax reform, and (4) banks that offer location-efficient mortgages (LEMs). With LEMs, mortgage rates and/or borrowing levels consider both the annual cost of commuting and the house cost. Thus, money saved on transportation is applied to a higher loan amount, improving the borrower's opportunity to qualify for a higher level mortgage or lower interest rate than a borrower with a long commute by car.

Neighborhood Transportation Cooperatives. Automobile transportation is a major need in semi-isolated residential communities. Cooperative efforts often spring up between parents to carpool children to their numerous activities. This often involves transportation for school, after school activities, shopping, and so forth. Neighborhood-focused transportation cooperatives are another way that neighborhoods can act to reduce the impact of urban sprawl.

Maximize Transportation Choices and Incentives. Make communities bicycle- and pedestrian-friendly, provide reliable mass transit, and encourage telecommuting. Numerous incentives and disincentives can be tried by communities to entice motorists out of their cars. A good way to begin is to give them an attractive way to walk or bike to their destinations.

The remainder of this chapter focuses on energy use by transportation modes, the environmental situation faced by many non-attainment areas, the possible advantages of using information technology to improve transportation, and several technologies whose application to transportation conveyances might improve transportation's role in sustainability.

13.2 TRANSPORTATION AND ENERGY

13.2.1 Oil Use in the United States

The oil embargo in 1973 woke America up to the immense amount of oil that its citizens were consuming. Since then, with the exception of transportation, every sector of the economy has been able to make significant changes to use alternative fuels, find other sources of energy, and improve system productivity. (See Table 13.3.) Transportation could only find ways to become more efficient in its use of gasoline and oil.

Since 1973, considerable progress has been made toward improving individual automobile and truck fuel economy. For example, increasing the average new automobile fuel economy from about 13 miles per gallon (mpg) in 1973 to 28.8 mpg today—a value that has remained constant since 1988—is significant. Despite these significant gains, the U.S. population and its economy continues to grow, resulting in many more automobiles on the road and each one being driven more miles.

TABLE 13.3   How Sectors of the U.S. Economy Consume Oil

Economic Sector	1973 MBPD*	1973 Percent	2001 MBPD	2001 Percent
Residential	1.49	8.6	0.85	4.3
Commercial	0.75	4.3	0.36	1.8
Industrial	4.48	25.9	4.86	24.7
Electric utilities	1.54	8.9	0.37	1.9
Transportation	9.05	52.3	13.26	67.3
Total	17.31	100.0	19.7	100.0

*MBPD = millions of barrels per day.

Source: ONRL, 2002, Table 1.13.

Table 13.4 adds to the message of Table 13.3 by showing the production and consumption of oil in million barrels per day (MBPD). Dependence on foreign oil has been estimated to cost the United States in the vicinity of $7 trillion (present value 1998 dollars) over the last 30 years, which is about as large as the sum total of payment on the national debt over the same period (Greene and Tishchishyna, 2001). The numbers of Table 13.4 might be a bit deceiving in that the level of U.S. production is down some so that our imports are up at the same time our transportation use is up.

The natural units of liquid petroleum (barrels of oil, each equal to 42 gallons) are converted to the heating value in terms of *British thermal units (BTU)* per gallon. Through the use of BTUs, we are able to compare modes and transportation consumption of oil energy by using the heating values from different parts of the barrel of oil as shown in Table 13.5. Because some vehicles use automotive gasoline achieving 125,000 BTU per gallon while others use diesel fuel at 138,700 BTU per gallon and so forth, heating values become more realistic than the actual number of gallons consumed.

The same units also allow the comparison of oil with other forms of energy, such as coal and electricity. Of particular note in Figure 13.3 is the increase in petroleum use by transportation from 17.83 quadrillion BTU (or simply "quads") to 25.84 quads in the 27 years since the embargo; an increase of 45 percent. Note that the use of petroleum has been growing more rapidly since 1985 than before.

TABLE 13.4   U.S. Production, Consumption, and Transportation's Role in Oil

Year	Domestic Production (MBPD)	Net Imports (MBPD)	US Petroleum Consumption* (MBPD)	Transportation Consumption (MBPD)	Imports as Percent of U.S. Petroleum Consumption	Transportation Petroleum Use Percent of Domestic Production
1960	7.99	6.03	17.31	5.15	16.5	64.5
1970	11.30	3.16	14.70	7.78	21.5	68.9
1980	10.24	6.38	17.10	9.57	37.3	93.5
1990	8.99	7.16	16.99	10.97	42.2	122.0
2001	8.05	10.90	19.65	13.26	55.5	164.8

Note: Domestic Production and Net Imports do not equal Consumption due to natural gas plant liquids, refinery processing gains, and stock changes.

Source: ORNL, 2002, Table 1.12.

TABLE 13.5   Approximate Gross Heat Content for Various Fuels

Fuel Type	Heat Content/Gallon
Automotive gasoline	125,000 BTU/gal
Diesel motor fuel	138,700 BTU/gal
Gasohol	120,900 BTU/gal
Aviation gasoline (high octane)	120,200 BTU/gal
Jet fuel (naptha)	127,500 BTU/gal
Jet fuel (kerosene + B39)	135,000 BTU/gal
Asphalt and road oil	158,000 BTU/gal
Residual fuel oil	149,700 BTU/gal

Source: ORNL, 2002, Table B-1.

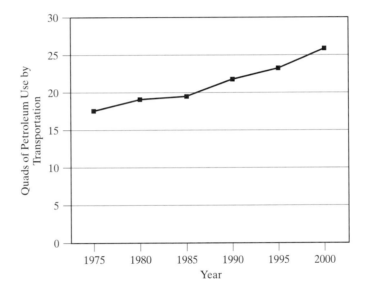

FIGURE 13.3
Plot of transportation use of petroleum since 1975. *Source*: ORNL, 2001.

13.2.2   Energy Use by Mode

The way transportation uses petroleum by mode is shown in Table 13.6. Examination of the data identify where our priorities should be focused, if we are going to participate in reducing the transportation dependence on petroleum. Table 13.6 shows clearly that it is automobiles and light duty trucks [including vans and sports utility vehicles (SUVs)] that account for most oil use among transportation modes. Heavy trucks are next, but they represent one of the strengths of the U.S. economy.

Even if we were able to reduce the use of all the non-highway modes (those that appear lower on the chart) by 50 percent, the transportation share of U.S. oil use would only decline from 68 to 64 percent. Thus, the non-highway modes do not represent areas for considerable reduction in oil use. It is the highway modes where the greatest potential for improved energy use is possible.

TABLE 13.6   Energy Use by Mode/Oil Use by Mode in 2000

Mode	Quads BTU	MBPD	Share of Oil (%)
Automobiles	9.126	4.599	34.15
Light trucks	6.617	3.335	24.76
Heavy trucks	4.563	2.300	17.08
Buses	0.208	0.143	1.06
Air	2.546	1.281	9.51
Water	1.300	0.654	4.86
Pipeline	1.009	0.508	3.78
Rail	0.607	0.305	2.26
Off-highway	0.680	0.342	2.54

Source: ONRL, 2001.

13.2.3   Highway Modes

Before looking at automobiles and light-duty trucks, a look at buses and heavy-duty trucks is in order. Increased transit ridership has been suggested as one means to help increase the level of sustainability. There are almost 75,000 transit buses in the United States, with annual oil consumption of about 0.1 quads. The buses carry 21,205 million passenger miles while traveling 2275 million vehicle miles. The average bus travels 30,662 miles per year and carries 9.32 persons per mile. Thus, the buses provide 0.212 passenger miles/1000 BTU. A single-occupancy automobile getting 22 miles per gallon in the city would transport 0.176 passenger miles per 1000 BTU.

Example 13.2

If the ridership on each bus were to increase to an average of 30 passengers with no increase in bus fuel economy and the number of buses in service were to double, how many single-occupancy automobile miles would be replaced? What reduction in transportation's BTUs would occur?

Solution to Example 13.1   The enlarged bus fleet would deliver 30 passengers/bus $\times$ 150,000 buses $\times$ 30,662 miles = 138,000,000 passenger miles for 0.2 quads of oil. The number of passengers moved would be 4,500,000 compared to 699,000, representing an increase of persons moved of 3,800,000.

The single driver automotive equivalent is 30,662 $\times$ 3,800,000 = 116,526,000 passenger miles. At 0.2 passenger miles per 1000 BTU, the single-occupancy autos would consume 0.5826 quads. Thus, the saving would be 0.5826 quads. Applying that saving to automobile use would reduce it from 9.082 quads to 8.449 quads, or 7.0 percent.

Example 13.2 indicates that even an extremely aggressive project to increase bus use will have benefits, but it would not make a very significant impact on the automobile mode.

Trucks offer a different situation. In 1997, there were about 72 million trucks in the United States. They are typically classified into three weight classes: light duty [less than 10,000 pound *Gross Vehicle Weight (GVW)*], medium duty (between 10,000 and 26,000 pounds GVW), and heavy duty (in excess of 26,000 pounds GVW). Table 13.7 indicates

TABLE 13.7   Percentage of Trucks by Size Ranked by Major Use in 2000

	Light Duty <10,000 lb (%)	Medium Duty 10,001 to 26,000 lb (%)	Heavy Trucks >26,000 lb (%)
Personal Use	74.6	7.0	0.3
Construction	7.6	20.2	17.6
Services	5.6	11.6	7.4
Agriculture	3.8	19.5	14.0
Local Delivery	4.1	16.6	13.5
For Hire	0.2	5.5	31.5
Other	4.2	19.6	15.7

Source: ORNL, 2002, Table 8.7.

the use to which the trucks are put. The uses of heavy-duty trucks in mainly for-hire, agriculture, and construction are important to the economy, and the uses of medium trucks in the local delivery function for wholesale and retail are also important.

Autos and Light-Duty Trucks.   The significant personal use of light-duty trucks (including SUV and vans) of 75 percent constitutes a problem. Table 13.8 indicates, from an operations view, the number of trucks and the fuel consumed in 1997. Light-duty trucks consume more than 70 percent of the gasoline and should be the focus of increased fuel economy and alternative fuel opportunities. Figure 13.4 shows that both automobiles and light-duty trucks are meeting their congressionally mandated CAFE (Corporate Average Fuel Economy) standards of 27.5 mpg and 20.7 mpg, respectively. Figure 13.4 also indicates that there has been little improvement since the mid-1980s.

Table 13.9 shows the fuel economy, the vehicle miles traveled, and the gasoline consumed by vehicle type. The change in vehicle-miles driven and fuel consumed that are attributable to light truck/van/SUV is significant. (See also Figure 13.5.) Although technology can improve the fuel efficiency of an individual vehicle, the type of vehicle that is driven and the amount that vehicle is driven can overwhelm the technological advances.

Car Buying.   The car companies compete in the market by producing what the customer wants. Table 13.10 shows how SUV sales have experienced an annual growth

TABLE 13.8   Use by Trucks of Different Weight Category

Operation	Light Duty <6000 lb	6000 to 10,000 lb	Medium 10,001 to 26,000 lb	Heavy Trucks >26,000 lb
Number of trucks in 1997	45.240 M	22.373 M	2.768 M	2.419 M
VMT per truck (miles per year)	13,328	12,952	13,675	47,020
Fleet fuel use in million gallons per year	35,184	21,226	4,410	18,523

Source: ORNL, 2001, Table 8.7.

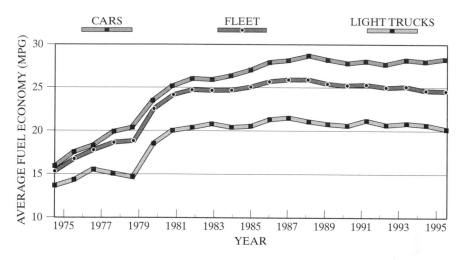

FIGURE 13.4
Average gains in fuel economy since 1975. *Source*: Pacific Institute, 1999.

TABLE 13.9   The History of Travel and Gas Consumed by Vehicle Type

	Automobiles			Light Trucks/Vans/SUVs			All	
Year	Fuel Economy (MPG)	Travel (% VMT)	Gas Consumed (MBPD)	Fuel Economy (MPG)	Travel (% VMT)	Gas Consumed (MBPD)	Gas Consumed (% Total)	Gas Consumed (MBPD)
1980	24.3	72.8	2.985	18.5	19.0	1.023	25.5	4.008
1985	27.6	70.2	2.945	20.7	22.0	1.230	29.5	4.175
1990	27.9	65.7	3.294	20.8	26.8	1.802	35.4	5.096
1995	28.6	59.4	3.282	20.5	32.6	2.513	43.4	5.795
2000	28.5	58.3	3.591	21.2	33.5	2.774	43.4	6.395

Source: ORNL, 2001.

TABLE 13.10   Annual Sales of Various Types of Light-Duty Highway Vehicles in 1980, 1990, and 2000

Type of Vehicle	1980	1990	2000	2000 Sales (%)	Fleet
Small pickup	516,412	1,135,727	1,071,730	6.2	
Large pickup	1,115,248	1,116,490	1,968,710	11.4	
Vans: small and large	341,714	1,331,570	1,640,890	9.5	
SUVs: small, medium, and large	243,163	930,838	3,625,623	21.0	
Light trucks/vans/SUVs	2,216,537	4,514,625	8,306,953	48.1	85,879,000
Subcompact, compact auto	4,685,213	5,433,870	4,326,667	25.0	
Mid size auto	3,073,103	2,211,503	3,352,198	19.4	
Large size auto	1,336,190	1,279,092	1,297,237	7.5	
Automobiles	9,094,506	9,224,465	8,978,102	51.9	127,721,000
Total light-duty vehicles	11,311,043	13,739,090	17,285,055		213,300,000

Source: ORNL 2001.

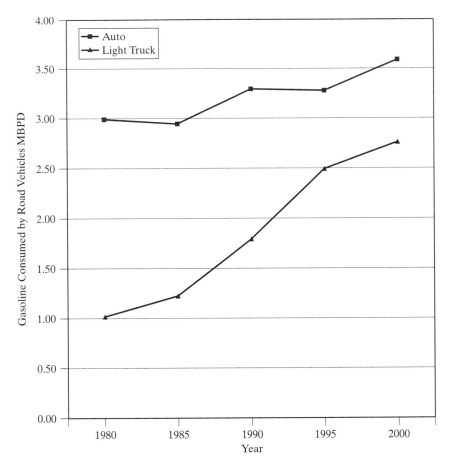

FIGURE 13.5
Gasoline use in the United States by automobiles vs. light-duty trucks, vans, and SUVs. *Source*:
Data from ONRL, 2002.

rate of almost 7 percent. Figure 13.6 illustrates the rapid growth in market share by
light-duty trucks, vans, and SUVs.

THINK ABOUT IT
Why are SUVs so popular? Is this a case in which the private market and
public policy clash? If so, how can the two positions be reconciled?

Auto Use Patterns. The last element of an automotive-oriented energy-saving
scenario would be to increase the productivity of the millions of autos already in the
fleet. A 1990 household survey determined average vehicle occupancy as a function of

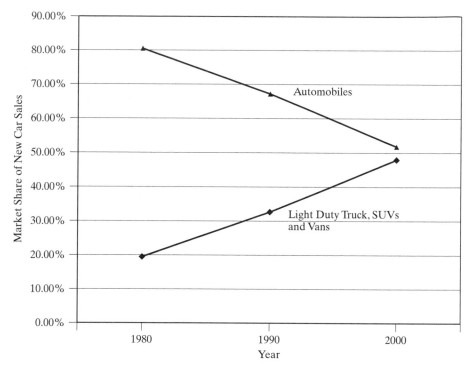

FIGURE 13.6
Market share of new sales of auto and light-duty trucks.

trip purpose. Table 13.11 shows that only 1 of 10 vehicles has two occupants or more; in the rest, the driver is the only occupant. The data from 1985 to 1999 shown in the bar graph of Figure 13.7 indicate that carpooling, which is a significant way to save fuel, is actually decreasing.

Unless we begin to ride with others, the nation is bound to continue to move in the direction of using more fuel and spending more income on transportation, especially as the costs of congestion also increase. Even if we can use new technology and alternative fuel sources to reduce oil consumption, relief from congestion will depend on the improved use of personal vehicles.

TABLE 13.11 Average Auto Occupancy by Trip Purpose		
Trip Purpose	1977	1995*
Home to work	1.3	1.1
Home to shopping	2.1	1.7
Other family/personal business	2.0	1.8
Social/recreation	2.4	2.0
All trips	1.9	1.6

Source: ONRL, 2002, Figure 11.2.

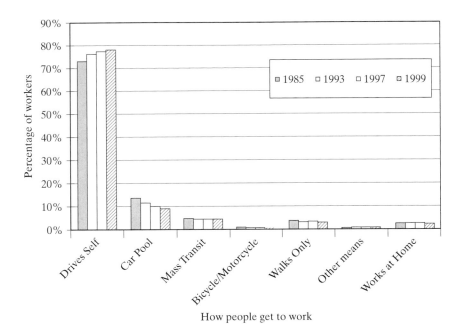

How people get to work

FIGURE 13.7
How people get to work 1985 to 1999. *Source:* Transportation Statistics Annual Report, 2000.

THINK ABOUT IT

If ridesharing is such an effective solution to congestion, why do so few people form or join a carpool?

Cost of Fuel. Americans pay far less for retail gasoline than do drivers in other countries. (See Figure 13.8.) The countries shown pay about the same for petroleum on the world market. A much higher tax is levied by most other countries than by federal and state governments in the United States. Some of the tax proceeds in other countries are used for improvement of rail and trolley service. The higher price means that drivers in these countries drive smaller, more fuel-efficient cars for economic reasons. Figure 13.8 shows that the price of fuel in real (inflation-adjusted) terms is *less* than it was under federal price control prior to 1978.

13.2.4   Energy Intensity

The concept of *energy intensity (EI)* is an attempt to place all of transportation under the same performance measure, namely, BTU per passenger mile. It identifies the amount of energy expended by each mode to move one passenger 1 mile. The calculations of EI shown in Table 13.12 reflect countrywide averages and do not account for changes in vehicle occupancy. The passenger car occupancy is assumed to be 1.6 persons per vehicle. From these figures, automobiles with an average occupancy of 1.6

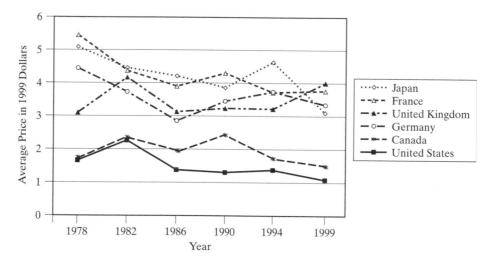

FIGURE 13.8
Gasoline prices for selected countries, 1978 to 1999, in constant 1999 U.S. dollars. *Source*: DOE International Energy Annual, 1999.

TABLE 13.12   Energy Intensity of Passenger Modes (BTU per Passenger-Mile)

	1975	1980	1985	1990	1995	1998
Air certificated air carrier						
Domestic operations	7,746	5,727	5,047	4,932	4,382	4,123
International operations	8,465	8,866	5,103	4,546	4,173	4,278
Highway						
Passenger car	4,743	4,348	4,267	3,812	3,721	3,672
Other 2-axle 4-tire vehicle	6,571	5,709	4,971	4,451	4,539	4,591
Motorcycle	2,354	2,125	1,896	1,990	2,227	2,330
Transit motor bus	N/A	2,742	3,396	3,723	4,155	3,729
Amtrak	2,383	2,164	2,094	2,064	1,838	2,138

Notes: To calculate total BTU, multiply fuel consumed by 120,900 BTU/gallon for air carrier, by 125,000 BTU/gallon for passenger car, other 2-axle 4-tire vehicle, and motorcycle, and by 138,700 BTU/gallon for transit motorbus and Amtrak. See Table 13.5. *Source*: ORNL, 2001.

show an EI slightly better than the transit bus. If EI is based on 1.1 persons per car, then the EI for automobiles increases to 5340 BTU/passenger-mile—poorer in terms of fuel use than the transit bus.

THINK ABOUT IT

Why do you think that transportation engineers have difficulty in finding the right number for average vehicle occupancy? If you had to estimate average vehicle occupancy for your area, what data would you collect? How would you collect it?

Example 13.3

A bus achieves 7 mpg of diesel fuel. How many passengers on the average will the bus have to carry to achieve 3729 BTU per passenger-mile?

Solution to Example 13.3   Because the fuel for the bus achieves 138,700 BTU/gal

$$\frac{138,700 \text{ BTU}}{\text{gal}} \times \frac{1 \text{ gal}}{7 \text{ mi}} \times \frac{\text{passenger-mile}}{3729 \text{ BTU}} = 5.31 \text{ passengers}$$

Example 13.4

Assuming that the average bus has space for 30 persons and that 30 persons were carried, what would the energy intensity of the bus be?

Solution to Example 13.4

$$\frac{\text{X BTU}}{\text{passenger-mile}} = \frac{138,700 \text{ BTU}}{\text{gal}} \times \frac{1 \text{ gal}}{7 \text{ mi}} \times \frac{1 \text{ bus}}{30 \text{ passengers}} = 660.5 \text{ BTU/passenger-mile}$$

Example 13.5

What are the energy intensities of a vanpool with seven riders (including the driver) and a car pool with two riders plus the driver? The fuel economy for the van is that of a medium size SUV at 20.3 mpg and the automobile is a midsize car at 28.7 mpg.

Solution to Example 13.5

$$\text{Vanpool} \frac{125,000 \text{ BTU}}{\text{gal}} \times \frac{\text{gal}}{20.7 \text{ mi}} \times 7 \text{ passengers} = 865 \text{ BTU/passenger-mile}$$

$$\text{Carpool} \frac{125,000 \text{ BTU}}{\text{gal}} \times \frac{\text{gal}}{28.7 \text{ mi}} \times 3 \text{ passengers} = 1452 \text{ BTU/passenger-mile}$$

Example 13.6

What overall energy intensity for light-duty highway vehicles (automobiles, vans, and SUVs) would reduce our consumption of gasoline by to a level equivalent to 70 percent of our domestic production in 2000 (Table 13.4), or about 4 MBPD? Use the value of 2,454,365,000,000 passenger-miles per year (ONRL, 2002).

Solution to Example 13.6   Convert 4 MBPD to BTU for vehicles using gasoline (125,000 BTU/gal in Table 13.5).

$$4 \times 10^6 \text{ bbl/day} \times 125000 \text{ BTU/gal} \times 42 \text{ gal/bbl} \times 365 \text{ days/year} = 7.665 \times 10^{15}$$
$$= 7.665 \text{ quads}$$

The value of annual passenger miles is 2.454×10^{12}

$$\text{Energy intensity EI} = \frac{7.665 \times 10^{15}}{2.454 \times 10^{12}} = 3153 \text{ BTU per person-mile}$$

13.3 ENVIRONMENTAL ISSUES

One of the major planning issues related to sustainability is the question of pollution. We want transportation systems that do not add contaminants to the air. We do not want noise from transportation activities to disturb our communities. This section briefly examines the question of pollution, especially air and noise, and what can be done to prevent, mitigate, or eliminate it.

13.3.1 Cars and Trucks on Congested Roads Pollute

In their normal operation, all vehicles that burn hydrocarbon fuels produce some form of emissions from their tail pipes. The emissions from internal combustion engines are carbon monoxide, nitric oxides, hydrocarbons, or volatile organic compounds, particulates and lesser amounts of sulfur dioxide and ammonia. When car engines are cold, they emit more pollutants. Vehicle emissions are also high when they are traveling slowly or are in stop-and-go traffic. When the road is jammed with vehicles—whether it is from daily traffic peaks, planned construction zones, or an unplanned freeway incident such as an accident or disabled vehicle—slower speeds mean more pollution.

Congestion is the result of an inequality of *supply* and *demand*—the *demand* for travel on a road at that time exceeds its *supply* (or capacity). It causes drivers and others considerable costs, in wasted gasoline and lost productivity, as well as causing health hazards. One solution is to reduce the demand for road space by putting more persons in larger vehicles (e.g., carpools or buses) or by charging a toll for the use of the road. Another solution would be to increase the supply by increasing the number of lanes or adding parallel high-speed roads, improving signal progression, or changing to one-way streets.

THINK ABOUT IT

How would "changing to one-way streets" increase the supply (or capacity) of a street system?

Figure 13.9 shows the relationship between uniform vehicle speed and emissions and fuel consumption. In most cases, higher uniform speeds means lower emissions.

13.3.2 Environmental and Mobility Concerns

Legislation called the Clean Air Act Amendments of 1990 (CAAA) was enacted by Congress in an attempt to get transportation planners, engineers, operators, and resource providers to begin to seriously attack a problem that was getting more severe. The CAAA legislation was followed by new legislation in transportation that set new goals and guidelines. It is the Intermodal Surface Transportation Act of 1991 (ISTEA), called "Ice Tea" by many. ISTEA had as its purpose to develop a national intermodal transportation system that is (US DOT, 1991).

1. Economically efficient.
2. Environmentally sound.

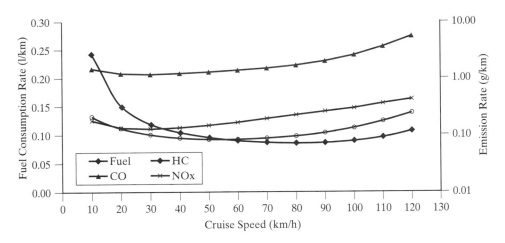

FIGURE 13.9
Emissions and fuel consumption vary with automobile speed. *Source*: Rakha and Ding 2003.

3. Provides the foundation for the nation to compete in the global economy.
4. Will move goods and people in an efficient manner.

Dr. Thomas Larson, Federal Highway Administrator (1988 to 1992) considered the Clean Air Act Amendments to be the most important recent legislation affecting transportation. He said:

> The Clean Air Act Amendments may have a greater effect on the Nation's transportation over the next 20 to 30 years than any non-highway law enacted since the 1960's. More than a decade in the making, the CAAA recast the planning function to ensure that, in areas experiencing air quality problems, transportation planning is geared to improved air quality as well as mobility. State and local officials have been challenged by the CAAA to find ways to reduce emissions from the vehicle fleet, to develop projects and programs that will alter driver patterns to reduce the number of single-occupant vehicles and to make alternatives such as transit and bicycles an increasingly important part of the transportation network. (Larson, 1992)

13.3.3 Transportation Under the Clean Air Act Amendments of 1990

The Clean Air Act Amendments (CAAA) of 1990 established National Ambient Air Quality Standards (NAAQS). Table 13.13 summarizes the NAAQS regulations for ozone. Ozone non-attainment areas are classified according to the second highest hourly level of ozone in the air during a year. Other NAAQ standards apply to carbon monoxide, hydrocarbons, oxides of nitrogen and sulfur, and particulates.

The table classifies urban areas according to the severity of the pollutant, usually along a corridor. Along with the classification program came a list of possible transportation changes that needed to be implemented to achieve attainment. For classifications of *severe* to *extreme*, the measures become increasingly strict. Each level includes its measures and those from the previous levels. Under the CAAA, each state is required to develop a State Implementation Plan (SIP). A SIP contains measures that will enable areas within the state that do not meet the NAAQS to improve their air quality. The SIP is used

TABLE 13.13 NAAQS Classifications and Requirements for Reducing Ozone

Classification	Requirements for Reducing Ozone in Non-Attainment Areas	Goal for 1 Hour Concentration of Ozone (PPM)	Date for Attainment
Marginal	Emissions inventory, implement SIP commitments, basic inspection and maintenance program	$0.121 \leq$ Ozone < 0.138	11/15/1993
Moderate	Must include inspection program; vapor recovery at gasoline stations; traffic control measures through traffic flow, reduced congestion: reduced vehicle use, etc.	$0.138 \leq$ Ozone < 0.160	11/15/1998
Serious	Enhanced inspection and maintenance program; clean fuel program for centrally fueled fleets of more than 10 cars	$0.160 \leq$ Ozone < 0.180	11/15/1999
Severe #1	Limitations on VMT; employer trip reduction programs (for companies with more than 100 employees); increase average occupancy to 25% of area average	$0.180 \leq$ Ozone < 0.190	11/15/2005
Severe #2	Reformulated gasoline in worst ozone areas, including Baltimore, Chicago, Hartford, Los Angeles, Milwaukee, New York City, Philadelphia, and San Diego. Others may participate.	$0.190 \leq$ Ozone < 0.280	11/15/2007
Extreme	Measures to reduce the use of heavy-duty vehicles during peak hours.	Ozone > 0.280	11/15/2010

Source: FHWA, 1992.

by the Environmental Protection Agency (EPA) to approve and monitor a state's air quality program. When communities generate transportation plans, programs, and projects, they must conform to the SIP. The CAAA provides for (a) greater integration of transportation planning and changes in transportation management and (b) the possibility for EPA to impose sanctions.

Transportation measures found in a typical SIP include HOV lanes for carpools, employer incentive programs to reduce the number of automobiles arriving at the workplace, and staggered work hours to shift some travel to off-peak hours. Along with specific projects for implementation, each state must engage in aggressive attempts to gather data by which the effects of improvements can be measured for compliance. Elements to be measured would include reductions in the growth rate of vehicle travel and congestion. The bottom line is, of course, a reduction in the concentration of pollutants. Some of the most stringent measures are being implemented in California and New York City. These include both the use of *reformulated gasoline (RFG)* and the required sale of electric automobiles.

The key issue for transportation engineers is how to avoid some of the draconian measures being suggested by the EPA. Some measures could easily cripple the level of transportation mobility. Air quality must be monitored along heavily traveled corridors, where there are concentrations of pollution. The answer does not lie in adding more and more lanes for single occupancy cars to use. Rather, the transportation engineer must find ways to improve *the movement of people* in the corridor without increasing the number of vehicles that produce the harmful emissions. Vehicle improvements, fuel changes, and increases in multiple vehicle occupancy are all possibilities.

13.3.4 Reformulated Gasoline and Reduction of Pollutants

The non-attainment areas for air pollution in the United States usually occur on heavily traveled corridors in and around the large cities. In severe non-attainment areas for ozone (see Table 13.14), the EPA has mandated non-polluting additives in gasoline to enhance cleaner burning. The three most likely additives are *methyl tertiary butyl ether (MTBE)*, *ethyl tertiary butyl ether (ETBE)*, and ethanol. Gasoline with these additives is called reformulated gasoline (RFG) and has been shown to provide considerable reduction in the level of ozone emissions. The pure ethanol additive, with its very high oxygen content, offers good capability for reducing ozone emissions. However, in hot weather, the fuel is so unstable that it evaporates before it can do much good. On the other hand, as was shown in an experiment in Colorado, ethanol did a marvelous job in winter when the ski resort traffic was very high.

Implementation of the RFG program has involved considerable controversy, because these additives alter the characteristics of combustion in the standard spark ignition engine, resulting in reduced efficiency. Although the use of RFG can result in a 20 percent reduction in *volatile organic compound (VOC)* emissions, there is much concern about fuel efficiency and how the minimum oxygen content standards are met in RFG blends. The most common oxygenate, MTBE, has been identified as a carcinogen that has been found in drinking water in some of the areas listed in Table 13.14. MTBE is being phased out in California (CGER, 1999). So, in the interest of improving public health, sometimes other health issues are created.

Pollutants have been steadily declining without RFG. Table 13.15 indicates the additional impact that RFG is having. In the table, the vehicle types are light-duty vehicles (passenger cars up to 6000 lb GVW); and light-duty trucks (pickups and minivans up to 8500 lb GVW). With RFG, the reductions in *hydrocarbons (HC)* in automobiles was about 50 percent; without RFG, other improvements reduced HC by 30 percent.

Emission reductions shown in Table 13.15 are estimated by MOBILE5b, the U.S. Environmental Protection Agency's (EPA) latest highway vehicle emissions factor model. (See USEPA Internet site http://www.epa.gov/otaq.) The effects of RFG are clearly positive from an environmental point of view. The major 10-year gains are 10 to 15 percent better with RFG than without.

TABLE 13.14   Major Areas Requiring Reformulated Gasoline

Sacramento	Milwaukee	Hartford
Los Angeles	Chicago	New York City
San Diego	Houston	Philadelphia
		Baltimore

Source: EPA, 1998.

TABLE 13.15 Estimated National Average Vehicle Emissions Rates by Vehicle Type Using Reformulated Gasoline (grams per mile)

Light-duty vehicles (autos)	1990 No RFG	1995 RFG	2000 RFG	10-Year Reduction (%)		2000 No RFG	Yr 2000 Compare
				With RFG	No RFG		
Exhaust HC	2.05	1.43	1.16	43.4	30.7	1.42	−18.3%
Non exhaust HC	1.04	0.69	0.37	64.4	28.8	0.74	−50.0%
Total HC	**3.09**	**2.12**	**1.53**	50.5	30.1	**2.16**	−29.2%
Exhaust CO	24.68	16.46	15.41	37.6	21.9	19.28	−20.1%
Exhaust NO_x	1.81	1.59	1.29	28.7	23.8	1.38	−6.5%
Light-duty trucks							
Exhaust HC	3.24	2.11	1.63	49.7	39.8	1.95	−16.4%
Non exhaust HC	1.44	0.85	0.47	67.4	37.5	0.90	−47.8%
Total HC	**4.68**	**2.96**	**2.10**	55.1	39.1	**2.85**	−26.3%
Exhaust CO	36.32	22.15	19.95	45.1	31.2	24.99	−20.2%
Exhaust NO_x	2.36	1.96	1.71	27.5	23.7	1.80	−5.0%

Source: National Transportation Statistics, 2001.

13.3.5 Alternate Fuels

Natural gas has emerged as a cost-effective, clean alternative fuel to refined gasoline. Its most common use is in fleets of vehicles owned by governments or transit agencies. Communities across the country are experimenting with buses that use *compressed natural gas (CNG)* or *liquified natural gas (LNG)*. Natural gas is produced from a readily available source and works in diesel engines with little modification. It provides somewhat less power than diesel but emits no particulates and about 30 percent of the HCs of the diesel. The Government Services Agency has a fleet of government vehicles fueled with CNG. Houston has a fleet of 250 buses that run on LNG. Figure 13.10 shows a bus that uses fuel cells and hydrogen gas. A fuel cell is an electrochemical device that produces DC electricity without combustion. The hydrogen is stored on the roof in tanks as a compressed gas. Toronto has CNG buses in its fleet. The fuel tanks for an equivalent range are much larger than diesel ones.

Chapter 13.4 discusses electric-powered vehicles and those with a battery assisted small internal-combustion engine called the *hybrid electric vehicle (HEV)*. Presently, electric cars have a limited range (about 100 miles) and require 4 to 6 hours to charge the battery. California and New York, working together, had mandated that sales of electric cars in their states be at 2 percent by 1998, growing to 10 percent in about 5 years. Its main competitor (the hybrid auto) is under development by automakers. The hybrid auto utilizes both an electric battery and a small internal combustion engine. Although its fuel efficiency is much higher (50 to 80 miles per gallon), it still falls under the rubric of an internal combustion engine. Two HEVs are now being marketed—the Toyota Prius (52 mpg in city driving) and the Honda Insight (56 mpg). Both are considered super low-emission vehicles (SULEVs).

FIGURE 13.10
Transit bus running on natural gas. Photo: Jon D. Fricker.

Compliance with the electric vehicle mandate has already been delayed once by Congress, and further delays can be expected unless the battery technology improves. Because there are no direct air emissions, electric cars provide the cleanest form of highway transport. When a technology for on-board energy storage is developed that can be competitive with gasoline, then the acceptance of the electric vehicle will grow. Until then, its future is limited. Regardless of regulations to the contrary, the United States is still a consumer-oriented economy, and at present there is no clear approach that will permit the electric car to be successfully marketed. However, the EPA has issued new pollution standards for the 2004 through 2007 model years. See Table 13.16.

13.3.6   Ridership Programs

Example 13.1 showed that ridership programs could have a large impact on fuel use and congestion. During the period 1973 to 1980, when transportation energy was a

TABLE 13.16   Summary of Emission Standards for Cars and Light Trucks Effective for the 2004 to 2007 Model Years After 120,000 Miles

Pollutant: Vehicle emission class	NMOG Non-Methane Organic Gases	CO Carbon Monoxide	NOx Nitrogen Oxide	PM Particulate Matter	HCHO Formaldehyde
LEV (low-emission vehicle)	0.156 (0.125)	4.2 (3.4)	0.6 (0.4)	0.08	0.018 (0.015)
ULEV (Ultra LEV)	0.090 (0.075)	4.2 (3.4)	0.07 to 0.30 (0.14 to 0.20)	0.01 to 0.02	0.018 (.015)
SULEV (Super LEV)	0.01 to 0.09	2.1	0.02 to 0.04	0.01	0.011
ZEV (zero-emission vehicle)	0.00	0	0.00	0.00	0.000

Numbers in () are after 50,000 miles.

Source: U.S. Federal Register, Vol. 65, No. 28, February 10, 2000, p. 6822–6870.

limited resource (the oil embargo was in 1973/4), several ridership programs were tested to determine their effectiveness to reduce automobile use. These include programs using:

- Vans and automobiles as shared ride vehicles (called carpools or vanpools).
- Outlying parking areas where travelers could park their vehicles (or be dropped off) and then travel to the Central Business district on a bus or other transit vehicle. These programs are called "park-and-ride" or "kiss-and-ride," respectively.
- Free or discounted bus passes offered by employers.

Other travel demand management techniques are aimed at reducing the number of automobiles on the road. One method is to increase parking prices or decrease the amount of parking. However, such strategies can also cause unforeseen consequences, such as changes in mode choices, destination choices, and land use development patterns. (Hamerslag, Fricker, Van Beek, 1995). Other strategies are:

- *Auto Measures.* Companies have gone to flexible hours to encourage travel at off-peak and, in some instances, have fostered programs encouraging employees to work at home.
- *Truck Measures.* Major truck measures could include prohibiting truck traffic during peak times on certain roads.
- *Congestion Pricing.* One way to manage congestion is to charge for highway use. Higher tolls are charged to those who drive during peak hours.

As transportation engineers work to build a program that will reduce pollution, they can consider policies or planning options such as those described below.

- *Point of production management* puts the onus for pollution reduction on the auto manufacturers. It permits automakers to make crucial trade-offs among safety, energy consumption, market appeal, and so forth. The results will take a long time to be achieved, because only about 10 to 12 percent of the fleet is replaced each year. The total automotive fleet is slowly moving toward 27.5 mpg, the value that new car sales were mandated to reach by 1992. Light-duty trucks, which include some of the heavier vans and SUVs, are at about 20.6 mpg, well below their automobile counterparts. With the improvement in fuel economy has come an improvement in tailpipe emissions.
- *Point of use control* focuses the problem on a given locale. The fleet that usually travels in that area will require time for integration of the capability to use other fuels. Either MTBE or ETBE can be used and, if the need is for winter ozone control only, pure ethanol can be used. Because emission control has a cumulative effect, regular inspection of individual automobiles is the only way to manage the system and determine its effectiveness.
- *Charges for pollution* have been proposed as a reasonable solution. One method is to charge motorists who travel in a single occupancy car during rush

hours. This approach helps to internalize the cost of congestion. Such a system requires that an alternative form of travel, such as adequate public transit, be available.

13.3.7 The Transportation Acts of 1991 (ISTEA) and 1998 (TEA 21)

In 1991, Congress showed its concern about the infrastructure, the environment, and the mounting traffic problems with landmark legislation that addressed congestion and a range of ground transport needs. The 1991 ISTEA legislation increased funding authorization levels by 63 percent for highways and 91 percent for transit. It also increased the flexibility with which state and local governments can shift funds among different modal programs. The follow-up TEA 21 legislation, passed in 1998, continued the same pattern.

ISTEA required each state to have a statewide congestion management plan (CMP), with special and immediate attention to those areas in the state that are non-attainment air quality areas. States that do not comply can expect that the federal government will hold back a portion of their road building/highway funds. The intent of the CMP section of the ISTEA legislation was to reduce congestion through a combination of transportation improvements, land use planning, trip reduction programs (travel demand management), transit service improvements, and the ITS program, partially funded under ISTEA. ITS attempts to address the problem of traffic congestion through the use of a variety of technological advances. It is intended also to entice local planning organizations to better integrate land use and transportation decision-making processes.

Trip Reduction and Travel Management. Planners were directed by ISTEA to consider ways to reduce the number of single-driver vehicles on congested corridors. Included are the establishment of HOV lanes and incentives for employers who implement ride-sharing programs. Because of its greater pollution problems, California requires all cities to adopt a *travel demand management (TDM)* ordinance. In this ordinance, all non-attainment corridors stress TDM programs that discourage the use of single-occupancy vehicles or encourage ways that travelers can forego the personal automobile as a commuting tool. Programs such as increased vehicle occupancy through ridesharing, increased transit use, and parking management are frequently advocated, because they encourage ridesharing.

Land Use Analysis Program.   Land use decisions made by local jurisdictions can have a significant impact on congestion. Planners must identify ways in which such decisions add to congestion through increased traffic or, alternately, serve to reduce traffic.

In addition to congestion management systems, ISTEA contained provisions related to privatization. Federal funds can now be used in the development of toll roads and funding for congestion-pricing pilot projects.

13.3.8 Transportation Noise

Transportation activities create noise. Noise mitigation can occur when a highway is depressed, when heavy groves of trees are planted along the highway, or when high

FIGURE 13.11
Highway noise barrier. Photo: Jon D. Fricker.

wooden or concrete barriers are erected. Figure 13.11 shows a barrier that reduces the highway noise heard by nearby residents.

Railroads are also noisy, but it is airport noise, for which barriers cannot be built, that is of far greater concern. As explained in section 11.5.4, noise from the engines of jet aircraft, in particular, have restrained development, constrained operations, and restricted the expansion of numerous airports in the United States. Because noise is also a function of the number of flights landing and taking off from an airport, the larger volume of flights offsets some of the gains made by implementing quiet engine programs.

Figure 13.12 shows the noise contours at Raleigh-Durham International Airport. Note the patterns with which the noise levels dissipate as the distance from the runways increase.

13.4   TECHNOLOGY IMPACTS SUSTAINABILITY

As was suggested earlier in this chapter, it is possible for new technology to impact sustainability. In Example 13.1, we showed the enormous impact that a doubling of automotive efficiency would make and how an improvement of flow by reducing congestion would improve traffic. In this concluding section, we will look more closely at the potential technology impacts on the horizon.

13.4.1   Use of Information and Communication Technologies

Transportation engineers have for many years been aware of the impact new information technologies, coupled with communications capabilities, might have on transportation productivity. The 1991 ISTEA created an initiative called Intelligent Vehicle Highway Systems (IVHS). A few years later, the program's name was changed to Intelligent Transportation Systems (ITS). Initial experiments showed that there were significant gains to be made in applying ITS technologies, especially

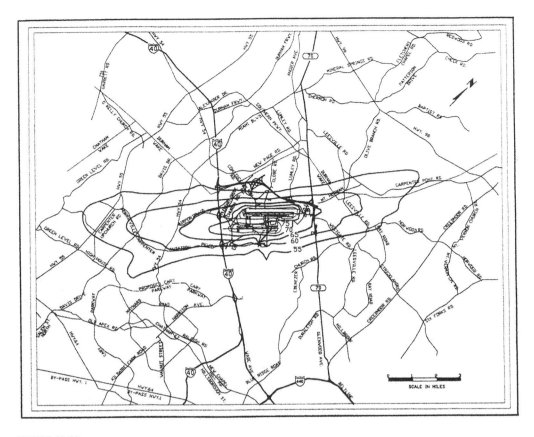

FIGURE 13.12
1996 annual average day noise contours for all operations on all runways at RDU. *Source:*
http://www.rduaircraftnoise.com/noise-info/96contour-report.pdf.

in reducing traffic congestion. Figure 13.13 shows a traffic operations center used to control traffic flow.

The heart of the ITS concept is the application of existing and emerging technologies to improve the efficiency of existing transportation facilities. Sensors, real-time computation, communication, and displays can provide valuable information to drivers, dispatchers, and traffic control personnel. There are many applications under study and testing. The ITS program motivated action by government and high-tech industry, much of it collaborative.

Building additional highway capacity is not always the most cost-effective solution to the congestion problem. Table 13.17 identifies 20 years of traffic growth without similar growth in highway facilities. This disparity provides much of the basis for ITS. The dramatic increase in traffic over the last 20 years led to increased congestion, higher fuel use and, perhaps most importantly, a decrease in public health through the pollutants emitted by the vehicles. The table, which gives national averages, does not reflect the increases in vehicle miles driven on some of the more heavily used corridors in the nation. For example, traffic in one corridor west of Indianapolis in those same 20 years

FIGURE 13.13
Traffic Operations Center. *Source:* Ada County Highway District Traffic Management Center, Boise, ID.

has increased by 130 percent. The preference for the automobile is so great in the United States, that more than 90 percent of all trips made in 1995 were with personal vehicles.

Goals of ITS. Transportation demand varies by time of day, as morning and evening peak hour travelers on urban freeways know well. Most daily congestion occurs during the rush period, but in most cities, it does not last for more than a few hours. Thus, it is difficult to argue that eliminating congestion for this short time by merely adding new lanes for traffic is the most cost-effective solution, especially if something else can be done. Three alternatives are possible:

1. Do nothing. Some feel that traffic control in cities is now effectively accomplished by the congestion, and adding more vehicular capacity will only encourage more traffic.

TABLE 13.17 Twenty Years of Change in Highway Use

Indicator	Percent Change from 1978 to 1998
Total road and street mileage	0.6% increase
Population of the nation	21.4% increase
Number of licensed drivers	31.3% increase
Number of motor vehicle registrations	42.6% increase
Total vehicle miles of travel	70.0% increase
Total motor use of fuel	23.8% increase
Gallons of motor fuel use per mile	27.1% decrease

Source: ITS Primer 1-5, ITE, 2000.

2. Use all the technology possible to make a traveler more aware of his options when congestion occurs and/or encourage use of a roadway where the flow is lower.

3. Increase the economic cost of using the highway during peak periods through tolls and/or heavy increases in parking prices, to provide an incentive for not using the highway.

The second alternative has led to research on transportation schemes and incentives for drivers and employers to improve the efficiency of the use of the existing infrastructure. Intelligent Transportation Systems (ITS) are intended to improve accessibility, safety, and productivity, while reducing travel delay, pollution, and fuel use.

Some ITS proponents look forward to the day when drivers will be able to enter an *automated highway* and relinquish control of their cars to the highway, which will move closely bunched vehicles at high speed. Collision avoidance systems will keep the occupants safe, even in heavy flow, and the capacity of such highways will be far greater than highways with autonomous drivers. Such highways and instrumented cars are already technically possible, but they have not yet been implemented as part of an operational system. There are still many things ITS can do before taking such a dramatic step. ITS is intended to be an integrated approach to transportation, linking all the elements of the transportation network—the vehicle, the infrastructure, and the traveler.

As generally conceived, Intelligent Transportation (Vehicle and Highway) Systems consist of four functional components (ITS Primer 3-1, ITE, 2000), not including the Advanced Traveler Information System (which can be considered a part of each component):

1. *Advanced Traffic Management Systems (ATMS)*
2. *Advanced Public Transportation Systems (APTS)*
3. *Advanced Vehicle Control Systems (AVCS)*
4. *Commercial Vehicle Operations Systems (CVO)*

Advanced Traffic Management Systems (ATMS). ATMS are generally systems that provide for optimum flow where traffic control signals are present. Sensors will provide the traffic management center (Figure 13.13) with information about traffic flow conditions. The center personnel can provide route guidance to drivers, if an incident or heavy flow causes delays on any segment. Figure 13.14 shows one concept of ITS. Traffic management is accomplished through messages on dynamic message signs, on the radio, and even on in-vehicle displays. Some automobiles already come equipped with map displays and advanced navigation system inputs like the global positioning systems (GPS). In an *Advanced Traveler Information System (ATIS)*, information about traffic conditions is used in real time to predict the shortest time route between a driver's current location and his/her intended destination.

FYI: A forerunner of ATIS was the helicopter traffic report provided by radio stations during rush hours in most large cities.

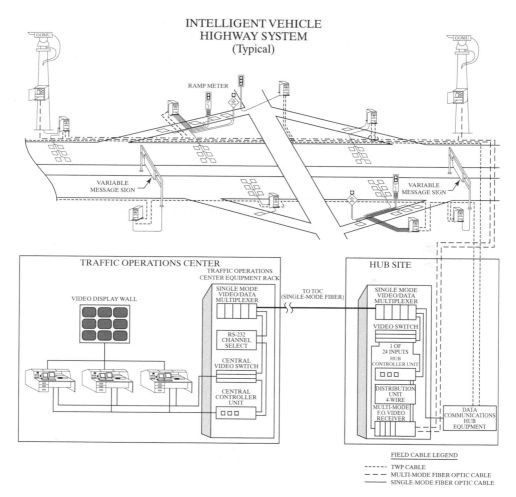

FIGURE 13.14

ITS highway system showing ATMS. *Source*: Halkias, 1997 Appendix I, Figure I.1.

Ramp Metering.   An early example of applying technology to a traffic problem was using a traffic signal to "ration" traffic onto an expressway. (See Figure 13.15.)

Using information about gap acceptance (see Chapter 8), the entrance ramp meter allows vehicles to enter the freeway when its detectors sense gaps in the mainline traffic stream. The major disadvantage of this approach is that each ramp is controlled independently, ignoring interdependencies with other ramps and, thus, may not be effective in dissipating traffic congestion. One proposed scheme is to calculate metering rates based on the predicted demands at the entrance ramps and the downstream capacity of the freeway. Further research has continued to apply improved measuring techniques and more sophisticated software to the subject. The positive outcome is improved flow for those drivers already on the freeway. This benefit, however, may be at the expense of longer waiting times for those waiting to enter.

One of the pioneers in ramp metering was the Minnesota DOT on I-35 in the Minneapolis-St. Paul Area. The Minnesota DOT had installed 430 ramp meters. A

FIGURE 13.15
Ramp meter. Photo by Jon D. Fricker.

subsequent study indicated that the system was having a positive effect. The Minnesota legislature mandated an evaluation of the traffic flow and safety impacts associated with turning off all 430-ramp meters for six weeks in Fall 2000. Results [Cambridge Systematics, 2001] presented to Minnesota DOT revealed that, *without* ramp meters, there was:

- A 9 percent reduction in freeway volume.
- A 7 percent reduction in freeway speeds.
- A 26 percent increase in crashes.
- A net annual increase in emissions of 1160 tons of pollutants.

Ramp meters made possible a 22 percent decrease in freeway travel times, which more than offset the ramp delays. The only category where ramp metering had a negative impact was in fuel consumption. A decrease in fuel use of 5.5 million gallons was possible without ramp metering. Clearly, ramp metering had made a difference.

Incident Detection and Response.   Another ITS example related to traffic management has to do with freeway incident detection and response. When an incident occurs on a heavily traveled freeway, a shock wave is generated. The phenomenon is best described by referring to Figure 13.16, which shows schematic "snapshots" of vehicles on a two-lane freeway when an incident occurs. If an incident occurs at time $t = t_0$, traffic can no longer move at the flow rate that existed previously. Traffic builds up and cars move closer to each other when traveling at the slower speed. The traffic buildup is a phenomenon that traffic engineers call the *shock wave*. Under this condition, the traffic flow rate will be whatever is allowed to pass by the location of the incident. Meanwhile, the shock from the incident continues to move farther back in the traffic stream. When the accident is cleared at $t = t_1$, traffic begins to move at the maximum flow possible on the roadway (V/C = 1). So now there are two waves— the shock wave, which continues to move farther back in the line of traffic, and the *clearing wave*, which permits traffic to move at maximum flow. Moving faster than the shock wave, the clearing wave eventually reaches the place in traffic where the

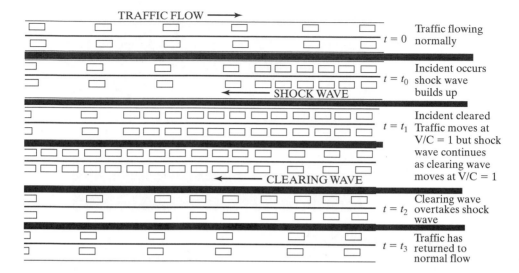

TRAFFIC FLOW ⟶

$t = 0$ Traffic flowing normally

Incident occurs
$t = t_0$ shock wave builds up

⟵ SHOCK WAVE

Incident cleared
$t = t_1$ Traffic moves at V/C = 1 but shock wave continues as clearing wave moves at V/C = 1

⟵ CLEARING WAVE

Clearing wave
$t = t_2$ overtakes shock wave

Traffic has
$t = t_3$ returned to normal flow

FIGURE 13.16
View of time sequence of freeway incident causing shock wave.

shock wave is $(t = t_2)$. Traffic will then thin out to return to the normal flow condition, as shown at t_3.

One cost-effective response to the effects of such incidents is the implementation of a highway assistance vehicle and personnel equipped to handle repairs or push a disabled car off the road, and equipped to call for more assistance, if needed. The motorist assistance program vehicle moves as quickly as possible to the place where an incident has been observed to have impeded or blocked traffic.

Queueing diagrams similar to the ones used in Chapter 3 can be used to analyze the effects of the incident and the ITS remedy. In the left diagram of Figure 13.17, the shaded portion shows the total delay experienced by all the motorists. When the *motorist aid vehicle (MAV)* arrives, the MAV personnel quickly analyze the situation and work to improve the flow by pushing the disabled vehicle to the shoulder, fixing a tire, calling repair service, notifying emergency response units, and/or do whatever is needed. As a result, the blockage is removed more quickly, allowing traffic to resume its normal flow sooner. The traffic buildup due to the shock wave is reduced and the benefit is the reduced delay time indicated by the striped portion in the right diagram of Figure 13.17.

Example 13.7

A freeway has three lanes in each direction and has a maximum flow of 6000 vph. It is operating at 3600 vph at $t = 0$. A collision occurs, blocking two lanes and restricting the flow of the third lane to 1800 vph. The freeway's constant speed is 60 mph and its 3-lane jam density is 600 vpm. The incident is completely cleared in 45 minutes and traffic returns to normal as soon as the backup is dissipated. How long does it take to dissipate the backup? What is the total delay?

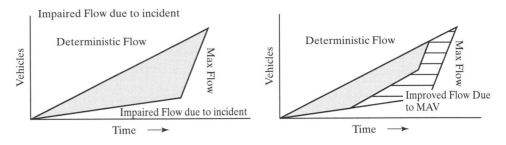

FIGURE 13.17
Queueing diagrams showing the advantage of the motorist assistance program.

Solution to Example 13.7 The length of time to clear begins at 45 minutes, when 60 cars per minute * 45 minutes = 2700 cars. The incident has blocked lanes, allowing only 30 cars per minute * 45 minutes or 1350 cars to get past the blockage. The queue is 2700 − 1350 = 1350 cars. The time at which the queue clears is where the arriving cars meet the freeway operating at capacity (100 cars/minute). See Figure 13.18.

Thus, $2700 + 60t = 1350 + 100t$; $t = 33.75$ minutes.

Total time to clear is $45 + 33.75 = 78.75$ minutes after the incident.

Number of cars affected $= 60 * 78.75 = 4725$ cars.

The delay time is the area under the curve

$$\text{Delay}_{before} = \frac{3600 * \dfrac{78.7}{60} \times \dfrac{78.7}{60}}{2} - \frac{1800 * 0.75 \times 0.75}{2} - \frac{\left(1350 + 3600 * \dfrac{78.7}{60}\right)}{2} * \frac{33.7}{60}$$

$$= 886 \text{ vehicle hours}$$

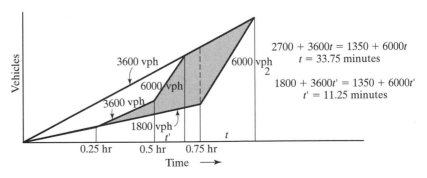

$2700 + 3600t = 1350 + 6000t$
$t = 33.75$ minutes

$1800 + 3600t' = 1350 + 6000t'$
$t' = 11.25$ minutes

FIGURE 13.18
Queueing diagram for Examples 13.7 and 13.8.

Example 13.8:   Motorist Assistance Program

An incident response program called the Mythaca Miracle Machine (MMM) is stationed along the freeway during peak traffic hours. Traffic is monitored by detectors, so that the location of an incident that causes the traffic to back up is known within a few minutes. As indicated in Figure 13.18, traffic backs up quickly, but the MMM vehicle is able to reach the scene of the incident in 15 minutes and, in 15 additional minutes, is able to push disabled cars to the shoulder, increasing the flow from 1800 vph to 3600 vph. Because this new service rate is the same as the deterministic arrival rate, the shock wave will not move farther back. The wrecker and ambulance come and clear things fully in 30 minutes, so that the freeway is opened up for full flow. What is the reduction in delay for cars on the freeway? At $15 per vehicle hour, what is the value of the time savings?

Solution to Example 13.8   From the queueing diagram, the time to clear the queue after $t = 30$ minutes is determined by where the arrival and service curves cross.

$$\text{Vehicles entering the queue after 30 minutes} = (60 \text{ vpm} * 30 \text{ min}) - (30 \text{ vpm} * 15 \text{ min})$$
$$- (60 \text{ vpm} * 15 \text{ min})$$
$$= 450 \text{ vehicles.}$$

Thus, t' is $450 + (60*t') = 100*t'$, which solves as $t' = 11.25$ minutes. The MMM vehicle allowed the full flow on the road to begin 41.25 minutes after the incident, compared to 78.75 minutes without the helper. That difference in delay translates into significant savings, just for the one incident.

$$\text{Delay}_{\text{withMMM}} = \frac{3600 * \dfrac{41.25}{60} \times \dfrac{41.25}{60}}{2} - \frac{1800 * 0.25 \times 0.25}{2} - \frac{450 + 1350}{2} \times 0.25$$
$$- \frac{(1350 + 3600) * \dfrac{41.25}{60}}{2} \times \frac{11.25}{60}$$
$$= 278.6 \text{ vehicle-hours}$$

The delay is thus reduced by more than a factor of 3. The saving, if every hour wasted costs $15.00, is equal to $(886 - 278.6) * \$15.00 = 607.4 * \$15.00 = \$9111$. It has been determined that incidents occur during rush hour about 100 times per year. Although each incident is different, if we assume that a saving of $9000 occurs 100 times, the total annual savings are $900,000. The cost of the van and the three people to operate it will be approximately 40 percent of the savings, for a benefit-cost ratio of 2.4. See Table 13.18.

TABLE 13.18   Annualized Costs for the MMM Vehicle

Vehicle ($100,000 @ 3 year life 6%)	$37,500
Labor	$150,000
Central control	$100,000
Surveillance equipment	$60,000
Fuel and maintenance (100,000 miles)	$25,000
Total	$372,500

Advanced Public Transportation Systems (APTS). A common complaint among bus riders is the uncertainty of bus arrivals. The APTS combines public transit vehicle location and vehicle tracking to inform the customers of bus arrival times. APTS consists of components indicated in Table 13.19. Figure 13.19 shows an arrival time and schedule kiosk for the bus service in Vail, Colorado.

Currently, the large amount of traffic data generated as the byproduct of ATMS activities are utilized in the operation of the traffic control system. When buses are a part of the traffic mix, they can benefit with improved priority in traffic flow as well as be beneficiaries of directly supplied traffic data and on-route guidance advice. The APTS system will offer improved service to the public. If the uncertainty regarding the arrival time of the next bus can be reduced, a major obstacle to transit customer satisfaction will have been achieved, and transit ridership may grow.

Advanced Vehicle Control Systems (AVCS). The multiple objectives of AVCS is to assist drivers in various tasks, to provide automatic warning of certain kinds of hazardous situations, and possibly to provide automatic control of vehicles. Specific systems include collision-warning devices, vision enhancement, automatic headway keeping, lateral control, and collision-avoidance braking. Technologies presently in widespread use are automatic cruise control and antilock brakes. Vehicle manufacturers will provide AVC systems whenever they can be provided at reasonable cost or whenever they increase the marketability of vehicles. One near-term future possibility is a collision-warning system. Some technical progress has been made in the more complex areas such as automatic steering and headway keeping, but formidable institutional and marketing obstacles will have to be overcome in these areas. For instance, it may be difficult or impossible to have automatically controlled vehicles on the same roadway with manually controlled ones. In addition, failures of automatic control systems involving many vehicles could have catastrophic consequences resulting in significant legal liability questions.

Commercial Vehicle Operations Systems (CVO). Most CVO applications are in the area of fleet control, telecommunications, and weight/toll priority. Similar systems are already used at various levels of sophistication by providers of emergency services, such as police, fire, and ambulance. Automatic vehicle location, automatic vehicle identification, and automatic paging are critical technical components of the CVO. Administrative improvements, such as one-stop shopping for permits and point-of-origin weight bypass sanctions (where stops at intermediate weigh stations are not required),

---

TABLE 13.19 Functions Provided by Equipment that Serves as Part of APTS

Sensors	Collision avoidance, door opening, automatic passenger counter
Vehicle tracking	Schedule adherence, route adherence, priority movement
Fare collection	On-board revenue
Emergency management	Covert audio monitoring, vehicle tracking, silent alarm, video surveillance
Vehicle maintenance data	Breakdowns, mechanical systems, fuel systems, electrical systems
Customer information	Kiosks, Internet, pagers, arrival signs, destination signs, next stop signs, audio information

Source: ITS Primer 7-2 through 7-6, ITE, 2000.

FIGURE 13.19
Showing variable message sign in Vail, Colorado. Advanced traveler information is a major feature of APTS. Here, a variable message sign tells the waiting passenger when the next bus will arrive at the stop. When the bus arrives, the sign reminds the traveler which route it is on. Here it is giving the time of day. Photos: Jon D. Fricker, July 2001.

also improve the system's management. For overweight trucks, weigh-in-motion devices are used to enforce truck weight limits, and electronic toll collection is offered on major toll roads.

One important aspect of ITS has been a program to eliminate the delays experienced by trucks at weigh stations. Under traditional procedures, trucks must stop at a weigh station every time they cross a state border. At each weigh station, they usually wait in line to get reweighed and have their credentials checked. Such activities waste much of their time and fuel. Six states (Florida, Georgia, Tennessee, Kentucky, Ohio, and Michigan), Canada's Ontario province, and the U.S. Federal Government joined in a partnership with the trucking industry to test a new system along Interstate Highway 75. I-75 goes from Miami, Florida, north to Ontario, Canada, via Detroit. It is an important truck corridor, with 25 weigh stations along the way.

In 1984, the Federal Highway Administration invested $8.4 million in an operational test to outfit each truck with a transponder and each station with the necessary sensors. The objective was to make I-75 a 1000-mile, one-stop trucking corridor from Miami to Ontario. The truck preclearance system begins with the truck being weighed (and its load secured) and identified at the first weigh station. After that, the truck needs to be in the proper lane as it approaches a weigh station. Using Automated Vehicle Identification (AVI) technology, the station reads the truck's vehicle identification number and weighs it with a weigh-in-motion (WIM) sensor. After checking on these pieces of information, the truck is signaled to bypass the station. There will be a series of changeable signs to notify the driver if he is to come into the weigh station.

For the 4500 trucks that participated, between 1 and 5 minutes were saved per weigh station (about 30 to 50 minutes per trip), depending on the station equipment and the truck traffic. Fuel savings were from 0.05 gallons to 0.18 gallons per weigh station (Iowa State University, 1998). A reasonable estimate of fuel saved would be 2.5 gallons per truck. The monetary value of savings could easily be more than $30 per trip. Other operational savings will accrue to the weigh station, which will need fewer personnel to operate the station, and to the state, which will have more time to spend on those trucks that need special attention (e.g., overweight trucks, trucks with bad paper work, and trucks with safety problems).

Example 13.9

The State of Florida has recorded 9900 heavy trucks per day in both directions at one station in Morgan County and 9300 at one further north in Hamilton County (Florida DOT, 1999). If 1000 loaded trucks per day are bypassing an average of 10 weigh stations on the way to Tennessee, Kentucky, Ohio, Michigan, and Canada, what annual savings will accrue if 30 minutes per trip and 2.5 gallons of fuel are saved per trip and time is valued at $15.00 per hour and fuel costs 1.25 per gallon?

Solution to Example 13.9   The fuel saved will be 2.5 gal/trip * 1000 trips per day * 365 days per year = 912,500 gallons of fuel.

The annual savings to truckers at 30 minutes per trip with this scenario are 182,500 hours of time.

$$\text{Cost Savings} = (912{,}500 * \$1.25) + (182{,}500 * \$15.00)$$
$$= \$1{,}140{,}625 + \$2{,}737{,}500$$
$$= \$3.878 \text{ million/day.}$$

The AVI Transponders on each vehicle not only signal the weigh station but permit carriers to know where their trucks are. The larger investment cost is on the part of the state for weigh-in-motion and AVI sensing equipment at each weigh station. More and more trucks are carrying the AVI transponders.

13.4.2   Automotive Technology

One of the hoped for advances toward sustainability is that new vehicle technologies will emerge that will be less harmful to the environment, use fewer non-renewable resources, and reduce congestion.

Electric Automobile.   The concept of the electric car has been around for a long time. In fact, early cars (1890 to 1910) were electric cars. They were touted as vehicles for women, because they didn't have to be crank-started like their internal combustion competitors. It is interesting that it was an electric component, the electric starter, that led to the downfall of the electric automobile. The internal combustion engine had much longer range and the ease of filling the gas tank in a few minutes.

The electric vehicle has electronics and a revolutionary new propulsion system to replace the internal combustion engine. With the limit on range that the present batteries can offer, auto design has placed a premium on reduced weight, lower aerodynamic resistance, and lightweight tires. During braking, energy is regenerated and sent back to the batteries. To replace the gas station, drivers of the electric vehicle have a 220-volt charging system installed in their homes. Should an extra charge ever be needed on the road, a portable charger in the trunk, which plugs into an ordinary 110-volt outlet, can be used to charge the lead acid battery.

In addition to the home charger, it is likely that an electric charger infrastructure similar to the present gas station infrastructure will become available to support the electric cars coming on the market. Technically, when electricity is used to power automobiles, such as electric cars, it is considered an alternate, pollution-free fuel. In this case, both the external power source (fuel is the electricity) and the drive power source is an electric motor rather than a combustion cylinder. Because there are no direct air emissions, electric cars provide the cleanest form of highway transport. (Note that air emissions related to the electric car all occur at the power plant, where the emissions are easier to control.) Car manufacturers have been working on electric automobiles to increase the range and reduce the battery charging time. Both Honda and General Motors have built EVs that could be driven, but the market is just not there at this time.

The second generation GM EV1 is an electric vehicle with two battery technologies: an advanced, high-capacity lead acid, and an optional nickel-metal hydride. Using a 220-volt charger, charging from 0 to 100 percent for the new lead-acid pack takes up to 5.5 or 6 hours. Charging for the nickel-metal hydride pack, which stores more energy, takes 6 to 8 hours. During braking, the electric motor generates electricity (regenerative), which is then used to partially recharge the battery pack. The EV1 has the lowest aerodynamic drag of the production vehicles on the road today, with a 0.19 drag coefficient. It is shaped like a teardrop when viewed from above.

The real problem with electric cars is the limited capacity of a reasonable weight battery. This limitation places limits on the range between "fill-ups" (battery charges) and the carrying capacity of the vehicle. When a technology for on-board energy storage is developed to be competitive with gasoline, then demand for the electric vehicle will grow.

Direct fuel cells (DFC), such as the hydrogen fuel cell technology, seem to be poised for a breakthrough soon. It will generate electricity with no combustion. DFCs are, in-effect, large continuously operating batteries that generate electricity, as long as fuel, such as natural gas, is supplied. DFC technology also eliminates external fuel processing to extract hydrogen from a hydrocarbon fuel. The schematic shown in Figure 13.20 indicates how the components of the fuel cell version of an electric vehicle would be laid out.

If the battery technology problems are solved, then the market for electric cars will grow and we will have a true *zero-emission vehicle (ZEV)*.

The Hybrid Electric Vehicle (HEV). The limited performance of batteries has led several companies, spurred on by the U.S. Department of Energy, to develop a vehicle whose gas mileage is about 80 to 100 miles per gallon. The idea is to develop a drive train where the gasoline engine runs at nearly constant speed and the battery operates to provide the acceleration and maneuvering and speed changes. The two devices serve

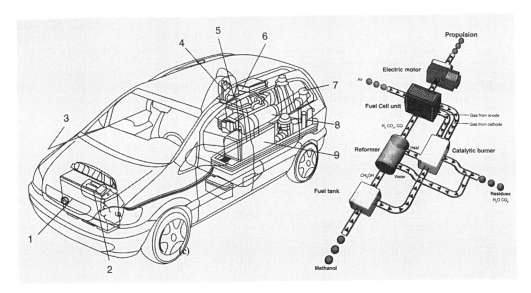

FIGURE 13.20
The principal components of an electric vehicle (fuel cell version). *Source*: Hodkinson and Fenton, 2001.

together to provide current to electric traction motors, which drive the wheels. By running at a constant speed, the gasoline engine is operating in its most efficient regime.

Development of the electric/hybrid drive train may reach a place where it will replace standard drive trains. In Figure 13.21, the heart of the system is in the "Energy Management & System Controls" and the "Hybrid Power Unit." The major difference between this technology and the straight electric vehicles is that the battery is smaller and is recharged during coasting and braking. The two basic HEV configurations are series and parallel.

Series Configuration. In the *series* hybrid, the internal combustion engine drives a generator that charges the batteries, which provide power to the electric motor that directly turns the driveshaft. In the series configuration, the gasoline internal combustion engine generates electric power, not direct torque to the drive wheels. The car functions very much like the diesel generator/traction motor system that drives today's railroad locomotives.

Parallel Hybrid. The *parallel* configuration is such that either the engine or the electric motor can provide torque to the driveshaft. A separate generator is not needed in the parallel HEV because, when the clutch is engaged, the internal combustion (IC) engine turns the driveshaft, but it also spins the motor's rotor, causing the motor to generate electricity to charge the batteries. Both the parallel and the series hybrid can be operated with power coming from either the internal combustion engine the battery or both.

One advantage of the parallel scheme is that a smaller engine and motor can be used, because the two hybrid components can work together. However if a parallel hybrid is running electrically, the batteries cannot be charged at the same time, because there is no generator and the internal combustion engine must be connected to the drive train.

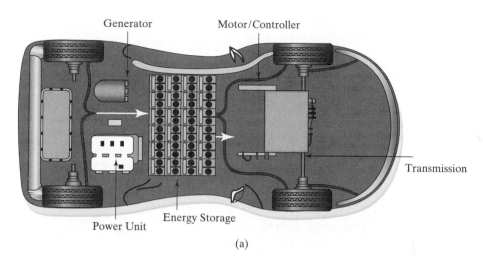

(a)

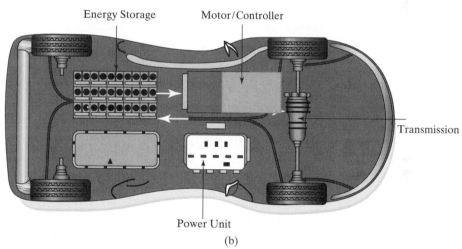

(b)

FIGURE 13.21

Configuration of the HEV. (a) Series Configuration for HEV. (b) Parallel configuration for HEV.
Source: Department of Energy website.

Vehicle Development.   Many believe the hybrid electric vehicle (HEV) will be the car of the near future. The HEV's IC engine provides about one fifth the power of a conventional automobile's IC engine. The design of this engine is optimized at its running speed, so although it produces some pollutants, it is very much cleaner than a conventional car. Toyota and Honda are already marketing HEVs in the United States. The Toyota and Honda HEVs are ranked as SULEVs, discussed in section 13.3, and achieve about 52 to 56 mpg in city driving.

In 1993, when the federal government announced the Partnership for a New Generation of Vehicles (PNGV) consortium, the "Big Three" U.S. automakers and a large number of smaller technical firms began looking at the HEV in greater earnest. The goal was to develop a midsize car that achieves 34 kilometers per liter (80 mpg) of gasoline without degrading driving performance achieved within a cost range that is competitive with other midsize autos.

13.4.3 High-Speed Rail Systems

There are two technologies that have been worked on mainly in other countries that offer high-speed rail opportunities. If higher speed were available in rail and the stations were convenient, such systems could be considered as operational potential in the United States; one is already achieving high speeds in Europe and Japan. The major reason that the United States is slow in embracing high-speed rail technology is the lack of sufficient ridership demand for a system that provides only a minor speed improvement over the automobile on the interstate highway. The low demand for high-speed rail in the United States is related to:

- The love of Americans for the automobile.
- The advances of the commuter airline industry.
- The size of the U.S. land mass.
- The poor "track record" of U.S. rail as a profitable venture.
- The lack of a passenger railroad culture.

High-Speed (Rail) Ground Transport (HSGT).   Railroad trains that use the trusted technology of steel-wheel-on-steel-rail have been able to operate at speeds of 300 to 400 kph (186 to 250 mph). With many of these trains in successful revenue service in France, Germany, Spain, and Japan, this is hardly a new technology. Ever since 1965, Japanese, French, Swedish, and German engineers have been developing and improving on high-speed rail operated as steel wheel on steel. Speeds have increased as engineers have found ways of improving track and ride. For example, the most recent French TGV train operating in revenue service between Paris and the cities of LeMans/Tours achieves peak speeds of 320 kph (200 mph) and has been tested at a sustained speed of more than 530 kph (330 mph).

HSGT in the United States is largely an economic question. One important drawback to implementation in the United States is the fact that 65 to 80 percent of the investment cost is for facilities (track, right-of-way, stations, catenaries, bridges, etc). Other areas of special concern are safety and noise.

The Amtrak Acela (pronounced "ah-SELL-ah") Express has been in service in the Northeast Corridor since December 2000. The Acela employs the "tilt train" technology shown in Figure 13.22. As the train enters a curve, the train's suspension adjusts the position of the cars to allow a higher speed through the turn without causing discomfort to the passengers. This is especially helpful on rail lines that are shared by freight and passenger services, where the degree of superelevation would have to be quite different for the two services.

In an HSGT system, average speed is usually considered the most important performance characteristic. The requirements for high-speed vehicles in a specific corridor arise from the need to decrease travel time relative to other modes of transport. Although a high *peak speed* is essential to reducing travel time, high *average speed* reflects the high-speed train's ability to negotiate curves and to accelerate and decelerate quickly at the station stops en route.

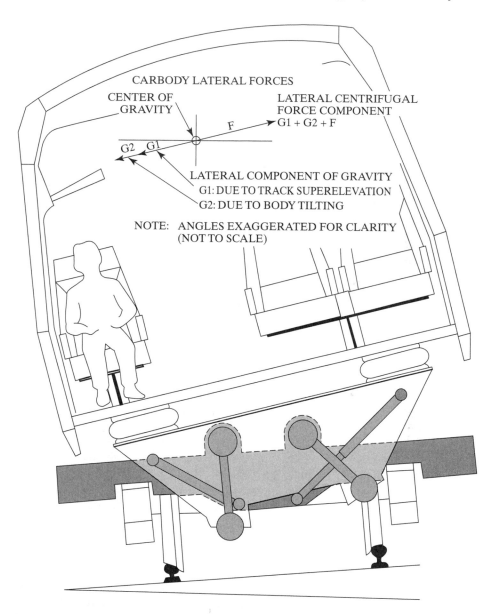

CARBODY LATERAL FORCES

CENTER OF
GRAVITY

LATERAL CENTRIFUGAL
FORCE COMPONENT
G1 + G2 + F

F

G2   G1

LATERAL COMPONENT OF GRAVITY
G1: DUE TO TRACK SUPERELEVATION
G2: DUE TO BODY TILTING

NOTE: ANGLES EXAGGERATED FOR CLARITY
(NOT TO SCALE)

FIGURE 13.22
Tilt train technology.

Magnetic Levitation.   Magnetic Levitation (MAGLEV) is an advanced technology in which magnetic forces lift, propel, and guide a vehicle over a guideway. Combining a configuration that has no guideway contact with state-of-the-art electric power and control system, the MAGLEV vehicle permits cruising speeds of up to 300 mph. A specially designed guideway, often elevated, is required. Because of its high speed,

MAGLEV provides competitive trip time savings over auto and airplane for trips less than 600 miles. Its acceleration capability is superior to high-speed rail.

Germany's Transrapid System.   Germany is taking a careful look at initiating a MAGLEV line on the Hamburg-Berlin route, to accommodate the considerable increase in passenger traffic after the 1991 unification of Germany. Interest from private industry in the technology is such that private capital will assist the construction of the MAGLEV line. The name of the German MAGLEV system is "Transrapid." As shown in Figure 13.23, the Transrapid MAGLEV vehicles are magnetically levitated and ride along a guideway. Synchronous linear motors provide the force to propel the vehicle along the guideway. Levitation results from energized magnets positioned on the undercarriage, or levitation frame. Lateral stability on the guideway is obtained by magnets, mounted on the undercarriage, facing the outer edges of the guideway. Suspension of the car comes from a series of levitation frames that align magnets and carry the levitation and guidance forces to the car body through pneumatic springs and links. When the levitation magnets are energized, the vehicle is lifted toward the guideway. On-board batteries energize the magnets during low-speed operation. The batteries are recharged when the vehicle is moving at speeds of 75 mph or more.

MAGLEV guideway structures are made up of steel or prestressed concrete beams. Important functions such as accuracy and stability of the guideway at the high speeds is achieved through the use of automated production techniques and construction of the piers and the foundations to appropriate specifications. The system is operated automatically, and its monitoring is achieved with the use of wave transmission between the vehicle and a central control center. Because the propulsion system is located in the guideway, MAGLEV has all-weather capability. The vehicle can also climb grades of 10 percent.

Because of the unique vehicle design that wraps around the guideway, derailment is not possible, making the Transrapid MAGLEV safe. Because the power is only turned on in the section of guideway that the vehicle occupies, a faster vehicle cannot run into the back of a slower vehicle. Head-on collisions are not possible, because the electromagnetic field in the guideway that propels the vehicles always moves the vehicles in the same direction and at the same speed. Separated guideway makes collision with other modes of transportation impossible.

Because it is electrically propelled and emits no pollutants, the MAGLEV system is not directly damaging to the environment. The electromagnetic system necessary for the vehicle propulsion, with magnets located beneath the guideway, results in minor magnetic fields in the vehicles or in the vicinity of the vehicles (10 to 30 mGauss at the sea level). Because the MAGLEV trains are elevated and there are no outside connections with the guideway, it is very quiet.

U.S. Demonstration.   In 1998, Congress passed the "Transportation Equity Act for the 21st Century," making available more than $218 billion over a 6-year period for surface transportation assistance to states and localities. Section 1218 of this Act created a national Magnetic Levitation Transportation Technology Deployment Program. The federal funds for planning and construction must be matched 1/3 to 2/3 by state, local, or private contributions. The first project that has been selected for this funding is a 45-mile system linking Pittsburgh Airport to Pittsburgh and its eastern

Operating Speed- 310 mph

Maximum Grade- 3.5% desired

Maximum Grade- 10% allowable

Empty Car weight- 90 tons

Cost varies
about 15–20 Million per mile

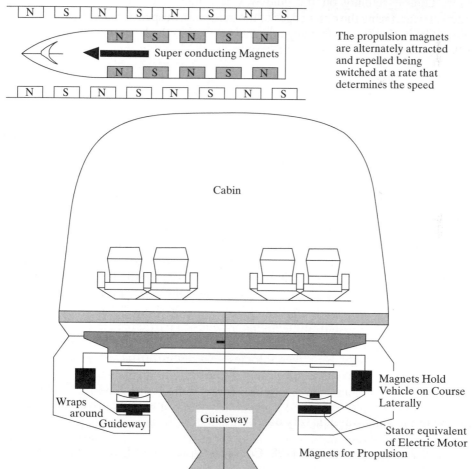

Super conducting Magnets

The propulsion magnets
are alternately attracted
and repelled being
switched at a rate that
determines the speed

Cabin

Magnets Hold
Vehicle on Course
Laterally

Wraps
around
Guideway

Guideway

Stator equivalent
of Electric Motor

Magnets for Propulsion

FIGURE 13.23
German Transrapid MAGLEV Train. *Source*: FRA Maglev website.

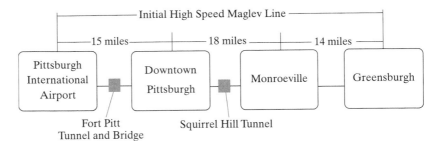

FIGURE 13.24
Initial route of Pittsburgh MAGLEV demonstration project. *Source*: FRA website.

suburbs. (See Figure 13.24.) The first segment is due for construction in 2004 or 2005, with revenue service beginning on July 1, 2005. The rugged terrain, a full four-season climate, and stops at an airport, downtown, and in the suburbs will demonstrate the full potential of MAGLEV technology to provide service in a variety of environments. The project would be the first stage of a system that eventually would provide high-speed service to Cleveland to the northwest and Philadelphia across the state to the east.

SUMMARY

Transportation is essential to our daily lives. We travel from place to place—to work, to shop, to go to school, to visit friends, to enjoy some recreational activity. To satisfy our need for food, clothing, and other essentials of life, goods and services are made accessible to us because of a transportation system. Although a nation's quality of life depends on an efficient, reliable, and safe transportation system, it is also true that the same transportation system can have negative impacts on the society it is designed to serve. Transportation systems are driven by energy—most of which are in the form of fossil fuels that cause a large amount of emissions that can affect air quality. Transportation modes can also cause unpleasant noise. The construction of transportation facilities can consume land that would otherwise be available for agricultural, commercial, or residential use, or left undeveloped.

The modern transportation engineer needs to participate in the search for ways to meet society's transportation needs, while minimizing the negative impacts of transportation activity. New technologies (e.g., more efficient vehicles, alternative fuels, ITS) and new strategies (e.g., better coordination between land use and transportation, travel demand management measures) are being proposed, evaluated, and tested as ways to maintain good accessibility without burdening society with excessive costs. In many ways, the techniques to analyze transportation problems that are presented in this book are the fundamentals of transportation engineering. In a larger sense, however, it is the ability of the transportation engineer to operate within the context of society that constitutes the true essentials of transportation engineering.

ABBREVIATIONS AND NOTATION

APTS	Advanced Public Transportation Systems (part of ITS)
ATMS	Advanced Traffic Management Systems (part of ITS)

ATIS	Advanced Traveler Information System (part of ITS)
AVCS	Advanced Vehicle Control Systems (part of ITS)
Bbl	Barrel (42 gallons)
BTU	British thermal unit
CAAA	Clean Air Act Amendments
CAFÉ	corporate average fuel economy
CMP	congestion management plan
CNG	compressed natural gas
CO	carbon monoxide
CVO	Commercial Vehicle Operations Systems (part of ITS)
E-90	mixture of 90 percent ethanol and 10 percent gasoline
EI	energy intensity
EPA	Environmental Protection Agency
ETBE	ethyl tertiary butyl ether
EV	electric vehicle
FRA	Federal Railroad Administration
GVW	gross vehicle weight
HC	hydrocarbons
HCHO	formaldehyde
HEV	hybrid electric vehicle
HOV	high-occupancy vehicle
HSR	high-speed rail
IC	internal combustion
ISTEA	Intermodal Surface Transportation Efficiency Act
ITS	intelligent transportation system
LEM	location-efficient mortgage
LEV	low-emission vehicle
LNG	liquified natural gas
MAV	motorist assistance vehicle
MAGLEV	magnetically levitated vehicle
MBPD	million barrels per day
MBPY	million barrels per year
MTBE	methyl tertiary butyl ether
NAAQS	National Ambient Air Quality Standards
NMOG	nonmethane organic gases
NOx	nitrogen oxide
NTPS	National Personal Transportation Survey
PPM	parts per million
PM	particulate matter
RFG	reformulated gasoline
SIP	State Implementation Plan (to meet the NAAQS)
STPP	Surface Transportation Policy Project
SULEV	super-low-emission vehicle
SUV	sports utility vehicle
TEA-21	Transportation Efficiency Act of the 21st Century
TDM	travel demand management
TOC	traffic operations center
TOD	transit-oriented design
TTI	Texas Transportation Institute at Texas A&M

ULEV   ultralow-emission vehicle
VOC   volatile organic compounds
ZEV   zero emission vehicle

GLOSSARY

Accessibility: The ability to reach a desired destination.

Acela: Swedish version of high-speed rail operating in the Northeast Corridor of the United States.

Barrel: 42 gallons.

Congestion pricing: Charge for highway use during rush hours.

Direct cost: Cost that the driver pays out of his pocket.

Emission standards: Standards set for automotive and truck emission from vehicles while driving and idling.

Energy efficiency: Inverse of energy intensity (miles per gallon).

Energy intensity: The amount of heating energy (BTUs) to move one passenger one mile.

Indirect costs: Costs from driving that are not paid directly by the driver.

MOBILE5b: EPA's driving cycle used to determine pollution and energy of automobiles and other vehicles.

Mobility: The ability to be mobile; to move from place to place.

Point of production management: Efforts to reduce fuel use and pollution by improving the vehicle during its production.

Point of use management: Efforts to reduce fuel use and pollution by conserving the manner in which vehicles are used.

QUAD: Quadrillion BTU or 10^{15} BTU.

Social costs: Costs borne by society such as health effects from pollution, those paid by taxes, etc.

Sustainability: Transportation that will sustain a high quality of life with low pollution at reasonable cost.

REFERENCES

[1] Bureau of Transportation Statistics, *Transportation Statistics Annual Report*, 2000, US DOT Report.

[2] Cambridge Systematics, Inc., *Twin Cities Ramp Metering Evaluation*, prepared for the Minnesota Department of Transportation, Oakland CA, February 2001.

[3] CGER, Commission on Geosciences, Environment, and Resources, "Ozone-Forming Potential of Reformulated Gasoline," National Research Council, 1999. http://books.nap.edu/books/0309064457/html/1.html#1

[4] Environmental Protection Agency, List of Federal Reformulated Gasoline Program Areas, Clean Air Act REQUIRED AREAS (as of July 2, 1998), Office of Mobile Sources, US Environmental Protection Agency, 1998. http://www.ofa.net/~DZC226.htm

[5] Federal Highway Administration, Transportation Programs and Provisions of the Clean Air Act Amendments of 1990, 1992.

[6] Florida Traffic Information CD for 1999.

[7] Greene, D.L. and N.I. Tishchishyna, *Costs of Oil Dependence, 2000 update*, Oak Ridge National Laboratory ORNL/TM 2000/152 and updates (2001).

[8] Hamerslag, R., J. D. Fricker, and P. Van Beek, "Parking Restrictions in Employment Centers: Implications for Public Transport and Land Use," *Transportation Research Record* 1499, p. 76–82, 1995.

[9] Halkias, John A., *Demonstration Project No. 105*, Publication FHWA SA-97-058, April 1997.

[10] Hodkinson, Ron and John Fenton, *Lightweight electric/hybrid vehicle design*, SAE International, Warrendale, PA, 2001.

[11] Institute of Transportation Engineers, *Intelligent Transportation Primer*, published by ITE in partnership with FHWA and ITS America, 2000.

[12] Iowa State University, Advantage I-75 Mainline Automated Clearance System, Center for Transportation Research and Education, Iowa State University, August 1998.

[13] Larson, Thomas, Administrator, U.S. Federal Highway Administration, 1992, in FHWA-PD-92-023.

[14] Oak Ridge National Laboratory (ORNL), *Transportation Energy Data Book, 2001*, U.S. Department of Energy.

[15] Oak Ridge National Laboratory (ORNL), *Transportation Energy Data Book, 2002*, U.S. Department of Energy.

[16] Pacific Institute, "Without Oil Crisis, Fuel Economy Declines", Global Change (Electronic Edition), 1999. http://www.globalchange.org/dataall/98dec9b.htm

[17] Rakha, Hesham and Yonglian Ding, "Impact of Stops on Vehicle Fuel Consumption and Emissions", *Journal of Transportation Engineering*, Vol. 129, Number 1, American Society of Civil Engineers, January/February 2003, p. 23–32.

[18] Surface Transportation Policy Project, *Ten Years of Progress*, STPP, January 2002.

[19] Texas Transportation Institute, *Urban Mobility Study 2002*, Texas A&M University.

[20] U.S. Department of Energy, *International Energy Annual*, 1999.

[21] U.S. Department of Energy, *Sustainable Transportation Introduction*, Smart Communities Network, Creating Energy Smart Communities, http://www.sustainable.doe.gov/transprt/trintro.shtml, as viewed 12 October 2002.

[22] U.S. Department of Transportation, *A Summary Intermodal Surface Transportation Act of 1991*.

[23] Vuchic, V., *Transportation for Livable Cities*, CUPR Press, Rutgers University, 1999.

OTHER REFERENCES NOT CITED IN TEXT

Sperling, D., M. Delucchi, P. Davis, and A. Burke, *Future Drie: Electric Vehicles and Sustainable Transportation*, Inland Press, 1995.

Society of Automotive Engineers, *Electric Hybrid Vehicles: Alternative Powerplants, Energy Management, and Battery Technology*, ISBN: 0768000408, Paperback, August 1997.

U.S. Department of Energy, *Electric & Hybrid Vehicles Program*, U.S. Department of Energy Business/Technology Books; ISBN: 0899342523, September 1995.

Federal Railroad Administration, *Tilt Train Technology: A State of the Art Survey*, 92/05, 1992.

Lynch, T., *High Speed Rail in the U.S.: Super Trains for Millennium*, edited by Thomas Lynch, G & B Science, ISBN: 9056996061, June 1998.

U.S. General Accounting Office, Report to the Chairman, Committee on Energy and Commerce, *High-Speed Ground Transportation—Issues Affecting Development in the United States*, House of Representatives, Nov. 1993.

EXERCISES FOR CHAPTER 13: TOWARD A SUSTAINABLE TRANSPORTATION SYSTEM

13.1 Large Federal Transit Program.   Many governments around the world impose a large gasoline tax to maintain a viable and well-used transit system. Because of the potential difficulties the United States has with imported oil, you are asked to evaluate a major new federal bus program. The project calls for 1 million buses to be added to communities around the country. The federal government will buy the buses and subsidize their operation. The cost to the government is estimated to be $700 billion over 3 years. Each bus rider will pay 25 cents per trip. The remainder of the payment for the bus operation would be subsidized by the program. Each bus is expected to deliver 1600 passenger miles on 150 miles of travel per day. Using 300 days per year,

- **(a)** How many automobile/SUV passenger miles will be replaced by bus? If the average auto/SUV occupancy is 1.5, how many vehicle miles traveled will be replaced?
- **(b)** How much fuel will no longer be used by automobile, SUVs, and vans, if their collective fuel economy is 25 mpg?
- **(c)** If the fuel that will continue to be used is taxed to pay for the program, what extra tax will have to be charged at the pump?
- **(d)** Because the tax increase is likely to raise the price per gallon considerably, there will be reduced driving. That will mean less gasoline will be purchased and less tax will be paid. If a 0.3 percent reduction in miles driven can be expected for every one percent increase in gasoline price above the current $1.50 per gallon, what will be the actual added tax revenues?
- **(e)** Transit bus fuel efficiency is 5 mpg on diesel. How many BTUs will the 1 million new diesel buses use per year and what will their energy intensity be?

13.2 CAFÉ Standards.   As a regulatory analyst, you are asked to evaluate a new level of CAFÉ standards. The goal is to reduce automobile and light-duty truck (LDT) fuel usage to a total of 5 MBPD 10 years from now. For the simplified evaluation, use the following assumptions:

- Ten percent (or more) of each vehicle type will be replaced each year, until the entire fleet will have the mix of the Year 2000 new vehicle sales. See Table 13.10.
- There will be no change in the number of automobiles and LDTs, and no change in the VMT by either vehicle type.
- Fifty percent of the automobile fleet in 10 years will be HEVs with a fuel efficiency of 55 mpg. The remaining 50 percent will be standard automobiles at their current fuel efficiency.

- **(a)** If the auto and LDT fuel consumption rates do not change from the Year 2000 values shown in Table 13.9, what will be the collective fuel consumption for the fleet of autos, HEVs, and LDTs 10 years from now? By how many MBPD will this fleet rate fail to meet the goal of 5 MBPD?
- **(b)** The remainder of the fuel economy requirement is to be made up by improving the LDT fleet from its present 21.2 mpg in ten years. What fuel economy standards for LDTs make total fuel consumption of no more than 5 MBPD possible?

13.3 Energy Intensity

- **(a)** If your automobile achieves 30 mpg and carries two persons, what is its energy intensity (EI)?
- **(b)** At 21 mpg, how many persons must you have occupying your SUV that gets only 17 mpg to achieve the same EI you obtained in part A?
- **(c)** The HEV gets 50 mpg and carries three passengers. What is its EI with three passengers? With three passengers? With two? With the driver alone?

13.4 Congestion Cost.   Your freeway has a capacity of 7200 vph and, on a given day, there is a constant flow of 4000 vph. A four-car crash occurs at $t = 0$, which reduces the flow past the crash site to 900 vph.

(a) It took the wrecker an hour to clear the roadway, before unimpeded flow could resume. How long will it take the freeway to return to normal flow? What is the total vehicle hours spent in delay? What is the total cost, if automotive costs include 0.05 gallons per hour ($1.50 per gal) and the lost time is valued at $15.00 per hour?

(b) The Mythaca Magical Machine is on patrol and reaches the scene within 5 minutes. MMM personnel assess the injuries and work to correct things and are able to open the flow to 1600 vph 15 minutes after arrival ($t = 20$). After the ambulance is loaded in another 15 minutes ($t = 35$), the flow increases to 2400 vph. The MM crew is able to get the last part of the accident cleared up at $t = 50$. What time delay was experienced now and what are the savings to motorists because of the MMM for this scenario?

13.5 Congestion Costs.   Your freeway has a capacity flow of 4800 vph. Today, it is running at a V/C of 0.75 when a vehicle blows a tire, swerves, and hits a truck.

(a) Traffic is totally stopped for 30 minutes, when it is cleared by a wrecker that was close by. What is the delay experienced by the motorists?

(b) The MMM comes along with the wrecker right behind it and the two of them get there in 5 minutes, clear one lane in 5 minutes ($t = 10$), and the second lane at $t = 25$ minutes. What amount of total delay did the MMM save in this instance?

13.6 Sustainability.   The report by the Surface Transportation Policy Project can be found at www.transact.org/report. Read the Preface, Introduction, and first two chapters (they are short) and then select five projects. Choose two projects from the first list:

Examples by Category

Enhancing Health, Safety, and Security
Conserving Energy and Enhancing the Environment
Creating Equitable and Livable Communities
Promoting Economic Prosperity

And choose three projects from the second list:

Bicycle Improvements
Coordinating Transportation and Land Use
Incentives for the Public
Intelligent Transportation Systems (ITS)
Pedestrian Improvements
Planning and Development
Public Participation
Road Improvements
Specialized Transit Service
Transit Improvement

For each project you chose, write a page discussing the project and its pros and cons from the perspective of sustainability. Consider how the topics and concepts learned in this course would apply to an analysis of moving the project to another location. What you would need to do to set it up? What goals you would have for it? How you would evaluate it?

Index

Note: t = table and f = figure